Advances in Intelligent Systems and Computing

Volume 1258

The series “Advances in Intelligent Systems and Computing” contains publications on theory, applications, and design methods of Intelligent Systems and Intelligent Computing. Virtually all disciplines such as engineering, natural sciences, computer and information science, ICT, economics, business, e-commerce, environment, healthcare, life science are covered. The list of topics spans all the areas of modern intelligent systems and computing such as: computational intelligence, soft computing including neural networks, fuzzy systems, evolutionary computing and the fusion of these paradigms, social intelligence, ambient intelligence, computational neuroscience, artificial life, virtual worlds and society, cognitive science and systems, Perception and Vision, DNA and immune based systems, self-organizing and adaptive systems, e-Learning and teaching, human-centered and human-centric computing, recommender systems, intelligent control, robotics and mechatronics including human-machine teaming, knowledge-based paradigms, learning paradigms, machine ethics, intelligent data analysis, knowledge management, intelligent agents, intelligent decision making and support, intelligent network security, trust management, interactive entertainment, Web intelligence and multimedia.

The publications within “Advances in Intelligent Systems and Computing” are primarily proceedings of important conferences, symposia and congresses. They cover significant recent developments in the field, both of a foundational and applicable character. An important characteristic feature of the series is the short publication time and world-wide distribution. This permits a rapid and broad dissemination of research results.

**** Indexing: The books of this series are submitted to ISI Proceedings, EI-Compendex, DBLP, SCOPUS, Google Scholar and Springerlink ****

More information about this series at http://www.springer.com/series/11156

Vera Murgul · Viktor Pukhkal
Editors

International Scientific Conference Energy Management of Municipal Facilities and Sustainable Energy Technologies EMMFT 2019

Volume 1

Springer

Editors
Vera Murgul
Peter the Great St.Petersburg Polytechnic
Saint Petersburg, Russia

Viktor Pukhkal
Saint Petersburg State University
of Architecture and Civil Engineering
Saint Petersburg, Russia

ISSN 2194-5357 ISSN 2194-5365 (electronic)
Advances in Intelligent Systems and Computing
ISBN 978-3-030-57449-9 ISBN 978-3-030-57450-5 (eBook)
https://doi.org/10.1007/978-3-030-57450-5

This Springer imprint is published by the registered company Springer Nature Switzerland AG
The registered company address is: Gewerbestrasse 11, 6330 Cham, Switzerland

Preface

This book presents a collection of the latest studies in the field of the sustainable development of urban energy systems and new strategies for the transportation sector.

The international scientific conference Energy Management of Municipal Facilities and Sustainable Energy Technologies EMMFT 2019 took place in Voronezh State Technical University on November 28–30, 2019 in the city of Voronezh.

This annual scientific event brought together guests and participants from throughout Russia and different foreign countries. As traditionally, the main topics to discuss were sustainable energy technologies, building energy modeling, energy efficiency in transport sector, electrical energy storage, energy management and life cycle assessment in urban systems and transportation.

The objective of the conference was the exchange of the latest scientific achievements, strengthening of academic relations with leading scientists of the European Union, creating favorable conditions for collaborative researches and implementing collaborative projects, encourage young scientists, doctoral and post-graduate students in their scientific and practical work related to the field of new energy technologies. The newest equipment and devices for HVAC-systems were demonstrated; the latest technologies of thermal protection of buildings were shared.

Over than 250 papers were submitted for the conference. All papers passed scientific and technical review. Finally, 136 papers were accepted.

Within the framework of technical review, all papers were thoroughly checked for the following attributes: compliance with the subject of the conference; plagiarism (acceptable minimum of originality was 90%); acceptable English language. At the same time, papers were checked by a technical proofreader (for the quality of images, absence of Cyrillic, etc.).

Scientific review of each paper was made by at least three reviewers. If the opinions of the reviewers were radically different, additional reviewers were appointed.

Live participation in the conference was an indispensable condition for the publication of a paper.

The book is intended for a broad readership: from policymakers tasked with evaluating and promoting key enabling technologies, efficiency policies and sustainable energy practices, to researchers and engineers involved in the design and analysis of complex systems.

All the participants and organizers express their gratitude to Springer publishing office and to the editing group of journal Advances in Intelligent Systems and Computing for publishing the proceedings of the conference.

Vera Murgul
Viktor Pukhkal

Organization

Scientific Committee

Samuil G. Konnikov	Full Member of the Russian Academy of Sciences, Ioffe Physical-Technical Institute of the Russian Academy of Sciences
Iurii Tabunschikov	Corr. Member of RAASN, Honorary Member of the International Ecoenergetic Academy of Azerbaijan, ASHRAE fellow member, REHVA Fellow Member, Corr. Member of VDI, Member of ISIAQ Academy, Winner of the 2008 Nobel Peace Prize as a Member of the Intergovernmental Panel on Climate Change
Antony Wood	Executive Director (CTBUH), Visiting Prof. of Tall Buildings, Tongji University, Shanghai, China, Studio Ass. Prof., Illinois Institute of Technology, Chicago, the USA
Viktor Pukhkal	Head of the Department of Heat and Gas supply and Ventilation, Saint Petersburg State University of Architecture and Civil Engineering
Sergey Anisimov	Wroclaw University of Science and Technology, Professor, Poland
Marianna M. Brodach	Moscow Architectural Institute (State Academy), Vice President of Russian Association of Engineers for Heating, Ventilation, Air-Conditioning, Heat Supply and Building Thermal Physics "ABOK", ASHRAE member, REHVA Fellow Member, Member of the Editorial Board of REHVA Journal
Igor Surovtsev	Head of the Department of Innovation and Building Physics Voronezh State Technical University
Daniel Safarik	Director (CTBUH China Office), Editor (CTBUH Journal), Chicago, the USA

Aleksander Szkarowski	Head of the Construction Networks and Systems Division Department of Civil & Environmental Engineering and Geodesy, Koszalin University of Technology, Koszalin, Poland
Alexander Solovyev	Head of the Research Laboratory of Renewable Energy Sources Lomonosov Moscow State University, Full Member of Russian Academy of Natural Sciences
Dietmar Wiegand	Technische Universität Wien TU Wien
Luís Bragança	Director of the Building Physics & Technology Laboratory, Guimaraes, University of Minho, Portugal
Zdenka Popovic	Belgrade University of Belgrade, Faculty of Civil Engineering, Serbia
Marco Pasetti	Università degli Studi di Brescia UNIBS, Italy
Valerii Volshanik	Moscow State University of Civil Engineering
Mirjana Vukićević	Faculty of Civil Engineering, University of Belgrade, Serbia
Sang Dae Kim	Chief Editor (International Journal of High-rise Buildings), Emeritus Professor, Department of Civil, Environmental and Architectural Engineering, Korea University, Seoul, South Korea
Alenka Fikfak	University of Ljubljana: Faculty of Civil and Geodetic Engineering (Department of Town & Regional Planning) Biotechnical Faculty (Department of Landscape Architecture), Slovenia
Milorad Jovanovski	Faculty of Civil Engineering, Ss. Cyril and Methodius University in Skopje, Macedonia
Škoda, Radek	Czech Technical University in Prague, Faculty of Mechanical Engineering, Department of Nuclear Energetics Technická
Paulo Cachim	Department of Civil Engineering, University of Aveiro, Portugal
Aires Camões	Director of the Materials of Construction Laboratory, Guimarães, University of Minho, Portugal
Michael Tendler	currently Professor of Fusion Plasma Physics at the Royal Institute of Technology, Stockholm (KTH) and Senior Science Expert and Member of the External Management Advisory Board of the ITER Organization, Kungliga Tekniska Högskolan, Sweden
Christoph Pfeifer	Professor of Process Engineering of Renewable Resources, University of Natural Resources and Life Sciences, Vienna, Austria
Antonio Andreini	The University of Florence, UNIFI, Italy
Pietro Zunino	DIME Universitá di Genova, Genoa, Italy

Olga Kalinina	Peter the Great St. Petersburg Polytechnic University, Russia
Tomas Hanak	Faculty of Civil Engineering, Brno University of Technology, Czech Republic
Vera Murgul	Peter the Great St. Petersburg Polytechnic University, Russia
Darya Nemova	Peter the Great St. Petersburg Polytechnic University
Norbert Harmathy	Budapest University of Technology and Economics, Department of Building Energetics and Building Services
Igor V. Ilyin	Peter the Great Saint-Petersburg Polytechnic University, Russia

Contents

Transportation Engineering and Traffic Engineering. Intelligent Transportation Systems

Solving the Multi-criteria Optimization Problem of Heat Energy Transport

Viktor Melkumov, Svetlana Tulskaya, Anastasiya Chuykina(✉), and Vladimir Dubanin

Voronezh State Technical University, 20-Letiya Oktyabrya Street, 84, Voronezh 394006, Russia
a.a.chuykina@mail.ru

Abstract. The work is dedicated to the improvement of technique of designing of the pipe network of heat supply systems based on the solution of a multi-criterial optimization problem. One of the most important stages of design, influencing future costs for construction and operation of heating systems is the choice of the pipeline route. The lack of baseline data at the initial design stage (in the absence of structural calculation) may lead to erroneous solution of the problem under consideration. This can be avoided by applying calculation methods based on aggregated parameters of a thermal network. However, individually they cannot describe all or even most significant characteristics of the system. In this regard, the improved method is proposed to solve the optimization problem of the trace pipeline network, which is based on the methods of system analysis using a number of aggregated parameters describing qualitative and quantitative characteristics of the heating system, which allows to increase the accuracy of selecting the best option (or group of options) of the pipe network. As a criterion of optimality, we suggest a generalized vector criterion which is a function of the material characteristics of the heat network, the moment the heat load, heat loss, reliability, and building-technological indicators. To determine the parameters of preferences (criteria weights), we selected ranking method, which allows us to reduce the time of the expert survey and increase the accuracy of the result. The obtained results can be used in the design and reconstruction of heating systems.

Keywords: Heat supply · Optimal route · Multi-criteria optimization · Aggregated parameters of the heating main · Transport problem · Heating networks

1 Introduction

The choice of the best option for tracing the pipeline network when transporting heat from a source to a consumer is a complex multifactor task. The solution of such problems can be solved using system analysis methods. The search for a solution to a multi-criteria problem can be carried out using a number of methods, for example, such as: the "ideal" point method, lexicographic ordering of criteria; highlighting the main criterion; folding a vector criterion, etc. The latter got the greatest spread for the type of optimization problem under consideration. This method takes into account the

V. Murgul and V. Pukhkal (Eds.): EMMFT 2019, AISC 1258, pp. 3–10, 2021.
https://doi.org/10.1007/978-3-030-57450-5_1

importance of particular parameters by constructing a scalar function, which is a generalized parameter of the vector criterion. The generalized parameter, in turn, can be converted into a function of various kinds, for example, into function (1) [1]

$$S = \sum_{k=1}^{n} x_k p_k, \quad (1)$$

where x_k – particular optimality criterion; p_k – weight of the particular criterion.

Thus, the further solution of the optimization problem is reduced to several stages, the selection of criteria by which the multi-criteria optimization will be carried out, the determination of their weight values and the solution of function (1).

2 Materials and Methods

Since the design of a transport pipeline system is a complex engineering task that requires significant labor and time costs, it is possible to reduce the volume of calculations of optimality criteria by using aggregated parameters that describe the main properties and characteristics of the designed heating network. The main aggregated indicators that exist at the moment are such as:

- the material characteristic of the heat network [2]

$$M = \sum_{i=1}^{n} D_{ini} l_i, \quad (2)$$

where D_{ini} – the inner diameter of the pipeline at a section of the heating network; l_i – the length of the heating network; n – the number of sections of the heating network.

At the initial design stage, when the diameters of the pipelines are unknown, formula (2) can be transformed according to [3] into the following form

$$M = E.\ G^{0.38} \cdot l, \quad (3)$$

$$E = \frac{A_d^a}{R_l^{0.19}}, \quad (4)$$

where G – the heat carrier flow rate in the line, kg/s; A_d^a – the coefficient related to the diameter of the pipeline, depending on the roughness; R_l – the specific linear pressure drop, kg/(m^{2}m);

- the annual heat loss [4]

$$Q_{t.l.} = q \cdot M_c, \quad (5)$$

where M_c – the conventional material characteristic of the heating system calculated on the outer surface of the insulation, m^2; q – the specific annual heat losses attributed to 1 m^2 of the conventional material characteristic of the heating network, Gcal/(year·m^2).

$$M_c = \mathrm{M} + 0.15 \sum l, \quad (6)$$

where $\sum l$ – the total length of the pipeline, m.

$$q = 3.6 \cdot \pi \cdot k \cdot (T_{av} - t_0)(1 + \beta) \cdot n \cdot 10^{-6}, \quad (7)$$

where k – the heat transfer coefficient of the heat conduit taking into account the thickness and material of the insulation, channel and type of soil, conventionally assigned to the outer surface of the insulation, W/(m^2 °C); T_{av} – the average annual heat carrier temperature, °C; t_0 – the average annual soil or ambient temperature, °C; β – the local heat loss coefficient; n – the number of hours of operation of the heating network per year;

- the thermal load moment [4]

$$Z_{ac} = \sum Z_i = \sum (Q_i^p \cdot l_{aci}), \quad (8)$$

where Z_i – the actual moment of heat load in the considered section, MW·m; Q_i^p – the estimated heat load in the considered section, MW; l_{aci} – the actual length of the considered section, m;

- the reliability of the heating network (it is customary to evaluate it by the reliability indicator, which should not be lower than the established level, the higher it is, the more reliable the system), the optimization criterion will take the form [4]

$$R_{syst}(t) = \frac{Q(t)}{Q_0} = 1 - \sum_{j=1}^{j=l} \frac{\Delta Q_j}{Q_0} \frac{\omega_i}{\sum \omega_i} \left(1 - e^{-\Sigma \omega_i t}\right), \quad (9)$$

where Q_0 – the estimated heat consumption; ΔQ_j – the lack of heat; $Q(t)$ – the mathematical expectation of the performance characteristics of the system; t – the time; ω_i – the failure flow parameter determined by the formula

$$\omega = \frac{\sum_{i=1}^{N} m_i}{N \Delta t} = \frac{m_{av}(t)}{\Delta t}, \quad (10)$$

where m_i – the failure rate; N – the number of identical sections of the heating network; Δt – the observation time; m_{av} – the average failure rate;

- the time spent on construction or reconstruction [5]

$$\mathrm{T}_{con} = \sum_{j=1}^{m} \sum_{k=1}^{n} \frac{\theta_{kj} v_{kj}}{N_{kj}}, \quad (11)$$

where N_{kj} – the workers; θ_{kj} – the labor costs per unit of construction work; v_{kj} – the volume of work; k – the sizes (i = 1, 2, …, n); j – the types of designs (j = 1, 2, …, m);

– the construction and technological indicators [5]

$$\theta_{con} = \sum\nolimits_{j=1}^{m} \sum\nolimits_{k=1}^{n} \theta_{kj} v_{kj}, \tag{12}$$

$$M_{con} = \sum\nolimits_{j=1}^{m} \sum\nolimits_{k=1}^{n} \mathrm{M}_{kj} v_{kj}, \tag{13}$$

where M_{kj} – the machine capacity per unit of construction work.

The search for weight values of optimality criteria can be carried out on the basis of various methods, for example, such as: the method of point estimates, the method of direct numerical estimates, the method of ranking criteria; frequency preference method; Churchman-Akof's method; Thurstone's method; method of linear folding of criteria. According to [6], the most appropriate method from the point of view of accuracy of the final result and the time spent on its processing is the ranking method. In this method, the relative frequencies of the converted ranks [6] are written in form (14) are taken as weight values of coefficients:

$$b_i = \sum j^m \cdot B_{ij} / \left(\sum i^n \sum j^m \cdot B_{ij} \right), \tag{14}$$

where B_{ij} – the rank of the converted criterion.

3 Results

Table 1 gives the ranks of the seven aggregated criteria for optimizing heating networks discussed above when interviewing ten experts.

Table 1. Ranks of heat network optimization criteria.

Experts	Converted criteria rank, B_{ij}						
	M	$Q_{t.l.}$	Z_{ac}	R_{syst}	T_{con}	θ_{con}	M_{con}
1	0	2	1	2	4	6	5
2	0	1	2	3	4	6	5
3	0	1	2	3	4	6	5
4	0	2	1	3	4	6	5
5	0	1	2	3	4	6	5
6	1	0	3	2	4	5	6
7	2	1	0	3	5	6	4
8	1	0	3	2	4	5	6
9	0	2	1	3	4	6	5
10	2	0	1	3	4	6	5

Table 2 shows the relative frequencies of the ranks of the considered aggregated criteria for optimizing heating networks determined by dependence (14).

Table 2. The relative frequency of the ranks of the aggregated criteria for the optimization of heating networks.

Relative frequency of the converted ranks	Aggregated optimality criteria						
	M	$Q_{t.l.}$	Z_{ac}	R_{syst}	T_{con}	θ_{con}	M_{con}
Expert 1	0.000	0.100	0.048	0.100	0.191	0.286	0.238
Expert 2	0.000	0.048	0.100	0.143	0.191	0.286	0.238
Expert 3	0.000	0.048	0.100	0.143	0.191	0.286	0.238
Expert 4	0.000	0.100	0.048	0.143	0.191	0.286	0.238
Expert 5	0.000	0.048	0.100	0.143	0.191	0.286	0.238
Expert 6	0.048	0.000	0.143	0.100	0.191	0.238	0.286
Expert 7	0.100	0.048	0.000	0.143	0.238	0.286	0.191
Expert 8	0.048	0.000	0.143	0.100	0.191	0.238	0.286
Expert 9	0.000	0.100	0.048	0.143	0.191	0.286	0.238
Expert 10	0.100	0.000	0.048	0.143	0.191	0.286	0.238
b_i	0.030	0.049	0.078	0.130	0.196	0.276	0.243

According to this method, the criterion with the lowest relative frequency value bi is the most important criterion.

The last step in determining the most optimal piping tracing during the transport of heat energy in heat supply systems, namely, solving Eq. (1), is possible if the values of the aggregated parameters are brought to a common view, since their dimension is the same. The simplest is the conversion option, in which the largest value of the aggregated parameter is taken equal to one, and smaller values are determined by compiling the proportion, thereby reduction of the parameters to a dimensionless form. The solution of function (1) for finding the best option for tracing the heating network shown in Fig. 1 is quite simple to implement using modern computing tools [7]. Graphically, the choice of the most optimal parameter can be seen in Fig. 2, which gives an example of the location of the five options for tracing the heating network relative to the weight vector, depending on the three most important, aggregated parameters, which are the coordinate axes.

Obviously, when considering a larger number of optimality criteria, the space from three-dimensional is transformed into n-dimensional. In our case, 7-dimensional.

The main disadvantage of using the method under consideration, as well as others, in which expert assessments are applied, is some subjectivity of the obtained research results. In addition, the accuracy of the solution to the optimization problem will be significantly affected by the qualification of the expert.

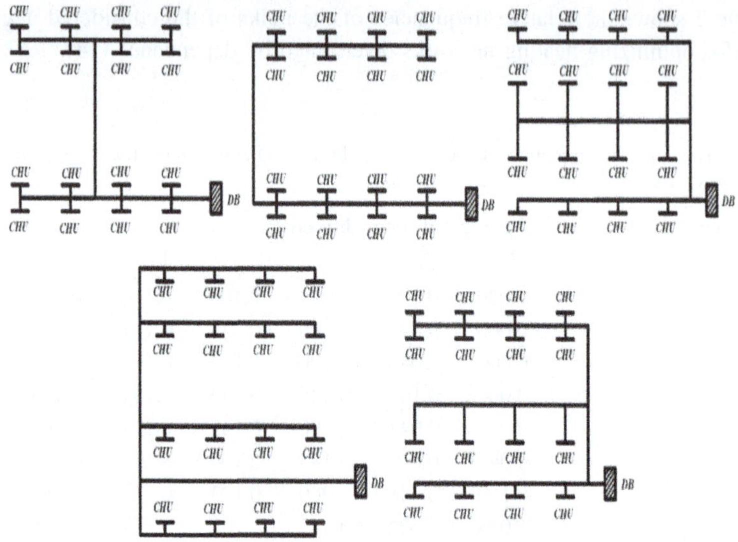

Fig. 1. Considered options for tracing pipelines of a heat supply system.

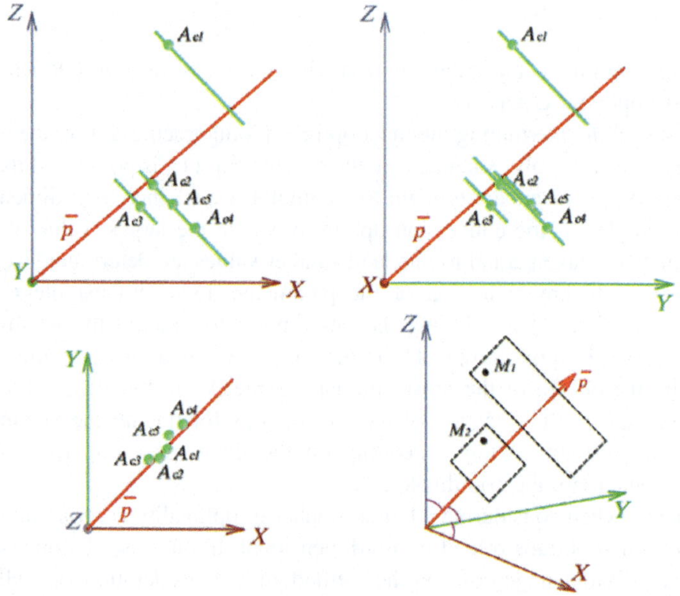

Fig. 2. The location of the five options for tracing the heating network relative to the weight vector according to three optimization parameters.

4 Discussion

In practice, when designing the route of the pipeline network of heat supply systems, the number of laying options can be estimated at hundreds or thousands, depending on the initial data and the required degree of development of the project. From this set, it is desirable to choose a small number of optimal or close to optimal options. In addition, according to studies [8], often, the optimal tracing option for one parameter may not coincide with the option for another parameter, which can lead to a contradiction when choosing the most suitable construction or reconstruction option.

In connection with the foregoing, the practical implementation of the considered methodology of the multi-criteria optimization problem of heat energy transport will allow us to find the most optimal option or a limited number of network trace options that take into account a number of basic parameters that reflect various properties of the system. In addition, the chosen method of searching for weight values of optimality criteria allows obtaining the most acceptable result with the least labor and time costs of experts. To increase the objectivity of the result, it seems possible to use the considered method together with the automated search methods for the most profitable options based on an analysis of the available data, in this case, the available design solutions for the heat supply systems.

5 Conclusions

The considered solution of the multi-criteria optimization problem of choosing the optimal route of the pipeline transporting thermal energy involves the search for the minimum of the function according to the vector criterion, which consists of a number of particular parameters and their weight values. The determination of their numerical values is conveniently carried out using the ranking method, which requires minimal time costs and is as close as possible to the most accurate solution. Using the considered aggregated parameters of the heating network as optimality criteria allows us to obtain a more accurate solution to the multifactor problem, taking into account both quantitative and qualitative characteristics of the heat supply system at the design, construction and operation stages.

References

1. Muravyeva, L., Vatin, N.: Application of the risk theory to management reliability of the pipeline. Appl. Mech. Mater. **635–637**, 434–438 (2014). https://doi.org/10.4028/www.scientific.net/AMM.635-637.434
2. Duda, M., Dobrianski, J., Chludzinski, D.: Analysis of the possibility of applications for a two-phase reverse thermosyphon in passive heat transport systems. In: E3S Web of Conferences, vol. 49, p. 00020 (2018). https://doi.org/10.1051/e3sconf/20184900020
3. Gorshkov, A.S., Vatin, N.I., Rymkevich, P.P., Kydrevich, O.O.: Payback period of investments in energy saving. Mag. Civ. Eng. **78**(2), 65–75 (2018). https://doi.org/10.18720/MCE.78.5

4. Muravyeva, L., Vatin, N.: Pipelines stability under extreme hydrodynamic conditions. Appl. Mech. Mater. **635–637**, 451–456 (2014). https://doi.org/10.4028/www.scientific.net/AMM.635-637.451
5. Muravyeva, L., Vatin, N.: Elaboration of the method for safety assessment of subsea pipeline with longitudinal buckling. Adv. Civ. Eng. (2016). https://doi.org/10.1155/2016/7581360
6. Muravyeva, L., Vatin, N.: Risk assessment for a main pipeline under severe soil conditions on exposure to seismic forces. Appl. Mech. Mater. **635–637**, 468–471 (2014). https://doi.org/10.4028/www.scientific.net/AMM.635-637.468
7. Muravyeva, L., Vatin, N.: The safety estimation of the marine pipeline. Appl. Mech. Mater. **633–634**, 958–964 (2014). https://doi.org/10.4028/www.scientific.net/AMM.633-634.958
8. Kolbin, V.V.: Generalized mathematical programming as a decision model. Appl. Math. Sci. **8**(70), 3469–3476 (2014). https://doi.org/10.12988/ams.2014.44231

Logistic Aspects of the Distribution of Electric Charging Stations on the Urban Road Network

Evgeny Makarov[1(✉)], Sergey Gusev[2], Elena Shubina[1], and Yulia Nikolaeva[1]

[1] Plekhanov Russian University of Economics Voronezh branch of PRUE. G. V. Plekhanov, Karl Marks Street, 67a, Voronezh 394030, Russia
ea_makarov@mail.ru

[2] Yuri Gagarin State Technical University of Saratov, Politehnicheskaja Street, 77, Saratov 410054, Russia

Abstract. Modern approaches to organization and management of an urban energy system deal with the optimal placement of charging stations for electric cars and other vehicles, that have dynamic charging while driving. The electric power flows are allocated between the consumers and must be taken into account in logistics network model under study in order to ensure sustainable and stable generation of energy for the urban transportation. Integration of different functional areas into the process of power supply of electric vehicles allows to optimize the total expenses for the city transportation service. The energy aspects of functioning of logistics systems are directly related to the problems of environmental protection in cities and locations of energy sources concentration. The distribution and redistribution of the energy power flows are an urgent applied research task, which is emerging as a promising work in the field of design of deployment models of electric charging stations on the city road network. The elaboration of the suggested approaches involves traditional and modern management models, including the models for estimating the entropy of the logistics system. The entropic model consists of the calculation of parameters of adaptation of the type, structure and properties of the available charging stations on the city road network. Planning of placement and layouts of charging stations is carried out after the final stage in calculating the parameter of organization of functioning of the analyzed resource-supplying logistics network. The total result and planned economic parameters of the charging stations should be measured and controlled.

Keywords: Urban energy system · Entropic model · Logistics network · Charging station

1 Introduction

The contemporary practice of tackling applied research tasks in the theory of logistics covers key states, problems and approaches to their solution, including the study of material flows and the ones that accompany in logistic systems and processes. In some cases, there are isolated states of the road and logistics network as well as systems that supply energy to keep the logistics systems functioning.

V. Murgul and V. Pukhkal (Eds.): EMMFT 2019, AISC 1258, pp. 11–23, 2021.
https://doi.org/10.1007/978-3-030-57450-5_2

The management practice of logistics systems involves a comprehensive analysis of industry components and detection of factors which affect the development of the logistics systems under study. The terminology is enriched with the definitions of the "first and last miles" and other categories, as well as the characteristics of the events taking place within the framework of goods distribution and accompanying processes. Consistency and process approach of logistics have many aspects and require taking into account not only well studied issues, but also the crossroads of the operational problems of information and energy power flows, their compatibility and practical experience, which is demonstrated by the above terms. The energy aspects of the functioning of logistics systems are directly related to the problems of environmental protection in cities and places of energy sources concentration. The distribution and redistribution of resource flows today are an urgent applied research task, to which this article is dedicated.

A logistics system of electric power supply for urban electric transport was chosen as the object of the study.

The efficient operation of the drive of any vehicle is due to a number of factors, both objective and subjective, the main of which are:

- state of the track infrastructure;
- technical condition of drive elements and rolling stock as a whole;
- weather conditions;
- qualifications of a driver, etc.

At the same time, one of the fundamental points that subsequently determines the efficiency of its work is the use of a traction motor with the optimal power parameters. Indeed, with insufficient engine power, acceptable dynamic index is usually achieved by forcing its operation modes, which is followed by increased energy consumption due to a decrease in drive efficiency.

The use of a high-power engine also leads to excess power demands, but in this case it is caused by the small performance values of a power unit which operates in a mode of under-utilization of capacity. This situation is typical for drives with both electric machines and thermal ones.

It is well known that there are three phases of a classic movement scheme for a passenger vehicle equipped with an electric drive: start, coasting, braking. At the same time, the energy necessary for movement on the entire haul is consumed at the start-up phase and is spent to overcome the resistance force. And it is also accumulated in the form of kinetic energy of the rolling stock.

An important factor affecting the energy consumption for movement is the haul length, the optimal value of which, according to the criterion of the minimum energy consumption for movement for ground-based electric transport is Lhaul = 550 m. This value became the corner stone for calculating the energy consumption for movement of a vehicle with an electric drive and with a heat engine [1, 2].

In accordance with the above, the maximum required value of power P to execute movement at point a of the curve is defined as (1):

$$P = (F_t - W)v_a = [(1+\gamma)m_{rs}a_s - W]v_a, \tag{1}$$

where Ft – traction on the wheel rim;
W – motion resistance force;
γ = 1.12…1.14 – inertia coefficient of the rotating masses of the vehicle;
mrs – mass of the rolling stock;
as – starting acceleration;
va – speed at point a.

The mechanical energy accumulated by the rolling stock is determined by the following expression (2):

$$A_t = (1+\gamma)m_{rs}v_a^2/2. \tag{2}$$

The power that is utilized by the engine to accelerate the rolling stock with the value as, taking into account the transmission efficiency, is determined by the expression (3):

$$P_{dv} = (P + Wv_a)\eta_d\eta_m, \tag{3}$$

where η_d – engine efficiency;
η_m – manual transmission efficiency.

Recuperation of energy by rolling stock in the modes of partial braking and emergency braking is possible subject to the following requirements:

- sufficient value of kinetic energy (a decrease in speed below a certain level leads to inefficiency of electrical braking and the need to replace it with mechanical one);
- availability of a consumer of electric energy;
- circuit support for the transition from the start mode to the regenerative braking mode.

The stored kinetic energy by the rolling stock when accelerating is sufficient to provide regenerative braking from the design speed to the minimum defined by the expression (4):

$$v_{\min} = k(U_p + I_t r) \Big/ c_e F, \tag{4}$$

where I_t – brake current providing a deceleration of 1.5 m/s^2;
r – recuperation of loop resistance;
$k = R_k/i_r$ – proportionality coefficient that depends on the radius of the wheel (R_k) and the gear reduction ratio (i_r).

Effective regenerative braking is possible up to velocity of 5–7 km/h when using pulse control for the rolling stock, and without - up to 15–20 km/h [3].

2 Materials and Methods

As known, human needs are unlimited, but resources are limited. The lack of resources and the potential for self-organization drive social progress. They equally affect the development of social systems, combining external and internal development factors. Commitment to control the distribution of resources is the fundamental of any social (scientific, technical) progress. Systemic contradictions are not initially defined, they dynamically change and correlate with the type of connections in the system and the structure of resource flows at different system levels of logistics. A feature of the exogenous approach of the General system theory is that the causes of systemic changes lie outside the social systems. They are static and fundamentally uncontrollable. A reaction to these causes can be controlled, but it is only possible under a certain systemic condition and subject to the availability of resources. It follows that the basis of any subjective social interest (consequence) is the objective need for something (reason). External factors are objective in social systems, when internal ones are subjective. The objective nature of external factors is due to the fact that external factors exist outside the system, and the environment generates their causes. And the subjective nature of internal factors is driven by the system itself, reflecting its own adaptive capacities. Therefore, the objectivity of systemic contradictions manifests itself at the social level in the form of subjective interests that reflect conflicting needs [4].

Under the influence of adverse external factors, the system organizes itself to adapt to the conditions of objective reality. Raymond Boudon rightly points out: "… an innovation can be perceived by the system only when the latter is capable of its perception. Acceptance (or rejection) of innovation is, therefore, a function of certain characteristics of the system." The resistance to innovation is based on an objective process which is described by Ilya Prigogine's theorem on the minimum level of production of entropy. Thus, the creation of information flows, that are independent of restructuring systems, is also necessary when introducing social innovations, as well as the formation of new external factors affecting the process.

There is nothing new about using electric traction to provide power to the vehicle propulsion system. However, researchers, both in Russia and abroad, has carefully scrutinized this question in recent years. Electric energy, as an alternative to solid fuel and liquid fuel resources, is used not only to service passenger flows, but is also actively being introduced into the cargo flows [5]. In this regard, one of the key problems in the development of this resource-supplying logistics system is the model of the distribution of charging stations on the road network of a city. If current collectors are installed on the vehicles, the energy supply is conducted via a contact high-voltage line. This type of systems is basic and supportive when using vehicles with dynamic recharging. Implementation of innovative technologies for wireless transmission of electric energy to recharge vehicles when driving is possible if the climate conditions imply the ambient temperature above 0C during the entire calendar period of use of this equipment. In addition, as it is indicated in the works of foreign researchers, these are expensive systems and their use is not always economically sound. Whereas, the situation is completely different for the energy supply of electric cars. It becomes problematic to charge vehicles, for instance, only at the end of routes of public transport, or to places of deployment. Therefore, it is important and relevant to optimize the distribution of charging stations and estimate their efficiency [6].

As discussed earlier, physical entropy is a measure of the functioning disorder of the system, and as one of the alternatives to solve the problem of placing a charging station in the city, a methodology is proposed based on the calculation of indicators characterizing the entropy of the system as a measure of adaptation of the charging station to the traffic flow [7].

Firstly, it is necessary to determine the reference and target values of the functioning parameters of the investigated resource-supply logistics network.

The methodology for determining the reference values for the parameters of the vectors of the intended structure and properties by the limit values is that, first of all, all parameters are divided into two groups.

The first group includes parameters, values of which are bounded from above. The maximum possible value equals to one and etc.

The second group includes all other factors except those that determine the composition sector.

An analysis of these parameters shows that their values are formed under the influence of a sufficiently large number of random variables of the same order of smallness and not dependent or little dependent on each other [8].

As is known, in the presence of these conditions, a normal law of distribution steps in, the probability density of which is (5):

$$f(x_j) = \frac{1}{\sigma\sqrt{2\pi}} e^{-\frac{(x-\bar{x})^2}{2\sigma^2}}, \tag{5}$$

where xj - random variable of one of the parameters of the vector of structure or property.

After testing the hypothesis of the normal distribution of any parameter using the Pearson's test, the numerical characteristics of the distribution are determined (6).

$$\bar{X} = \int_{-\infty}^{\infty} xf(x)dx, \tag{6}$$

$$\sigma^2 = \int_{-\infty}^{\infty} (x - \bar{x})^2 f(x)dx. \tag{7}$$

After that, using the three sigma rule, the reference or target value of parameter is determined (8).

$$X_{\text{эт}} = \bar{X} \pm 3\sigma\,. \tag{8}$$

The methodology for economic justification of the reference or target values of the parameters of the target vectors is based on minimizing the total cost of distribution of 1 kW/h of electric energy and identifying a set of indicators that ensures this cost.

The methodology for determining the reference or target values of the parameters of the target vectors from the average values consists in the following operations.

First of all, the goals and objectives of the analysis are discussed. The hierarchy of time series will depend on this.

If a comparative analysis of the adaptive properties of charging stations within a large city needs to be performed, then the analysis should be based on sampling parameters that set the target vector for the same type enterprises. If it is required to analyze data in a region or a single territorial entity, but within several months' period, variational series are selected respectively.

The expected value of each parameter from the selected variational series will be the reference value of the corresponding parameter.

Due to the fact that there is no possibility to obtain the general population, it is suggested to use statistical mathematical expectation in order to estimate the mathematical expectation. At the same time, the numerical characteristics obtained from experience should have a good approximation to the characteristics of the general population, i.e. be of good quality: consistent, unbiased, efficient [9].

Let us see how much the statistical estimate of mathematical expectation complies with these requirements.

Statistical mathematical expectation is calculated by the formula (9):

$$M^*(x) = \frac{\sum_{i=1}^{n} x_i}{n} = \frac{\sum_{i=1}^{n} x_i m_i^*}{n} = \frac{x_1 + x_2 + \ldots + x_n}{n}, \tag{9}$$

As an estimate of the mathematical expectation, its statistical value (M (x)) is taken and the feasibility of the condition is checked (10):

$$\tilde{M}(x) = \sum_{\substack{i=1 \\ n \to \infty}}^{n} \frac{x_i}{n} \to \bar{M}(x). \tag{10}$$

According to the law of large numbers, with a large number of observations, the arithmetic average of the observed values of a random variable converges in probability to its mathematical expectation (P) (11):

$$\underset{n \to \infty}{P}\left\{\left|\frac{x_1 + x_2 + \ldots + x_n}{n} - \bar{M}(x)\right| < \varepsilon\right\} = 1, \tag{11}$$

whereby (12)

$$P\{|M^*(x) - \bar{M}(x)| < \varepsilon\} = 1. \tag{12}$$

Therefore (13)

$$\frac{x_1 + x_2 + \ldots + x_n}{n} = M^*(x). \tag{13}$$

Consequently, if statistical expectation is taken as an estimate for the mathematical expectation, then the consistency requirement for n $\to \propto$ will be feasible.

If the accepted estimate is unbiased, the following condition must be satisfied for mathematical expectation (14):

$$M\left[\tilde{M}(x)\right] = \bar{M}(x) \quad n \to \infty \quad . \tag{14}$$

The invariable is put outside the expectation sign, and the mathematical expectation of the sum is equal to the sum of the mathematical expectations, therefore (15)

$$M\left[\frac{\sum_{i=1}^{n} x_i}{n}\right] = \frac{1}{n} M \sum_{i=1}^{n} x_i = \frac{1}{n} \sum_{i=1}^{n} M(x_i) = \frac{n}{n} \bar{M}(x) = \bar{M}(x). \tag{15}$$

A set of indicators is suggested to be used in this example for experimental calculations.

Given the analogy in the location of gas stations, their fixed and variable costs per month, the intensity on the street-road network of the city in places where it is planned to place charging stations are taken into account.

The speed parameters of the traffic flow will be also considered in the calculations, since this will be significant in the case of several charging stations and in the condition of ensuring competition.

The technological charging time, as well as the total residence time of the vehicle (preliminary estimates), are determined on the basis of the technical documentation (Table 1).

Table 1. Indicators and their numerical values for 2018 to test this methodology and use it as an initial stage of the estimate.

Month of the year	Fixed costs, thousand rubles	Variable costs, thousand rubles	Intensity of the traffic flow, unit/month	Number of stations	Charging time, hours	Time spent at the charging station, hours	Speed of the traffic flow, km/h
January	6712.7	4521	10773	6	1.2	3.3	40
February	15000.64	4823.86	12661	6	1.3	4.6	43
March	8846.29	4646	11775	6	1.4	3.5	46
April	12487	3712.95	12344	6	0.95	5.3	34
May	14014.54	2714.69	10834	6	0.9	3.5	35
June	13976	2987.7	14044	6	1	2.7	45
July	14838.52	1209.84	15892	6	1.3	4	56
August	18671.8	2330.26	14125	6	1.25	5.5	57
September	19198.18	3792.06	14600	6	1.05	5.4	60
October	19110.26	5406	11177	6	1.1	5.9	50
November	18418.45	5276.24	10564	6	1.35	5.3	45
December	32046.83	7990.08	10124	6	1.4	4.2	40
Target values	32046.83	7990.08	15892	10	Basics of indicators	2.7	40

According to the given sequence of calculations, let's determine the mismatch indicators and their normalized values (Tables 2 and 3).

Table 2. Mismatch matrix.

January	25334.13	3469.08	5119	4	1.4	5.9	60
February	17046.19	3166.22	3231	3	0.2	2.6	20
March	23200.54	3344.08	4117	3	0.1	1.3	17
April	19559.83	4277.13	3548	4	0	2.4	14
May	18032.29	5275.39	5058	4	0.45	0.6	26
June	18070.83	5002.38	1848	4	0.5	2.4	25
July	17208.31	6780.24	0	4	0.4	3.2	15
August	13375.03	5659.82	1767	4	0.1	1.9	4
September	12848.65	4198.02	1292	4	0.15	0.4	3
October	12936.57	2584.08	4715	4	0.35	0.5	10
November	13628.38	2713.84	5328	4	0.3	0.6	15
December	0	0	5768	4	0.05	1.7	20
Average value	15936.72917	3872.52333	3482.583	3,833333	0.236364	1.981818	15.36364
Mean square deviation	6398.876575	1760.77113	1866.439	0,389249	0.373761	0.98387	7.513624

Table 3. Normalized mismatches.

January	1.468601671	0.22912878	0.876759	0.428174	0.097291	0.628317	0.617061
February	0.17338369	0.40113296	0.134793	2.140872	0.364842	0.692996	0.217786
March	1.135169705	0.3001204	0.339908	2.140872	0.632393	0.425038	0.181489
April	0.566208895	0.22978947	0.035049	0.428174	0.571586	1.404472	1.41561
May	0.327488866	0.79673425	0.844076	0.428174	0.705362	0.425038	1.282519
June	0.333511798	0.64168287	0.875777	0.428174	0.437811	1.238153	0.048397
July	0.198719387	1.6513882	1.865897	0.428174	0.364842	0.08316	1.512404
August	0.400335768	1.01506473	0.919175	0.428174	0.231067	1.607751	1.645496
September	0.482597083	0.18486029	1.17367	0.428174	0.304035	1.506112	0.713855
October	0.468857171	0.73174947	0.660304	0.428174	0.17026	1.404472	0.048397
November	0.36074288	0.65805448	0.988737	0.428174	0.498618	0.286438	0.617061
December	2.490551112	2.19933373	1.22448	0.428174	0.632393	2.014309	2.04477

The probability of a mismatch is 0.0833333.

Thus, if the statistical expectation is taken to estimate the mathematical expectation, then the requirement of unbiasedness will be fulfilled [9].

3 Results

The development of mathematical models of the adaptive properties of charging stations in the electric vehicles power supply network [9] is carried out on the basis of methods for the formation of vectors of composition, structure and properties, target vectors, and also quasi-ordered zones.

Power networks of regions, cities, towns can be treated as elements, depending on the purpose of the analysis.

The parameters of vectors of the state and targets will be considered as a situation of disorder with respect to the composition, structure, and properties of the system.

Generalization of time intervals means that disorganization is generalized over a period of time, divided into intervals (decade, month, year, etc.).

In accordance with these considerations, in order to assess the functioning disorganization of the network of charging stations, the models of the target entropy H_t and maximum H_{max} will be the following (16–17):

$$Ht = \sum_{i=1}^{I}\sum_{j=1}^{J}\sum_{k=1}^{K} P_{ijk} \times ln((q_{ijk} - \xi(X_{ijk})w_{ijk} + 1), \tag{16}$$

$$H_{\max} = \ln\left\{(\sum_{i=1}^{I}\sum_{j=1}^{J}\sum_{k=1}^{K} \max((Q_{ijk} - E(x_{ijk}))) \times \omega_{ijk} + 1\right\}, \tag{17}$$

where I, J, K - the number of elements, target parameters and time intervals correspondingly over which the entropy of the system is generalized;
P_{ijk} - probability of a mismatch;
q_{ijk} - module of the mismatch vector;
$\varepsilon(x_{ijk})$ - boundary of the quasi-ordered zone;
w_{ijk} - logic variable.

The purpose of the logic variable w_{ijk} is to reset the value of the disorder parameter, and hence the value of entropy, if $q_{ijk} - \varepsilon(x_{ijk}) < 0$.

Therefore, the following mathematical notation is valid (18):

$$w_{ijk} = \{1, \ \ if \ \text{qijk} - \varepsilon \geq 0, \tag{18}$$

$$\{0, \ \ if \ \text{qijk} - \varepsilon < 0,$$

Calculations of the target entropy in the system of the distribution of electric charging stations are given in Table 4.

Table 4. Target entropy.

January	0.075304322	0.01719213	0.052462	0.0297	0.007737	0.040629	0.040051
February	0.013324301	0.02810676	0.010538	0.095375	0.02592	0.043875	0.01642
March	0.063212178	0.02187141	0.024383	0.095375	0.040837	0.029517	0.013898
April	0.037388165	0.01723692	0.002871	0.0297	0.037674	0.073111	0.073496
May	0.023607424	0.04883089	0.050998	0.0297	0.044481	0.029517	0.068773
June	0.023984659	0.04131015	0.052419	0.0297	0.03026	0.067138	0.003939
July	0.015104484	0.08125695	0.08774	0.0297	0.02592	0.006657	0.07677
August	0.028059337	0.05838761	0.054325	0.0297	0.017323	0.079874	0.081072
September	0.032816278	0.01413541	0.064701	0.0297	0.022122	0.076561	0.044895
October	0.032040389	0.04576101	0.04225	0.0297	0.013102	0.073111	0.003939
November	0.025669232	0.04213708	0.057292	0.0297	0.033712	0.02099	0.040051
December	0.104171636	0.09691188	0.066627	0.0297	0.040837	0.091948	0.092785

The organization of functioning in terms of composition (Ocomp), structure (Ost) and properties (Oprop) will be the following (19–21):

$$Ocomp = 1 - \frac{\sum_{i=1}^{I}\sum_{n=1}^{N}\sum_{k=1}^{K} P_{ink} \times \ln((q_{ink} - \varepsilon(X_{ink}))\omega_{ink} + 1)}{\ln\{(\sum_{i=1}^{I}\sum_{n=1}^{N} \max_k((Q_{ink} - E(X_{ink}))\omega_{ink} + 1)\}}, \tag{19}$$

where N is for parameters of the composition vector.

$$O_{st} = 1 - \frac{\sum_{i=1}^{I}\sum_{M=1}^{M}\sum_{k=1}^{K} P_{imk} \times \ln((q_{imk} - \varepsilon(X_{imk}))\omega_{imk} + 1)}{\ln\{(\sum_{i=1}^{I}\sum_{m=1}^{M} \max_k((Q_{imk} - E(X_{imk}))\omega_{imk} + 1)\}}, \tag{20}$$

where M is for parameters of the structure vector.

$$O_{prop} = 1 - \frac{\sum_{i=1}^{I}\sum_{\bar{k}=1}^{\bar{K}}\sum_{k=1}^{K} P_{i\bar{k}k} \times \ln((q_{i\bar{k}k} - \varepsilon(X_{i\bar{k}k}))\omega_{i\bar{k}k} + 1)}{\ln\{(\sum_{i=1}^{I}\sum_{\bar{k}=1}^{K} \max_k((Q_{i\bar{k}k} - E(X_{i\bar{k}k}))\omega_{i\bar{k}k} + 1)\}}, \tag{21}$$

where $\overline{K}$ is for parameters of the property vectors.

With (3.31), (3.32), (3.33) and taking into account Figure 2.14, the vector C is determined (22):

$$\bar{C} = \sqrt{B_{comp}^2 + B_{st}^2 + B_{prop}^2}. \tag{22}$$

It should be noted that, depending on the choice of the method for substantiating the reference or target values of the parameters of the target vectors, the interpretation of these estimates will have its own characteristics.

The calculation results are presented in Table 5.

Table 5. Results of the calculations.

Target general entropy							
0.513138199	0.54106579	0.487748	0.339926	0.632926	0.556088	0.554213	0.474682
Maximum entropy							
1.250059635	1.16294258	1.052882	1.1445	0.533777	1.103371	1.113426	1.096437
Organization of the system by indicators							
0.620272192	0.55875878	0.461853	0.573834	0.363168	0.426371	0.500561	0.494532

Target entropy general on all indicators – 5.24247.
Maximum entropy general on all indicators – 10.74818.
Organization general on all indicators – 0.51224.

Application of the above-mentioned method, using the structural dependence between random variables, allows to separate the initial set of factors characterizing the resource system of transport and its elements into three subsets.

Such a classification is conducted on the basis of the nature of the information which is drawn from the indicators as composition, structure and property. The set consists of the indicators that define each of the three components of C.

Of course, the set of indicators that is used to calculate the level of adaptation of, for example, the traffic flow and the charging station, is not finite and can be supplemented with other values or vice versa.

4 Discussions

A feature of the adaptation of innovative transport (innovation stands by electric vehicles (cars) in this case) is that it will operate all day, while buses will not have to go to the depot to recharge - they will be charged wirelessly through induction coils built in the road surface.

Milton Keynes is a city where a similar system is used. This is yet the only city in the UK that has implemented wireless charging via induction coils. This technology has already been successfully tested in Italy, South Korea, Germany and the Netherlands.

3 coils are used to recharge the buses, 2 of which are installed at the end points of the route, and the third one is in the middle. Transport can make up for two-thirds of the energy spent on the full route with a 10-min stop.

This project will cut the amount of carbon dioxide emissions in the city by 500 tons. This, in turn, will reduce air pollution. According to open sources, the city council of Milton Keynes plans to transfer all city routes to electricity in the coming years and, in addition, to establish 50 fast charging stations that can be used by electric taxis. This option requires a prepared network infrastructure for power supply and electric grids, as well as the availability of a well-developed and adapted system of tenders for servicing urban passenger flows. When forming the network and taking into account the mileage of the rolling stock, in our opinion, it is advisable to take into account the amount of energy received from the recovery under different operating conditions. In the authors views, the values of the regenerated energy under various operating conditions should be taken into account when designing the network and considering the travel distance of a rolling stock.

In order to quantitatively estimate the theoretically possible return of the energy to the source, the data on the distance of the hauls will be used. The minimum haul is 200–250 m, and the maximum is 800 or more, when the optimal distance for metropolises is 500–600 m. Traction and energy calculations show that, for example, for a single-body trolley with a DK-210AZ traction motor, the speed does not exceed 60 km/h for hauls over 250 m during start-up. The velocity of the initiation of braking depends on the haul distance insignificantly and can be taken equal to 30 km/h. By the moment the rolling stock begins to brake, it retains only 25% of the kinetic energy, which has been accumulated during coasting, if the vehicle is hauled in usual mode and for optimal distance. Part of this energy (up to 10%) is spent on overcoming resistance to motion during coasting and braking to a complete stop. Another part of the energy (up to 3%) goes to mechanical brakes when braking at a speed of 5–7 km/h. Thus, in the classic traffic pattern, it is theoretically possible to recuperate up to 20% (with account of the efficiency of the pulse regulator and motor - up to 17%) of the kinetic energy which has been accumulated during acceleration of the vehicle to 60 km/h.

5 Conclusion

The contemporary practice of organizing transport systems includes various aspects in the planning and management of urban space, including trends in changes in the urban environment, but it is also accompanied by a whole range of contradictions, including territorial, economic, environmental, social and other problems [9]. The focus on the enlargement of cities and the accumulation of cultural and social life in them breed the problems of mobility of citizens. Arising transport issues become vital components and come into the picture of work organization in almost any city. A model of free, quick and convenient movement is a necessary form of interaction between the municipal government and each resident of the city, what is the bedrock of the modern development of a city as "the environment with open skies and no borders". Such a point of view should have as its basis a network of power generating plants adapted to the existing traffic network and the correspondence matrix of city residents. Logistics of energy flows and aspects of its manifestation are associated with cost optimization in the process of functioning of charging stations by adaptive distribution on the urban road network. The layout can be determined by calculating the entropy of the system, as the resulting value of the adaptation of the network of charging stations to the functioning traffic flow on the road network of the city. As it is known, the logistical aspects become seen in the integral consideration of the problem of effective resource supply and conservation. It is the integration of elements that will make it possible to fully utilize the potential of logistics in solving the above-mentioned problem.

The main objective of functioning of logistics networks and their power supply is the completeness, quality and promptness of services for the emerging flows. It is required to ensure equal access to quality service for all participants, regardless of their status in the system under study [9].

The discrepancy between the expected level of service quality and the actual one should be minimized, and in the ideal case, eliminated. Therefore, it is necessary to monitor the service indicators in order to ensure the smooth work of the charging stations.

References

1. Sizova, E., Zhutaeva, E., Gorshkov, R., Smirnov, V., Kochetkova, E.: Methodical bases for forming the structure of management of innovative activity of large building holdings. In: MATEC Web of Conferences (2018). https://doi.org/10.1051/matecconf/201817001126
2. Golov, R., Narezhnaya, T., Voytolovskiy, N., Mylnik, V., Zubeeva, E.: Model management of innovative development of industrial enterprises. In: MATEC Web of Conferences (2018). https://doi.org/10.1051/matecconf/201819305080
3. Lukmanova, I., Golov, R.: Modern energy efficient technologies of high-rise construction. In: E3S Web of Conferences (2018). https://doi.org/10.1051/e3sconf/20183302047
4. Nezhnikova, E.: The use of underground city space for the construction of civil residential buildings. Proc. Eng. **165**, 1300–1304 (2016). https://doi.org/10.1016/j.proeng.2016.11.854
5. Safronova, N., Nezhnikova, E., Kolhidov, A.: Sustainable housing development in conditions of changing living environment. In: MATEC Web of Conferences, vol. 106, p. 08024 (2017). https://doi.org/10.1051/matecconf/201710608024
6. Marques, A.C., Fuinhas, J.A., Pires Manso, J.R.: Motivations driving renewable energy in European countries: a panel data approach. Energy Policy **38**(11), 6877–6885 (2010). https://doi.org/10.1016/j.enpol.2010.07.003
7. Gusev, S., Makarov, E., Vasiliev, D., Marosin, V.: Smart management and power consumption forecasting in passenger transport. In: Advances in Intelligent Systems and Computing. Springer, Cham (2019). https://doi.org/10.1007/978-3-030-19868-8_9

Improving the Experimental Technique of Asynchronous Single-Phase Motors Equivalent Circuits Research

Dmitry Tonn(✉), Sergey Goremykin, Nikolay Sitnikov, Alexander Mukonin, and Alexander Pisarevsky

Voronezh State Technical University, Moscovsky Prospect, 14, 394026 Voronezh, Russia
tonnda@yandex.ru

Abstract. The paper considers an improved experimental technique for determining the parameters of low-power single-phase and capacitor asynchronous motors equivalent circuits. The proposed technique differs from the classical one because it takes into account the branch of the equivalent circuit containing the active and inductive scattering resistance of the rotor winding in various operating modes. A system of nonlinear algebraic equations has been obtained based on the single-phase asynchronous motor equivalent circuits in ideal idling speed and short circuit modes. It has been shown that the resulting system can be reduced to a system of two equations. It is proposed to solve the indicated systems of equations by applying various iterative methods implemented in universal mathematical software of symbolic mathematics. The proposed technique makes it possible to determine such single-phase asynchronous motor replacement circuit parameters as the active and inductive scattering resistance of the rotor winding, and the inductive resistance of mutual induction more precisely in comparison with the classical one. The way of application of the obtained improved technique for determining the capacitor asynchronous motors parameters has been given. More precise experimentally determined values of parameters allow to increase the calculation accuracy of single-phase and capacitor asynchronous low power motors transients.

Keywords: Experimental technique · Single-phase asynchronous motors · Capacitor asynchronous motors · Equivalent circuit · Parameters · Transients

1 Introduction

Single-phase AC motors that convert electrical energy into mechanical one are widely used in automation systems, small industrial enterprises, agriculture, and everyday life. Such motors have specific properties.

Nowadays low-power asynchronous single-phase motors (ASM) which are commercially produced, for example, for household appliance, are the most widely used, [1]. Such production volume is characterized by high material costs and a significant energy resources expenditure. In recent years, the electricity consumption by various household appliances has become equal to the needs of industrial production, and the

V. Murgul and V. Pukhkal (Eds.): EMMFT 2019, AISC 1258, pp. 24–34, 2021.
https://doi.org/10.1007/978-3-030-57450-5_3

consumption of structural, active, and insulating materials has reached the volumes of both turbo and hydro generators production.

Therefore, it is necessary to pay close attention to the developments and research focused on optimization and cost reduction in production, and most importantly, when operating such electric machines in various operating modes.

There is a wide variety of low-power ASM designs. ASM, in which the auxiliary winding with a phase-shifting capacitor remains on in the operating mode, are called asynchronous capacitor motors (ACM) [2]. Such engines have good performance, but are usually used in electric drives with not heavy starting conditions, since they produce a low moment when turned on.

ACM are the most common of all ASM since such electric machines have a number of outstanding characteristics. They have a power factor close to unity and smaller dimensions than other ASM, so the material utilization ratio in ACM is much higher. ACM are reliable and easy to use. Such engines reach high net power on the shaft and have the maximum efficiency of all ASM types. The idle speed of the ACM is characterized by a large current value in the auxiliary phase of the winding. The considered ACM have a distributed stator winding, which makes it possible to obtain a rotating magnetic field, and a short-circuited rotor in the form of a "squirrel cage".

Dynamic and transient processes are real modes of ASM and ACM functioning. Therefore, even the stationary or static operation mode of such electric machines seems to be a quasi-steady non-stationary mode. In this case, the rotor speed fluctuates, that is, it rotates extremely unevenly. It is difficult to calculate and study transients in the ASM, and in the ACM in particular, in comparison with the established modes of their operation. For a clear understanding and analysis of the entire spectrum of physical phenomena occurring during the operation of this class of electrical machines, it is necessary to study the transient processes that arise in them [3].

To study the electromechanical transients in ASM and ACM, it is desirable to use a system of differential equations, which is a mathematical model of such machines. Now it is possible to solve such systems of differential equations with accuracy acceptable for engineering problems by various numerical methods using computer facilities [4].

Scientific research carried out in the field of transient studies in asynchronous electric machines at the present stage pays close attention to various issues of mathematical modeling and taking into account the whole spectrum of physical phenomena and processes that arise during the ASM and ACM operation in various modes [5]. At the same time, new types of ASM and ACM are being developed and introduced into production [6].

The described mathematical model allows one to calculate both static (quasi-steady) and dynamic (transient) modes of ASM and ACM operation, as well as to determine their characteristics. The coefficients of the variables in the model are the parameters of the ACM equivalent circuit, such as the stator windings active resistance, the stator windings inductive resistance, the inductive resistances of mutual induction, the active and inductive resistances of the rotor winding. The multiphase short-circuited rotor winding in this model is converted to the equivalent two-phase one.

The calculation accuracy largely depends on the ACD equivalent circuit parameters. Based on this, accurate determination of the ACD parameters is a prerequisite for the study and calculation of transients in such engines [7]. A number of publications by

Russian and foreign scientists are devoted to the exact determination of the equivalent circuits parameters in asynchronous motors [7].

In the well-known classical technique for the experimental determination of the ASM and ACM parameters some assumptions have been made. If these assumptions are remoted, it is obvious that the technique will becomes more accurate, and, consequently, the calculation accuracy of transients in such electric motors increases.

2 Experimental

First, let's pay attention to the ASM which is obtained from a two-phase symmetrical motor when one phase of the winding is broken. In all operating modes, the equivalent circuit of such an electric machine has the form shown in Fig. 1 [8].

In this diagram: r_1 is the resistance of stator main phase winding; $x_{\sigma 1}$ is the resistance of the stator main phase windings scattering; r_2' is the active resistance of the equivalent rotor winding phase; $x_{\sigma 2}'$ is the inductive reactance of the scattering equivalent rotor winding phase; x_m is the inductive mutual induction resistance; S is the slide.

In this equivalent circuit, as in the mathematical model, we neglect the resulting mechanical losses, as well as steel losses. The equivalent circuit shown in Fig. 1 is applicable for all slip values.

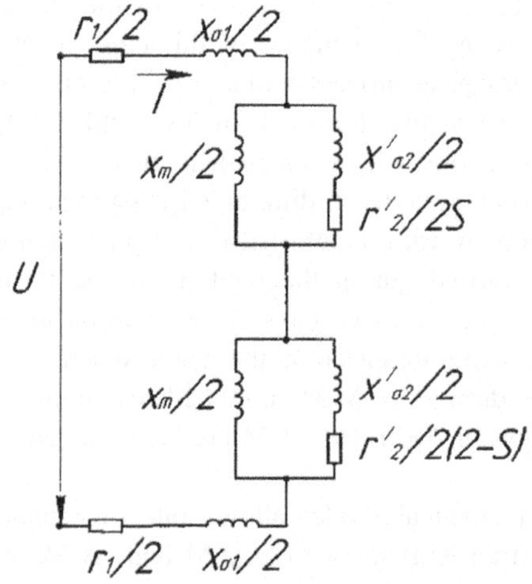

Fig. 1. The ASM equivalent circuit.

At the time of start-up (i.e., during a short circuit), when $S = 1$, the equivalent circuit will take the form shown in Fig. 2.

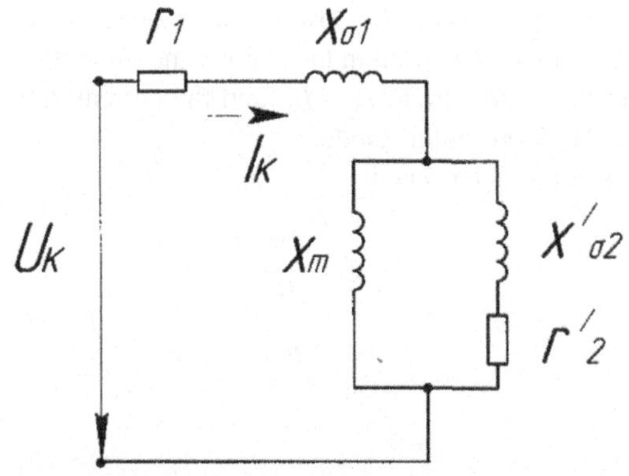

Fig. 2. The ASM equivalent circuit in short circuit mode.

The resistances r_1 and $x_{\sigma 1}$ are determined by the removed rotor method. In this case, the electric motor is disassembled, and the rotor is removed from the stator bore. A voltage lower than the rated voltage is supplied to the stator phase winding, such that the current in the stator winding is less than or equal to the rated current $I_1 \leq I_H$, and the voltage U_1, current I_1, and stator winding power P_1 are measured in the removed rotor mode.

The following quantities are determined:

$$Z_1 = \frac{U_1}{I_1}, \tag{1}$$

$$r_1 = \frac{P_1}{I_1^2}, \tag{2}$$

$$x_{\sigma 1} = \sqrt{Z_1^2 - r_1^2}, \tag{3}$$

where Z_1 is the total resistance of the phase (main phase) of the stator winding in the removed rotor mode.

Also, the active resistances of the stator windings can be measured using a DC bridge.

The expressions required to determine the resistances of the equivalent rotor winding r_2' and $x_{\sigma 2}'$ can be obtained from the short-circuit experiment results. In the existing technique it is assumed that $x_m \gg \left|(r_2' + jx_{\sigma 2}')\right|$ (see Fig. 2), where $j = \sqrt{-1}$ is the imaginary unit. However, the values x_m and $\left|(r_2' + jx_{\sigma 2}')\right|$ have comparable values, although x_m is more than the complex resistance module $\left|(r_2' + jx_{\sigma 2}')\right|$. We do not make this assumption, which is the basis of the classical technique for determining the ASM and ACM parameters, but we take into account the branch of the equivalent circuit containing the resistances r_2' and $jx_{\sigma 2}'$.

A short circuit test is carried out with a fixed (braked) rotor, when slip $S = 1$. Engine braking is carried out by fixing the output end of the motor shaft in a bench vise. In this experiment, the voltage U_K lower than the rated voltage is applied to the phase stator winding so that the current in the stator winding in the short circuit mode is less than or equal to the rated current $I_K \leq I_H$, and the current value I_K and the power P_K are measured in the short circuit mode.

Then the quantities are determined:

$$Z_K = \frac{U_K}{I_K}, \quad (4)$$

$$r_K = \frac{P_K}{I_K^2}, \quad (5)$$

$$x_K = \sqrt{Z_K^2 - r_K^2}, \quad (6)$$

where Z_K is the total resistance in short-circuit mode;
r_K is the active resistance in short-circuit mode;
x_K is the inductive resistance in short-circuit mode.

From the equivalent circuit shown in Fig. 2, we have

$$Z_K = r_1 + jx_{\sigma 1} + \frac{jx_m\left(r_2' + jx_{\sigma 2}'\right)}{r_2' + j\left(x_m + x_{\sigma 2}'\right)}. \quad (7)$$

On the other hand, according to the short circuit test results it is known that

$$Z_K = r_K + jx_K. \quad (8)$$

Having selected the real and the imaginary parts of expression (7) by elementary transformations, we obtain the following expressions:

$$r_K = \frac{r_1\left[\left(r_2'\right)^2 + \left(x_m + x_{\sigma 2}'\right)^2\right] + x_m^2 r_2'}{\left[\left(r_2'\right)^2 + \left(x_m + x_{\sigma 2}'\right)^2\right]}, \quad (9)$$

$$x_K = x_{\sigma 1} + \frac{x_m\left[\left(r_2'\right)^2 - x_{\sigma 2}'\left(x_m + x_{\sigma 2}'\right)\right]}{\left[\left(r_2'\right)^2 + \left(x_m + x_{\sigma 2}'\right)^2\right]}. \quad (10)$$

In these expressions, the resistances r_2', $x_{\sigma 2}'$, and x_m are the unknown value. At the same time, according to the results of removed rotor and short circuit experiments the values of the resistances r_1, $x_{\sigma 1}$ и r_K, x_K are known respectively. The specified resistance values are known in absolute units of measurement, which can be converted to relative units if necessary in the system of basic values used in the calculations.

Note. During the short circuit test, it is necessary to understand whether the instrument readings depend on the spatial position of the rotor. In that case, if such a dependence exists, then we find $Z_{K\min}$ and $Z_{K\max}$, and therefore $r_{K\min}$, $r_{K\max}$, $x_{K\min}$, and $x_{K\max}$. For the calculated values we take

$$r_K = \frac{r_{K\min} + r_{K\max}}{2}, \tag{11}$$

$$x_K = \frac{x_{K\min} + x_{K\max}}{2}. \tag{12}$$

The expressions necessary to determine the resistance x_m can be obtained from the results of the idle test. We start the engine without load on the shaft, then turn off the starting (auxiliary) winding. Then the engine will continue to operate in single-phase mode. Knowing that the losses in the idle mode are not large, then we consider the slip to be equal to zero $S = 0$, i.e. we consider the ideal idle mode. In this case the equivalent circuit of the engine under consideration will take the form shown in Fig. 3 [8].

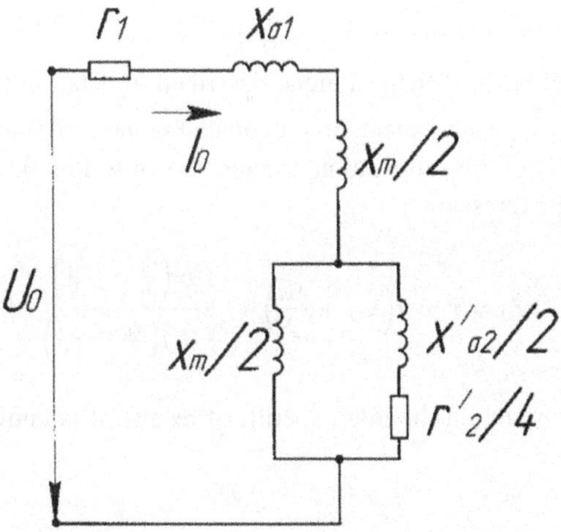

Fig. 3. The ASM equivalent circuit in the idle mode.

In this experiment, a voltage U_0 equal to the nominal voltage ($U_0 = U_H$) is applied to the stator phase winding. With this voltage, we measure the idle current I_0 and the idle power P_0.

Next, we find the following quantities:

$$Z_0 = \frac{U_0}{I_0}, \tag{13}$$

$$r_0 = \frac{P_0}{I_0^2}, \tag{14}$$

$$x_0 = \sqrt{Z_0^2 - r_0^2}. \tag{15}$$

where Z_0 is the total resistance in the idle mode;
r_0 is the active resistance in the idle mode;
x_0 is the inductive resistance in the idle mode.

In the existing classical technique for the experimental determination of parameters in this case, it is taken into account that the following inequality is valid $\frac{x_m}{2} \gg \left|\left(\frac{r_2'}{4} + \frac{jx_{\sigma 2}'}{2}\right)\right|$ (see Fig. 3), i.e., the branch of the equivalent circuit containing resistances $\frac{r_2'}{4}$ and $\frac{jx_{\sigma 2}'}{2}$ is neglected, and this makes it possible to simplify the determination of parameters, while reducing the study accuracy. However, the magnitudes $\frac{x_m}{2}$ and modulus of the complex resistance $\left|\left(\frac{r_2'}{4} + \frac{jx_{\sigma 2}'}{2}\right)\right|$ are comparable, although the magnitude $\frac{x_m}{2}$ is greater than the modulus of the complex resistance $\left|\left(\frac{r_2'}{4} + \frac{jx_{\sigma 2}'}{2}\right)\right|$. We do not make such an assumption, which is used as the basis of the technique described in detail in [2], but we take into account the branch of the equivalent circuit containing the resistances $\frac{r_2'}{4}$ and $\frac{jx_{\sigma 2}'}{2}$.

Next, an analysis of the substitution scheme shown in Fig. 3, makes it possible to write the following expression

$$Z_0 = r_1 + j\left(x_{\sigma 1} + \frac{x_m}{2}\right) + \frac{j\frac{x_m}{2}\left(\frac{r_2'}{4} + j\frac{x_{\sigma 2}'}{2}\right)}{\frac{r_2'}{2} + j\left(\frac{x_m}{2} + \frac{x_{\sigma 2}'}{2}\right)}. \tag{16}$$

On the other hand, based on the idle experiment results, it is known that

$$Z_0 = r_0 + jx_0. \tag{17}$$

Having selected the real and imaginary parts of the complex expression (14) by simple transformations, we obtain the following expressions:

$$r_0 = r_1 + \frac{x_m^2 r_2'}{\left[(r_2')^2 + 4\left(x_m + x_{\sigma 2}'\right)^2\right]}, \tag{18}$$

$$x_0 = x_{\sigma 1} + \frac{x_m}{2} + \frac{2x_m\left[(r_2')^2 - x_{\sigma 2}'\left(x_m + x_{\sigma 2}'\right)\right]}{\left[(r_2')^2 + 4\left(x_m + x_{\sigma 2}'\right)^2\right]}. \tag{19}$$

In these expressions, as in the previous experiment, the resistances r_2', $x_{\sigma 2}'$, and x_m are unknown values. At the same time, according to the results of the removed rotor and idling experiments, the values of the resistances r_1, $x_{\sigma 1}$ and r_0, x_0 are known respectively. The specified resistance values, as in the previous case, are known in absolute units of measurement, which, if necessary, can be converted to relative units in the system of basic values adopted in the calculations.

3 Evaluation

Expressions (9), (10), (16), (17) represent a system of four nonlinear algebraic equations, where the parameters of the ASM equivalent circuit r'_2, $x'_{\sigma 2}$, x_m act as unknown quantities. One of the necessary and sufficient conditions for solving systems of equations is the correspondence of the number of unknown quantities to the number of equations of the system. Therefore, to determine the unknown parameters of the ASM equivalent circuit, it suffices to use any three equations from the expressions (9), (10), (18), (19). Since the expression (17) is the most cumbersome it is proposed to solve a system that includes expressions (9), (10), and (18):

$$\begin{cases} r_K = \dfrac{r_1\left[\left(r'_2\right)^2 + \left(x_m + x'_{\sigma 2}\right)^2\right] + x_m^2 r'_2}{\left[\left(r'_2\right)^2 + \left(x_m + x'_{\sigma 2}\right)^2\right]}, \\ x_K = x_{\sigma 1} + \dfrac{x_m\left[\left(r'_2\right)^2 - x'_{\sigma 2}\left(x_m + x'_{\sigma 2}\right)\right]}{\left[\left(r'_2\right)^2 + \left(x_m + x'_{\sigma 2}\right)^2\right]}, \\ r_0 = r_1 + \dfrac{x_m^2 r'_2}{\left[\left(r'_2\right)^2 + 4\left(x_m + x'_{\sigma 2}\right)^2\right]}. \end{cases} \tag{20}$$

To minimize the amount of computation and reduce the complexity of calculating parameters, this system of three nonlinear algebraic equations can be reduced to a system of two equations by expressing a quantity r'_2 from the second equation (expression (10)). Then we have:

$$r'_2 = \sqrt{\frac{\left(x_m + x'_{\sigma 2}\right)\left(\left(x_1 - x_K\right)\left(x_m + x'_{\sigma 2}\right) - x_m x'_{\sigma 2}\right)\right)}{x_m + x_1 - x_K}}. \tag{21}$$

Considering the above and the remaining equations of the system (20), we obtain the following system of two equations with respect to the parameters x_m and $x'_{\sigma 2}$:

$$\begin{cases} r_K = \dfrac{r_1\left[\left(r'_2\right)^2 + \left(x_m + x'_{\sigma 2}\right)^2\right] + x_m^2 r'_2}{\left[\left(r'_2\right)^2 + \left(x_m + x'_{\sigma 2}\right)^2\right]}, \\ r_0 = r_1 + \dfrac{x_m^2 r'_2}{\left[\left(r'_2\right)^2 + 4\left(x_m + x'_{\sigma 2}\right)^2\right]}, \end{cases} \tag{22}$$

where r'_2 is determined by the expression (21).

Solving the system of equations (20) for unknown quantities r'_2, $x'_{\sigma 2}$, x_m or the system of equations (22) for unknown quantities $x'_{\sigma 2}$ and x_m taking into account the expression (21), we determine the desired parameters of the ASM equivalent circuit.

It is possible to solve the above system of equations for determining unknown quantities r_2', $x_{\sigma 2}'$, x_m both in absolute units of measurement and in relative units in the system of basic quantities adopted in the calculations, depending on in which quantities the data of experimental studies are taken.

Direct and iterative methods are used to solve the systems of equations. Direct methods are far from always applicable for solving practical problems. To solve the indicated systems of equations it is proposed to use iterative methods, which make it possible to obtain a solution to the system by performing successive approximations.

To perform the calculations, it is needed to specify a certain approximate solution - the initial approximation, and then the first cycle of calculations, called iteration, is carried out using the selected algorithm. To determine the approximate initial values, the sought-after parameters of the equivalent circuits of ASM and ACM, you can use the data of experimental studies performed by the classical technique, or take them from the experience of designing and calculating such electrical machines, and you can also use reference materials.

The following approximation is found as a result of a series of iterations. Iterations are performed until a solution with the necessary accuracy is obtained. Thus, when applying iterative methods, the error in calculating the final results does not accumulate since the accuracy of the calculation depends only on the result of the previous iteration.

To solve these systems of equations, it is proposed to use the Levenberg-Marquardt, Newton, Seidel methods, the method of simple iterations, and any other iterative methods implemented in mathematical universal software of symbolic mathematics, such as Mathcad, MATLAB, Maple, Mathematica, Derive, and some others.

There are two phase windings on the ACM stator: main (working) and auxiliary (starting). The active resistances and inductive scattering resistances of both phases of the stator are determined by the removed rotor method. A short circuit test is carried out for each of the phase windings. Its results allow us to write down the expressions necessary to determine r_2' and $x_{\sigma 2}'$ which are the resistances of the equivalent rotor winding. Based on the results of the idle mode test, the expressions necessary to determine the resistance x_m can be obtained. The resistance of mutual induction x_m is determined only for the main winding. Thus, it is necessary to calculate the system of equations (20) or the system (22) taking into account the expression (21), for each of the phase windings separately, as a result of which the parameter values r_{2a}', r_{2B}', x_{2a}', x_{2B}' are obtained.

The ACM is considered as a reduced motor, when the auxiliary winding and the equivalent rotor winding are brought to the main stator winding. Thus, the experimental studies described above can also determine the coefficient of reduction to the main winding. It will be equal to:

$$k^2 = \frac{x_{2a}'}{x_{2B}'}, \tag{23}$$

$$k = \sqrt{\frac{x_{2a}'}{x_{2B}'}}, \tag{24}$$

or

$$k^2 = \frac{r'_{2a}}{r'_{2B}}, \tag{25}$$

$$k = \sqrt{\frac{r'_{2a}}{r'_{2B}}}. \tag{26}$$

4 Conclusions

The following conclusions can be drawn from the arguments and expressions given above.

1. As a result of the studies, it is possible to obtain new systems of equations, the solution of which allows us to determine the parameters of the ASM equivalent circuits r'_2, $x'_{\sigma 2}$, x_m with higher accuracy compared to the classical technique for the experimental ASM and ACM parameters determination.
2. The study made it possible to obtain new systems of equations, the solution of which allows one to determine the parameters of the ACM equivalent circuits r'_{2a}, r'_{2B}, x'_{2a}, x'_{2B}, x_m with higher accuracy compared to the classical technique for the experimental ASM and ACM parameters determination.
3. More accurate values of the parameters determined experimentally allow one to increase the calculation accuracy of the low power ASM and ACM transients, and hence the degree of correspondence of the mathematical model to a real engine.

References

1. Pan, Z.H., Zhou, J.L., Jiang, X.: Investigating the effects of steel slag powder on the properties of self-compacting concrete with recycled aggregates. Constr. Build. Mater. **200**, 570–577 (2019). https://doi.org/10.1016/j.conbuildmat.2018.12.150
2. Nedeljkovic, M., Ghiassi, B., Laan, S., Li, Z.M., Ye, G.: Effect of curing conditions on the pore solution and carbonation resistance of alkali-activated fly ash and slag pastes. Cem. Concr. Res. **116**, 146–158 (2019). https://doi.org/10.1016/j.cemconres.2018.11.011
3. Yan, X.C., Jiang, L.H., Guo, M.Z., Chen, Y.J., Song, Z.J., Bian, R.: Evaluation of sulfate resistance of slag contained concrete under steam curing. Constr. Build. Mater. **195**, 231–237 (2019). https://doi.org/10.1016/j.conbuildmat.2018.11.073
4. Pu, L., Unluer, C.: Durability of carbonated MgO concrete containing fly ash and ground granulated blast-furnace slag. Constr. Build. Mater. **192**, 403–415 (2018). https://doi.org/10.1016/j.conbuildmat.2018.10.121
5. Farhan, N.A., Sheikh, M.N., Hadi, M.N.S.: Investigation of engineering properties of normal and high strength fly ash based geopolymer and alkali-activated slag concrete compared to ordinary Portland cement concrete. Constr. Build. Mater. **196**, 26–42 (2019). https://doi.org/10.1016/j.conbuildmat.2018.11.083

6. Samchenko, S., Kozlova, I., Zemskova, O.: Model and mechanism of carbon nanotube stabilization with plasticizer. In: MATEC Web of Conferences, vol. 193, p. 03050 (2018). https://doi.org/10.1051/matecconf/201819303050
7. Smirnov, V., Dashkov, L., Gorshkov, R., Burova, O., Romanova, A.: Methodical approaches to value assessment and determination of the capitalization level of high-rise construction. In: E3S Web of Conferences (2018). https://doi.org/10.1051/e3sconf/20183303030
8. Artyushina, G.G., Sheypak, O.A., Golov, R.S.: Podcasting as a good way to learn second language in e-learning. In: ACM International Conference Proceeding Series (2017). https://doi.org/10.1145/3026480.3029590

Reinforcing a Railway Embankment on Degrading Permafrost Subgrade Soils

Sergey Kudryavtcev[1], Tatiana Valtceva[1(✉)], Zhanna Kotenko[1], Aleksey Kazharsrki[1], Vladimir Paramonov[2], Igor Saharov[3], and Natalya Sokolova[4]

[1] Far Eastern State Transport University, 47 Serishev Street, 680021 Khabarovsk, Russia
vtu25@mail.ru
[2] Emperor Alexander I St. Petersburg State Transport University, 8 Moskovski Street, 190000 Saint-Petersburg, Russia
[3] Saint-Petersburg State University of Architecture and Civil Engineering, 2-Krasnoarmeskya Street, 190000 Saint-Petersburg, Russia
[4] Financial University Under the Government of the Russian Federation, 49 Leningradsky Prospect, 125993 Moscow, Russia

Abstract. The article considers options for stabilizing the thawing process of permafrost soil of railways during the reconstruction period. The analysis of the engineering and geological conditions made it possible to design rock cooling structures at this facility, which are berms and cover slopes of the subgrade with fractionated rocky soil. The technical characteristics of fractionated rock soil structures have been developed and tested in this cryological area and have shown their effective operation for more than 30 years. As a result of the operation of the railway embankment, permafrost degrades and its boundary is at different depths depending on local conditions and the condition of the drainage systems from the subgrade. The position of the upper permafrost boundary should be established during surveys, if it is not advisable to restore the frozen base to a depth of 10 m, it is necessary to strengthen the thawed weak base and create conditions for the consolidation of thawed soils.

Keywords: Permafrost soil · Geological conditions · Reinforcement · Modeling · Deformations · Frozen and thawed soils

1 Introduction

Currently, experience has been accumulated in the design, construction and operation of buildings and structures on the Trans-Siberian and Baikal-Amur highways (the Eastern training ground of Russian Railways) in the conditions of permafrost and freezing soils.

To ensure the quality and reliability of buildings and structures at the Eastern Range of Russian Railways, the most modern and high quality materials, advanced construction technologies and effective methods of calculation and operation are used.

V. Murgul and V. Pukhkal (Eds.): EMMFT 2019, AISC 1258, pp. 35–44, 2021.
https://doi.org/10.1007/978-3-030-57450-5_4

Areas with the spread of frozen soils are characterized by a continental climate with large annual amplitudes. The annual amplitude of the air temperature is the difference in the average monthly temperatures of the warmest and coldest months. Large amplitude in the continental climate is created by lowering winter temperatures. Annual amplitudes in the continental climate are 25–40 °C, where extreme values are observed after solstices. So, the minimum temperature is observed in January, and the maximum - in July [1, 2].

2 Influence of Insolation on Road

Geographic latitude determines zoning in the distribution of climate elements. Solar radiation enters the upper boundary of the atmosphere, depending on geographic latitude. It determines the midday height of the Sun and the duration of the radiation. The absorbed radiation is distributed more difficultly, since it depends on cloud cover, the albedo of the earth's surface, and the degree of transparency of the air.

Zoning also underlies the distribution of air temperature. The temperature depends not only on the absorbed radiation, but also on the circulating conditions. Zoning in the temperature distribution leads to zoning of other meteorological climate values. The influence of geographical latitude on the distribution of meteorological values becomes more noticeable with height when the influence of other climate factors associated with the earth's surface weakens.

The temperature of the outside air changes in the diurnal course following the temperature of the earth's surface; therefore, the temperature course averaged over many years is taken into account. The average daily temperature amplitude depends on the latitude of the area (with increasing latitude, the daily air temperature decreases, since the midday sun decreases above the horizon), the nature of the soil cover (the larger the temperature amplitude of the surface itself, the greater the amplitude of air temperature), the proximity of water basins, forms terrain (on convex forms - peaks and slopes of mountains and hills - the daily amplitude of air temperature is reduced in comparison with flat terrain, and concave - valleys, ravines, ins - increased).

In summer, air temperature is distributed depending on latitude: the lowest temperature is set on the Arctic coast and the highest is at the southern borders of permafrost distribution. The exceptions are the coasts of the Bering and the Okhotsk seas, where the temperature in summer is lower than at the same latitudes in the continental regions [3, 4].

Solar radiation or the radiant energy of the sun is the main source of heat and light for the surface of the Earth and its atmosphere. A quantitative measure of solar radiation entering the surface is the energy illuminance or radiation flux density, expressed in watts per square meter (W/m^2), that is, 1 J of radiant energy is supplied per 1 m^2/s [4, 5].

The energy illuminance of solar radiation is expressed by the solar constant So, which is determined by the emissivity of the Sun and the distance between the Earth and the Sun. Using satellites and rockets, it was found that So = 1367 W/m^2 with an amendment of 0.3%, while the average distance between the Earth and the Sun was taken into account as 149.6 106 km [4, 5].

The solar constant experiences fluctuations from year to year due to the constant change in solar activity. The amount of energy per year 5.49 · 1024 J, approximately equal to the heat from the combustion of 400 thousand tons of coal, enters the illuminated space of the Earth at the upper boundary of the atmosphere. The energy received from the Sun in 1.5 days is approximately equal to the energy of all power plants of the Earth generated during the year [4, 5].

The intensity of direct solar radiation depends on the height of the Sun, the transparency of the atmosphere, season of the year, geographical latitude and exposure of the place. The intensity of direct solar radiation in summer is greatest - in the East at 7–8 a.m., in the West - at 4–5 p.m.

Part of the direct radiation due to the influence of molecules of atmospheric gases and aerosols goes into diffuse, which reaches the Earth's surface, is partially reflected from it and partially absorbed by it (about 20–23%). The higher the sun and the pollution of the atmosphere are, the more diffuse radiation is, which increases cloud cover. It contributes to its increase in snow cover, which has a large negative ability [4].

Different climatic conditions in the northern regions of the Far East and Siberia are due to the difference in heat input. In areas with the spread of frozen soils, the reduced heat input is explained by the low standing of the sun over the horizon for a long night, and therefore a large amount of heat generated by radiation is lost to radiation [2–4]. The decrease in heat input is also associated with the high reflectivity of snow and ice, and the heat consumption for their melting.

The effect of solar radiation on the temperature of the elements is taken into account in the form of additional heating at 100 °C of a surface layer illuminated by the sun with a thickness of 15 cm [5]. The total solar radiation (direct and diffuse) to a vertical surface with a cloudless sky, MJ/m^2. SP 131.13330.2012 Construction climatology. Updated edition of Construction Regulations 23-01-99*.

3 Thermophysical Model of the Process of Freezing, Frozen Burning and Thawing

In the Russian Federation, the "Termoground" mathematical model was developed, which makes it possible to analyze the processes of freezing, frost heaving and thawing according to established temperature and humidity fields. The software module "Termoground" [6] was implemented in the FEM models software package.

Freezing-thawing processes are described by the heat equation for unsteady thermal conditions in three-dimensional soil space by the following equation [7].

$$C_{th(f)}\rho\frac{\partial T}{\partial t} = \lambda_{th(f)}\left(\frac{\partial^2 T}{\partial x^2} + \frac{\partial^2 T}{\partial y^2} + \frac{\partial^2 T}{\partial z^2}\right) + q_V \qquad (1)$$

where is the specific heat of soils (frozen or thawed); ρ - soil density; T is the temperature; t is the time; $\lambda_{th(f)}$ - thermal conductivity of soils (frozen or thawed); x, y, z - coordinates; qv is the power of internal heat sources. This equation allows us to determine the values of the incoming and outgoing heat flux from the elementary volume of the soil, leaving the main flow of the soil volume at a point in time equal to the change in the value of heat rotations. Under steady conditions, the flow entering and leaving the elementary volume of soil is the same at any time. In this case, the left side of the equation is reduced, and the equation will look like:

$$\lambda\left(\frac{\partial^2 T}{\partial x^2}+\frac{\partial^2 T}{\partial y^2}+\frac{\partial^2 T}{\partial z^2}\right)+q_V=0 \tag{2}$$

The heat capacity function consists of two parts. The first part is the volumetric heat capacity of the soil (thawed or frozen), the second part is the latent heat of phase transitions in the negative temperature range, absorbed or given away by the soil due to changes in the groundwater phase, presented in the form:

$$C_{(f)}=C_{(f)}+L_0\frac{\partial W_W}{\partial T} \tag{3}$$

where L_0 = 335·106 J/m^3 = 335·103 kJ/m^3 = 8975 Btu/ft^3 = 79760 kcal/m^3 is the heat of water-ice phase transformations; Ww - humidity of frozen soil due to unfrozen water.

When the function of unfrozen water content in soils is determined, the total content of unfrozen water can be expressed as

$$W_W=K_W\cdot W_P \tag{4}$$

where Wp - humidity at the rolling border; Kw - coefficient of unfrozen water content in frozen clay soils, taken according to [2, 7]. Substituting relation (3) into expression (1), we obtain the complete differential equation:

$$\rho_в\left(C_{th(f)}+L_0\frac{\partial W_W}{\partial T}\right)\frac{\partial T}{\partial t}=\lambda_{th(f)}\left(\frac{\partial^2 T}{\partial x^2}+\frac{\partial^2 T}{\partial y^2}+\frac{\partial^2 T}{\partial z^2}\right)+q_V \tag{5}$$

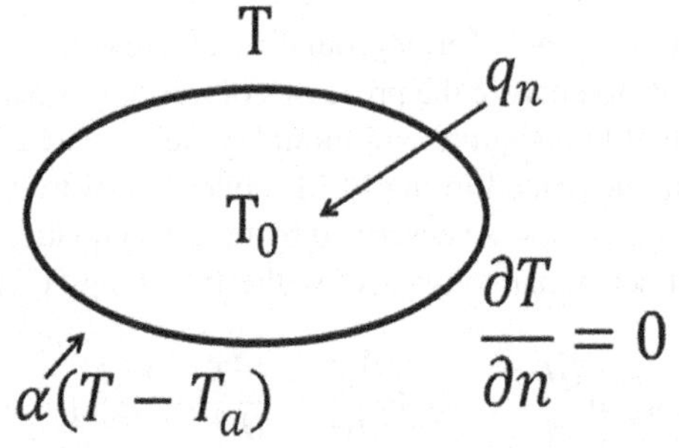

Fig. 1. Boundary conditions of the heat conduction problem.

The initial condition for Eqs. (1) and (5) is the given value of the temperature field in the studied region T (x, y, z) of the soil at time t = T0 (Fig. 1). The boundary conditions can be of four types.

1. If the soil temperature at surface S is known, then T = T0(S, t)
2. If a heat flux is specified inside the region Sq, then
   where n is the direction vector of the external normal to the surface; qn is the density of the heat flux, which is considered positive if the soil loses heat. Physical examples of heat flux sources are heat supply pipes, water vapor, or power or communication cables laid in the ground. In each of these cases, the cross-sectional area of the pipe or cable is small compared to the size of the surrounding soil.
3. If convective heat transfer occurs on the surface of the soil Sα, then

$$\lambda\left(\frac{\partial T}{\partial n}\right)+\alpha(T-T_{\alpha})=0 \tag{6}$$

   where α is the heat transfer coefficient; Ta is the temperature of the surrounding atmosphere.
4. If a heat flux is given at the boundaries of the region under consideration, then

$$\lambda=\left(\frac{\partial T}{\partial n}\right)=0 \tag{7}$$

   Heat flux qn and convective heat loss do not occur on the same section of the boundary surface. If there is heat loss due to convection, then there is no heat removal or influx due to heat flow and vice versa.

4 Numerical Modelling of the Freezing and Thawing Process Taking into Account the Influence of Solar Radiation on the Roadbed

The Baikal-Amur Mainline passes through the territory with severe climatic conditions at 52–56 latitudes, at different angles of light and through permafrost areas with a depth of 1–3 to hundreds meters and a high seismicity of up to 9 points. In the process of the highway construction and the ongoing modernization, the latest designs are used, new methods of construction and operation of facilities in difficult engineering conditions are developed and patented.

To numerically simulate the effect of solar insolation depending on the latitude and direction of the cardinal directions, thermotechnical calculations of the railway section from Hani to Tynda were performed.

Having considered the materials of eight transverse profiles of engineering and geological surveys on the railway section, the results of the analysis of the stability of the railway track for warm and cold periods of the year, as well as overview information from space images, it is advisable to design rock cooling structures on this object, which are berms and cover slopes of earthen canvases by fractionated rocky soil. The technical characteristics of the fractionated rock soil structures were developed and tested by the permafrost station in this cryological area, and have shown their effective operation for more than 20 years.

Figure 2 shows the design diagram of the transverse profile of a railway embankment for numerically modeling the effect of solar insolation depending on the latitude and direction of the cardinal directions for a period of two years.

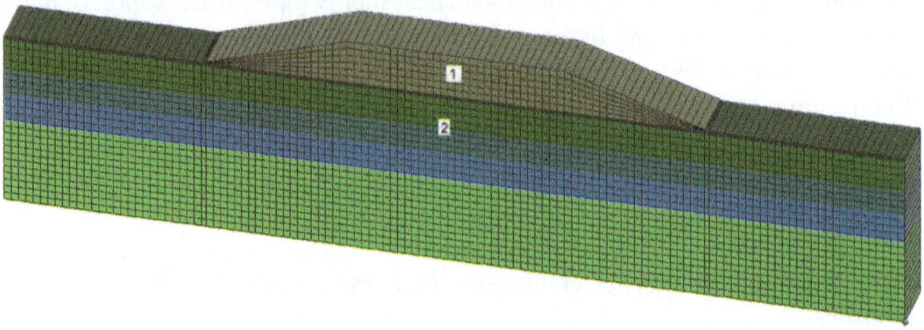

Fig. 2. The design diagram of the transverse profile of the railway embankment. 1 - embankment; 2 - frozen base.

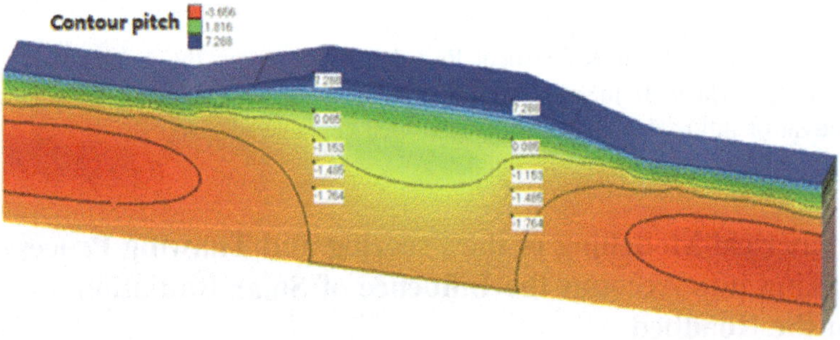

Fig. 3. The contours of the temperature of the embankment and the base without taking into account the influence of solar insolation.

Figure 4 demonstrates zones of thawed and frozen soils are given without taking into account the effects of terminal insolation. The value of thawed soil is up to 1.3 m. Under the embankment, the soil remains frozen.

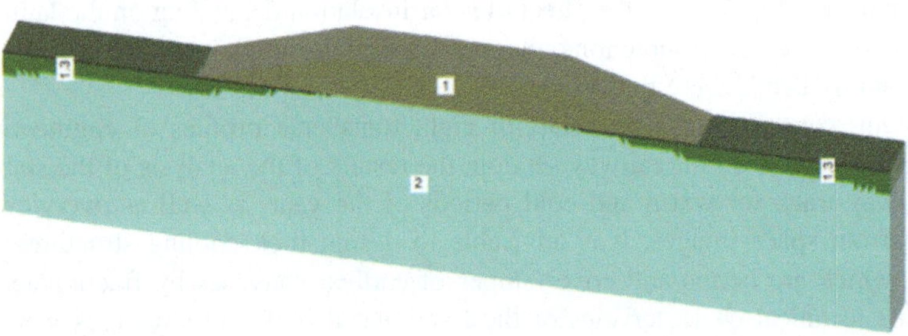

Fig. 4. Zones of thawed and frozen soils of the embankment and base, excluding the impact solar insolation.

Figure 5 shows deformation of thawing of frozen embankment soils without taking into account the influence of solar insolation.

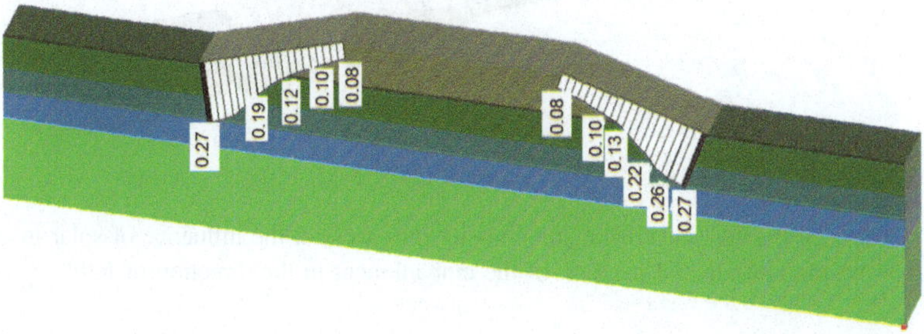

Fig. 5. Defrosting of the main site and slopes of the embankment without taking into account the influence of solar insolation.

Excluding the effect of solar insolation (Fig. 3), deformations of the main site of the embankment for a period of two years are up to 8 cm, and up to 27 at the base of the slope. Figure 6 illustrates the isolines of the temperature of the embankment and the base, taking into account the effect of solar insolation, depending on the latitude of the embankment in the direction of light.

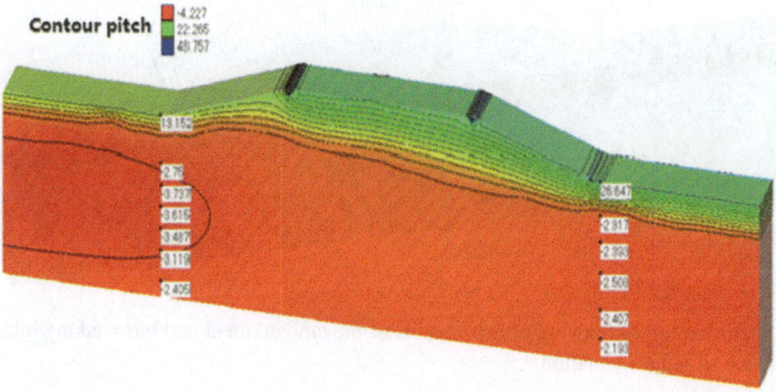

Fig. 6. Temperature contours of the embankment and base, taking into account the influence of solar insolation and the direction of the cardinal directions for a period of two years.

In Fig. 7 zones of thawed and frozen soils are given without taking into account the effect of final insolation, taking into account the war of final insolation depending on the latitude and location of the embankment in the direction of light. The value of thawed soil is more than 4 m in the field and up to 1.8 m under the embankment on the south side of the world, and on the north side, thawing is 1.3 m.

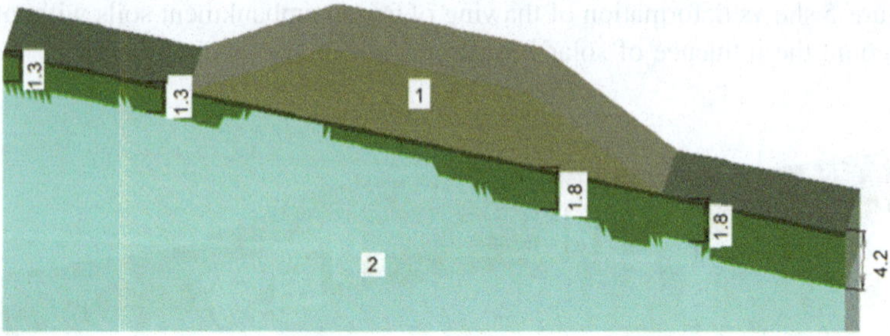

Fig. 7. Zones of thawed and frozen soils, taking into account the influence of solar insolation, depending on the latitude and location of the embankment in the direction of light.

Figure 8 demonstrates deformation of thawing of frozen soils of the embankment is given taking into account the influence of solar insolation depending on the latitude and location of the embankment in the direction of light. The amount of deformation of the embankment during thawing of the soil is from 34 cm at the base of the slope and up to 30 cm edge of the main platform of the embankment from the south side of the world, and from the north side from 21 cm to the base of the slope and up to 12 cm edge of the main platform of the embankment.

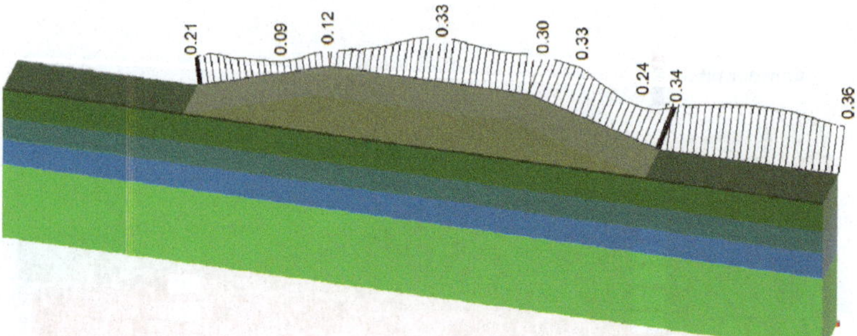

Fig. 8. Deformation of thawing of frozen soils of the embankment and base, taking into account the influence of solar insolation.

As can be seen from the deformation diagram, that the most deformable section of the road is on the south side with a strain of up to 34 cm and the base is thawed under the embankment to 1.8 m over a two-year period.

5 Conclusions

1. At present, experience has been accumulated in the design, construction and operation of buildings and structures on permafrost and freezing soils.

2. Solar radiation or the radiant energy of the Sun is the main source of heat and light for the surface of the Earth and its atmosphere, therefore, it is advisable to use the total solar radiation (direct and scattered) presented on the vertical surface as the source data for the normative documents.
3. When installing cooling structures, all drainage systems must be brought in good condition, and strengthened in case of necessity, for example, ditches must be strengthened, or equipped with waterproofing, or replaced with composite trays. In places of significant failures where transverse filtration occurs, antifiltration screens should be provided or additional culverts should be arranged.
4. Thermophysical calculations were carried out in the Termoground software package, which allowed us to analyze the processes of freezing, frost heaving and thawing according to steady-state temperature and humidity fields.
5. Without taking into account the influence of solar insolation, deformations of the main site of the embankment for a period of two years are up to 8 cm, and at the base of the slope up to 27 cm. Under the body of the embankment, the base soil is frozen.
6. Taking into account the influence of solar insolation, depending on the latitude and location of the embankment in the direction of light, deformation of thawing of frozen soil of the embankment to a depth of 1.8 m, the amount of deformation of the embankment during thawing of soil to 34 cm at the base of the slope and to 30 cm of the edge of the main embankment site from the south cardinal points. On the north side, the amount of deformation of the embankment during thawing of the soil is from 21 cm at the base of the slope and up to 12 cm of the edge of the main platform of the embankment, since the base soil under the slope is frozen.
7. When designing structures on permafrost soils, it is advisable to carry out calculations using boundary conditions when the heat flux is set at the boundaries of the area under consideration, taking into account the influence of solar insolation and direction to the cardinal points for a long period of time.

References

1. Paramonov, V.N., Sakharov, I.I., Kudryavtsev, S.A.: Forecast the processes of thawing of permafrost soils under the building with the large heat emission. In: MATEC Web of Conferences, vol. 73, p. 05007. EDP Sciences, France (2016). https://doi.org/10.1051/matecconf/20167305007
2. Kudryavtsev, S., Paramonov, V., Sakharov, I.: Strengthening thawed permafrost base railway embankments cutting berms. In: MATEC Web of Conferences, vol. 73, p. 05002. EDP Sciences, France (2016). https://doi.org/10.1051/matecconf/20167305002
3. Kudryavtsev, S.A., Arshinskaya, L.A., Valtseva, T.U., Berestyanyy, U.B., Zhusupbekov, A.: Developing design variants while strengthening roadbed with geomaterials and scrap tires on weak soils. In: Proceedings of the International Workshop on Scrap Tire Derived Geomaterials - Opportunities and Challenges, IW-TDGM 2007, Yokosuka, pp. 171–178 (2008)
4. Kudriavtcev, S., Berestianyi, I., Goncharova, E.: Engineering and construction of geotechnical structures with geotechnical materials in coastal arctic zone of Russia. In: Proceedings of the 23rd International Offshore and Polar Engineering Conference, ISOPE 2013, pp. 562–566 (2013)

5. Kudruavtsev, S.A., Valtseva, T.Y.: The use of geosynthetic materials in special engineering geological conditions of the Far East. In: Proceedings of the 11th ICG - International Conference on Geosynthetics, Seoul, Korea, 16–21 September, pp. 321–326 (2018)
6. Kudryavtsev, S.A., Berestyanyy, Y.B., Valtseva, T.Y., Goncharova, E.D., Mikhailin, R.G.: Geosynthetical materials in designs of highways in cold regions of Far East. In: Proceedings of the International Conference on Cold Regions Engineering. "Cold Regions Engineering 2009: Cold Regions Impacts on Research, Design, and Construction", p. 546 (2009)
7. Kudriavtcev, S., Valtseva, T., Berestianyi, I., Goncharova, E., Mihailin, R.: Motorway structures reinforced with geosynthetic materials in polar regions of Russia. In: Proceedings of the International Offshore and Polar Engineering Conference. "Proceedings of the 24th International Ocean and Polar Engineering Conference, ISOPE Busan", Busan, pp. 1141–1143 (2014)

Competition Development on the Ground Passenger Transportation Market in Krasnodar Krai, Russia

Svetlana Grinenko(✉), Lyudmila Prikhodko, Ekaterina Belyakova, and Margarita Tatosyan

Sochi State University, Plastunskaya str., 94, Sochi, Russia
sveta.grinenko@gmail.com

Abstract. This article analyzes the competition development in the passenger transport market. The purpose of the study is to assess the state and development of the competitive environment in the ground passenger transportation market of Krasnodar Krai. The object of the study is a competitive environment in the ground passenger transportation market of Krasnodar Krai. The study covered several areas, such as identification of the main consumers of transport services, competitiveness factors, ways to improve competitiveness, barriers assessment. Objectives of the study are: the assessment of customers' satisfaction with the service quality of ground passenger transportation in Krasnodar Krai, the assessment of the level of competition and administrative barriers by transport organizations of Krasnodar Krai. The study used a questionnaire method. Data were collected by using a quota sampling method based on standardized questionnaires for service consumers and for service producers in on-line mode. In order to address the low level of competition between public and private carriers, it is necessary to create a competitive environment in the market of ground passenger transportation. It can be done due to changing the existing intermunicipal route network, routes with regulated and unregulated tariffs.

Keywords: Passenger transportation · Transport services market · Carriers · Passengers · Krasnodar Krai

1 Introduction

Modern research in the field of transport service should be considered from various perspectives, representing both technical, technological and organizational aspects.

The authors R. Yashiro, H. Kato, E. Vitvitskii, M. Simul, S. Porkhacheva, A. Novikov, I. Novikov, A. Katunin, A. Shevtsova, A. Kostsov in their studies present methods of modeling traffic flows, estimates the capacity of the transport infrastructure, the criteria for the formation of passenger transportation systems, which directly affects the quality of transport [1–4].

Researchers Y. Averianov, K. Glemba, A. Gritsenko, Y. Baranov, A. Bodrov, D. Lazarev consider the quality of transport services from the standpoint of training as a factor in improving traffic safety, reducing the number of road accidents, which is especially important for passenger transport [5, 6].

V. Murgul and V. Pukhkal (Eds.): EMMFT 2019, AISC 1258, pp. 45–59, 2021.
https://doi.org/10.1007/978-3-030-57450-5_5

S. Guidon, M. Wicki, T. Bernauer, K. Axhausen are devoted to issues of innovation, quality characteristics of the fleet, its updating and impact on the service [7]. The study of transport systems as a service sector affecting society and territorial management is discussed in publications by Yale Z. Wong, David A. Hensher, C. Mulley, K. Pangbourne, Miloš N. Mladenović, D. Stead, D. Milakis, and Caitlin D. Cottrill [8–10].

Next, we should single out the works devoted to the problems of vehicle routing, the formation of an optimal passenger transportation structure, which are described in the works of Y. Shao, M. Dessouky, Y. Chao, M. Zishan, Y. Ma, Y. Gao [11–13].

We should also single out a pool of works in which the authors propose a solution to the problems of developing a market mechanism for regulating transport services, creating flexible transport services in the public transport market, creating a system of interaction between government and business in the process of market transformation in the transport sector, developing a model of competition between independent transport services operators Are authors M. Egan, M. Jakob, C. Mulley, John D. Nelson, M. Taylor, A. Hallsworth, Huw CWL Williams, J. Abdulaal [14–17].

It should be noted that despite a significant amount of work in the field of transport and transport systems, the formation of market mechanisms in the process of economic transformation, the development of competition in the transport services market remain relevant and served as the basis for this study.

Passenger transportation plays a significant role in the transport services market due to its high socio-economic importance. The population's need for transportation is connected with production activities (trips to workplace and business trips) and with cultural and household needs (leisure trips, tourism, and excursions).

Intercity transportation is provided by rail, bus, and personal vehicles. The demand for these services is elastic in terms of price and income, which determines its dependence on non-price competition factors, such as reliable and convenient schedules, comfortable rolling stock, that would make the population's final choice of a certain type of transport. In this regard, the competitiveness of railway transport compared to intercity bus passenger transport is related to its high carrying capacity, reliability and regularity [1, 2].

The social role of railway transport should be noted in the transportation of the urban population to suburban areas and to places of mass recreation. It's about 3.5 million passengers transported by Russian Railways daily. According to opinion polls, more than 41% of commuters make trips to work and school, and about 29% - to the country house. More than 45% of passengers use rail transport almost daily.

Intercity transportation varies considerably in the distance of passengers' trips. There is local transportation, which is carried out mainly by rail or buses and long-distance transportation, carried out by rail or air transport and a small share of bus and water routes [2–4].

2 Materials and Methods

The objects of the study are organizations that provide ground transportation services for passengers in Krasnodar Krai.

Organizations of various structural and legal forms took part in the survey (Fig. 1). 64% of respondents have a private form of ownership and the majority of them - 40% are individual entrepreneurs.

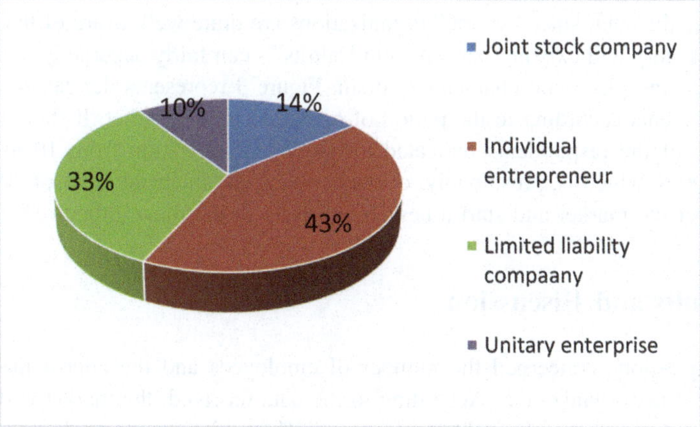

Fig. 1. Structural and legal form of the organization.

The main consumers of public transport services are residents of the region. The main consumers of charter transport services are city residents within the region and other subjects of the Russian Federation.

The respondents include organizations that provide various types of services (Fig. 2), among which 50% are carriers that provide services, 29% are service organizers who act as intermediaries between carriers and consumers, 11% are operators and 8% are organizations that provide information services in the field of ground passenger transportation. Organizations engaged in the operation of roads, referred to as "other".

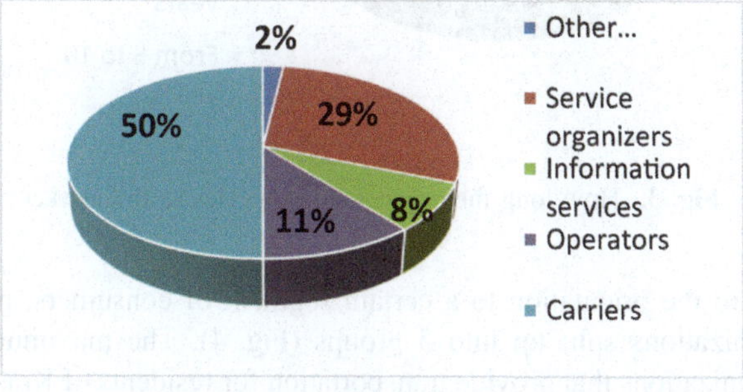

Fig. 2. Type of economic entities.

Respondents of the survey included heads of departments (18%), low level employees (28%), owners (28%), directors and deputy directors (26%). This sample composition provides a sufficiently well-founded and diversified view of the ground passenger transportation market. The survey includes organizations with work experience in the market of several months up to 15 years or more, which also confirms the validity of the data, since "young" organizations are quite well aware of the problems of market entry and existing barriers, and "adults" - can fairly accurately represent the changes taking place and characterize them. Figure 3 represents the categorization of the respondents according to the period of operation in the market. It should be noted that none of the respondents indicated the period of operation from 10 to 15 years (2004–2009), which is, presumably, connected with the financial crisis of 2007–2008, when enter the market and start a new enterprise was the most difficult [5, 6].

3 Results and Discussion

Several questions concerned the number of employees and the approximate annual turnover of the organization. According to the data received, the market is dominated by organizations with up to 100 employees (82%), which are divided almost equally into enterprises with up to 15 employees (40%) and from 15 to 100 employees (42%). In terms of annual turnover, there are organizations with an approximate turnover of up to 800 million rubles (88%). They represent two unequal groups – those with the turnover that does not exceed 120 million rubles (59% of the total) and the organizations with a turnover of 120 to 800 million rubles per year (29% of the total).

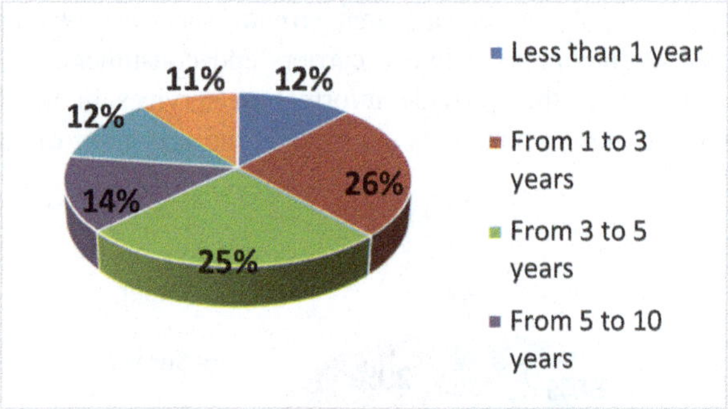

Fig. 3. How long the organization operates in the market.

According to the orientation to a certain segment of consumers, passenger transportation organizations split up into 5 groups (Fig. 4). The maximum share - 44% belongs to organizations that provide transportation for residents of Krasnodar Krai and 37% - for residents of municipalities where these organizations operate.

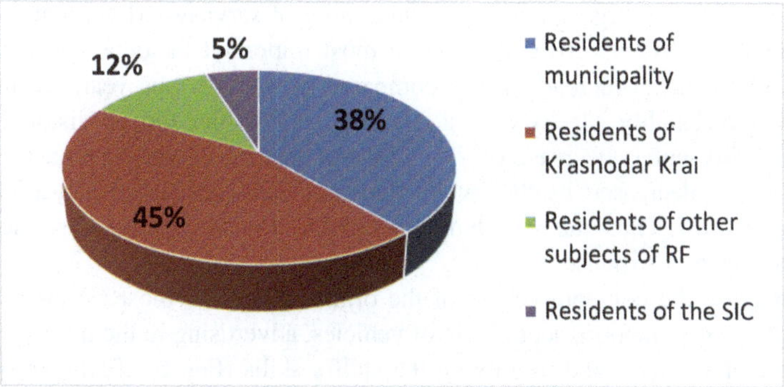

Fig. 4. The main consumers of services.

12% of organizations carry out transportation for other subjects of the Russian Federation and only 7% - for residents of the CIS and foreign countries (5% and 2% respectively).

3.1 Results of the Survey of Transport Organizations on the State of the Competitive Environment in the Ground Passenger Transportation Market

The main factors of competitiveness are the quality of services and the price. As the price of public transport services is regulated by the regional and municipal authorities, it cannot act as the main competitive factor for public transport enterprises. The main competitive factor is the price of charter transport services.

The quality of passenger transport services is determined primarily by the timeliness and reliability of services, and, then by the comfort of passengers [7].

The remaining competitive factors, such as cost of transportation, related services and seasonality, increase economic efficiency and diversify the risks of enterprises (Fig. 5).

Fig. 5. Competitive factors.

In the first place, organizations put the quality of services (61%), that presented a 10% growth in 2019 vs 2018; the second most important factor is the service price (54%), which had an increase of 3% compared to the previous year, the third place shared the seasonality (28%) with a decrease of 2% and the provision of related services (26%) with a decrease of 7%. The significance of the location of the organization (17%) decreased by 8%, while transport expenses (18%) had a 3% growth compared to 2017. Thus, the growth dynamics was shown by factors of service quality, price and transport expenses.

To increase the competitiveness of the organization, the most common measures used were: costs reduction, acquisition of vehicles, advertising in the media, enhancing the quality of services, and improving the staff's skills (Fig. 6). Technology acquisition, R & D, marketing strategies, and online advertising were used less frequently. Price reduction, enhancement of related services and development of new modifications of the services provided were not practically applied.

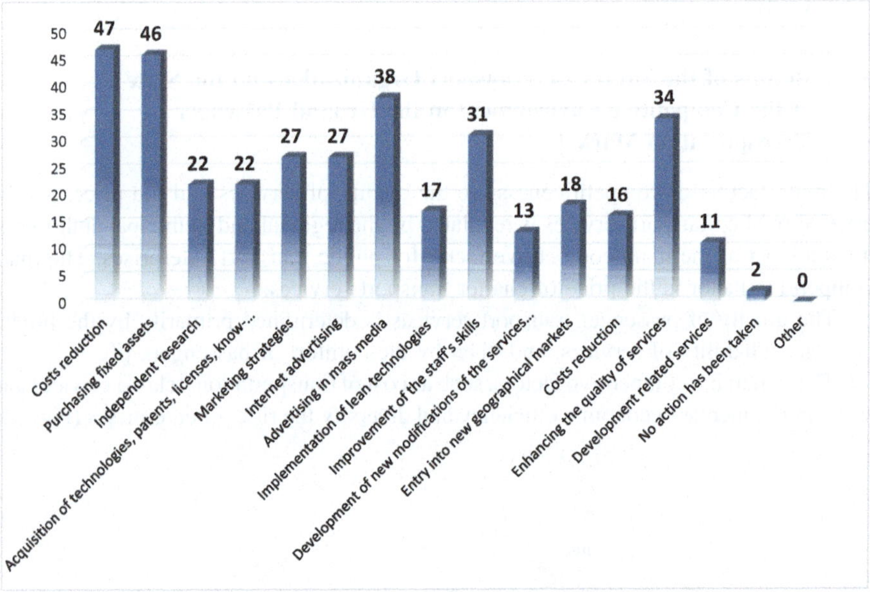

Fig. 6. Measures to improve competitiveness.

Costs reduction, purchasing fixed assets, improving the quality of services, and improving the staff's skills have the leading position. Enterprises began to actively apply marketing strategies and Internet advertising, though little attention is paid to the development of new modifications of services and related services.

The competition in the transport services market is evaluated as weak or moderate by most carriers, especially those in the field of commercial public transport (Fig. 7). Municipal public transport enterprises assess the level of competition as weak [8, 9].

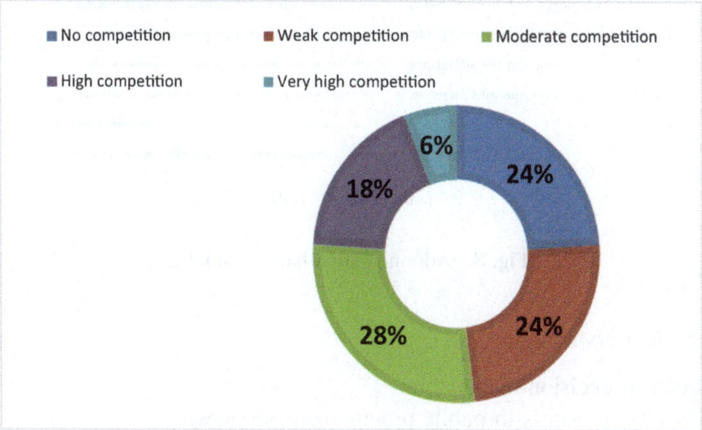

Fig. 7. The level of competition.

The level of competition has increased for charter bus transportation. Carriers assess the level of competition as very high and undertake significant efforts to maintain market position.

As for public transport enterprises, the level of competition between commercial carriers for maintaining their position increased due to the policy of municipal and regional authorities to reduce the number of carriers in the market (consolidation of carriers). Thus, the largest enterprises operating in the market for more than 10 years have been able to maintain their positions by now.

In 2018, commercial carries showed a high level of competition and economic efficiency compared to municipal ones. This is because commercial carriers are guided by economic priorities, while municipal ones are financed from municipal budget and are guided mainly by political priorities.

Commercial carriers, unlike municipal ones, rely on their own economic resources and try not to attract credit funds. Municipal carriers often purchase rolling stock on credit or lease, which due to interest on the deferred payment reduces their economic efficiency in the future. But the availability of credit funds for municipal carriers is higher due to budget support. This explains why municipal carriers often need subsidies from the budget, while commercial carriers maintain their profitability.

Administrative barriers assessment. According to the weighted average assessment the indicators are ranked as follows: in the first place – the existing legal framework, in the second place- the high tax burden, in the third – the procedures for obtaining permits/licenses, in the fourth – the dialogue with the authorities (Fig. 8). For non-governmental organizations, the high tax burden is in the first place, the existing legal framework is in the second place, and other types of barriers are the same for all enterprises.

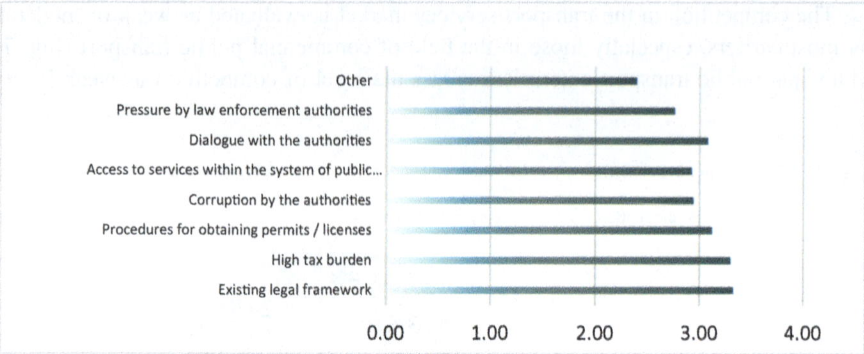

Fig. 8. Administrative barrier ranking.

"Other" barriers include:

1. the timing of decision-making;
2. the difficulty of access to public procurement services;
3. high level of taxes;
4. difficulty of attracting credit;
5. high fuel prices.

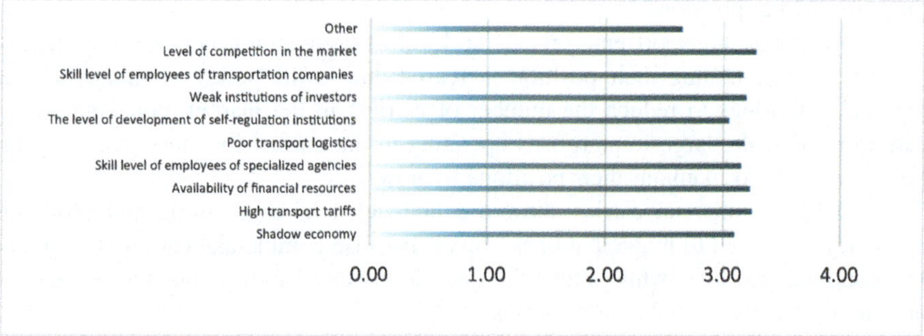

Fig. 9. Non-administrative barrier ranking.

According to the weighted average assessment of non-administrative barriers the indicators are ranked as follows: in the first place – high transport tariffs, in the second place – availability of financial resources (loans), in the third – weak institutions of investors, in the fourth – poor transport logistics (Fig. 9).

The assessment of the activity of authorities in the transport services market is shown in Fig. 10.

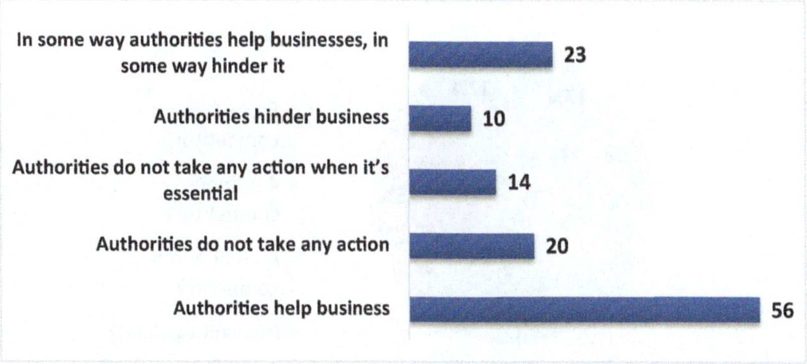

Fig. 10. The assessment of the activity of authorities in the transport services market.

Table 1 presents the structure of the responses given by state and private organizations as well as individual entrepreneurs in respect of the evaluation of authorities' activity in the market of transport services. It shows that municipal organizations are assisted by the authorities, which is due to the importance of public transport, but at the same time it has a negative impact on the competition development.

Table 1. Evaluation of authorities' activity in the market of transport services.

Actions on the part of authorities	The response rate (%)		
	All types of organizations	Government organizations	Private organizations
Authorities help business	46	52	42
Authorities do not take any action	16	13	18
Authorities do not take any action when it's essential	11	16	9
Authorities hinder business	8	3	11
In some way, the authorities help businesses, in some way they hinder it	19	16	20

In 2018, all organizations claimed that the introduction of the resort fee wouldn't have any impact on their competitiveness. In 2019, only 56% of respondents claimed the same, 24% believed that the competition would increase and 20% - that the competition would decrease.

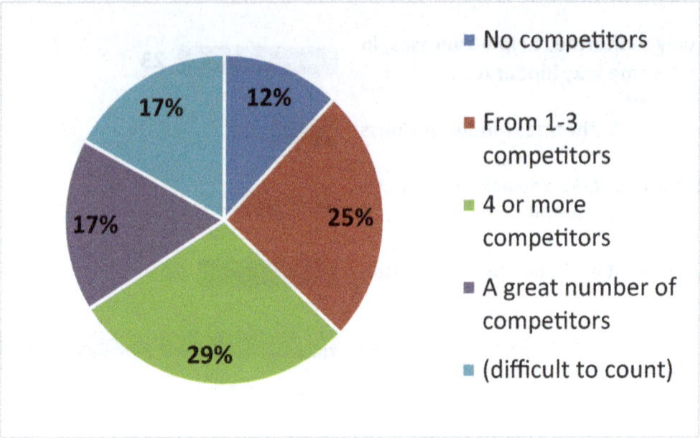

Fig. 11. The number of competitors offering similar or alternative services in the transport services market.

The estimation of the number of competitors offering similar or alternative services in the transport services market according to transport organizations is presented in Fig. 11. 17% of respondents do not have any information about their competitors operating in the market, 12% believe that they offer unique services and the majority-29%, are confident that there is a significant number of competitors on the market [7, 8].

The results highlight the active formation of a competitive environment in the passenger transport market. Answering a question about competitive environment changes, the majority of respondents noted that the number of competitors has increased over the past three years. They also mentioned that each segment in the transport services market has competitors offering similar services, which provides a choice for consumers.

In general, there has been a decline in administrative and economic barriers, along with an increase in the level of assistance provided by the authorities, and this also indicates additional opportunities to mitigate existing restrictions.

3.2 The Monitoring Results of Consumer Satisfaction with the Quality of Ground Passenger Transportation Services

To estimate the consumer environment in the market of transport services of Krasnodar Krai, by the order of the Ministry of Economy of the region, the research team of Federal State Budget Educational Institution of Higher Education "Sochi State University" conducted a survey of transport services consumers. Respondents were asked to complete a feedback survey on the services provided by passenger transportation companies of Krasnodar Krai.

The procedure of the survey was regulated. Some questions were formulated in such a way as to clarify the substantive aspects of the answer.

In total, 410 people from different regions and cities of Krasnodar Krai took part in the study, among them respondents from Armavir, Goryachy Klyuch, Krasnodar, Novorossiysk, Sochi, Tuapse, Abinsky District, Beloglinsky District, Yeisk District,

Kalininsky District, Krasnoarmeysky District, Mostovsky District, Novokuban District, Tikhoretsky District.

The survey provided a set of sociological and transport data to determine the satisfaction with the characteristics of the transport services market.

A number of criteria have been chosen to conduct a survey of respondents' views on improving the quality of transport services provided by transportation companies of Krasnodar Krai. The most important criteria were social status, age, education, average monthly income per family member, public transport assessment, the number of users of public transport, preferences for modes of transport and others. Based on the survey data, dependency graphs were made [5].

Most of the respondents were students (59%), then come workers of different groups (15%), followed by specialists and engineers (10%). A considerable percentage of the total number of respondents - were pensioners (5%). It should also be noted that among the respondents there were civil servants (2%), senior/ middle managers (5%), business owners and entrepreneurs (1%) as well as unemployed citizens (1%).

The significant number of students can be explained not only by the fact that the survey was conducted on the initiative of the educational institution, but also by the fact that this is the biggest user segment due to social and economic factors.

The gender composition of the respondents (women 68%, men 32%) is due both to a greater willingness of women to participate in surveys and to the fact that women are more likely to use public transport. The structure of respondents using public transport is shown in Fig. 12.

The representativeness of the survey is confirmed by the fact that 54%, of the total number of respondents, use public transport almost every day to get to their workplace. Business and shopping trips got 16%. So, the total number of respondents using public transport daily is 70%. And only 15% of respondents practically do not use land public transport.

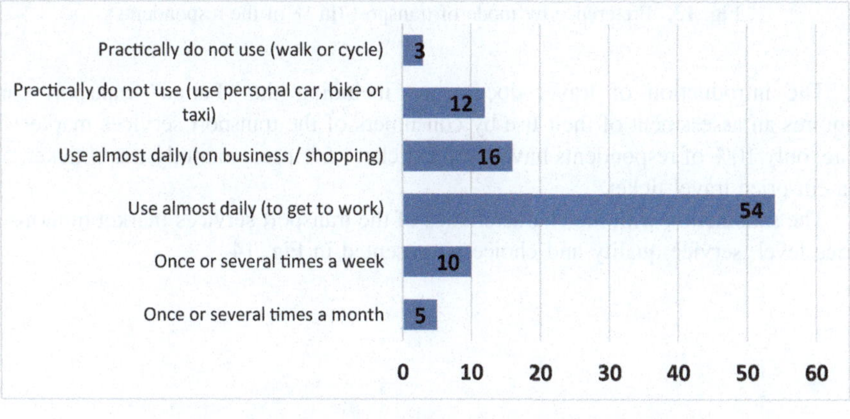

Fig. 12. Structure of respondents using public transport.

The main role in the research of the transport services market is played by such factors as performance assessment, satisfaction of the characteristics of transport services market in terms of price level, quality of service, choice of services [7]. Therefore, these criteria were chosen to assess the quality of public transport services. The majority (38%) of the respondents believe that public transport in cities and districts of Krasnodar Krai works “rather good”, while 25% of the respondents evaluated its work as – “good”, 6.5% - “bad”, and 19% - “rather bad”. It is worth noting that there were respondents who found it difficult to answer this question (6%).

The most common mode of transport in Krasnodar Krai is still public transport. The majority of respondents prefer buses/trolleybuses (i.e. 40% of the total number of respondents). Personal transport is used by 22% of the respondents; the least preferred transport is rail (3% of the total number of respondents). This is mostly because there are a lot of cities and districts in Krasnodar Krai with no rail transport (tram) at all. It should also be mentioned that 10% of respondents use taxi services. Figure 13 clearly shows the distribution of preferences by mode of transport among the people surveyed.

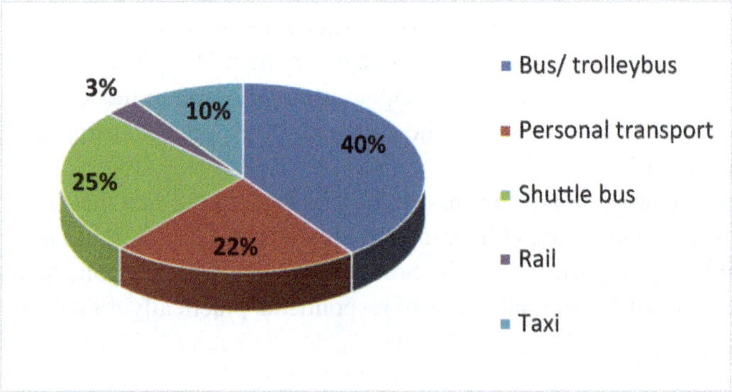

Fig. 13. Preference by mode of transport (in % of the respondents).

The introduction of travel documents, including the “Palma” transport card, requires an assessment of their use by consumers of the transport services market. To date, only 10% of respondents have a travel ticket - 5% use a monthly travel ticket, 5% - a cut-price travel ticket.

The satisfaction with the characteristics of the transport services market in terms of price level, service quality and choice is presented in Fig. 14.

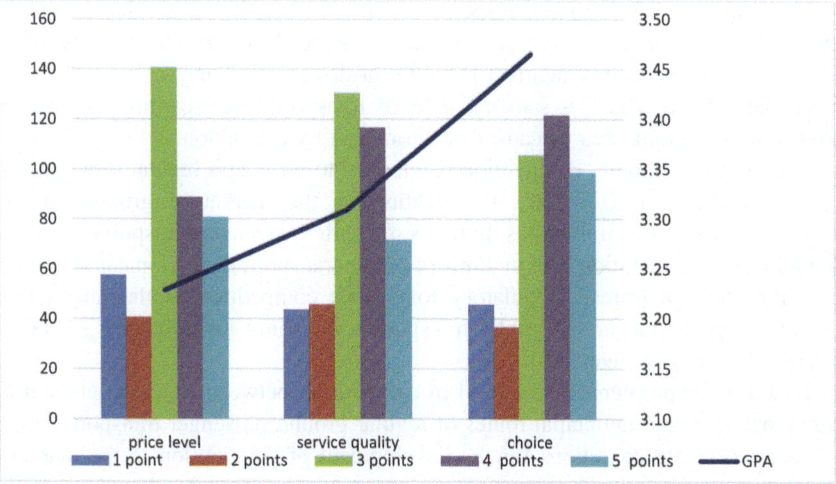

Fig. 14. Satisfaction with the characteristics of the transport services market.

The satisfaction with the characteristics of the transport services market in terms of the price level is estimated on a 5-point scale, ranging from satisfactory (5 points) to unsatisfactory (1 point). The vast majority of respondents (34.4%) gave the price level 3 points, and the least number of respondents (10.0%) gave it 2 points. The weighted average score is 3.23 points, which is below the 2018 level.

The satisfaction with the characteristics of the transport services market in terms of quality of service is as following: most of the respondents (31.9%) gave it 3 points, and the least number of respondents (10.73%) gave it only 1 point. The weighted average score is 3.31 points, which is also below the 2018 level.

As for the satisfaction with the characteristics of the transport services market in terms of choice, the vast majority of respondents (29.8%) estimated the market by 4 points; 25.8% gave it 3 points; the least number of respondents (9.0%) gave it 2 points. The weighted average score is 3.47 points.

An important issue of our time is the environment, and the question of whether environmental transport is important for consumers was included in the survey. 51% of respondents answered "Yes", 38% - "Yes, if it does not affect convenience/cost" and only 11% answered "No".

4 Conclusions

The level of service, as well as the desire of the customer use these services again affects the success of the enterprise. In a highly competitive environment, the quality of staff service plays a crucial role in building relationships between consumers and producers. In this respect, the personnel of transport services organizations are of great importance.

The majority (75%) of the respondents rate the quality of service provided by personnel of the organizations as satisfactory, while 25% consider it to be unsatisfactory. The rudeness and incorrect behavior of drivers, smoking and use of a mobile phone behind the wheel, a sanitary state of transport, overcrowding of buses and electric trains were named as causes of unsatisfactory evaluation.

In a market economy, competition is intended to serve as a regulator of consumer-producer interaction. However, the situation on the market of ground passenger transportation is not unambiguous. In terms of charter passenger transportation, there is a high level of competition. But in terms of public transport, the regional and municipal authorities have a primary regulatory role; thus, competition in the interaction of consumers and producers of public transport occurs only in the choice of mode of transport by the passenger [10].

To solve the problem of low level of competition between state carriers and non-state carriers on intermunicipal routes of regular ground passenger transportation, it is necessary to create conditions for the development of competition in the market, by reformatting the existing intermunicipal route network into a combination of routes with regulated and unregulated tariffs [11–13].

The way to solve the problems mentioned above is cooperation of the regional and municipal authorities with scientific organizations for effective management of regular passenger transport on the basis of the use of strategic management and marketing tools, analytical support for decision-making, technical and economic analysis of the route network and modeling of transport systems [14–17].

References

1. Yashiro, R., Kato, H.: Success factors in the introduction of an intermodal passenger transportation system connecting high-speed rail with intercity bus services. Case Stud. Transp. Policy **7**(4), 184–186 (2019). https://doi.org/10.1016/j.cstp.2019.10.001
2. Vitvitskii, E., Simul, M., Porkhacheva, S.: Innovative technology for evaluation of capacity of thoroughfares. Transp. Res. Proc. **20**, 88–102 (2017). https://doi.org/10.1016/j.trpro.2017.01.109
3. Novikov, A., Novikov, I., Katunin, A., Shevtsova, A.: Adaptation capacity of the traffic lights control system (TSCS) as to changing parameters of traffic flows within intellectual transport systems (ITS). Transp. Res. Proc. **20**, 56–66 (2017). https://doi.org/10.1016/j.trpro.2017.01.074
4. Kostsov, A.: Results of studies on traffic volume at left-turn exits of grade-separated intersections. Transp. Res. Proc. **36**, 26–28 (2018). https://doi.org/10.1016/j.trpro.2018.12.106
5. Averianov, Y., Glemba, K., Gritsenko, A.: Research results of the professional training process for mobile machines operators as a factor of improving traffic safety. Transp. Res. Proc. **36**, 65–70 (2018). https://doi.org/10.1016/j.trpro.2018.12.036
6. Baranov, Y., Bodrov, A., Lazarev, D.: Methods for investigating road accidents. Transp. Res. Proc. **36**, 122–128 (2018). https://doi.org/10.1016/j.trpro.2018.12.038
7. Guidon, S., Wicki, M., Bernauer, T., Axhausen, K.: Transportation service bundling – for whose benefit? consumer valuation of pure bundling in the passenger transportation market. Transp. Res. Part A: Policy Pract. **131**, 75–78 (2020). https://doi.org/10.1016/j.tra.2019.09.023

8. Wong, Y.Z., Hensher, D.A., Mulley, C.: Mobility as a service (MaaS): charting a future context. Transp. Res. Part A: Policy Pract. **131**, 84–88 (2020). https://doi.org/10.1016/j.tra.2019.09.030
9. Pangbourne, K., Mladenović, M.N., Stead, D., Milakis, D.: Questioning mobility as a service: unanticipated implications for society and governance. Transp. Res. Part A: Policy Pract. **131**, 96–101 (2020). https://doi.org/10.1016/j.tra.2019.09.033
10. Cottrill, C.D.: MaaS surveillance: privacy considerations in mobility as a service. Transp. Res. Part A: Policy Pract. **131**, 53–56 (2020). https://doi.org/10.1016/j.tra.2019.09.026
11. Shao, Y., Dessouky, M.: A routing model and solution approach for alternative fuel vehicles with consideration of the fixed fueling time. Comput. Ind. Eng. **142**, 142–145 (2020). https://doi.org/10.1016/j.cie.2020.106364
12. Chao, Y., Zishan, M.: System dynamics model of shanghai passenger transportation structure evolution. Proc. – Soc. Behav. Sci. **96**, 100–105 (2013). https://doi.org/10.1016/j.sbspro.2013.08.127
13. Ma, Y., Gao, Y.: Passenger transportation structure optimization model based on user optimum. Proc. Eng. **137**, 68–70 (2016). https://doi.org/10.1016/j.proeng.2016.01.251
14. Egan, M., Jakob, M.: Market mechanism design for profitable on-demand transport services. Transp. Res. Part B: Method **89**, 90–94 (2016). https://doi.org/10.1016/j.trb.2016.04.020
15. Mulley, C., Nelson, J.D.: Flexible transport services: a new market opportunity for public transport. Res. Transp. Econ. **25**(1), 26–30 (2009). https://doi.org/10.1016/j.retrec.2009.08.008
16. Taylor, M., Hallsworth, A.: Power relations and market transformation in the transport sector: the example of the courier services industry. J. Transp. Geogr. **8**(4), 55–58 (2000). https://doi.org/10.1016/S0966-6923(00)00014-4
17. Williams, H.C.W.L., Abdulaal, J.: Public transport services under market arrangements, part I: a model of competition between independent operators. Transp. Res. Part B: Method **27** (5), 65–71 (1993). https://doi.org/10.1016/0191-2615(93)90023-4

Numerical Modeling of a Vertical Steel Tank Differential Settlement Development

Aleksandr Tarasenko[1], Petr Chepur[1], and Alesya Gruchenkova[2](✉)

[1] Industrial University of Tyumen, Volodarskogo str., 38, 625000 Tyumen, Russia
[2] Surgut Oil and Gas Institute, Entuziastov str., 38, 628405 Surgut, Russia
alesya2010-11@yandex.ru

Abstract. The object of this study is a vertical aboveground steel storage tank with a floating roof and a capacity of 50,000 m^3, which has an area of heterogeneity of the soil base under its bottom and foundation. The magnitude of the heterogeneity area is considered in a wide range of possible values in accordance with the diagnostic data of real facilities. The main cause of tank accidents is differential settlement with subsequent destruction of the metal structure. The determination of the actual stress-strain state of the tank during the development of differential settlement is an important task for determining the values of permissible deformations. In the study, numerical methods were used to solve the problem. To model the heterogeneity zone, the Drucker-Prager model of a linear elastoplastic material with implementation in the ANSYS finite element software package was used. A finite element model of the tank was developed with high detail of its metal structures: walls, bottoms, annular plate, stiffening rings. As a result of the calculations, the maximum possible settlement values of the RVSPK-50000 base were determined in the presence of a heterogeneity zone due to the stiffness of its metal structure. The dependences of the maximum stresses on the value of differential settlement for the accepted range of values from 10 to 95 m were obtained.

Keywords: Numerical models · Tank · Drucker-Prager model

1 Introduction

Current trends in the development of oil main transportation can be characterized as a constant increase in the unit nominal storage volume in oil storage facilities and construction areas are located in more remote territories with complex engineering and geological conditions. Despite the fact that vertical steel tanks with a capacity of more than 50,000 m^3 fall within facilities of hazard class I and the requirements for design solutions and the quality of construction and installation works are at a very high level, there are cases of the development of differential settlement of such facilities [1–4].

Studies devoted to the analysis of the stress-strain state of tanks during the development of base settlement, as a rule, are limited to the strength analysis of the wall and bottom structures [5, 6]. The properties of the soil base are either not taken into

V. Murgul and V. Pukhkal (Eds.): EMMFT 2019, AISC 1258, pp. 60–70, 2021.
https://doi.org/10.1007/978-3-030-57450-5_6

account at all, or are set according to simplified models that do not correctly reflect the real nature of its deformation under the influence of operational and off-design loads. Ultimately, this approach leads to greater errors in determining the stress-strain state of the structure as a whole, in particular, its lower assembly.

It is also necessary to take into account the fact that many of the tanks described were built on pilot projects. Thus, any project includes elements of research and calculations for new working conditions of the reconstructed elements of a reinforced concrete tank. In world practice, there are cases of trouble-free operation of vertical steel tanks (VSTs) with differential settlements of more than 1 m [7]. However, according to the current design standards in the Russian Federation, permissible differential settlement of the external bottom contour is limited to 40 mm for new VSTs and 80 mm for those operated for more than 20 years. Permissible settlement values in storage tank construction have significant differences in different countries, which is due to the type of structures, materials and design approaches used in regulatory documents.

Errors at the stages of surveys and design, violations of the technology of construction and installation works, deviations in the mode of hydraulic testing, changes in the technological scheme of pumping oil after commissioning of the facility (as a result, the appearance of off-design loads) may be the causes of the appearance of zones of heterogeneity and the development of differential tank settlements.

Therefore, the aim of the present paper is to develop a numerical model of a vertical steel tank, which would allow us to determine the ultimate deformation parameters of the RVSPK-50000 tank in the presence of the soil base heterogeneity zone. Moreover, based on the results of diagnostic inspections and statistical reports, it is necessary to take into account all possible in practice arc intervals of heterogeneous zones under the external bottom contour. The authors also set the task of obtaining an array of parameters of the stress-strain state of the structure, taking into account the joint work of the soil base and metal structures of the tank as a thin-walled shell structure with finite bending stiffness.

2 Methods

An analysis of the diagnostic results of more than 40 tanks of the RVSPK-50000 type showed that any settlement of the external bottom contour can be represented by a combination of the magnitude of the arc of the external bottom contour L, along which the settlement develops, and the vertical component of the settlement u. The amount of settlement for L with a combination of operational loads is finite and will be determined by the elastic-plastic properties of the tank [8, 9]. Knowing the entire interval of settlement values encountered in practice and changing the magnitude of the arc on which it occurs, one can obtain the interval of possible values of deformations of the external tank bottom contour. Settlement cannot develop beyond these values. For research the authors propose to use the ANSYS software product tools, namely, using a foundation with minimal strength properties in the zone of differential settlement development. An analysis of the interaction of the foundation and structure is proposed to be performed using a soil model based on the Drucker-Prager yield criterion.

This model uses the Drucker-Prager yield criterion with either an associated or nonassociated flow rule. The yield surface does not change with progressive yielding, hence there is no hardening rule and the material is elastic - perfectly plastic. The equivalent stress for Drucker-Prager is:

$$\sigma_{e} = 3\beta\sigma_{m} + \left[\frac{1}{2}\{s\}^{T}[M]\{s\}\right]^{\frac{1}{2}} \tag{1}$$

σ_m – mean or hydrostatic stress; $\{s\} = \{\sigma\} - \sigma_m \, [1\ 1\ 1\ 0\ 0\ 0]^T$ – deviatoric stress; β – material constant; [M] – as defined with (4.1–24) [10].

This is a modification of the von Mises yield criterion that accounts for the influence of the hydrostatic stress component: the higher the hydrostatic stress (confinement pressure), the higher the yield strength. β is a material constant which is given as:

$$\beta = \frac{2\sin\phi}{\sqrt{3}(3 - \sin\phi)} \tag{2}$$

φ – input angle of internal friction.

The material yield parameter is defined as:

$$\sigma_y = \frac{6c\cos\phi}{\sqrt{3}(3 - \sin\phi)} \tag{3}$$

c – input cohesion value.

With the advent of powerful finite element analysis packages, it became possible to obtain reliable numerical solutions taking into account the real geometry of the system and the high detail of its interacting elements.

The authors propose to take advantage of the capabilities of the ANSYS finite element software package and analyze the structural behavior of RVSPK-50000, taking into account the joint work of the annular reinforced concrete foundation, the metal structure of the tank and the base with heterogeneity zones composed of weak highly compressible soils. Our experience [2, 11] suggests that for the correct solution of such a problem it is necessary to develop the most accurate tank model that takes into account not only the stiffness of the structure, but also the contact interaction at the boundaries of the elements of the “soil – foundation – bottom – wall” system. When solving the contact problem, it is necessary to account for the possibility of separation of contact surfaces when modeling settlement, for example, disconnection of the storage tank metal structure from the foundation ring, and the foundation sagging above the heterogeneity zone.

The finite element model of the tank was constructed in accordance with the model design of RVSPK-50000 by Melnikov Central Research and Design Institute of Steel Structures; its verification was considered in [1]. The diameter of the tank is 60.7 m; its height is 17.95 m; wall thicknesses for belts I–XII vary from 17 to 8 mm with alignment along the inner surface; the stiffening rings from a bent profile—a corner 100 × 300 with a thickness of 8 mm—are located on the V and VIII wall belts; the wind girder is an L-shaped design of paired sheet and beam elements reinforced with spacers with an interval of 2.5 m, welded to the XII wall belt through intermediate mounting plates; the foundation is circular, from reinforced concrete, of a rectangular profile with dimensions 1.5 × 0.4 m; the lower nine wall belts and annular plates are made of steel 16G2AF (manganese-vanadium alloy steel with nitrogen) with a yield strength σt = 440 MPa, other structures are made of steel 09G2S (low-alloyed silicon-manganese steel) with a guaranteed yield strength σ_t = 325 MPa. To model walls, bottom, annular plate, stiffness rings, four-node SHELL181 finite elements with six degrees of freedom in each node are used, taking into account membrane tension – compression and bending; for the ring foundation, a 20-node SOLID186 element is used which has three degrees of freedom in each node. The soil base is modeled using 10-node SOLID187 finite elements with three degrees of freedom in each node and support for large deformations.

Figure 1 presents the proposed design scheme of RVSPK-50000, taking into account the current loads and fixing conditions.

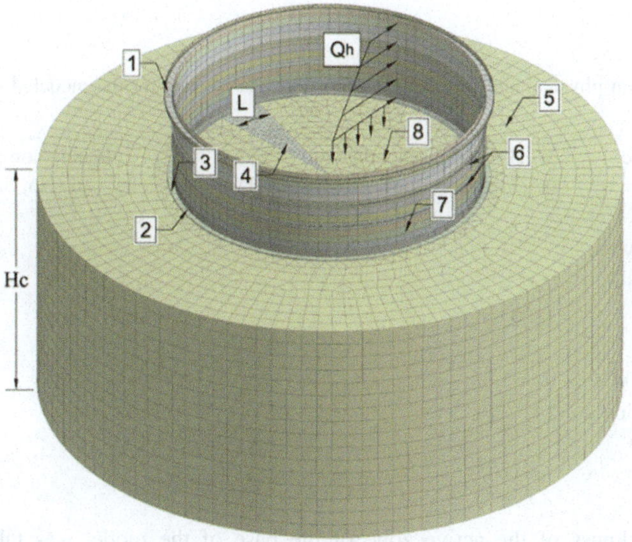

Fig. 1. Design scheme: 1 – wind girder; 2 – tank bottom edge; 3 – reinforced concrete ring foundation; 4 – highly compressed soil in the heterogeneity zone; 5 – soil base with design characteristics; 6 – stiffness rings at wall courses V and VIII; 7 – wall; 8 – bottom; Q_h – hydrostatic load; L – length of the sector in the heterogeneity zone (along the external bottom contour); H_c – compression depth.

Diagnostic surveys [12, 13] confirm that the heterogeneity zone under the bottom and annular plate of the VST, as a rule, has the shape of a triangular sector. The heterogeneity zone in this study is defined by a triangular sector with an arc size L along the external contour of the VST. Slepnev I.V. proposed such a design scheme for the first time in his work [3], where the structural loading was created by cutting a segment of the foundation ring of the VST. However, in [1], a section of the wall, annular plate and bottom "sagged" over the heterogeneity zone, i.e. there was no soil and foundation within the given sector. This was justified by the fact that such a loading scheme reflects the most unfavorable working conditions of the VST metal structure in the presence of a heterogeneity zone.

When performing test calculations, as well as analyzing the works [14, 15] and the data of diagnostic surveys [12, 13], the intervals were set that determine the minimum and maximum size of the sector of the heterogeneity zone of the RVSPK-50000 base (along the external bottom contour). The minimum value was L_{min} = 10 m, the maximum L_{max} = 95.3 m (which corresponds to the roll of RVSPK-50000). The properties of highly compressible soil of the heterogeneity zone were set using the Drucker-Prager model, taking into account the most unfavorable case that has ever been encountered according to diagnostic surveys of real storage tanks. Table 1 shows the physical and mechanical characteristics of soils for the modeled sections of the base.

Table 1. Main physical and mechanical characteristics of soils for the modeled sections of the base.

Characteristic	Weak fluid-plastic clay soil of the heterogeneity zone (No. 4, Fig. 1)	Artificially compacted sand-clay soil (No. 5, Fig. 1)
Elasticity modulus E, MPa	5	30
Poisson ratio μ	0.43	0.3
Density ρ, kg/m^3	1800	1650
Cohesion c, kPa	0	4
Internal friction angle φ, deg	14	36

The thickness of the active zone of the base of the model was taken equal to H_c = 40 m. This value was determined in accordance with the recommendations of Standard 653 American Petroleum Institute from the results of calculating the foundation plate on an elastic base for highly compressible soil of the heterogeneity zone (4, Fig. 1). The area of the active zone of the base is assumed to be 11,304 m^2 and has a diameter of 120 m.

The boundary conditions of the model are determined by the rigid fixing of the bottom face of the soil base at the "–40 m" mark and also by the restriction of lateral soil movement around the perimeter of the computational domain. A non-trivial task of the developed model is to take into account the contact interaction of loaded structures.

Thus, the contacts of the metalwork of the bottom, walls and annular plates (surface – surface, edge – surface) connected by welds are modeled as bonded deformable, but without the possibility of separation, a bonded contact. However, the contact of the bottom, annular plate and the foundation with the soil base, and the contact of annular plates with the foundation ring cannot be set as bonded, because with large deformations in the heterogeneity zone of the base, the contact area can change. Detachment of metal structures from the foundation, rising and lowering of the foundation ring and the bottom can occur, as a result of which interpenetration of the contact surfaces (reinforced concrete foundation into the ground) and their separation at any point are possible. Therefore, the authors used an algorithm - an extended Lagrange method, which takes into account the sliding friction force proportional to the normal reaction in the contact model [10]. The contact area in this case can be changed and contains both sections of adhesion, sliding, and complete separation. This method allows controlling the amount of penetration of the contact surfaces, while calculating the value of contact stiffness in the normal direction based on the Young's modulus - E and the size of adjacent elements [16]. The calculation results confirm the appropriateness of the applied approach. So, with large values of the heterogeneity zone, the RVSPK-50000 corner weld joint actually "hangs" over the base section composed of highly compressible soil. The model consists of 301,647 finite elements and takes into account physical and geometric nonlinearities.

3 Results and Discussions

Figure 2 and 3 show diagrams of deformations of the soil base and metalwork of RVSPK-50000 with the minimum considered heterogeneity zone L = 10 m and the maximum L = 95.3 m (along the external bottom contour), in which case a roll of the tank with a bent fracture occurs at the transition boundary of a weak soil and soil with design characteristics. In the design scheme of RVSPK-50000 with the presence of a heterogeneity zone, the hydrostatic load Q_h = 144.2 kPa and the height of the oil innage level H = 17 m (oil density ρ = 865 kg/m^3) are taken into account. For visualization, the diagrams present soil and metalwork deformations using a scale factor of $\times$10.

The authors interpreted the data obtained as a result of finite element modeling. Figure 4 presents a graph that shows the dependences of the maximum vertical component of differential settlement u on the length of the heterogeneity zone sector, and the dependences are presented for three key zones - bottom ring contour, bottom and wall. Because at a given ratio of thickness to diameter bottom behaves like a membrane, its deformations have maximum values and exceed 900 mm in the vertical direction with a sector L = 95 m. With a maximum filling of the tank, the bottom ring contour settlement with the same sector L is 325 mm. As an element having the greatest cylindrical stiffness with such a design scheme, the wall is deformed by 110 mm. For the differential settlement zone with a sector length L = 10, the maximum deformations are 205, 80, and 18 mm, corresponding to the previous listing. The greatest settlement values are observed in the middle of the heterogeneity zone sector under the bottom, regardless of the length of the sector.

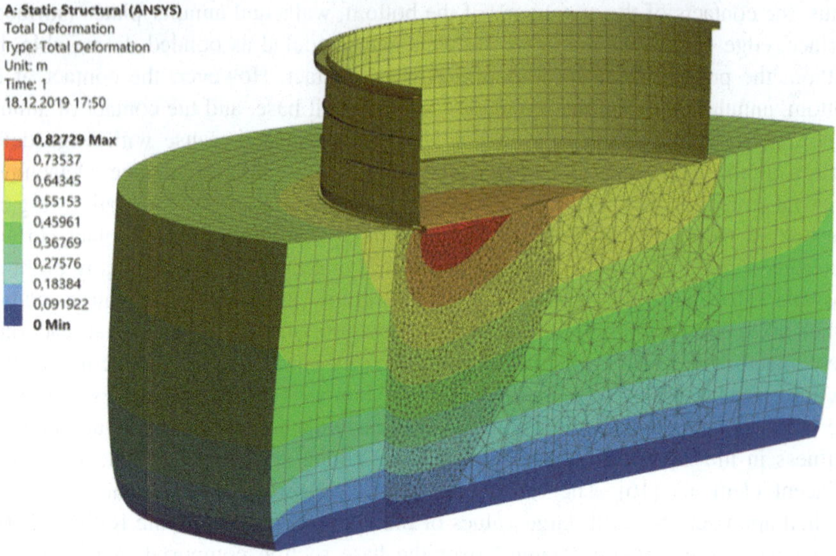

Fig. 2. Diagram of displacements for the base and foundation of the RVSPK-50000 storage tank: with minimum L = 10 m (in section view).

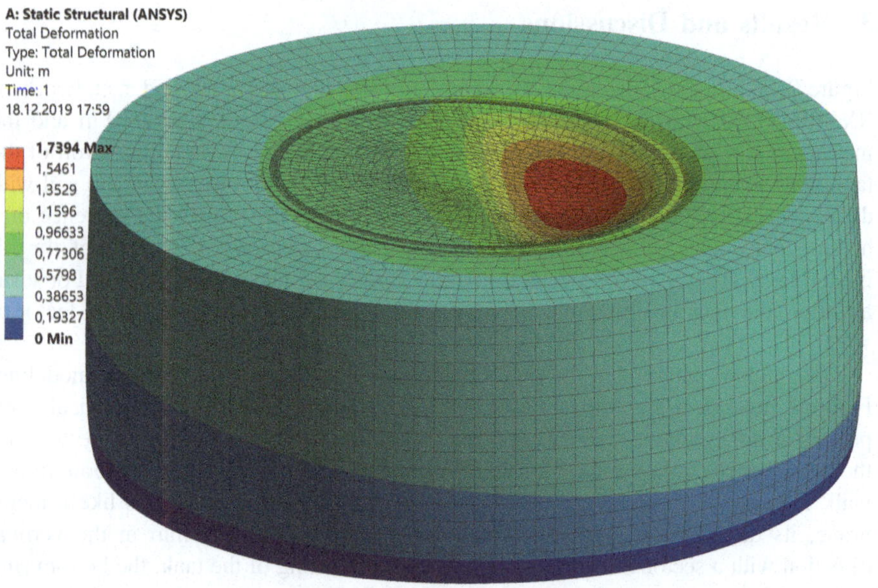

Fig. 3. Diagram of displacements for the base and foundation of the RVSPK-50000 storage tank: with maximum L = 95.3 m of the heterogeneity zones.

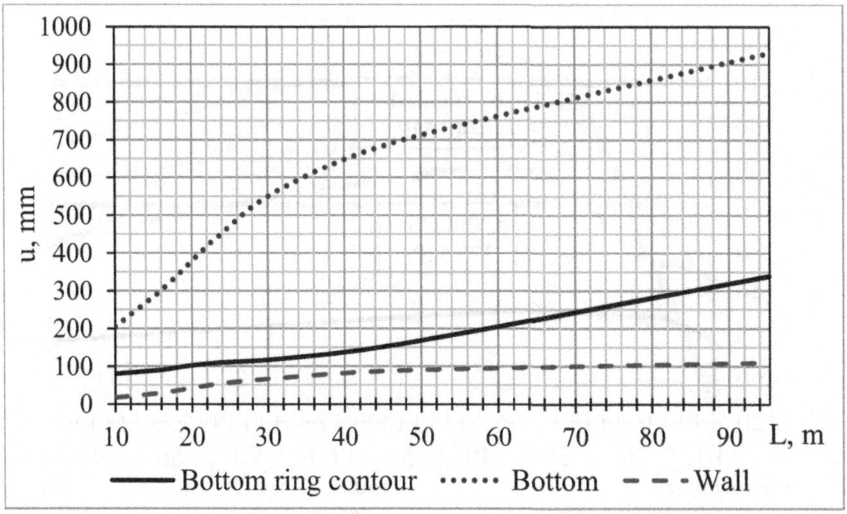

Fig. 4. Dependence of the maximum calculated settlement of RVSPK-50000: a) bottom ring contour b) bottom c) wall on the sector length of the heterogeneity zone.

Note that structures of this type have great cylindrical stiffness, which prevents the development of differential settlement. However, it must be borne in mind that when the external bottom contour settles by more than 40 mm with a 20-m length of the heterogeneity zone sector, a sharp increase in stresses occurs in the tank metal structures, which confirms the requirements of the current regulatory documents.

The developed model with the contact interaction of the base with the structure made it possible to establish the acting stresses in the tank metalwork and determine the most unfavorable cases of deformation during differential settlement (Fig. 5).

With a sector length of the heterogeneity zone from 30 to 50 m, the maximum stresses in all the load-bearing elements of RVSPK-50000 exceed the yield strength of steel. However, at maximum values of the sector of the heterogeneity zone (70–95.3 m), the stress-strain state level decreases. This is due to the stiffness characteristics of the structure and the forms of possible deformations under non-axisymmetric loads. Note that the greatest stresses occur in three zones: the wind ring above the area of heterogeneity, in the places of wall fracture at the boundary of two types of base soils, as well as in an additional stiffening ring on the V belt of the wall.

Figure 5 shows the dependences of the acting equivalent stresses in the metal structures of RVSPK-50000 on the sector length of the heterogeneity zone for metal structures of belts I–XII of the wall, stiffening rings in belts V and VIII, as well as the wind girder.

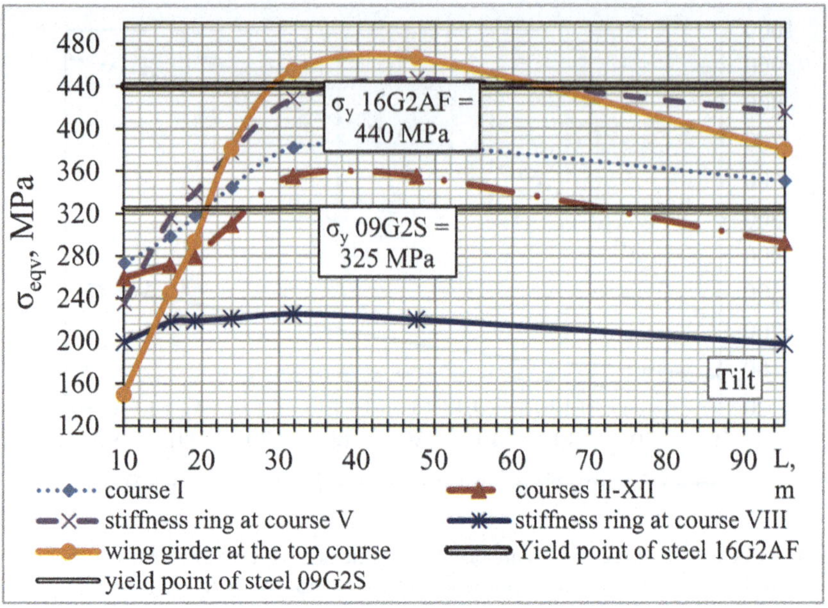

Fig. 5. Dependence of acting equivalent stresses in the metalwork of RVSPK-50000 tanks on the length of the heterogeneity zone sector L.

4 Conclusions

1. The finite-element model of RVSPK-50000 was developed which allows determining the stress-strain state of the structure and soil mass with the existing heterogeneity zone taking into account the Drucker-Prager physical model, contact interaction of the "base-foundation-bottom-tank" system.
2. It was established that with the development of differential settlement, the separation of the reinforced concrete ring with the annular plate and the bottom leads to a sharp increase in the stress-strain state of the entire structure, which indicates the need to develop and improve solutions to strengthen the design of the foundation ring.
3. The dependences of the maximum calculated settlement value of the bottom, annular plate and walls of RVSPK-50000 on the length of the heterogeneity zone sector (in the interval L from 10 to 95 m) were obtained 95% of all cases of differential settlement encountered in practice according to the inspection of 40 vertical steel storage tanks fit into this interval [12, 13].
4. The dependences of the maximum acting stresses on the size of the heterogeneity zone L for various structural elements of the tank are obtained: wall belts 1–12, stiffening rings on belts 5 and 8, and the wind girder.
5. The proposed approach can be extended to the stress-strain state analysis during the development of differential base settlement for other tank sizes constructed both according to Russian and foreign designs.

References

1. Tarasenko, A.A., Konovalov, P.A., Zekhniev, F.F., Chepur, P.V., Tarasenko, D.A.: Effects of nonuniform settlement of the outer bottom perimeter of a large tank on its stress-strain state. Soil Mech. Found. Eng. **53**(6), 405–411 (2017). https://doi.org/10.1007/s11204-017-9420-1
2. Tarasenko, A., Chepur, P., Gruchenkova, A.: Determining deformations of the central part of a vertical steel tank in the presence of the subsoil base inhomogeneity zones. In: AIP Conference Proceedings, vol. 1772, p. 060011 (2016). https://doi.org/10.1063/1.4964591
3. Slepnev, I.V.: Stress-Strain Elastic-Plastic State of steel Vertical Cylindrical Tanks with Inhomogeneous Base Settlement. Moscow Engineering and Building Institute, Moscow (1988)
4. Gorelov, A.S.: Heterogeneous Soil Bases and Their Influence on Work Vertical Steel Tanks. Nedra, Saint Petersburg (2009)
5. Korobkov, G.E., Zaripov, R.M., Shammazov, I.A.: Numerical Modeling Stress-Strain State and Stability of Pipelines and Tanks in Difficult Operating Conditions. Nedra, Saint Petersburg (2009)
6. Gorban, N.N., Vasiliev, G.G., Leonovich, I.A., Salnikov, A.P.: Study of the functioning models of tank farms of marine terminals in the Russian Federation. Oil Ind. **1**, 77–80 (2020). https://doi.org/10.24887/0028-2448-2020-1-77-80
7. Tarasenko, A.A., Chepur, P.V.: Aspects of the joint operation of a ring foundation and a soil bed with zones of inhomogeneity present. Soil Mech. Found. Eng. **53**(4), 238–243 (2016). https://doi.org/10.1007/s11204-016-9392-6
8. Lukyanova, I.E., Mikhailova, V.A., Kantemiov, I.F., Yakshibaev, I.N.: Study of ignition of binding substances used in foundations of tanks. In: IOP Conference Series Earth and Environmental Science, vol. 1, no. 378, p. 012019 (2019). https://doi.org/10.1088/1755-1315/378/1/012019
9. Terzeman, J.V., Teregulov, M.R.: Analysis of the stress-strain state of the toroidal transition connecting the wall and the bottom of the tank. In: Journal of Physics: Conference Series, vol. 1425, p. 012001 (2020). https://doi.org/10.1088/1742-6596/1425/1/012001
10. Bruyaka, V., Fokin, V., Soldusova, E., Glazunova, N., Adeyanov, I.: Engineering Analysis in ANSYS Workbench. Samara State Technical University, Samara (2010)
11. Tarasenko, A., Chepur, P., Gruchenkova, A.: The use of a numerical method to justify the criteria for the maximum settlement of the tank foundation. In: AIP Conference Proceedings, vol. 1899, p. 060003 (2017). https://doi.org/10.1063/1.5009874
12. Gorban, N.N., Vasiliev, G.G., Leonovich, I.A.: Analysis of existing approaches to modeling cyclic loading of the oil tank wall of marine terminals. Oil Ind. **3**, 110–113 (2019). https://doi.org/10.24887/0028-2448-2019-3-110-113
13. Yudakov, V.A., Fan, S.D., Fan, I.A., Teregulov, M.R., Bagdasarova, Y.A.: Improving the operational reliability of vertical steel tank bottoms for oil and petroleum products. Oil. Bus. **8**(608), 59–65 (2019). https://doi.org/10.30713/0207-2351-2019-8(608)-59-65
14. Vasiliev, G.G., Salnikov, A.P., Katanov, A.A., Likhovtsev, M.V., Ilin, E.G.: Optimizing the desktop processing of the terrestrial laser scanning data in assessing the stress-strain state of tanks. Pip. Sci. Tech. **3**(2), 112–117 (2019). https://doi.org/10.28999/2514-541X-2019-3-2-112-117

15. Latypova, L.A., Lukyanova, I.E.: Calculation of the stress-strain state of a vertical steel tank with a volume of 5000 m^3 under horizontal seismic impact in the ANSYS software complex. Oil Gas Bus. **1**(17), 113–119 (2019). https://doi.org/10.17122/ngdelo-2019-1-113-119
16. Korobkov, G.E.: Numerical modeling of stress-deformed state and stability of pipelines and tanks in complicated operating conditions. Nedra, Saint Petersburg (2009)

New Methods for Determining Poisson's Ratio of Elastomers

Viktor Artiukh[1(✉)], Vladlen Mazur[2], Yurii Sagirov[3], and Arkadiy Larionov[4]

[1] Peter the Great St. Petersburg Polytechnic University, Polytechnicheskaya, 29, 195251 St. Petersburg, Russia
artiukh@mail.ru
[2] LLC 'Saint-Petersburg Electrotechnical Company', Pushkin, Parkovaya, 56, 196603 St. Petersburg, Russia
[3] Pryazovskyi State Technical University, Universytets'ka, 7, Mariupol 87500, Ukraine
[4] Moscow State University of Civil Engineering, Yaroslavskoe Shosse, 26, Moscow 129337, Russia

Abstract. Nowadays, coefficient of transverse deformation of material (Poisson's ratio) can be determined by experimental methods very approximately. The reasons for the low accuracy are in the method itself, by which deformations are measured in local areas and very approximately. These deformations can be unevenly distributed over cross section and volume of a sample. It is proposed in this paper to determine Poisson's ratio through bulk modulus. Various new schemes of experiments with complex stress state were implemented. Features of determining Poisson's ratio of elastomers are shown. Any theoretical formula that includes Poisson's ratio is suitable for this; for example, Hooke's law or formula that establishes relationship between three elastic constants or formula of bulk modulus. From these formulas interesting and original experiments follow that make it possible to determine Poisson's ratio with high accuracy. All elastomers have almost the same bulk modulus, which turned out to be 3000 MPa despite different values of Young's modulus. Error for elastomers is approximately 2%; the error increases for harder materials. Thus, it becomes possible to determine Poisson's ratio with sufficient accuracy for any material.

Keywords: Elastomer · Deformation · Poisson's ratio · Elasticity · Energy-efficiency · Stress state

1 Introduction

Practice of designing and operating machines for various purposes is characterized by widespread usage of new structural materials, namely plastics, ceramics, elastomers. These materials significantly differ from usual steel grades. Point is not only that any new material is more durable or less durable than steel, these are fundamentally different materials with different elastic characteristics, different specific energy-efficiency and other time dependences [1–8].

V. Murgul and V. Pukhkal (Eds.): EMMFT 2019, AISC 1258, pp. 71–80, 2021.
https://doi.org/10.1007/978-3-030-57450-5_7

2 Formulation of Task

At the same time, long-term practice of working with different steel grades has some negative aspects associated with fact that elastic constants of steel, namely Young's modulus, shear modulus and Poisson's ratio are structurally insensitive and they are not analyzed during transition from one steel grade to another. However, this approach is unacceptable in relation to other materials. In reference literature lack of information is often seen in chapter 'Poisson's ratio' or 'shear modulus'. It is not at all because they are not defined but rather because secondary importance is given to these characteristics [9–15].

3 Objectives of This Paper

There is opinion that strength of structure or specific detail is determined by strength characteristics of material from which this detail is made. In fact, this is true only in relation to standard samples of materials. For most details and devices it is much more complicated. It is enough to take any statically indeterminate structure or case of impact loading to make sure that all elastic constants significantly affect strength, stiffness and energy-efficiency of structure because distribution of forces inside considered system depends on them [16–21].

4 Materials and Methods

Buffer device shown in Fig. 1 can be considered as example. Energy of impact is absorbed during volume compression using plunger pos. 2 of elastomer pos. 1 installed in housing pos. 3. It is easy to show from generalized Hooke's law that strength of the housing depends on the elastomer/filler of inside volume of the housing. Even with static loading of such device value μ determines pressure on the housing walls. At $\mu = 0.5$ the pressure is maximum and at $\mu = 0$ the pressure is completely absent.

During impact loading application its value depends on initial value of energy and rigidity of the system receiving the impact. Rigidity of the buffer device shown on Fig. 1 is determined by bulk modulus of the elastomer (filler) which is related to other elastic constants by equation

$$K = \frac{E}{3(1 - 2\mu)} \tag{1}$$

From Eq. (1) it can be seen that Poisson's ratio of the filler significantly affects rigidity of the buffer device and, consequently, value of dynamic loads. This effect becomes crucial if elastomer is used as filler. Reference data for different rubber grades gives value equal to $0.47 \leq \mu \leq 0.50$. These data must be treated with great caution. Value $\mu = 0.5$ is meaningless because it means that incompressible material is used but there are no such materials. Device shown on Fig. 1 becomes inoperative with such filler.

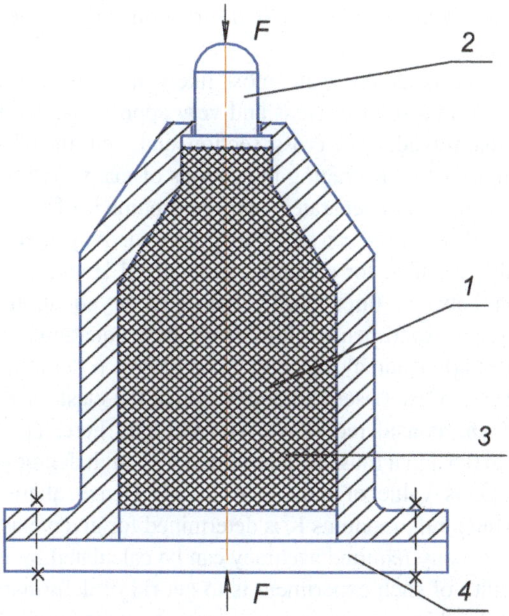

Fig. 1. Buffer device with elastomer: pos. 1 is elastomer; pos. 2 is plunger; pos. 3 is housing; pos. 4 is bottom.

Two values of μ that differ by 0.01 can be taken, for example, $\mu_1 = 0.48$ and $\mu_2 = 0.49$. Both of these values are in range of possible ones. From Eq. (1) it can be obtained

$$K_1 = \frac{E}{3(1-2\mu_1)} = \frac{E}{3(1-2\cdot 0.48)} = \frac{E}{0.12};$$
$$K_2 = \frac{E}{3(1-2\mu_2)} = \frac{E}{3(1-2\cdot 0.49)} = \frac{E}{0.06}, \quad (2)$$

it means that stiffness changes by 2 times. This emphasizes importance of accurately determining Poisson's ratio. Value of μ for elastomers must be known with accuracy of 0.001 and sometimes even higher and it can be achieved.

Well-known methods of determining μ can be considered because there is the same situation here as in reference data [11–14]. Poisson's ratio is found as per equation

$$\mu = \left|\frac{\varepsilon'}{\varepsilon}\right|, \quad (3)$$

transverse deformation ε′ and longitudinal deformation ε are measured at any place of sample using mechanical or electrical tensometers. It provides very low accuracy. For different steel grades range of $0.25 \leq \mu \leq 0.33$ can be obtained and for different rubber grades often $\mu > 0.5$ is obtained which makes no sense (if material is not biological, energized from outside and if destruction has not yet begun, i.e. micro cracks did not occur).

Reasons of such low accuracy are most likely in methodology itself by which deformations are measured in local areas and very approximately. These deformations can be unevenly distributed over cross section and volume of sample. Therefore, accuracy can be improved by averaging indications of many vertical sensors located in different places of sample and separately the same number of horizontal sensors.

New experimental schemes can also be proposed. Any theoretical formula that includes μ is suitable for this; for example, generalized Hooke's law or equation that establishes relation between three elastic constants or equation of bulk modulus. Interesting and original experiments follow from these equations that make it possible to determine μ with high accuracy. Specific examples of μ definition are given below.

Example number 1. Case described above (when the elastomer is compressed in a closed volume) can be considered. There is specificity here, i.e. Poisson's ratio for elastomers is close to 0.5 and it must be determined with high degree of accuracy because this significantly affects value of bulk modulus K. Hence, there is conclusion that experiment from which bulk modulus K is determined for given elastomer can be made and then μ with almost any required accuracy can be calculated by usage of Eq. (1).

The main difficulty of such experiment is to get rid of deformation of the housing, plunger, and testing machine during measurements; in other words, to isolate only bulk deformation of the elastomer from total deformation. Such experiment was done in laboratory 'Resistance of Materials' of 'Peter the Great St. Petersburg Polytechnic University'. Bulk modulus K and Poisson's ratio μ were determined for polyurethanes of grades 'SCU-PFL-70' (CIS) and 'Adipren L-167' as well as for rubber grade 'B-14' (CIS). Tested cylindrical sample was under compression.

Tested device is shown in Fig. 2 and Fig. 3. Different volumes of elastomer pos. 3 were compressed by plungers pos. 2 in the same metal housing pos. 1 (while maintaining loading scheme). In other words, two experiments were carried out:

1) with volume of the elastomer V_1;
2) with volume of the elastomer $V_2 = V_1 - V_0$, where V_0 is volume of metal detail pos. 4.

Searched stiffness characteristic was obtained as difference of two characteristics. Rigidity of the testing machine and the device itself was automatically subtracted in this case. Steel detail was considered absolutely rigid because bulk modulus of the elastomer is $K \approx 3000$ MPa [1, 3] which in comparison with steel gives an error of no more than 2%. For value of K this is quite acceptable; however, there are no technical obstacles to take into account rigidity of the steel detail.

The main results of these experiments are:

1. All elastomers, despite different values of E, showed almost the same modulus K which turned out to be equal to 3000 MPa.
2. Error for elastomers is approximately 2%; error increases for harder materials.
3. Following Poisson's ratio values were obtained for the above given elastomers:
   a. polyurethane 'SCU-PFL-70' $\mu = 0.4984$;
   b. polyurethane 'Adipren L-167' $\mu = 0.4970$;

c. rubber 'B-14' $\mu = 0.4993$.

4. All obtained values turned out to be larger than those given in reference literature.

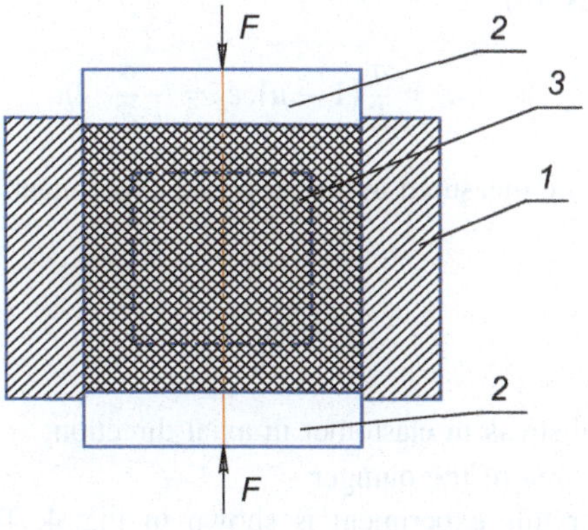

Fig. 2. First loading variant: pos. 1 is housing; pos. 2 is plunger; pos. 3 is elastomer.

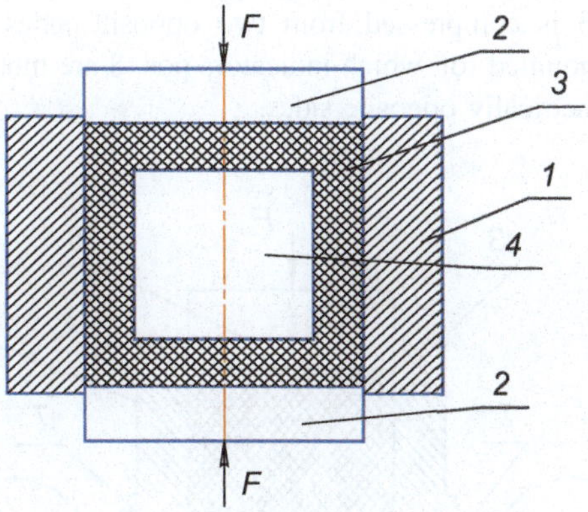

Fig. 3. Second loading variant: pos. 1 is housing; pos. 2 is plunger; pos. 3 is elastomer; pos. 4 is steel detail.

Example number 2. Equations of generalized Hooke's law:

$$\begin{cases} \varepsilon_x = \frac{1}{E}\left[\sigma_x - \mu(\sigma_y + \sigma_z)\right]; \\ \varepsilon_y = \frac{1}{E}\left[\sigma_y - \mu(\sigma_x + \sigma_z)\right]; \\ \varepsilon_z = \frac{1}{E}\left[\sigma_z - \mu(\sigma_x + \sigma_y)\right]. \end{cases} \quad (4)$$

Experiment in which the sample is loaded so that

$$\sigma_x = \sigma_y = \sigma; \sigma_z = 0. \tag{5}$$

It is obtained from Eqs. (4)

$$\varepsilon_x = \varepsilon_y = \frac{\sigma}{E}(1 - \mu); \varepsilon_z = -\frac{\sigma}{E} \cdot 2\mu. \tag{6}$$

E is known, σ and εz are measured in the proposed experiment. It allows to determine μ by equation

$$\mu = \frac{E \cdot \varepsilon_z}{2\sigma}. \tag{7}$$

where $\sigma = \frac{F}{A}$ is normal stress in elastomer in axial direction;

A is cross-section area of the plunger.

Device scheme for this experiment is shown in Fig. 4. The device consists of cylindrical housing pos. 1 with through transverse cylindrical hole pos. 2. In the housing pos. 1 there is elastomer cylindrical block pos. 3 with a transverse hole pos. 4 matching the hole pos. 2. Cylindrical sample pos. 5 is installed in the holes pos. 2 and pos. 4. Block pos. 3 is compressed from two opposite sides by plungers pos. 6. Brackets pos. 7 are mounted (on which indicators pos. 8 are mounted) on the housing pos. 1 from two diametrically opposite sides.

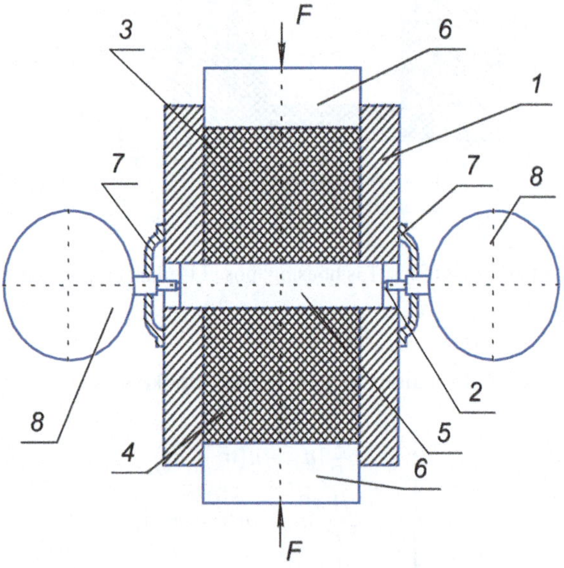

Fig. 4. Scheme of the device for determining μ: pos. 1 is housing; pos. 2 is hole in housing; pos. 3 is elastomer cylindrical block; pos. 4 is hole in the block; pos. 5 is sample; pos. 6 is plunger; pos. 7 is fixing bracket; pos. 8 is indicator.

The device is installed in the testing machine and loaded axially. Compressive force F is fixed using force meter of the machine and stress σ is determined from it; εz is measured by indicator; Poisson's ratio μ is determined by Eq. (7).

The most acceptable test parameters are: σ = 50…100 MPa; d = 10…20 mm; E = 10^3…10^4 MPa. It means that the method is suitable for plastics, glasses, concrete, granite, etc.

Advantage of the method is that entire volume of material is involved in the experiment, disadvantage of the method (which, however, can be significantly reduced) is in presence of friction between the elastomer and the sample. It should also be noted that for hard materials axial deformations of the sample are very small and difficult to measure with the indicator.

Example number 3. Experimental equipment shown in Fig. 5 can be considered. Cylindrical sample pos. 1 is rigidly fixed at one end and loaded at the other end by force F transverse to axis causing bending and torsion in the sample sections.

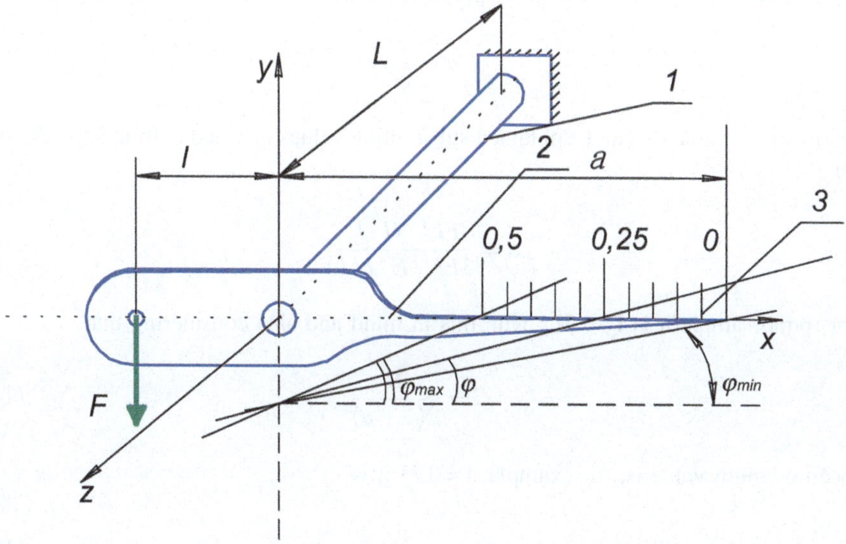

Fig. 5. Scheme of experimental equipment for determination of μ: pos. 1 is sample; pos. 2 is bar; pos. 3 is scale of fixed ruler.

Movement of end section of the sample can be considered. From bend of the sample it shifts down by value

$$y = \frac{F \cdot L^3}{3EI_x}, \tag{8}$$

and it rotates from torsion by angle

$$\varphi = \frac{Fl \cdot L}{G \cdot I_p}. \tag{9}$$

Distance a at which the bar pos. 2 (its right end) crosses horizontal (ruler pos. 3) can be found

$$tg\varphi = \frac{y}{a}, \tag{10}$$

it is from (10)

$$a = \frac{y}{tg\varphi}. \tag{11}$$

$tg\varphi \approx \varphi$ can be considered at first approximation because angle φ is small, then

$$a = \frac{y}{\varphi}. \tag{12}$$

Below given equation can be obtained substituting values of y and φ from Eqs. (8) and (9)

$$a = \frac{FL^3 \cdot G \cdot I_p}{3EI_x \cdot F \cdot l \cdot L}. \tag{13}$$

For round sample it is $I_p = 2I_x$; with this in mind and also considering that

$$G = \frac{E}{2(1+\mu)} \tag{14}$$

specific length value is, for example, l = L/3 it is

$$a = \frac{L}{1+\mu}, \tag{15}$$

it means that intersection with horizontal occurs at distance depending on desired value μ. Boundary values of a are equal to

$$a_{\min} = \frac{L}{1+0.5} = 0.667\,\mathrm{L};\, a_{\max} = \frac{L}{1+0} = 1.0\,\mathrm{L}. \tag{16}$$

This gap can be calibrated within range of μ that is from 0 to 0.5.

Measurement of μ on the experimental equipment is as follows:

1. Force F should be applied in the right place (value of the force does not play role but deformations should be small and elastic).

2. Line has to be drawn by pencil along the right side of the bar to intersection with axis x.
3. Scale on axis x should be read.

This method does not require special measuring devices and is universal, i.e. suitable for all materials.

5 Conclusion

These examples could be continued, and on the basis of the three described experiments, it can be concluded that value of μ can be determined for any material and with sufficient accuracy.

Acknowledgments. The reported study was funded by RFBR according to the research project №19-08-01241a. The authors declare that there is no conflict of interest regarding the publication of this paper. This research work was supported by the Academic Excellence Project 5-100 proposed by Peter the Great St. Petersburg Polytechnic University.

References

1. Al-Quran Firas, M.F., Matarneh, M.E., Artukh, V.G.: Choice of elastomeric material for buffer devices of metallurgical equipment. Res. J. Appl. Sci. Eng. Technol. **4**(11), 1585–1589 (2012)
2. Artiukh, V.G., Karlushin, S.Yu., Sorochan, E.N.: Peculiarities of mechanical characteristics of contemporary polyurethane elastomers. Procedia Eng. **117**, 938–944 (2015). https://doi.org/10.1016/j.proeng.2015.08.180
3. Artiukh, V.G., Galikhanova, E.A., Mazur, V.M., Kargin, S.B.: Energy intensity of parts made from polyurethane elastomers. Mag. Civil Eng. **81**(5), 102–115 (2018). https://doi.org/10.18720/MCE.81.11
4. Balalayeva, E., Artiukh, V., Kukhar, V., Tuzenko, O., Glazko, V., Prysiazhnyi, A., Kankhva, V.: Researching of the stress-strain state of the open-type press frame using of elastic compensator of errors of "Press-Die" system. In: Advances in Intelligent Systems and Computing, vol. 692, pp. 220–235. Springer (2018). https://doi.org/10.1007/978-3-319-70987-1_24
5. Artiukh, V., Mazur, V., Kukhar, V., Vershinin, V., Shulzhenko, N.: Study of polymer adhesion to steel. In: E3S Web of Conferences, vol. 110, p. 01048 (2019). https://doi.org/10.1051/e3sconf/201911001048
6. Ishchenko, A., Artiukh, V., Mazur, V., Poberezhskii, S., Aleksandrovskiy, M.: Experimental study of repair mixtures as glues for connecting elastomers with metals. In: MATEC Web of Conferences, vol. 265, p. 01016 (2019). https://doi.org/10.1051/matecconf/201926501016
7. Efremov, D.B., Gerasimova, A.A., Gorbatyuk, S.M., Chichenev, N.A.: Study of kinematics of elastic-plastic deformation for hollow steel shapes used in energy absorption devices. CIS Iron Steel Rev. **18**, 30–34 (2019)
8. Sotnikov, A.L., Rodionov, N.A., Ptukha, S.V.: Analysis of mechanical loading of the hinges and supports of the mold vibration mechanism on a continuous caster. Metallurgist **58**(9), 883–891 (2015)

9. Pestryakov, I.I., Gumerova, E.I., Kupchin, A.N.: Assessment of efficiency of the vibration damping material «Teroson WT 129». Constr. Unique Build. Struct. **5**(44), 46–57 (2016)
10. Yakovlev, S.N., Mazurin, V.L.: Vibroisolating properties of polyurethane elastomeric materials, used in construction. Mag. Civil Eng. **6**, 53–60 (2017). https://doi.org/10.18720/MCE.74.5
11. Datta, J.: Synthesis and investigation of glycolysates and obtained polyurethane elastomers. J. Elastomers Plast. **42**, 117–127 (2010)
12. Zhang, H., Chen, Y., Zhang, Y., Sun, X., Ye, H., Li, W.: Synthesis and characterization of polyurethane elastomers. J. Elastomers Plast. **40**, 161–177 (2008)
13. Rek, V.: Kinetic parameters estimation for thermal degradation of polyurethane elastomers. J. Elastomers Plast. **38**, 105–118 (2006)
14. Valero, M.F.: Preparation and properties of polyurethanes based on castor oil chemically modified with yucca starch glycoside. J. Elastomers Plast. **41**, 223–244 (2009)
15. Maniak, I., Melnikov, B., Semenov, A., Saikin, S.: Experimental investigation and finite element simulation of fracture process of polymer composite material with short carbon fibers. Appl. Mech. Mater. **725–726**, 943–948 (2015)
16. Gonella, L.B.: New reclaiming process of thermoset polyurethane foam and blending with polyamide-12 and thermoplastic polyurethane. J. Elastomers Plast. **41**, 303–322 (2009)
17. Yokoyama, N.: Properties of polyurethane containing new phenolic additives. J. Elastomers Plast. **39**, 347–369 (2007)
18. Belkin, A.E., Narskaya, N.L.: Raschet elastomernogo tsilindricheskogo amortizatora s uchetom vyazkih svoistv materiala [Analysis of an Elastomer Cylindrical Shock-Absorber with Regard to Viscous Properties of the Material]. Univ. News: Eng. **8**(665), 12–18 (2015). https://doi.org/10.18698/0536-1044-2015-8-12-18. (in Russian)
19. Jayasree, T.K.: Effect of fillers on mechanical properties of dynamically crosslinked styrene butadiene rubber. J. Elastomers Plast. **40**, 127–146 (2008)
20. Yakovlev, S.N.: Self-oscillation of an elastic polyurethane coating in polishing. Russ. Eng. Res. **34**(5), 295–298 (2014)
21. Yakovlev, S.N.: Dynamic hardening of structural polyurethanes. Russ. Eng. Res. **36**(4), 255–257 (2016)

Regularities of City Passenger Traffic Based on Existing Inter-district Links

Oleksandr Stepanchuk[1(✉)], Andrii Bieliatynskyi[1,2], and Oleksandr Pylypenko[1]

[1] National Aviation University, Kiev 03058, Ukraine
olstnau@gmail.com
[2] North Minzu University, 204 Nort-Wenchang Street, Xixia District, Yinchuan, Ningxia, People's Republic of China

Abstract. The study is aimed at solving the problem of creating and ensuring the conditions for optimal intensity of vehicles on the elements of the street network of cities. The basis of the work is experimental and theoretical studies of the principles of city passenger traffic and needs of city inhabitants in use of the street network. The results of a questionnaire survey of Kyiv residents regarding their commute to work are considered and analyzed, taking into account the administrative-territorial division of city territory. The survey of the population traffic in Kyiv was conducted, which revealed the principles of its traffic between the districts, the average travel distance within every district was determined, and the percentage of use of different types of vehicles was calculated as well. Such approach enables to determine the effectiveness of the use of transport links between city districts, which allows estimating the size of inter-district connections, determining the most priority and problematic traffic directions.

Keywords: City · Traffic system · Traffic flow · Passenger traffic · Traffic route · Transports links

1 Introduction

The everyday transportation of a large number of people on the city road network forms usually high-intensity vehicle and pedestrian flows that require extra time for large number of people to get to their destination points.

From this it follows that transport system of any modern city is one of the most important components affecting significantly the level of whole city infrastructure functioning.

The transport system is the most complex element of major and largest cities and its development and operation effectiveness determines the quality level of living conditions of the whole city.

As of today the observations of transport system operation condition in major and largest cities of Ukraine, and especially the city of Kyiv, proves its ineffectiveness. The vehicle delays occur frequently and time of such delays exceeds usually the time necessary for free traveling within the city by public transport or by individual cars.

V. Murgul and V. Pukhkal (Eds.): EMMFT 2019, AISC 1258, pp. 81–93, 2021.
https://doi.org/10.1007/978-3-030-57450-5_8

This requires the development and implementation of effective measures improving urban road traffic conditions, which depend on the technical state and the number of vehicles on the city road network. It should be noted also that over the last 25 years, the construction of new streets and roads in Ukrainian cities has been very slow (less than 1% over the whole period) against the background of an increase in the number of vehicles, especially cars, which is accompanied by an annual increase - from 3% to 5% [1]. This indicates that development of the city road networks and its elements is far behind of both the annual increase in the number of vehicles and the constant improvement of road technical and operational characteristics.

Thus, the issue of ensuring the necessary conditions for the efficient vehicle traffic on the road network of cities in the current situation is an urgent problem.

2 Materials and Methods

The transport problem of cities is that usually from 50% to 90% of their population is concentrated in relatively limited territory (2–5%) [2]. For example, the city of Kyiv has 2.76 million registered residents, while in fact 3.1 million people live in it and about 3.5 million people is the daily number of this city population. In the largest city of Ukraine exists essential disproportion between living places and workplaces: there are 1.35 million workplaces in the economic complex of the city, when 36.2% of city residents live on the Left Bank of city having only 19.7% of workplaces, while 63.8% of residents live on the Right Bank having 81.3% of jobs [2]. This disproportion problem between living places and workplaces has not been solved yet.

The disproportion of such type in urban planning decisions is the main reason why people are forced to travel across the city territory. Moreover, 70–90% of the traffic flow on the streets of urban settlements of Ukraine is formed by passenger transport and motor cars account for 84% of their total number. More than 95% of all motor cars are individual cars [2].

In recent years, many domestic and foreign scientists have been studying the traffic flows and the use of models to solve problems of traffic flows distribution on the city street network. These are the works of H. Barselo, J. Vardrop, D. Witham, O. Gasnikov, B. Grinschilds, V. Hooke, V. Dolly, D. Drew, V. Zhivogladov, H. Kassas, O. Lobashov, M. Ossetrin, V. Polishchuk, S. Ramming, E. Reitsen, V. Semenov, V. Silianov, V. Filipov, F. Heit, R. Herman, Y. Tsibenko, V. Sheshtokas, J. Sheffi, and others [3–8].

To find the efficient ways for solving the problems of traffic conditions improvement on the road network of cities, it is necessary to analyze possible methods that allow improving its transport and operational qualities, enhancing traffic safety conditions and increasing the traffic capacity of the road network elements. In the same time, to increase the vehicles traffic efficiency in urban conditions, the speed of goods delivery and passenger traffic, comfort and safety conditions of passengers, as well as to reduce the transportation cost, it is necessary to improve traffic conditions on arterial roads and city streets as well.

3 Results

In general, all vehicles moving on the city road network are divided into three groups:

- the vehicles moving according to a clear public transport schedule;
- vehicles moving according to production schedules of enterprises and organizations;
- vehicles moving free without any stable schedule.

The presence in the transport flow of such three types of vehicles, in fact, creates a disorganized nature of traffic on the street network. It is obvious that the implementation of general management decisions concerning planning the routes for individual cars on the streets of cities is very difficult, because it is impossible to predict the probable directions of their travelling.

The presence of significant percentage of individual cars on the road network creates various obstacles for traffic and impairs the public passenger transport operation. The traffic conditions of public passenger transport moving in a traffic flow are determined by the conditions of current traffic flow, which are characterized by two main indicators: traffic intensity level of road network and travel speed. The problem of increasing the speed and safety of public passenger transport while increasing the intensity of traffic flows becomes extremely urgent in all cities of Ukraine and at the same time very difficult to solve [9].

This problem has to be solved by using urban planning and traffic management methods beginning even at the design stage to develop functional zoning patterns of the city so as to take into account the effects of further motorization and possible problems of city transport service [10].

The main task of urban transport is to provide comfortable transportation of inhabitants and goods within city territory with minimal time spent [11, 12]. But, today, as it was mentioned above, the arterial roads of the largest cities of Ukraine are extremely "saturated" by individual cars throughout the working day and this causes congestions in the main directions of traffic [13].

As stated in [14], the formation of congestions on the city arterial roads is the result of combination of three main factors:

- organizational-managerial one, where the scheme of organization and management of traffic is designed without taking into account the peculiarities of formation and distribution of traffic flows on a given section of the road network;
- deficient one, where there is insufficient size of the roadway;
- unpredictable one, when there are possible accidents with severe consequences, adverse weather conditions, natural disasters;
- significant repair and construction work on the road network without the necessary organization and management of traffic.

All these factors are interrelated between each other. The causes of traffic congestions on city streets can be interconnected between each other in different combinations: external causes (traffic accidents, road works, and weather conditions); the level of transport demand (daily fluctuations of traffic intensity, fluctuations associated with various activities); physical parameters of roads (technical means of traffic organization, changes in traffic capacity).

It is clear that solution of problems dealing with traffic improvement on city streets and travel time reduction require making a few necessary actions, which include a set of planning, engineering, management and organizational measures.

The road network is a very expensive and difficult-to-alter element of urban infrastructure. Its designing process is one of the complex issues of urban transport planning theory. Making any city planning decision related to solving the city's socio-economic problems will, in most cases, lead to an increase and change of traffic intensity on road network of cities.

The increase of urban traffic volume requires an increase of capital investment in the construction of new and reconstruction of existing transport facilities, improvement of the quality of design works and more detailed feasibility study of them.

But, as practice shows, the measures dealing with construction or reconstruction of sections or elements of the city road networks not always lead to a positive result. It should be noted that the construction of new sections of streets, roads, overpasses, bridges, etc. solves the transport problem in the city for a very short time period. After the construction work completion, a significant number of vehicles begin to include such new sections of the street network into their routes immediately and with time the total number of vehicles using the new sections increases again leading to a further deterioration of road conditions in this traffic direction.

The issue of ensuring the rational development of the city street network is generally reduced to determining the following parameters:

- network capacity E, which represent the network ability to insure the certain volumes of urban traffic;
- network saturation level G, which represent the maximum expected volume of urban traffic.

The condition $G > E$ means that city road network does not perform its functions, it does not provide necessary traffic for growing passenger and vehicle traffic flows. The condition $G < E$ means that city has an excessive road network, which leads to irrational investment in road construction. Thus, in order to substantiate the road network development, it is necessary to find parameters that correspond to the maximum values of G and E characteristics.

The established rational relations between these parameters provide the most expedient development of the city road network. It should be noted that in many cities, even when the parameters G and E correspond each other, there are congested places on certain sections of the road network and transport hubs, which lead to significant time losses, traffic speed reduction, etc.

Therefore, it is urgent to question the effective use of the potential of the existing road network, the ability to pass the maximum number of vehicles and the proper satisfaction of the city inhabitants needs in their transportation. Today, the increase of vehicle traffic volume has further focused on the need to assess the correspondence level of the existing street network to current volumes of vehicle and pedestrian traffic and to evaluate how reliable the network will be in the future.

The potential of the road network enables to reveal the proportion of the network sections being used efficiently for vehicle and pedestrian traffic and the proportion of idle sections of the road network in the course of their intensive operation. Each route should have in its potential sufficient reserves and, likewise, duplicate traffic options that will ensure the traffic flows without delay.

In [15], it is stated that the proportion of efficient use of the road network is obtained by comparing the actual transport capacity of the certain sections with the potential of the whole network, where operation efficiency of the road network's elements (bridges, interchanges, intersections, etc.) are analyzed additionally.

$$Q_{rnw} = \sum l_{ef} / \sum l_{rnw} \tag{1}$$

where Q_{rnw} - proportion of efficient use of the road network;
l_{ef} - length of the road network section used efficiently, m;
l_{rnw} - total length of city road network, m

In [1] it is noted that tens of thousands of cars, buses, trolleybuses form the traffic flows structure on road network of cities every day, and, at first glance, it seems that there is no regularity in these traffic flows, but, in fact, each city has only its own rhythm of traffic. The city traffic rhythm is an objective spatial-temporal regularity in the integrated traffic flow conditions. It depends on urban, planning, economic, social aspects and on methods and means of traffic management.

Therefore, in this study, we are interested in the influence of the city's living mechanism on its street network, in particular, the correspondence of urban traffic demand to the city's transport supply.

Hence, in order to achieve our goal, it is necessary to pay attention precisely to the city population traffic, taking into account the volume and direction of population traffic in the city territory, and determine the indicator of the average travel distance of the city inhabitants. This will enable to identify the directions of correspondence to its inhabitants and the main traffic routes of vehicles, as well as to identify the busiest transport hubs and road sections on each route and predict alternative traffic routes [16].

To make a reasonable forecast of the prospects for the development of the city road network and its transport, it is necessary to determine, by theoretical calculations, the nature of passenger traffic and its volume depending on different type of transport and traffic routes. The theoretical methods of passenger flows calculation are based on the objective laws of urban traffic within the city, the nature and intensity of which are directly related to current city planning, its planning structure and principle of population settling.

As a result of the analysis of the obtained experimental observations, it was found that the average travel distance from place of living to the workplaces in the city of Kyiv is 11.5 km (Table 1). Each city district has its own indicator. This take place due to location of districts (central, middle or peripheral zone) and the availability of adequate number of workplaces.

Table 1. Distribution of average travel distance (commute to work) for Kyiv residents between administrative districts.

The city administrative districts	Travel distance, km									
	Holosiivskyi	Darnytskyi	Desnyanskiy	Dniprovskyi	Obolonskiy	Pecherskyi	Podilskyi	Svyatoshinskiy	Solomenskiy	Shevchenkivskyi
Holosiivskyi	3.3	15.4	20.1	19.5	17.7	8.2	13.8	11.9	9.7	10.2
Darnytskyi	13.7	1.7	7.7	6.2	20.8	12.6	15.0	25.7	18.5	15.1
Desnyanskiy	22.0	14.1	6.0	3.7	11.1	13.5	12.6	20.0	19.0	15.2
Dniprovskyi	12.5	6.5	6.3	3.1	9.1	10.6	12.8	22.0	16.6	11.6
Obolonskiy	15.4	20.0	15.2	11.6	2.6	12.0	5.7	15.7	13.2	8.4
Pecherskyi	4.8	11.1	12.1	6.8	12.6	2.4	5.9	15.2	8.6	4.5
Podilskyi	7.7	15.2	15.6	8.6	6.1	10.5	1.6	9.2	11.5	6.6
Svyatoshinskiy	12.9	23.4	22.7	21.9	12.6	14.0	13.2	4.5	7.3	10.8
Solomenskiy	4.7	16.8	2.6	15.5	11.3	10.5	9.6	7.8	2.8	5.4
Shevchenkivskyi	12.9	16.0	17.6	12.5	7.6	6.6	4.1	5.4	6.3	2.8

The smallest distance of travels to workplaces is typical for the central districts of the city, namely for residents of Pechersk, Shevchenkivsky and Podilsky districts. Based on the data obtained, it is possible to determine such traffic routes, which are the shortest ones for inhabitants of every administrative district.

In order to create and ensure the required conditions for efficient street network operation, it is necessary to determine the portion of passenger traffic and population number using the above-ground transport and how such type traffic routes are distributed over the city road network.

Choosing the type of transport and traffic route to travel in the city is based on the following factors: the desire to save time; reduction of the travel length; preference to use the most popular and attractive routes (for example, through the city center, main street, etc.), as well as the routes having the highest traffic capacity and safety level due to their straightness property.

Therefore, taking into account the conditions and purpose of our study, it would be rational to determine the quantitative indicators characterizing the use of the appropriate types of vehicles by the residents of each district (Table 2).

The data collected by interviewing shows that about 61% of all inhabitants use public passenger transport when traveling for workplaces. Also, based on the data obtained, it can be argued that approximately 25.5% of Kyiv residents use their own cars to get to workplaces, 33.8% use the subway (the highest degree is for Obolonskyi and Darnytskyi districts). Approximately 9.6% of Kyiv residents get to workplaces on foot and 1.7% by bicycle.

The study found that when traveling more than 6 km, a time-saving indicator is always preferable. But it should be noted also that nowadays the time saving factor for cities where there are significant traffic queues is dominant for the routes that are less congested.

The choice of transport type (by car or public transport) by population depends on the convenience degree and the transportation cost. The convenience degree means the comfort level of a trip performed by the chosen type of transport vehicle during the minimum time spent.

The results represented by Table 3 indicate that only 66.2% of the population who owns a car uses it for all kinds of city travels (almost every day), and the remaining 33.8% - use their own vehicles periodically, only in specific cases (shopping, going on vacation, etc.).

Table 2. Particularities of traffic of residents of the city of Kiev to their workplaces.

The name of city district	Transport use proportion, %								
	On foot	By bicycle	Car	Bus or trolleybus	Fixed-route taxi	Street car	Subway	Urban train	Taxi
Holosiivskyi	5.0	0.8	27.5	12.8	34.5	1.5	41.5	0.08	3.3
Darnytskyi	9.8	5.0	25.8	9.1	38.3	4.5	42.8	0.55	2.5
Desnyanskiy	5.3	1.3	22.8	24.5	30.5	9.5	32.4	1.8	1.3
Dniprovskyi	3.5	1.5	24.5	20.5	35.0	5.6	36.5	1.3	1.5
Obolonskiy	12.3	1.2	23.5	17.5	21.5	5.0	41.5	0.6	1.1
Pecherskyi	11.3	0.5	28.5	7.5	26.5	0.8	33.0	0.05	3.8
Podilskyi	14.8	1.3	21.8	25.0	15.5	5.0	32.5	0.2	2.3
Svyatoshinskiy	3.3	2.5	28.8	16.0	30.0	16.3	32.5	0.1	1.8
Solomenskiy	17.8	1.8	26.3	21.8	28.0	7.5	15.0	0.5	2.0
Shevchenkivskyi	13.0	1.5	25.3	14.8	26.0	5.0	30.0	0.4	2.3

The coefficient of individual cars use for the city of Kiev distributed by administrative districts as follows: Holosiivskyi - 0,786; Darnytskyi - 0.714; Desniansky - 0.75; Dniprovskyi - 0.714; Obolonskiy - 0,563; Pecherskyi - 0,733; Podilsky - 0.60; Svyatoshynskyi - 0.688; Solomenskiy - 0,558; Shevchenkivskyi - 0,556.

The coefficient of two or more transport vehicles use for passengers traveling by one route, depending on their living place, is also represented in Table 3 and is 21% for the whole city.

Almost 69% of residents use only one type of transport while traveling to workplace and back. It should be also noted that 71.0% of residents use above ground street transport vehicles when traveling to workplaces, and 45.5% of them use only public passenger transport.

It is clear that in addition to permanent city inhabitants, it is necessary to take into account the daily flows of visitors to the city and workers coming from other settlements and regions. These additional passengers and traffic flows must be provided for the peripheral districts of the city. The distribution of such additional flows between districts of the city is performed according to the same pattern as for permanent residents. It depends on the number of workplaces and the location of cultural objects that form the city gravity centers.

To determine the number of vehicles that run on the city streets during the day, we use the division of vehicles into three groups, depending on their operation schedule. That is, we consider traffic of separate types of vehicles by dividing the whole traffic into individual cars, public passenger vehicles, freight transport and transit transport. It should also be noted that today there are a significant number of vehicles in the city of Kyiv that have legal registration in other settlements, but are permanently situated on the city territory. Today the percentage of such individual motor cars account for 11.2% according to the surveys conducted at the department of airports and highways reconstruction of NAU. Therefore, this category of vehicles should also be taken into account [9].

It should be noted that all types of vehicles, except the transit one, are in motion for a maximum of t hours from the active hours of day T, and the rest of the time they are at the parking lots.

Hence, the number of vehicles (Nv) that are both located and driving along the city road network can be found by the formula:

Table 3. Use of vehicles by residents of Kyiv during their travelling to workplaces.

The district of living	Use of own car everyday, %	Use of only one type of public transport, %	Use of two or more types of public transport, %	Use of street public transport, %	Use of street public transport including individual cars, %
Holosiivskyi	78.6	61.5	33.3	47.3	74.8
Darnytskyi	71.4	75.0	17.5	47.4	73.2
Desnyanskiy	75.0	70.0	25.0	55.0	77.8
Dniprovskyi	71.4	72.5	25.0	55.5	80.0
Obolonskiy	56.3	67.5	17.5	39.0	62.5
Pecherskyi	73.3	75.0	12.5	34.0	62.5
Podilskyi	60.0	60.0	22.5	40.5	62.3
Svyatoshinskiy	68.8	72.5	25.0	46.0	74.8
Solomenskiy	58.8	60.0	20.0	49.8	76.1
Shevchenkivskyi	55.6	75.0	12.5	40.8	66.1

$$N_v = ((N_{ic} + N_{icn}) \times \eta_{ic}) \times \frac{t_{ic}}{T_{ic}} + ((N_{pv} + N_{pvn}) \times \eta_{pv}) \times \frac{t_{pv}}{T_{pv}} + ((N_{fv} + N_{fvn}) \times \eta_{fv}) \times \frac{t_{fv}}{T_{fv}} + N_T \times \frac{t_T}{T_T} \quad (2)$$

where N_{ic} - the number of individual cars registered in the city, unit;
N_{icn} - the number of individual cars that are not registered but are located permanently in the city, unit;
η_{ic} - the individual car use factor;
t_{ic} - the maximum possible number of hours during which an individual car is in motion, hour;
T_{ic} - the number of “active hours” per day for individual cars, hour;
N_{pv} - the number of passenger vehicles registered in the city, unit;
N_{pvn} - the number of passenger vehicles which are not registered but permanently are in the city, unit;
η_{pv} - the passenger vehicle use factor;
t_{pv} - the maximum possible number of hours during which the passenger vehicle is in motion, hour;
T_{pv} - number of “active hours” of the day for passenger vehicles, hour;
N_{fv} - the number of freight vehicles registered in the city, unit;
N_{fvn} - the number of freight vehicles which are not registered but are permanently located in the city, unit;
η_{fv} - the freight vehicle use factor;
t_{fv} - the maximum possible number of hours during which the freight vehicle is in motion, hour;

T_{fv} - number of "active hours" of the day for freight vehicles, hour;
N_T - the number of vehicles entering the city (external and transit transport) daily, unit;
t_T - the maximum number of hours during which the transport entered in the city is in motion, per hour;
T_T - the number of "active hours" of a day for transit transport commuting to the city, hour

Analyzing the possible population traffic through the city, it clearly traces the main directions of population traffic flows through the territory of every district of the city during the inter-district transportations.

Based on the data obtained above on the traffic flow volumes during peak periods of city transport system operation for all districts of the city, the main routes of population traffic were determined, that to reveal regularities of passenger traffic in the city of Kyiv (commutes to work) and to determine ways of their distribution according to traffic directions, as well as to find the volumes of intra-district, external and transit traffic for each district of the city (Table 4).

The results show that inter-district traffic are much higher than intra-district ones. This indicates that not only the street network with its characteristics must meet the needs of traffic, but also the number and capacity of its elements, which ensure proper connections between the city districts.

Therefore, it is necessary to determine quantitative and qualitative indicators of the places providing transport links between the districts and to specify their functional importance in ensuring the city connectivity.

Considering the possible passenger traffic through the city territory, it is possible to find the gravity centers, where the main passenger flows in the territory of each city district are directed to during the inter-district traffic, as well as to reveal the presence of complex sections and elements of the street network on the territory of each district and their capabilities, and in such way to identify traffic volumes and congestion level of complex sections of the street network. This will also enable to estimate roughly the traffic volume, above ground traffic density and find possible alternative routes to redistribute the congested traffic flows to less "saturated" sections of the network and determine the required number of additional elements (bridges, junctions at different levels, signalized intersections, etc.), which will ensure the reliability of the operation of both the separate elements and the whole street network of the city.

The main influence on the traffic flows formation has a population density, planning features and geometric parameters of traffic networks. To further analyze the city traffic and to reveal the possibilities of its road network in performing its functions of ensuring the continuous traffic of vehicles, it's suggested to determine the number of available connections between the city districts, considering them as their problematic places. It is such places that the main transport flows are directed through to travel between city districts.

For any city it's possible to find a certain number (n) of such places through which you can travel from one city district to another. This enables to form the schematic diagram of traffic connections for any city (Fig. 1).

Considering the level of traffic flow density running through every street element linking city districts, it is possible to set the importance level (coefficient) of such connection that characterizes the weight of use of such element when forming the inter-district traffic. The importance coefficient of the linking element k_I is the ratio of the number of vehicles entering or exiting a certain district during a certain time through a particular element of the street network to the number of vehicles entering or exiting this district through all possible elements providing inter-district traffic during the same period of time.

$$k_{im}^{j} = \frac{N^{j}}{N_{all}} \tag{3}$$

where k_{im}^{j} - importance coefficient of j-th linking element of city street network;
N^{j} - the number of vehicles entering or exiting the certain district through j-th element of city street network;
N_{all} - the number of vehicles entering or exiting the certain district through all possible elements providing its links with another districts.

Table 4. The volumes of intra-district, external and transit traffic for districts of the city of Kyiv (commutes to work).

The name of city district	The intra-district traffic, vehicles	The traffic between neighboring districts, vehicles	The transit traffic, vehicles	Total number of vehicles which exit the district, vehicles	Total number of vehicles which enter the district, vehicles
Holosiivskyi	4874	18425	6508	24933	32277
Darnytskyi	3827	6556	25053	31609	6400
Desnyanskiy	4207	6346	25136	31482	7292
Dniprovskyi	5082	12516	20367	32883	8106
Obolonskiy	6116	8413	17862	26275	6836
Pecherskyi	5445	9623	3838	13461	44291
Podilskyi	4071	11290	3449	14739	33171
Svyatoshinskiy	4325	18414	20170	38584	7363
Solomenskiy	12562	12897	16456	29353	25951
Shevchenkivskyi	7346	15099	2923	18022	75153

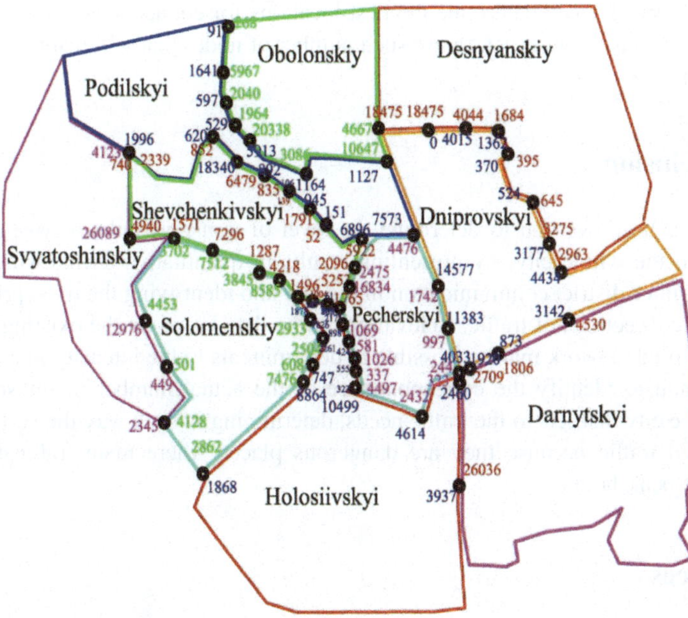

Fig. 1. The number of individual cars (commute to work), crossing the elements of the street network providing links between the districts of Kyiv.

4 Discussion

The results obtained from the observations show that congestion places coincide with the element of city street network having the heaviest traffic. It should also be noted that additional sections of heavy traffic often occur in places adjacent to the street network elements that provide inter-district traffic links. Such places are characterized by the maximum number of routes that pass through them on the basis of the shortest route condition.

The issue of ensuring the connectivity of the city street network is one of the key questions in the problem of transportation service of inhabitants. There are n elements (nodes) in the city that provide street traffic links with all its districts. It should be noted that the effectiveness of such node depends on its location and its ability to serve the district needs:

$$e_i = 1 - g_i \tag{4}$$

where e_i - the effectiveness indicator of the node operation in the i-th district; g_i - coefficient of inefficient use of the node in the i-th district (possibility of unsatisfactory user request).

In such way we can determine the best locations for connections between districts of the city or to be more exact, to use such number of nodes that will insure the greatest total effect.

5 Conclusion

This approach allows you to determine the level of transport links between any city district and the whole city, by presenting results in quantitative terms, which allows evaluating inter-district communication in the city and identifying the most priority and problematic directions of traffic. Relevant baseline data based on the existing structure of the city road network make it possible to determine its traffic intensity in a particular direction and to identify the correspondence of the actual number of transport links between the city districts to the traffic needs, determining in such way the sections with complicated traffic because they are dangerous places where main traffic flows are directed at peak hours.

References

1. Reitsen, E.: Organization and Safety of Urban Traffic. CIK GROUP Ukraine, Kyiv (2014)
2. Master plan for development of Kyiv and its suburban area by 2025 (project). http://drive.google.com/file/d/0BxbGBoNdb1j6TTRuS3RMQjFINTA. Accessed 01 Apr 2020
3. Kesuma, P., Rohman, M., Prastyanto, C.: Risk analysis of traffic congestion due to problem in heavy vehicles: a concept. In: IOP Conference Series: Materials Science and Engineering, vol. 650 (2019). https://doi.org/10.1088/1757-899X/650/1/012011
4. Lizbetin, J., Bartuska, L.: The influence of human factor on congestion formation on urban roads. Procedia Eng. **187**, 206–211 (2017). https://doi.org/10.1016/j.proeng.2017.04.366
5. Kiunsi, R.B.: A review of traffic congestion in Dar es Salaam city from the physical planning perspective. J. Sustain. Dev. **6**(2) (2013). https://doi.org/10.5539/jsd.v6n2p94
6. Agyapong, F., Ojo, T.K.: Managing traffic congestion in the Accra Central Market, Ghana. J. Urban Manag. **7**(2), 85–96 (2018). https://doi.org/10.1016/j.jum.2018.04.002
7. Hossain, M.T., Hasan, M.K.: Assessment of traffic congestion by traffic flow analysis in Pabna Town. Am. J. Traffic Transp. Eng. **4**(3), 75–81 (2019). https://doi.org/10.11648/j.ajtte.20190403.11
8. Žiliūte, L., Laurinavičius, A., Vaitkus, A.: Investigation into traffic flows on high intensity streets of Vilnius city. Transport **25**(3), 244–251 (2010). https://doi.org/10.3846/transport.2010.30
9. Stepanchuk, O., Bieliatynskyi, A., Pylypenko, O., Stepanchuk, S.: Peculiarities of city street-road network modelling. Procedia Eng. **134**, 276–283 (2016). https://doi.org/10.1016/j.proeng.2016.01.008
10. Stepanchuk, O., Bieliatynskyi, A., Pylypenko, O., Stepanchuk, S.: Surveying of traffic congestions on arterial roads of Kyiv city. Procedia Eng. **187**, 14–21 (2017). https://doi.org/10.1016/j.proeng.2017.04.344
11. Ben-Dor, G., Ben-Elia, E., Benenson, I.: Assessing the impacts of dedicated bus lanes on urban traffic congestion and modal split with an agent-based model. Procedia Comput. Sci. **130**(C), 824–829 (2018). https://doi.org/10.1016/j.procs.2018.04.071

12. Nguyen-Phuoc, D.Q., Young, W., Currie, G., De Gruyter, C.: Traffic congestion relief associated with public transport: state-of-the-art. Public Transp. 1–27 (2020). https://doi.org/10.1007/s12469-020-00231-3
13. Timkina, S., Stepanchuk, O., Bieliatynskyi, A.: The design of the length of the route transport stops' landing pad on streets of the city. In: IOP Conference Series: Materials Science and Engineering, vol. 708 (2019). https://doi.org/10.1088/1757-899X/708/1/012032
14. Bakhtina, O.: Development of methods of calculation and estimation of congestion conditions of traffic flow on the street-road network of cities (on the example Krasnodar). Dis. cand. tech. of sciences. Armavir (2006)
15. Zhivoglyadov, V.: Methodology of traffic management efficiency improvement. Dis. doctor of Engineering Sciences. Armavir (2008)
16. Stepanchuk, O., Bieliatynskyi, A., Pylypenko, O.: Modelling the bottlenecks interconnection on the city street network. In: VIII International Scientific Siberian Transport Forum. TransSiberia 2019. Advances in Intelligent Systems and Computing, vol. 1116, pp. 889–898. Springer, Cham (2020). https://doi.org/10.1007/978-3-030-37919-3_88

Geosynthetic Reinforced Interlayers Application in Road Construction

Valerii Pershakov[1], Andrii Bieliatynskyi[1,2(✉)], and Oleksandra Akmaldinova[1]

[1] National Aviation University, Kiev 03058, Ukraine
beljatynskij@ukr.net
[2] North Minzu University, 204 Nort-Wenchang Street, Xixia District, Yinchuan, Ningxia, People's Republic of China

Abstract. The article is devoted to the analysis of theoretical and experimental data of the geosynthetic layers reinforcing functions and their functional interaction with other layers in the top dressing structure. The materials and results are presented in accordance with the basis of a project on a highway section reconstruction by means of geosynthetics. An optimal method for solving the main problem in road construction, being the impact of negative external factors causing the destruction of the pavement structure is studied in the paper. The essence of the method is impregnation of the synthetic material with a binding solution, in this case, a bitumen emulsion, which will ensure its good adhesion to asphalt concrete. Results of the study: The theoretical and experimental data of geosynthetic layers reinforcing functions and their functional interaction with other layers in the structure of road covering were analyzed. The materials and results found on the basis of highway section reconstruction project with the use of geosynthetics are presented. Given the current trends in the road construction providing for the optimization of all processes in order to improve their operational properties and achieve maximum economic efficiency of the decisions made, the research on geosynthetic materials is relevant today.

Keywords: Geosynthetics · Technologies · Construction · Road surface covering · Geosynthetic layers · Geotextiles

1 Introduction

The rapid increase in the number of heavy road transport, increase in traffic intensity and, as a result, increase in axial loading of the road surface contribute to developing deformations of asphalt concrete roads based on conventional bitumen. The main problem here is the deformation of road pavements. Bitumen can no longer fully satisfy today's requirements. All over the world, work is constantly being carried out to create new modern road materials and technologies, to adjust the regulatory requirements to their physical and mechanical properties. This is aimed at increasing the road pavement durability in modern operating conditions. The main task is to analyze the road technology construction using geosynthetic layers, to review their characteristics, identify disadvantages and advantages.

V. Murgul and V. Pukhkal (Eds.): EMMFT 2019, AISC 1258, pp. 94–103, 2021.
https://doi.org/10.1007/978-3-030-57450-5_9

Nowadays, non-rigid road surfacing with asphalt concrete layers predominates on the highways of Ukraine, as well as in the whole world. Such road dressing is often quickly destroyed by transport loads and requires early repairs.

Destruction is manifested depending on the loading conditions and nature. Cracks are a quite common type of destruction of asphalt concrete layers. That's why, in recent years, increase the durability of such layers, reinforcing interlayers in the form of synthetic materials are widely used in practice. Though, their application is not sufficiently supported by a theoretical base that would allow to calculate the asphalt concrete layers of road dressing taking into account the specificity of reinforcing synthetic materials operation.

In recent decades, many scientific works of domestic and foreign scientists have been devoted to investigation of the fracture resistance of asphalt concrete layers, but not enough attention has been paid to the study of the reinforcing synthetic layers effect on the overall stress and strain state of these layers.

So, there are practically no theoretical papers in this field that would allow to develop a single computing technique for asphalt concrete layers reinforced with synthetic materials. It is known that road-construction materials and road bed soils exhibit viscous-elastic properties under loading, which must be taken into account when determining the stress and strain state of road dressing reinforced asphalt concrete layers under the action of transport loads and in assessing their limiting state.

The interaction of asphalt concrete as a material with a synthetic layer is a complex process that requires a careful study in the process of theoretical and practical research.

Review of recent publications on the topic: geosynthetic materials in road construction are reviewed and analyzed according to national standards (DSTU EN 14030: 2006, SOU 45.2-00018112-025: 2007, VBN B.2.3-218-544: 2008, supplement to VBN B.2.3-218-544: 2008, EN 12224, P B.2.3-218-21476215-795: 2011); national [1–7] and foreign publications [8, 9].

2 Materials and Methods

Application of reinforcing materials in highways designing, construction, reconstruction and repair. In recent years, structures using tens of thousands of geotextile materials have become widespread throughout the world, and systematic surveys indicate their high reliability, durability and technical and operational indices. Geotextile material is being increasingly used in the ground bed construction and road covering though, the specifics of its work in road covering is less studied than in the ground bed.

A number of foreign publications are devoted to the issue of road dressing reinforcement, describing the experience of construction of road structures including the "asphalt concrete - reinforcing layer – asphalt concrete" structural elements. Geotextile, fiberglass, various polymeric nets, metal gauzes and metal pins are used as layers [2].

Research and observation results have shown that geotextile enhances the drainage properties and has reinforcing effect, thereby slowing down the process of pavement fracturing.

Geosynthetics are extensively used to strengthen and monitor the development of cracks in road surfaces in Australia, Belgium, Canada, Italy, Spain, Slovenia, The Netherlands, Czech Republic, Germany, France [2].

Geotextiles make it possible to limit and reduce fracturing of the asphalt concrete covering. It is necessary to use low-compressible geotextile materials that do not increase the pavement deformation under loading and are not too rigid. They must be compatible with bitumen and heat-resistant in the temperature range of laying the asphalt concrete mixtures.

From this review we can conclude that:

- the use of geotextile interlayers in or between the road dressing structural layers and the conditions of their joint work are of great practical importance for road construction;
- geotextile used for reinforcement of road structures has now become an independent construction material, which in many cases cannot be replaced by a traditional material;
- reinforcement of road dressing structural layers with geotextile materials allows to increase their strength, to prevent of formation of broken cracks in covering asphalt concrete layers, to reduce the materials consumption of the road structure;
- data of systematic reinforced structure surveys indicate their reliability, durability and high technical and operational performance;
- layered road structures increase the culture of production, by reducing the production cycle, technological effectiveness of operations and their quality indices.

The main types of synthetic materials and their general characteristics.

It is relevant to consider the classification of geosynthetic materials according to the British methodology, which is designed to review, analyze and systematize specific samples. This classification, shown in Table 1, is essential in the process of selecting rational types of road structures, depending on the desired properties, engineering and geological, soil and climatic conditions.

The area, efficiency and feasibility of using synthetic rolled materials (SM) are determined by their properties, depending on the composition of the raw material and production technology. To manufacture SM different polymers are used: polyamide (PA), polyester (PET), polyether (PETh), polypropylene (PP), polyethylene (PE) and others (Table 2). Mixtures of polypropylene and polyethylene are referred to as polyolefins. Additives may be added to get special properties. Polyvinyl chloride (PVC), polyethylene, bitumen are used as coatings.

The properties of non-woven geotextiles depend on the method of strengthening the road bed:

- mechanical (needle-punching) – material, anisotropic in two mutually perpendicular directions, differs in low tensile strength, permeability and high deformability;
- chemical - hardening is achieved by putting in binding glue into the material, which fixes the fibers at the points of contact;
- thermal - the road bed is heat-calendered with fibers sintering.

Table 1. Classification of geosynthetic materials.

Name	Material, polymer	The area of application	Physico-mechanical indices
Non-woven: needle punched, thermally bonded	Polypropylene	Substrate for composites, anti-clogging and reducing filtration properties, for arranging drainage structures and separating layers	Tensile strength, load-bearing capacity, relative elongation for nominal strength, modulus of elasticity, resistance to light and exposure to chemical agents, porosity
Knitted and woven	Polyester and propylene	Reinforcement of bases belonging to the category of weak, slopes with steepness above average value, retaining walls	Relative elongation for strength and tensile strength. Chemical resistance and light resistance. Modulus of elasticity, flowability boundary and cone puncture strength
Geogrids: - woven, - Extrusive	Polypropylene, glass, polyamide, polyester, polyethylene	Reinforcement: Ground and natural foundations, arrangement of rigid and flexible piles, pileworks, reinforcement of asphalt coverings	Relative elongation for strength, modulus of elasticity, creep deformation, resistance to light, density, external friction coefficient
Bulk geograte s: - modular, - Gabion grid, - honeycomb	Polypropylene	Reinforcement of slopes, cones and embankments. Reinforcement of foundations of increased steepness slopes	Tensile strength, frost resistance and resistance to chemical effect
Compositional: - porous, - fibrous - multilayer with plastic frame and protective layers, non-woven low density materials	Polypropylene, polyethylene, polyester	Strengthening of slopes and arrangement of drainage in places with difficult geological and climatic conditions	Moisture and water resistance, relative deformation and tensile strength
Geomembranes	Polypropylene and polyethylene	Reduction of active stresses by reducing friction with soil components	Water tightness, elongation at rupture. Ultimate strength, thickness, material density

(continued)

Table 1. (*continued*)

Name	Material, polymer	The area of application	Physico-mechanical indices
Dampproof material	Polypropylene and bentonite	Arrangement of completely waterproof elements	Protection against adverse effects of the aquatic environment

Table 2. Fibers used for manufacturing synthetic materials.

Index	Fiber-forming polymers			
	Polyester, polyether	Polypropylene	Polyamide	Polyethylene
Density, g/cm	1.36–1.38	0.90–0.92	1.14	0.95–0.96
Water absorption, at 21 °C and relative humidity 65%	0.2–0.5	0	3.5–4.5	0
Fracture tensile strength, MPa	35–90	22–55	45–70	32–65
Elongation at rupture, %	15–40	15–30	30–80	15–30
Creeping ability	Insignificant	High	Insignificant	Very high

3 Results

Due to the structure thus formed, SM has excellent water permeability and high tensile and fracture strength characteristics. Thermo-bonded SMs are characterized by high marginal elongation (up to 70%) and increased durability.

Secondary raw materials, including those containing non-synthetic components, may be used to manufacture roadway SMs, provided that their properties meet the requirements. Depending on the manufacturing method, the SMs are divided into woven and non-woven. Woven SMs have a regular structure, high strength, high modulus of elasticity. But they do not have sufficient water permeability in the road bed plane. Such materials should be used in case the SM layers perform the functions of reinforcement, protection, but not drainage.

Asphalt concrete covering reinforcement polyester grids are used, having high strength and low deformation characteristics, chemical and biological stability, as well as good compatibility - adhesion to bitumen and heat resistance in the working temperatures range when laying asphalt concrete.

4 Discussion

Reinforcement of road asphalt concrete coverings. When applying reinforcing grids in road covering construction with asphalt concrete layers, no additional demands for structural layer materials are made; road-construction materials must meet the requirements of the State Standards of Ukraine.

Reinforcement of road asphalt concrete coverings is aimed at increasing the durability of road structure. It can be accomplished by introducing a reinforcing layer of woven or non-woven type, if their characteristics meet the requirements (Table 3).

Table 3. Requirements for reinforcing synthetic materials.

Surface density, g/m	Tensile strength at breaking, min. kN/m	Elongation, max, %	Melting point, °C
100–400	20	20	+180–200

When reinforcing the road covering asphalt concrete layers, the SM is placed directly under the asphalt concrete layer, which will be considered reinforced, that is, it will have increased strength and deformation properties as compared to conventional one. In this case, the reinforcing layer is placed only within the width of the roadway. At strengthening the asphalt concrete coverings with synthetic materials the minimum thickness of the asphalt concrete layer above the interlayer should be not less than 5 cm.

The number of asphalt concrete layers and their thickness is assigned depending on the required load bearing capacity of the road. When arranging two- and three-layer coverings, it is advisable to lay the reinforcing mesh to increase the fracturing resistance in the area of high tensile stresses.

Reinforcing grid is laid over the entire width of the roadway apon the existing covering or between the second and third layers of asphalt concrete, in the areas of concentrated high shear stresses:

- in places of intensive transport braking and at stops;
- under the top asphalt concrete layer.

During current repair of asphalt concrete covering, the reinforcing grid is used in places of irreversible deformations: cracks, subsidence, potholes, etc. The reinforcing grid is laid onto a leveled and binder treated surface. In case of a significant damage to the existing covering milling is applied. The existing covering layer is partially removed, increasing the thickness of the reinforcement layer over the damaged area to the depth of milling. The reinforcing grid is laid symmetrically with respect to the axis of the destroyed section or crack [8].

When reinforcing rigid foundations (cement concrete, layers treated with mineral binders) with asphalt concrete layers, the reinforcing grid is placed only over transverse and longitudinal cracks if the distance between cracks and joints is more than 3 m.

With the spacing between the transverse cracks less than 3 m, it is advisable to reinforce the asphalt concrete covering along the entire roadway width.

When the roadway is extended, there is a problem of ensuring the reliable joint work of the existing design having a formed structure and structural binding. The different nature of the designs operation causes new binding and a new construction, where the processes of deformation and consolidation have not yet been completed.

It is possible to reduce the amount of tangent stresses in the contact zone of the two structures by increasing the contact zone, creating a so-called leaning effect. It is recommended to reinforce the contact areas with synthetic grids to provide the fracturing resistance of the covering.

Methods of calculating the reinforced asphalt concrete covering of non-rigid road dressing for the strength under the action of vehicles.

Asphalt concrete covering is calculated for the fracture resistance to the transport action, as well as other monolithic layers under the current regulatory and technical document. According to this regulatory method, it is required that in monolithic layers of road dressing stresses resulting from the deflection under repeated short-term loading do not cause structural rupture of the material or fracturing, i.e. the condition

$$K_{mts} \leq \frac{R_{zg}}{\sigma_r} \tag{1}$$

must be ensured, where

K_{mts} is the strength factor, given the specified level of reliability;
R_{zg} are the maximum allowable tensile stresses of the layer material with regard to fatigue;
σ_r is the highest tensile stress in the layer calculated.

The asphalt concrete calculation characteristics, modulus of elasticity and fatigue ratio are determined at loading time of 0.1 s. and T - 0 °C. The allowable tensile stress is determined by the dynamic flexural strength test at a strain rate of 100 mm/min.

Thus, according to the current calculation method, only one type of loading mode at a time of 0.1 s is used, when only the tensile stresses in the lower part of the covering are taken into account when it is bent by the action of the transport estimated load, transmitted in the form of a distributed vertical load over the area of the circle, equidimensional to the wheel imprint.

According to the national normative document for the reinforced asphalt concrete layers the following condition must be fulfilled

$$K_{mts} \leq \frac{k_{ef}^{a} \cdot R_{zg}}{\sigma_r} \tag{2}$$

where, K_{mts}, R_{zg}, σ_r are the values given in the above formula;

k_{ef}^{a} is the reinforcement efficiency coefficient determined by a special methodology.

An analysis of the world experience in the construction and operation of road surfacing with layers increasing the fracturing resistance and strength of asphalt concrete pavements shows that the layers can be divided into two groups: “soft” and “stiff”. Each group has a particularity in the mechanism of their impact on road covering fracturing resistance and performance [7].

The choice of the strength theory and the criterion of the local boundary condition is usually somewhat conventional. The theory of determining the stress-strain state must correspond to the experimentally obtained characteristics of the materials.

Several strength theories are currently used to predict asphalt pavement fracturing. Some theories require special methods for determining the strength characteristics of materials; others use traditional characteristics or their modifications.

The determination of stress-strain state is carried out by means of the method of finite elements. The elements are prismatic, two-dimensional with the crack model, its environment included into the general model of the road pavement. This method takes into account the effect of annual changes in temperatures.

The methodology developed in SoyuzdorNII and KADI takes into account not only the effect of transport means and annual temperature fluctuations, but daily temperature fluctuations as well [7].

In setting the problem, the road top dressing is presented as an elastic multilayer package laid onto a rigid base. Given that the elasticity theory equation is considerably simplified when solving a plane problem, we have restricted ourselves to studying the plane deformation. But even with this approach, the boundary conditions reflecting the presence of a series of vertical rectangular sections in one of several layers, do not allow to solve the problem analytically. Therefore, we had to use also the method of finite elements. As a basis the isoparametric quadratic element was adopted, as it is the most effective in terms of accuracy and time of calculation.

When constructing a finite element grid, thickening was done to increase the accuracy of calculations near the vertices of the sections. In addition to the quadrangular elements triangular elements have been used. Depending on the material used as a layer, one of two calculation schemes was used. For example, geotextile was modeled by presenting it as an elastic thin layer, characterized with a thickness, a very low modulus of elasticity, and finite Poisson coefficient. The grid was modeled by a set of deformable elementary plates, interconnected by hinged joints. This approach takes into account the fact that the real grid layer does not perceive the compressive stresses in the horizontal direction. The model of a set of elementary plates is characterized by the modulus of elasticity obtained at stretching Poisson’s ratio equal to 0, and thickness.

Given the reliability of the technique and a great experience in determining the tensile strength in bending, in the first phase of research, it was decided to evaluate the road asphalt pavement fracturing resistance by a stretch arising over the crack under the action of road transport. In this case, we may assume that the integrity of the covering will not be broken, if the tensile stress during repeated bending won’t exceed the admissible limits for asphalt concrete established, taking into account the fatigue phenomena.

The thermal stressed state of the fracture-block covering was considered, given the asphalt concrete relaxation ability. At the same time, when evaluating the temperature stresses, the Volterr-Boltzmann ratio of linear viscosity and elasticity theory, and the principle of temperature-time analogy were used. As a result, the relationship between the nature of temperature fluctuations (daily and annual) and the degree of danger of temperature fracturing have been established and subsequently used. At the same time, And a simple method for experimental determination of the parameters of the road covering material long-term durability at different temperatures has been developed. To take into account the interlayers, their impact on the process of cracks formation when determining the long-term strength parameters, the composition “asphalt - reinforcing layer - fractured - block base (initiator of cracks)” has been researched.

5 Conclusion

Analysis of existing approaches to increasing the asphalt concrete layer stability by means of geosynthetic reinforcing layers has shown the disparity of current theoretical and experimental studies. So far, there has been no single technique optimally ensuring maximum results of the interaction and compatibility of road pavement materials and reinforcement layers.

The interaction of asphalt concrete as a material with a synthetic interlayer is a complex process that requires careful theoretical and practical research. These materials are completely different in the origin, composition and properties, which complicates the mechanism of their interaction. In the process of laying the asphalt concrete layer onto the interlayer, the “reinforcing layer - asphalt layer” system is made up and an inseparable contact between them is formed due to the adhesive ability of the asphalt binder and the mechanical adhesion and catching of the interlayer individual parts with the asphalt concrete mineral components. If geogrid is used as a reinforcing layer, its joints together with links work as anchors in asphalt concrete, being a support for coarse filler (crushed stone, gravel) [10].

The coupling of the reinforcing layer with asphalt concrete is provided by: the increase in asphalt concrete resistance to flexural stresses, caused by armature irregularities and periodic profile, i.e. mechanical adhesion of the reinforcing layer with asphalt concrete; formation of friction on the reinforcing layer surface due to its compression by the asphalt concrete during rolling; bonding of the asphalt concrete with the reinforcing interlayer due to the presence of bitumen [11].

When arranging a reinforced asphalt concrete layer during construction, the reinforcing interlayer comes into close contact with the asphalt concrete. But in the process of operation due to the action of various factors, at the points where the displacement of the interlayer relative to the asphalt layer is not possible (absolute contact) the area of contact is reduced as a result of partial separation of the reinforcing layer from the asphalt layer. Therefore, in this case it is possible to distinguish two contact areas:

- he area with absolute contact;
- he area where the displacement of the intermediate layer relative to the asphalt layer is possible.

Tensile stresses are transmitted from the asphalt concrete layer to the reinforcing layer only due to their mutual friction and mechanical adhesion. This case is characterized by the fact that between the interlayer and the asphalt layer there arise external friction forces, internal friction forces and accompanying friction, as well as the phenomenon of adhesion of macroscopic particles with the surface [11].

To ensure a reliable contact between the interlayer and the asphalt layer, it is necessary that the friction and adhesion forces are maximized. This is possible only under several conditions:

- the modulus of elasticity of the grid must not be lower than the asphalt concrete modulus;
- the size of the grid cells should be sufficient to allow the mixture to penetrate to ensure good catching and adhesion between the upper and lower layers of the covering (base);
- to transfer the tensile force, the grid adhesion with the asphalt concrete layer must be sufficiently strong;
- the grid material must have high temperature resistance without impairing its basic physical and mechanical characteristics.

An optimal method for solving the main problem in road construction has been investigated, namely, the impact of negative external factors contributing to the destruction of the road pavement design. The essence of the method is compulsory impregnation of the synthetic material with a binding solution, in this case, a bituminous emulsion, which will ensure its good adhesion with the asphalt concrete [11].

References

1. Kostrytskyi, V., Kolomiets, A., Artemenko, L., Gamelak, I.: Investigation of the performance characteristics of geographers intended for reinforcing asphalt concrete. KNUDT Bull. **6**, 46–50 (2007)
2. Gamelak, I., Kostritsky, V., Artemenko, L.: Problems of using geosynthetic materials in road construction and ways of solving them. KNUDT Bull. **6**, 17–27 (2009)
3. Shevchuk, V., Zhurba, G.: Modern areas of development. In: Avtoshahovik Ukrainy. International Conference on Geosynthetics, vol. 6, pp. 38–40 (2006)
4. Mozgovy, V., Onishchenko, A., Garkusha, M., Aksyonov, S.: Modern aspects of increasing the resistance of non-rigid road wear. Avtoshahovik Ukrainy **5**, 25–30 (2012)
5. Gamelak, I., Raykovsky, V.: Analysis of transport and operational indicators of the state of highways of national importance. Highways **1**(237), 24–28 (2014)
6. Mironchuk, S.: A method for determining the stability of asphalt-concrete road pavements to the accumulation of residual deformations under the influence of dynamic loads. Cand. Thesis. Voronezh (2015)
7. Stefashina, N.: The use of geosynthetic reinforcing layers in road construction. Master's thesis. NAU, Kyiv (2020)
8. Koerner, R.: Designing with Geosynthetics, 5th edn. (2005)
9. Jones, C.J.F.P.: Developments and Innovations in Geosynthetic Material Technology. University of Newcastle upon Tyne, UK
10. Krayushkina, K., Khymeryk, T., Bieliatynskyi, A.: Basalt fiber concrete as a new construction material for roads and airfields. In: IOP Conference Series: Materials Science and Engineering, vol. 708, no. 1, pp. 1–9 (2019). https://doi.org/10.1088/1757-899X/708/1/012088
11. Onishchenko, A., Stolyarova, L., Bieliatynskyi, A.: Evaluation of the durability of asphalt concrete on polymer modified bitumen. In: E3S Web Conference, vol. 157 (2019). https://doi.org/10.1051/e3sconf/202015706005

Research of the Properties of Bitumen Modified by Polymer Latex

Artur Onishchenko[1(✉)], Artem Lapchenko[1], Oleh Fedorenko[1], and Andrii Bieliatynskyi[2,3]

[1] National Transport University, 1, Mykhaila Omelianovycha - Pavlenka Str., Kyiv 01010, Ukraine
artur_onish@bigmir.net
[2] National Aviation University, Kyiv 01010, Ukraine
[3] North Minzu University, 204 Nort-Wenchang St. Xixia District, Yinchuan, Ningxia, People's Republic of China

Abstract. There are given the investigation results of bitumen modified with the polymer latex Butonal NS 104 depending on temperature, time and the modifying agent amount, taking into account the manufacturing company BASF recommendations. The carried research has shown the possibility to produce bituminous polymer on the polymer latex Butonal NS basis that meets the Ukraine technological normative documents requirements. On the conducted investigations basis, there have been established the rational parameters of bituminous polymer binding agent production process thus allowing determining the conditions of the polymer latex Butonal NS 104 application under the domestic production conditions. The study found that bituminous polymer bending agent physical-mechanical properties change in the wide limits (by 50–100%) depending on the polymer amount, and the preparation process parameters show the possibility of active regulation of its properties under the particular operating conditions. It is established that bituminous polymer bending agent properties resistance and stability under the high process temperatures influence show the possibility of its sufficiently long-term storage under the production conditions with the original properties keeping.

Keywords: Paving bitumen · Polymer latex Butonal NS · Bituminous polymer preparation process · Binding agent properties

1 Introduction

In road-building industry, the bitumen modification with the polymers is one of the most effective ways to extend service life of the road-building materials produced on the organic binding agents [1–6].

The most effective and recognized are the styrene- butadiene-based polymers [3–6]. Among the polymers of this type the polymer latexes of the Butonal NS 104 series (BASF production, USA) advantageously differ due to their properties [4].

Taking into consideration the fact that domestic paving bitumens differ from those used in the USA, it emerged necessity to study influence of such modifying agents on

V. Murgul and V. Pukhkal (Eds.): EMMFT 2019, AISC 1258, pp. 104–116, 2021.
https://doi.org/10.1007/978-3-030-57450-5_10

the properties of the paving bitumens used in Ukraine. Cationic latex Butonal NS 198 became one of the first modifying latexes used in Ukraine. On basis of the previous investigations performed by the Derzhdor research institute and KNARU research teams it was determined the polymer latex Butonal NS 198 on the paving bitumen properties [4]. The laboratory test results confirmed the enhancement of the binding agent physical-mechanical properties depending on the polymer amount. This research have allowed to apply the Butanol NS 198 polymer at the following highways construction: Kyiv – Odessa (km 217–236, km 247–236, km 247–252); Kharkov – Simferopol (km 536–km 538); Kyiv – Chop (km 331–km 335), and the other objects.

Recently one more type of the polymer latex appeared in Ukraine – Butonal NS 104 designed to enhance the paving bitumens modifying efficiency. However, it is obvious that to achieve the maximum benefit of the polymer use and to study its influence on the bitumen physical-mechanical properties the preparation process rational parameters should be determined taking into account operating conditions. In this paper there is investigated the influence of the modified with polymer latex Butonal NS 104 the bitumen preparation parameters on the paving bitumen properties. There was conducted the experimental investigation to determine the following:

- the polymer rational consumption for bitumen modifying;
- influence of the bituminous polymer preparation temperature and time on its main characteristics;
- influence of the preparation temperature and time on the bituminous polymer homogeneity.

To perform this work, the samples preparation procedures were developed. The experiments were carrying out using the samples of bitumen with the polymer latex Butonal NS 104 provided by the BASF representatives – "International chemical production, Ltd.". Research was performed in the "Transport construction materials and designs" prof. G.K. Sunya laboratory of the road building materials and chemistry chair of the National transport university.

2 Materials and Methods

Procedure of the bituminous polymer preparation before testing. To determine the influence of the polymer latex Butonal NS 104 on the bituminous polymer physical-mechanical properties during these investigations, the petroleum paving bitumen 90/130 was used as one of the most regnant in Ukraine. The polymer latex has being added into this bitumen to determine its amount and other process-dependent parameters which are relevant to production of the bituminous polymer according to domestic requirements. Output paving bitumen 90/130 had the following output data (Table 1).

There were prepared the bituminous polymer samples containing 2%, 4% and 6% of the Butonal NS 104. Temperature during preparation varied from 160 °C to 200 °C. The bituminous polymers preparation duration varied from 1 to 8 h. Such extended range of the modifying parameters is used also to determine their limit values under the operating conditions.

Table 1. Requirements and results of physical and technical properties of investigated bitumen.

Parameters description	Unit	UNSt 4044 requirements to petroleum paving bitumen 90/130	Parameter value
Needle penetration depth at 25 °C	0.1 mm	91 to 130	97
Softening temperature according to KiK	°C	47 to 53	49.0
Stretchability (shortness) at temperature of 25 °C	cm	Minimum 55	82.6
Properties change after heating:			
Residual penetration	%	Minimum 60	95
Softening temperature change	°C	Maximum 6	2
Fragility temperature	°C	Maximum −10	−22
Flash temperature	°C	Minimum 240	240
Adhesion	%	–	90

To modify bitumen binding agent it was developed the O-1 installation (Fig. 1) which allows ensuring the process with the specified technological modes.

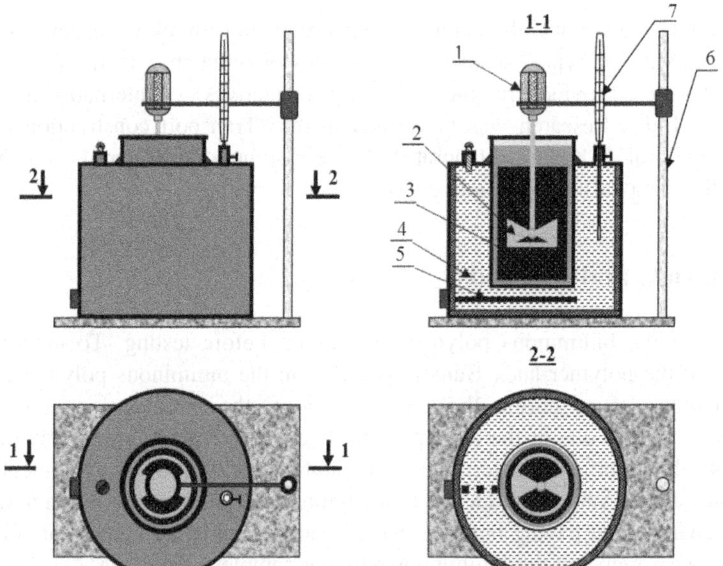

1 – electric motor; 2 – mixer; 3 – binding agent; 4 – high-temperature liquid (commercial glycerin or industrial oil); 5 – heating element; 6 – laboratory stand; 7 – electric contact thermometer

Fig. 1. Diagram of reactor for bituminous polymer preparation.

The bituminous polymer preparation operation sequence was the following:

- bitumen heating to operating temperature;
- introducing into bitumen the necessary amount of the polymer latex Butonal NS 104;
- heating of the binding agent to preparation operating temperature with continuous mixing;
- hold-up of the bituminous polymer in reactor at operating temperature with continuous mixing during the modifying specified time period necessary to get the required physical-mechanical properties.

3 Results

Laboratory research results. The output bitumen and modified with polymer bitumen tests were carried out according to applicable regulations [4]. The output and modified bitumen test results are shown in Fig. 2, 3 and 4. Results of determining the penetration of bitumen modified with the Butonal NS 104 polymer various amounts and prepared at various temperatures have shown (Fig. 2, 3) that the viscosity significant change occurs during the initial three hours of the bitumen modifying.

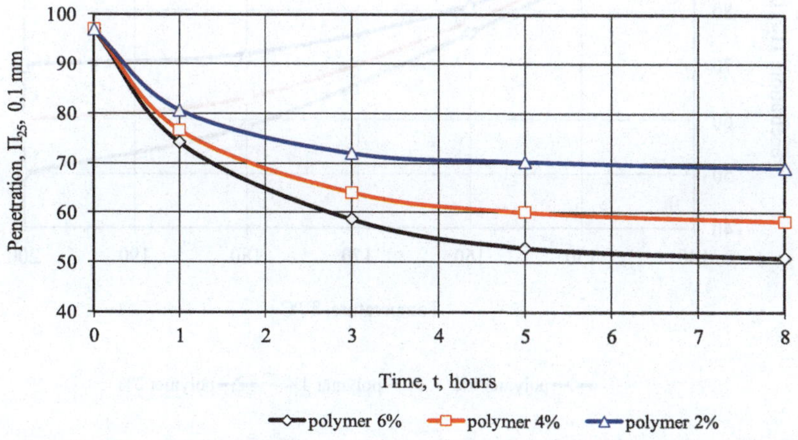

Fig. 2. Bituminous polymer penetration dependence (at 25 °C) upon the preparation time at the polymer various amount.

While analyzing the bituminous polymer viscosity change results in a case of the preparation temperature change from 160 °C to 200 °C, it is seen that at its increase, the viscosity value increases too (Fig. 4).

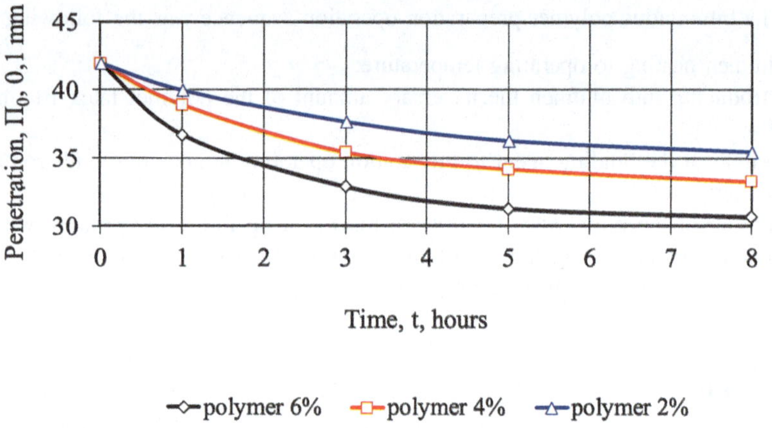

Fig. 3. Bituminous polymer penetration dependence (at 0 °C) upon the preparation time at the polymer various amount.

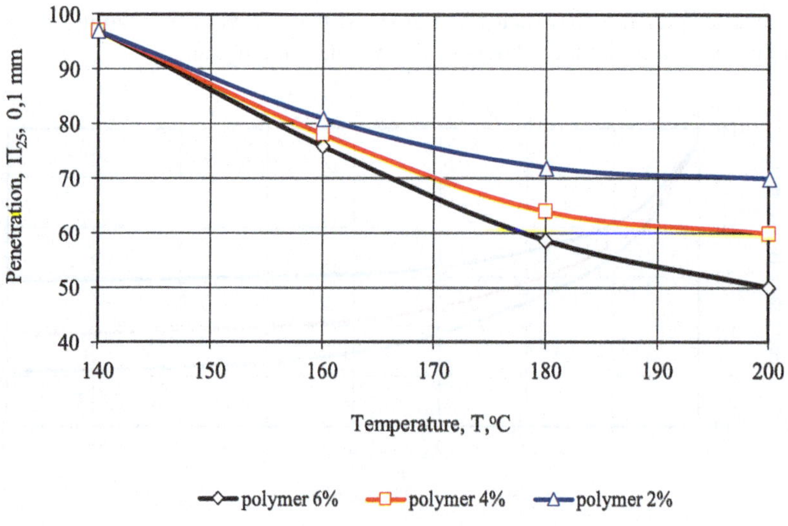

Fig. 4. Bituminous polymer penetration dependence (at 25 °C) upon the preparation temperature at the polymer various amount.

The results of determination of the bituminous polymer softening temperature dependence upon the preparation temperature and time showed its change character (Fig. 5, 6) similar to penetration change.

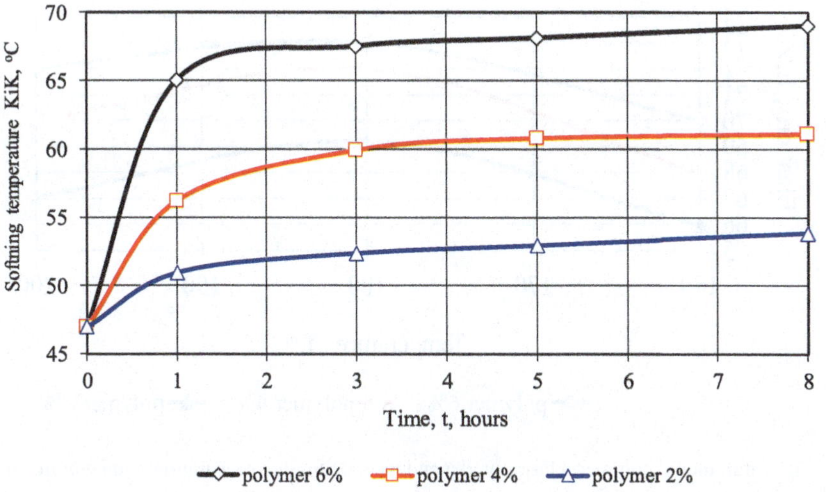

Fig. 5. Bituminous polymer softening temperature dependence upon the preparation time at the polymer various amount.

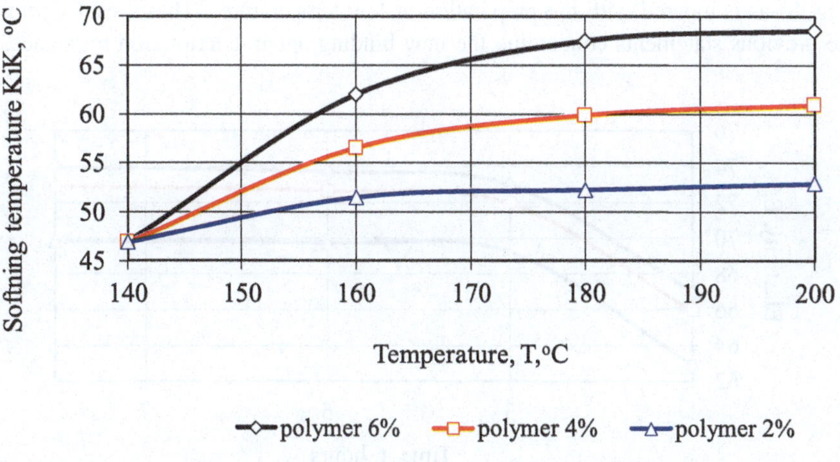

Fig. 6. Bituminous polymer softening temperature dependence upon the preparation temperature at the polymer various amount.

The elasticity occurrence even at the small amount of polymer and modifying time (at 2%, after 1 h of preparation (Fig. 7)) confirms the possibility of the bituminous polymer bending agent elastic properties considerable enhancement.

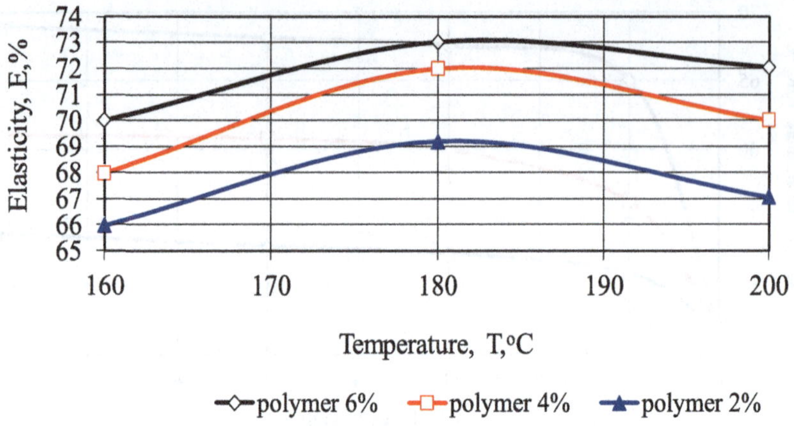

Fig. 7. Bituminous polymer elasticity dependence upon the preparation temperature at the polymer various amount.

It is especially important to note that the elasticity value at the high temperatures o the bituminous polymer bending agent preparation virtually remained invariable (Fig. 8) as compared with the preparation at low temperatures. These results confirm the previous statements concerning the new binding agent deterioration resistance.

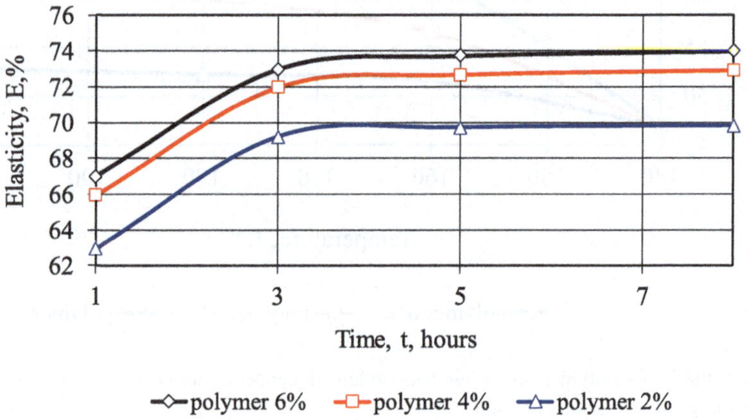

Fig. 8. Bituminous polymer elasticity dependence upon the preparation time at the polymer various amount.

It may be seen that the bituminous polymer bending agent heat resistance factor change after heating at various polymer amounts and time of its modifying is very slight. After the first hour of modifying at preparation temperature of 180 °C this parameter was higher by 2–4 °C as compared with the output bituminous polymer bending agent, that is explained by the non-modified bending agent, while after 3 h of modifying this parameter didn't exceed 3 °C (Fig. 9), i.e. met the applicable requirements.

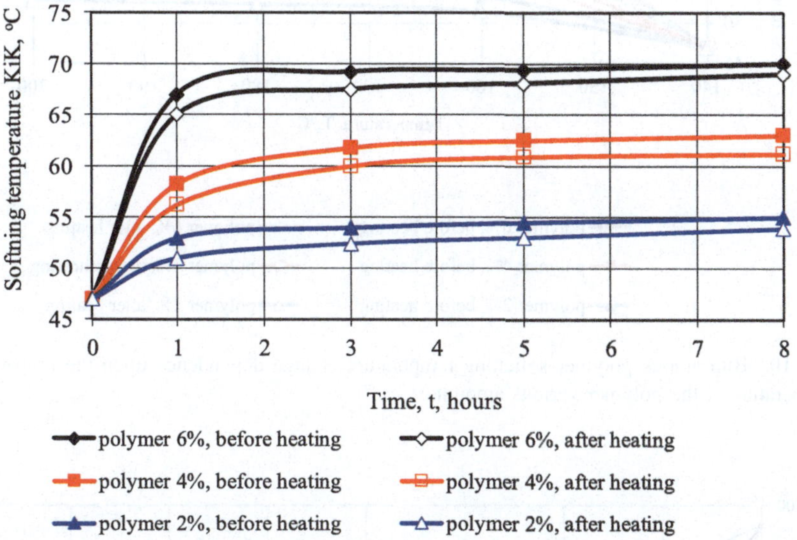

Fig. 9. Bituminous polymer softening temperature change dependence upon the preparation time at the polymer various amounts.

At lowing the bituminous polymer bending agent preparation temperature to 160 °C, this parameter remained the same, and at the temperature increase up to 200 °C, its maximum value was 5 °C (Fig. 10), that also meets the requirements [4–6].

Dependence of the bituminous polymer bending agent disintegration factor during storage (according to penetration factor) upon the polymer amount, the preparation time and temperature showed its slight change during modifying both at low temperatures and at high ones (5–7 degrees of penetration) (Fig. 11, 12).

At the polymer concentration of 2% the KiK factor difference depending on the preparation time varied from 1 to 4 °C, while at the polymer concentration of 6% this difference was equal to 3–5 °C (Fig. 13, 14). Such results are observed in the entire range of the bituminous polymer bending agent preparation temperature change.

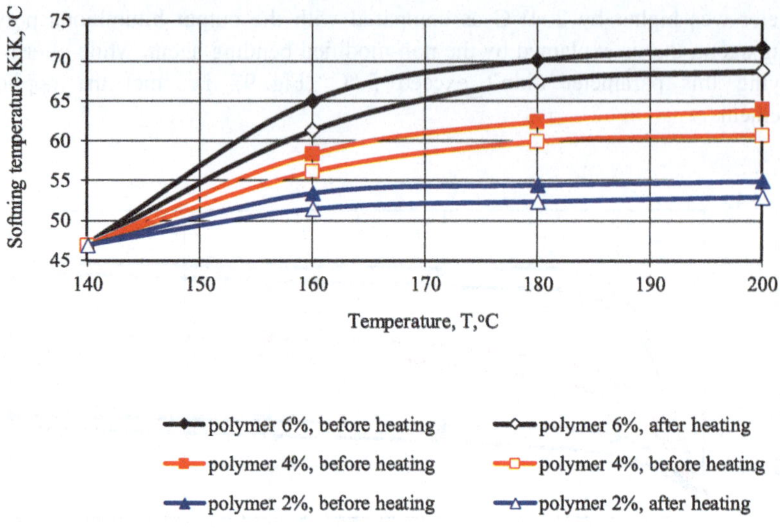

Fig. 10. Bituminous polymer softening temperature change dependence upon the preparation temperature at the polymer various amounts.

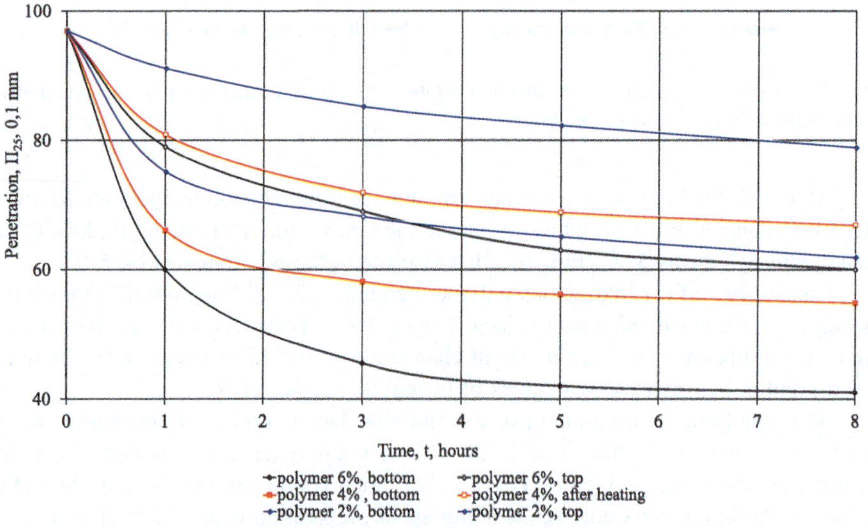

Fig. 11. Bituminous polymer disintegration during storage (according to penetration factor) dependence upon the preparation time at the polymer various amounts.

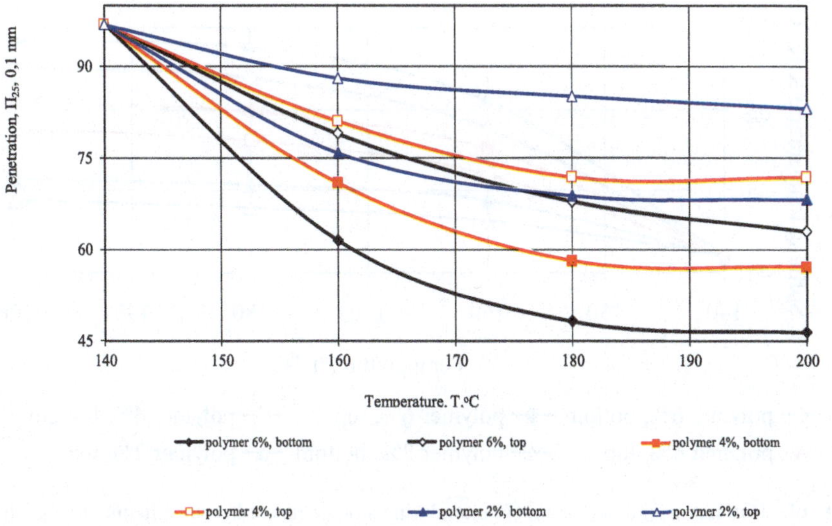

Fig. 12. Bituminous polymer disintegration during storage (according to penetration factor) dependence upon the preparation temperature at the polymer various amounts.

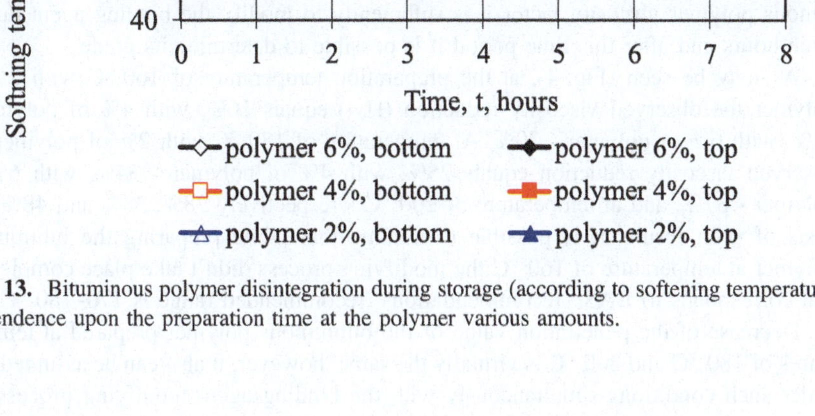

Fig. 13. Bituminous polymer disintegration during storage (according to softening temperature) dependence upon the preparation time at the polymer various amounts.

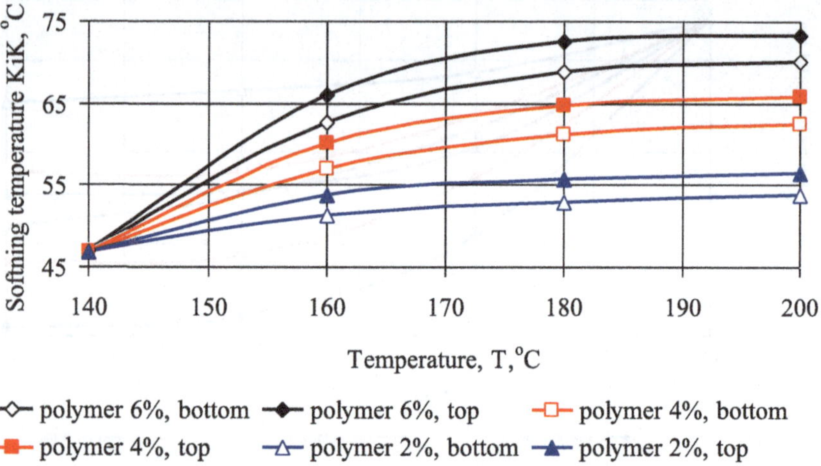

Fig. 14. Bituminous polymer disintegration during storage (according to softening temperature) dependence upon the preparation temperature at the polymer various amounts.

4 Discussion

As may be seen, at the polymer concentration of 2% and the bituminous polymer preparation temperature of 180 °C in 3 h of modifying the penetration value ($П_{25}$) reduced by 25%; at the polymer concentration of 4% – by 34%; and at the polymer concentration of 6% - by 44%; while in 8 h this reduction was respectively 31%, 40% and 47% (Fig. 2). I.e. the penetration value becomes stable after 3 h of the modifying bitumen. Similar penetration change of the bituminous polymer takes place at testing temperature of 0 °C as well (Fig. 3).

On basis of these results it can be made the conclusion that to establish the bituminous polymer viscosity factor it is sufficiently to modify the binding agent during three hours and after this time period it is possible to determine its grade.

As may be seen (Fig. 4), at the preparation temperature of 160 °C with 2% of polymer the observed viscosity reduction ($П_{25}$) equals 16%, with 4% of polymer - 18%, with 6% of polymer - 20%. At temperature of 180 °C with 2% of polymer the observed viscosity reduction equals 25%, with 4% of polymer - 34%, with 6% of polymer - 37%; and at temperature of 200 °C - respectively 28%, 38% and 48%. On basis of such results it is possible to assume that while preparing the bituminous polymer at temperature of 160 °C the modifying process didn't take place completely, that corresponds to BASF recommendations (recommended range is 170–180 °C).

Decrease of the penetration value of the bituminous polymer prepared at temperatures of 180 °C and 200 °C is virtually the same; however, it also can be assumed that under such conditions simultaneously with the binding agent modifying process the intensive processes of its deterioration occur. It is important also to note that at the preparation temperature increase and especially at polymer amount increase (up to 6%) it is possible the bituminous polymer transfer to more viscose grade. This feature of the produced bituminous polymer should be taken into consideration during preparation, placement and consolidation of polymer road concrete mix.

In case of 2% polymer concentration already after the first hour of bituminous polymer bending agent preparation (Fig. 5) the heat resistance value increased by 5 °C and remained virtually constant at its further maturing in reactor (Fig. 1).

In case of 4% polymer concentration the main increase in the heat resistance value took place during the first hour of bituminous polymer bending agent preparation and corresponded to 10 °C. After 8 h of the bituminous polymer bending agent preparation this increase was equal to 14 °C.

In case of 6% polymer concentration these parameters were equal to 15 °C and 21 °C. It is necessary to note that the bituminous polymer bending agent was prepared at temperature of 180 °C according to BASF recommendations. In case of the bituminous polymer bending agent preparation temperature reduction from 200 °C to 160 °C (Fig. 6) the softening temperature value respectively increased by 1 °C and decreased by 3–5 °C. It confirms the previous assumption that the bituminous polymer bending agent preparation temperature of 160 °C is insufficient for quick polymer integration to get the homogeneous mixture, and the heat resistance slow increase at the preparation temperature of 200 °C confirms that the deterioration processes don't develop.

To examine the possible bituminous polymer bending agent deterioration at too high temperatures there was studied the bituminous polymer bending agent elasticity change in this range of the temperatures change and preparation time periods. The following results were obtained.

In case of 2% polymer (Fig. 7) concentration after the first hour of preparation the elasticity factor was equal to 62%, after 3 h – 69%, and it was remaining constant at the maximum time of the bituminous polymer bending agent – 8 h. Similar increase of the elasticity factor was found at modifying 4% of polymer respectively by 66% and 73%, and at 6% of polymer – 67% and 73%. Thus, the polymer amount increase even up to 6% virtually doesn't influence on the increase of the elasticity factor which is sufficiently high at the smaller amount of the polymer too.

The bituminous polymer bending agent important parameters which characterize its processability, usability in production conditions, and determine the produced binding agent quality as well, are the bituminous polymer bending agent properties change after heating and disintegration during storage. Therefore, in this work there have been conducted investigations and obtained the results which confirmed the very slight change of such properties at given modifying temperatures (Fig. 9).

Although this parameter is not rated the obtained results show the sufficient homogeneity of produced binding agent after modifying. Dependence of the bituminous polymer bending agent disintegration factor during storage (according to softening temperature) upon the polymer amount and its preparation time showed slight change of this parameter that additionally confirmed the obtained bituminous polymer bending agent homogeneity and the processability as well. However, it is necessary to note that at the polymer amount increasing the disintegration factor value increases too, although remaining the allowable limits (Fig. 11, 12).

5 Conclusion

Conducted investigations of bitumen modified with the Butanol NS 104 polymer latex allowed making the following conclusions.

1. The bituminous polymer bending agent on basis of the Butanol NS 104 polymer latex in compliance with all standard parameters meets the requirements to bitumen modified with polymers.
2. The bituminous polymer bending agent physical-mechanical properties change in the wide limits (by 50–100%) depending on the polymer amount and the preparation process parameters shows the possibility of its properties active regulation under the particular operating conditions.
3. The bituminous polymer bending agent properties resistance and stability under the high process temperatures influence shows the possibility of its sufficiently long-term storage under the production conditions with the original properties keeping.
4. Short time of this bituminous polymer bending agent preparation (on average 2–3 h) allows saving the considerable energy resources and advantageously distinguishes it from the bituminous polymer bending agents produced with the other modifiers.
5. The Butanol NS 104 polymer latex rational amount for modifying the bitumen of this grade constitutes 2–3%, the preparation time is within the limits of 2–3 h, and the optimal preparation temperature is close to 170–180 °C.

References

1. Kang, Y., Song, M., Pu, L., Liu, T.: Rheological behaviors of epoxy asphalt inder in comparison of ase asphalt inder and SBS modified asphalt inder (Реологічна поведінка епоксидного асфальтового в'яжучого у порівнянні із звичайним асфальтовим в'яжучим та модифікованим асфальтовим в'яжучим SBS). Constr. Build. Mater. Shaanxi **76**, 343–350 (2015). https://doi.org/10.1016/j.conbuildmat.2014.12.020
2. Esfahani, M.A., Jamaloei, M.H., Torkaman, M.F.: Rheological and mechanical properties of bitumen modified with Sasobit, polyethylene, paraffin, and their Mixture. J. Mater. Civil Eng. **31**(7), 04019119 (2019). https://doi.org/10.1061/(ASCE)MT.1943-5533.0002664
3. Yin, H., Zhang, Y., Sun, Y., Xu, W., Yu, D., Xie, D.: Performance of hot mix epoxy asphalt binder and its concrete. Mater. Struct. **48**(11), 3825–3835 (2015)
4. Solouki, A., Muniandy, R., Hassim, S., Kheradmand, B.: Rheological property investigation of various Sasobit-modified bitumen. Pet. Sci. Technol. **33**(7), 773–779 (2015). https://doi.org/10.1080/10916466.2015.1010040
5. Yu, J., Cong, P., Wu, S.: Laboratory investigation of the properties of asphalt modified with epoxy resin (Laboratory determination of asphalt concrete properties modified with epoxy resin). J. Appl. Polym. Sci. **12**(6), 3557–3563 (2009). https://doi.org/10.1002/app.30324
6. Simnofske, D., Mollenhauer, K.: Effect of wax crystallization on complex modulus of modified bitumen after varied temperature conditioning rates. In: IOP Conference Series: Materials Science and Engineering, vol. 236 (2017). https://doi.org/10.1088/1757-899x/236/1/012003

Formation of a Soil Wedge by a Bulldozer with a Controlled Blade

Gennadiy Voskresenskiy(✉) and Evgeniy Kligunov(✉)

Pacific National University, Tihookeanskaya str., 136,
Khabarovsk 680035, Russia
004179@pnu.edu.ru, kes70@bk.ru

Abstract. The formation process of a soil wedge performed by a bulldozer with a controlled blade is considered. The most effective trench earthwork method for soil or rocks was implemented. The volume of the soil wedge moving through the trench increases by 15–20% compared with the layer by layer excavation. The use of a hydraulic drive allows improving the capabilities of working equipment. The operating process of the soil wedge formation and transportation performed by a hydraulic bulldozer with blade tilt in the transverse plane and a variable blade installation angle (back and forth) is analyzed. The blade tilt and installation angle variability can increase the bite depth of the blade when tilting forward and provide an increased productivity. The movement of the bulldozer on a horizontal track section when operating using a trench method is adopted as assumption. As a result, the calculated correlations for determining the soil wedge volume of the bulldozer were obtained, and the forces of resistance to the wedge transportation in the trench method were determined. The soil volume in front of the blade increases due to the increase in the wage base length, and the soil volume located above the blade decreases with a general increase in the soil wedge volume. New design also allows increasing the volume of the soil wedge and increasing the productivity of the bulldozer with a slight increase in the energy consumption of the wedge moving process.

Keywords: Excavation of soils and rocks · Bulldozer · Controlled blade · Trench method · Soil wedge · Resisting forces of the soil wedge

1 Introduction

Bulldozers are one of the main machines used in road construction, in the construction of residential and industrial complexes, for transporting rocks. The excavation of soil and rocks with a bulldozer starts with the operation of cutting and assembling a wedge in front of the blade and continues with transporting the prism to the discharge site [1–13].

The trench method for soil or rocks excavation is accepted as the most effective. The volume of the soil wedge moving along the trench increases by 15–20% compared with the layer by layer excavation [14, 15].

V. Murgul and V. Pukhkal (Eds.): EMMFT 2019, AISC 1258, pp. 117–126, 2021.
https://doi.org/10.1007/978-3-030-57450-5_11

2 Methods

The use of a hydraulic drive allows improving the capabilities of working equipment. Caterpillar manufactures bulldozers with a blade tilt in the transverse plane and a variable blade installation angle (back and forth) that can increase the bite depth of the blade when tilted forward and provide increased productivity. Studying the operating process of the formation and movement of the soil wedge shows that determining the soil wedge volume in front of the bulldozer blade in known methods is carried out according to a simplified model and does not take into account the inclination angle of the blade [2, 5, 9, 10]:

$$V_{pr} = \frac{BH^2}{2tg\varphi} \tag{1}$$

where B, H – width and height of the blade; φ – slope of response.

3 Results

In order to determine the effectiveness of using a bulldozer with a controlled blade, the design scheme for the soil wedge volume formation is considered, taking into account a possible change in the blade installation angle.

According to the design scheme presented in Fig. 1, it follows that the longitudinal wedge section and, consequently, the wedge volume increase with an increase in the inclination angle of the bulldozer blade.

The movement of the bulldozer on a horizontal section of the track when working in a trench way is accepted as an assumption. The volume of the soil wedge is as follows:

$$V = F_{\Sigma} \cdot B, \tag{2}$$

where $F_{\Sigma} = F_0 + F_1$ – area of the longitudinal wedge section; F_0 – area limited by the curved part of the blade and the line L connecting the top of the blade with the cutting edge; F_1 – area of a triangle OAK_1 (Fig. 1).

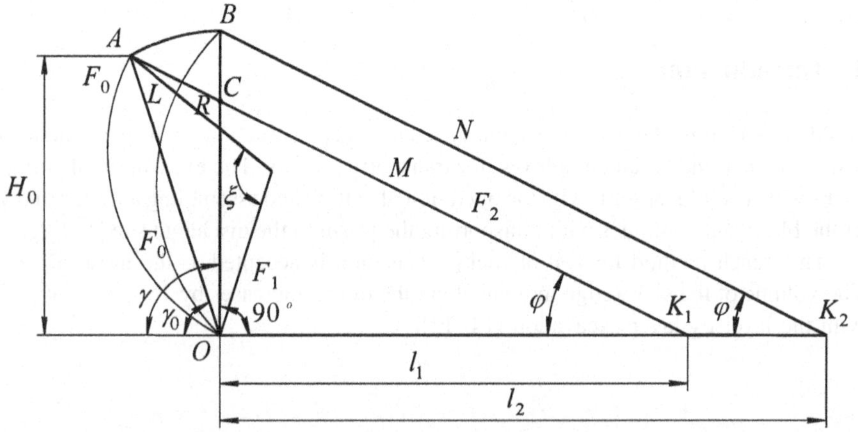

Fig. 1. The design scheme for determining the soil wedge volume of the bulldozer with a variable blade angle γ.

Area F_0 can be defined as follows, according to Fig. 1:

$$F_0 = \frac{\pi R^2}{360^o}\xi - LR\cos\frac{\xi}{2}, \tag{3}$$

where R – radius of the curved part of the blade; ξ – sector angle.

The area F_1 can be obtained as following using the design scheme (Fig. 1).

$$F_1 = \frac{H_A \times l}{2}. \tag{4}$$

The total volume of the soil wedge will be as follows:

$$V_\Sigma = (\frac{\pi R^2}{360^o}\xi - LR\cos\frac{\xi}{2})^B + \frac{L^2 B}{2}(\frac{\sin^2\gamma}{tg\varphi} - \cos\gamma \cdot \sin\gamma). \tag{5}$$

It can be assumed that an increase in the volume of the soil wedge leads to an increase in motion resistance forces.

The motion resistance forces of the soil wedge can be determined according to the design scheme (Fig. 2, 3).

It is assumed that a force W_x acts from the side of the soil wedge, a horizontal force acting on the base of the soil wedge $W_x = G_1 \cdot \mu$

$$W_x = V_1 \cdot \rho \cdot \mu \cdot g, \tag{6}$$

where V_1 – soil volume within the OCD limits, m^3; ρ – soil density, kN/m^3; μ – coefficient of friction between two soils, μ = 0.5…0.7.

The gravity force V_2 of the soil wedge volume acting above the blade of the bulldozer is transmitted to the chassis of the tractor and resists the movement of the tractor

$$W_{mp} = G_2 \times f, \tag{7}$$

where $G_2 = V_2 \cdot \rho$ is the gravity force of the soil; f – rolling resistance coefficient of the crawler attachment, f = 0.1.

Volume V_1 will be as follows

$$V_1 = \frac{H_c \cdot l}{2} \cdot B. \tag{8}$$

Volume V_2 can be determined as follows

$$V_2 = V_\Sigma - V_1. \tag{9}$$

Positions of points A and C the length of the soil wedge base l can be calculated for γ from 75° to 90° (Fig. 2).

$$H_A = L \cdot \sin\gamma; \tag{10}$$

$$l = \frac{H_A}{tg\varphi} - L\cos\gamma, \tag{11}$$

$$H_C = l \cdot tg\varphi. \tag{12}$$

In case when $\gamma = 90°$, $H_C = H_A$. At the angle $\gamma > 90°$ the positions H_A, H_C, l can be obtained as following

$$H_A = L \cdot \sin\gamma; \tag{13}$$

$$H_C = L \cdot \sin\gamma; \tag{14}$$

$$l = \frac{H_A}{tg\varphi} - L\cos\gamma. \tag{15}$$

The influence of the inclination angle γ on the soil wedge volume and on the positions of points A and C will be determined for the blade dimensions of the D-9R Caterpillar bulldozer: H = 1.93 m; B = 4.35 m; R = 1.9 m; $\xi = 65°$; $\gamma = 75°$; $\gamma_{max} = 75° (95°)$, $\varphi = 35°$, L = 2.0 m, $\rho = 16$ kN/m^3.

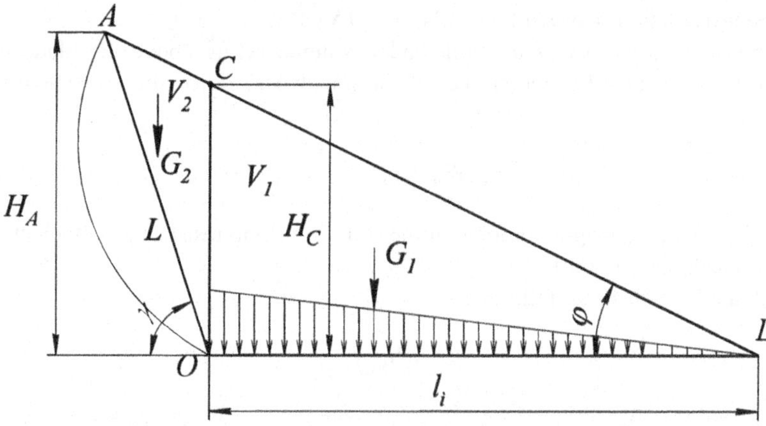

Fig. 2. The design scheme for determining the motion resistance forces of soil (side view of the blade).

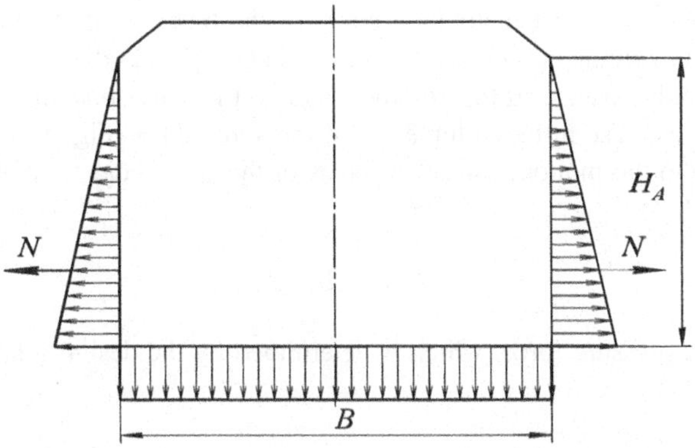

Fig. 3. The design scheme for determining the motion resistance forces of the soil (front view of the blade).

The main results of calculating the parameters of the soil wedge are given in Table 1 and are presented in Fig. 4.

Table 1. The results of calculating the parameters of the soil wedge.

γ, deg	75	80	85	90	95
H_A, m	1.93	1.97	1.99	2.00	1.99
H_C, m	1.57	1.72	1.86	2.0	1.99
l_1, m	2.24	2.46	2.66	2.85	3.01
V_2, m^3	3.86	3.42	2.87	2.17	2.17
V_1, m^3	7.64	9.2	10.76	12.39	13.22
V_Σ, m^3	11.5	12.62	13.63	14.56	15.4

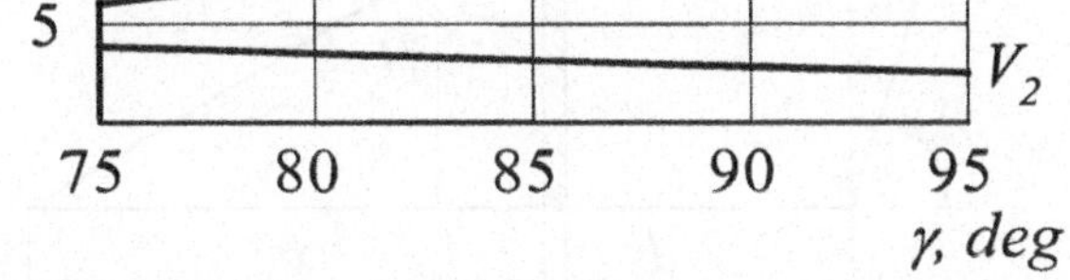

Fig. 4. The influence of the blade inclination angle of the on the volume of parameters of the soil prism V_Σ, V_1, V_2.

With an increase in the inclination angle of the blade γ, the volumes V_1 and V_2 redistribute. The volume of soil in front of the blade increases due to an increase in the length of the wedge base, and the volume V_2 of soil located above the blade decreases with a general increase in the volume V_Σ of the soil wedge (Fig. 4).

In addition to the motion resistance forces of the soil wedge, lateral friction forces also act (Fig. 2, 3).

$$2F_{fr} = 2N \cdot \mu, \tag{16}$$

where N – side pressure force, which is determined by the design scheme (Fig. 5, 6).

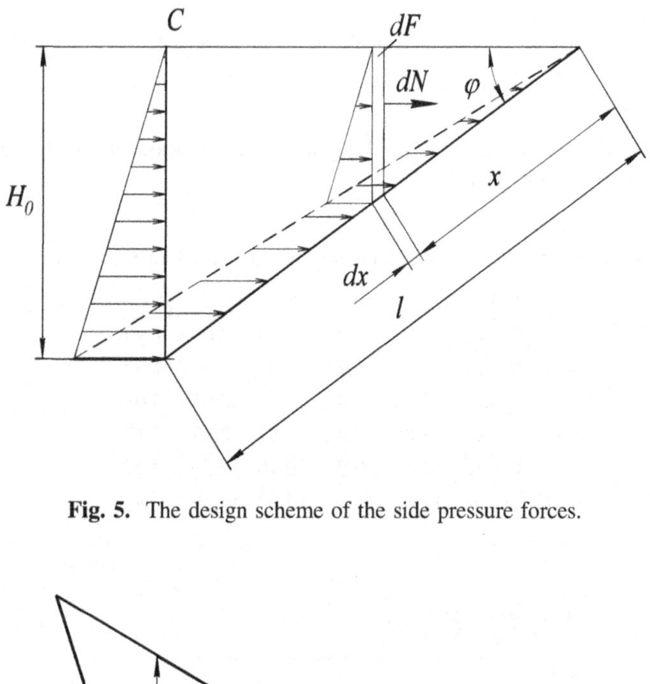

Fig. 5. The design scheme of the side pressure forces.

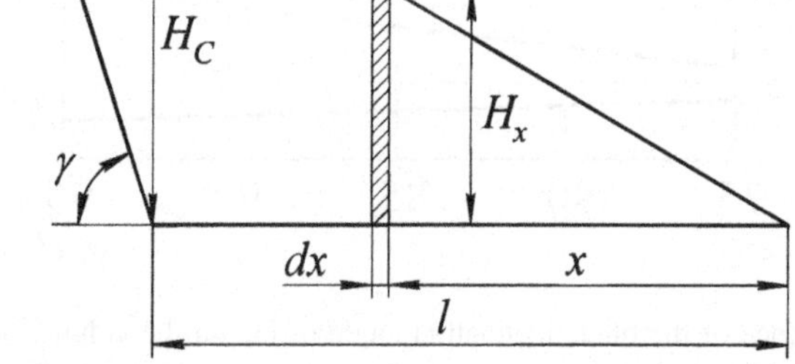

Fig. 6. The model for determining the position of the elementary area.

Elemental strength dN can be determined as following

$$dN = G_{mid} \cdot dF, \tag{17}$$

where $dF = H_x \cdot dx = x \cdot tg\varphi \cdot dx$ – elementary area; $G_{mid} = \rho \cdot g \frac{H_x}{2}$ – average pressure, then dN can be defined as follows

$$dN = \rho \cdot g \frac{H_x}{2} \cdot x \cdot tg\varphi \cdot dx, \tag{18}$$

or, substituting its value instead of H_x, the following can be obtained

$$dN = \rho \cdot g \frac{x \cdot tg\varphi}{2} \cdot x \cdot tg\varphi \cdot dx. \tag{19}$$

Pressure force is calculated

$$N = \int_0^l \rho \cdot g \frac{tg^2\varphi}{2} \cdot x^2 dx, \tag{20}$$

$$N = \frac{\rho \cdot g}{2} \cdot tg^2\varphi \frac{l^3}{3}. \tag{21}$$

Substituting its value $l = \frac{H_C}{tg\varphi}$ instead of l, the following is obtained

$$N = \frac{\rho \cdot g}{6} \cdot \frac{H_c^3}{tg\varphi}. \tag{22}$$

The friction force will be

$$2F_{mp} = \frac{\rho \cdot g}{3} \cdot \frac{H_C^3}{tg\varphi} \cdot \mu \tag{23}$$

The total motion resistance force of the soil wedge

$$W_\Sigma^f = W_x + W_{fr} + 2F_{fr}. \tag{24}$$

The motion resistance force of the bulldozer

$$W_\Sigma^B = W_\Sigma^f + G_P \cdot f, \tag{25}$$

where G_p – tractor gravity with bulldozer and cultivator attachments, G_p = 484 kN.

The results of calculating the power characteristics are given in Table 2 and in Fig. 7.

Table 2. The influence of the blade inclination angle on the motion resistance forces of the soil wedge and caterpillar tractor.

γ, deg	75	80	85	90	95
W_x, kN	59.9	72.1	84.3	97.1	103.6
$2F_{fr}$, N	14.4	18.9	24.0	29.8	30.7
W_f^{pr}, N	6.05	5.36	4.5	3.4	3.4
W_Σ^{pr}, kN	80.35	96.36	112.8	130	137.7
S_d, kN/m^3	6.98	7.63	8.27	8.92	8.94
W_Σ^B, kN	128.75	144.76	161.2	178.4	186.1

With an increase in the blade inclination angle, the volume of the transported soil wedge and the total tractor motion resistance increase. Taking into account the tractor's drawbar performance, it can be concluded that the tractor's speed is slightly reduced compared to the blade position with $\gamma = 75°$, as the gearbox provides traction resistance up to 190 kN in the steeply falling traction section at a speed of 3.5…3.6 km/h.

Changes in traction resistance are presented in Fig. 5.

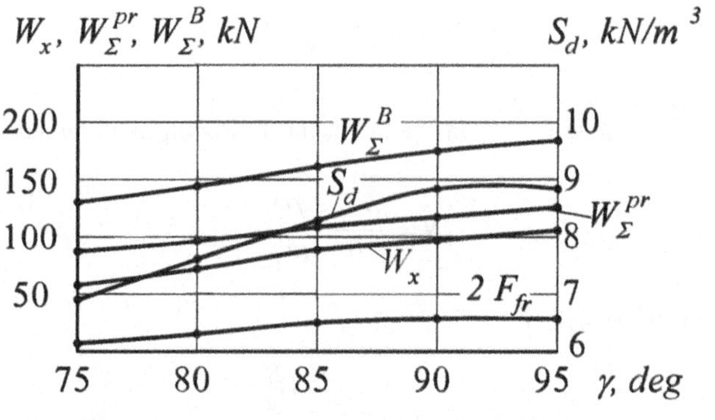

Fig. 7. The influence of the blade inclination angle γ on the resistance forces.

It is possible to evaluate the energy intensity of the wedge transportation process performed by a bulldozer with a controlled blade by introducing the resistivity index

$$S_d = \frac{W_\Sigma^{pr}}{V_{pr}}. \tag{26}$$

According to the graph $S_d(\gamma)$, the energy intensity increases from 6.98 kN/m^3 to 8.94 kN/m^3, or by 28%, with an increase in the blade inclination angle γ, while the soil wedge increased by 34% for angles γ from 75° to 95°.

4 Conclusions

The new design of a bulldozer with a controlled blade allows for more intensive cutting of knives into the soil at the collecting section due to the increased blade inclination angle. It also has the ability to increase the soil wedge volume and increase the machine productivity with a slight increase in the energy consumption of the wedge moving process.

At the same time, an increase in the energy intensity of the soil wedge transportation by 28% is offset by an increase in the volume of the soil wedge by 34% for angles γ from 75° to 95° and contributes to an increase in productivity.

References

1. Ren, L.Q., Han, Z.W., Li, J.Q., Tong, J.: Experimental investigation of bionic rough curved soil cutting blade surface to reduce soil adhesion and friction. Soil Tillage Res. **85**, 1–12 (2006). https://doi.org/10.1016/j.still.2004.10.006
2. Kim, S.-H., Lee, Y.-S., Sun, D.-I., Lee, S.-K., Yu, B.-H., Jang, S.-H., Kim, W., Han, C.-S.: Development of bulldozer sensor system for estimating the position of blade cutting edge. Autom. Constr. **106** (2019). https://doi.org/10.1016/j.autcon.2019.102890
3. Hirayama, M., Guivant, J., Katupitiya, J., Whitty, M.: Path planning for autonomous bulldozers. Mechatronics **58**, 20–38 (2019). https://doi.org/10.1016/j.mechatronics.2019.01.001
4. Zhou, W., Cai, Q.-X., Chen, S.-Z.: Study on dragline-bulldozer operation with variations in coal seam thickness. J. China Univ. Min. Technol. **17**, 464–466 (2007). https://doi.org/10.1016/S1006-1266(07)60126-6
5. Qinsen, Y., Shuren, S.: A soil-tool interaction model for bulldozer blades. J. Terrramech. **31**, 55–65 (1994). https://doi.org/10.1016/0022-4898(94)90007-8
6. Ito, N.: Bulldozer blade control. J. Terrramech. **28**, 65–78 (1991). https://doi.org/10.1016/0022-4898(91)90007-S
7. Muro, T.: Tractive performance of a bulldozer running on weak ground. J. Terrramech. **26**, 249–273 (1989). https://doi.org/10.1016/0022-4898(89)90039-6
8. Osinenko, P., Streif, S.: Optimal traction control for heavy-duty vehicles. Control Eng. Pract. **69**, 99–111 (2017). https://doi.org/10.1016/j.conengprac.2017.09.010
9. Schott, D.L., Lommen, S.W., van Gils, R., de Lange, J., Kerklaan, M.M., Dessing, O.M., Vreugdenhil, W., Lodewijks, G.: Scaling of particles and equipment by experiments of an excavation motion. Powder Technol. **278**, 26–34 (2015). https://doi.org/10.1016/j.powtec.2015.03.012
10. Ucgul, M., Saunders, C., Fielke, J.M.: Comparison of the discrete element and finite element methods to model the interaction of soil and tool cutting edge. Biosys. Eng. **169**, 199–208 (2018). https://doi.org/10.1016/j.biosystemseng.2018.03.003
11. Ucgul, M., Fielke, J.M., Saunders, C.: Three-dimensional discrete element modelling (DEM) of tillage: accounting for soil cohesion and adhesion. Biosys. Eng. **129**, 298–306 (2015). https://doi.org/10.1016/j.biosystemseng.2014.11.006
12. Shmulevich, I., Asaf, Z., Rubinstein, D.: Interaction between soil and a wide cutting blade using the discrete element method. Soil Tillage Res. **97**, 37–50 (2007). https://doi.org/10.1016/j.still.2007.08.009

13. Atkins, T.: Burrowing in soils, digging and ploughing. In: The Science and Engineering of Cutting, pp. 327–351 (2009). https://doi.org/10.1016/b978-0-7506-8531-3.00014-6
14. Li, G., Wang, W., Jing, Z., Zuo, L., Wang, F., Wei, Z.: Mechanism and numerical analysis of cutting rock and soil by TBM cutting tools. Tunn. Undergr. Space Technol. **81**, 428–437 (2018). https://doi.org/10.1016/j.tust.2018.08.015
15. Maciejewski, J., Jarzębowski, A., Trąmpczyński, W.: Study on the efficiency of the digging process using the model of excavator bucket. J. Terramech. **40**, 221–233 (2003). https://doi.org/10.1016/j.jterra.2003.12.003

On the Impact of Metrological Support on Efficiency of Special Equipment

Rustam Khayrullin(✉)

Moscow State University of Civil Engineering, 129337 Moscow, Russia
zrkzrk@list.ru

Abstract. The problem of construction of mathematical model of special technics or special objects and its metrological maintenance is extremely difficult. The basic complexity is connected by difficulties formalization of such system. The system is difficult to formalize primarily because of the large number of goals and sub-goals of metrological support. Moreover, the goals and sub-goals are often contradictory. This leads to the need for search of compromises between sub-goals. The article is devoted to the construction of mathematical models of operation of special equipment with metrological support. The new approach to construction of semi-Markov models of maintenance both the special equipment and its metrological support is presented. Semi-Markov models are proposed that describe a consistent change in the degree of achievement of the goals of operating special equipment and objects with metrological support. The influence of the interdependence of costs, losses and probabilities of being in various states of special equipment and objects on their quality is investigated. Two-dimensional losses and costs surfaces are constructed depending on the parameters of metrological support. The structure of these surfaces is studied. The possibility of classifying special equipment and objects for specifying the requirements for their metrological support was considered.

Keywords: Metrological support · Semi-Markov model · Special equipment and objects

1 Introduction

A huge amount of work has been devoted to the problem of constructing and studying large-scale and poorly formalized systems [1–6]. Annually, in the Institute of Management Problems named after V.A. Trapeznikov of the Russian Academy of Sciences, international conferences are held devoted to the management of the development of large-scale systems (for example MLSD-2018, MLSD-2019). One of the main directions of the scientific conference is the modeling of large-scale and difficult for formalization systems [7–10].

One of the oldest problems of optimization of metrological support (MS) of special equipment and objects (SEO) at all periods of their life cycle is the lack of models of interdependence of the quality of MS and SEO [11–15]. There is a fairly large number of theoretical and practical works in which the impact of MS on SEO readiness is

V. Murgul and V. Pukhkal (Eds.): EMMFT 2019, AISC 1258, pp. 127–135, 2021.
https://doi.org/10.1007/978-3-030-57450-5_12

evaluated, on individual indicators of their effectiveness for intended use or on consumption of various types of resources. Attempts to develop a universal integrated model for assessing the impact of MS on "ensuring the potential readiness of SEO, readiness for use, effective use for the intended purpose, failure-free operation of SEO, the health of personnel and saving all types of resources [4] have not yet been made. There are two main reasons for the rejection of the universal integrated research model of MS and SEO.

The first reason is that the system of goals for the operation of the SEO is an extremely complex, poorly formalized structure. Any goal regulated in [4] is decomposed into sub-goals. For example, "the effectiveness of the application of the destination aircraft - interceptor," is the sum of the interception rate, interception range, the probability of hitting the target, etc. That is, the a priori set of goals for the operation of an SEO can be considered infinitely large. In practical problems, the total number of special objects and their goals reaches hundreds of thousands.

The second reason is that costs and losses associated with the achievement of various SEO goals (with the transition of SEO to different states) are distinguished by their physical essence. For example, achieving the required level of availability is modeled by time and loss, while minimizing resource consumption takes into account the amount of fuel consumed (volume or weight), energy losses (kilowatt/hour), consumables (monetary units or pieces), etc.

In [14] the formulation of the problem of creating the theoretical foundations of the operation of SEO with MS is given. In [15] the method for distribution of controlling volumes of metrological support the objectives of complex organizational and technical systems with the use of semi-Markov models are suggested.

This article provides the formulation and solution of a fundamentally new problem of developing a mathematical model of the operation of large scale system: SEO with MS. The approach developed in the article differs from classical approach in that in classical approach each vertex of the graph is characterized by the probability of being in a given state, and the edges of the graph - by the time of transition from one state to another. In classical approach the transition time can be both as random as deterministic quantity. In the approach proposed in the article each vertex of the graph (state) corresponds to the total losses and costs associated with being in this state.

2 Materials and Methods

2.1 General Formulation of the Problem and the Main Suggestions

Let the operation of SEO is described by a stationary random process of a sequential transition from the initial state in which all the goals are achieved, to the states that exclude the possibility of achieving one, two or more goals. Transitions are interpreted by the occurrence of faults (failures) in the SEO or the manifestation (detection) of faults (failures) with the subsequent restoration of the SEO. There are classes of functions of interdependencies of the probabilities of a change in the state of an SEO, losses in these states, and the cost of returning to the initial state. It is necessary to model and classify the change in total costs and losses during the operation of the SEO.

Such formulation of the problem allows to circumvent the problems of uncertainty of the initial data and their composition. The need to justify the nomenclature of the studied SEO states and the probabilities of transitions between them, reducing costs and losses with different physical essence to one universal equivalents disappears. Instead of quantitative characteristics of random processes, one can use qualitative estimates: "relatively high or low probability", "equal probability", "incommensurate losses", etc. In the early stages of designing the MS and SEO systems, as well as modeling the stages of the life cycle of the SEO with MS, there can be no other estimates.

Even the most qualified and authoritative specialist - expert is very difficult in developing tactical and technical tasks for the development of SEO to predict quantitative estimates of the probability of reducing the effectiveness of SEO, accidents, catastrophes and corresponding losses. It is quite another thing to designate the "order" of these assessments or their correlation.

Obviously, the complexity of solving the formulated problem will be directly proportional to the number of simulated SEO states and possible transitions between them. Therefore, to illustrate the solution, it is advisable to introduce additional restrictions and assumptions:

1. The operation process of the SEO is considered without division into the procedures for application, storage of maintenance, etc. That is, all procedures are performed in a timely manner, with proper quality and ensure the achievement of specified goals. This process in the model should be described by one state: - "perfect" state of SEO (all goals are achieved, all costs for its operation correspond to the given, losses are zero).
2. The degradation of an SEO during operation is represented by a sequence of transitions to other states in which one, two, or more of the objectives of exploitation are not achieved $(S_0, S_1, \ldots, S_n)$.
3. Degradation processes are stationary.
4. Any subsequent state is more "grave" than the previous one.
5. Certain costs and losses correspond to each SEO state $(c_0, c_1, \ldots, c_n)$. All costs for the operation of SEO, except for restoration (repair), are assumed to be zero.
6. Transitions are possible only in more "severe" states $(S_i \rightarrow S_j, \ldots j > i)$. Each transition is characterized by a certain probability (p_{ij}). Allowed the transition to the initial state $(S_i \rightarrow S_0)$, which simulates the situation of the manifestation of an apparent failure (apparent failure) with subsequent recovery.
7. Transition probabilities and losses are interrelated by some functions: $p_{ij} = F(i, j, \mathbf{a})$, $c_i = \Phi(i, \mathbf{b})$, where a and b are the vectors of controlled variables.
8. Total costs and losses are estimated by the indicator $L = \sum_{i=0}^{n} c_i \pi_i$ (π_i - the stationary probability of SEO being in - state). It can be considered an indicator of the quality of SEO.

2.2 Model 1

The process of operating SEO without MS [7] can be described by the semi-Markov model shown in Fig. 1a).

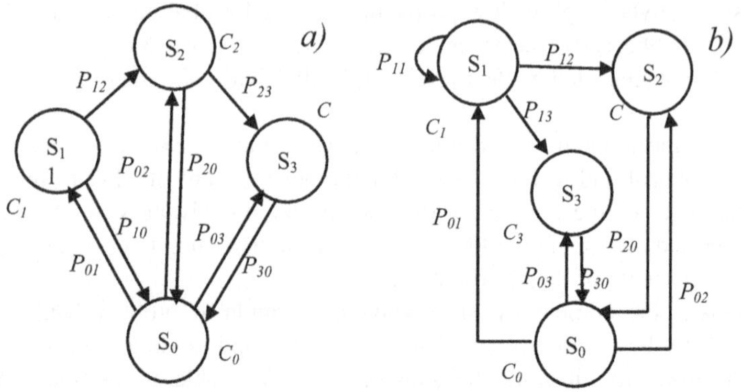

Fig. 1. The simplest models of operation SEO without MS and with MS.

The model describes a fairly idealized process of operating SEO:

1. SEO is basically in the initial state Fig. 1b: ($p_{00} > 0$, $p_{00} + p_{01} + p_{02} + p_{03} = 1$).
2. Gradually degrading, the SEO enters the first state (for example, the state of increased consumption of resources, in particular fuel and lubricants). This condition is characterized by additional losses c1.
3. When the increase in resource consumption becomes noticeable "with the naked eye", repairs are carried out and SEO returns to its original state S_0. Repair costs are negligible compared to losses c_1). The probability of an apparent failure is given by the vector of the controlled variables a.
4. If the over-expenditure of resources has not manifested itself, the SEO goes to the second state (for example, the state of loss of readiness for use, in particular the difficulty of starting the engine or its failure). Losses in this state may be higher than in the first, and vice versa. It all depends on the duration of the operation of SEO with increased consumption of resources. In the proposed model, the ratio c_1 and c_2 is determined by the vector of controlled variables b.
5. Lack of readiness may occur (if attempts are made to use SEO for its intended purpose, performance check during maintenance, etc.) and then SEO is restored, or it may not manifest, SEO becomes more "serious", for example, non-returnable losses. The mechanism for setting transition probabilities and losses in all states is the same.
6. From the third state, SEO with a probability equal to one returns to the initial state (compensation of losses).

2.3 Model 2

In [8] the simplest operation model of an SEO with MS was described. The SEO can be in one of four states: S_0 - fully operational, S_1 - practically operational, but with increased loss of resource, S_2 - faulty, state S_3 - simulates a generalized state and includes monitoring, checking and repair states (Fig. 1b).

3 Results

Figure 2 shows (according the model 1) a part of the results of modeling the quality of SEO based on the control of the functions of the interdependencies of the probabilities of a change in the state of the SEO, losses in these states and the cost of returning to the initial state. For clarity, two-dimensional surfaces of the total costs of the two control variables (vectors a and b were replaced with scalars a and b) are shown.

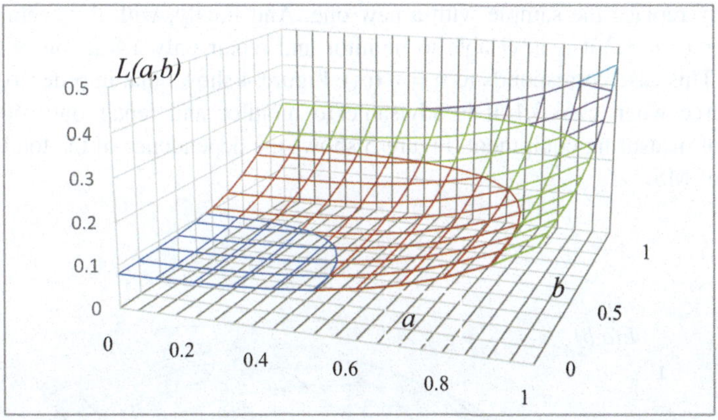

Fig. 2. Dependence of total losses and probabilities L on the controlled parameters a and b.

The specific choice of vectors of controlled variables a and b is ambiguous. In the simplest case, the range of permissible values of the variables a and b is a square $0 \leq a \leq 1$, $0 \leq b \leq 1$. The parameter a provides the ability to change the transition probabilities p_{ij} by 1, 2 or 3 orders of magnitude. For example, from $p_{00} > 0,01p_{23}$ to $p_{00} \approx p_{23}$. Similarly, b it provides for the variation of unit costs from $c_1 > 0,01c_3$ to $c_1 \approx c_3$.

The total costs and losses are used as an indicator of efficiency, and the volume of control η is used as a controlled parameter (the percentage of SEO samples tested from the total number of SEO in the state S_1). Vector a replaced by a scalar η (Fig. 3). In Fig. 5 shows two-dimensional surface of the total costs of the two control variables (vectors a and b replaced with scalars a and b) are shown.

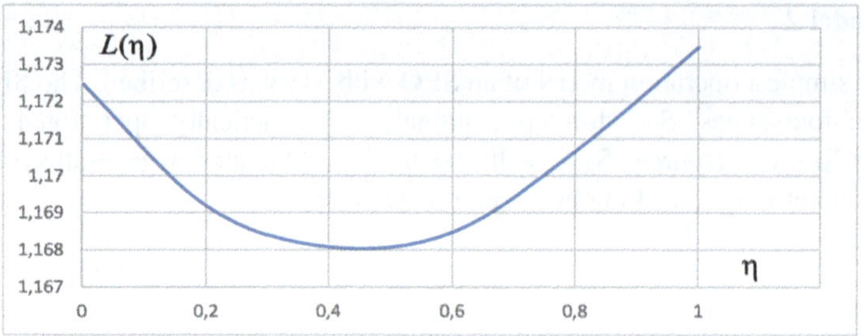

Fig. 3. The minimum value of the criterion is achieved with an optimal amount of MS.

The results of calculations showed that if the cost of MS is small, then to minimize L it is advisable to subject all samples to control, and, if necessary, repair. This case corresponds to $\eta = 1$. If the cost of MS is large, then it is advisable not to implement the MS in general, that is $\eta = 0$. In case of failure or accident, it is advisable to completely replace the sample with a new one. And finally, with the average cost of MS: $2.9 < c_3 < 3.2$ it is advisable to monitor and repair only a fraction of the faulty samples. This case corresponds to $\eta = \eta\ (c_3)$. Figure 4 shows that in order to minimize the resource when $c_3 = 3.1$ it is advisable to monitor and repair only 40% of the samples of measuring equipment. Figure 5 shows the dependence of the total losses on the cost of MS.

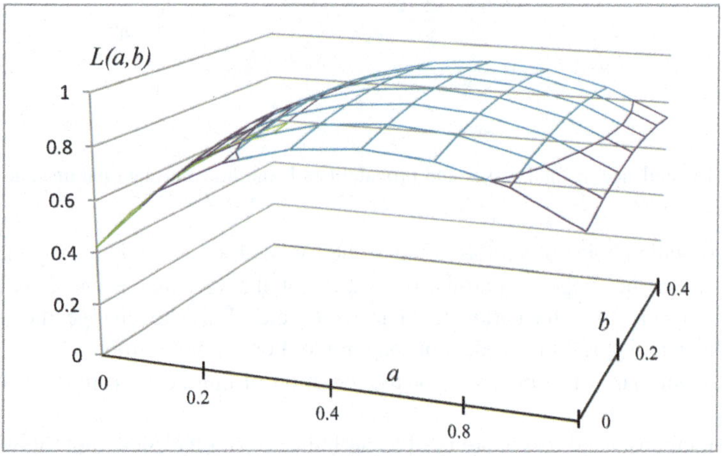

Fig. 4. Dependence of total losses and probabilities L on the controlled parameters a and b.

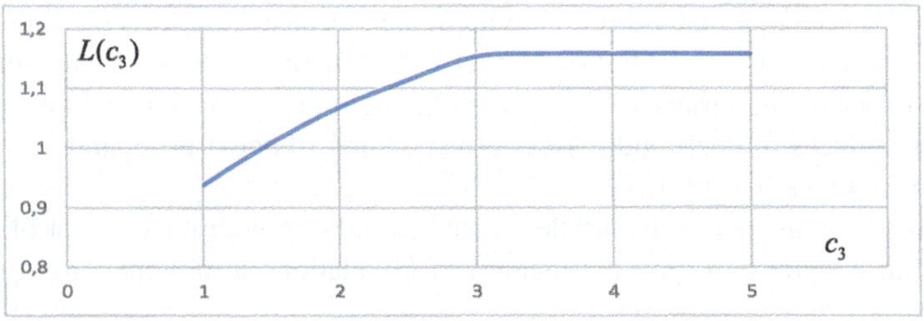

Fig. 5. The dependence of the total losses and probabilities of the cost of MS.

4 Discussions

The obtained results allow us to classify SEO based on the choice of discrete combinations of controlled variables a and b. For example:

Class A: The SEO that characterized by a sharp decrease in transition probabilities (the "heavier" the state, the less likely the transition to it) and the same increase in losses (the "heavier" the state, the higher the loss).
Class B: The SEO that characterized by a sluggish decrease in transition probabilities and a "sluggish" increase in losses.
Class C: The SEO that characterized by a sluggish decrease in the probability of transitions and a "sharp" increase in losses.
Class D: The SEO that characterized by a sharp decrease in transition probabilities and a "sluggish" increase in losses.

If we examine more states of SEO, as well as a wider class of functions for changing transition probabilities, costs and losses, then the SEO classification can be more detailed.

Obviously, by changing the vectors of the controlled variables a and b, it is possible to estimate the a priori sensitivity of the model to the introduction of MS into the operation of an SEO in order to timely identify the prerequisites for the appearance of failures (malfunctions) and then eliminate them. Note that in this case there will be additional costs for MS.

Thus, the solution of the problem under consideration allows to classify SEO in order to further shape its appearance when designing an SEO with MS system. Classification can be carried out, for example, on the basis of an analysis of the structure of the total cost hyper-surface L(a, b). Varying the vector of the controlled variables a and b allows both to expand the surfaces of the total losses relative to the "diagonals" and to change the positions of the local extremes of the hyper-surfaces L(a, b) and K_G(a, b). The determining criterion for attributing SEO to a particular class is the presence of characteristic points on the surfaces of total costs L(a, b), such as the corner point of the domain of controlled variables, in which the extremes of total costs are reached, the local point of the function extremum L(a, b), points and lines of inflection of functions, etc.

The most effective method for studying hyper-surfaces L(a, b) is the finite element method for studying semi-Markov models. This method allows as to analyze the structure of the hyper-surfaces on the basis of a preliminary analysis of finite elements - the hyper-surfaces of costs and losses corresponding to being in separate states of the operation of an SEO with MS.

The results obtained in the article are zin accordance with the concept of [14], can enter, as an integral part, into the theoretical foundations of the operation of an SEO with MS.

5 Conclusions

The scientific and methodical approach to the construction of semi-Markov models for the operation of special equipment has been developed, which makes it possible to form models having an arbitrary finite dimensionality of the space of technical states. The constructed models can be effectively investigated and solved by means of standard algorithms and programs for solving systems of linear algebraic equations, as well as developed by the author the finite element method for the study of stationary semi-Markov models.

The dependences of the total losses and probabilities on the controlled parameters are constructed. The basic principles of the classification of SEO with MS are proposed.

References

1. Belov, M.V., Novikov, D.A.: General-system approach to the development of complex activity technologies. Procedia Comput. Sci. **112**, 2076–2085 (2017). https://doi.org/10.1016/j.procs.2017.08.220
2. Novikov, L.S., Baranov, D.G., Gagarin, Yu.F., Dergachev, V.A., Voronina, E.N.: Measurements of microparticle fluxes on orbital space stations from 1978 until 2011. Adv. Space Res. **59**(1215), 3003–3010 (2017). https://doi.org/10.1134/s0020441211010180
3. Burkov, V.N., Korgin, N.A., Novikov, D.A.: Control mechanisms for organizational-technical systems: problems of integration and decomposition. IFAC-PapersOnLine **49**(32), 1–6 (2016). https://doi.org/10.1016/j.ifacol.2016.12.180
4. Kandoba, I.N., Koz'min, I.V., Novikov, D.A.: Admissible controls in a nonlinear time-optimal problem with phase constraints. IFAC-PapersOnLine **51**(32), 251–255 (2018). https://doi.org/10.1016/j.ifacol.2018.11.390
5. Khayrullin, R.Z., Zubkov, S.I., Lutskova, T.A.: To constructing strategies for purchasing, repairing and retiring of control and measuring equipment. Syst. Methods Technol. **4**(40), 85–90 (2018). https://doi.org/10.18324/2077-5415-2018-4-85-90
6. Zedgenizov, A., Burkov, D.: Methods for the traffic demand assessment based on the quantitative characteristics of urban areas functioning. Transp. Res. Procedia **20**, 724–730 (2017). https://doi.org/10.1016/j.trpro.2017.01.117
7. Ereshko, F.I., Medennikov, V.I.: Features of large-scale events in the field of industrial infrastructure and evaluation of their effectiveness. In: Proceedings of 2018 Eleventh International Conference "Management of Large-Scale System Development". https://doi.org/10.34706/de.2018.04.03

8. Valuev, A.M.: Modeling of the transport flow through crossroads with merging and divergence points. In: Proceedings of 2018 Eleventh International Conference "Management of Large-Scale System Development". https://doi.org/10.1109/mlsd.2018.8551915
9. Efremenko, V.F., Pashchenko, F.F.: Regional innovation system as an instrument of social-economic development of the territorial entity. In: Proceedings of 2018 Eleventh International Conference "Management of Large-Scale System Development". https://doi.org/10.1109/mlsd.2018.8551912
10. Arakelyan, E., Mezin, S.V., Sabanin, V.R.: Technical opportunities to improve the intelligence of modern automatic control system of large power plants. In: Proceedings of 2018 Eleventh International Conference "Management of Large-Scale System Development". https://doi.org/10.1109/mlsd.2018.8551935
11. Chunovkina, A.G., Pokhodun, A.I., Sulaberidze, V.Sh.: The problem of determining and adjusting the inter-calibration intervals of measuring instruments. Meas. Tech. **62**, 86–868 (2020). https://doi.org/10.1007/s11018-020-01706-2
12. Francisco, S., Guzmán, J., Rosa, B., Rodríguez, C., Doimeadios, M., Ángel, R.: Analytical metrology for nanomaterials: present achievements and future challenges. Analytica Chimica Acta **1059**, 1–15 (2019). https://doi.org/10.1016/j.aca.2019.02.009ISBN:0003-2670
13. Gao, W., Haitjema, H., Fang, F., Leach, R., Cheung, C., Savio, E., Linares, J.: On-machine and in-process surface metrology for precision manufacturing. CIRP Ann. **68**(2) (2019). https://doi.org/10.1016/j.cirp.2019.05.005. ISBN 0007-8506
14. Kostoglotov, A.A., Andrashitov, D.S., Kornev, A.S., Lazarenko, S.V.: Method synthesis algorithms ratings dynamic software for measuring systems and measuring instruments based on the combined maximum principle. Meas. Tech. **6**, 20–24 (2019). https://doi.org/10.32446/0368-1025it.2019-6-20-24
15. Popenkov, A.Ya., Khayrullin, R.Z.: Distribution of controlling volumes of metrological support for the objectives of complex organizational and technical systems with the use of semi-Markov models. In: Proceedings of 2018 Eleventh International Conference "Management of Large-Scale System Development". https://doi.org/10.1109/mlsd.2018.8551917

Assessment of the Conditions for Allocating Independent Road Safety ITS Subsystem

Elena Pechatnova[1(✉)] and Vasiliy Kuznetsov[2]

[1] Altai State University, 61, Lenina Ave., Barnaul 656049, Russia
phukcia@yandex.ru
[2] Altai State Agricultural University, 98, Krasnoarmeysky Ave., Barnaul 656049, Russia

Abstract. Intelligent transport systems (ITS) are developing dynamically in many countries around the world, including Russia. One of the main stated goals is to improve road safety (RS). The paper considers the prospects and conditions for the allocation of an independent RS subsystem within the framework of an intelligent transport system. A structural model is obtained that is easy to implement in the existing ITS. The transition from the closed type of ITS functioning is proposed: for the successful functioning of the RS subsystem, the use of dispatching information, forces and means of state services is additionally recommended. The results of evaluation of the ability to use the accessible tool of ITS components in the proposed subsystem that preliminarily grouped in two blocks: block "monitoring" and block "management". The study was conducted using the example of the Russian Federal road A-322. Based on the use of the risk theory and "Driver-Car–Road–Environment" approach to assessing the level of potential danger, an algorithm for the functioning of a complex RS subsystem has been developed, the main idea of which is to compare the current risk to the established limit value. A multiplicative form is proposed as a basic model for assessing the risk of road traffic accident. Based on the results of the study, recommendations were made on the conditions and opportunities for implementing the RS subsystem.

Keywords: Intelligent transport systems · Road safety · ITS subsystems · ITS functioning algorithms

1 Introduction

Intelligent transport systems (ITS) as well as other computer technologies and automated systems are being developed and improved in many countries of the world [1–3].

Large differences in the structure of ITS are recorded in cities and roads outside of human settlements. The main reason for differences in different traffic conditions, the amount of traffic flow and the speed difference [4].

V. Murgul and V. Pukhkal (Eds.): EMMFT 2019, AISC 1258, pp. 136–145, 2021.
https://doi.org/10.1007/978-3-030-57450-5_13

One of the leading goals of ITS is to improve road safety (RS). Many studies have shown that using ITS and informing drivers of hazards leads to a significant reduction in road traffic accident rates [5–8].

The greatest severity of consequences is caused by road traffic accidents that occurred on roads outside human settlements [9]. However, the task of increasing the RS on highways in Russia is solved mainly by increasing the means of automatic registration of traffic regulations violations. This is not insufficient, since the objects are installed locally and drivers observe the speed limit only on a certain section of road [1]. In addition, there is no real-time response to the offense. Other ways to prevent road traffic accidents (the work of state services, informing drivers) are practically not associated with ITS. Thus, a comprehensive work of the existing ITS and RS enhancement tools is needed, which can be implemented using an independent RS subsystem within ITS.

Separate studies are devoted to the study of the role of ITS and the design of ITS components for ensuring RS. Work [10] is devoted to the development of key performance indicators (KPIs) for traffic management and ITS, one of the four strategic topics is occupied by RS. Cognitive mechanisms are proposed as a basis for the operation of the ITS security system in research [11]. The paper [12] presents methods for designing and developing an intelligent multimodal transport system in order to improve safety.

The purpose of the work was to assess the conditions and prospects for the implementation of the RS subsystem on roads outside human settlements as an independent component of ITS.

2 Defining the Structural Model of the RS Subsystem

2.1 Architecture of the Existing ITS

One of the leading conditions for allocating a RS subsystem is its simple integration into the existing ITS system, so the consideration of the current ITS scheme is an important part of the study.

The forms of ITS organization and architecture are determined by state regulations and depend on the local characteristics of the road network.

The physical architecture of ITS in Russia is organized in a hierarchical way and is shown in Fig. 1. In general, ITS consists of subsystems [13]. According to GOST R 56294-2014, 5 levels are currently allocated.

There is an integration platform at level I, which manages all complex subsystems. There are several complex subsystems at level II: an automated traffic management system, a road condition management subsystem, a traffic regulations control and vehicle control subsystem, and a user services subsystem. Each of the complex subsystems is based on one or more tool subsystems that are located at level III. Currently, there are 18 tool subsystems in use, most of which are shown in Fig. 1. There are elements of subsystems at level IV, which are executive elements of tool subsystems. Level V is an equipment.

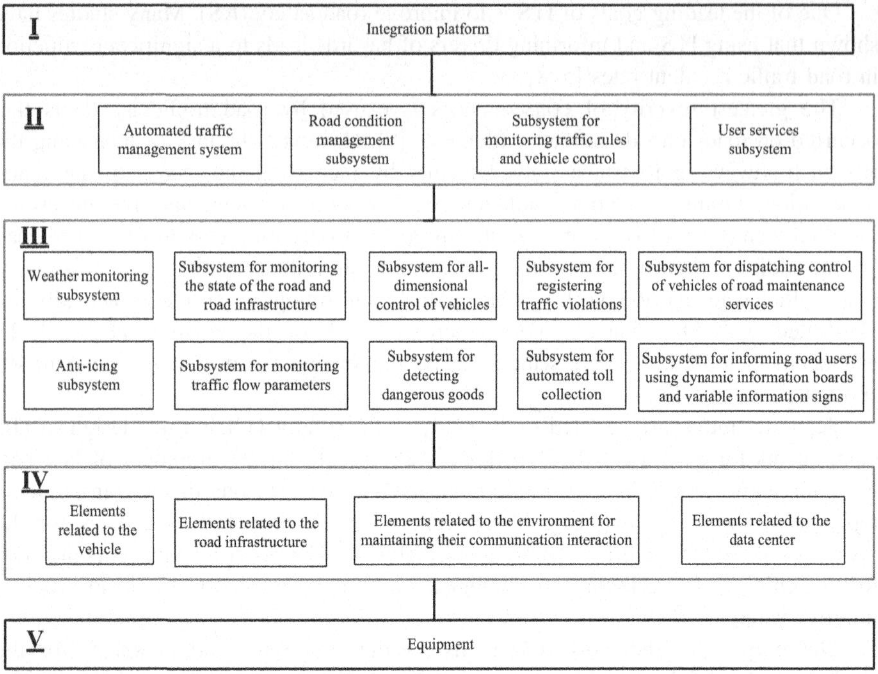

Fig. 1. Physical architecture of ITS.

2.2 Components and Architecture of the Proposed RS Subsystem

For the best implementation of the RS subsystem in the current ITS, the use of existing components of levels III-V is proposed; the subsystem itself is represented at level II, i.e. it is a complex subsystem. An abbreviated block diagram of the proposed subsystem is shown in Fig. 2.

It is proposed to group ITS subsystems (components) at the tool level into 2 blocks: “monitoring” and “management”. Thus, the instrumental block has two main functions: control over the necessary indicators and execution of decisions on RS management.

Based on the information received from levels IV and V, on the tool level in the “monitoring” block collects, transmits, processes and stores the received data. Based on data processing using a given mathematical model, the complex RS subsystem makes decisions on control actions that are passed to the tool subsystems (the “management” block), and are implemented using ITS elements and equipment. This algorithm is most common in existing ITS, but it is closed. For more efficient operation of the RS subsystem, additional use of dispatching information as input data and the use of forces and means of state services as execution of decisions is proposed.

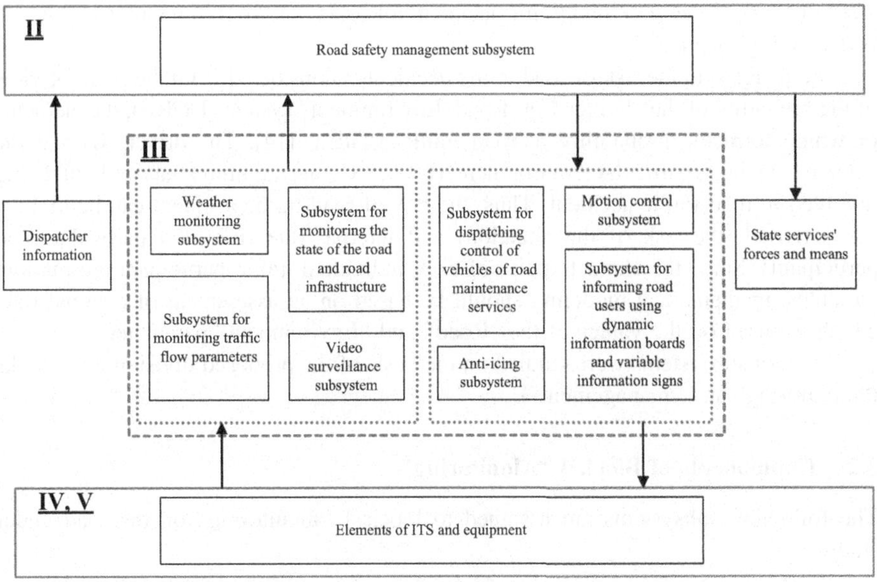

Fig. 2. Proposed block diagram of the RS subsystem.

In accordance with the adopted administrative division of territories in the Russian Federation, it is proposed to place a control center in each municipal formation (region) of the subject of the Russian Federation.

3 Materials and Methods

3.1 Research Subject and Theoretical Basis

To develop proposals for the implementation of the RS subsystem on the basis of the existing ITS, an analysis of the operating components of the ITS was carried out. The work of tool subsystems has been studied and the directions and prospects for the development of a complex RS subsystem have been determined. The study was carried out on the road outside the human settlements A-322 Barnaul-Rubtsovsk – the state border with the Republic of Kazakhstan, the length of which is 321 km.

With regard to the development of the complex RS subsystem, it was determined that its main focus should be the assessment of the current and projected risk of road traffic accidents on road sections. To determine the basic mathematical model, the risk theory is used, the basic position of which is the concept of acceptable risk. Despite the concept of "VisionZero", which is common in European countries and implies zero risk of death, this position is not adequate in Russia in the short perspective: at the end of

2018, the risk of death in road traffic accident in Russia is 1.24·10-4 people/year, which is a very high value.

The increase in the risk of road traffic accident is considered from the point of view of the reliability of the "Driver-Car–Road–Environment" system (DCRE), the elements of which form the probability of road traffic accident [14]. The driver (B) and the vehicle (A) in a particular environment (D and C) are at immediate risk of being involved in road traffic accident. Thus, the risk of road traffic accident can be divided into external (the risk of the situation) and internal (the risk of a particular road participant). Since the characteristics of each individual traffic participant are random variables, mathematical modeling should be based on an assessment of external risk, which is based on the values of the "Road" and "Environment" parameters.

The tool subsystems were studied on the basis of the proposed division into blocks ("monitoring" and "management").

3.2 Components of Block 1 "Monitoring"

The following subsystems are assigned to block 1 "monitoring" on the road under study:

1. Traffic flow parameters monitoring subsystem – traffic intensity monitoring points (TIMP).
2. Weather monitoring subsystem-automated road and weather stations (ARWS).
3. Video surveillance subsystem.

Their location on the A–322 road is shown in Fig. 3.

To assess the performance of the traffic flow parameters monitoring subsystem, the number of failures in operation was analyzed, for this purpose, data on hourly intensity values were obtained for three TIMP located on 38, 128 and 166 km of the analyzed road. Information is available for 2018.

3.3 Components of the Block of 2 "Management"

The components of block 2 "management" on the selected road are the following:

1. Subsystems of dispatching control of the vehicle for road maintenance and de-icing services.
2. Subsystem of informing of participants of traffic using dynamic information boards.

The perspective role of the second component was evaluated by analyzing the number and location of its elements.

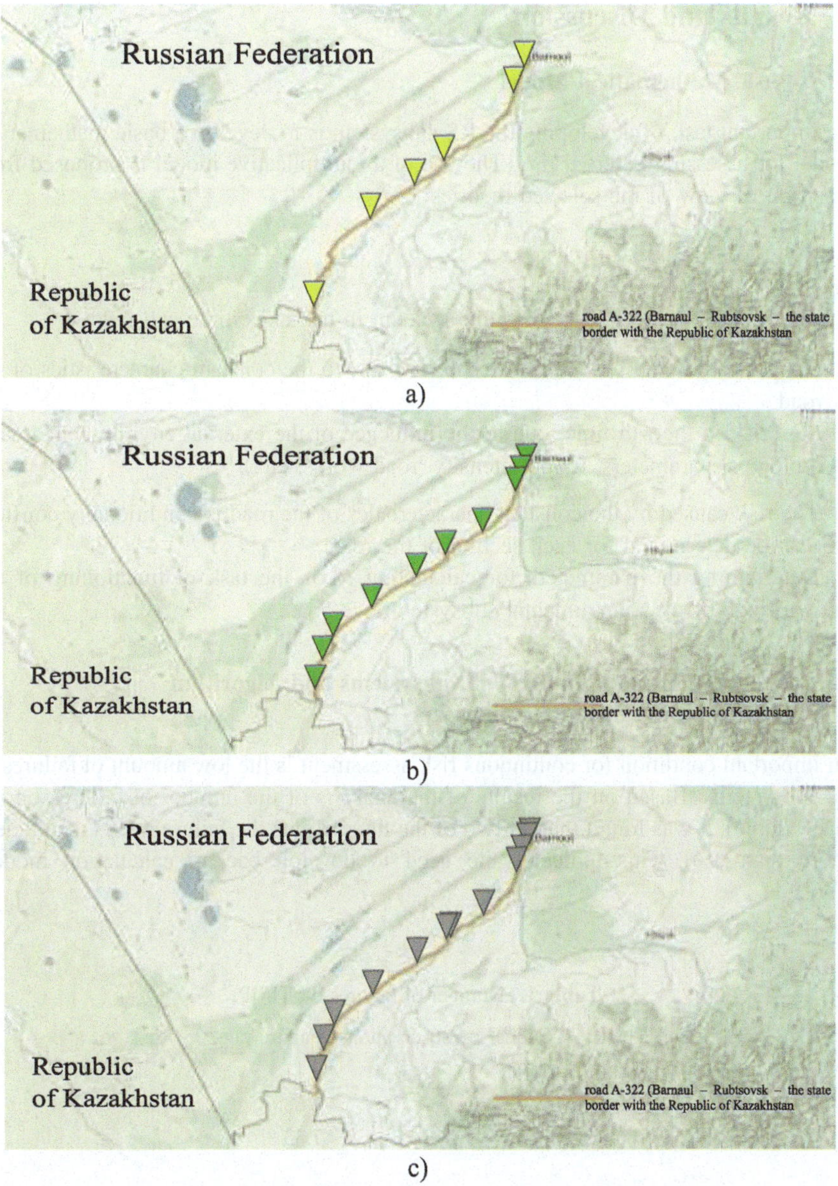

Fig. 3. Layout of traffic intensity monitoring points (a), automated road stations (b), and video complexes (c).

4 Results and Discussion

4.1 Basic Mathematical Model

An important task of developing the RS subsystem is to develop a basic mathematical model for assessing external risk. The use of a multiplicative model is proposed from the point of view of the selected theories:

$$R^{i \times t} = R^{i}_{road} \cdot R^{t}_{e} \tag{1}$$

where $R^{i \times t}$ – is the risk of road traffic accident in the space-time cell i × t,

R^{i}_{road} – risk on the i-th section of the road due to the constant characteristics of the road,
R^{t}_{e} – risk in the t-th time, due to the influence of the external environment (meteorological parameters, traffic intensity, road works, etc.).

The risk caused by the constant characteristics of the road is conditionally constant and can be determined for each section of the road.

Determining the parameters for calculating R^{t}_{e} is the task of functioning of the "monitoring" block's instrumental subsystems.

4.2 Results of Evaluation of Tool Subsystems and Algorithm of Functioning f the RS Subsystem

An important condition for continuous risk assessment is the low amount of failures in the subsystems. Based on the results of the analysis of the number of failures on the TIMP (table), it was found that on two of the three points the failure rate is insufficient (more than 3%). This indicates the need to develop backup calculation models (Table 1).

Table 1. Number of failures on TIMP.

TIMP	Number of observations	Failure rate, %
38 km	4169	8.11
128 km	8760	2.68
166 km	8760	7.33

The location of the TIMP on the road does not allow estimating the traffic intensity on many sections located between them: the distance between some TIMP is 90 km. This proves the need to obtain appropriate calculation models or install additional TIMP.

The ARWS analysis showed that most of the leading meteorological parameters are monitored, but there are no sensors for meteorological visibility distance. According to regulatory documents, such sensors should only be installed for dangerous places, but this parameter is important for determining the overall impact of weather conditions on road traffic accidents.

The video surveillance system can play an important role in RS in the future [15]. The assessment of the video surveillance subsystem showed that the distribution of complexes is fairly uniform, they are stationed in places of high road traffic accident rate. However, video surveillance can only serve as additional information in the monitoring control block.

The information obtained using data on the constant road risk on the road section (R^i_{road}) is converted to the value of the risk of road traffic accident on the i-th road section based on the formula (1). This value is then compared with the established acceptable risk. Its value can be accepted at the level of the target value defined in the regulatory documents.

The general algorithm of the complex subsystem is shown in Fig. 4.

If the acceptable level is exceeded, measures are implemented to increase the RS. They can be implemented using ITS tools (components of block 2 "management") and using the forces and means of state services (Fig. 2).

The road under study has a subsystem for dispatching control of the vehicle for road maintenance and anti-icing services. Its work is based on data from on-board navigation and communication GLONASS/GPS terminals installed on vehicles.

The use of dynamic information boards and road signs on roads is a promising and effective way to inform road users about the danger of the situation [1]. Only 2 such objects function on the road under study, which is extremely insufficient for the effective operation of the management system of RS.

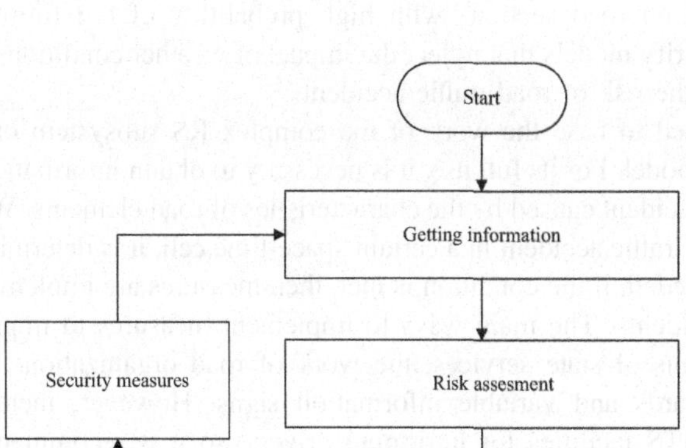

Fig. 4. Algorithm of complex subsystem operation.

Thus, the tool subsystems of block 1 “monitoring” are more developed and adapted for the operation of a complex RS subsystem than the components of block 2 “management”.

5 Conclusions

The study revealed that currently, the task of improving RS within the operation of ITS is one of the leading in Russia, but really solved only indirectly: via the traffic control system, monitoring traffic regulations violations. An independent RS subsystem within ITS is practically not implemented, and is a promising area of ITS development. The problem is most acute on roads outside human settlements. The main condition for the development of the RS subsystem is its affordable implement-ability in the existing ITS system, so the physical architecture is proposed based on the analysis of the existing ITS. At the stage of architecture development, it is proposed to abandon the closed operation of the subsystem and supplement it with the components “dispatcher information” - for receiving input data and “forces and means of state services” - for implementing the decisions made. In addition, the functional division of tool subsystems into 2 blocks is proposed: block 1 “monitoring” and block 2 “management”.

The analysis of the tool subsystems used on the A–322 road revealed that they can be used for the effective operation of the complex RS subsystem. However, additional works are required: installation of additional TIMP, development of backup mathematical models for calculating traffic intensity at a given time, as well as models for determining traffic intensity between TIMP; installation of sensors for meteorological visibility range on road sections with high probability of fog formation. It is also necessary to clarify models that reflect the impact of weather conditions and traffic flow parameters on the risk of road traffic accident.

It is proposed to base the work of the complex RS subsystem on the basis of a multiplicative model. For its full use, it is necessary to obtain information about the risk of road traffic accident caused by the characteristics of road elements. When calculating the risk of road traffic accident in a certain space-time cell, it is determined whether the set limit is exceeded, if the condition is met, then measures are implemented to prevent road traffic accidents. The main ways to implement measures to improve RS are the forces and means of state services, the work of road organizations, and the use of information boards and variable information signs. However, there are extremely insufficient of ITS facilities for informing drivers, so it is recommended to increase their number. Information boards are recommended to be installed near exits from cities or at the intersection of major traffic flows.

The results obtained indicate that the RS subsystem can function independently within ITS framework, but additional technical and research works are required for its effective operation.

References

1. Jarašūnienė, A., Batarlienė, N.: Lithuanian road safety solutions based on intelligent transport systems. Transport **28**(1), 97–107 (2013). https://doi.org/10.3846/16484142.2013.782895
2. Wang, X., Zhang, F., Li, B., Gao, J.: Developmental pattern and international cooperation on intelligent transport system in China. Case Stud. Transp. Policy **5**(1), 38–44 (2017). https://doi.org/10.1016/j.cstp.2016.08.004
3. Agureev, I., Elagin, M., Pyshnyi, V., Khmelev, R.: Methodology of substantiation of the city transport system structure and integration of intelligent elements into it. Transp. Res. Procedia **20**, 8–13 (2017). https://doi.org/10.1016/j.trpro.2017.01.003
4. Pechatnova, E., Kuznetsov, V.: Study of the relationship between time and traffic flow on motorways. J. Phys.: Conf. Ser. **1333**, 032063 (2019). https://doi.org/10.1088/1742-6596/1333/3/032063
5. Szczepanik, T., Besta, P.: Impact of intelligent transportation systems on road traffic safety. Zeszyty Naukowe Politechniki Częstochowskiej Zarządzanie **29**, 208–216 (2018). https://doi.org/10.17512/znpcz.2018.1.17
6. Pauer, G.: Development potentials and strategic objectives of intelligent transport systems improving road safety. Transp. Telecommun. J. **18**(1), 15–24 (2017). https://doi.org/10.1515/ttj-2017-0002
7. Mfenjou, M.L., Ari, A.A.A., Abdou, W., Spies, F.: Methodology and trends for an intelligent transport system in developing countries. Sustain. Comput.: Inform. Syst. **19**, 96–111 (2018). https://doi.org/10.1016/j.suscom.2018.08.002
8. Khorasani, G., Tatari, A., Yadollahi, A., Rahimi, M.: Evaluation of intelligent transport system in road safety. Int. J. Chem. Environ. Biol. Sci. (IJCEBS) **1**(1), 110–118 (2013)
9. Pechatnova, E., Sergeeva, Ja.: Assessment of influence of meteorological parameters on the risk of road traffic accidents on roads outside settlements. In: IOP Conference Series Earth and Environmental Science, vol. 272, p. 022175 (2019). https://doi.org/10.1088/1755-1315/272/2/022175
10. Kaparias, I., Bell, M.G.H., Eden, N., Gal-Tzur, A., Komar, O., Prato, C.G., Tartakovsky, L., Aronov, B., Zvirin, Y., Gerstenberger, M., Tsakarestos, A., Nocera, S., Busch, F.: Key performance indicators for traffic management and intelligent transport systems. CONDUITS Deliv. **3**, 5 (2011)
11. Malygin, I., Komashinskiy, V., Korolev, O.: Cognitive technologies for providing road traffic safety in intelligent transport systems. Transp. Res. Procedia **36**, 487–492 (2018). https://doi.org/10.1016/j.trpro.2018.12.134
12. Asaul, A., Malygin, I., Komashinskiy, V.: The project of intellectual multimodal transport system. Transp. Res. Procedia **20**, 25–30 (2017). https://doi.org/10.1016/j.trpro.2017.01.006
13. Zhankaziev, S.: Current trends of road-traffic infrastructure development. Transp. Res. Procedia **20**, 731–739 (2017). https://doi.org/10.1016/j.trpro.2017.01.118
14. Galkin, A., Davidich, N., Filina-Dawidowicz, L., Davidich, Y.: Improving the safety of urban freight deliveries by organization of the transportation process considering driver's state. Transp. Res. Procedia **39**, 54–63 (2019). https://doi.org/10.1016/j.trpro.2019.06.007
15. Bommes, M., Fazekas, A., Volkenhoff, T., Oeser, M.: Video based intelligent transportation systems – state of the art and future development. Transp. Res. Procedia **14**, 4495–4504 (2016). https://doi.org/10.1016/j.trpro.2016.05.372

Change of Geometric and Dynamic-Strength Characteristics of Crosspieces in the Operation

Irina Shishkina(✉)

Russian University of Transport (MIIT), Chasovaya str. 22/2,
125190 Moscow, Russia
shishkinaira@inbox.ru

Abstract. The paper discusses the changes in the geometric shape and size of the crosses examined on the main tracks of the Russian Railways, and analyzes the effect of these changes on the dynamic forces and other indicators of the work of the crosspieces, with the goal of differentiating their performance, establishing operating conditions, and developing measures to increase the reliability of the crosses. Intensive wear and need for frequent change of crosspieces due to their insufficient reliability is one of the reasons for their shortage and speed limits on railways. Irregularities in the longitudinal profile increase rapidly as the crosses with fixed elements wear in the zone of rolling the wheels through the gutter. The transverse shape of the crosses also undergoes changes. By the time the cross reaches wear of 6–8 mm, the transverse profiles are stabilized. The values of the dynamic pressure forces of the wheels of the rolling stock increase with increasing wear of the crosspieces, as well as the magnitude of the vibration accelerations of the crew units and the crosspieces itself.

Keywords: Crosspiece · Outlines · Dimensions · Dynamic forces · Operating conditions · Reliability

1 Introduction

A crosspiece with fixed elements is one of the most vulnerable nodes of turnouts. Its service life is 2.5–3 times less than the service life of the switch as a whole. This is due to the difficult conditions for rolling the wheels of the rolling stock through the troughs of the cross, the sharp drops of vertical and horizontal rigidity, the presence of angles of impact of the wheels in the guardrails and other features of the interaction of the crosses with the rolling stock. Intensive wear and the need for frequent change of crosspieces due to their insufficient reliability is one of the reasons for their shortage and speed limits on railways. Therefore, it is necessary to analyze the changes in the geometric shape and size of the crosses and the effect of these changes on the dynamic forces and other performance indicators of the crosses, in order to differentially evaluate their performance, establish operating conditions, as well as develop measures to increase the reliability of crosses [1, 2].

V. Murgul and V. Pukhkal (Eds.): EMMFT 2019, AISC 1258, pp. 146–155, 2021.
https://doi.org/10.1007/978-3-030-57450-5_14

2 Materials and Methods

The contact of the wheel and the cross in the zone of rolling from the core to the guardrail and vice versa occurs along narrow contact pads, which leads to the formation of local unevenness. Wheel rolling paths in this zone are distorted, increasing the dynamic forces of interaction and accelerating the wear of the crosspieces [3, 4].

A survey of 211 crosses on the main tracks of the Russian Railways shows that about 50% of the crosses have a wear exceeding 6 mm. The greatest wear in the areas where it is regulated by the current rules was observed on the guardrails in core sections of 12 and 20 mm wide and on the core in section 40 mm. Most often, maximum wear occurs on the guardrail in a section of 20 mm, i.e., in the section where the wheel rolls from the core to the guardrail and vice versa, and where the average width of the contact strips is 5–6 mm. The greatest wear of the entire sample of examined crosses was observed: on the guardrail in the cross section of 12 mm (in 25.2% of cases), on the guardrail in the cross section of the core 20 mm (59.3%) and on the core in the cross section 40 mm (15.5% of cases).

Core sections with a width of less than 20 mm are gradually included in the work as the guardrail wear increases, and with a further increase in wear, sections with a core width of more than 30 mm are included [5, 6]. A rolling zone with a maximum depth of unevenness is located on heavily worn crosses between sections with a core width of 10–12 to 40–50 mm and is up to 55 cm in length (Fig. 1).

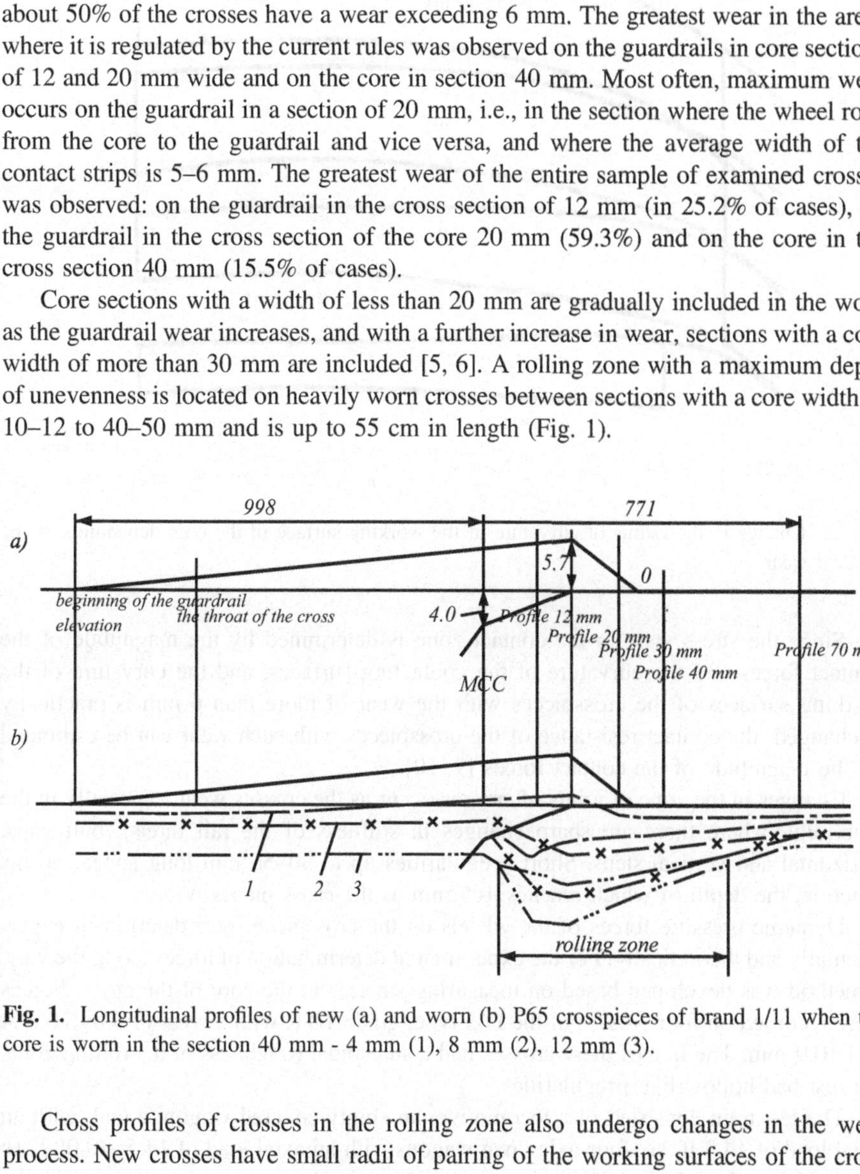

Fig. 1. Longitudinal profiles of new (a) and worn (b) P65 crosspieces of brand 1/11 when the core is worn in the section 40 mm - 4 mm (1), 8 mm (2), 12 mm (3).

Cross profiles of crosses in the rolling zone also undergo changes in the wear process. New crosses have small radii of pairing of the working surfaces of the cross with side faces in accordance with OST 32-11-78. At the beginning of operation, the radii of the cross sections of the working surfaces of the core of the crosspiece increase

(Fig. 2). The running-in of various sections ends when the cross reaches a wear of 6–8 mm [7, 8]. In the future, the cross-sectional shape of the working surfaces of the cross remains quite stable.

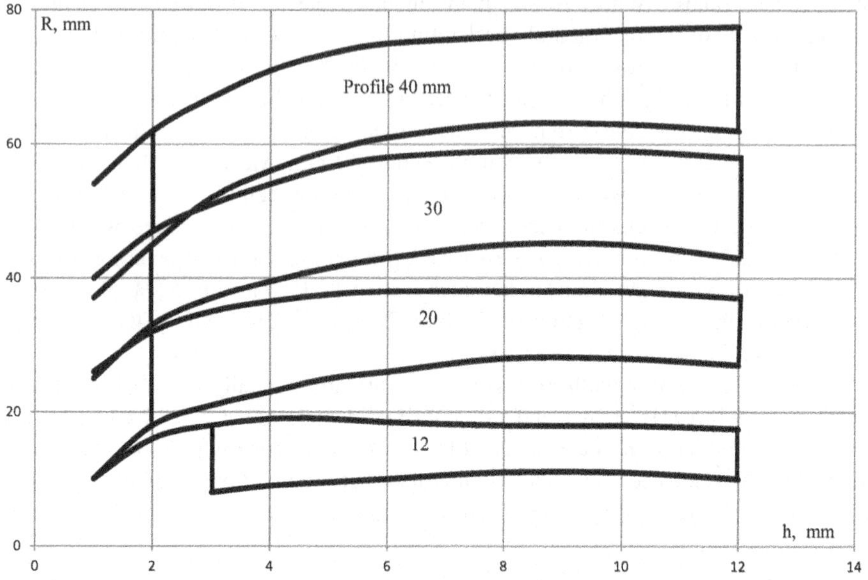

Fig. 2. Change in the radius of curvature of the working surface of the core depending on the vertical wear.

Since the stress state in the contact zone is determined by the magnitude of the contact forces and the curvature of the contacting surfaces, and the curvature of the working surfaces of the crosspieces with the wear of more than 6 mm is practically unchanged, the contact resistance of the crosspieces with such wear can be estimated by the magnitude of the contact forces [9, 10].

Changes in the zone of joints of crosses occur as the crosses wear, especially in the rear joint, where there are sharp changes in stiffness of the rail thread, butt gaps, horizontal and vertical steps. Short irregularities up to 30–50 mm long appear at this junction, the depth of which reaches 4–5 mm as the cross pieces wear.

Dynamic pressure forces of the wheels on the crosspiece were determined experimentally and theoretically. For the experimental determination of forces along the way, a method was developed based on measuring stresses in the core of the cross. Forces were recorded on four crosses of the P65 type, grade 1/11, with a wear of 3.0; 6.0; 7.4 and 10.0 mm. The first of these crosses had a sinusoidal roughness in the rolling zone, the rest had hollow-like irregularities.

The test train consisted of a locomotive, an empty covered wagon, a tank with an axial load of 14.3 tf, six four-axle open wagons with an axial load of 14.5; 20.0; 22.0; 23.0; 24.0 and 25.0 tf. Arrivals were carried out in the anti-wool (AW) and in the woolly (W) directions along the direct path of the tested turnouts at speeds of 25, 40, 60 and 80 km/h. The test results showed (Fig. 3) that as the wear of the crosspieces and the speeds of movement of the pressure forces of the wheels of the rolling stock also increase [11, 12].

The values of axle box accelerations of a freight open wagon with a load of 23 tf/axle were also measured in order to assess the dynamic qualities of worn crosses. The measurements were carried out in summer using accelerometers. The dependence of the maximum measured accelerations on the amount of wear of the crosses and the speeds of movement (Table 1) turned out to be less stable than the dependence obtained for dynamic forces, which is associated with the steepness of the trajectories of the studied crosses.

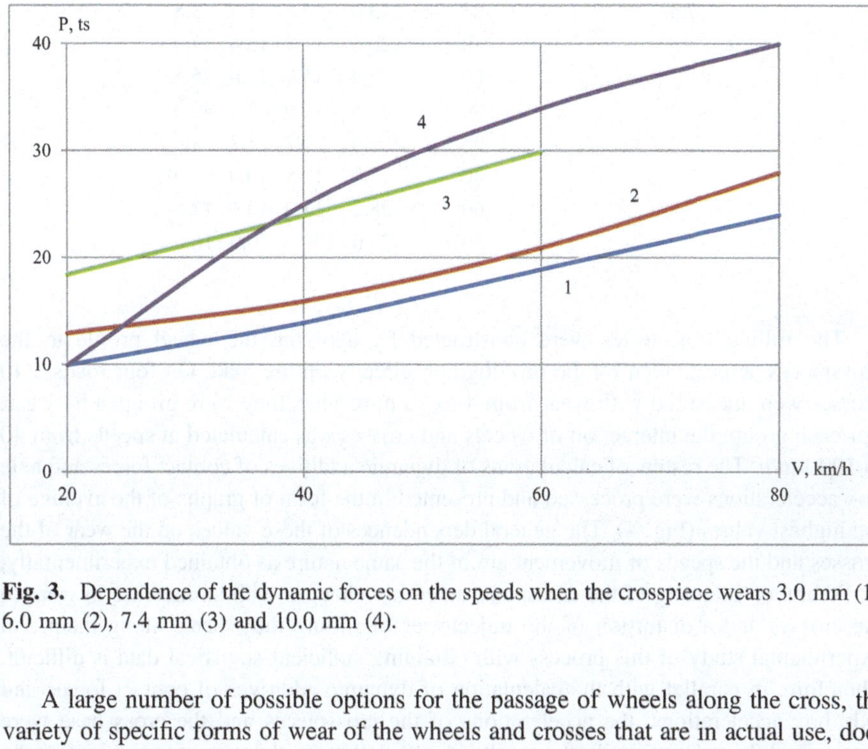

Fig. 3. Dependence of the dynamic forces on the speeds when the crosspiece wears 3.0 mm (1), 6.0 mm (2), 7.4 mm (3) and 10.0 mm (4).

A large number of possible options for the passage of wheels along the cross, the variety of specific forms of wear of the wheels and crosses that are in actual use, does not allow experimentally to study the whole variety of effects of wheels on the cross in the process of wear. Therefore, to determine the statistical characteristics of the effects of wheels on the crosses, theoretical calculations were carried out. The dynamic system "crew - path" was modeled as a single system with four degrees of freedom with elastic-viscous nonlinear bonds between the elements. The calculated parameters of the path in the area of the crosspiece node are determined experimentally, by a method based on the analysis of vibrations of the path elements in the process of shock loading. In the calculations, the trajectories of rolling the wheels of a freight car with a medium-network form of a rim rental, which is in the most probable position in the gutter of the crosspiece - pressed back to the guardrail were used [13, 14].

Table 1. Dependence of the maximum measured accelerations on the amount of wear of the crosspieces and speeds.

Crosspieces wear, mm	V, km/h	Acceleration of axle boxes, g			
		Passenger cars		Freight cars	
		W	AW	W	AW
6.0	25	7.5	6.0	4.9	3.5
	40	11.0	8.3	5.8	6.9
	60	13.0	13.2	9.4	14.1
	80	14.1	16.9	12.0	18.9
7.4	25	13.0	15.1	11.5	8.8
	40	20.2	17.4	15.4	17.0
	60	27.9	15.8	23.0	26.5
	80	41.5	45.0	43.8	40.0
12.2	25	9.0	9.7	5.2	4.8
	40	13.0	10.5	10.0	7.0
	60	28.2	13.0	15.1	12.5
	80	24.0	139	20.9	16.0

The rolling trajectories were constructed by applying the wheel profile to the crosspieces across, taken by the profilograph directly on the road. On four roads, 240 crosses were measured with wear from 4 to 12 mm. Then they were grouped by wear. For each group, the interaction of wheels and crosses was calculated at speeds from 40 to 100 km/h. The results of calculations of dynamic additives of contact forces and axle box accelerations were processed and presented in the form of graphs of the average of the highest values (Fig. 4). The general dependences of these values on the wear of the crosses and the speeds of movement are of the same nature as obtained experimentally.

Vibration intensity of the crosspiece assembly changes with an increase in wear of the crosses and a distortion of the trajectories of rolling the wheels along them. An experimental study of this process with obtaining sufficient statistical data is difficult. Therefore, in parallel with the calculation of dynamic additives of contact forces and axle box accelerations, the accelerations of the crosspieces and the cross base were determined. In order to verify the results of calculations in the area of crosspieces, type P65, grade 1/11 with wear 3.0; 7.4; 10.0 mm under the mentioned test train vibration measuring equipment recorded acceleration cross-node. The experimental results were compared with the calculated ones obtained for the same crosses and rolling stock wheels. As perturbing factors, the trajectories of wheel rolling along the crosses were introduced, obtained by superimposing the profiles of the wheel rims of the experimental composition on the corresponding diameters of the crosses participating in the experiment [15, 16].

3 Results

As a result, it was found that both in the experiment and according to the calculated data, an increase in the wear of the cross pieces led to an increase in the accelerations of the cross pieces and the cross base. Thus, the accelerations of the crosspieces and the cross base due to the influence of gondola cars of the test train with a wear of 7.4 mm compared to a wear of 3 mm in the speed range of up to 80 km/h increased 1.37–1.86 times according to the calculated data and 1.46–1.84 times during the experiment. With a cross wear of 10 mm compared to a wear of 3 mm, accelerations increased 1.59–2.00 times as calculated and 2.0–2.47 times in the experiment.

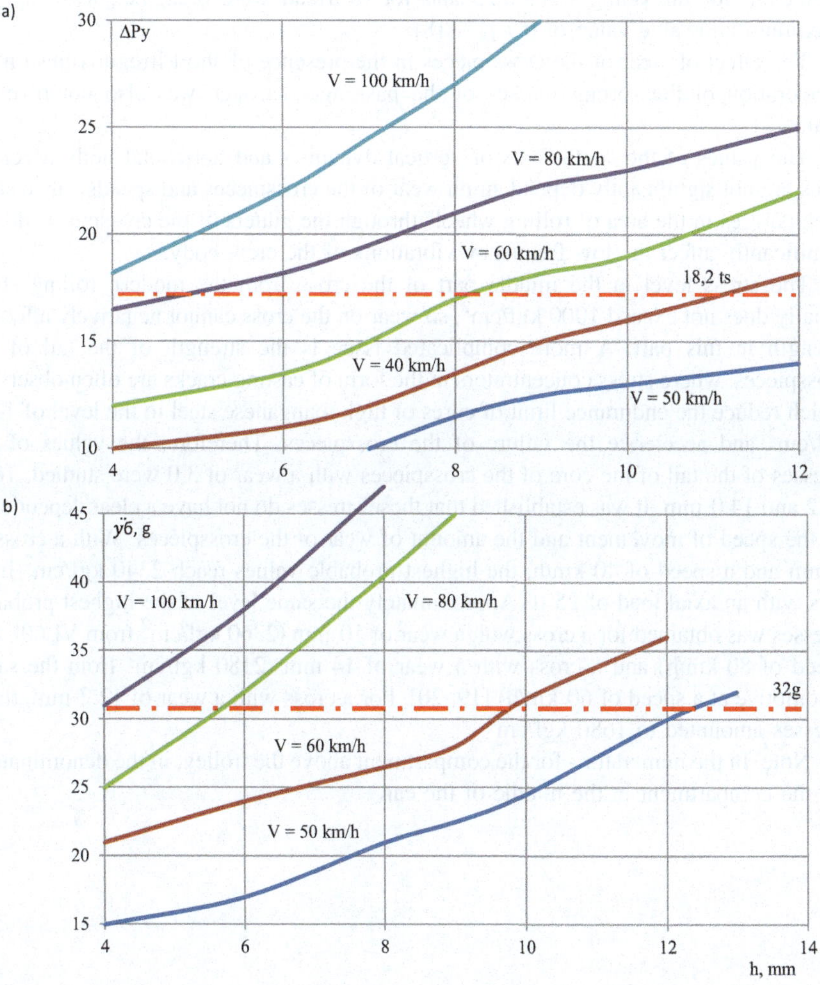

Fig. 4. Dependence on the wear of the crosses of the averages of the largest values of the dynamic additives of contact forces (a) and axle box accelerations of a loaded gondola car (b).

A great influence on the growth of vibration accelerations is exerted by the speed of movement. Under experimental conditions, the vibration acceleration of a wooden beam increased by 2.4–4.0 times with an increase in speed from 40 to 80 km/h and reached 160–170 g with a 10 mm cross.

Thus, one should expect an increase in the intensity of accumulation of residual deformations in the ballast and a breakdown of the fastening elements with an increase in the wear of the crosses.

In the zone of crosses with different wear according to stresses in the thrust beam of a loaded four-axle gondola car, the values of the coefficients of the vertical dynamics of the sprung mass are determined. It has been established that these values are not appreciably dependent on the values of wear and speed. At speeds from 25 to 80 km/h and cross wear from 3 to 10 mm, the maximum values of the vertical dynamics coefficient for unloading were 0.25 and for overload were 0.22, i.e., less than the maximum allowable value of 0.7 [17, 18].

The effect of wear of the cross pieces in the presence of short irregularities on the acceleration of the sprung masses of the passenger carriage was also not revealed (Table 2).

The values of the coefficients of vertical dynamics and horizontal body accelerations are not significantly dependent on wear of the crosspieces and speeds, since short irregularities in the area of rolling wheels through the gutters of the crosspieces do not significantly affect the low-frequency vibrations of the crew body.

The stress level in the middle part of the cross from the modern rolling stock usually does not exceed 1000 kgf/cm^2, so wear on the cross cannot negatively affect its strength in this part. A more complicated issue is the strength of the tail of the crosspieces, where stress concentrators in the form of casting cracks are often observed, which reduce the endurance limit of cores of high manganese steel to the level of 1200 kgf/cm^2 and accelerate the failure of the crosspieces. Therefore, the values of the stresses of the tail of the core of the crosspieces with a wear of 3.0 were studied; 10.0; 12.2 and 14.0 mm. It was established that these stresses do not have a clear dependence on the speed of movement and the amount of wear of the crosspieces. With a cross of 3 mm and a speed of 40 km/h, the highest probable values reach 2340 kgf/cm^2 from cars with an axial load of 25 tf. Approximately the same level of the highest probable stresses was obtained for a cross with a wear of 10 mm (2360 kgf/cm^2 from VL60k at a speed of 80 km/h) and a cross with a wear of 14 mm (2180 kgf/cm^2 from the same locomotive at a speed of 60 km/h) [19, 20]. For a cross with a wear of 12.2 mm, these stresses amounted to 1680 kgf/cm^2.

Note. In the numerator - for the compartment above the trolley, in the denominator - for the compartment in the middle of the car.

Table 2. Dependence of horizontal body accelerations on the amount of wear of the crosses and speeds.

Speed, km/h	Crosspieces wear, mm	Values of horizontal body accelerations, m/s^2	
		W	AW
25	3	0.83/0.24	0.69/0.24
	10	0.89/0.48	0.66/0.24
40	3	1.32/0.72	1.09/0.36
	10	0.76/0.48	1.32/0.31
60	3	1.22/0.48	1.15/0.48
	10	1.32/0.31	1.32/0.41
80	3	1.32/0.24	1.48/0.72
	10	1.42/0.41	1.16/1.04

Consequently, the stress state of the tail of the core, located at a distance of 1.5–2.0 m from the rolling zone, is noticeably independent of the wear of the crosspieces in this zone, since this part is closer to the junction of the core with adjacent rails, in which there are irregularities caused by the presence of steps, saddles and sudden changes in stiffness, predetermining a high level of force on the cross. The obtained stress values indicate the need to strengthen the tail of the core of the crosses [21, 22].

The stress values in the counter rail according to the test results also do not depend on the amount of wear of the crosspieces and the speeds of movement. This is due to the fact that the stress level in the counter rail to a greater extent depends on the gauge and grooves of the cross piece [23], as well as on the observance of the distance between the working faces of the counter rail and guardrail, which should not exceed 1435 mm according to the PTE.

4 Conclusions

1. As crosses with fixed elements wear in the zone of rolling the wheels through the trough, irregularities in the longitudinal profile intensively grow. Changes and the transverse outlines of the crosses undergo. By the time the cross reaches wear of 6-8 mm, the transverse profiles are stabilized.
2. With the increase in wear of the crosses, the values of the dynamic pressure forces on them of the wheels of the rolling stock increase, as well as the magnitude of the vibration accelerations of the crew nodes and the crosspiece assembly.

References

1. Gluzberg, B., Korolev, V., Loktev, A., Shishkina, I., Berezovsky, M.: Switch operation safety. E3S Web Conf. **138** (2019). https://doi.org/10.1051/e3sconf/201913801017
2. Korolev, V.: Switching Shunters on a slab base. Advances in Intelligent Systems and Computing, vol. 1116, pp. 175–187 (2020). https://doi.org/10.1007/978-3-030-37919-3_17
3. Glusberg, B., Savin, A., Loktev, A., Korolev, V., Shishkina, I., Alexandrova, D., Loktev, D.: New lining with cushion for energy efficient railway turnouts. Advances in Intelligent Systems and Computing, vol. 982, pp. 556–570 (2020). https://doi.org/10.1007/978-3-030-19756-8_53
4. Korolev, V.: Guard rail operation of lateral path of railroad switch. Advances in Intelligent Systems and Computing, vol. 1115, pp. 621–638 (2020). https://doi.org/10.1007/978-3-030-37916-2_60
5. Loktev, A.A., Korolev, V.V., Shishkina, I.V., Basovsky, D.A.: Modeling the dynamic behavior of the upper structure of the railway track. Procedia Eng. **189**, 133–137 (2017). https://doi.org/10.1016/j.proeng.2017.05.022
6. Savin, A., Kogan, A., Loktev, A., Korolev, V.: Evaluation of the service life of non-ballast track based on calculation and test. Int. J. Innov. Technol. Explor. Eng. **8**(7), 2325–2328 (2019)
7. Glusberg, B., Korolev, V., Shishkina, I., Loktev, A., Shukurov, J., Geluh, P., Loktev, D.: Calculation of track component failure caused by the most dangerous defects on change of their design and operational conditions. MATEC Web Conf. **239** (2018). https://doi.org/10.1051/matecconf/201823901054
8. Loktev, A., Korolev, V., Shishkina, I., Chernova, L., Geluh, P., Savin, A., Loktev, D.: Modeling of railway track sections on approaches to constructive works and selection of track parameters for its normal functioning. Advances in Intelligent Systems and Computing, vol. 982, pp. 325–336 (2020). https://doi.org/10.1007/978-3-030-19756-8_30
9. Loktev, A.A., Korolev, V.V., Gridasova, E.A.: Influence of high-frequency cyclic loading on mechanical and structural characteristics of rail steel under extreme conditions. IOP Conf. Ser.: Mater. Sci. Eng. **687** (2019). https://doi.org/10.1088/1757-899x/687/2/022036
10. Savin, A., Suslov, O., Korolev, V., Loktev, A., Shishkina, I.: Stability of the continuous welded rail on transition sections. Advances in Intelligent Systems and Computing, vol. 1115, pp. 648–654 (2020). https://doi.org/10.1007/978-3-030-37916-2_62
11. Gridasova, E., Nikiforov, P., Loktev, A., Korolev, V., Shishkina, I.: Changes in the structure of rail steel under high-frequency loading. Advances in Intelligent Systems and Computing, vol. 1115, pp. 559–569 (2020). https://doi.org/10.1007/978-3-030-37916-2_54
12. Shishkina, I.: Determination of contact-fatigue of the crosspiece metal. Advances in Intelligent Systems and Computing, vol. 1115, pp. 834–844 (2020). https://doi.org/10.1007/978-3-030-37916-2_82
13. Loktev, A., Korolev, V., Shishkina, I., Illarionova, L., Loktev, D., Gridasova, E.: Perspective constructions of bridge crossings on transport lines. Advances in Intelligent Systems and Computing, vol. 1116, pp. 209–218 (2020). https://doi.org/10.1007/978-3-030-37919-3_20
14. Loktev, A.A., Korolev, V.V., Shishkina, I.V.: High frequency vibrations in the elements of the rolling stock on the railway bridges. IOP Conf. Ser.: Mater. Sci. Eng. **463** (2018). https://doi.org/10.1088/1757-899x/463/3/032019
15. Loktev, A.A., Korolev, V.V., Poddaeva, O.I., Chernikov, I.Y.U.: Mathematical modeling of antenna-mast structures with aerodynamic effects. IOP Conf. Ser.: Mater. Sci. Eng. **463** (2018). https://doi.org/10.1088/1757-899x/463/3/032018

16. Korolev, V., Loktev, A., Shishkina, I., Zapolnova, E., Kuskov, V., Basovsky, D., Aktisova, O.: Technology of crushed stone ballast cleaning. IOP Conf. Ser.: Earth Environ. Sci. **403** (2019). https://doi.org/10.1088/1755-1315/403/1/012194
17. Savin, A.V., Korolev, V.V., Shishkina, I.V.: Determining service life of non-ballast track based on calculation and test. IOP Conf. Ser.: Mater. Sci. Eng. **687** (2019). https://doi.org/10.1088/1757-899x/687/2/022035
18. Savin, A., Korolev, V., Loktev, A., Shishkina, I.: Vertical sediment of a ballastless track. Advances in Intelligent Systems and Computing, vol. 1115, pp. 797–808 (2020). https://doi.org/10.1007/978-3-030-37916-2_78
19. Lyudagovsky, A., Loktev, A., Korolev, V., Shishkina, I., Alexandrova, D., Geluh, P., Loktev, D.: Energy efficiency of temperature distribution in electromagnetic welding of rolling stock parts. E3S Web Conf. **110** (2019). https://doi.org/10.1051/e3sconf/201911001017
20. Glusberg, B., Loktev, A., Korolev, V., Shishkina, I., Alexandrova, D., Koloskov, D.: Calculation of heat distribution of electric heating systems for turnouts. Advances in Intelligent Systems and Computing, vol. 982, pp. 337–345 (2020). https://doi.org/10.1007/978-3-030-19756-8_31
21. Loktev, A.A., Korolev, V.V., Poddaeva, O.I., Stepanov, K.D., Chernikov, I.Y.: Mathematical modeling of aerodynamic behavior of antenna-mast structures when designing communication on railway transport. Vestn. Railw. Res. Inst. **77**(2), 77–83 (2018). https://doi.org/10.21780/2223-9731-2018-77-2-77-83. (in Russia)
22. Loktev, A.A., Korolev, V.V., Loktev, D.A., Shukyurov, D.R., Gelyukh, P.A., Shishkina, I. V.: Perspective constructions of bridge overpasses on transport main lines. Vestn. Railw. Res. Inst. **77**(6), 331–336 (2018). https://doi.org/10.21780/2223-9731-2018-77-6-331-336. (in Russia)
23. Glusberg, B., Savin, A., Loktev, A., Korolev, V., Shishkina, I., Chernova, L., Loktev, D.: Counter-rail special profile for new generation railroad switch. Advances in Intelligent Systems and Computing, vol. 982, pp. 571–587 (2020). https://doi.org/10.1007/978-3-030-19756-8_54

Selecting a Turnout Curve Form in Railroad Switches for High Speeds of Movement

Vadim Korolev(✉)

Russian University of Transport (MIIT), Chasovaya str. 22/2, 125190 Moscow, Russia
korolevadim@mail.ru

Abstract. When designing railroad switches for high speeds of movement, it is necessary to provide passengers with a comfortable ride when the train moves on a side track by limiting the values of the so-called centrifugal acceleration and increment (change) of centrifugal acceleration per unit time (second). If it is necessary to realize the movement speed of the railroad switch on the side track over 50 km/h, the main factor in determining the radius of the turnout curve according to the conditions of driving comfort is the limitation of the increment (change) value of centrifugal acceleration per unit time (second). Based on this, when designing railroad switches for high speeds of movement, it is advisable to use curves of variable radius as a turnout curve. Curves of variable curvature are considered: cubic parabola; fourth-degree parabola and a sinusoid, recommended as turnout curves, which are given a comparative assessment of the conditions for their use in railroad switches of flat grades. Of the considered curves of variable radius, the curve that varies according to the law of a sinusoid is the most suitable for use as a turnout curve in railroad switches of flat grades for high speeds of movement.

Keywords: Railroad switch · Side track · Speed of movement · Radius of the turnout curve · The increment of the centrifugal acceleration per unit time · Curves of variable radius · Sinusoidal curve

1 Introduction

The continuously increasing volume of freight and passenger traffic on the railways of our country necessitates a significant increase in the movement speeds of both freight and passenger trains. The solution of these issues, in turn, puts forward a number of tasks to improve the design and condition of the switch economy of our railways.

Already, there is a need to speed up the design and implementation of railroad switches on the operated railway network, which can significantly increase the permissible train speeds, especially on the side track [1, 2].

One of the starting points in the design of railroad switches for high speeds of movement is to provide passengers with a comfortable ride when the train moves on a side track by limiting the values of the so-called centrifugal acceleration a and the increment (change) of centrifugal acceleration per unit time (second) ψ [3, 4].

V. Murgul and V. Pukhkal (Eds.): EMMFT 2019, AISC 1258, pp. 156–172, 2021.
https://doi.org/10.1007/978-3-030-57450-5_15

Based on the conditions for providing passengers with a comfortable ride when the crew moves along the railroad switch to the side track, the turnout curve must satisfy the requirements that the centrifugal accelerations that appear, as well as their increment (change) per unit time (second), do not exceed some established values [5, 6]:

$$a \leq a_{add} \tag{1}$$

$$\psi \leq \psi_{add} \tag{2}$$

where a is the magnitude of the outstanding centrifugal acceleration, m/s^2;

ψ is the magnitude of the change in the outstanding centrifugal acceleration per unit time (second), m/s^3.

Based on the studies, to verify the conditions of ride comfort when moving a passenger train along the railroad switch to the side track, it is possible to consider acceptable in the car body: centrifugal acceleration $a = 1.0\ m/s^2$, and its increment (change) per unit time $\psi = 0.8\ m/s^3$, since the effect of these values on passengers within the turnout curves will be short-term and repeated at relatively large intervals [7, 8].

On the railway network, railroad switches with turnout curves of constant radius were widely used.

In the railroad switch of the last construction made of P50 rails with a 1/18 crossing, the radius of the tongue to a section of 40 mm is 1698.0 m, and then along the tongue and the turnout curve is 960.0 m.

The permissible movement speed of trains on the side track of the indicated railroad switch from the conditions of providing passengers with comfortable ride according to experimental data can be set within 80–85 km/h [9, 10].

Professor G. M. Shakhunyants, when considering the question about speed of movement at railroad switches on the side track, pointed out that permissible speeds of movement should be assigned in such a way that as a result of the interaction of the track and the rolling stock, unacceptable effects were not caused on both the rolling stock and for passengers.

2 Materials and Methods

With the established allowable values of outstanding centrifugal acceleration, changes in centrifugal acceleration per unit time and kinetic energy loss per stroke when entering the switch, the train speed and radius of the turnout curve are interdependent and determine the main parameters of the railroad switch [11, 12].

In the calculations described below, a given train speed is taken, and the radius of the turnout curve is determined as a function of speed.

As it known, the magnitude of the outstanding centrifugal acceleration when the crew moves in a curve of radius R having an elevation of the outer rail h can be determined by the formula:

$$a_{out} = \frac{V^2}{R} - \frac{gh}{S_1} \tag{3}$$

In the absence of elevation of the outer rail, which usually takes place in conversion curves, the magnitude of centrifugal acceleration will be determined by the formula:

$$a_{out} = \frac{V^2}{R} \tag{4}$$

Having taken the a = 1.0 m/s^2 and conversion factor 3.62, we obtain the expression for determining R_{min} from the conditions of limiting the magnitude of centrifugal acceleration:

$$R_{min} = \frac{V^2}{3.6^2 \cdot 1.0} = 0.0772\, V^2 \mathrm{m} \tag{5}$$

Professor P. G. Koziychuk suggested to determine the increment of centrifugal acceleration at the entrance to the switch (at R = const) by the formula:

$$\psi = \frac{V^3}{Rb} \tag{6}$$

Having taken ψ = 0.8 m/s^3, b = 17.0 m (the distance between the axes of rotation of the trolleys in all-metal railway carriages) and also introducing a conversion factor 3.63 for the convenience of using the formula, we obtain the expression for determining R_{min} from the conditions for limiting the magnitude of the increment (change) of centrifugal acceleration per unit time (second)

$$R_{min} = \frac{V^3}{3.6^3 \cdot 0.8 \cdot 17} = 0.00158\, V^3 \tag{7}$$

Thus, the minimum magnitude of the turnout curve radius (in the case R = const) at the set speed of movement according to the conditions of ride comfort will be determined:

1. from the conditions for limiting the magnitude of the centrifugal acceleration from the expression R_{min} = 0.0772 V^2 m;
2. from the conditions of limiting the increment (change) of centrifugal acceleration per unit time (second) from the expression R_{min} = 0.00158 V^3 m.

The Table 1 shows the results of calculating the minimum radii depending on the speed of the train movement on the side track.

Table 1. The results of the calculation of the minimum radii depending on the speed of the train movement on the side track.

The speed of the train movement on the side track, km/h	Magnitudes	
	From the conditions of limiting the magnitude of centrifugal acceleration (a = 1.0 m/s^2)	From the conditions for limiting the magnitude of the increment (change) of centrifugal acceleration per unit time (ψ = 0.8 m/s^3)
1	2	3
40	123.52	101.12
50	193.00	197.12
60	277.92	341.28
70	378.28	541.94
80	494.08	808.96
90	625.32	1151.82
100	772.00	1580.00
110	934.12	2102.98
120	1111. 68	2730.24
130	1304.68	3471.26
140	1513.12	4335.52
150	1737.00	5332.50
160	1976.32	6471.68

According to Table 1, a graph of magnitudes R_{min} depending on the speed of the train movement along the turnout curve for a = 1.0 m/s^2 and ψ = 0.8 m/s^3 is constructed, from which it is established that when the speed of movement along the side track is up to 50 km/h, the possible minimum radius of the turnout curve is determined by limiting the magnitude of centrifugal acceleration, and at a speed of more than 50 km/h - by limiting the increment (change) of centrifugal acceleration per unit time. From this it can be concluded that in railroad switches for high speeds of movement, the parameters of the turnout curve will be primarily determined by limiting the magnitude of the change in centrifugal acceleration per unit time. In this case, it is especially advisable to use variable curvature curves as turnout curves, in which the change in curvature meets the requirements of a monotonic increment (change) in the magnitude of centrifugal acceleration.

In a variable curvature curve that meets these requirements, a less drastic effect is exerted than in a constant radius curve by factors (changes in accelerations over time) that are especially unpleasant for the human body, which is very important when realizing high speeds of the passenger train movement [13, 15].

The feasibility of using variable radius curves in railroad switches for high speeds of movement is indicated in their studies by prof. P.G. Koziychuk, prof. S.V. Amelin, PhD in tech. sciences G.I. Ivaschenko and other authors.

It is known that a variable curvature turnout curve in railroad switches of flat grades can consist of one or two turnout curve branches, or it can be composed of two turnout curves and of a constant radius curve insert between them.

In the turnout curve of variable radius, varying from $\rho = 0$ to $\rho = R$, the minimum radius magnitude from the condition for limiting the magnitude of centrifugal acceleration will be determined, as for the constant radius curve, from expression (5).

That's why the magnitudes R_{min} for curves of variable radius, determined from the conditions for limiting the magnitude of centrifugal acceleration, will correspond at various speeds to the data given in Table 1 (column 2).

Prof. B.N. Vedenisov suggested defining the change in centrifugal acceleration per unit time ψ in curves of variable radius, within the centrifugal acceleration a increases from $a = 0$ to some magnitude $a = V^2/R$, as follows if $a = V^2/\rho$, and $\psi = da/dt$, then

$$\psi = \frac{d\left(\frac{V^2}{\rho}\right)}{dt} \tag{8}$$

When $\rho = R$, the magnitude of the change in centrifugal acceleration per unit time will be determined from the expression:

$$\psi = \frac{V^2}{Rt} \tag{9}$$

where $t = l/V$, here l is the theoretical length of the turnout curve branch.

Then $\psi = V^3/Rt$, whence $l = V^3/\psi R$.

Taking $\psi = 0.8$ m/s^3 and introducing a conversion factor of 3.6^3 for the convenience of using the formula, we obtain:

$$l = \frac{V^3}{3.6^3 \cdot 0.8 \cdot R} = 0.0268\frac{V^3}{R} \tag{10}$$

where V is the speed of the train movement, km/h,

R is the smallest magnitude of the turnout curve radius, m.

For various speeds of train movement and corresponded to them R_{min} from the conditions of limiting the magnitude of centrifugal acceleration, adopted according to Table 1 (column 2), the branch length of the turnout curve was calculated from the conditions of limiting the magnitude of the increment (change) of centrifugal acceleration per unit time. The results of the calculations are presented in Table 2.

Table 2. The results of the calculations of branch length of the turnout curve.

The speed of the train movement on the side track, km/h	The magnitude R_{min} from the conditions of limiting the magnitude of centrifugal acceleration (a = 1.0 m/s^2)	The theoretical branch length of the turnout curve from the conditions for limiting the magnitude of the increment (change) of centrifugal acceleration per unit time (ψ = 0.8 m/s^3)
1	2	3
40	124	13.83
50	193	17.36
60	278	20.82
70	378	24.32
80	494	27.78
90	625	31.26
100	772	34.71
110	934	38.19
120	1112	41.65
130	1305	45.12
140	1513	48.60
150	1737	52.07
160	1976	55.55

The Table 2 shows that at the eccepted magnitudes R_{min} from the conditions for limiting the magnitude of centrifugal acceleration, the theoretical length of the branch of the turnout curve, determined from the conditions for limiting the increment (change) of centrifugal acceleration per unit time, increases significantly less with increasing speed of movement than the radius magnitude of the circular curve (Table 1 column 3). This indicates the appropriateness of the use in railroad switches for high speeds of movement as a turnout curve of variable curvature curves [16, 17].

Crossing marks in ordinary and symmetrical railroad switches for high speeds of movement are recommended to be taken at the stages: 1/18, 1/22 and 1/36.

The variable curvature curves are considered below: cubic parabola $y = x^3/6C$; fourth-degree parabola $y = x^4/12C$; and a sinusoid $y = a_1 \sin \varphi$; recommended as turnout curves, which are given a comparative assessment of the conditions for their use in railroad switches of flat grades.

The question of the design of railroad switches with a cubic parabola as a turnout curve is presented in the work of professor P. G. Koziychuk.

The railroad switch scheme with the turnout curve, divided according to the law of the parabola of the third and fourth degree, is presented in Fig. 1.

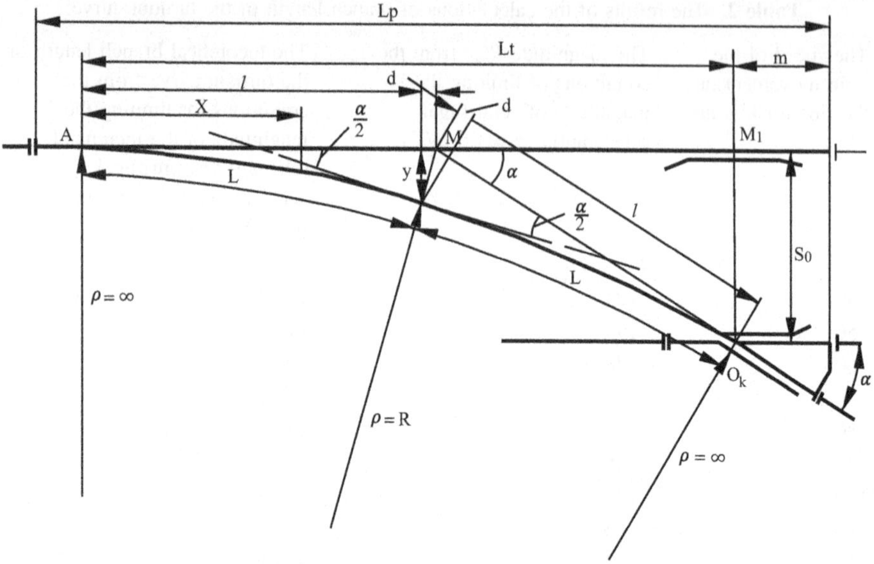

Fig. 1. The railroad switch scheme with the turnout curve.

The turnout curve consists of two symmetrically located turnout curves AB and O_kB.

The equation for determining the magnitude of the increment (change) of centrifugal acceleration per unit time ψ at any point in the curve, divided according to the law of the cubic parabola, has the form:

$$\psi = \frac{V^3}{Rl}\left[\frac{1}{\left(1+\frac{x^4}{4R^2l^2}\right)^{3/2}} - \frac{3x^4}{2R^2l^2\left(1+\frac{x^4}{4R^2l^2}\right)^{5/2}}\right] \tag{11}$$

The expressions for determining the magnitude of the change in centrifugal acceleration per unit time in the curves divided according to the laws of parabolas of the third and fourth degree can be significantly simplified by making the assumption, as a result of which the final result is determined with sufficient accuracy for practical purposes.

It is known that the initial equation in the derivation of the formulas for determining ψ was the differential equation of curvature, which is represented by the expression:

$$\frac{1}{\rho} = \frac{\frac{d^2y}{dx^2}}{\left[1+\left(\frac{dy}{dx}\right)^2\right]^{3/2}} \tag{12}$$

It was supported that dy/dx = tg(α/4) = const.

The indicated assumption is taken from the conditions that the tangent angle of the tangent straight at the points of the turnout curve branch and axis X will change from 0 to α/2 (Fig. 2).

In the view of small angles of the crossings of considered railroad switches (1/18—a = 3°10′12.5″; 1/22—a = 2°35′50″ и 1/36—a = 1°35′6.25″), the possible error in the assumption noted, as numerical calculations showed, in determining the value of ψ in the curves divided according to the laws of the parabola of the third and fourth degree, does not exceed hundredths of a percent [18, 19].

The equation of the third-degree parabola has the form:

$$y = \frac{x^3}{6C} = \frac{x^3}{6Rl} \tag{13}$$

The first and second derivatives of Eq. (13) will be:

$$\frac{dy}{dx} = \frac{x^3}{2Rl} = tg\left(\frac{\alpha}{4}\right) \tag{14}$$

After substituting the magnitudes of the derivatives in the equation of curvature (12), we obtain:

$$\frac{1}{\rho} = \frac{x}{Rl} \cdot \frac{1}{\left[1 + tg^2 \frac{\alpha}{4}\right]^{3/2}} \tag{15}$$

After conversion of expression (15), having known that $\frac{1}{1+tg^2\frac{\alpha}{4}} = \cos^2\frac{\alpha}{4}$ we rewrite formula (15) in the form:

$$\frac{1}{\rho} = \frac{x}{Rl} \cdot \cos^3\frac{\alpha}{4} \tag{16}$$

Multiplying both sides of the Eq. (16) on V^2 we obtain the expression of centripetal (or, which is one and the same, in magnitude of centrifugal) acceleration:

$$\frac{V^2}{\rho} = \frac{V^2 x}{Rl} \cdot \cos^3\frac{\alpha}{4} \tag{17}$$

Taking V = const, we have x = Vt, then expression (17) can be rewritten in the form:

$$\frac{V^2}{\rho} = \frac{V^3 x}{Rl} \cdot \cos^3\frac{\alpha}{4} \tag{18}$$

Bearing in mind that the increment of centrifugal acceleration per unit time is determined by Eq. (6), and differentiating expression (18), we obtain:

$$\frac{d\left(\frac{V^2}{\rho}\right)}{dt} = \frac{d}{dt}\left(\frac{V^3 x}{Rl} \cdot \cos^3\frac{\alpha}{4}\right) \tag{19}$$

or

$$\psi = \frac{V^3}{Rl} \cdot \cos^3\frac{\alpha}{4} \tag{20}$$

From the obtained expression (20) it follows that in the curve, which varies according to the law of the third-degree parabola, change of centrifugal acceleration per unit time ψ is a constant value in practice.

To compare the curves under consideration, we restrict ourselves to calculating the projection (abscissa) of the length l of the branch of the turnout curve, the value of the finite radius R and the length of the segment d. For a cubic parabola, the magnitudes of l, R and d will be determined by the formulas:

$$l = \frac{3S_0}{\sin\alpha\left(3 + tg^2\frac{\alpha}{2}\right)} \tag{21}$$

$$R = \frac{1}{2tg\frac{\alpha}{2}} \tag{22}$$

$$d = \frac{l}{3}tg^2\frac{\alpha}{2} \tag{23}$$

The question of the design of railroad switches with a fourth-degree parabola as a turnout curve was considered in the work of professor S.V. Amelin.

The increment (change) of centrifugal acceleration per unit time at any point in the curve, divided according to the law of the fourth-degree parabola, is determined by the formula:

$$\psi = \frac{2V^3 x}{Rl^2}\left[\frac{1}{\left(1 + \frac{x^6}{9R^2l^4}\right)^{3/2}} - \frac{x^6}{2R^2l^4\left(1 + \frac{x^6}{9R^2l^4}\right)^{5/2}}\right] \tag{24}$$

For a curve that varies according to the law of the fourth-degree parabola, as well as for the cubic parabola, we obtain the equation for determining ψ in a simplified form [20, 21].

The fourth-degree parabola equation has the form:

$$y = \frac{x^4}{12C} = \frac{x^4}{12Rl} \quad (25)$$

After the performed calculations, as well as for the cubic parabola, we will have an expression for determining ψ in the considered curve in the following form:

$$\psi = \frac{2V^3x}{Rl^2}\cos^2\frac{\alpha}{4} \quad (26)$$

From Eq. (23) it follows that in a curve that varies according to the law of the fourth-degree parabola, the magnitude of the change in centrifugal acceleration per unit time when entering the curve (at point A in Fig. 2) and leaving the curve (at point O_k) at x = 0 is equal to zero and reaches its maximum magnitude at the end of the branch of the turnout curve, i.e., in the middle of the turnout curve at x = l.

In the translated curve, divided according to the law of the fourth-degree parabola, the projection (abscissa) of length l of one branch of the transition curve, the value of the finite radius R and the length of the segment d will be determined by the formulas [1]:

$$l = \frac{4S_0}{\sin\alpha\left(4 + tg^2\frac{\alpha}{2}\right)} \quad (27)$$

$$R = \frac{1}{3tg\frac{\alpha}{2}} \quad (28)$$

$$d = \frac{l}{4}tg^2\frac{\alpha}{4} \quad (29)$$

The question of the design of railroad switches using a sinusoid as a turnout curve is described in the work of professor P. G. Koziychuk.

A curve that changes according to the law of the sinusoid, when it is rotated at the angle of the crossing α, is shown in Fig. 2.

The equation of sinusoid in accordance with the notation adopted in Fig. 3, will have the form:

$$y = \alpha_1 \sin\varphi \quad (30)$$

where α_1 is the radius of the circle forming a sinusoid; he is the largest ordinate of a sinusoid at its peak.

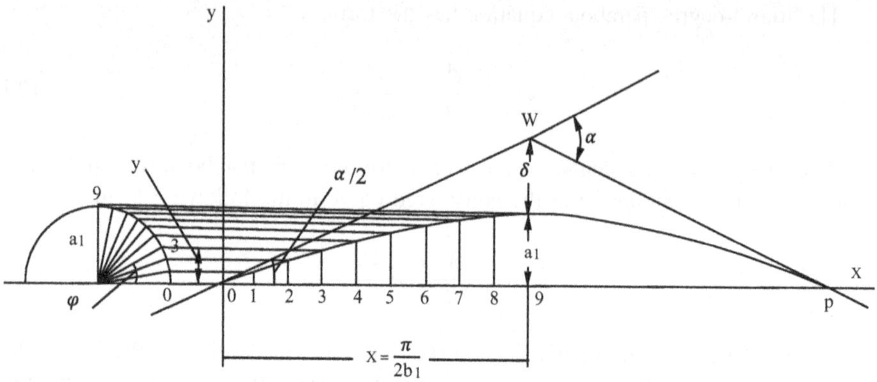

Fig. 2. The curve that changes according to the law of a sinusoid, when it is rotated at an angle of a crossing α.

Abscissas x are connected with angular magnitudes by the dependence:

$$b_1 x = \varphi \tag{31}$$

Substituting the dependence (31) in the expression (30), we obtain the equation of the sinusoid of the form:

$$y = \alpha_1 \sin b_1 x \tag{32}$$

Performing the calculations below we obtain the expression of the increment (change) of centrifugal acceleration per unit time for a curve that varies according to the law of a sinusoid.

The first and second derivatives of the sinusoid Eq. (32) will be:

$$\frac{dy}{dx} = \alpha_1 b_1 \cos b_1 x \tag{33}$$

$$\frac{d^2y}{dx^2} = -\alpha_1 b_1^2 \sin b_1 x \tag{34}$$

Since the minus sign of the second derivative characterizes the convexity of the curve, and in this case only the absolute magnitude is of interest, therefore, in subsequent calculations, the second derivative is omitted.

Substituting the values of derivatives (33) and (34) in the differential equation of curvature (12), we will have:

$$\frac{1}{\rho} = \frac{\alpha_1 b_1^2 \sin b_1 x}{(1 + \alpha_1^2 b_1^2 \cos^2 b_1 x)^{3/2}} \tag{35}$$

Multiplying both sides of Eq. (35) by V^2, we obtain the expression for centripetal (or, which is the same, in magnitude of centrifugal) acceleration:

$$\frac{V^2}{\rho} = \frac{V^2 \alpha_1 b_1^2 \sin b_1 x}{(1 + \alpha_1^2 b_1^2 \cos^2 b_1 x)^{3/2}} \tag{36}$$

It was previously noted (6) that the increment (change) of centrifugal acceleration per unit time (second)

$$\psi = \frac{d\left(\frac{V^2}{\rho}\right)}{dt} \tag{37}$$

In accordance with the dependence (8), the expression (36) for determining ψ of a curve that changes according to the law of a sinusoid can be rewritten in the form:

$$\frac{d\left(\frac{V^2}{\rho}\right)}{dt} = \frac{d}{dt}\left(\frac{V^2 a_1 b_1^2 \sin b_1 x}{\left(1 + a_1^2 b_1^2 \cos^2 b_1 x\right)^{3/2}}\right) \tag{38}$$

Multiplying and dividing the right side of Eq. (38) by dx we will have:

$$\psi = \frac{dx}{dt} \cdot \frac{d}{dx}\left(\frac{V^2 a_1 b_1^2 \sin b_1 x}{\left(1 + a_1^2 b_1^2 \cos^2 b_1 x\right)^{3/2}}\right) \tag{39}$$

Sinse dx/dt = V, after transformations, the expression (39) takes the form:

$$\psi = a_1 b_1 V^3 \cdot \frac{d}{dx}\left(\frac{\sin b_1 x}{\left(1 + a_1^2 b_1^2 \cos^2 b_1 x\right)^{\frac{3}{2}}}\right) \tag{40}$$

Differentiating the right side of Eq. (40), we obtain:

$$\psi = \frac{a_1 b_1^3 V^3}{2}\left[\frac{2\cos b_1 x}{\left(1 + a_1^2 b_1^2 \cos^2 b_1 x\right)^{\frac{3}{2}}} + \frac{3 a_1^2 b_1^2 \sin 2b_1 x \sin b_1 x}{\left(1 + a_1^2 b_1^2 \cos^2 b_1 x\right)^{\frac{5}{2}}}\right] \tag{41}$$

Studying Eq. (41) at the characteristic points of the curve, changing according to the law of the sinusoid, we will have:

$$\text{at } b_1 x = 0; \ \cos b_1 x = 1; \ \sin b_1 x = 0 \tag{42}$$

$$\psi = \frac{a_1 b_1^3 V^3}{\left(1 + a_1^2 b_1^2\right)^{3/2}} \tag{43}$$

$$t\, b_1x = \pi/2; \cos b_1x = 0; \;\sin b_1x = 1; \;\sin\; 2b_1x = 0 \tag{44}$$

$$\psi = 0 \tag{45}$$

The railroad switch scheme with the turnout curve, divided according to the law of a sinusoid, is presented in Fig. 3.

Coordinates x_1 and y_1 for practical use, when dividing the turnout curve, will be determined by the formulas:

$$x_1 = x\cos\frac{a}{2} + y\sin\frac{a}{2}; \tag{46}$$

$$y_1 = x\sin\frac{a}{2} - y\cos\frac{a}{2}. \tag{47}$$

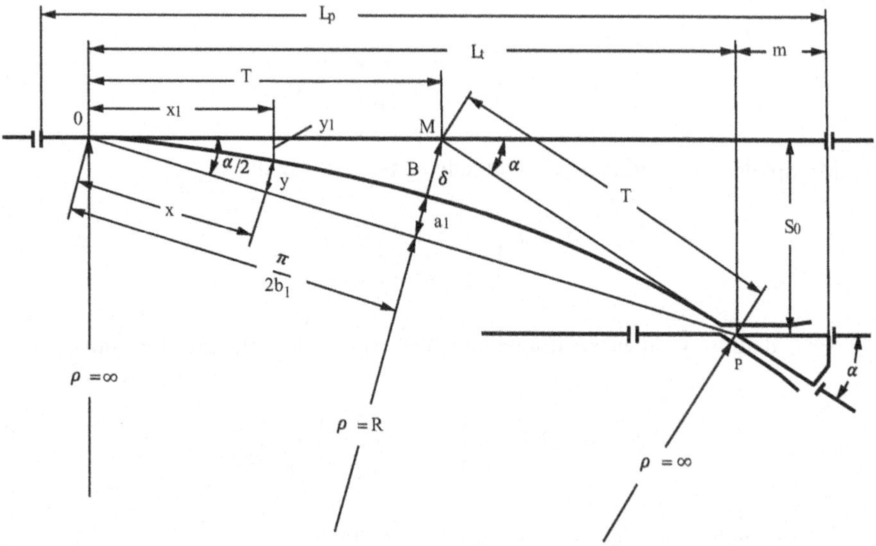

Fig. 3. The railroad switch scheme with the turnout curve, divided according to the law of a sinusoid.

Expressions (43) and (45) make it possible to establish that a change in centrifugal acceleration per unit time (second) ψ, in a curve divided by the law of a sinusoid, when entering the curve (at point 0) and when leaving the curve (at point P) for $b_1x = 0$ has a maximum value. For $b_1x = \pi/2$, i.e., in the middle of the turnout curve, $\psi = 0$.

The magnitudes of a_i and b_i, as mentioned above, and the length of the tangent T can be determined from the expressions [5]:

$$a_1 = R \cdot tg^2\frac{a}{2} \tag{48}$$

$$b_1 = \frac{1}{R \cdot tg\frac{a}{2}} \tag{49}$$

$$T = \frac{\pi R\, tg\frac{a}{2}}{2cos\frac{a}{2}} \tag{50}$$

The minus sign of the formulas for determining a_1 and b_1 is omitted, since for the case under consideration the absolute value of the values a_1 and b_1 is of interest.

The magnitude of the smallest radius R will be determined as follows. We take for the theoretical end of the sinusoid OBP the mathematical center of the crossing—point P. This can be admitted, given that practically the curve moves away from the tangential line begins to affect only at a distance of 6 m [22, 23].

From Fig. 3 establishes that

$$S_0 = T \text{ sin a} \tag{51}$$

Substituting in the expression (51) the magnitude of T from formula (50), we will have:

$$S_0 = \frac{\pi R\ tg\frac{\alpha}{2}}{2\cos\frac{\alpha}{2}} \sin\alpha \tag{52}$$

whence

$$R = \frac{2S_0 \cos\frac{\alpha}{2}}{\pi \cdot tg\frac{\alpha}{2}\sin} \tag{53}$$

When comparing, we consider railroad switches of the above grades with turnout curves, divided according to the laws of a cubic parabola, fourth-degree parabola and sinusoid.

When calculating the basic geometric characteristics of the turnout curves of variable curvature l, d, T, and R_{min} (Figs. 2 and 3), the following formulas were used: for a cubic parabola - (21), (22) and (23); for a fourth-degree parabola - (27), (28) and (29); for a sinusoid - (50) and (53).

3 Results

The main parameters of the considered curves of variable curvature, which are included in the calculation equations when determining permissible speeds, in the case of movement along the railroad switch to the side track, are presented in Table 3.

Table 3. The main parameters of the considered curves of variable curvature when determining the permissible speeds in the case of movement on the railroad switch on the side track.

Curve type	Magnitude ψ at the characteristic points of the turnout curve			Crossing grade	Projection lengths of one branch of the turnout curve with additional length l + d or T, m	Magnitude of the final radius R, m
	At the beginning	At the middle	At the end			
Cubic parabola	Constant			1/18	27.558	497.814
				1/22	33.632	741.665
				1/36	55.096	1999.162
Fourth-degree parabola	0	Max	0	1/18	27.558	331.900
				1/22	33.631	494.464
				1/36	55.096	1327.768
Sinusoid	Max	0	Max	1/18	27.561	633.832
				1/22	33.633	944.289
				1/36	55.091	2535.532

Based on the data Table 3 sets the following:

The curve that most fully meets the requirements of a smooth increase in the magnitude of the change in centrifugal acceleration per unit time is a curve that varies according to the law of the fourth-degree parabola. The main disadvantage of this curve is the significantly lower magnitudes of R_{min} for the same grades of railroad switches in comparison with the other curves, as a result of which the maximum speeds of train movement under the conditions of ride comfort are the smallest.

Curves that vary according to the laws of the cubic parabola and sinusoid have approximately the same properties as the centrifugal acceleration per unit time changes in the range of the turnout curve. At the same time, the turnout curve, divided according to the law of a sinusoid, compared to the turnout curve, divided according to the law of a cubic parabola, has other conditions being equal, has large values of R_{min}, which determine the maximum speeds of train movement according to the conditions of ride comfort.

4 Conclusions

1. If it is necessary to realize the speed of movement at the railroad switch to the side track over 50 km/h, the main factor when determining the radius of the turnout curve according to the conditions of ride comfort is the limitation of the increment (change) of centrifugal acceleration per unit time (second). Based on this, when designing railroad switches for high speeds of movement, it is advisable to use variable radius curves as a turnout curve.
2. Of the considered curves of variable radius (cubic parabola, fourth-degree parabola, sinusoid), the curve that varies according to the law of the sinusoid is the most acceptable for use as a turnout curve in railroad switches of flat grades for high speeds of movement.

References

1. Korolev, V.: switching shunters on a slab base. Advances in Intelligent Systems and Computing, vol. 1116, pp. 175–187 (2020). https://doi.org/10.1007/978-3-030-37919-3_17
2. Gluzberg, B., Korolev, V., Loktev, A., Shishkina, I., Berezovsky, M.: Switch operation safety. E3S Web Conf. **138** (2019). https://doi.org/10.1051/e3sconf/201913801017
3. Savin, A., Suslov, O., Korolev, V., Loktev, A., Shishkina, I.: Stability of the continuous welded rail on transition sections. Advances in Intelligent Systems and Computing, vol. 1115, pp. 648–654 (2020). https://doi.org/10.1007/978-3-030-37916-2_62
4. Loktev, A., Korolev, V., Shishkina, I., Chernova, L., Geluh, P., Savin, A., Loktev, D.: Modeling of railway track sections on approaches to constructive works and selection of track parameters for its normal functioning. Advances in Intelligent Systems and Computing, vol. 982, pp. 325–336 (2020). https://doi.org/10.1007/978-3-030-19756-8_30
5. Savin, A., Kogan, A., Loktev, A., Korolev, V.: Evaluation of the service life of non-ballast track based on calculation and test. Int. J. Innov. Technol. Explor. Eng. **8**(7), 2325–2328 (2019)
6. Loktev, A.A., Korolev, V.V., Shishkina, I.V.: High frequency vibrations in the elements of the rolling stock on the railway bridges. IOP Conf. Ser.: Mater. Sci. Eng. **463** (2018). https://doi.org/10.1088/1757-899x/463/3/032019
7. Glusberg, B., Savin, A., Loktev, A., Korolev, V., Shishkina, I., Alexandrova, D., Loktev, D.: New lining with cushion for energy efficient railway turnouts. Advances in Intelligent Systems and Computing, vol. 982, pp. 556–570 (2020). https://doi.org/10.1007/978-3-030-19756-8_53
8. Korolev, V.: Guard rail operation of lateral path of railroad switch. In Advances in Intelligent Systems and Computing, vol. 1115, pp. 621–638 (2020). https://doi.org/10.1007/978-3-030-37916-2_60
9. Loktev, A.A., Korolev, V.V., Shishkina, I.V., Basovsky, D.A.: Modeling the dynamic behavior of the upper structure of the railway track. Procedia Eng. **189**, 133–137 (2017). https://doi.org/10.1016/j.proeng.2017.05.022
10. Glusberg, B., Korolev, V., Shishkina, I., Loktev, A., Shukurov, J., Geluh, P., Loktev, D.: Calculation of track component failure caused by the most dangerous defects on change of their design and operational conditions. MATEC Web Conf. **239** (2018). https://doi.org/10.1051/matecconf/201823901054
11. Glusberg, B., Savin, A., Loktev, A., Korolev, V., Shishkina, I., Chernova, L., Loktev, D.: Counter-rail special profile for new generation railroad switch. Advances in Intelligent Systems and Computing, vol. 982, pp. 571–587 (2020). https://doi.org/10.1007/978-3-030-19756-8_54
12. Loktev, A.A., Korolev, V.V., Gridasova, E.A.: Influence of high-frequency cyclic loading on mechanical and structural characteristics of rail steel under extreme conditions. IOP Conf. Ser.: Mater. Sci. Eng. **687** (2019). https://doi.org/10.1088/1757-899x/687/2/022036
13. Gridasova, E., Nikiforov, P., Loktev, A., Korolev, V., Shishkina, I.: Changes in the structure of rail steel under high-frequency loading. Advances in Intelligent Systems and Computing, vol. 1115, pp. 559–569 (2020). https://doi.org/10.1007/978-3-030-37916-2_54
14. Korolev, V., Loktev, A., Shishkina, I., Zapolnova, E., Kuskov, V., Basovsky, D., Aktisova, O.: Technology of crushed stone ballast cleaning. IOP Conf. Ser.: Earth Environ. Sci. **403** (2019). https://doi.org/10.1088/1755-1315/403/1/012194
15. Shishkina, I.: Determination of contact-fatigue of the crosspiece metal. Advances in Intelligent Systems and Computing, vol. 1115, pp. 834–844 (2020). https://doi.org/10.1007/978-3-030-37916-2_82

16. Loktev, A., Korolev, V., Shishkina, I., Illarionova, L., Loktev, D., Gridasova, E.: Perspective constructions of bridge crossings on transport lines. Advances in Intelligent Systems and Computing, vol. 1116, pp. 209–218 (2020). https://doi.org/10.1007/978-3-030-37919-3_20
17. Savin, A.V., Korolev, V.V., Shishkina, I.V.: Determining service life of non-ballast track based on calculation and test. IOP Conf. Ser.: Mater. Sci. Eng. **687** (2019). https://doi.org/10.1088/1757-899x/687/2/022035
18. Lyudagovsky, A., Loktev, A., Korolev, V., Shishkina, I., Alexandrova, D., Geluh, P., Loktev, D.: Energy efficiency of temperature distribution in electromagnetic welding of rolling stock parts. E3S Web Conf. **110** (2019). https://doi.org/10.1051/e3sconf/201911001017
19. Glusberg, B., Loktev, A., Korolev, V., Shishkina, I., Alexandrova, D., Koloskov, D.: Calculation of heat distribution of electric heating systems for turnouts. Advances in Intelligent Systems and Computing, vol. 982, pp. 337–345 (2020). https://doi.org/10.1007/978-3-030-19756-8_31
20. Loktev, A.A., Korolev, V.V., Poddaeva, O.I., Chernikov, I.Y.U.: Mathematical modeling of antenna-mast structures with aerodynamic effects. IOP Conf. Ser.: Mater. Sci. Eng. **463** (2018). https://doi.org/10.1088/1757-899x/463/3/032018
21. Loktev, A.A., Korolev, V.V., Poddaeva, O.I., Stepanov, K.D., Chernikov, I.Y.: Mathematical modeling of aerodynamic behavior of antenna-mast structures when designing communication on railway transport. Vestn. Railw. Res. Inst. **77**(2), 77–83 (2018). https://doi.org/10.21780/2223-9731-2018-77-2-77-83. (in Russ.)
22. Savin, A., Korolev, V., Loktev, A., Shishkina, I.: Vertical sediment of a ballastless track. Advances in Intelligent Systems and Computing, vol. 1115, pp. 797–808 (2020). https://doi.org/10.1007/978-3-030-37916-2_78
23. Loktev, A.A., Korolev, V.V., Loktev, D.A., Shukyurov, D.R., Gelyukh, P.A., Shishkina, I. V.: Perspective constructions of bridge overpasses on transport main lines. Vestn. Railw. Res. Inst. **77**(6), 331–336 (2018). https://doi.org/10.21780/2223-9731-2018-77-6-331-336. (in Russ.)

Image Blurring Function as an Informative Criterion

Alexey Loktev[1(✉)] and Daniil Loktev[2]

[1] Russian University of Transport (MIIT), Chasovaya str. 22/2, 125190 Moscow, Russia
aaloktev@yandex.ru
[2] Bauman Moscow State Technical University (National Research University), 2-nd Baumanskaya, 5, 105005 Moscow, Russia

Abstract. The paper considers the issue of modeling the blurring of the object image on the primary image in an automated monitoring and control system for various categories of moving and stationary objects. The blurring function takes into account the value of the environment parameters between the monitoring system and the object under study, the color components of the object image and background, the movement parameters of the object and the detector, the parameters of tools of detection and primary image processing. The proposed model of the image blurring function makes it possible to present blurring as an important informative criterion that can be used to determine the parameters of movement and state of the desired object, as well as to more accurately assess the possibility of using algorithms and procedures for detecting and recognizing individual objects. The image blurring model allows taking a new look at the border between different objects, the object and the background. The boundary blurred layer can determine not only the noise components of the image and be a quality criterion for determining the sharpness, contrast and clarity of the photo, but also carry important information about the state and behavior of the object itself, as well as generally characterize the operation of an automated monitoring and control system, taking into account the color gradient in the visible spectrum range; the movement of the research object and the hardware of the monitoring system; primary processing of the object image.

Keywords: The primary image · Object of study · Parameters of the photo detector · Characteristics of the environment · Optical spectrum · Color components · Reduced blurring function

1 Introduction

One of the most popular and promising methods for obtaining primary information about the object under study is the analysis of its image on a picture or a series of pictures. The process of computer perception of visible images is quite complex, requires significant computing power, multi-level algorithms that formalize the presented information and compliance with the performance parameters of the entire system, in which the delay in processing each frame will be fixed. Despite significant achievements in this area, there are currently no universal methods, algorithmic,

V. Murgul and V. Pukhkal (Eds.): EMMFT 2019, AISC 1258, pp. 173–183, 2021.
https://doi.org/10.1007/978-3-030-57450-5_16

mathematical, software and hardware provision that can solve the problems of detecting, capturing, recognizing objects of various classes and detecting geometric, kinematic and dynamic parameters of their state and behavior in a wide range of possible values.

2 Materials and Methods

To operate an automated monitoring and control system in real conditions, it is necessary to meet the requirements of information security; scalability; data representation in a convenient form for analysis; intelligent interface; integration with other systems that process monitoring results; network interaction with other complexes and systems [1, 2].

When using such monitoring systems, the tasks of determining the contours of images of objects, improving contrast, sharpness, clarity of images, reducing noise, etc. are solved. In the first stage from a sequence of images, selected individual pictures, suspected of containing the image of the desired object, then the picture by reducing the blurring, color correction is improved, the third stage is the object detection and determination of characteristic coordinates, in the fourth stage the contour of the image is completely determined or a set of cascaded classifiers to identify the object, i.e. assign it to one of the specified classes or to establish that such objects in the database do not exist [3–5].

Blurring can determine the minimum distance between detected objects, the maximum speed of movement of objects for their recognition, and the parameters of photo and video detectors necessary for remote monitoring, monitoring, or diagnostics in specific conditions and for certain groups of control objects [4–6].

Technical vision systems based on obtaining and processing graphic information in photo and video formats are associated with the need to solve a number of problems of image analysis that show a fragment of three-dimensional space at a certain time: various brightness and geometric parameters of images of individual objects in the general image (brightness gradients, colors, textures, shapes, sizes, the presence of shadow effects, etc.) [6–8]; changes in the overall disposition of objects in each individual image from a series or sequence (changes in brightness, background, interference between the observer and the object under study, changes in the parameters of the state and behavior of objects, the presence of precipitation, suspended solid particles, various noise effects, etc.) [5, 8, 9]; time delay from the moment of image acquisition, object detection, capture, recognition and subsequent processing operations, this factor is associated with the speed of software and hardware provision. The operation of the entire automated control and monitoring system largely depends on the primary images obtained. To ensure their required quality, algorithms are used to improve contrast and sharpness, as well as carry out measures to compensate for lighting and control the direction of shooting [10, 11].

A number of parameters that characterize the quality of the resulting image, such as contrast, sharpness, and clarity, are associated with determining the boundaries of individual objects, which always has a certain degree of blurring. The nature of the appearance of such a boundary layer between objects is different (Figs. 1, 2), but most often it is assumed that the width of the boundary layer (the blur layer) depends on the lens focus parameters (Fig. 1); hardware parameters that affect the display of different colors and the brightness transition between them (Figs. 1, 2); from the initial processing of the original image in a photo or video detector and the format of recording the file to memory (Figs. 1, 2) [12, 13].

Figure 1 shows the primary images of a stationary object under study (book) with a stationary photo detector and different distances between them: the distance between the object and the detector is 1 m in Fig. 1a, in Fig. 1b – 0.8 m, in Fig. 1c – 0.6 m, in Fig. 1d – 0.4 m. The parameters of the photo detector before and after its movement remain the same. We can note different blurring of images of objects such as a book (the object under study, which was focused on by the lens for the photo shown in Fig. 1a), the arms of a chair, the seat of a chair, the door of a room in the photos in Fig. 1a and Fig. 1b.

Blurring due to the movement of the object under study (car) when the photo detector is stationary is shown in Fig. 2, but in general case of monitoring, the photo detector can also move [14–16].

Fig. 1. Primary images of a stationary object with the same photo detector settings and different distances to it.

The parameters of the boundary layer can not only indicate the image quality [4, 17], but also be a full-fledged informative criterion that determines the totality of parameters of the entire monitoring and control system, as well as parameters of the state and behavior of the object under study.

The blurring function is proposed to describe by seven main factors: the characteristics of the environment from the monitoring system to the object under study (blurring due to the environment, σ_{mat}); the dependence of the blurring from color, in fact, this dependence on wavelength, the corresponding range of the visible spectrum λ (blurring due to the color, σ_{col}); features of the motion of the detection system, i.e. movement of the object of research and the hardware of the monitoring system (blurring due to the motion of an object and monitoring system, σ_{mov}) (Fig. 2); a particular state of the background (blurring due to the background σ_{bg}); parameters of detection tools (blurring due to the detector, σ_{det}) (Fig. 1); primary processing of the object image (blurring due to primary processing, σ_{pp}) (Figs. 1, 2); features of the state and behavior of the object of study (blurring due to the state of the surface of the object of study, σ_{sc}).

a)

b)

Fig. 2. Primary images of a moving object with the same photo detector settings.

Let's consider the effect on the blurring function of the parameters of the environment located between the object of study and the photo detector. It is fundamentally related to the different transparency of the environment for different wavelengths and the nonlinear dependence of the phase coefficient $\beta(\lambda)$ within the spectrum of available wavelengths (colors present in the image) [7, 18, 19]. The nonlinearity of the function $\beta(\lambda)$ depends on the blurring for each of the available colors and the change in the transparency function for different wavelengths at the distance between the detector and the object of study.

In general, the blurring due to environmental parameters from the monitoring system to the object under study is proposed to be determined by the formula

$$\sigma_{mat} = \Delta\lambda \frac{L}{H} \frac{\lambda}{v_f} \frac{dP_{tr}^2}{d\lambda^2} \quad (1)$$

here $\Delta\lambda$ - the range of wavelengths corresponding to the color components available in the image, P_{tr} - is a function of the transparency of the environment depending on the refractive index (P_{ref}), light (P_{ill}), the presence of a temperature fronts (P_{tf}) and various noises (P_{ns}), taking into account these characteristics we can write the expression: $P_{tr} = P_{ref} + P_{ill} + P_{tf} + P_{ns}$; L – given the distance to the object; H – given the typical size of the object; λ – wavelength for a specific color; v_f – the group given speed of propagation of the radiation used wavelengths in the environment.

Often, the environment transparency function can be described by a dependence that smoothly changes from the central axis (the central axis of the photo detector, which is the shortest distance from the detector to the trajectory of the movement of object under study) to the image border and depends on the reduced distance from the detector to the object [17, 20], the characteristic size of the object itself and the size of the primary image

$$P_{tr}(L) = P_{tr}^0\left[1 + K_{sp}\frac{P_{tr}^0}{P_{tr}^m}\left(\frac{L}{H_m}\right)^n\right], \quad (2)$$

here n - is a non-linearity indicator that shows the degree of change in the environment parameters between the edges of the primary image; P_{tr}^0 - transparency function on the central axis of the photo detector; K_{sp} - coefficient that takes into account changes in space metrics when it is displayed as a flat image; P_{tr}^m - averaged transparency function for the reference shooting situation; H_m - the reduced width of the space at the level of the object under study, falling on the primary image.

The dependence of the blurring from the color σ_{col} (the dependence on the wavelength corresponding to the range of the visible spectrum λ) has a nature similar to the nature of the previous component of the blurring function and is related to the features of propagation in space of radiation of different wavelengths in accordance with the expression $c/v_\Phi = \beta/k_0$.

When choosing an object detection system, it is important that the time of obtaining a photo-fixation image should be commensurate with the time of moving the object to the distance between two points in space, the trajectory of movement between which is close to a straight line. Since the process of obtaining the primary image takes some time, it is proposed to consider the blurring □mov, as a function that determines the relationship between the state parameters of the detection system and the object under study, obtained during the time between the beginning and end of the initial image formation [8, 9, 21].

The resulting blurring due to the movement of the object and the photo detector can be represented as one of the following formulas:

$$f(\sigma_{ob}) = \frac{\sigma'_{ob} + \sigma''_{ob}}{2}, \; f(\sigma_{sy}) = \frac{\sigma'_{sy} + \sigma''_{sy}}{2} \qquad (3)$$

or

$$f(\sigma_{ob}) = \sigma'_{ob} + \sigma''_{ob}, \; f(\sigma_{sy}) = \sigma'_{sy} + \sigma''_{sy}, \qquad (4)$$

or

$$f(\sigma_{ob}) = \max_{\sigma'_{ob},\sigma''_{ob}} (\sigma_{ob}), \; f(\sigma_{sy}) = \max_{\sigma'_{sy},\sigma''_{sy}} (\sigma_{sy}). \qquad (5)$$

Here $f(\sigma_{ob})$ and $f(\sigma_{sy})$ - are functions that link blurring by moving the object or detector relative to the metric basis at the beginning and end of obtaining the primary image.

The specific choice of the function that links the final blurring of the object image in the picture and the movement of the object under study and the control system's photo detector, taking into account their different locations at the time that determines the beginning and end of the primary image acquisition process, significantly depends on the detector parameters and can be determined empirically [11, 12, 22]. If the image size of an object is larger than a certain threshold value, we must take into account the aberrations that occur due to the linear size of the object.

The appearance of blurring of image fragments is also possible due to their distance from the axis of the photo detector and is associated with the angle between the axis of the photo detector and the line connecting the detector (as a point) and the extreme point of the object under study.

$$L^2 + \frac{H^2}{4} = \frac{L^2}{\cos^2 \theta_1}, \qquad (6)$$

where θ_1 - is the angle between the axis of the photo detector and the line connecting the detector and the most distant point of the object under study.

Due to the curvature of the display of the spatial object on the plane, the blurring at the points of the image of the object on the axis of the photo detector and more distant points will differ

$$\sigma_{sz} = f(\sigma_{ax}, \sigma_{ex}), \qquad (7)$$

where σ_{sz} - is the blurring associated with the size of the object image, $f(\sigma_{ax}, \sigma_{ex})$ - is a function that links the blurring of image points on the axis of the photo detector (σ_{ax}) and at the greatest distance from it (σ_{ex}).

If the blurring on the axis of the photo detector is taken as a conditional minimum, then we can assume the following ratio, which is valid for linear changes in the parameters of the environment over the entire width of the object

$$\sigma_{sz} = \sigma_{ex} - \sigma_{ax} = \frac{LP_{tr}^{ex}}{v_f \cos\theta_1} - \frac{LP_{tr}}{v_f}, \tag{8}$$

here P_{tr}^{ex} -is the transparency function of the environment at the level of the most remote point of the object under study.

If the object size is greater than a certain threshold value corresponding to the distance from the detector L_d, the following relationship is assumed for this type of blurring

$$\sigma_{sz} = \begin{cases} \frac{LP_{tr}^{ex}}{v_f \cos\theta_1} - \frac{LP_{tr}}{v_f} \\ \frac{\sqrt{LL_d}P_{tr}^{ex}}{v_f \cos\theta_1} - \frac{LP_{tr}}{v_f} \end{cases} \text{at } L < L_d, \text{ at } L \geq L_d \tag{9}$$

The L_d distance is determined experimentally for a specific photo or video detector, a specific object under control, and specified shooting conditions.

If we need to calculate a blurring σ_{sz}, it is added to the blurring σ_{mov} and then taken into account in the calculations along with it, i.e. it is actually part of it.

Blurring due to the features of the background state σ_{bg} depends on the image, color, color gradient in different directions, the parameters of the environment between the object and the background, distance and movement parameters (ripples, flickering, movement of individual background elements) [18, 20, 23], the aggregate state of individual elements that make up the background layer of the image, anisotropic properties of the environment.

An important assumption is how to define the background practically and theoretically, this study proposes to consider the background of everything around the image of the desired object, without regard to the depth of an arrangement of individual background elements.

Blurring the image of an object due to the background σ_{bg} is suggested to be represented as the following formula:

$$\sigma_{bg} = K_g \int\limits_{l_{dis}} \int\limits_{l_{mov}} \oint\limits_{S} \left(G\frac{\partial U}{\partial l} - U\frac{\partial G}{\partial l} \right) dl_{dis} dl_{mov} ds \tag{10}$$

where K_g - is a vector of coefficients taking into account the curvature of metric of the image relative to the metric of the space with distance from the axis of the photo detector, G - is the Green's function for the spatial contour of the investigated object; S - flat contour of the image of the investigated object; l_{dis} - is the distance from the object to conventional boundaries background; l_{mov} - the direction of movement of the object; l - is the direction from the observer to the object at the shortest distance; U (r, φ, z) - the function of changing the color components from the border of the image of the object in three directions of anisotropy.

The calculation of the blurring caused by the parameters of the detection tools (σ_{det}) and the primary processing of the object image (σ_{pp}) in this study was carried out on the basis of field experiments with specified photo detectors and cameras, with their supported data presentation formats. The influence of the state and behavior of the surface of the object of study on blurring (σ_{sc}) was not considered, since objects were detected in a solid aggregate state, with the shape and color not changing during observing them, situations of dynamic contact between objects with their subsequent deformation were also not considered in this study [15, 18, 20].

Since the visible range has a wavelength range from 0.37 to 0.75 microns, the spectrum will change when propagated in a non-absolutely transparent space. The overall blurring can be represented as a function that links all the above mentioned types of blurring,

$$\sigma_{\Sigma} = f\left(\sigma_{mat}, \sigma_{col}, \sigma_{mov}, \sigma_{bg}, \sigma_{\det}, \sigma_{pp}, \sigma_{sc}\right), \tag{11}$$

most often in engineering applications of tasks of monitoring and control can be based on the sum of individual blurring components

$$\sigma_{\Sigma} = \sigma_{mat} + \sigma_{col} + \sigma_{mov} + \sigma_{bg} + \sigma_{\det} + \sigma_{pp} + \sigma_{sc}. \tag{12}$$

For specific conditions use of information-measuring system of control and monitoring, and for given classes of studied objects, for certain intervals of the measured values of the parameters of state and behavior of objects, the ratio between the individual components of the blurring function can vary significantly, thus simplifying the model to determine the blurring of an object image, and may not take into account the individual components in view of their smallness, relative to other terms.

The presented mathematical model of blurring elements of the object image on the picture takes into account many different parameters. This makes it difficult to determine the impact of each factor. For the convenience of comparative analysis for some classes of applied tasks, it is proposed to use the blurring function reduced to a unit of distance and a unit of color range

$$\sigma_{\Delta\lambda L} = \frac{K_{sp}}{v_f}\left[\lambda^2 \frac{d^2\beta}{d\lambda^2} + 2\lambda \frac{d\beta}{d\lambda}\right], \tag{13}$$

where $\sigma_{\Delta\lambda L}$ - is the reduced blurring function, K_{sp} - is the coefficient that determines the geometric parameters of the space near the object, and β - is the characteristic of displaying color components (colors) in space.

To solve the tasks of detecting, capturing, recognizing objects and detecting of parameters of their behavior and state, it is proposed to use a characteristic such as the slope coefficient of the blurring function $\sigma_{\Sigma}(\lambda)$ for each value of the color component

$$S_{\sigma\lambda} = \frac{d\sigma_{\Sigma}(\lambda)}{d\lambda}. \tag{14}$$

If the object under study has one or a limited number of color components on its surface, the slope coefficient of the blurring function can be represented as a straight line that is valid for a given range of colors and the tangent of the angle of inclination to the wavelength axis is equal to:

$$S_{\sigma\lambda} = tg\varphi = \frac{\Delta\sigma_{\Sigma}(\lambda)}{\lambda_1 - \lambda_2}, \tag{15}$$

where λ_1 and λ_2 - are the wavelengths corresponding to the colors of the left and right border of the optical range used in the image.

In general, it can be noted that the blurring function does not significantly depend on the wavelength (color components), the dependence of the transparency function from the color components can be ignored, in general, it can be argued that the blurring is proportional to the range of colors available in the composition, i.e. the width of the emission spectrum.

The characteristic of displaying colors in space can be represented as

$$\beta^2 = \frac{1}{H_m^2}\left[\frac{N^2}{P_{tr}^{pr}} - M^2\right], \tag{16}$$

here $N = kH_m\sqrt{P_{tr}/P_{tr}^{ax}}$, P_{tr}^{ax} - is the transparency function of the environment on the detector axis in the direction of the object under study $k = 2\pi/\lambda$, k - is the wave number in free space, M - is the parameter for transmitting a separate color component in space, which can be determined using the Gaussian elimination

$$M^2 = \frac{k_1}{H_0^2}\int_0^{\infty}\left[\left(\frac{dF_0(H)}{dH}\right)^2 + N^2 f(H)(F_0(H))^2\right] HdH, \tag{17}$$

here f (H) - is the dependence that determines the gradient of the transparency function in the direction perpendicular to the segment “detector – object under study”, most often, the polynomial form of this dependence is assumed, based on the actual conditions for obtaining the primary image, the linearization procedure f (H) can also be used for small linear dimensions of space; H_0 - is the reduced transverse size of the object image.

The Gaussian method is based on the representation of the blurring distribution field in the direction transverse to the detector axis by the Gaussian function

$$F_0(H) = \exp\left[-\frac{1}{2}\left(\frac{H}{H_0}\right)^2\right]. \tag{18}$$

3 Conclusions

The proposed model of blurring images of objects in the pictures allows taking a new look at the border between different objects, the object and the background. The border blurred layer can determine not only the noise components of the image and be a quality criterion for determining the sharpness, contrast and clarity of the photo, but also carry important information about the state and behavior of the object itself, as well as generally characterize the operation of an automated monitoring and control system, taking into account the color gradient in the visible spectrum range; movement of the research object and hardware of the monitoring system; primary processing of the object image.

References

1. Korolev, V.: Switching shunters on a slab base. In: Advances in Intelligent Systems and Computing, vol. 1116, pp. 175–187 (2020). https://doi.org/10.1007/978-3-030-37919-3_17
2. Gluzberg, B., Korolev, V., Loktev, A., Shishkina, I., Berezovsky, M.: Switch operation safety. In: E3S Web of Conferences, vol. 138 (2019). https://doi.org/10.1051/e3sconf/201913801017
3. Savin, A., Suslov, O., Korolev, V., Loktev, A., Shishkina, I.: Stability of the continuous welded rail on transition sections. In: Advances in Intelligent Systems and Computing, vol. 1115, pp. 648–654 (2020). https://doi.org/10.1007/978-3-030-37916-2_62
4. Loktev, A., Korolev, V., Shishkina, I., Chernova, L., Geluh, P., Savin, A., Loktev, D.: Modeling of railway track sections on approaches to constructive works and selection of track parameters for its normal functioning. In: Advances in Intelligent Systems and Computing, vol. 982, pp. 325–336 (2020). https://doi.org/10.1007/978-3-030-19756-8_30
5. Savin, A., Kogan, A., Loktev, A., Korolev, V.: Evaluation of the service life of non-ballast track based on calculation and test. Int. J. Innov. Technol. Exploring Eng. **8**(7), 2325–2328 (2019)
6. Loktev, A.A., Korolev, V.V., Shishkina, I.V.: High frequency vibrations in the elements of the rolling stock on the railway bridges. In: IOP Conference Series: Materials Science and Engineering, vol. 463 (2018). https://doi.org/10.1088/1757-899x/463/3/032019
7. Glusberg, B., Savin, A., Loktev, A., Korolev, V., Shishkina, I., Alexandrova, D., Loktev, D.: New lining with cushion for energy efficient railway turnouts. In: Advances in Intelligent Systems and Computing, vol. 982, pp. 556–570 (2020). https://doi.org/10.1007/978-3-030-19756-8_53
8. Korolev, V.: Guard rail operation of lateral path of railroad switch. In: Advances in Intelligent Systems and Computing, vol. 1115, pp. 621–638 (2020). https://doi.org/10.1007/978-3-030-37916-2_60
9. Loktev, A.A., Korolev, V.V., Shishkina, I.V., Basovsky, D.A.: Modeling the dynamic behavior of the upper structure of the railway track. Procedia Eng. **189**, 133–137 (2017). https://doi.org/10.1016/j.proeng.2017.05.022
10. Glusberg, B., Korolev, V., Shishkina, I., et al.: Calculation of track component failure caused by the most dangerous defects on change of their design and operational conditions. In: MATEC Web of Conferences 239 (2018). https://doi.org/10.1051/matecconf/201823901054

11. Glusberg, B., Savin, A., Loktev, A., Korolev, V., Shishkina, I., Chernova, L., Loktev, D.: Counter-rail special profile for new generation railroad switch. In: Advances in Intelligent Systems and Computing, vol. 982, pp. 571–587 (2020). https://doi.org/10.1007/978-3-030-19756-8_54
12. Loktev, A.A., Korolev, V.V., Gridasova, E.A.: Influence of high-frequency cyclic loading on mechanical and structural characteristics of rail steel under extreme conditions. In: IOP Conference Series: Materials Science and Engineering, vol. 687 (2019). https://doi.org/10.1088/1757-899x/687/2/022036
13. Gridasova, E., Nikiforov, P., Loktev, A., Korolev, V., Shishkina, I.: Changes in the structure of rail steel under high-frequency loading. In: Advances in Intelligent Systems and Computing, vol. 1115, pp. 559–569 (2020). https://doi.org/10.1007/978-3-030-37916-2_54
14. Korolev, V., Loktev, A., Shishkina, I., Zapolnova, E., Kuskov, V., Basovsky, D., Aktisova, O.: Technology of crushed stone ballast cleaning. In: IOP Conference Series: Earth and Environmental Science, vol. 403 (2019). https://doi.org/10.1088/1755-1315/403/1/012194
15. Shishkina, I.: Determination of contact-fatigue of the crosspiece metal. In: Advances in Intelligent Systems and Computing, vol. 1115, pp. 834–844 (2020). https://doi.org/10.1007/978-3-030-37916-2_82
16. Loktev, A., Korolev, V., Shishkina, I., Illarionova, L., Loktev, D., Gridasova, E.: Perspective constructions of bridge crossings on transport lines. In: Advances in Intelligent Systems and Computing, vol. 1116, pp. 209–218 (2020). https://doi.org/10.1007/978-3-030-37919-3_20
17. Savin, A.V., Korolev, V.V., Shishkina, I.V.: Determining service life of non-ballast track based on calculation and test. In: IOP Conference Series: Materials Science and Engineering, vol. 687 (2019). https://doi.org/10.1088/1757-899x/687/2/022035
18. Lyudagovsky, A., Loktev, A., Korolev, V., Shishkina, I., Alexandrova, D., Geluh, P., Loktev, D.: Energy efficiency of temperature distribution in electromagnetic welding of rolling stock parts. In: E3S Web of Conferences, vol. 110 (2019). https://doi.org/10.1051/e3sconf/201911001017
19. Glusberg, B., Loktev, A., Korolev, V., Shishkina, I., Alexandrova, D., Koloskov, D.: Calculation of heat distribution of electric heating systems for turnouts. In: Advances in Intelligent Systems and Computing vol. 982, pp. 337–345 (2020). https://doi.org/10.1007/978-3-030-19756-8_31
20. Loktev, A.A., Korolev, V.V., Poddaeva, O.I., Chernikov, I Y.U.: Mathematical modeling of antenna-mast structures with aerodynamic effects. In: IOP Conference Series: Materials Science and Engineering, vol. 463 (2018). https://doi.org/10.1088/1757-899x/463/3/032018
21. Loktev, A.A., Korolev, V.V., Poddaeva, O.I., Stepanov, K.D., Chernikov, I.Y.: Mathematical modeling of aerodynamic behavior of antenna-mast structures when designing communication on railway transport. Vestnik Railway Res. Inst. **77**(2), 77–83 (2018). https://doi.org/10.21780/2223-9731-2018-77-2-77-83. (in Russian)
22. Savin, A., Korolev, V., Loktev, A., Shishkina, I.: Vertical sediment of a ballastless track. In: Advances in Intelligent Systems and Computing, vol. 1115, pp, 797–808 (2020). https://doi.org/10.1007/978-3-030-37916-2_78
23. Loktev, A.A., Korolev, V.V., Loktev, D.A., et al.: Perspective constructions of bridge overpasses on transport main lines. Vestnik Railway Res. Inst. **77**(6), 331–336 (2018) https://doi.org/10.21780/2223-9731-2018-77-6-331-336. (in Russian)

Deformations and Life Periods of the Switch Chairs of the Rail Switches

Boris Glusberg[1], Alexey Loktev[2], Vadim Korolev[2], Irina Shishkina[2](✉), Mikhail Berezovsky[3], and Pavel Trigubchak[3]

[1] Railway Research Institute (JSC VNIIZhT), 3rd Mytischinskaya Str., 10, Moscow 129626, Russia
[2] Russian University of Transport (MIIT), Chasovaya Str. 22/2, 125190 Moscow, Russia
shishkinaira@inbox.ru
[3] JSC "Novosibirsk Switch Factory", Aksenova Str., 7, 630025 Novosibirsk, Russia

Abstract. Observations of the operation of the switch chairs show that they are deformed during operation. In some cases, their deformations appear in the form of cracks due to the high level of bending stresses. The switch chair is a plate of variable stiffness lying on an elastic foundation. The paper considers the built probabilistic model of the operation of switch chairs of the rail switches on reinforced concrete bars, well consistent with the results of the trial operation of such switches, which allows calculating the failure distribution of switch chairs depending on the characteristics of the metal from which they are made, the initial tightening of the fasteners and the stress spectrum arising from train load. The graphs of failures of switch chairs in transit at various axial loads, obtained using the developed model, make it possible to plan the replacement of switch chairs of the rail switchers with reinforced concrete beams when changing the main metal parts. With an average network axial load of 140 kN, at the time of the first change of metal parts, it is necessary to prepare a replacement of 2–3% of switch chairs. With increased axial loads of 250 kN, the number of replaceable switch chairs should be about 16%.

Keywords: Rail switch · Lining with a cushion (switch chair) · Probabilistic model · Dependence of the distribution of failures of the switch chairs · Axial loads · Planning for the replacement of switch chairs

1 Introduction

Observations of the operation of the switch chairs show that are deformed during operation they. In some cases, their deformations appear in the form of cracks due to the high level of bending stresses [1, 2].

The switch chair is a plate of variable stiffness lying on an elastic foundation.

V. Murgul and V. Pukhkal (Eds.): EMMFT 2019, AISC 1258, pp. 184–196, 2021.
https://doi.org/10.1007/978-3-030-57450-5_17

2 Materials and Methods

Professor V.F. Yakovlev proposed the option of calculating a switch chair of the rail switch with a base in the form of wooden beams.

In the model of V.F. Yakovlev switch chair is a plate with a stepwise change in stiffness, lying on an elastic foundation.

When calculating the switch chair, two schemes should be considered that characterize its operation (Fig. 1).

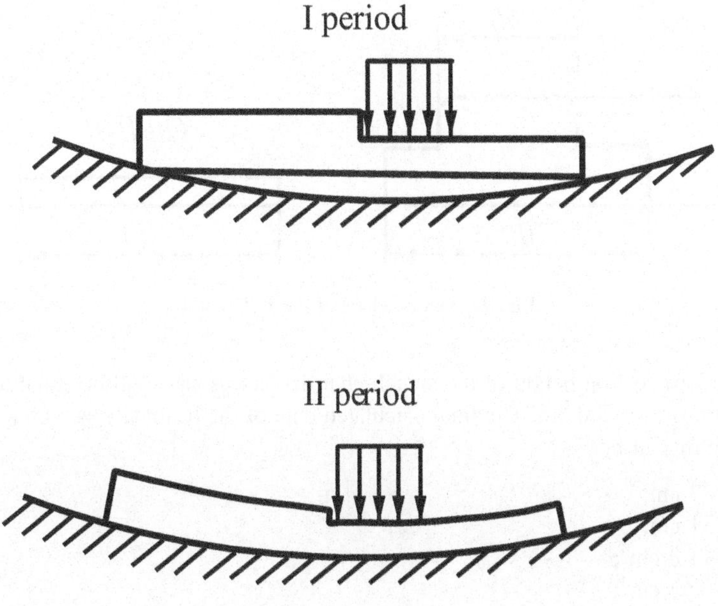

Fig. 1. Schemes of a switch chair operation.

Practically both schemes of operation of the plate will be implemented within one loading cycle [3, 4].

Tension in the switch chair is determined by the formula

$$\sigma = \sigma' + \sigma'', \quad (1)$$

where σ'' - stress arising in a switch chair due to bending of the beam;

σ' - stress arising in the switch chair due to deformations of the elastic base of the beam.

Characteristics of calculation elements. The moment of inertia of a switch chair in cross Sect. 1—I (Fig. 2) is equal to 216.28 cm^4. Based on the accepted calculation scheme, we replace Sect. 1-1 with a fictitious rectangular section of the same stiffness (Fig. 2).

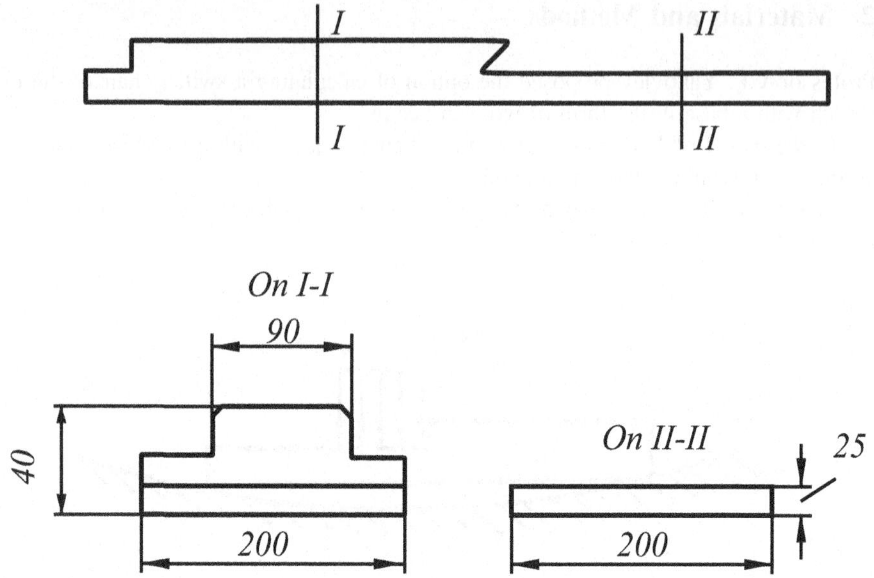

Fig. 2. Switch chair of the rail switch.

Fictitious section height of the switch chair h1 in this case will be equal to 5.2 cm and width b—to 18.0 cm. The final calculated data of the beam (strips with a width of b1 = 1 cm) will be:

$l_1 = 33$ cm;
$l_1 = 31$ cm;
$I_1 = 11.8\ cm^4$;
$I_2 = 0.67\ cm^4$;
$w_1 = 4.45\ cm^3$;
$w_2 = 0.67\ cm^3$;
$C = 250\ kg/cm^3$;
$u = 250\ kg/cm^3$;
$k_1 = 0.035\ cm^{-1}$;
$k_2 = 0.080\ cm^{-1}$;
$a = 13.0$ cm;
$q = 427.35 \times 10^{-5}P$.

1. The calculation of the stresses arising in the switch chair due to the bending of the transverse beam σ″, shows that they are equal

$$\sigma'' = \frac{M(x)}{w(x)} = \frac{0.0254P}{0.67} = 0.0379\text{P} \tag{2}$$

2. The stresses arising in the switch chair due to deformations of the elastic base are calculated by the method of initial parameters according to three design schemes [5, 6]:

first: the beam is taken of variable (step) stiffness (Fig. 3a); It is assumed that the lining and a cushion attached to it work as a whole cross-section;
second: beam stiffness is assumed constant for the case when rivets of cushion are weakened, and the lining and cushion work separately (Fig. 3b);
third: beam assumed of variable (speed) stiffness as at the first scheme, loaded with a concentrated force.

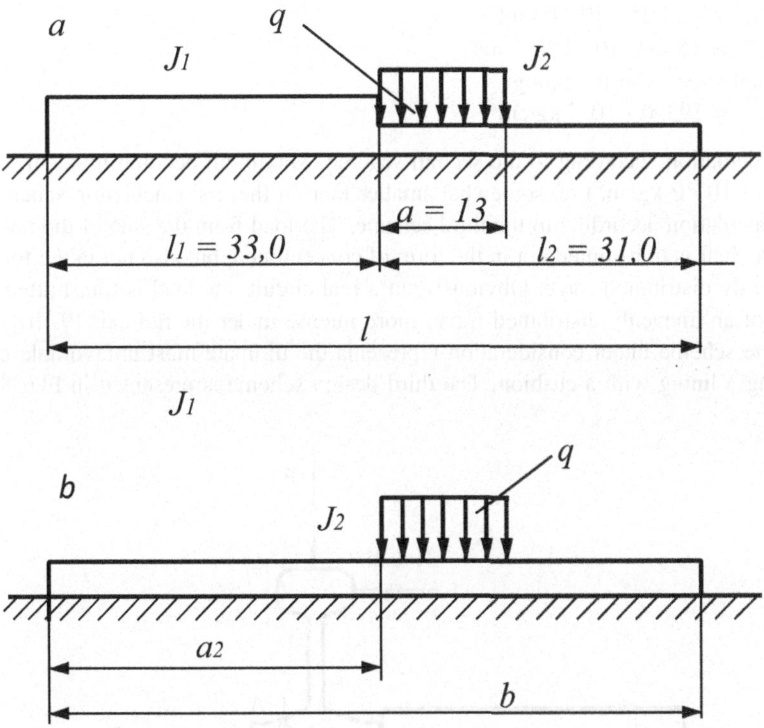

Fig. 3. - Design schemes: a) - beam of variable stiffness; b) - beam of constant stiffness.

The point of application of force is determined by the line of action of the resultant vertical load P and the horizontal transverse force H = 0.2P.

In the first two design schemes, the load is assumed to be uniformly distributed on the contact area of the lining and the sole of the frame rail [7, 8].

The final calculation results are presented on the diagrams of bending moments, deflections and stresses (Fig. 4).

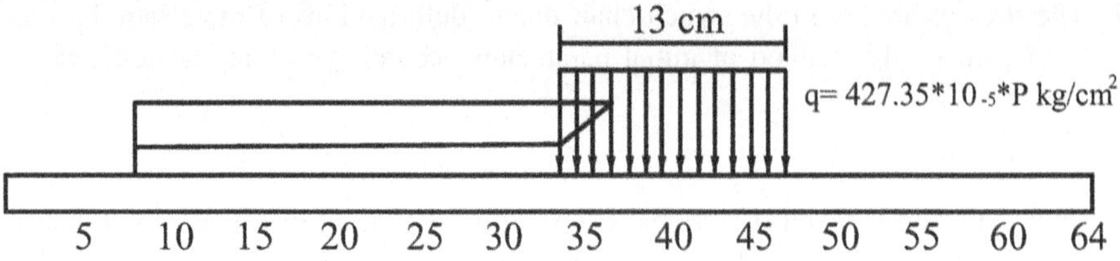

Fig. 4. Diagrams of bending moments, deflections and stresses.

The maximum bending moment, deflection and stress occur in the section near the axis of the frame rail:

$M_{max} = 10470 \cdot 10^{-5}P$ kg cm;
$y_{max} = -0.805 \cdot 10^{-5}P$ cm;
$\sigma'_{max} = 15000 \cdot 10^{-5}P\,kg/cm^2$.
Total stresses in the lining:
$\sigma_{max} = 19390 \cdot 10^{-5}$ kg/cm^2.

Calculation according to the second scheme. Maximum bending moment M_{max} is $10470 \cdot 10^{-5}P$ kgcm, t. e. somewhat smaller than in the first calculation scheme.

Calculation according to the third scheme. The load from the sole of the rail to the switch chair is transmitted not in the form of concentrated, but also not in the form of a uniformly distributed force. Obviously, in a real circuit, the load is transmitted in the form of an unevenly distributed force, more intense under the rail axis [9, 10].

The scheme under consideration represents the ultimate most unfavorable case of loading a lining with a cushion. The third design scheme is presented in Fig. 5.

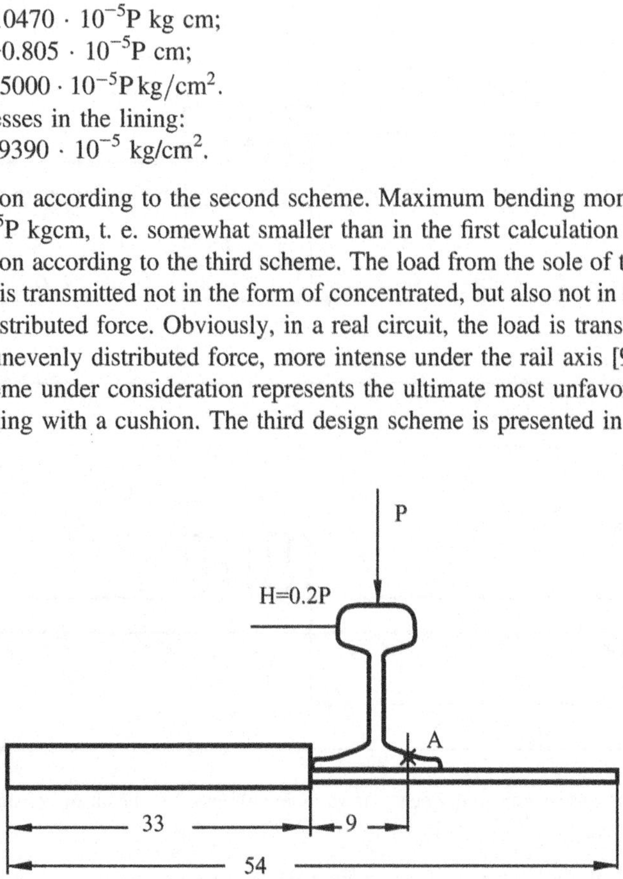

Fig. 5. Design scheme III.

The maximum bending moment in this case will be

$$M_{max} = +17860 \cdot 10^{-5} P\,kg \cdot cm,$$

and the maximum tensions are

$$\sigma'_{max} = 26665 \cdot 10^{-5} P\,kg/cm^2.$$

Total maximum stresses caused by the bending of the beam and the deflection of the lining on the beam (elastic foundation)

$$\Sigma\sigma_{max} = 30455 \cdot 10^{-5} P\,kg/cm^2.$$

In studies tensions in the indicated section at P = 10000 kg are 2200 kg/cm^2. It should be noted that in laboratory experiments, a piece of beam laid on a rigid foundation.

According to the third design scheme σ = 3045 kg/cm^2 (at P = 10 t).

The calculation of the lining according to the first two schemes (distributed load) shows that the magnitude of the stresses from the practically realized pressures on the lining (up to 8000 kg) is lower than the allowable.

The calculation of the lining with the cushion (switch chair) according to the third scheme (concentrated force) shows that the stresses from the loads of the same order on the lining are higher than the yield strength.

Considering the data on the practice of operating the linings with the cushion (switch chairs) and the results of the calculations above, we can recommend that in the future, when selecting sections of the linings, they should be calculated according to the third scheme, which gives a certain margin of safety [11, 12]. According to this scheme, we obtain that at the yield strength of steel grade St.3, σ_t = 2200 kg/cm^2 the lining thickness of 20 mm can be safely allow a pressure of about 7100 kg. Under existing loads, pressure on the sleeper of a larger size can be realized, therefore, the dimensions of the lining are insufficient. At the same time, the calculation does not take into account the uneven wear of the bed of the beam under a moving load, and this factor also adversely affects the working conditions of the lining [13, 14].

Thus, an approximate calculation of the lining can be recommended according to the following scheme:

1. determination of stresses in the lining arising in connection with the bending of the beam, σ'';
2. determination of stresses in the lining arising in connection with its deflections on the beam (solid elastic base), σ';
3. determination of total stresses $\Sigma\sigma$.

In this case, in the calculation scheme, it is necessary to take into account the variability of the cross section of the lining with the cushion, and the load from the rail on the lining should be taken in the form of concentrated force (third design scheme).

When calculating these linings, we took the dynamics coefficients for: locomotive—1.45; coaches—1.3. Permissible stresses for a lining with a cushion are equal to the yield strength of steel grade St.3 excluding safety factor—2200 kg/cm^2.

From the graphs, for static axle loads up to 23 t, the thickness of the lining should be about 24 mm. It should be noted that in the calculation a slightly lower value of the dynamics coefficients and an overestimated value of the permissible stresses were taken.

The experience of landfill tests of rail switches showed that part of the metal switch chairs breaks down along the transverse fractures arising from the high level of bending stresses. So, in the P65 railroad switch on reinforced concrete bars, after passing 760 million tons of gross cargo, 28% of switch chairs were broken. Since when replacing the main metal elements of the translation, the broken linings are also replaced, it is necessary to know in advance the number of linings to be replaced [15, 16].

The results of dynamic tests of rail switches on reinforced concrete beams showed that the main operational factor determining the stress state of the linings is the axial loads of the rolling stock. In this regard, the calculations of the failure of the pads were carried out for various axial loads (up to 250 kN). The initial data for the calculation were the results of the fatigue tests of the linings with a cushion and the data of dynamic tests of the switch chairs of the rail switches on reinforced concrete beams. The calculation model is built taking into account the work of the linings in transit [17, 18].

When attaching the linings with cushion to the switch beam tightening tensile stresses appear in the upper edge of the lining σ_t. When the rolling stock passes through the transfer, the cyclic stresses from the train load are superimposed on the tightening tensile stress σ_c of the opposite sign. General stresses at the moment of passage of the i-th wheel above the lining σ_i are defined as the algebraic sum of stresses from tightening of embedded bolts and train load:

$$\sigma_i = \sigma_t - \sigma_{ci} \tag{3}$$

Thus, in the cycle of passage of the wheel over the switch chair, the stresses vary from σ_t to σ_i. The asymmetry coefficient of the cycle is determined as $r_i = \sigma_i/\sigma_t$, or taking into account the (3)

$$r_i = 1 - \sigma_{ci}/\sigma_c \tag{4}$$

The number of loading cycles that the lining with the pillow can withstand before breaking can be determined from the fatigue relation

$$N_i = \begin{cases} (\sigma_{ri}/\sigma_i)^m . N_0 & at\, \sigma_i \geq \sigma_{ri}\,; \\ \infty & at\, \sigma_i < \sigma_{ri}\,, \end{cases} \tag{5}$$

where σ_{ri} - the longevity limit of the lining at a given asymmetry of the loading cycle;

N_0 - test base;
m = 2.35 - the fatigue curve indicator obtained when testing switch chairs in laboratory conditions.

Taking into account that σr can be expressed through the limit of long endurance with a symmetrical cycle σ − 1 using the cycle asymmetry coefficient r:

$$\sigma_r = \sqrt{2/(1-r)\sigma_{-1}} \tag{6}$$

and substituting formulas (4) and (6) into (5), we finally obtain for damaging cycles:

$$N_i = \left[\sqrt{2\sigma_t/\sigma_{ci}/(\sigma_t - \sigma_{ci})}\right]^m (\sigma_{-1})^m N_0, \tag{7}$$

If material properties are known σ_{-1} and the results of fatigue tests are m and N0, then, using expression (7), it is possible to determine the number of cycles until the linings are broken, depending on the initial stress arising from bolt tightening, and the change in stress under the train, which is directly recorded by the sensor in transit. When the train load passes over the lining, various stress cycles can occur in it, so the total number of cycles before the break of the lining must be determined taking into account the accumulation of damage for each cycle [19, 20].

According to the theory of summation of damage the expression [1] can be written for the total number of cycles n_0 before the lining is broken:

$$n_0 = 1/\int_{\sigma_{dam}} \frac{dn_0}{N}, \tag{8}$$

where σ_{dam} - damaging stresses.

Integration is carried out for all cycles, where tensions σ_{dam} are damaging.

Differential dn0, representing the probability density of the occurrence of stresses of a certain level for the period before the break, can be determined using the probability density of stresses from the train load $P(\sigma_c)$, obtained experimentally by shifting the tension values by σ_t:

$$dn_0 = -P(\sigma_t - \sigma_c)d\sigma_c \tag{9}$$

Substituting formulas (9) and (7) into expression (8) and taking into account that the metal properties do not depend on the train load, we obtain:

$$n_0 = (\sigma_{-1})^m N_0/\int_{\sigma dam} (0.5\sigma_t\sigma_c - \sigma_c^2 + 0.5\sigma_c^3/\sigma_t) \times P(\sigma_t - \sigma_c)d\sigma_c \tag{10}$$

If all cycles are damaging (for example, taking into account corrosion and other factors that reduce durability), then the second factor in expression (10) does not depend on the properties of the material. Marking it by α, we get:

$$n_0 = \alpha(\sigma_{-1})^m \tag{11}$$

In this case, the number of cycles that the lining can withstand before breaking is completely determined by the long-term endurance limit σ_{-1} of the linings. Knowing the distribution of the long-life endurance limit of the linings, one can find the distribution of the running time of the linings before breaking, using the usual methods of transforming random variables [21–23]:

$$P[n_0] = P[\sigma_{-1}(n_0)] \mid J \mid^{-1}, \tag{12}$$

where P $[n_0]$ - the desired density of the distribution of the number of cycles to break the linings;

P $[\sigma_{-1}(n_0)]$—distribution density σ_{-1} of linings for function $\sigma_{-1}(n_0)$;
$\mid J \mid$—Jacobian of the conversion.

It is known that the long-term endurance limit of standard carbon steels has a distribution close to normal. With this in mind, by expressing $\sigma_{-1}(n_0)$ from (11) and substituting all the quantities in (12), we finally obtain:

$$P[n_0] = \frac{1}{\sqrt{2\pi S_{\sigma_{-1}}}} \text{x } exp\left\{ -\frac{1}{2}\left[\frac{(n_0/\alpha)^{1/m} - \overline{\sigma_{-1}}}{S_{\sigma_{-1}}}\right]^2 \right\} \frac{1}{m\alpha^{1/m} n_0^{1-1/m}}, \tag{13}$$

where $\overline{\sigma}_{-1}$ - average lining long-term endurance limit;

$S\sigma_{-1}$ - standard deviation σ_{-1}.

By substitution in (13) various values of the number of cycles to the break of the linings, we can obtain the probability density corresponding to them, i.e. build distribution for n_0.

Integrating expression (13) from 0 to n_0, we determine the number of linings (their share of the total) that will be broken by kinks after the rail switch passage $n0P_{ax}$ mln tons of gross cargo (at this point P_{ax} – is an axial load).

If not all cycles are damaging, then the expression for P $[n_0]$ is obtained cumbersome, since the integral in (10) is not taken in closed form. In this case, it is advisable to carry out the calculation according to the formula (10) by numerical methods, since the calculation according to the formula (13) will lead to an overestimation of the probability of the linings failure (calculation to the stock).

The constructed probabilistic model allows using formulas (10) and (13) to calculate the distribution of lining failures depending on the operating time, taking into account the properties of the metal, the initial tightening of the embedded bolts and a given stress spectrum from the moving units. These formulas were the basis for further calculation. To use them, it is necessary to know the characteristics of the distribution of the longevity limit of the lining $\overline{\sigma}_{-1}$ and $S\sigma_{-1}$.

Switch chairs for rail switches on reinforced concrete bars are made of steel St3. According to GOST 380-60 for laboratory samples, the tensile strength of this steel is σ_{ts} = 380–470 MPa. It is known that for samples there is a dependence

$$^{-}\sigma_{-1}^{sam} = (0.55 - 0{,}0001\sigma_{ts})\sigma_{ts}, \tag{14}$$

where $^{-}\sigma_{-1}^{sam}$—average long-term endurance of samples.

When passing from laboratory samples to real details, the long term endurance limit changes k times. According to the data for the linings of St3 k = 0.67. Substituting the values $\overline{\sigma}_{ts}$ and k to (14), we get $\overline{\sigma}_{-1}$ = 130 MPa.

For machine parts, the coefficients of variation of the endurance limits vary in the range 0.05–0.15. Taking the coefficient of variation equal to 0.10, we get $S\sigma_{-1}$ = 13 MPa.

Due to the fact that data for σ_{-1} $S\sigma_{-1}$ were taken on the basis of dependencies generalized for machine parts, before performing basic calculations, a verification calculation was made for operating conditions in which the normal stress distribution was accepted: $\overline{\sigma}_c$ = 180 MPa; $S\sigma_c$ = 22 MPa; P_{ax} = 210 kN. The calculation results are shown in Table 1.

Table 1. The calculation results.

σ_c, MPa	r	σ_r, MPa	$N_i \cdot 10^{-4}$ cycles	Probability P
0	+1.0	$>\sigma_c$	∞	–
180	−0.16	$>\sigma_c$	∞	–
195	−0.26	155	∞	–
201	−0.30	152	9.7	0.17519
223	−0.44	144	8.7	0.06066
244	−0.58	138	7.9	$59774 \cdot 10^{-6}$
266	−0.72	132	7.3	$2292 \cdot 10^{-7}$
287	−0.85	127	6.9	$3385 \cdot 10^{-10}$
310	−1.00	122*	6.5	$45 \cdot 10^{-11}$

* σ_{-1} is chosen with probability ($\overline{\sigma}_{-1} - 0.58S\sigma_{-1}$), corresponding to 28% of failed linings.

In the verification and main calculations σ_t = 155 MPa was accepted, which corresponds to the tightening torque of bolts equal to 160 Nm.

Since after passing of 750 million tons gross 28% of the switch chairs failed, the check consisted in determining the estimated operating time until 28% of the linings failed.

Using the data Table 1 and formula (10), we obtain n_0 = 38.4 million of cycles, which, with an axial load of 210 kN, corresponds to passage of 806 million tons of gross. Thus, the difference is 6.1%, i.e., the accuracy is quite satisfactory for fatigue calculations.

A similar check was carried out for the operation of the rail switch with switch chairs under the lining of rubber 6 mm thick. As a result of the calculation, it was established that the failures of switch linings under average network axial loads and the indicated conditions should not be observed. This coincided with the results of the pilot operation of switches on reinforced concrete bars with 6 mm linings on the network.

It should be noted that with a thickness of rubber linings of 6 mm, an increased yield of the main metal parts of rail switches on reinforced concrete bars is observed, therefore, in modern designs of such switches, linings with a thickness of 10 mm are used. The main calculation was carried out in a similar way. The stress data σc for various axial loads are given in Table 2.

Values given in the Table 2 are taken from the test results of the P65 type switch, as the weakest construction of switches on reinforced concrete bars.

Table 2. Values of stresses for various axial loads.

Indicators	The magnitude of the stress, MPa, under axial load P_{ax}, kN			
	56	140	210	250
$\overline{\sigma}_c$	60	120	165	190
$S\ \sigma_c$	80	26	30	32

The graphs of failures of switch chairs, obtained using the developed model, in transit at various axial loads make it possible to determine how many linings with a cushion will need to be replaced after skipping a certain amount of cargo by the switch (up to a passage of 1200 million tons gross). With axial loads up to 120 kN, the fatigue life of the linings is so that they should not fail practically.

The output of switch linings with an average network axial load of 140 kN is small and at the time of the first change of metal parts (after passing about 300 million tons of gross cargo) it is about 2–3%.

At the time of the second change of metal parts (after passing about 600 million tons of gross), it will be necessary to replace 4–5% of the linings with a cushion. With increased axial loads, the yield of the linings along the kinks increases sharply. So, with an axial load of 250 kN, about 16% of the switch linings will have to be replaced already at the first change of metal parts, and about 39% at the second, i.e. almost for a half.

3 Conclusions

1. The built probabilistic model of the operation of switch chairs on reinforced concrete bars is in a good agreement with the results of the trial operation of such switches, and expressions (10) and (13) allow us to calculate the failure distribution of switch linings depending on the characteristics of the metal from which they are made, the initial tightening of the fasteners and the stress spectrum arising from the train load.
2. The graphs of failures of switch chairs in transit at various axial loads, obtained using the developed model, make it possible to plan the replacement of switch chairs on switches with reinforced concrete beams when changing the main metal parts.

3. With an average network axial load of 140 kN, at the time of the first change of metal parts, it is necessary to prepare a replacement of 2–3% of the linings with a cushion. At the increased axial loads of 250 kN, the number of replaceable switch linings with a cushion should be about 16%.

References

1. Glusberg, B., Korolev, V., Shishkina, I., Loktev, A., Shukurov, J., Geluh, P., Loktev, D.: Calculation of track component failure caused by the most dangerous defects on change of their design and operational conditions. In: MATEC Web of Conferences, vol. 239 (2018). https://doi.org/10.1051/matecconf/201823901054
2. Savin, A., Kogan, A., Loktev, A., Korolev, V.: Evaluation of the service life of non-ballast track based on calculation and test. Int. J. Innov. Technol. Exploring Eng. **8**(7), 2325–2328 (2019)
3. Loktev, A.A., Korolev, V.V., Poddaeva, O.I., Chernikov, I Y.U.: Mathematical modeling of antenna-mast structures with aerodynamic effects. In: IOP Conference Series: Materials Science and Engineering, vol. 463 (2018). https://doi.org/10.1088/1757-899x/463/3/032018
4. Loktev, A., Korolev, V., Shishkina, I., Chernova, L., Geluh, P., Savin, A., Loktev, D.: Modeling of railway track sections on approaches to constructive works and selection of track parameters for its normal functioning. In: Advances in Intelligent Systems and Computing, vol. 982, pp. 325-336 (2020). https://doi.org/10.1007/978-3-030-19756-8_30
5. Gluzberg, B., Korolev, V., Loktev, A., Shishkina, I., Berezovsky, M.: Switch operation safety. In: E3S Web of Conferences, vol. 138 (2019). https://doi.org/10.1051/e3sconf/201913801017
6. Loktev, A.A., Korolev, V.V., Poddaeva, O.I., Stepanov, K.D., Chernikov, I.Y.: Mathematical modeling of aerodynamic behavior of antenna-mast structures when designing communication on railway transport. Vestnik Railway Res. Inst. **77**(2), 77–83 (2018). https://doi.org/10.21780/2223-9731-2018-77-2-77-83. (in Russian)
7. Loktev, A., Korolev, V., Shishkina, I., Illarionova, L., Loktev, D., Gridasova, E.: Perspective constructions of bridge crossings on transport lines. In: Advances in Intelligent Systems and Computing, vol. 1116, pp. 209-218 (2020). https://doi.org/10.1007/978-3-030-37919-3_20
8. Loktev, A.A., Korolev, V.V., Loktev, D.A., Shukyurov, D.R., Gelyukh, P.A., Shishkina, I. V.: Perspective constructions of bridge overpasses on transport main lines. Vestnik Railway Res. Inst. **77**(6), 331–336 (2018) https://doi.org/10.21780/2223-9731-2018-77-6-331-336. (in Russian)
9. Loktev, A.A., Korolev, V.V., Shishkina, I.V.: High frequency vibrations in the elements of the rolling stock on the railway bridges. In: IOP Conference Series: Materials Science and Engineering, vol. 463 (2018). https://doi.org/10.1088/1757-899x/463/3/032019
10. Loktev, A.A., Korolev, V.V., Gridasova, E.A.: Influence of high-frequency cyclic loading on mechanical and structural characteristics of rail steel under extreme conditions. In: IOP Conference Series: Materials Science and Engineering, vol. 687 (2019). https://doi.org/10.1088/1757-899x/687/2/022036
11. Glusberg, B., Savin, A., Loktev, A., Korolev, V., Shishkina, I., Alexandrova, D., Loktev, D.: New lining with cushion for energy efficient railway turnouts. In: Advances in Intelligent Systems and Computing, vol. 982, pp. 556-570 (2020). https://doi.org/10.1007/978-3-030-19756-8_53

12. Korolev, V.: Guard rail operation of lateral path of railroad switch. In Advances in Intelligent Systems and Computing, vol. 1115, pp. 621-638 (2020). https://doi.org/10.1007/978-3-030-37916-2_60
13. Glusberg, B., Loktev, A., Korolev, V., Shishkina, I., Alexandrova, D., Koloskov, D.: Calculation of heat distribution of electric heating systems for turnouts. In: Advances in Intelligent Systems and Computing, vol. 982, pp. 337-345 (2020). https://doi.org/10.1007/978-3-030-19756-8_31
14. Lyudagovsky, A., Loktev, A., Korolev, V., Shishkina, I., Alexandrova, D., Geluh, P., Loktev, D.: Energy efficiency of temperature distribution in electromagnetic welding of rolling stock parts. In: E3S Web of Conferences, vol. 110 (2019). https://doi.org/10.1051/e3sconf/201911001017
15. Savin, A.V., Korolev, V.V., Shishkina, I.V.: Determining service life of non-ballast track based on calculation and test. In: IOP Conference Series: Materials Science and Engineering, vol. 687 (2019). https://doi.org/10.1088/1757-899x/687/2/022035
16. Korolev, V.: Switching shunters on a slab base. In: Advances in Intelligent Systems and Computing, vol. 1116, pp. 175-187 (2020). https://doi.org/10.1007/978-3-030-37919-3_17
17. Savin, A., Suslov, O., Korolev, V., Loktev, A., Shishkina, I.: Stability of the continuous welded rail on transition sections. In: Advances in Intelligent Systems and Computing, vol. 1115, pp. 648-654 (2020). https://doi.org/10.1007/978-3-030-37916-2_62
18. Savin, A., Korolev, V., Loktev, A., Shishkina, I.: Vertical sediment of a ballastless track. In: Advances in Intelligent Systems and Computing, vol. 1115, pp. 797-808 (2020). https://doi.org/10.1007/978-3-030-37916-2_78
19. Gridasova, E., Nikiforov, P., Loktev, A., Korolev, V., Shishkina, I.: Changes in the structure of rail steel under high-frequency loading. In: Advances in Intelligent Systems and Computing, vol. 1115, pp. 559-569 (2020). https://doi.org/10.1007/978-3-030-37916-2_54
20. Shishkina, I.: Determination of contact-fatigue of the crosspiece metal. In: Advances in Intelligent Systems and Computing, vol. 1115, pp. 834-844 (2020). https://doi.org/10.1007/978-3-030-37916-2_82
21. Korolev, V., Loktev, A., Shishkina, I., Zapolnova, E., Kuskov, V., Basovsky, D., Aktisova, O.: Technology of crushed stone ballast cleaning. In: IOP Conference Series: Earth and Environmental Science, vol. 403 (2019). https://doi.org/10.1088/1755-1315/403/1/012194
22. Glusberg, B., Savin, A., Loktev, A., Korolev, V., Shishkina, I., Alexandrova, D., Loktev, D.: New lining with cushion for energy efficient railway turnouts. In: Advances in Intelligent Systems and Computing, vol. 982, pp. 556-570 (2020). https://doi.org/10.1007/978-3-030-19756-8_53
23. Loktev, A.A., Korolev, V.V., Shishkina, I.V., Basovsky, D.A.: Modeling the dynamic behavior of the upper structure of the railway track. Procedia Eng. **189**, 133–137 (2017). https://doi.org/10.1016/j.proeng.2017.05.022

Wear Peculiarities of Point Frogs

Irina Shishkina(✉)

Russian University of Transport (MIIT), Chasovaya str. 22/2,
125190 Moscow, Russia
shishkinaira@inbox.ru

Abstract. According to the study, operation of the point frogs was divided into four stages based on their wear conditions. First stage is the settlement of cast part relative to the counter-rails due to the selection of joint backlash. Next is crushing of the metal without (second stage) and with a change in the position of the regulated wear points (third stage) and galling without crushing (fourth steady-state wear stage). It was established that the wear rate is determined mainly by the level of dynamic forces at the stage of steady-state wear, in the contact of the wheels and the frog. The intensity of wear in frog's cross-section is determined under various operating conditions. The geometry of the frogs at the steady-state wear stage is determined as well. Formulas allowing determining the specific wear of the frogs at different axial loads and speeds are obtained as a function of dynamic forces in the contact of the wheels and frogs. It was found that at the crushing stage, the wear rate is mostly determined by the initial hardness of the metal and its change during the frog's operation. It has been established that the loss of the regulated height of the frog's elements due to metal's crushing—hardening stage is about 42% in all sections of the core, and about 56% (12–20 mm) in the sections of the counter-rail. The rest height of the elements (58% and 44%, respectively) is lost at the crashing stage as a result of galling.

Keywords: Point frog · Wear · Four operating stages · Dynamic forces · Hardness of metal · Crushing of metal

1 Introduction

The first main process forming the regulated wear of the frog is the settlement of the cast part due to selecting the joint backlashes for the rail counter-rails. The second process is the loss of the metal from the initial contour cross section as a result of galling by wheels and metal slipout due to plastic deformations determined by insufficiently high yield point values for 110G13L steel [1, 2].

2 Materials and Methods

The frog's railway operation be divided into four main stages. At the first stage, the construction properties and the frog's manufacturing quality determine the possibility of mutual movements of the counter-rails and the cast part. According to the

V. Murgul and V. Pukhkal (Eds.): EMMFT 2019, AISC 1258, pp. 197–206, 2021.
https://doi.org/10.1007/978-3-030-57450-5_18

measurement results on railways and the experimental ring, the settlement of the cast part in modern designs of typical crosses is on average 0.5 mm in the core and 0.8 mm in the most intensively operating counter-drail. The main settlement part occurs under the first trains. After passing of 3–5 million tons of gross cargo along the frog, the settlement is realized almost completely [3, 4].

After settlement, the crushing process begins (second and third stages). Crushing in the rolling zone begins at the junctions of the working surfaces with the lateral ones. Therefore, the position of the regulated wear points remains unchanged at the beginning (second stage) [5, 6]. After the crushing process reaches the points where wear is regulated, their position also starts changing (the third stage). During the second and third stages of wear, there is an intense hardening of the metal, accompanied by large plastic deformations.

Plastic deformations decrease as the frog is riveted [7, 8]. The galling part starts to prevail in wear, i.e., the frog switches to the fourth stage of wear, which is characterized by a linear correlation between the regulated height loss by the operating frog's surfaces running time.

To date, there is no calculation method that takes into account all the features of the wear process. Therefore, the dependences obtained from operation and the test results for galling of the friction pair "steel wheel rim - 110G13L steel" were used as the basic ones.

The dimensionless characteristic is used for calculating the wear of machine elements is the wear rate [9, 10]. The wear rate is the loss of height on a single friction path and is expressed by the following formula:

$$\mathrm{fh}I_h = \Delta H / L_{fr} \tag{1}$$

where ΔH is loss of height of the abraded element on the friction path L_{fr} (the amount of wheel sliding along the frog when rolling over a given section).

By using a variation of this formula for the case when the lost mass of the abraded element and its density are known, we obtain as follows $I_h = 4.52 \cdot 10^{-8}$.

The wear rate is a universal characteristic of the wear for a given material in a given friction pair and characterizes its operation under any contact conditions. Therefore, the value I_h obtained during the tests can be used for the case of operation of frogs in transit, taking into account changes in contact parameters (forces and geometry) [11, 12].

As a result of a large number of tests for various materials, it was determined that the wear rate has a nonlinear dependence on contact pressure [2]:

$$I_h = c \cdot p^{\alpha} \tag{2}$$

where c is the coefficient taking into account contact geometry and elastic properties of contacting bodies; p is the contact pressure; α is an exponent characterizing the effect of contact pressure on wear rate.

At the fourth stage, the correlation between the frog wear and the run time is linear. The wear rate is constant and is characterized by specific wear, which is the ratio of the regulated wear increment to the run time [13, 14]. The Department of Transport Construction in RUT (MIIT) has accumulated a large amount of data on the wear of frog in various operating conditions [15–17]. Table 1 represents the data on the specific wear dh/dN of the frogs after finished crushing at various axial loads, which were recounted from [3].

Table 1. Specific wear of the frogs after the finished crushing at various axial loads.

Axial load, kN	Specific wear, mm/mil cycles	
	of counte-rail in the most worn out place	of 40 mm core
75	0.375	0.278
100	0.500	0.350
150	0.570	0.510
170	–	0.476
200	0.620	0.460

Specific wear is almost the average wear of the entire frog's cross section in one wheel passage. Passing along the frog, the wheels come into contact with it at various points [18, 19]. The actual metal loss for each wheel at each contact is localized on the width of the contact pad, which is only a part of the working surface. The metal loss in one wheel pass from the contact pad can be expressed in terms of specific wear as follows:

$$h_1 = \frac{dh}{dN}\frac{L}{2b} \tag{3}$$

where L - the operating part width of the frog's surface; 2b—average width of the contact pad.

According to the definition of wear rate from formula (3), the following could be obtained:

$$I_h = \frac{h_1}{L_{mp}} = \frac{1}{L_{mp}}\frac{dh}{dN}\frac{L}{2b} \tag{4}$$

The slipping value can be defined as a function of the coefficient of specific sliding and the length of the contact ellipse: $L_{fr} = 2\alpha\delta$, where $2\alpha\delta$ is the length of the contact area in the movement direction. δ is the specific sliding coefficient, depending on the geometry of the contact and the elastic constant elements.

The width of the contact pad can be expressed via length using the eccentricity of the contact ellipse e: $b = a\sqrt{1 - e^2}$.

After substituting the expressions for L_{fr} and b in Eq. (4), the following can be obtained:

$$I_h = \frac{dh}{dN}\frac{L}{4a^2\sqrt{1 - e^2}\delta} \tag{5}$$

The eccentricity of the contact ellipse does not depend on the contact force. Therefore, for two different axial loads, the ratio of wear intensities has the following form [20, 21]:

$$\frac{I_{h(p1)}}{I_{h(p2)}} = \frac{\left({}^{dh}/_{dN}\right)_1 \alpha_2^2}{\left({}^{dh}/_{dN}\right)_2 \alpha_1^2} \quad (6)$$

The change in contact pressure leads to a change in the size of the contact pads, proportional to the cubic root of the dynamic contact force value P_d:

$$ {}^{a_2}/_{a_1} = \sqrt[3]{P_{d2}/P_{d1}} \quad (7)$$

Taking into account that in accordance with the theory of contact stresses, the highest contact pressure is also proportional to the cubic root of the contact force. After substituting formulas (7) and (2) into Eq. (6) and, the following can be obtained:

$$\frac{\left({}^{dh}/_{dN}\right)_1}{\left({}^{dh}/_{dN}\right)_2} = \left(\frac{P_{d1}}{P_{d2}}\right)^{\frac{1}{3}(\alpha+2)} \quad (8)$$

Equation (8) allows calculating the exponent α, which determines the correlation between the wear rate and the dynamic forces acting on the frog. The operational wear rates for the frogs have a spread determined by the specific operating conditions, so the coefficient α was determined by all possible combinations of axial loads, for which wear rate were obtained in Table 1. 16 calculations were carried out in total. With that, dynamic additives of contact forces were taken from calculations performed on a computer. The average value of the α coefficient is 2.85.

Since the average value of the wear rate is 2/3 of the intensity corresponding to the highest pressure on the contact area, Eq. (2) can be exposed in the following form:

$$I_h^{av} = \frac{2}{3}cp^{\alpha} \quad (9)$$

After substituting the test results on the Amsler machine into this equation, $c = 4.15 \cdot 10^{-13}$ is obtained. Thus, the final correlation between the wear rate and the highest pressure in the contact has the following form

$$I_h = 4.15 \cdot 10^{-13} p^{2.85} \quad (10)$$

The wear theory of elements of higher pairs, to which a frog can be attributed, was developed with reference to the calculating the wear of cam control gears [22, 23]. In accordance with this theory, a formula is derived for determining the surface wear of a pair element in one loading cycle:

$$h_1 = \frac{2}{3} c p^{\alpha} \alpha \delta \tag{11}$$

In accordance with the theory of contact stresses:

$$\alpha = m_{\alpha} \sqrt[3]{P_d/(2AE)}; \quad p = m_p \sqrt[3]{P_d A^2 E^2}, \tag{12}$$

where m_a and m_p are coefficients depending on the geometry of contact and the elastic characteristics of the contacting bodies; A is a parameter depending on the geometry of the contacting bodies; E is elasticity modulus of 110G13L steel. The following is obtained after substituting relations (12) into Eq. (11) and giving numerical coefficients:

$$h_1 = 4.51 \cdot 10^{-6} m_p^{2.85} m_{\alpha} P_d^{1.28} A^{1.57} \delta \tag{13}$$

As per the Eq. (13), the metal loss from the frog's surface by each passing wheel is determined by the geometric characteristics of the contact and the magnitude of the contact force.

Calculating the metal loss from the frog's surface by each particular wheel is not of great practical interest, but the correlations for the specific wear of the frogs can be obtained using Eq. (13). By substituting the most probable values of the contact geometrical parameters into Eq. (13) and considering Eqs. (3) and (10), the correlations could be obtained for the 40 mm core and the counter-rail with the cross section of 12–20 mm:

$$\left(\frac{dh}{dN}\right)_c = 0.0250 \left(\frac{P_{ax}}{2} + \Delta P_d\right)^{1.61}$$

$$\left(\frac{dh}{dN}\right)_y = 0.0356 \left(\frac{P_{ax}}{2} + \Delta P_d\right)^{1.61} \tag{14}$$

where P_{ax} is the axial load, kN; ΔP_d—dynamic addition of contact forces, kN.

Formulas (14) allow determining the frog's specific wear under different operating conditions at various axial loads, speeds, various rail tracks. In the case when it is necessary to assess the propositions, which imply changes in the frog's geometry, Eq. (13) should be used.

Let us consider the height loss of the working surfaces due to crushing. This part of the frog's regulated wear is due to the fact that the impacts exerted by the rolling stock wheels exceed the maximum permissible regarding the condition of transition through the first limit state (from elastic to plastic deformations). In order to study the frog's operation in detail at these wear stages on the experimental ring, measurements of the frogs of types P50 and P65 of grade 1/11 were carried out. In addition to the standard measurement methodology, cross-sectional profiles were measured as well. As a result, the parts of the cross-sectional areas removed from the original cross-sectional profile were determined using an electronic device measuring the area. The areas that were outside the original contour were determined as well. Moreover, sectional areas were defined, by which the original sectional area was reduced.

The metal areas displaced beyond the initial contour correspond to wear due to crushing, and the decrease in the initial cross-sectional area corresponds to wear due to the metal loss caused by passing wheels (galling). The graphs of the correlations of wear and run time are presented in Fig. 1.

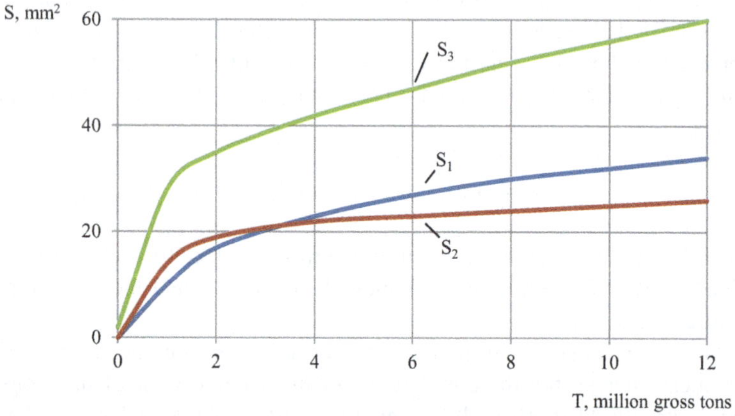

Fig. 1. Depreciation of the frog counter-rail's cross-section at the stage of predominant crushing (12 mm cross-section): S1 is worn-out area; S2 is the area remaining outside the primary contour; S3 is the area lost from the initial contour.

Analyzing the measurement results showed that for the rolling area, the parts of the worn-out metal and the one removed by crushing significantly depend not only on the operating time, but also on the initial geometry of the frog. The duration of the second wear stage, i.e. the stage at which the magnitude of the regulated height loss of the frog's cross-sectional element remains unchanged, also depends on the initial geometry. The analysis of the measurement results showed that the second stage is practically absent for the counter-rails in the sections, according to which their work is usually evaluated. For the core with the 40 mm section, the average duration of the second stage corresponds to the run time of 1.2 million tons gross.

The total duration of the crushing stage varies due to differences in the geometry of specific frogs. Relatively stable results are obtained only for sections, in which all wheels are in contact with only one of the frog's elements, either with the core or with the counter-rail.

Let us consider the data presented in Table 2 regarding the metal loss due to crushing and galling in the second and third stages. It can be seen that for all sections of the rolling area, the loss of core height at crushing stage is about 42% of the total loss of area and section height. On the counter-rail the height loss is 56% as a result of crushing in sections, in which the wear is limiting (12–20 mm sections). The data obtained indicate that even under such conditions as an experimental ring, which are severe for the crushing stage, about half the height loss of the frog's elements at this stage occurs due to metal galling. Metal's fractional losses can be easily obtained from the data in Table 2.

Table 2. Loss of the regulated height of the frog's elements due to crushing of metal.

Core's cross-sections	Loss of the regulated height of the frog's elements, %, due to crushing of metal	
	Core	Counter-rail
MCK	–	0.63
12 mm	–	0.56
20 mm	0.41	0.56
30 mm	0.43	0.43
40 mm	0.42	–

Let us consider the galling of the frog's metal in the third wear stage. According to current theoretical concepts of wear during plastic contact, the wear rate in this case depends on the contact pressure. This is the complex characterizing the physical and mechanical properties of the material, and the microgeometry of the contact surface. Since the microgeometry of the contacting bodies remains unchanged at the third wear stage of, the expression for the wear rate can be converted as follows:

$$I_h = k(p/HB)^{\frac{1+\beta t_1}{1-\beta}}\delta_0 - t_1 \tag{15}$$

where k is coefficient taking into account contact microgeometry and frictional properties of wearing bodies; t_1 is low-cycle fatigue curve parameter; δ_0 is the elongation at break; β is the coefficient depending on material properties.

Taking into account that there is a correlation between the values of δ and HB for steel 110GCHZL, the dependence (15) can be rewritten as follows:

$$I_h = mp^{\frac{1+\beta t_1}{1-\beta}} HB^{\frac{1+2t_1\beta-t_1}{1-\beta}} \tag{16}$$

Coefficient β is determined according to data obtained for the fourth wear stage, when the elastic contact $\alpha = 1 + \beta t$ takes place. Here t is the parameter of the high-cycle bulk fatigue curve.

For 110G13L steel, the indicator t ranges between 6.5—8.6 according to the results of bending tests. T the average value $t_{av} = 7.6$, then $\beta = (\alpha - 1)/t = 0.11$.

Thus, m и t_1 remain unknown in Eq. (16). According to the data in available researches, the values of the t_1 coefficient vary from 2 to 3 for different materials. After substituting β and t_1 in Eq. (16), the exponent at contact pressure will be from 1.37 to 1.48 at extreme t_1 values. Due to the fact that the contact pressure depends on the dynamic force to the degree of 1/3, the exponent at R_d will fluctuate at various possible values of only 0.46–0.49 range, depending on the wear rate.

The exponent for HB may vary from—0.79 to—1.59. This shows that at the crushing stage the wear rate depends on the dynamic forces in the contact to a much lesser extent than at the steady wear stage, and is mainly determined by the initial hardness of the metal and its change.

3 Conclusions

1. It is advisable to divide the work of the frogs on the railway into four stages according to their wear conditions. First stage is the settlement of cast part relative to the counter-rails due to the selection of joint backlash. Next is crushing of the metal without (second stage) and with a change in the position of the regulated wear points (third stage) and galling without crushing (fourth steady-state wear stage). At the stage of steady wear, the wear rate is determined mainly by the level of dynamic forces in the contact of the wheels and the cross.
2. At the steady wear stage, the wear rate is determined mainly by the level of dynamic forces in the contact of the wheels and the frog.
3. The wear rate in the cross sections under different operating conditions and various geometry of frogs at the steady wear stage can be quantified based on correlation (13). In particular, it was used to obtain formulas (14), which allow determining the specific wear of the frogs at different axial loads and speeds, as a function of the dynamic forces in the contact of the wheels and frogs.
4. At the crushing stage, the wear rate is mostly determined by the initial hardness of the metal and its change during the operation of the frog.
5. It has been established that the loss of the regulated height of the frog's elements due to metal's crushing at the crushing—hardening stage is about 42% in all sections of the core, and about 56% (12–20 mm) in the sections of the counter-rail. The rest height of the elements (58% and 44%, respectively) is lost at the crashing stage as a result of galling.

References

1. Shishkina, I.: Determination of contact-fatigue of the crosspiece metal. In: Advances in Intelligent Systems and Computing, vol. 1115, pp. 834–844 (2020). https://doi.org/10.1007/978-3-030-37916-2_82
2. Gridasova, E., Nikiforov, P., Loktev, A., Korolev, V., Shishkina, I.: Changes in the structure of rail steel under high-frequency loading. In: Advances in Intelligent Systems and Computing, vol. 1115, pp. 559–569 (2020). https://doi.org/10.1007/978-3-030-37916-2_54
3. Loktev, A.A., Korolev, V.V., Gridasova, E.A.: Influence of high-frequency cyclic loading on mechanical and structural characteristics of rail steel under extreme conditions. In: IOP Conference Series: Materials Science and Engineering, vol. 687 (2019). https://doi.org/10.1088/1757-899x/687/2/022036

4. Korolev, V.: Guard rail operation of lateral path of railroad switch. In: Advances in Intelligent Systems and Computing, vol. 1115, pp. 621–638 (2020). https://doi.org/10.1007/978-3-030-37916-2_60
5. Loktev, A., Korolev, V., Shishkina, I., Chernova, L., Geluh, P., Savin, A., Loktev, D.: Modeling of railway track sections on approaches to constructive works and selection of track parameters for its normal functioning. In: Advances in Intelligent Systems and Computing, vol. 982, pp. 325–336 (2020). https://doi.org/10.1007/978-3-030-19756-8_30
6. Savin, A., Kogan, A., Loktev, A., Korolev, V.: Evaluation of the service life of non-ballast track based on calculation and test. Int. J. Innov. Technol. Exploring Eng. **8**(7), 2325–2328 (2019)
7. Glusberg, B., Loktev, A., Korolev, V., Shishkina, I., Alexandrova, D., Koloskov, D.: Calculation of heat distribution of electric heating systems for turnouts. In: Advances in Intelligent Systems and Computing, vol. 982, pp. 337–345 (2020). https://doi.org/10.1007/978-3-030-19756-8_31
8. Lyudagovsky, A., Loktev, A., Korolev, V., Shishkina, I., Alexandrova, D., Geluh, P., Loktev, D.: Energy efficiency of temperature distribution in electromagnetic welding of rolling stock parts. In: E3S Web of Conferences, vol. 110 (2019). https://doi.org/10.1051/e3sconf/201911001017
9. Korolev, V., Loktev, A., Shishkina, I., Zapolnova, E., Kuskov, V., Basovsky, D., Aktisova, O.: Technology of crushed stone ballast cleaning. In: IOP Conference Series: Earth and Environmental Science, vol. 403 (2019). https://doi.org/10.1088/1755-1315/403/1/012194
10. Savin, A.V., Korolev, V.V., Shishkina, I.V.: Determining service life of non-ballast track based on calculation and test. In: IOP Conference Series: Materials Science and Engineering, vol. 687 (2019). https://doi.org/10.1088/1757-899x/687/2/022035
11. Loktev, A.A., Korolev, V.V., Shishkina, I.V., Basovsky, D.A.: Modeling the dynamic behavior of the upper structure of the railway track. Procedia Eng. **189**, 133–137 (2017). https://doi.org/10.1016/j.proeng.2017.05.022
12. Glusberg, B., Korolev, V., Shishkina, I., Loktev, A., Shukurov, J., Geluh, P., Loktev, D.: Calculation of track component failure caused by the most dangerous defects on change of their design and operational conditions. In: MATEC Web of Conferences, vol. 239 (2018). https://doi.org/10.1051/matecconf/201823901054
13. Loktev, A.A., Korolev, V.V., Shishkina, I.V.: High frequency vibrations in the elements of the rolling stock on the railway bridges. In: IOP Conference Series: Materials Science and Engineering, vol. 463 (2018). https://doi.org/10.1088/1757-899x/463/3/032019
14. Gluzberg, B., Korolev, V., Loktev, A., Shishkina, I., Berezovsky, M.: Switch operation safety. In: E3S Web of Conferences, vol. 138 (2019). https://doi.org/10.1051/e3sconf/201913801017
15. Korolev, V.: Switching shunters on a slab base. In: Advances in Intelligent Systems and Computing, vol. 1116, pp. 175–187 (2020). https://doi.org/10.1007/978-3-030-37919-3_17
16. Glusberg, B., Savin, A., Loktev, A., Korolev, V., Shishkina, I., Chernova, L., Loktev, D.: Counter-rail special profile for new generation railroad switch. In: Advances in Intelligent Systems and Computing, vol. 982, pp. 571–587 (2020). https://doi.org/10.1007/978-3-030-19756-8_54
17. Glusberg, B., Savin, A., Loktev, A., Korolev, V., Shishkina, I., Alexandrova, D., Loktev, D.: New lining with cushion for energy efficient railway turnouts. In: Advances in Intelligent Systems and Computing, vol. 982, pp. 556–570 (2020). https://doi.org/10.1007/978-3-030-19756-8_53
18. Loktev, A.A., Korolev, V.V., Poddaeva, O.I., Chernikov, I Y.U.: Mathematical modeling of antenna-mast structures with aerodynamic effects. In: IOP Conference Series: Materials Science and Engineering, vol. 463 (2018). https://doi.org/10.1088/1757-899x/463/3/032018

19. Loktev, A.A., Korolev, V.V., Poddaeva, O.I., Stepanov, K.D., Chernikov, I.Y.: Mathematical modeling of aerodynamic behavior of antenna-mast structures when designing communication on railway transport. Vestnik Railway Res. Inst. **77**(2), 77–83 (2018). https://doi.org/10.21780/2223-9731-2018-77-2-77-83. (in Russian)
20. Savin, A., Suslov, O., Korolev, V., Loktev, A., Shishkina, I.: Stability of the continuous welded rail on transition sections. In: Advances in Intelligent Systems and Computing, vol. 1115, pp. 648–654 (2020). https://doi.org/10.1007/978-3-030-37916-2_62
21. Savin, A., Korolev, V., Loktev, A., Shishkina, I.: Vertical sediment of a ballastless track. In: Advances in Intelligent Systems and Computing, vol. 1115, pp. 797–808 (2020). https://doi.org/10.1007/978-3-030-37916-2_78
22. Loktev, A., Korolev, V., Shishkina, I., Illarionova, L., Loktev, D., Gridasova, E.: Perspective constructions of bridge crossings on transport lines. In: Advances in Intelligent Systems and Computing, vol. 1116, pp. 209–218 (2020). https://doi.org/10.1007/978-3-030-37919-3_20
23. Loktev, A.A., Korolev, V.V., Loktev, D.A., Shukyurov, D.R., Gelyukh, P.A., Shishkina, I. V.: Perspective constructions of bridge overpasses on transport main lines. Vestnik Railway Res. Inst. **77**(6), 331–336 (2018). https://doi.org/10.21780/2223-9731-2018-77-6-331-336. (in Russian)

Change of Geometric Forms of Working Surfaces of Turnout Crosspieces in Wear Process

Vadim Korolev(✉)

Russian University of Transport (MIIT), Chasovaya Street 22/2, 125190 Moscow, Russia
korolevadim@mail.ru

Abstract. The paper discusses the study of changes in the geometric shapes of the working surfaces of the crosspieces of turnouts in the wear process, since the wear resistance and defect resistance of the crosspieces are determined by the level of dynamic forces, as well as by the contact conditions that are associated with the shapes of the crosspiece profiles of the zone of rolling crosses undergoing major changes in the process of wear. The studies are based on surveys of the main tracks of the roads. The research results show that the process of changing the profile shapes of the working surfaces of the crosspieces in the wear is converging, the shapes of the working surfaces remain practically stable after wear to a value of about 4 mm, the final stabilization of the shapes of the profiles of the rolling zone is achieved (up to half the measurement point) during wear 6 ÷ 8 mm, stabilized crosspiece profile shapes wear evenly. It is advisable to design new and repair profiles of the crosspieces so that the profiles across the roll-in zone are as close as possible to the stabilized ones. This will reduce the stabilization period, and thereby improve the working conditions of the crosspieces along the way.

Keywords: Crosspiece · Wear · Cross profile · Cross profile shape · Stabilized forms · Improving the working conditions of the crosspiece

1 Introduction

Great attention is currently being paid to improving the geometry of working surfaces from the point of view of increasing the efficiency of crosspieces [1]. Work is underway to improve standards for crosspieces, create crosspieces with an allowance for riveting, and create rational repair profiles.

As a rule, the main attention is paid to improving the longitudinal profile of the crosspieces from the position of reducing dynamic forces in the contact of the wheels and the crosspiece [2, 3]. However, the wear resistance and defect resistance of the crosspieces are determined by the level of dynamic forces, as well as by the contact conditions, which are associated with the shapes of the cross profiles of the rolling zone of the crosspieces, which undergo large changes in the wear.

V. Murgul and V. Pukhkal (Eds.): EMMFT 2019, AISC 1258, pp. 207–218, 2021.
https://doi.org/10.1007/978-3-030-57450-5_19

2 Materials and Methods

A study was conducted on the change in the geometric shapes of the working surfaces of crosspieces of the type P65 of the 1/11 grade in the wear, carried out according to the results of surveys of the main tracks on the roads. The total number of samples was more than 200 cross pieces with wear from 0 to 12 mm.

We studied eight cross profiles of the crosspieces in rolling - the front insert, neck, profile at the position of the mathematical point of the cross, profiles with a core width of 12, 20, 30, 40, and 70 mm (Fig. 1).

To study the shape of the cross profiles, the wear of the crosspieces was measured using special pallets according to the profilogram at four points of each cross profile of the core and four points of each cross profile of the guardrail (Fig. 2 and 3).

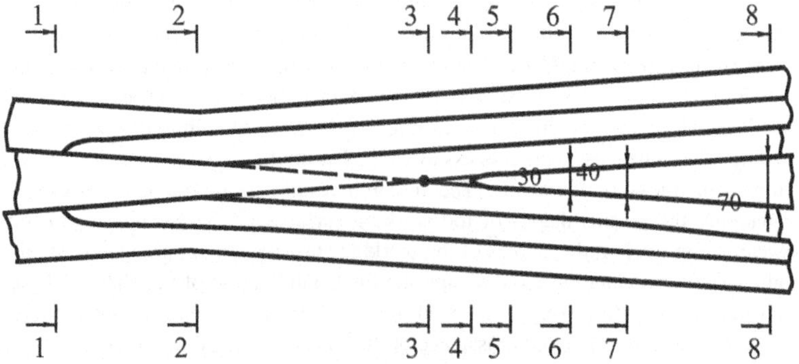

Fig. 1. Cross-profile of the rolling zone of the crosspiece, where the diameters of the working surfaces were removed: / 1-1 / - front insert; / 2-2 / - neck; / 3-3 / - math center of the crosspiece.; / 4-4 / - / 8-8 / - profiles with a core width of 12, respectively; 12; 20. 30; 40 and 70 mm.

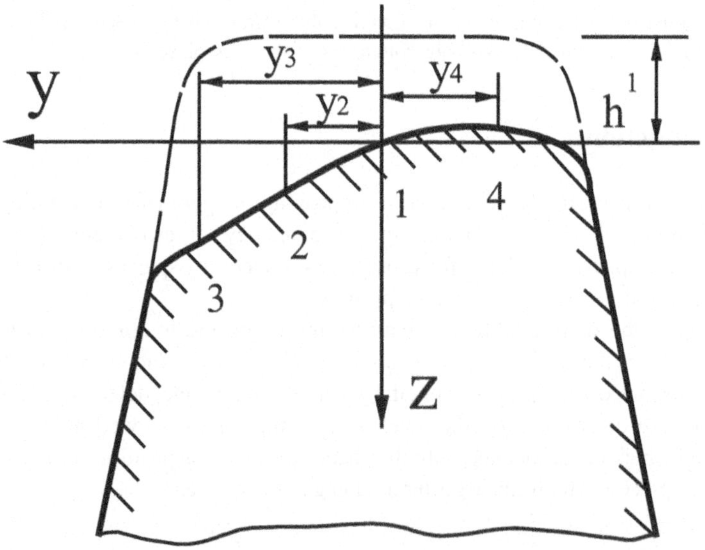

Fig. 2. Measurement of core wear.

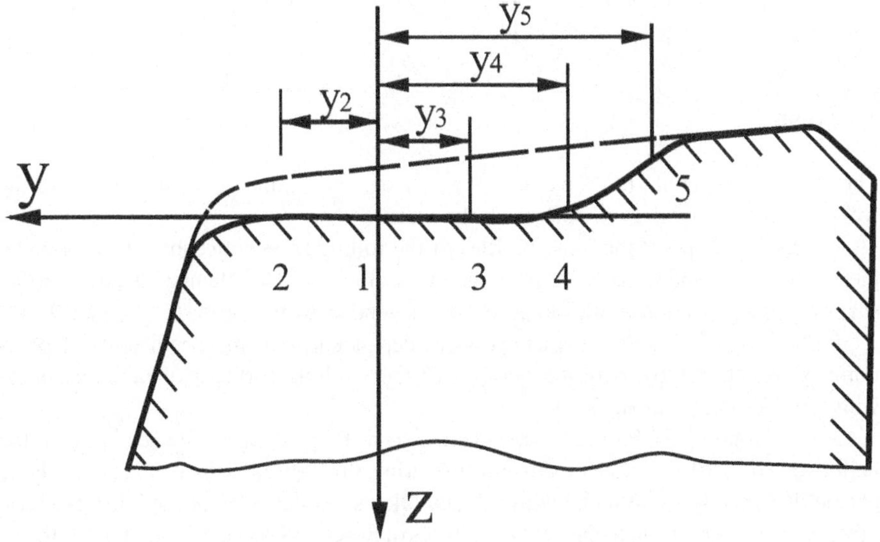

Fig. 3. Measurement of guardrail wear.

The distance between the points yi is given in Table 1. Designation of the ordinates of the measurement points in the Table 1 corresponds to the notation in Fig. 2 and 3.

The obtained depreciation values were statistically processed.

The required number of research objects to obtain statistically reliable data in each profile, for each wear, was determined on the basis of a confidence interval of ± 0.1 mm, which corresponds to the measurement accuracy when examining the diameter according to the profilogram.

Table 1. Coordinates of measuring points of diameters, mm.

Cross profile	Core			Guardrail			
	y_3	y_2	y_4	y_3	y_2	y_4	y_5
Neck		–	–	16	8	−8	−17
Math center of the crosspiece	–	–	–	16	8	−8	−17
Profile 12 mm	5	2,5	−3	14	7	−7	−17
Profile 20 mm	8	3	−4	12	6	−9	−19
Profile 30 mm	12	6	−6	12	6	−9	−19
Profile 40 mm	18	9	−9	10	5	–	−19
Profile 70 mm	32	16	−16	–	–	–	–

The statistical forms of cross profile wear were approximated by power polynomials by the least squares method [4, 5].

3 Results

The resulting material allows us to analyze the changes in the geometric shapes of the working surfaces of the crosspieces in the wear.

Geometric shape of the cross profiles of the rolling zone noticeably changes under the influence of the wheels of passing crews (Fig. 4). The change in cross-profile shapes in the operation is illustrated by the dependences in Fig. 5, 6, 7, 8 and 9.

Figure 5, 6, 7, 8 and 9 depict the wear dependences of the coefficients of polynomials, which approximate the outlines of the working surfaces of the crosspieces with the statistically most likely form of wear.

A common pattern for these dependencies is a sharp change in values in the initial period of work (the period of crushing). During this period, which undergoes large plastic deformations, the metal "settles" and "floats" to the sides of the cross grooves, forming characteristic flows on the lateral faces of the cores and guardrails (Fig. 4) [6, 7].

The loss of height of the core and guardrails during the crushing period reaches 50% or more of the wear value normalized by the Instructions.

Due to the uneven load in different profiles of the rolling zone, this process does not end simultaneously. Guardrail profiles, opposite the core with a width of 55 ÷ 50 mm, come into operation later than the remaining profiles of the rolling zone. Initially, only crew wheels having a saddle-shaped wear form pass through these profiles, then, as the core wears in these profiles, an ever larger group of wheels begins to roll along the guardrail. Different wheels contact the cross in these profiles in different places, so the crushing process gradually spreads along the diameter [8, 9].

After the completion of the crushing and the formation of a deformation-resistant riveted layer, the process of shape change slows down, the curves in Fig. 5, 6, 7, 8 and 9 become more gentle [10, 11].

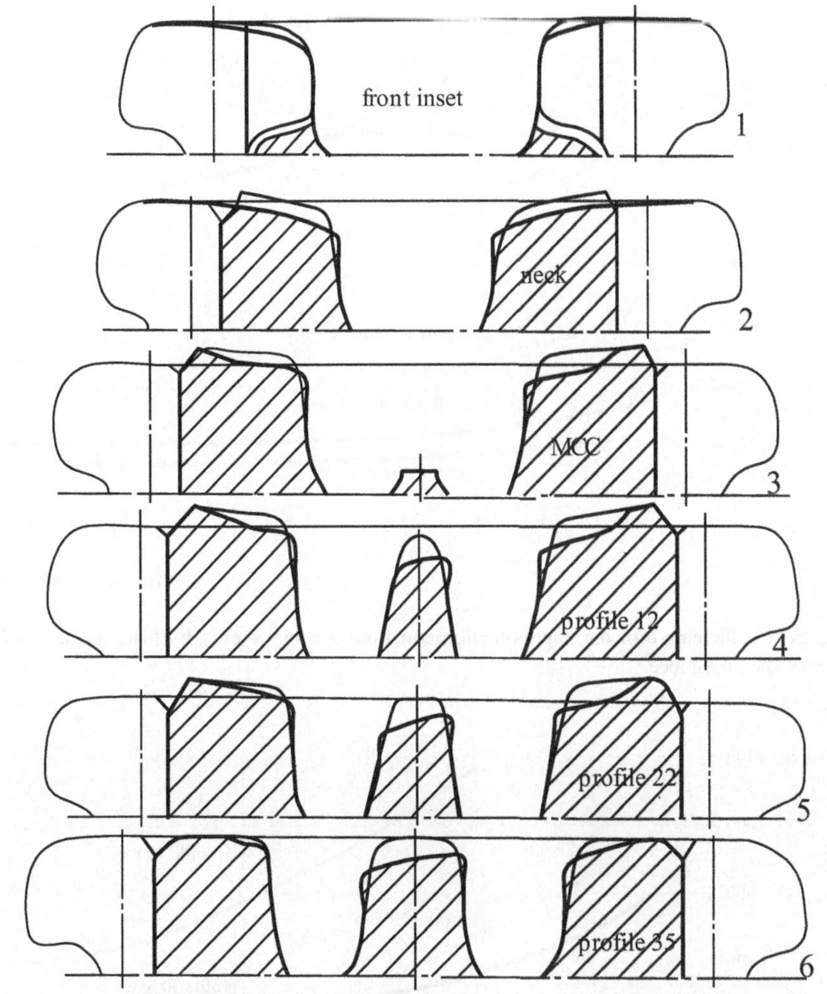

Fig. 4. Cross profiles of the rolling zone of the crosspiece P65 grade 1/11.

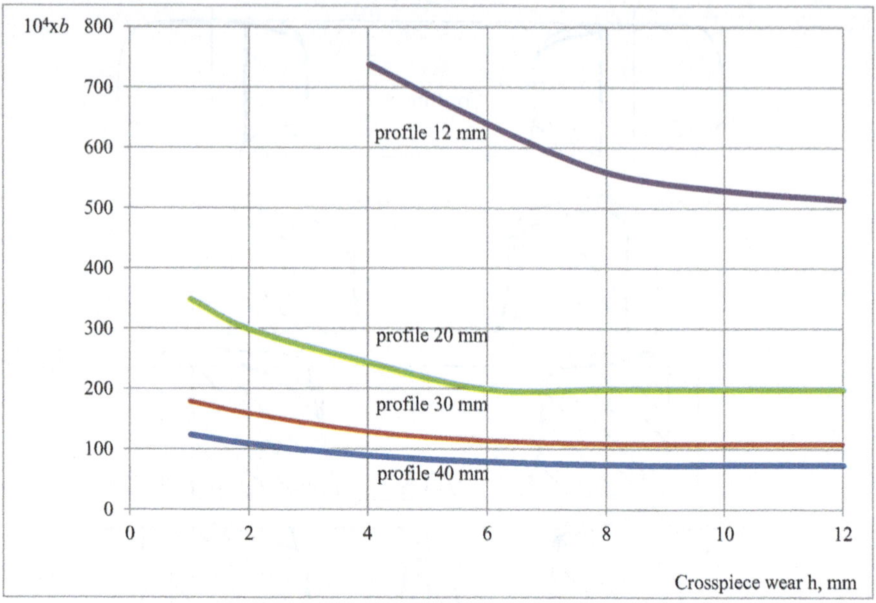

Fig. 5. Coefficients b of the approximation equations $z = by^2 + cy + h$ of the diameter of the core of the crosspiece.

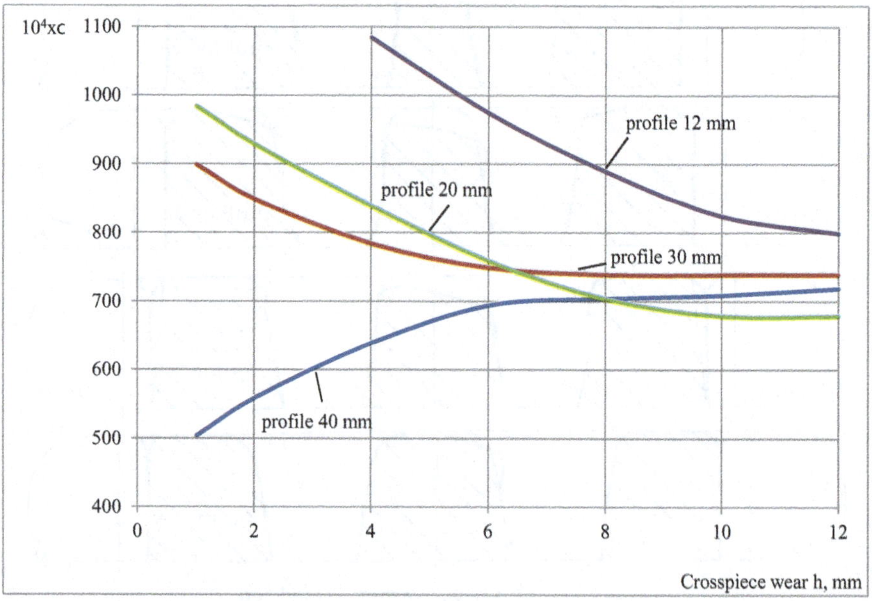

Fig. 6. Coefficients c of the approximation equations $z = by^2 + cy + h$ of the diameter of the core of the crosspiece.

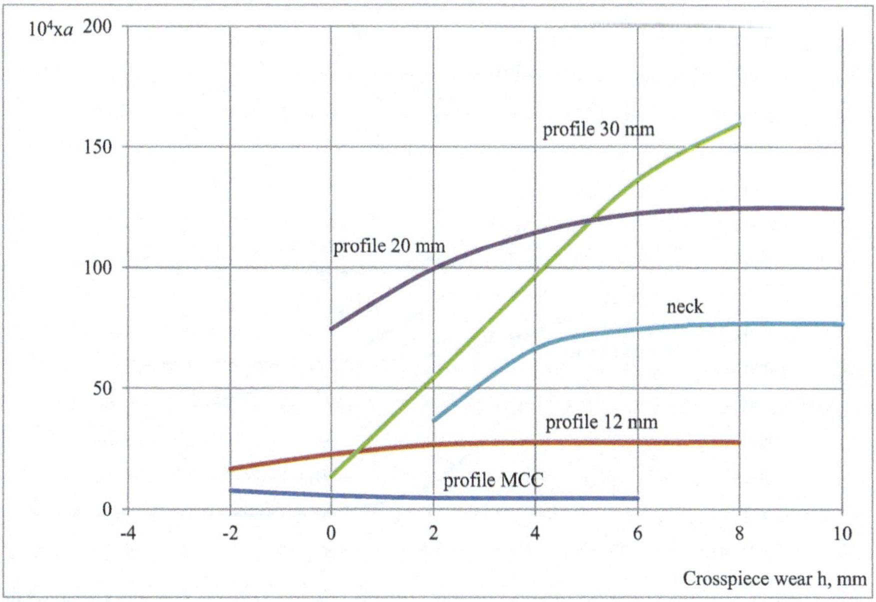

Fig. 7. Coefficients a of the approximation equations $z = ay^3 + by^2 + cy + h$ of guardrail widths.

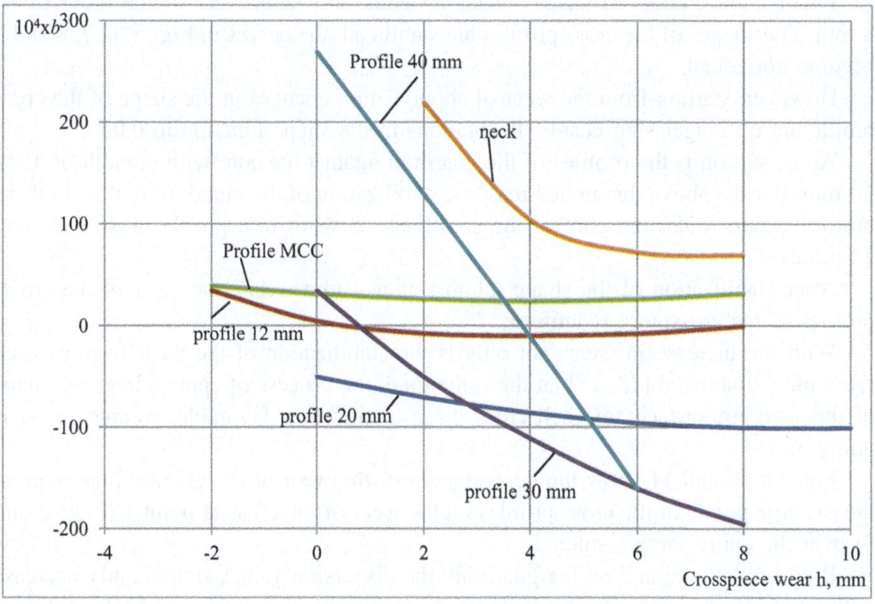

Fig. 8. Coefficients b of approximation equations $z = ay^3 + by^2 + cy + h$ of guardrail widths.

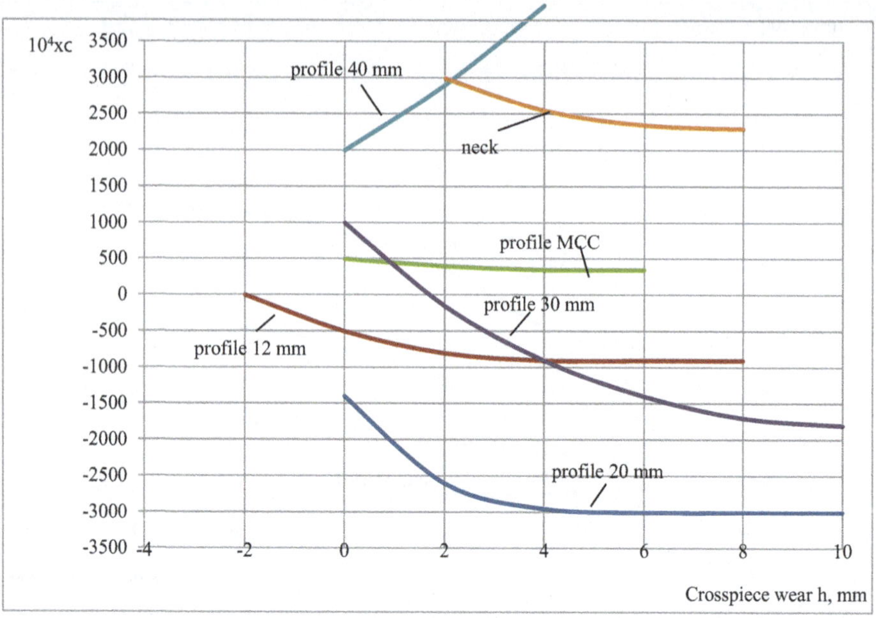

Fig. 9. Coefficients c of approximation equations $z = ay^3 + by^2 + cy + h$ of guardrail width.

Finally, the process of shapes changing ends only with wear of the order of 6–8 mm. The shapes of the cross profiles are stabilized, the curves in Fig. 5, 6, 7, 8 and 9 become horizontal.

However, starting from the wear of about 4 mm, changes in the shape of the cross profile are no longer significant - fluctuations in the shape function to 0.05.

An exception is the profiles of the guardrail against the core with a width of 35 ÷ 50 mm. For the above mentioned reasons, stabilization of the guardrail outline in these profiles occurs when the core of the crosspiece is worn in a profile of 40 mm over 12 mm.

After stabilization of the shape (almost after 4 mm wear), the wear of the cross-profiles of the crosspiece is uniform.

With the increase in wear, not only is the stabilization of the wear form of each cross piece stabilized [12, 13], at the same time, the process of approaching the shape of the working surfaces of each cross piece to the most favorable average shape is going on.

Figures 10 and 11 show the dependences of the wear of the guardrail and core at the extreme points of the cross profiles on the wear of the cross at point 1 (Figs. 2 and 3) over the entire sample studied.

Both on the core and on the guardrail, the dispersion values significantly decrease with an increase in wear. In the considered wear limits, this decrease is 2.2–6.0 times.

The analysis shows that the series of average functions of the shapes of the working surfaces absolutely converge to the functions of the shapes with wear of 8 mm. Therefore, the forms of wear of the working surfaces (except the guardrail in profiles

35 ÷ 50 mm) with wear of 8 mm are stable for these operating conditions of the cross, and the average across the entire set of cross pieces of the shape of the working surfaces with wear of 8 mm is stable for crosses in medium network conditions.

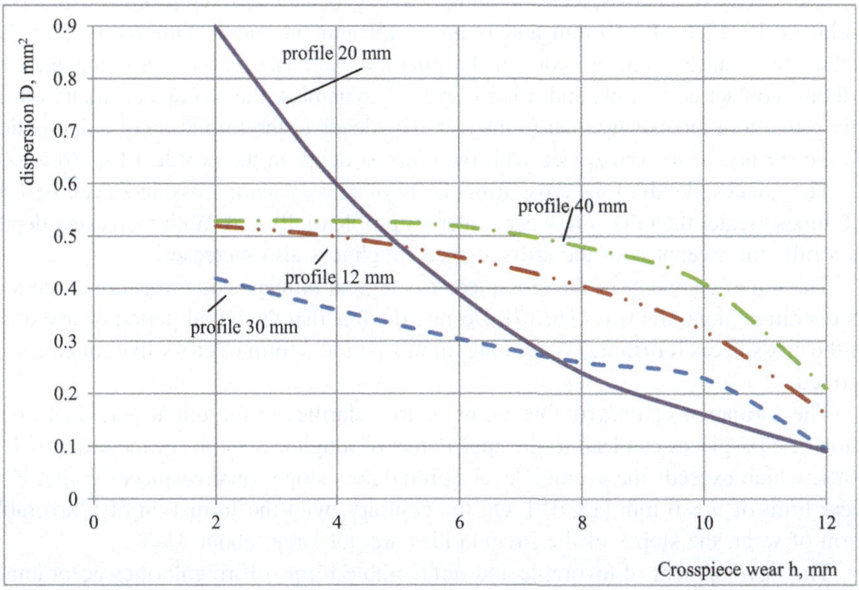

Fig. 10. The dependence of the dispersion of the wear of the cores at the extreme points of the profiles (points 3 and 4 of Fig. 2) on the wear in the middle of the profile (t. 1).

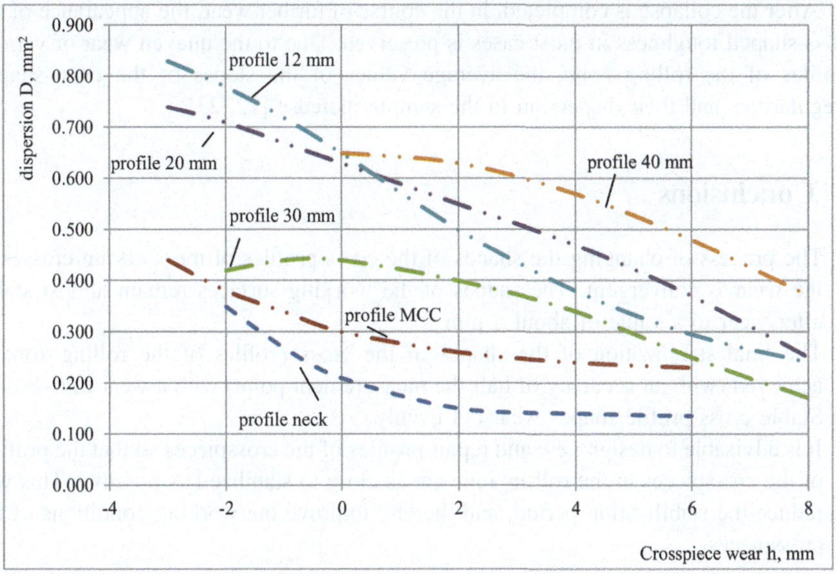

Fig. 11. Dependence of the guardrail wear dispersion at the extreme points of the cross profiles (points 2 and 5 of Fig. 3) on the ordinate in t. 1.

The stabilized cross-profile shapes wear out evenly, so they are the most rational in terms of metal performance.

The process of changing the longitudinal profile of the crosspieces has its characteristics [14, 15].

On heavily worn crosspieces, a rolling wave is located between profiles with a core width of 10–12 to 40–50 mm and is up to 450 mm in length. Due to the fact that rolling the wheels from the core to the guardrail and vice versa is accompanied by difficult contact conditions and a high level of dynamics, the crosspiece wears out in this zone more intensively than in the rest. This leads to the formation of a depression on the surface of the crosspiece with the greatest depth in the profile 12 ÷ 20 mm.

The greatest depth of the cross-irregularity of heavily worn crosspieces can be 1.5–1.8 times greater than the core wear at a cross profile of 40 mm. With increasing depth, as a rule, the steepness of the cross-shaped roughness also increases.

The initial geometry of the crosspiece has a great influence on the entire process of its operation along the way [16, 17]. Despite the fact that the initial period of operation of the crosspieces (crushing) is not long, in this period a form of cross-like roughness is formed.

The formation of unfavorable forms of irregularities in the initial period of operation of crosspieces can lead to the appearance of roughness slopes of the order of 18–20‰, which exceeds the average level of roughness slopes on crosspieces with a PTE wear limit of h = 6 mm [18, 19]. On the contrary, with the formation of a favorable form of wear, the slopes of the irregularities are not large (about 4‰).

The development of favorable and unfavorable forms of irregularities accordingly affects formation of the trajectories of the wheels along the crosspiece, and thereby the level of dynamic impact, which, in turn, affects the intensity of wear, creating a positive feedback between the dynamics and wear [20, 21].

After the collapse is completed, in the course of further wear, the appearance of the cross-shaped roughness in most cases is preserved. Due to the uneven wear of various profiles of the rolling zone, the average values of the slopes of the cross-shaped irregularities and their dispersion in the sample increase [22, 23].

4 Conclusions

1. The process of changing the shapes of the cross profiles of the working crosses in the wear is convergent. The shapes of the working surfaces remain almost stable after wear to a value of about 4 mm.
2. The final stabilization of the shapes of the cross profiles of the rolling zone is achieved (with an accuracy of half the measurement point) with a wear of 6–8 mm. Stable cross-profile shapes wear out evenly.
3. It is advisable to design new and repair profiles of the crosspieces so that the profiles of the crosspieces in the roll-in zone are as close to stabilized as possible. This will reduce the stabilization period, and thereby improve the working conditions of the crosspieces.

References

1. Shishkina, I.: Determination of contact-fatigue of the crosspiece metal. In: Advances in Intelligent Systems and Computing, vol. 1115, pp. 834–844 (2020). https://doi.org/10.1007/978-3-030-37916-2_82
2. Glusberg, B., Savin, A., Loktev, A., Korolev, V., Shishkina, I., Alexandrova, D., Loktev, D.: New lining with cushion for energy efficient railway turnouts. In: Advances in Intelligent Systems and Computing vol. 982, pp. 556–570 (2020). https://doi.org/10.1007/978-3-030-19756-8_53
3. Glusberg, B., Savin, A., Loktev, A., Korolev, V., Shishkina, I., Chernova, L., Loktev, D.: Counter-rail special profile for new generation railroad switch. In: Advances in Intelligent Systems and Computing, vol. 982, pp. 571–587 (2020). https://doi.org/10.1007/978-3-030-19756-8_54
4. Loktev, A.A., Korolev, V.V., Shishkina, I.V., Basovsky, D.A.: Modeling the dynamic behavior of the upper structure of the railway track. Procedia Eng. **189**, 133–137 (2017). https://doi.org/10.1016/j.proeng.2017.05.022
5. Korolev, V.: Guard rail operation of lateral path of railroad switch. In Advances in Intelligent Systems and Computing, vol. 1115, pp. 621–638 (2020). https://doi.org/10.1007/978-3-030-37916-2_60
6. Savin, A., Suslov, O., Korolev, V., Loktev, A., Shishkina, I.: Stability of the continuous welded rail on transition sections. In: Advances in Intelligent Systems and Computing, vol. 1115, pp. 648–654 (2020). https://doi.org/10.1007/978-3-030-37916-2_62
7. Savin, A., Korolev, V., Loktev, A., Shishkina, I.: Vertical sediment of a ballastless track. In: Advances in Intelligent Systems and Computing, vol. 1115, pp. 797–808 (2020). https://doi.org/10.1007/978-3-030-37916-2_78
8. Gluzberg, B., Korolev, V., Loktev, A., Shishkina, I., Berezovsky, M.: Switch operation safety. In: E3S Web of Conferences, vol. 138 (2019). https://doi.org/10.1051/e3sconf/201913801017
9. Loktev, A.A., Korolev, V.V., Gridasova, E.A.: Influence of high-frequency cyclic loading on mechanical and structural characteristics of rail steel under extreme conditions. In: IOP Conference Series: Materials Science and Engineering, vol. 687 (2019). https://doi.org/10.1088/1757-899x/687/2/022036
10. Glusberg, B., Korolev, V., Shishkina, I., Loktev, A., Shukurov, J., Geluh, P., Loktev, D.: Calculation of track component failure caused by the most dangerous defects on change of their design and operational conditions. In: MATEC Web of Conferences, vol. 239 (2018). https://doi.org/10.1051/matecconf/201823901054
11. Gridasova, E., Nikiforov, P., Loktev, A., Korolev, V., Shishkina, I.: Changes in the structure of rail steel under high-frequency loading. In: Advances in Intelligent Systems and Computing, vol. 1115, pp. 559–569 (2020). https://doi.org/10.1007/978-3-030-37916-2_54
12. Savin, A.V., Korolev, V.V., Shishkina, I.V.: Determining service life of non-ballast track based on calculation and test. In: IOP Conference Series: Materials Science and Engineering, vol. 687 (2019). https://doi.org/10.1088/1757-899x/687/2/022035
13. Loktev, A., Korolev, V., Shishkina, I., Chernova, L., Geluh, P., Savin, A., Loktev, D.: Modeling of railway track sections on approaches to constructive works and selection of track parameters for its normal functioning. In: Advances in Intelligent Systems and Computing, vol. 982, pp. 325–336 (2020). https://doi.org/10.1007/978-3-030-19756-8_30

14. Glusberg, B., Loktev, A., Korolev, V., Shishkina, I., Alexandrova, D., Koloskov, D.: Calculation of heat distribution of electric heating systems for turnouts. In: Advances in Intelligent Systems and Computing, vol. 982, pp. 337–345 (2020). https://doi.org/10.1007/978-3-030-19756-8_31
15. Lyudagovsky, A., Loktev, A., Korolev, V., Shishkina, I., Alexandrova, D., Geluh, P., Loktev, D.: Energy efficiency of temperature distribution in electromagnetic welding of rolling stock parts. In: E3S Web of Conferences, vol. 110 (2019). https://doi.org/10.1051/e3sconf/201911001017
16. Loktev, A.A., Korolev, V.V., Poddaeva, O.I., Chernikov, I Y.U.: Mathematical modeling of antenna-mast structures with aerodynamic effects. In: IOP Conference Series: Materials Science and Engineering, vol. 463 (2018). https://doi.org/10.1088/1757-899x/463/3/032018
17. Loktev, A.A., Korolev, V.V., Poddaeva, O.I., Stepanov, K.D., Chernikov, I.Y.: Mathematical modeling of aerodynamic behavior of antenna-mast structures when designing communication on railway transport. Vestnik Railway Res. Inst. **77**(2), 77–83 (2018). https://doi.org/10.21780/2223-9731-2018-77-2-77-83. (in Russian)
18. Loktev, A.A., Korolev, V.V., Shishkina, I.V.: High frequency vibrations in the elements of the rolling stock on the railway bridges. In: IOP Conference Series: Materials Science and Engineering, vol. 463 (2018). https://doi.org/10.1088/1757-899x/463/3/032019
19. Loktev, A., Korolev, V., Shishkina, I., Illarionova, L., Loktev, D., Gridasova, E.: Perspective constructions of bridge crossings on transport lines. In: Advances in Intelligent Systems and Computing, vol. 1116, pp. 209–218 (2020). https://doi.org/10.1007/978-3-030-37919-3_20
20. Savin, A., Kogan, A., Loktev, A., Korolev, V.: Evaluation of the service life of non-ballast track based on calculation and test. Int. J. Innov. Technol. Exploring Eng. **8**(7), 2325–2328 (2019)
21. Korolev, V., Loktev, A., Shishkina, I., Zapolnova, E., Kuskov, V., Basovsky, D., Aktisova, O.: Technology of crushed stone ballast cleaning. In: IOP Conference Series: Earth and Environmental Science, vol. 403 (2019). https://doi.org/10.1088/1755-1315/403/1/012194
22. Loktev, A., Korolev, V., Shishkina, I., Illarionova, L., Loktev, D., Gridasova, E.: Perspective constructions of bridge crossings on transport lines. In: Advances in Intelligent Systems and Computing, vol. 1116, pp. 209–218 (2020). https://doi.org/10.1007/978-3-030-37919-3_20
23. Korolev, V.: Switching shunters on a slab base. In: Advances in Intelligent Systems and Computing, vol. 1116, pp. 175–187 (2020). https://doi.org/10.1007/978-3-030-37919-3_17

Optimization Model of the Transport and Production Cycle in International Cargo Transportation

Valery Zubkov[1(✉)] and Nina Sirina[2]

[1] JSC Federal Freight Company, OAO "FGK", Kuibysheva Str., 44, Letter D, 620026 Yekaterinburg, Russia
Zubkovvvl973@gmail.com

[2] Ural State University of Railway Transport (USURT), ul. Kolmogorova, 66, 620034 Yekaterinburg, Russia

Abstract. Organization of international cargo transportation in the interaction space of different modes of transport is a multi-level transport and production cycle, which consists of many technological processes. For effective realization of transport and production cycle of cargo delivery in international traffic, it is necessary to solve problems of its optimization at all functional levels. The solution of optimization problems is achieved by applying an economic and mathematical model, on the basis of which the optimal variant of international cargo transportation in the conditions of interaction of several modes of transport is determined. The paper presents an economic and mathematical model of optimization of international cargo transportation in the interaction space of transport modes. The given model takes into account both economic indicators, and influence of internal and external space on a transport and production cycle. The developed economic and mathematical model of optimization of the international cargo correspondences provides a choice of an optimum variant of realization of transport services, including in space of interaction of modes of transport. The task of optimization of transport and production cycle of international cargo transportations is solved in the context of consecutive performance of technological processes, subprocesses, operations, sub-operations and norms.

Keywords: Transport and production cycle · Technological processes · International cargo transportation · Optimization · Optimal solution

1 Introduction

Definition and choice of optimum model of the international cargo transportation, consists in realization of such variant of transportation at which observance of the basic indicators is reached, namely: quantity of used modes of transport at performance of a cycle of delivery of cargoes, the organization of interaction of subjects of transport

V. Murgul and V. Pukhkal (Eds.): EMMFT 2019, AISC 1258, pp. 219–228, 2021.
https://doi.org/10.1007/978-3-030-57450-5_20

services in space of interaction of modes of transport, specificity of correspondence of cargoes, cost availability of transport services, the general expenses connected with a transport and production cycle.

The model chosen should also take into account economic indicators, in particular the payment of possible penalties in the event of failure to meet the conditions of carriage and delivery of goods. In addition, the model should take into account the impact of internal and external factors on the quality of transport services when implementing the transport and production cycle.

The bulk of export-import cargo in Russia is transported by several modes of transport, with over 70% of this share being transported by rail.

In the category of international transport, all transport models using railways are mixed. These models differ from others in that cargo, or its consignment, is delivered to a transshipment or logistics terminal by one mode of transport. Further after technological operations (sub-operations) directly on an overload or operations (sub-operations) of storage of cargo, lots of cargoes, are directed to the addressee by other type of transport. In existing communications of international cargo transportation, the subjects of a single transport and production cycle, interact at all its stages, using different transport and accompanying documents and tariff rates.

In the process of international cargo transportation using a mixed model, an important task is to ensure quality and effective technological process (sub-process) of the transfer of goods in the interaction space of modes of transport. In this case, the target parameter of the effectiveness of the model is the method of loading and unloading operations or the method of transshipment of goods [1].

When organizing the delivery of goods in international transport categories and applying a mixed model, it is appropriate to consider the following key criteria:

- optimality of the modes of transport used and number of vehicles;
- optimality of the cargo route;
- optimality of the mode of cargo transportation (choice of the place of loading or transshipment of goods, choice of the place of change of vehicles and modes of transport);
- period of validity of each stage of transportation and period of validity of the whole transport and production cycle;
- the delimitation of the areas of responsibility of the subjects, for each stage of the transport process.

Determining the optimal mode of transport for a specific international freight transport, the consumer of transport services analyzes the characteristics of the modes of transport, transported goods, as well as the cargo characteristics of terminals of departure, transshipment or trans-shipment and destination [2].

When monitoring the possibility of organizing the international transportation of goods, it is necessary to take into account the interstate legal norms and rules of the transport and production cycle.

Besides the basic transport and logistics processes (subprocesses), the transport and production cycle of cargo delivery supposes granting of a great number of additional transport services which at use of traditional methods and mechanisms of the organization and management are realized by intermediaries in many respects. As a result, as the range of transportation increases, the number of transport services subjects (representatives of the customs sector, providers, operators of transport and logistics services, etc.) increases proportionally, which leads to higher prices for transport services and, ultimately, to higher prices for finished products. The solution of such economic problems requires optimization and improvement of transport complex management models. For these purposes, an organizational multi-agent model of transport and logistics system management has been developed and partially implemented, which is considered in [3], and the technology of interaction of subjects of transport services and the delimitation of their areas of responsibility is presented in [4].

Transportation of international goods is a multilevel transport and production cycle, the implementation of which at all stages requires the solution of the following optimization tasks:

- rational formation of consignments with dimensions and weight for all modes of transport used in international cargo transportation;
- optimal choice of transport means of transport;
- optimal choice of means of loading and unloading, as well as cargo transshipment or transshipment terminals;
- definition of rational and effective variants of routes of the mixed model, ways of cargo transportation taking into account border and customs control of the states-participants of the international cargo transportation;
- selection of the optimal variant of international cargo transportation.

The solution of optimization problems is achieved by the application of economic and mathematical model on the basis of which the optimal variant of international cargo transportation in conditions of interaction of several types of transport is determined. For application of such model, it is necessary to know the basic parameters of the unified transport and production cycle [5].

2 Methods of Model Building

In connection with variety of functioning of uniform transport and industrial cycle, coordination and stability of its subprocesses on all life cycle of delivery of cargoes, is a subject for perfection and optimization (search of the best value), provided the minimum cumulative financial expenses [6].

The scheme for determining the optimal delivery model in an international multimodal transport is shown in Fig. 1.

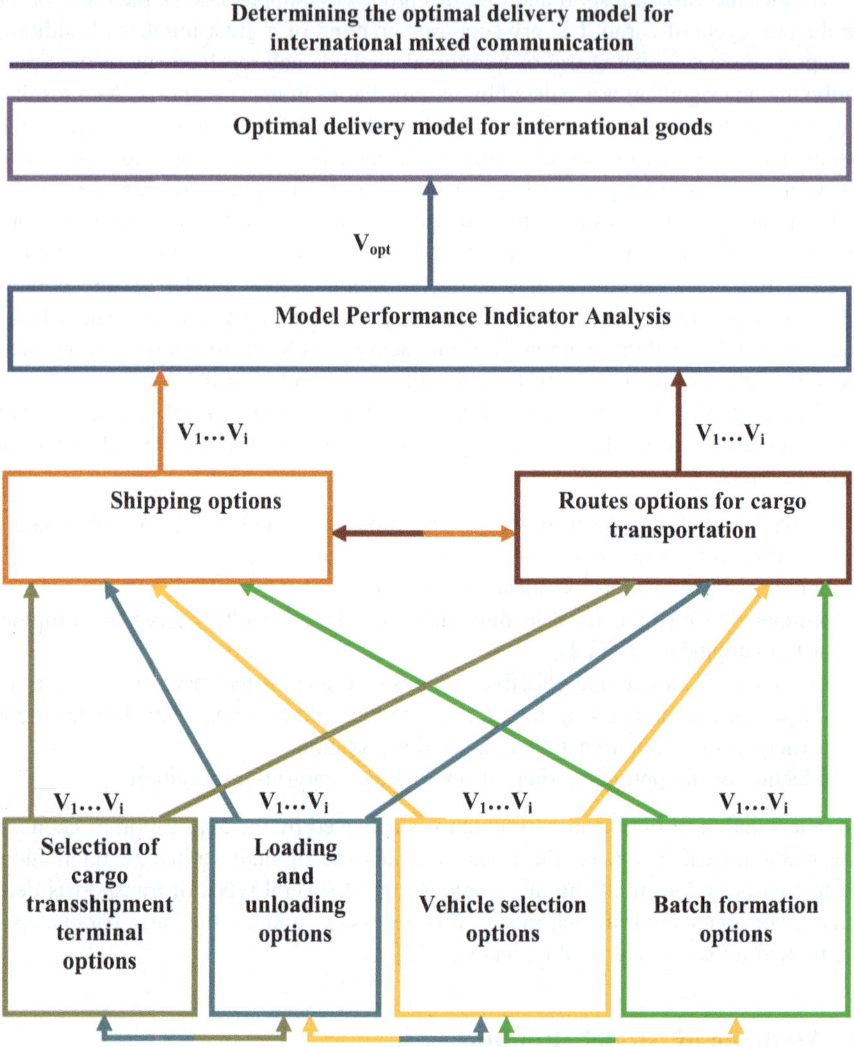

Fig. 1. Scheme for determining the optimal delivery model.

Figure 1 shows that in order to determine the optimal delivery model for international multimodal transport V_{opt}, is necessary to identify options for possible freight criteria $V_1 \ldots V_i$ and analyze the performance of the production model [7].

Figure 2 shows the model of optimization of transport and production cycle of international cargo transportation in the space of interaction of modes of transport and presents values:

A_e, A_g - the total number of consignments at the entrance, exit;
G_e, G_g - inbound, outbound cargo traffic;

W_e, W_g the mass of cargo on the inlet, the outlet;
D_1, D_2, D_i - transport distance of the goods in possible variants;
T_1, T_2, T_i - contractual delivery periods for possible options;
C_1, C_2, C_i - the cost of delivery according to possible options.

As shown in Fig. 2, the transport and production cycle optimization model includes the following main technological processes:

- cargo transportation from the consignor to the transshipment and logistics terminal or cargo handling terminal;
- transfer of cargo from one type of transport to another within the boundaries of the transport and logistics terminal transshipment or cargo handling terminal;
- cargo transportation from a transshipment terminal or cargo terminal to its destination;
- cargo transportation from destination to consignee.

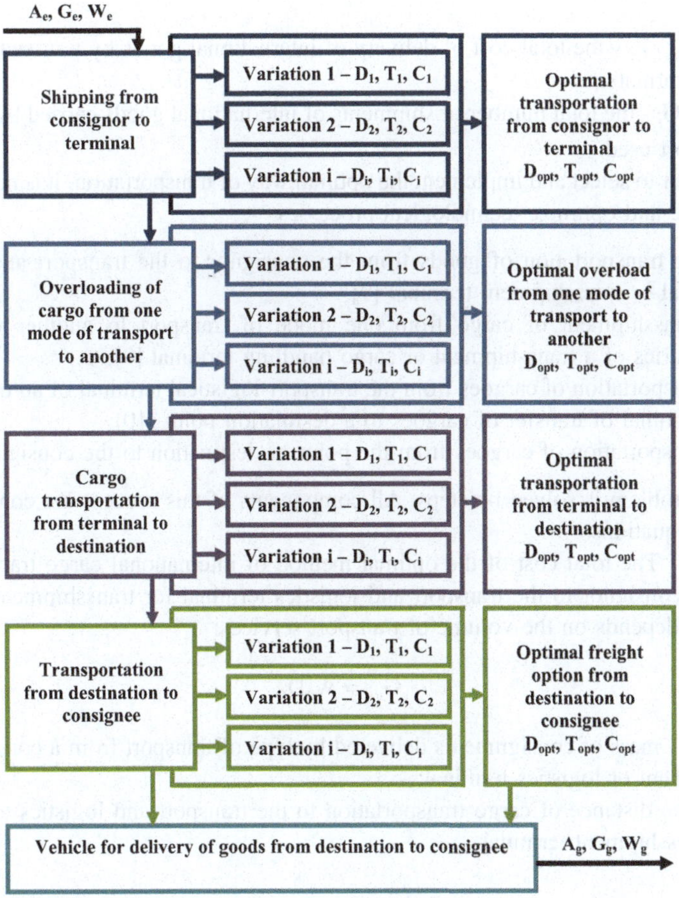

Fig. 2. Model for optimization of transport and production cycle of international cargo transportation in the interaction space of transport modes.

On the basis of set of variants of possible criteria of planned international cargo transportation optimum variants on each basic technological process which include efficiency criterion, that is the minimum total expenses at realization of technological processes are defined [8].

The decision of a problem of optimization in the given model is possible at the expense of a substantiation of probable ways of optimization of a transport and industrial cycle of transportation of the international cargoes in the mixed traffic.

3 Economical-Mathematical Model

We define the criterion of effectiveness, that is, the minimum total cost of delivery of one consignment of international goods, by the formula:

$$\frac{\sum_{i=1}^{k} Z_i}{\sum_{j=1}^{a} G_j} \to \min, \tag{1}$$

where $\sum_{i=1}^{k} Z_i$ - the total cost of delivery of international goods by a mixed model on possible variants;

$\sum_{j=1}^{a} G_j$ - the total number of shipments of international goods carried by the mode of transport used.

In order to select and implement the optimal way of transportation, it is necessary to investigate and optimize technological processes:

- for the transportation of goods from the consignor to the transport and logistics terminal or transshipment terminal [9];
- for transshipment of cargo from one mode of transport to another within the boundaries of a transshipment or cargo handling terminal [9];
- on transportation of cargoes from the transport-logistical terminal of an overload or the terminal of transfer of cargoes to a destination point [10];
- on transportation of cargoes from the point of destination to the consignee [11].

The problem is solved by steps. All components of this process are considered as separate equations.

Step 1. The total cost of the optimal method of international cargo transportation from the consignor to the transport and logistics terminal for transshipment or transshipment depends on the volume of transport services:

$$G_{ai} = w_p D_i, \tag{2}$$

where w_p - mass of consignments delivered by type of transport from a consignor to a transshipment or logistics terminal;

D_i - the distance of cargo transportation to the transport and logistics terminal or cargo transshipment terminal.

Optimum total costs for cargo transportation services to the transportation and logistics terminal for transshipment or transshipment of cargo, depending on the contractual period of their delivery T_i is determined as follows:

$$optZ_{ai} = w_p D_i F_{di}, \quad (3)$$

where F_{di} - agreed fee for the delivery of goods i type of transport from the consignor to the transport and logistics terminal for transshipment or transshipment.

Step 2. Definition of total expenses of the best way of an overload or transfer of cargoes from one type of transport on another in borders of the transport and logistical terminal of an overload or the terminal of transfer of cargoes [12].

The minimum total cost of loading and unloading operations at transport-logistic transshipment terminals or cargo transshipment terminals are calculated as:

$$optZ_{appi} = w_p D_{ppi} F_{ppi}, \quad (4)$$

Where D_{ppi} - the distance travelled by cargo technical means during transshipment or reloading of cargoes from one mode of transport to another for the contractual period T_i;

F_{ppi} - agreed fee for handling operations at a transshipment or transshipment terminal.

Step 3. Determining the total cost of the optimal way of transportation of cargoes from the transport and logistics terminal of transshipment or cargo terminal to the destination. Transportation costs taking into account the contractual period T_{i2} of cargo delivery are calculated by expression:

$$optZ_{ai2} = w_p D_{i2} F_{di2}, \quad (5)$$

where D_{i2} - the distance of cargo transportation from the transshipment or transshipment terminal to its destination;

Fd_{i2} - agreed fee for cargo transportation services i type of transport from a transshipment terminal or cargo transshipment to the destination.

Step 4. Determination of the total costs of providing transport services at the optimal destination.

At possible storage of cargoes on specialized cargo platforms of the destination point, the costs are determined as follows:

$$optZ_{si} = w_p t_{si} F_{si}, \quad (6)$$

where t_{si} - contractual cargo storage period;

F_{si} - agreed fee for additional cargo storage services i.

Costs related to loading and unloading operations are calculated by expression:

$$optZ_{api} = w_p (D_{pi} + D_{i3}) F_{pi}, \quad (7)$$

where D_{pi} - the distance traveled by cargo technical means in the production of loading from a specialized cargo area to the mode of transport;

D_{i3} - transport distance from the destination to the consignee;
F_{pi} - agreed fee for loading and unloading services at destination.

As a result, the total cost associated with transport services at destinations is defined as:

$$optZ_{mn} = w_p t_{si} F_{si} + w_p (D_{pi} + D_{i3}) F_{pi} \tag{8}$$

The target function of the economic and mathematical model using the efficiency criterion (minimum total costs) is as follows:

$$\sum_{i=1}^{k} Z_i = \min(optZ_{ai} + optZ_{appi} + optZ_{ai2} + optZ_{mn}) \tag{9}$$

After transformation, we get:

$$\sum_{i=1}^{k} Z_i = \min[w_p \sum (D_i F_i)] \tag{10}$$

In expression (10) the criterion of optimality which provides check of efficiency of the accepted decisions on each technological process and estimates their influence on full calculation of a transport and industrial cycle is accepted [13].

4 Results

Parameters of the carried out calculations show that application of economic and mathematical model of optimization of transport and production cycle of international cargo transportation in the space of interaction of modes of transport, allows raising efficiency of the given transportations, namely, it is admitted reduction of total expenses for delivery of cargoes in the mixed traffic (railway - automobile), more than on 8 percent from earlier planned expenses [14, 15]. The positive dynamics of decrease in the added value of the transported finished goods have been received.

The model of optimization of the transport and production cycle is applied in the Russian Federation in the organization of international cargo transportation using mixed models of transportation: railroad-sea and railroad-auto.

5 Conclusion

Nowadays, the issues of qualitative forecasting, planning, monitoring and modeling of transport and production cycle of international cargo transportation in the area of interaction of modes of transport, are of great scientific and practical interest for scientists, engineers and specialists of transport branch, dealing with the problems of interaction of different modes of transport. Justification, development or improvement of transport and production cycle of international cargo transportation management, the task is very difficult. The economic and mathematical model of transport and production cycle of international cargo transportation considered in the article provides an opportunity to determine and choose the optimal variant of rendering this transport service in conditions of interaction of different types of transport.

References

1. Müller, S., Wolfermann, A., Huber, S.: A nation-wide macroscopic freight traffic model. Procedia – Soc. Behav. Sci. **54**, 221–230 (2012). https://doi.org/10.1016/j.sbspro.2012.09.741
2. Sirikijpanichkul, A., Ferreira, L., Lukszo, Z.: Optimizing the location of intermodal freight hubs: an overview of the agent based modelling approach. J. Transp. Syst. Eng. Inf. Technol. **7**(4), 71–81 (2007). https://doi.org/10.1016/S1570-6672(07)60031-2
3. Zubkov, V., Sirina, N.: Improvement of cargo transportation technology in rail and sea traffic. In: Popovic, Z., Manakov, A., Breskich, V. (eds.) TransSiberia 2019. AISC, vol. 1116, pp. 1110–1119. Springer, Cham (2020). https://doi.org/10.1007/978-3-030-37919-3_109
4. Zubkov, V.V., Sirina, N.F.: Advanced technologies of international cargo correspondence in railway transport. In: IOP Conference Series: Materials Science and Engineering, International Conference on Transport and Infrastructure of the Siberian Region (SibTrans-2019), Moscow, Russian Federation, vol. 760 (2019). https://doi.org/10.1088/1757-899x/760/1/012056
5. Stokoe. M.: Space for Freight – managing capacity for freight in Sydney – a CBD undergoing transformation. Transp. Res. Procedia **39**, 488–501 (2019). https://doi.org/10.1016/j.trpro.2019.06.051
6. He, Z., Rayman-Bacchus, L., Wu, Y.: Self-organization of industrial clustering in a transition economy: a proposed framework and case study evidence from China. Res. Policy **40**, 1280–1294 (2011). https://doi.org/10.1016/j.respol.2011.07.008
7. Sirina, N., Yushkova, S.: Operation of infrastructure and rolling stock at railway polygon. In: Popovic, Z., Manakov, A., Breskich, V. (eds.) TransSiberia 2019. AISC, vol. 1115, pp. 367–383. Springer, Cham (2020). https://doi.org/10.1007/978-3-030-37916-2_36
8. Johansson, I., Jin, J., Ma, X., Pettersson, H.: Look-ahead speed planning for heavy-duty vehicle platoons using traffic information. Transp. Res. Procedia **22**, 561–569 (2017). https://doi.org/10.1016/j.trpro.2017.03.045
9. Sirina, N.F., Yushkova, S.S.: Integrative management of infrastructure and traction equipment at the railway area. Vestnik Railway Res. Inst. **78**(6), 328–339 (2019). https://doi.org/10.21780/2223-9731-2019-78-6-328-339. (in Russian)
10. Pimentel, C., Alvelos, F.: Integrated urban freight logistics combining passenger and freight flows – mathematical model proposal. Transp. Res. Procedia **30**, 80–89 (2018). https://doi.org/10.1016/j.trpro.2018.09.010
11. Pangbourne, K., Mladenović, M.N., Stead, D., Milakis, D.: Questioning mobility as a service: unanticipated implications for society and governance. Transp. Res. Part A: Policy Pract. (2019). https://doi.org/10.1016/j.tra.2019.09.033
12. Jarašūnienė, A., Sinkevičius, G., Mikalauskaitė, A.: Analysis of application management theories and methods for developing railway transport. Procedia Eng. **187**, 173–184 (2017). https://doi.org/10.1016/j.proeng.2017.04.363

13. Wang, X., Meng, Q.: Discrete intermodal freight transportation network design with route choice behavior of intermodal operators. Transp. Res. Part B: Methodol. **95**, 76–104 (2017). https://doi.org/10.1016/j.trb.2016.11.001
14. Nabais, J.L., Negenborn, R.R., Carmona Benítez, R.B., Botto, M.A.: Achieving transport modal split targets at intermodal freight hubs using a model predictive approach. Transp. Res. Part C: Emerg. Technol. **60**, 278–297 (2015). https://doi.org/10.1016/j.trc.2015.09.001
15. Matteis, T., Liedtke, G., Wisetjindawat, W.: A framework for incorporating market interactions in an agent based model for freight transport. Transp. Res. Procedia **12**, 925–937 (2016). https://doi.org/10.1016/j.trpro.2016.02.044

Dam Failure Model and Its Influence on the Bridge Construction

Artur Onishchenko[1(✉)], Andrii Koretskyi[1], Iryna Bashkevych[1], Borys Ostroverkh[2], and Andrii Bieliatynskyi[3,4]

[1] National Transport University, 1, Mykhaila Omelianovycha - Pavlenka Street, Kiev 01010, Ukraine
artur_onish@bigmir.net
[2] Institute of Hydromechanic of NAS Ukraine, 54, Volodymyrska, Kiev 01030, Ukraine
[3] National Aviation University, Kiev 01010, Ukraine
[4] North Minzu University, 204 North-Wenchang Street Xixia District, Yinchuan, Ningxia, People's Republic of China

Abstract. When drawing up a feasibility study of designing estimates for the repair of a bridge across the Kunka River of the M-12 Stryi-Ternopil-Kropyvnytskyi-Znamianka motorway section (via Vinnytsia) due to the proximity of a number of reservoirs formed by obsolete dams with increased pressure and a high probability of damage to the bridge from the formation and passage of a breakthrough wave, the problem of developing measures to ensure security moving through bridge arose. To simulate and analyze the parameters of the breakthrough wave in the bridge area, the available cartographic data with bottom surface marks and the determination of the coastal zone at the level of the maximum supported horizon were processed. The magnitude of the bridge opening was calculated and the total overflow through the bridge was estimated. It allowed making conclusions regarding the structural solutions of the bridge structure. The results of the development of the site calculating model and its use for calculating the costs, velocities and flood zones are presented in the form of initial data. The final values of the flow velocity of the breakthrough wave near the supports, frontal and end slopes of the bridge bulk dam, splash marks, the need to erect protective structures on the pressure slope of the dam were determined using mathematical modeling of the dam failure. Also, recommendations on the necessary layout and structural protective measures of the motorway and bridge were developed.

Keywords: Hydromorphodynamics · Numerical simulation of a dam failure · Bed forms and processes · Kinematics of channel flows · General analysis of GIS data

V. Murgul and V. Pukhkal (Eds.): EMMFT 2019, AISC 1258, pp. 229–237, 2021.
https://doi.org/10.1007/978-3-030-57450-5_21

1 Introduction

Factors of hydrodynamic danger of disturbance of the state of hydraulic structures for the motorway section and bridge in the downstream pool of the dam can be both natural and man-made (for example, destruction of the dam due to a decrease in its strength) and other factors. The destruction (failure) of the hydraulic structure is a multifactorial process and arises as a result of various forces of nature (earthquakes, hurricanes, floods, showers and other hydrometeofactors, even erosion due to concentrated filtration through animal burrows, etc.), or human activities (transport, large-scale bombing, sabotage), or due to structural defects (low-quality materials, cracks), or design errors. The methodology for calculating the parameters of a breakthrough wave is not sufficiently developed and is determined by engineering formulas according to Ukrainiane State Standart DBN V.2.4-3: 2010 "Hydrotechnical, energy and land reclamation systems and structures, underground mining. Hydraulic structures. Key Points." or Russian State Standard SNiP 2.05.03-84 "Bridges and pipes" for the survey and design of railway and road bridges through waterways".

In its physical essence, a breakthrough wave is an uncontrolled movement of the flow of water and silt mixture, at which the depth, width, surface slope and flow velocity change over time [1–7]. In this case, the parameters of the breakthrough (skipping) wave are determined at a given distance L from the dam, depending on the topography of the terrain and other obstacles.

To draw up a spatial model of the propagation of a breakthrough wave, one should use the drawings of the planned location of the dam and the motorway with the use of high-altitude survey, which is indicated by marks and contours.

The height of the breakthrough wave and the speed of its propagation depends on the volume and depth of the reservoir, the area of the mirror of the water basin, the size of the passage, the difference in water levels in the upper and lower pools, hydrological and topographic conditions of the river channel and its floodplain. In the region of zero alignment (dam body) the height of the breakthrough wave Hlp is determined by the application formulas 5.3 according to guide to Russian State Standard SNiP 2.05.03-84 "Bridges and pipes" for the survey and design of railway and road bridges through waterways".

2 Breakthrough Wave Motion Simulation Theory

Calculation of a dam failure under the forces of natural or technogenic catastrophic loads is a difficult mathematical task, since initially a contact medium consisting of a reservoir, bottom sludge and soil body of a dam having different physicomechanical properties, after a catastrophic dam failure turns into an inhomogeneous water-ground mixture. In case of introducing some simplifications that do not affect the final result of breakthrough flows, it is possible to formulate a problem statement for a numerical solution.

For this, it is possible to use modern software systems, the Open FOAM in CDF software package, e.g., which is freely available [8], but requires additional transformations to satisfy the boundary conditions.

As part of the problem, the movement of the soil mixture is considered as a viscous fluid, which is described by the differential equations of continuity and motion, the equations of conservation of mass, momentum and energy. In a three-dimensional system, the control equations of fluid mechanics can be written in differential form, which is compiled in such a system of equations:

– conservation of mass:

$$\frac{\partial \rho}{\partial t} + \nabla \cdot (\rho \mathbf{u}) = 0 \tag{1}$$

– conservation of momentum:

$$\frac{\partial \rho \mathbf{u}}{\partial t} + \nabla \cdot (\rho \mathbf{u}\mathbf{u}) = \rho g + \nabla \cdot \sigma \tag{2}$$

– conservation of energy:

$$\frac{\partial \rho e}{\partial t} + \nabla \cdot (\rho e \mathbf{u}) = \rho g \mathbf{u} + \nabla \cdot (\sigma \mathbf{u}) - \nabla \cdot \mathbf{q} + \rho Q \tag{3}$$

where ρ - fluid density, u - three-dimensional velocity field, σ - displacement stress tensor, e - total specific energy, Q - volume energy source, q - heat flow, g - gravity acceleration vector.

This system of three equations is uncertain, because the number of unknown variables is greater than the number of equations. For a Newtonian, incompressible (ρ constant) and isothermal fluid, system of Eqs. (1), (2) and (3) can be simplified to the form of:

$$\nabla \cdot \mathbf{u} = 0 \tag{4}$$

$$\frac{\partial \mathbf{u}}{\partial t} + \nabla \cdot (\mathbf{u}\mathbf{u}) = \mathbf{g} - \nabla p + \nabla \cdot (\upsilon \nabla \mathbf{u}) \tag{5}$$

where υ is the kinematic viscosity and p is the kinematic pressure.

Multiplying the momentum equation by the liquid density, we obtain the final form of the continuity and momentum equations for a homogeneous liquid field in the form of:

$$\nabla \cdot \mathbf{u} = 0 \tag{6}$$

$$\frac{\partial \rho \mathbf{u}}{\partial t} + \nabla \cdot (\rho \mathbf{u}\mathbf{u}) = -\nabla P + \nabla \cdot \tau + \rho \mathbf{g} + F \tag{7}$$

where P is the pressure (P p ρ), η is the viscous stress tensor, F is the momentum source relative to the surface tension:

$$F = \int_{S(t)} \sigma \kappa' \mathbf{n}' \delta(\mathbf{x} - \mathbf{x}')\, dS \tag{8}$$

where κ is the curvature and n is the normal vector of the contact surface.

The viscous stress term can be changed using the formula of Newton's law to get a more convenient look. The final form of this term is as follows:

$$\nabla \cdot \tau = \nabla \cdot \left(\mu\left[\nabla \mathbf{u} + (\nabla \mathbf{u})^T\right]\right) = \nabla \cdot (\mu \nabla \mathbf{u}) + (\nabla \mathbf{u}) \cdot \nabla \mu \tag{9}$$

The modified pressure gradient is defined as:

$$\begin{aligned} \nabla p^* &= \nabla P - \nabla(\rho \mathbf{g} \cdot \mathbf{x}) \\ &= \nabla P - \rho \mathbf{g} - \mathbf{g} \cdot \mathbf{x} \nabla \rho \end{aligned} \tag{10}$$

In order to determine which part of the cell is a viscous fluid and which part is air:

$$\alpha(x, y, z, t) = \begin{cases} 1 & \text{(for the zone (x, y, z, t)occupied by the liquid 1)} \\ 0 < \alpha < 1 & \text{(for the zone (x, y, z, t)of contact)} \\ 0 & \text{(for the zone (x, y, z, t)occupied by the liquid 2)} \end{cases} \tag{11}$$

At the interface, the contact conditions are satisfied.

The transfer of the indicator function in time is determined by the advection equation:

$$\frac{\partial \alpha}{\partial t} + \nabla \cdot (\alpha \mathbf{u}) = 0 \tag{12}$$

Equation (12) shows that in an incompressible fluid the conservation of mass is equivalent to the conservation of volume and, therefore, the conservation of the function α is observed. The properties of a local liquid (ρ and μ), which is a mixture of the physical properties of both liquids (indices 1 and 2 denote different liquids) are determined by the equations:

$$\rho = \alpha \rho_1 + (1 - \alpha) \rho_2 \tag{13}$$

$$\mu = \alpha \mu_1 + (1 - \alpha) \mu_2 \tag{14}$$

The preservation of the phase function is important, especially in the case of liquids with a high density, where small errors in the volumetric part create significant errors in the physical properties. We introduce an additional term in the phase differentiation function, namely, the artificial compression term in the form of the transportation equation:

$$\frac{\partial \alpha}{\partial t} + \nabla \cdot (\alpha \bar{\mathbf{u}}) + \underbrace{\nabla \cdot [\mathbf{u}_r \alpha (1 - \alpha)]}_{\text{Artifial compression term}} = 0 \quad (15)$$

where $\mathbf{u}_r = \mathbf{u}_1 - \mathbf{u}_2$ is the relative velocity vector between two fluids, which is also called the compression rate, and is the average velocity calculated by the average velocity between two fluids:

$$\bar{\mathbf{u}} = \alpha \mathbf{u}_1 + (1 - \alpha) \mathbf{u}_2 \quad (16)$$

3 Statement of the Problem and Initial Data

Within this study, a test calculation of the catastrophic dam failure was carried out according to the mathematical model of instantaneous failure in the framework of a two-dimensional plane wave. To carry out the calculations, a program system was created based on the Open FOAM in CDF complex [8], which is freely available by adapting the computational grid, limit and initial conditions for implementing the use of the interFoam multiphase resolver.

The distance between the motorway route and the dam is 30 m, which is unacceptable according to Ukrainian State Standard DBN V.2.4-3: 2010 "Hydrotechnical, energy and land reclamation systems and structures, underground mining. Hydraulic structures. Key Points., since there is the possibility of erosion of the embankment of the motorway and bridge in the area of extinguishing the energy of the breakthrough wave.

Modeling the movement of water-soil flow during a dam failure is described by the movement of a viscous fluid with a free surface with the following parameters:

- dynamic viscosity of the water-soil mixture ν = 1 Pa*sec;
- viscosity of the water-soil mixture ν = 10^{-3} Pa*sec;
- viscosity of air ν = 10^{-6} Pa*sec;
- dam mark 199 m, road mark 197 m;
- the width of the dam of the pond L = 100 m, the distance from the dam to the motorway lr = 33 m,
- the width of the embankment of the motorway Br = 40 m.

At present, there are no sufficient topographic, hydrological, and geological data for calculating the bridge opening.

A study of the topography of the area showed that the motorway crosses the permanent watercourse of the Kunka River within the village of Kunka with a sufficiently large catchment area to the bridge cross section. Determination of the catchment area of the Kunka River to the dam of the village of Kunka on the M-12 motorway (Vinnytsia-Gaysin section) picket 475 + 034 km was carried out using the QGIS software [9], which gave the catchment area of S = 133.23 km^2. On analogy with the example given in Appendix 2 according to guide to Russian State Standard SNiP 2.05.03-84 "Bridges and pipes" for the survey and design of railway and road bridges through waterways", the reservoir volume is approximately V = 1.2 * 106 m^3, and the flow rate of the breakthrough wave of the pond dam can be Q = 740 m^3/s.

The results of mathematical modeling within a flat problem are shown in Figs. 1, 2, 3, 4, where the phase color method indicates: water-ground mixture - red color, air - blue color, embankment of the motorway - gray color. Linear dimensions are shown in 1:10 scale.

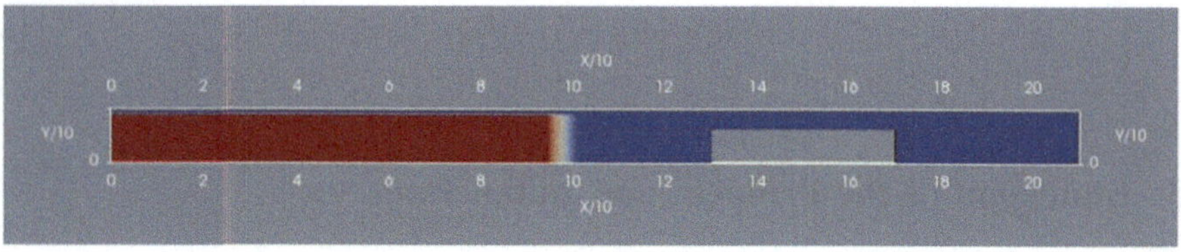

Fig. 1. Model of a dam failure at the initial moment t = 0 s.

It is shown that in the conditions of close proximity of the motorway and bridge, when the flow of breakthrough materials can be disturbed through the bridge opening due to the formation of soil congestion, a breakthrough wave overflows up to 2 m high through the motorway and bridge with the formation of a catastrophic zone of erosion of the motorway section and the bridge. About the same height of the breakthrough wave will pass through the floodplain of the motor-way on both banks of the river. More precisely, the parameters of the breakthrough wave can be obtained within the more complex three-dimensional formulation of the problem.

Also, mathematical modeling of the flow movement of water-soil mixture was carried out in the case when part of the flow passes on the bridge crossing section with an opening under the bridge up to 3 m high. The simulation results are shown in the Fig. 5, 6, 7, 8, 9. According to the calculation results, we cannot also exclude in this case the overflow of a breakthrough wave through the embankment of the motorway, if we take into account the spatial compression of the stream by the channel of the floodplains. In post-processing data processing mode, flow rates and pressure on the bridge structure can be obtained.

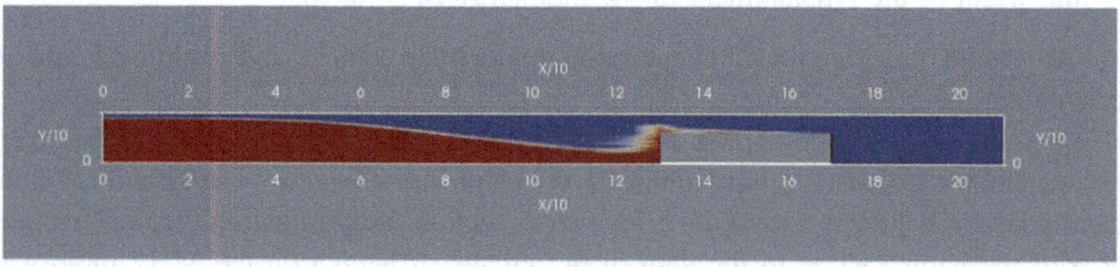

Fig. 2. Model of a dam failure in t = 5 s.

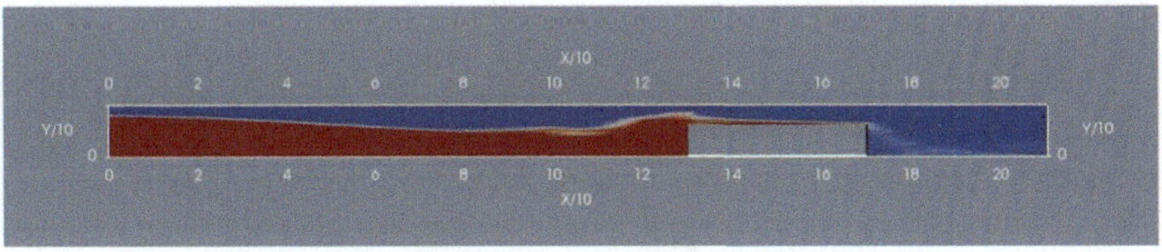

Fig. 3. Model of a dam failure in t = 10 s.

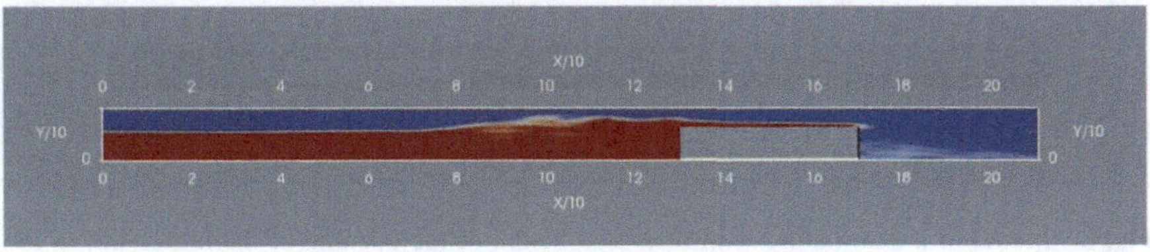

Fig. 4. Model of a dam failure in t = 15 s.

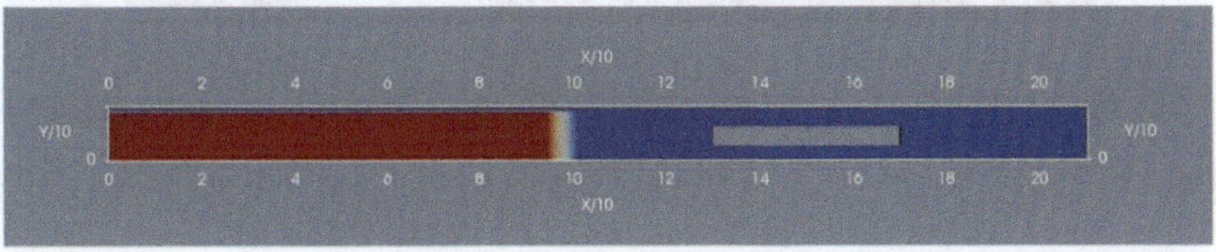

Fig. 5. Model of a dam failure with increased height of the opening of the bridge at the initial moment t = 0 s.

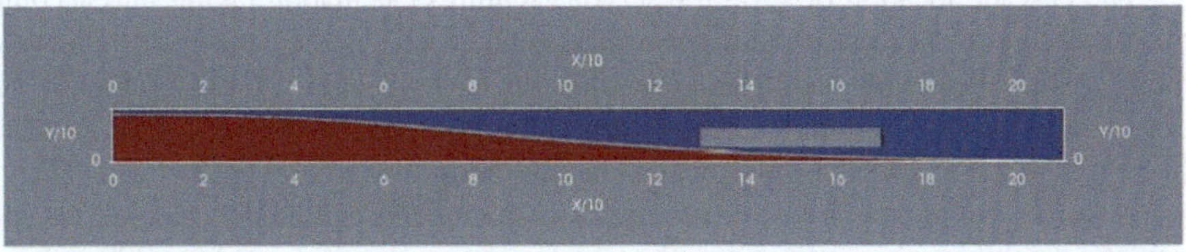

Fig. 6. Model of a dam failure with increased height of the opening of the bridge in t = 5 s.

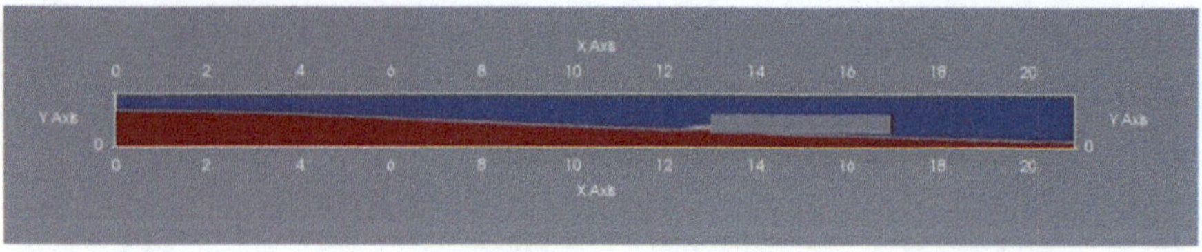

Fig. 7. Model of a dam failure with increased height of the opening of the bridge in t = 10 s.

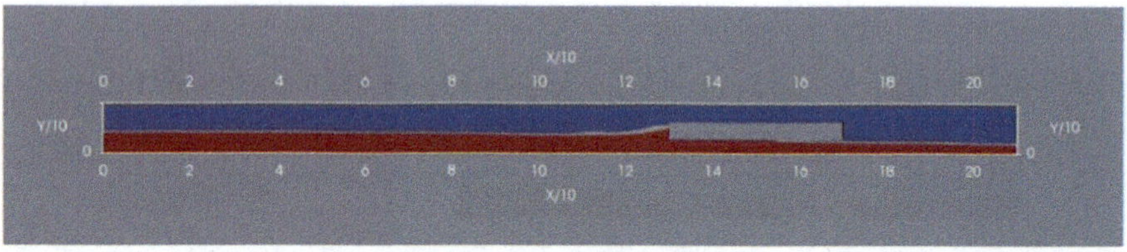

Fig. 8. Model of a dam failure with increased height of the opening of the bridge in t = 15 s.

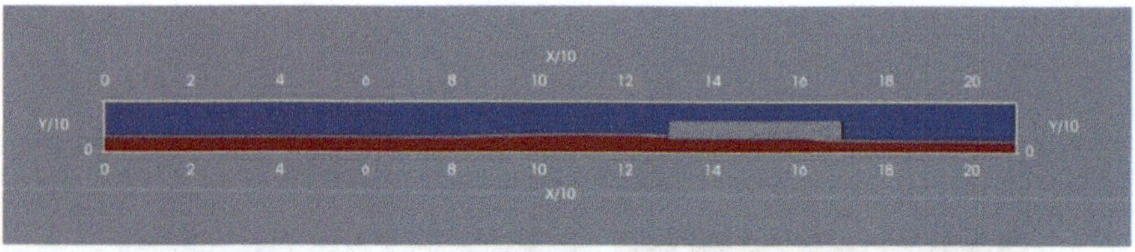

Fig. 9. Model of a dam failure with increased height of the opening of the bridge in t = 20 s.

4 Conclusions and Recommendations

At present, calculations of the bridge opening cannot be carried out in sufficient volume and accuracy due to the lack of initial data on the available cascade locations of several reservoirs above the motorway.

The cascade location of several reservoirs requires calculation taking into account the possibility of successive failure of several dams, based on the created mathematical model. The calculation results indicate that the cross section of the bridge in the downstream pool should be located at a distance of more than 20 * Nb from the dam (the pressure of the breakthrough wave Hb can be up to 9 m).

According to the requirements of regulatory documents and model calculations, the risk of a breakthrough wave requires the transfer of a motorway for at least 200 m from the dam.

To protect the bridge, additional measures to protect the embankment of the motorway at the dam and bridge section with plates and flow dams at the opening can be recommended.

5 Discussion

An analysis of the existing methods for calculating the breakthrough wave parameters showed that the techniques are underdeveloped and determined by engineering formulas. As part of this study, a test calculation of a catastrophic breakthrough of a dam was performed using a mathematical model of instantaneous destruction within a plane two-dimensional wave. For calculations, a system of programs was created based on the Open FOAM complex. Model calculations show that the breakthrough wave poses a significant threat to the road embankment, bridge and roadway and requires the transmission of the road at least 200 m from the dam.

References

1. Chanson, H.: Hydraulics of Open Channel Flow, 2nd edn. Butterworth heinemann, Australia (2004). https://doi.org/10.1016/b978-0-7506-5978-9.x5000-4
2. Chmielowski, W.Z.: Fuzzy controllers operating flood wave passage through a hypothetical cascade of retention reservoirs. In: Fuzzy Control in Environmental Engineering, pp. 109–119 (2016). https://doi.org/10.1007/978-3-319-19261-1_5
3. Mazumder, S.K.: Flow Transition Design in Hydraulic Structures. CRC Press, Boca Raton (2020). https://doi.org/10.1201/9780429280979
4. Moglen, G.: Fundamentals of Open-Channel Flow, 1st edn. CRC Press, Boca Raton (2015). https://doi.org/10.1201/b18359
5. Zhang, L., Peng, M., Chang, D., Xu, Y.: Dam Failure Mechanisms and Risk Assessment. Wiley, Hoboken (2016). https://doi.org/10.1002/9781118558522
6. Akan, O.A.: Fundamentals of Open-Channel Flow, 1st edn. Butterworth Heinemann, USA (2006). https://doi.org/10.1016/b978-075066857-6/50002-3
7. Stepanov, K.A.: Simplified methodology for modeling the propagation of a breakthrough wave to protect lands from flooding. Sci. J. Russ. Res. Inst. Land Reclam. Prob. **4**(12), 130–140 (2001)
8. OpenFOAM®: Open source CFD documentation. Extended Code Guide. http://www.openfoam.com/documentation/cpp-guide/html/
9. QGIS Training Manuals. https://docs.qgis.org/testing/en/docs/training_manual/

Simulation of Traffic Flows Optimization in Road Networks Using Electrical Analogue Model

Viktor Danchuk[1(✉)], Olena Bakulich[1], Serhii Taraban[2], and Andrii Bieliatynskyi[3,4(✉)]

[1] National Transport University, Mykhaila Omelianovycha - Pavlenka Street 1, 01010 Kiev, Ukraine
vdanchuk@ukr.net
[2] State Enterprise "State Road Transport Research Institute", Peremohy Avenue 57, 03113 Kiev, Ukraine
[3] National Aviation University, Lubomyra Gusara Avenue 1, 03058 Kiev, Ukraine
beljatynskij@ukr.net
[4] North Minzu University, 204 Nort-Wenchang Street, Xixia District, Yinchuan, Ningxia, People's Republic of China

Abstract. The electrical analogue model for the analysis and optimization of traffic flows in the urban road network has been improved. Implementation method for the simulation model, which is aimed at the increase of throughput capacity of the road network by means of redistribution of the congestion levels of different sections, has been proposed. Key aspects of the electrical simulation and optimization of the traffic flows in the urban environment were explored, having taken a real fragment of Kyiv urban road network as an example for verification. Optimization is achieved by redistributing traffic flows within the road network sections in order to unload congested areas and ensure uniform load of the road network of the city as a whole. It is shown that the proposed optimization method using a simulation electrical analogue model can be an important component for determining the conditions for the rational organization of the functioning of traffic flows on the urban road networks.

Keywords: Traffic flow · Urban road network · Electrical analogue simulation model · Optimization

1 Introduction

In the context of ongoing rapid urbanization and urban development, public needs call for a permanent expansion and improvement of the transport links system, improvement of their efficiency, reliability, safety and quality. Improvement of road infrastructure to transform it into an efficient multi-level transport and logistics system, in turn, requires a significant inflow of capital investment and material resources. At the same time, it should be noted that the financial risks increase significantly, if development laws of urban transportation systems and distribution of traffic congestion over

V. Murgul and V. Pukhkal (Eds.): EMMFT 2019, AISC 1258, pp. 238–254, 2021.
https://doi.org/10.1007/978-3-030-57450-5_22

the road network components are ignored, which leads to a number of negative social, economic and environmental effects. Such effects include, in particular, irregularity of traffic congestion of urban road network (overload or underutilization of individual network components), occurrence of traffic queues, higher accident rates and air pollution levels.

Making optimal decisions when designing road network and organizing traffic, as well as developing efficient traffic management strategies in the city requires consideration of the whole range of parameters of time-space distribution of traffic flows over the road network. Given the stochastic nature of traffic and intermittent volatility of its time-space parameters, adequate evaluation and forecasting of key operational parameters of the traffic flows over road network is quite challenging. Today this problem is solved by mathematical simulation methods.

In general, mathematical simulation of traffic flows involves development of models that use different simulation techniques - from simulation of each individual vehicle movement to description of the traffic density function within the road section/space interval. Current transport flow models can be differentiated by functionality, mathematical tools, input data, level of detail etc. For example, by functionality traffic flow models can be divided into two classes - simulation models (for traffic flow dynamics) and predictive models (for traffic congestion).

Mathematical simulation of traffic flows, in terms of the object's level of detail, can be divided into macroscopic, mesoscopic and microscopic (including submicroscopic).

Macroscopic simulation methods and models describe the traffic flow condition by correlation of key parameters of traffic flows - speed, intensity, and density. Such models do not take into account the size of individual vehicles, they are derived from the analogy of flows in continuous mediums.

The basic model is hydro-dynamic (known as Lighthill-Whitham-Richards, LWR model), where equation of the transport flow continuity has been introduced and, accordingly, the speed is presumed to be a function of the flow density [1]. One of the major shortcomings of the LWR model is the assumption that average flow q at any specific time is related to a certain equilibrium value at the given density of vehicles k. This leads to the inadequate representation of traffic flows in the road irregularity areas and when moving in "stop-and-go" mode which is inherent for heavy traffic flow.

From the viewpoint of the level of detail in the description of the traffic flows operation, mesoscopic simulation is an interim level, where traffic flow and driver behavior are described in aggregate terms, in particular by means of the probabilistic function [2]. Namely, traffic intervals distribution models [3], cluster model [4], and kinetic model [5] fall into this classification. One of the major shortcomings of the mesoscopic models is their negligence of the traffic flow dynamics.

A detailed description of the vehicle movement in a traffic flow can be obtained using microscopic simulation methods [6]. Models of this class study the instant connection between two vehicles in terms of distance between them and their speed. Microscopic simulation allows to mathematically describe the vehicle movement and the driver behavior. Here the driver behavior is described by a certain set of rules formulated mainly as "if .... then ...." condition. Exact location, speed and acceleration of the vehicle are determined at any moment of time depending on the driver's behavior and dynamic response of the vehicle. In general, microscopic simulation allows to analyse the traffic flow at the intersection, impact of the traffic light mode on the traffic flow dynamics, efficiency of traffic management measures in the individual road network components. However, this class of models requires a considerable set of input data, completeness and validity of which largely determine the adequacy of the simulation results.

To-date the increasing use of wireless technologies, such as GPS technologies and mobile telephone networks, has become a frequent practice and facilitates acquisition of large data arrays on population mobility and development of new tools for road network analysis. Application of such technologies allows, in particular, obtaining of a traffic congestion map, assessment of the current condition of traffic flows distribution during vehicle routing, short-term forecasting of the traffic flows in the individual sections of the city's road network, etc.

Up to 2017, one of the market leaders in traffic monitoring services in Ukrainian cities was "Yandex traffic jams" web-service - an extended edition of "Yandex maps" service. Using Yandex allowed obtaining information about the level of road network congestion by processing user data from navigation applications. Such web services display actual traffic congestion on the road network, allow making optimal routes based on network congestion and predictive assessment of traffic situation on the streets and roads for the next 15, 30, 45 and 60 min. Such web services however cannot be used as a simulation tool. It is explained by the fact that they operate historical data and real-time user data which are characteristic for already implemented scenarios of traffic management.

Methods and models based by analogy on the processes occurring in electrical circuits can also be applied to simulation traffic flows in logistic applications. However, an analysis of the studies known today indicates that solutions for complex networks (using this analogy) have been made only for individual transport tasks, which for the most part relate to finding the shortest path or finding the maximum flow in the network.

The transport problem for the shortest path, in the context of the use of an electrical analogue model, is to find such a path between two nodes of a branched electrical circuit in which the sum of the weights of individual sections (branches) of the electrical circuit is minimal. The features of this approach to the simulation of traffic flows, as well as its possibilities for practical applications, are considered in detail and described in [7]. However, in our opinion, this approach to simulation is not without a number of significant shortcomings, the main of which are as follows. The road network model, presented in the form of a branched electric circuit, does not fully (or not enough) reproduce the parameters of the roads, as well as the features of the organization of traffic on sections of the road network (including traffic direction, width of the carriageway of road sections, the number of traffic lanes and other equally important parameters).

It should be noted here that in recent years, the use of evolutionary programming methods can be considered more promising for solving the problems of finding the minimum route of various physical nature, primarily there are the genetic algorithms [8, 9???], and a method based on the effects of self-organization of ant agents [10]. In particular, the modified method of ant's self-organization [11] has found its application for solving the problems of finding the minimum route in the conditions of non-linear non-stationary dynamics of traffic flows on sections of the urban street-road network. Here, it must be recognized that in recent years there has been a rapid development of theoretical concepts, models, methods for describing non-linear non-equilibrium dissipative processes and phenomena occurring in systems of various physical nature, for example, in the field of phase transitions of molecular crystals, self-organization of the enterprise under the influence of external stochastic and deterministic changes in factors of a non-linear market environment, etc. [12–14]. One of the main approaches that are used to describe such processes is a synergistic approach based on the principles of self-organization.

At the same time, in recent years, an approach is applying based on studying the distribution of traffic flows by modelling a network of roads in the form of a branched electric circuit consisting of non-linear models of the cost functions of communication lines [15, 16]. This approach is designed to analyse the state of traffic load in a road network. Compared with the classical solution to find the distribution of traffic flows (based on the application of the well-known Frank-Wulf algorithm), the simulation results by the proposed approach are almost identical. At the same time, despite the positive results of such simulation, it should be noted that this approach has limitations on application, since, as you can see, it is focused on the description of the road network in a macroscale section, without taking into account its internal structure to a sufficient extent. This does not allow us to correctly solve the problem of optimizing the functioning of traffic flows in the network (including finding optimal solutions for individual sections and traffic interchanges of the network).

It should also be noted that the use of electrical analogies has found its application, including in solving optimization problems aimed at maximizing network bandwidth. In 2014, a group of American scientists in the city of Portland (Oregon), led by Kelner and Lorenzo Orechchia, used analogue electrical models to study information networks in terms of finding the maximum flow on the network [17]. In the general case, the results showed that analogue electrical networks can be used to solve this class of optimization problems. Moreover, these networks can be quite large dimension. This makes it expedient to search for various ways to implement such an approach for the correct solution of the class of optimization problems of the functioning of transport networks and traffic flows.

Therefore, currently available methods and models for description of traffic flows operation on road network are far from perfection. In particular, it concerns analysis and evaluation of road network congestion and optimization of traffic flows in cities, which is aimed at determining conditions and guidelines of rational traffic management on the road network to increase its throughput capacity.

The purpose of this paper is to improve the electrical analogy simulation model for the analysis of traffic flows in the urban road network, since implementation of such model will enable to determine pre-conditions for rational traffic management on the road network in order to increase its throughput capacity.

To achieve this purpose, it is necessary to solve the following problems:

- to find and generalize analogies in the laws of traffic flows behavior on the road network and processes behavior in electrical circuits;

- to develop implementation method for the electrical analogy simulation model, capable to analyse and optimize the traffic flows behavior on road network by means of redistribution of the congestion levels of different sections of relevant road network;
- to perform electrical simulation and optimization of the traffic flows, having taken a real fragment of Kyiv road network as an example for verification of the electrical analogy simulation model and the proposed method of its implementation.

2 Analysis of Analogies in the Regularities of Traffic Flows Behavior on the Road Network and Processes Occurring in Electrical Circuits

On the basis of the analogies found in the laws of traffic flows behavior on the road network and processes occurring in electrical circuits, an advanced electrical analogy simulation model for the analysis of traffic flows in the urban road network has been developed.

The proposed analogies include, in particular:

- point electrical charge is taken as a single unit (vehicle) of the traffic flow;
- current flowing through the electrical resistor reflects the intensity of the traffic flow at the relevant section of road network.
- road network sections are represented by electrical resistors, conductivity of which determines the throughput capacity of the respective section;
- field strength across the resistor reflects the density of the traffic flow in the road network section;
- voltage drop across the resistor reflects the integral flow in the relevant elementary section;
- electromotive force of the power source is the maximum transport flow in the corresponding road network section.

The systematization of the characteristics of the electric circuit and their compliance with the characteristics of the traffic stream and the road network in the most complete, generalized form is described and justified in the work [18]. In the paper [18] is has been also proven that the fundamental law of electrical engineering, the Ohm's law, which describes the relation of such characteristics as current, voltage, and resistance, is an isomorphic reflection of the basic equation of a traffic flow, which interrelates the major traffic parameters - namely, intensity, density and speed.

The revealed analogies between the electrical characteristics and parameters of road network and traffic flows are shown in Table 1 in a generalized form.

The process of loading and unloading of road network in time (from zero to maximum Nmax.) and vice versa can be represented as a quasi-stationary current arising in the RC circuit connected to the current power supply. Analogies between the respective processes are presented in Table 2.

Thus, it is proposed that the behavior of the traffic flow in road network sections to be reflected as quasi-stationary electrical processes occurring in the conductors of

Table 1. Analogies between the electrical characteristics and parameters of road network and traffic flows.

Electrical characteristics	Parameters of road network and traffic flows
Current: $I = \frac{dQ}{dt}$, where $Q = \sum q$ – conductor charge; q - point charge	Intensity of traffic flow through the road network cross-section: $J_t = \frac{dN_t}{dt}$, where dN_t – number of traffic units (vehicles) per hour dt; $[J_t]$ = hour^{-1}
Current density: $j = \frac{I}{S} = \frac{1}{S} \cdot \frac{dQ}{dt}$, where S – cross-sectional area of conductor	Intensity of traffic flow through the unit road network cross-sectional area: $J_s = \frac{1}{S} \cdot \frac{dN_t}{dt}$, where S – road network cross-sectional area; $[S]$ = km
Resistor resistance: $R = \frac{1}{\gamma} \cdot \frac{l}{S}$, where l – conductor length; γ – specific conductivity	Flow resistance determined by the driving time of vehicle (flow) within the relevant road network section: $R = \frac{1}{\gamma} \cdot \frac{l}{S}$, where l – road section length, γ – throughput capacity of the unit cross-sectional area; $[R]$ = hour
Voltage drop: $U = E \cdot l$, where E – electrical field strength	Number of units (vehicles) within the road network section: $N_l = \rho \cdot l$, where ρ – traffic flow density; $[\rho]$ = km^{-1}
Conductor capacitance: $C = \frac{Q}{U} = \frac{\int_t I(t)dt}{\int_l E(l)dl}$.	Traffic flow per unit time /traffic flow per unit length ratio: $C = \frac{Q}{U} = \frac{\int_t J(t)dt}{\int_l \rho(l)dl} = \frac{N_t}{N_l}$, $[C]$ = non-dimensional value

Table 2. Analogy between electrical and traffic processes.

Electrical processes	Traffic processes
Ohm's law in a differential form: $j = \gamma \cdot E$, where $\gamma = \frac{1}{\rho}$ – specific conductivity; ρ– specific resistivity	Major equation of motion: $J_t = \upsilon \cdot \rho$, where $\upsilon = \frac{l}{R}$ - traffic flow speed; ρ – traffic flow density
Capacitor voltage at charging: $U = U_{\max}(1 - e^{-\frac{t}{\tau}})$, where $\tau = RC$ – capacitor charge relaxation time	Process of road network loading with traffic flows: $N_l = N_l^{\max}(1 - e^{-\frac{t}{\tau}})$, where $\tau = RC$ – traffic flow relaxation time. Traffic flow intensity (in line with the Ohm's law in differential form) at time moment t: $J_t = \frac{N_{\max}}{R}(1 - e^{-\frac{t}{\tau}})$
Capacitor voltage at discharging: $U = U_{\max} e^{-\frac{t}{\tau}}$, where $U_{\max.}$ – initial voltage at capacitor	Process of road network unloading: $N_l = N_l^{\max} e^{-\frac{t}{\tau}}$, Traffic flow intensity at time moment t: $J_t = \frac{N_{\max}}{R} e^{-\frac{t}{\tau}}$.

electrical circuits. If there is a change in current or voltage on any part of the circuit, then after a certain time these changes will take place in other parts of the circuit. Since the disturbance propagation time is significantly, by the order of magnitude, shorter than the duration of these processes, it can be assumed that at any time the current value is the same in all series circuits. Similar processes in electrical circuits are explored on the basis of electrical current laws when applied to the instantaneous values of currents and voltages in the circuit sections.

3 Methodology of Electrical Simulation and Optimization of Traffic Flows on the Road Network

It is feasible to perform electrical simulation, studies and optimization of traffic flows on the road network in a software environment of circuitry simulators, such as Electronics WorkBench or NI Multisim. Using such simulators ensures easy and convenient measurement of the circuit parameters, as well as structural and parametric modification of electrical circuits.

Each section of the road network is modelled in electrical circuit by a series connection of diode and resistor, so in general the circuit will reflect the configuration of the structural elements and traffic management on the road network. Here, diode is responsible for setting the appropriate current direction, and therefore the traffic on a certain section of the road network, and the resistor is responsible for the throughput capacity. Simulation within such an electrical model allows description of the traffic in real conditions, since it takes into account the traffic direction, effect of intersections, etc.

This representation of the traffic flow dynamics on road and street sections allows simulation of custom-configured elements (multi-lane roads, junctions, intersections, etc.) in the form of single-line electrical diagrams. In this case, absolutely all (without exception) vehicle movement directions, allowed at the intersections (straight forward, right and left turns, U-turn) can be taken into consideration. It is also important to note that the two-way streets and roads can be considered as elements with separate lanes (separated by a central strip). Finally, the proposed representation of the model enables evaluation of traffic congestion of road network, including evaluation of individual lines and nodes of network in terms of their overloading or underloading.

A prerequisite for electrical simulation is that the total inbound and total outbound traffic flows for the road network (in the model, total input and total output currents) shall be the same.

Since the differences between total inbound and total outbound transport flows clearly depend on the size of network, it is obvious that when simulation a macro-objects, these difference can be significant. In order to improve the correct application of the model, it is advisable to make the electrical diagram of road network more sophisticated by connecting internal current supply sources, which will imitate the main centers of gravity for the sources of additional vehicles. Residential buildings, institutions, parking lots, service shops and malls, and other public buildings and structures can be centers of gravity.

The proposed model can be used for optimization of traffic flows on the road network. Optimization is achieved by redistributing traffic flows within the road network sections in order to unload congested areas and ensure better and uniform loading of the road network of the city as a whole. The main effect of this consists of increased traffic speed in congested areas and in general on the network, reducing the cost of transportation and accident rate on the streets and roads of the city.

Such redistribution of traffic flows will enable lower traffic delays in road network sections. It is obvious, however, that the delays reduction will take place on partially unloaded but still congested areas, but on under loaded sections there could be even no delays due to the available capacity reserves of these network elements. Given this, it can be assumed that in entire network the total traffic delays would be reduced.

Then we can state that the optimization goal (target function) of traffic flows is to maximize the throughput capacity of the road network, which depends on the level of traffic congestion in its sections. Variable parameters - congestion of road network sections. Mathematical model - electrical analogue simulation model for the analysis of traffic flows in the road network.

Problem to increase the throughput capacity of road network is a classical maximum network flow problem, solved by means of redistribution of congestion level of its sections [18].

In this context, the optimization goal problem for traffic flows on the road network can be presented as below:

$$\begin{cases} f(z_1, z_2, \ldots, z_n) \rightarrow \max \\ J_{in} - J_{out} = 0 \\ 0 \leq z_i \leq 0.7. \end{cases} \quad (1)$$

Here $f(z_1, z_2, \ldots, z_n)$ – throughput capacity of road network, which depends on the levels of congestion in sections $z_1, z_2, \ldots, z_n$; J_{in}, J_{out} – traffic intensity at the inbound and outbound sections respectively.

In this case:

$$z_i = \frac{J_i}{P_i} \in (0, 1), \quad (2)$$

where J_i – simulated traffic flow intensity on i-th section; P_i– throughput capacity of the driveway on i-th section.

As is commonly known [19], at $z \rightarrow 1$ a high density of traffic flows is observed, the speed slows down, and traffic conditions become complicated. At $z > 0.7$, the maximum saturation of the flow is achieved, and further use of streets and roads at this level of congestion is considered non-feasible.

Optimization goal presumes the following assumptions:

- road network with a finite number of elementary sections and intersections can be represented by an electrical diagram, each branch of which consists of a resistor and a series ideal diode. Electrical diagram is built according to the predefined design of the road network and allowed traffic directions on sections and intersections;

- in the electrical diagram, two-way roads can be considered as elements with separate lanes separated by a central strip;
- the total inbound and total outbound traffic flows for the road network (in the model, total input and total output currents) shall be the same, i.e. the conservation law is met;
- optimization task is solved with the exception of force-majeure events on the sections and intersections of the relevant road network, in particular, traffic collisions and vehicle accidents;
- in the electrical simulations of the intraday traffic flows distribution on the road network, change in the traffic intensity of the inbound and outbound traffic flows is implemented with a time discretion that is equivalent or better than the duration of vehicles existence in the respective network. This helps avoid the impact of traffic control devices (in particular, traffic lights) on the traffic intensity behavior at sections of the road network and ensures simulation adequacy;
- in the simulation of large road networks (or network fragments), the main centers of gravity for the sources of additional vehicles are represented in the analog electrical circuit as internal current supply sources of negative or positive sign, depending on the characteristic conditions for the period under review.

In order to introduce disturbances or impact of other factors occurring on road network, in particular, such as vehicle accidents in certain sections of the road network, it is necessary to recalculate the resistor values in the relevant branches of the simulation circuit, as applicable. In this case, the distribution of traffic flows on the road network according to the simulation results will be fundamentally different as the circumstances require.

4 Results of Electrical Simulation and Optimization of the Traffic Flows, Implemented for a Real Fragment of Road Network (Case of Kyiv City)

Key aspects of the electrical simulation and optimization of the traffic flows, were explored, having taken a real fragment of Kyiv road network as an example for verification. For simulation, a fragment was chosen whose road network is quite saturated and branched. The fragment under review consists of forty elementary sections, thirteen sections out of them are inbound and outbound. An elementary section is understood as part of a street or road between two nearest intersections.

For the relevant road network fragment, in a software environment of circuitry simulator NI Multisim, an electrical circuit was built. Its elements fully reflect the topology of the simulated road network, configuration parameters of the structural elements of network, specific features of the traffic management on the sections and intersections of streets and roads.

The constructed electrical circuit consisted of the following elements:

- the input current source which is an analogue of the traffic flow intensity at the corresponding input section of the studied fragment of the road network (J_{in_i}, where i = 1, 2, …, n, is the identifier of the i-th input current source);
- the output current source which is an analogue of the traffic flow intensity in the corresponding output section of the studied fragment of the road network (J_{out_i}, where i = 1, 2, …, n, is the identifier of the i-th source of the output current);
- the electrical resistance of the branch of the electric circuit, which is an analogue of the resistance of the corresponding section of the road crossing and is determined by the transit time of the traffic flow in the corresponding section of the investigated road network (R_i, where i = 1, 2, …, n, is the identifier of the i-th section of electric scheme)
- ammeters (A_i, where i = 1, 2, …, n, is the identifier of the ammeter in series in the i-th branch of the electrical circuit that is analogue of the studied fragment of the road network).

Series ammeters in the branches of analogy electrical circuit for the same fragment ensure control over current value through different branches of this circuit. Value of the ammeter current is an analogy to the intensity of the traffic flow on the respective sections of the network.

To perform electric simulation of distribution of traffic flows (in terms of traffic intensity) on the sections of fragment under review, information on the traffic flow intensity in the inbound (J_{in}) and outbound (J_{out}) sections of a given network fragment is required. In view of this, field observations of the dynamic behavior of the intensity of traffic flows were made on the respective road network sections. Observations were made during the intraday period with a constrained and dense traffic flow (between 16-00 and 20-00). Figure 1 shows traffic flow intensity in the inbound (J_{in}) and outbound (J_{out}) sections of a given network fragment for the above-mentioned intraday period, which correlates to the throughput capacity of these sections.

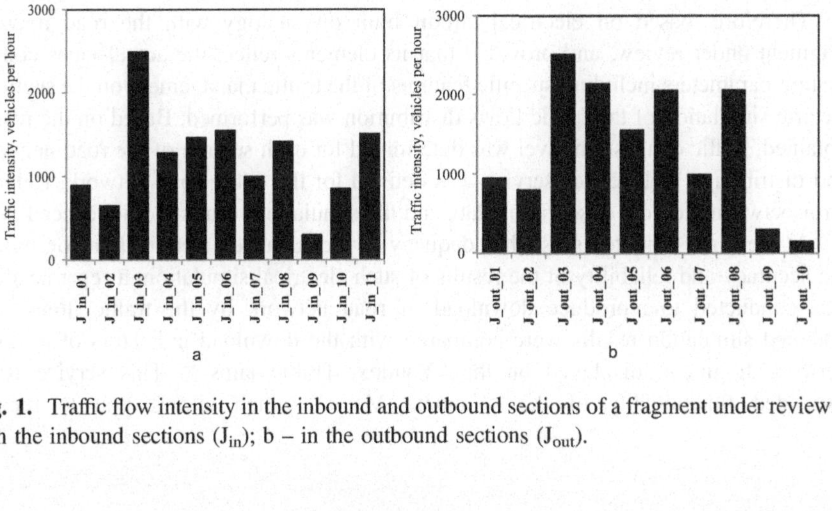

Fig. 1. Traffic flow intensity in the inbound and outbound sections of a fragment under review: a – in the inbound sections (J_{in}); b – in the outbound sections (J_{out}).

It was established posteriori that for the fragment under review the difference between the total inbound and total outbound traffic flow for the analysed intraday period does not exceed 10%. This assumption allowed, on the basis of the electrical circuit, to carry out correct electric simulation of the traffic flows distribution (in terms of traffic intensity) on the network sections under review.

For a qualitative description of the dynamic behavior of the traffic flow, relative indices of the basic parameters of the traffic flow - intensity, density and average speed - are often considered. Depending on the specific threshold values of these indicators, all possible traffic flow conditions are classified into certain categories, which, according to [19], have been called "levels of service" (Table 3). It should be noted that there is more than one classification of the levels of service in addition to those given in [19], but for the interpretation of the simulation results in this study it can be considered quite plausible.

Table 3. Classification of levels of service.

Level of service	z	c	ρ	Traffic condition
A	<0,2	>0.9	<0.1	Free-flow traffic
B	0.20–0.45	0.70–0.90	0.10–0.30	Stable traffic flow
C	0.45–0.70	0.55–0.70	0.30–0.70	Restricted flow
D	0.70–1.0	0.40–0.55	0.70–1.00	High-density flow
E	0.00; 1.00	<0.40	–	Unstable flow, forced traffic

"Level of congestion." In view of this, in this paper only one indicator was chosen for the levels of service at the road network sections, namely the level of congestion *(z)*, which was determined for each of the sections according to (2). In this case, the values of the traffic intensity J_i on the corresponding section of the road network were obtained from the electric simulation of the distribution of traffic flows on the network. It should also be noted that due to the highly correlated nature of such classification (see Table 3), if z_i is known, we can presume what limits (range) the relevant speed and density of traffic flow falls within.

Therefore, based on electrical circuit built on analogy with the road network fragment under review, and provided that its elements reflect the actual lanes configuration parameters including specific features of the traffic management on the sections, electric simulation of the traffic flows distribution was performed. Based on the results obtained, traffic congestion level was determined for each section of the road network, and distribution of levels of service was defined for the entire road network. Relative error between the field observation data and the simulation results did not exceed 20%.

At the same time, to assess the adequacy of the advanced electrical analog model, the accuracy and reliability of the results of such electrical simulation, a separate study was conducted to reproduce download of road network by the traffic flows. The obtained simulation results were compared with the download indicators of a model network fragment, displayed on the "Yandex. Traffic jams ". This service is an expanded functionality of the "Yandex. Maps", which allows you to receive information about the level of congestion of sections of the road network by processing information from users of navigation programs. Similar functionality for some cities (including Kiev city) is provided by the Google Maps and Nokia Maps services. During the study, the Pearson correlation coefficient between the indicators of download of network sections obtained by the results of electrical simulation and indicators according to the web service was calculated, which in the general case was 0.86. This indicates a strong relationship between the respective indicators. In addition, the reliability of the results of electrical simulation is confirmed by observations of the intensity of traffic flows, which were carried out in separate sections of the road network (discrepancy of results was not more than 20%).

It should also be noted that this error is not intrinsic for the model, since it can be mainly explained by the neglect of additional sources of traffic intensity increase within the road network sections under review. For example, in a certain period of time in the morning, vehicles start leaving residential area for work locations. It becomes a source of additional vehicles that generate respective traffic flows. This is very much supported by the fact that during the day simulation error on each specific section is volatile and caused by the intermittent changes in the traffic intensity on certain sections, which are not covered by the model. From a simulation standpoint, this may be compensated by inclusion of an additional current supply source (even variable) in the electrical circuit. Variable current supply source can be adjusted to cover additional sources of traffic intensity on sections of the road network in the model.

In general, if we compare the proposed approach for simulation the behavior of traffic flows with other methods known today that allow us to solve similar problems, we can confidently say that such an approach from the point of view of reliability, validity and accuracy demonstrates not worse, and sometimes better results due to the presence of obvious advantages to it. First of all, these advantages are the accessibility, versatility and simplicity of applying this approach to the simulation of road networks and traffic flows. In addition, the fact that in order to ensure electric simulation of the functioning of traffic flows in a road network with sufficiently high accuracy and reliability, it is necessary to have a relatively small set of input data.

Figure 2 shows distribution profile of "levels of service" for the road network fragment under review depending on traffic congestion level (z) at the inbound (J_{in}) and outbound (J_{out}) sections of the respective road network. Here elements of the analogue electrical circuit built for simulation of traffic flows distribution in road network fragment under review (see Fig. 2), reflect the actual lanes configuration parameters and specific features of the traffic management on these sections.

According to Fig. 2, it was found that:

- at $z\ (J_{in}, J_{out}) = \mathbf{0.00–0.20}$ share of road sections with "A" level of service varies within 100-42.9%, with "B" level of service – 2.4-57.1%, and sections with "C", "D", "E" level of service are not existent;
- at $z\ (J_{in}, J_{out}) = \mathbf{0.20–0.45}$ share of road sections with "A" level of service varies within 33.3-7.1%; with "B" level of service – 76.2-31.0%; with "C" level of service – 2.4-59.5%, with "D" level of service equals to 2.4% (at $z\ (J_{in}, J_{out}) = 0.38$), and sections with "E" level of service are not existent;

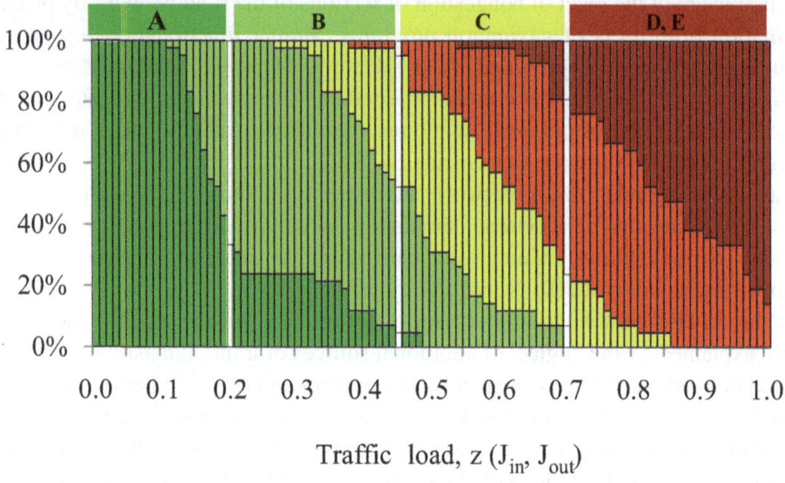

Fig. 2. Distribution profile of "levels of service" for the road network fragment under review (based on analogue electrical simulation of the distribution of traffic flows on the network).

- at $z\ (J_{in}, J_{out})$ = ***0.45–0.70*** share of road sections with "A" level of service equals to 4.8%, only at $z\ (J_{in}, J_{out}) = 0.47$, with "B" level of service varies within 28.6-4.8%, with "C" level of service – 61.9-23.8%, with "D" level of service – 4.8-59.5%, with "E" level of service – 2.4-19.0%.
- at $z\ (J_{in}, J_{out})$ = ***0.70–1.00*** road sections with "A", "B" level of service are not existent, share of road sections with "C" level of service varies within 21.4-4.8%, with "D" level of service –59.5 - 14.3%, with "E" level of service – 19.0-85.7%.

According to the study results, throughput capacity of the road network fragment under review (at the level of network congestion 0.70) is ≈ 4900 vehicles /hour.

At the same time, based on the results of electrical simulation optimization of traffic flows on the road network was performed according to (1). To achieve optimization goal, resistance values in certain circuit branches were accordingly recalculated. Such recalculation reflect changes in the actual lanes configuration parameters, namely - number of lanes in the driveway actually available for vehicles, as well as changes in the throughput capacity of those lanes according to (2). From a standpoint of changes in traffic management, such recalculation of resistance means, in particular, a ban for vehicle parking on the respective sections of road network. Given that, a simulation rerun was performed, which revealed that the throughput capacity of the road network fragment under review can be increased by 32%, up to 6800 vehicles /hour.

Figure 3 shows distribution profile of "levels of service" for the road network fragment under review, which was built based on results of analogue electrical simulation and relevant calculations (2) based on optimized electrical circuit.

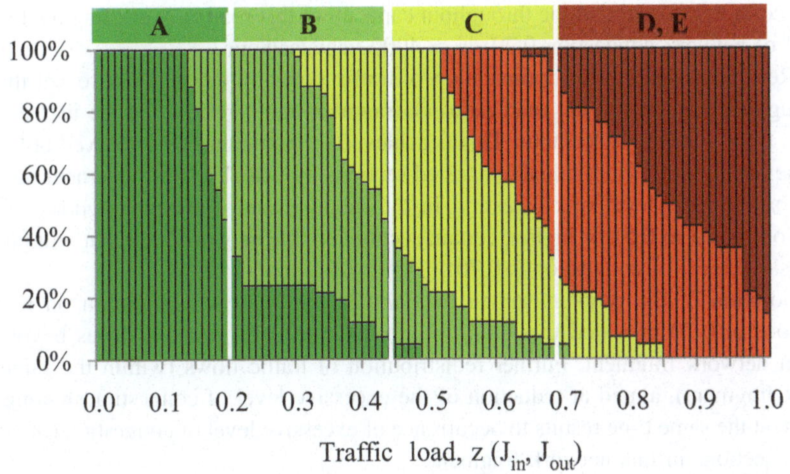

Fig. 3. Distribution profile of "levels of service" for the road network fragment under review (based on results of analogue electrical simulation after optimization of electrical circuit according to (1), (2)).

According to Fig. 3, it was found that:

- at z (J_{in}, J_{out}) = ***0.00–0.20*** share of road sections with "A" level of service varies within 100-45.2%, with "B" level of service – 11.9-54.8%, and sections with "C", "D", "E" level of service are not existent;
- at z (J_{in}, J_{out}) = ***0.20–0.45*** share of road sections with "A" level of service varies within 33.3-7.1%, with "B" level of service – 76.2-31.0%, with "C" level of service – 2.4-61.9%, and sections with "D", "E" level of service are not existent;
- at z (J_{in}, J_{out}) = ***0.45 – 0.70*** share of road sections with "A" level of service equals to 4.8%, with "B" level of service varies within 31.0-4.8%, with "C" level of service – 78.6-2149%, with "D" level of service – 9.8-66.7%, and sections with "E" level of service – 2.4–7.1%.
- at z (J_{in}, J_{out}) = ***0.70–1.00*** sections with "A" level of service are not existent, share of road sections with "B" level of service equals to 4.8%, only at z (J_{in}, J_{out}) = 0.70, sections with "C" level of service varies within 21.4-4.8%, sections with "D" level of service – 64.3-14.3%, and sections with "E" level of service – 14.3–85.7%.

5 Discussion of Results

The results of electric simulation of traffic flows allow for the conclusion that at the level of congestion in the inbound and outbound sections of road network equal to 0.38 or higher, level of congestion reaches 0.70 on certain sections in the network, which corresponds to "D" level of service. Accordingly, at a level of congestion equal to 0.69, which corresponds to the upper limit of "C" level of service, about 80% of the sections of the entire network under review can be attributed as areas with "D", "E" level of service. It was found that the throughput capacity of this road network fragment (at the level of network congestion 0.70) is ≈ 4900 vehicles/hour.

Results obtained after optimization of traffic flows on the network reveal that the throughput capacity of the road network fragment under review can be increased by 32%, up to 6800 vehicles/hour. This is achieved, in particular, by improved utilization of the lane width of individual congested sections by vehicles (for movement only). In this case, at the level of congestion in the inbound and outbound sections of road network equal to 0.52 or higher, certain sections in this network fragment will have an excessive level of congestion (>0.70).

Solution of the problem aimed to maintain the level of congestion on network sections at 0.70 provided that $z\ (J_{in},\ J_{out})$ equals to 0.52, probably goes beyond the given network fragment. Further redistribution of traffic flows (within the same network fragment), aimed at reduction of the excessive level of congestion in congested areas, at the same time results to occurrence of excessive level of congestion (>0.70) on other sections in this network fragment.

It should be therefore noted that the proposed analogue electrical simulation model can become a meaningful tool for analysis of urban transportation system condition. Use of this model will allow, in particular, to find a solution of such fundamental interrelated problems as:

- definition of urban road network congestion (setting distribution of traffic flows by traffic intensity on the sections of road network, including those based on predictive assessment of network saturation by traffic flows);
- definition of road network sections with excessive level of congestion and development of mitigation measures;
- parametric optimization of traffic light control on the network sections and nodes with excessive level of congestion;
- evaluation of effect from commissioning of new infrastructure facilities (roads, bridges, tunnels, etc.), including identification of priority facilities and urban development plans;
- efficiency assessment of investment into road network development and improvement projects (identification and implementation of priority projects);
- development of measures for mitigation of adverse environmental impact of motor vehicles in the cities;
- etc.

6 Conclusions

In a generalized form, existence of an analogy between the parameters and functionality of the process behavior in the electrical circuit and traffic flows on the road network was found, described, systematized and justified. Based on the results obtained, the electrical analogy simulation model for the analysis of traffic flows in the urban road network has been improved. Analysis of the road network condition within this model allows finding the most efficient solutions to ensure rational organization of the traffic flows on urban road network.

Implementation method for the advanced electrical analogue simulation model was proposed and justified, which enables to optimize the traffic flows behavior on road network by means of redistribution of the congestion levels of different sections. Such optimization is achieved by redistributing traffic flows within the road network sections in order to unload congested areas and ensure better and uniform loading of the road network of the city as a whole. The main effect of this consists of increased traffic speed in congested areas and in general on the network, reducing the cost of transportation and accident rate on the streets and roads of the city, etc.

Key aspects of the electrical simulation and optimization of the traffic flows in urban environment were explored, having taken a real fragment consisting of forty elementary sections, thirteen sections out of them are inbound and outbound, of road network in one of administrative districts of Kyiv as an example. The advanced simulation model and proposed implementation method have been duly tested and assessed. Relative error between the field observation data and the simulation results did not exceed 20%. Optimization of traffic flows, which is aimed at increase of throughput capacity of the road network, was performed within the fragment under review. The obtained simulation results indicate that due to improved utilization of the lane width of individual congested sections by vehicles (for movement only), the throughput capacity of the road network fragment under review can be increased by 32% compared to its actual condition, namely up to 6800 vehicles/hour.

Acknowledgment. Grateful acknowledgment of the authors is due to Viktor I. Kryvenko, Professor of electronics and computing department of the National Transport University, in particular, for his kind advice on selection of simulation software by NI Multisim (electrical circuit simulator) for performing a number of simulations and studies.

References

1. Lighthill, M.H., Whitham, G.B.: On kinematic waves II. a theory of traffic flow on long crowded roads. Proc. Royal Soc. Lond **229**, 317–345 (1955). https://doi.org/10.1098/rspa.1955.0089
2. van Wageningen-Kessels, F., van Lint, H., Vuik, K., Hoogendoorn, S.: Genealogy of traffic flow models. EURO J. Transp. Logistics **4**(4), 445–473 (2014). https://doi.org/10.1007/s13676-014-0045-5
3. Oyala, C.O., Otumba, E.O.: Modelling of distribution of the "Matatu" traffic flow using Poisson distribution in a highway in Kenya. Int. Math. Forum **13**(8), 385–392 (2018). https://doi.org/10.12988/imf.2018.8636
4. Zhang, Y., Ye, N., Wang, R., Malekian, R.: A method for traffic congestion clustering judgment based on grey relational analysis. Int. J. Geo-Inf. **5**(71), 1–15 (2016). https://doi.org/10.3390/ijgi5050071
5. Borsche, R., Kimathi, M., Klar, A.: A class of multi-phase traffic theories for microscopic, kinetic and continuum traffic models. Comp. Math. Appl. **64**, 2939–2953 (2012). https://doi.org/10.1016/j.camwa.2012.08.013
6. Maciejewski, M.: A comparison of microscopic traffic flow simulation systems for an urban area. Transp. Prob. **5**(4), 27–38 (2010)

7. Bullinghamand, J., Matthews, P.: Electronic simulator for quickest routes in a road network. Traffic Eng. Control **12**(5), 240–243 (1970)
8. Glover, F., Kochenberger, G.: Handbook of metaheuristics. In: International Series in Operations Research & Management Science, vol. 57, p. 570 (2003)
9. Rejer, I., Lorenz, K.: Classic genetic algorithm vs. genetic algorithm with aggressive mutation for feature selection for a brain-computer interface. Przegląd Elektrotechniczny **91** (2), 98–102 (2015). https://doi.org/10.15199/48.2015.02.24
10. Dorigo, M., Gambardella, L.M.: Ant colonies for the travelling salesman problem. BioSystems **43**(2), 73–81 (1997). https://doi.org/10.1016/S0303-2647(97)01708-5
11. Danchuk, V., Bakulich, O., Svatko, V.: Building optimal routes for cargo delivery in megacities. Transp. Telecom. **20**(2), 142–152 (2019). https://doi.org/10.2478/ttj-2019-0013
12. Puchkovska, G.O., Danchuk, V.D., Makarenko, S.P., Kravchuk, A.P., Kotelnikova, E.N., Filatov, S.K.: Resonance dynamical intermolecular interaction in the crystals of pure and binary mixture n-paraffins. J. Mol. Struct. **708**(1–3), 39–45 (2004). https://doi.org/10.1016/j.molstruc.2004.02.010
13. Puchkovska, G.O., Makarenko, S.P., Danchuk, V.D., Kravchuk, A.P.: Temperature study of resonance intermolecular interaction in normal long-chain carboxylic acid crystals using IR absorption spectra. J. Mol. Struct. **744–747**, 53–58 (2005). https://doi.org/10.1016/j.molstruc.2005.01.002
14. Danchuk, V.D., Kozak, L.S., Danchuk, M.V.: Stress testing of business activity using the synergetic method of risk assessment. Actual Probl. Econ. **171**(9), 189–198 (2015)
15. Cho, H.-J., Huang, H.: A circuit simulation technique for congested network traffic assignment problem. AIP Conf. Proc. **963**, 993–996 (2007). https://doi.org/10.1063/1.2836261
16. Huang, K., Cheng, C.-C.: A solution algorithm based on circuit simulation for the traffic assignment problem. In: Proceedings of the 40th International Conference on Computers and Industrial Engineering: Soft Computing Techniques for Advanced Manufacturing and Service Systems, pp. 1–6. NCTU Academic Hub, Shanghai (2010). https://doi.org/10.1109/iccie.2010.5668306
17. Knight, H.: New algorithm can dramatically streamline solutions to the "max flow" problem. MIT News **4**, 21–26 (2014)
18. Danchuk, V., Kryvenko, V., Oliinyk, R., Taraban, S.: Electrotechnical model for research of traffic flows. Bull. Nat. Transp. Univ. **21**(2), 28–32 (2010)
19. Lobanov, E.M.: Transport urban planning: textbook for university students. Transport, Moscow, 240 (1990)

Automation of the Solution to the Problem of Optimizing Traffic in a Multimodal Logistics System

Julia Poltavskaya[1(✉)], Olga Lebedeva[1], and Valeriy Gozbenko[1,2]

[1] Angarsk State Technical University, 60, Street Chaykovskogo, Angarsk 665835, Russia
juliapoltavskaya@mail.ru
[2] Irkutsk State Transport University, 15, Street Chernyshevskogo, Irkutsk 664074, Russia

Abstract. The development of methods for automating the distribution of finished products is possible using mathematical methods. Algorithms based on these methods will improve the efficiency of the transport industry and the logistics component in the network. Transportation costs account for a large part of the total logistics costs, so optimizing a multimodal logistics system increases the productivity of vehicles. The use of a mathematical model for optimizing the multimodal logistics chain allows controlling the functioning of transportation processes and combining incoming cargo information into a single data collection and processing system, which will improve the quality and efficiency of all links in the transport and logistics chain as a whole. The simulation results showed that in order to optimize a multi-modal logistics network in an urban agglomeration, many conditions, such as cost reduction, delivery time limits and the introduction of fines for exceeding exhaust gas emissions should be taken into account. The influence of the above factors on the optimal solution of the problem is studied. To solve it, the use of a heuristic algorithm for automating the design of new logistics systems was proposed taking into account changing demand and changes in transportation tariffs.

Keywords: Multimodal logistic system · Transport optimization · Minimization of transportation cost · Heuristic algorithm

1 Introduction

With the development of globalization and information technology in the transport industry, the relevance of tasks with a logistics component in the development of transport and logistics chains is increasing. The work of logistics centers allows to optimize the distribution processes of finished products using mathematical methods, which increases the competitiveness and efficiency of the industry as a whole. Thus, measures for modernization in the field of freight transportation are planning and managing the processes of moving goods from departure point to the destination in the transport network with minimal costs. Transportation takes a third of the total cost of

V. Murgul and V. Pukhkal (Eds.): EMMFT 2019, AISC 1258, pp. 255–261, 2021.
https://doi.org/10.1007/978-3-030-57450-5_23

logistics, and the efficient organization of transport routes increases the productivity of the entire system [1–3].

Freight transportation in a multimodal logistics network can be carried out by all types of transport: road, rail, water and air. Transportation options depend on the number of modes of transport used. The use of the combined method of transportation allows to increase the efficiency of transport services. Multi-modal transportation not only allows to use the advantages of each type of transport, but also eliminate their disadvantages.

The increase in freight traffic determines the economic development of the region, but there are also negative consequences. Climate change is associated with various emissions of pollutants, including from freight vehicles. Freight transport contributes about 5.5% of all global greenhouse gas emissions [4]. Carbon dioxide emissions during transportation account for 93% of the total volume, while storage accounts for only 7% [5]. In view of this, the creation of an environmentally sustainable logistics system is a relevant area of research. Improving the quality of logistics services, reducing transportation costs and reducing external impacts on the environmental component of the region will allow to achieve a stable balance between the economic, environmental and social goals of the development strategy of the country and its subjects [1, 6–9].

Optimization of the logistics network is a strategic planning, including the planning of combined transport and the organization of interaction of logistics facilities. Therefore, optimization of a multimodal logistics network, taking into account time and cost costs and environmental emissions, is an important and difficult task.

A literature review showed that earlier studies [2, 10–13] focused on the design of the logistics network and include optimization of the supply chain at the regional level, taking into account the environmental situation.

The traditional model for optimizing a multimodal logistics network is based on minimizing the total cost of transportation or the efficiency of the goods distribution process, and does not take into account the environmental impact on the environment [14]. The model under consideration includes all of the above factors. The study aims to optimize the multimodal logistics network in order to minimize the total cost of transportation and carbon emissions.

2 Materials and Methods

Figure 1 shows a diagram of a logistics network in which cargo is transported from a departure point (O) to a destination (D) through a certain number of intermediaries (I_1... I_n).

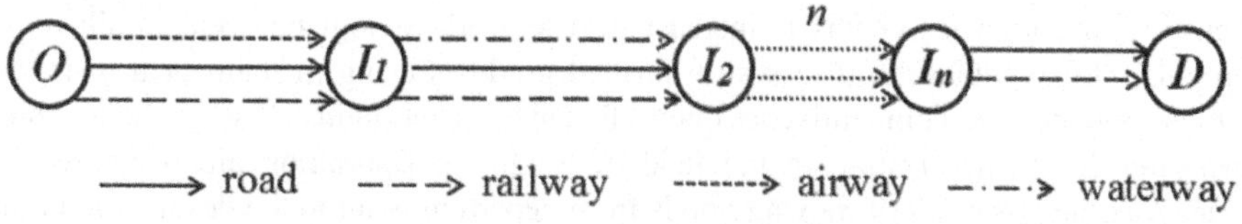

Fig. 1. Types of routing messages in a multimodal logistics system.

For each carrier, an urgent and cost-effective transportation scheme is required, which allows to deliver over long distances for a given time period set by the sender, with minimal cost. Thanks to the development of logistics technologies and the possibility of automation of transportation management processes, multimodal logistics networks can be used to optimize the entire supply system. The figure also reflects alternative transportation methods by various modes of transport (for example, automobile, rail, air), which are available between pairs of points. The time and cost of transportation, throughput, and carbon emissions vary depending on the mode of transport selected for a particular pair of points. The total time of arrival at the destination cannot exceed the set time. Transshipment from one type of transport to another is characterized by different durations in time, and is carried out in the transport node of the network. Thus, the choice of optimal congestion nodes affects the possibility of increasing network bandwidth. Choosing the best options allows to minimize the total cost of delivery, taking into account permissible restrictions on the time of transportation and volume of transportation [15, 16].

2.1 Formulation of the Model

When formulating the model conditions, the following restrictions are established:

1. For the transportation of goods between departure points and destination, only one of the alternative options may be selected.
2. Transshipment from one type of transport to another is carried out once in each transport node of the network in question.
3. The option of transportation using the same type of transport between different points of departure and destination is characterized by the same speed.
4. A linear relationship is observed between the total transportation costs, distance and volume of transportation.

The objective function (Eq. (1)) minimizes the total costs, which include five components, namely: transportation costs, fixed costs for transshipment of goods in transport nodes, fines for exhaust emissions [1]:

$$minZ = \sum_{i \in N/D} \sum_{m \in M} c_{ij}^m d_{ij}^m Q + \sum_{i \in T} \sum_{m \in M} \sum_{l \in M} f_i^{ml} Q + \sum_{i \in T} K_i z_i \lambda \left(\sum_{i \in N/D} \sum_{m \in M} e_{ij}^m d_{ij}^m Q + \sum_{i \in T} \sum_{m \in M} \sum_{l \in M} e_i^{ml} Q \right) \tag{1}$$

$$t_{ij}^m = \frac{d_{ij}^m}{v^m}, \forall i \in T, m \in M, \tag{2}$$

$$\sum_{i \in N/D} \sum_{m \in M} t_{ij}^m x_{ij}^m + \sum_{i \in T} \sum_{l \in M} \sum_{m \in M} y_i^{ml} t_i^{ml} = w, \tag{3}$$

$$\sum_{m \in M} x_{ij}^m = 1, \forall i \in T, \tag{4}$$

$$\sum\nolimits_{m \in M} \sum\nolimits_{l \in M} y_i^{ml} = 1, \forall i \in T, \quad (5)$$

$$x_{ij}^{m} + x_{ij}^{l} \geq 2y_i^{ml}, \forall i \in T, \quad (6)$$

$$x_{ij}^{m} Q \leq u_{ij}^{m}, \forall i \in T, m \in M, \quad (7)$$

where N – many items on the net; N/D – many points except destination; T – many departure points; M – many transportation options between i-points; c_{ij}^{m} – transportation cost from i to j point m by means of transport; d_{ij}^{m} – transportation distance from i to j point for m type of transport; Q – transportation volume; f_i^{ml} – transshipment cost from m mode of transport to l at the i-point; K_i – fixed overload costs at i-point; z_i – boolean variable, takes the value 1 if overload is performed at the i-point, otherwise 0; □ – penalty for exhaust emissions per unit of emissions; e_{ij}^{m} – volume of emissions during transportation of goods from i to j point m by mode of transport; e_i^{ml} – volume of emissions during transshipment from m mode of transport to l at the i-point.

Restriction (2) sets the duration of transportation between two points. Restriction (3) reflects that the total duration of transportation includes the movement time between points and the time of transshipment in the distribution center. Equation (4) indicates that only one type of transport is used to transport goods between two points. Restriction (5) indicates that overloading can only be carried out in the distribution center. Restriction (6) provides the possibility of transit transportation through point i. Constraint (7) ensures that the throughput capacity of the transport network m between points i and j is not exceeded.

2.2 Heuristic Algorithm for Solving the Problem

Consider the use of the heuristic algorithm to solve the problem of optimizing a multimodal logistics network [1, 4].

The essence of the heuristic algorithm is as follows: the shortest route is searched for, taking into account the time limit on transportation in a virtual network. Then an alternative transportation option is selected from the last pair of points to the first, so that the total service time limit is respected.

Each communication line represents a certain type of transportation for a given network from point of departure O to destination D through (K + 1) points that are included in the virtual logistics network (Fig. 2).

The following notation is used in the algorithm:

Dist[i] – shortest way from departure point O to point i;

T[i] – total transportation time from point O to point i;

T_f – actual value of the duration of transportation to destination D;

t_1 and t_2 – upper and lower bounds on the time for which transportation to destination D can be carried out;

C[i][j] – transportation cost between points i and j;

V_0 – many points; set of virtual items V = {O, 1, 2, 3, ···, (8 * k + 1), ···, (8 * k + 8), D};

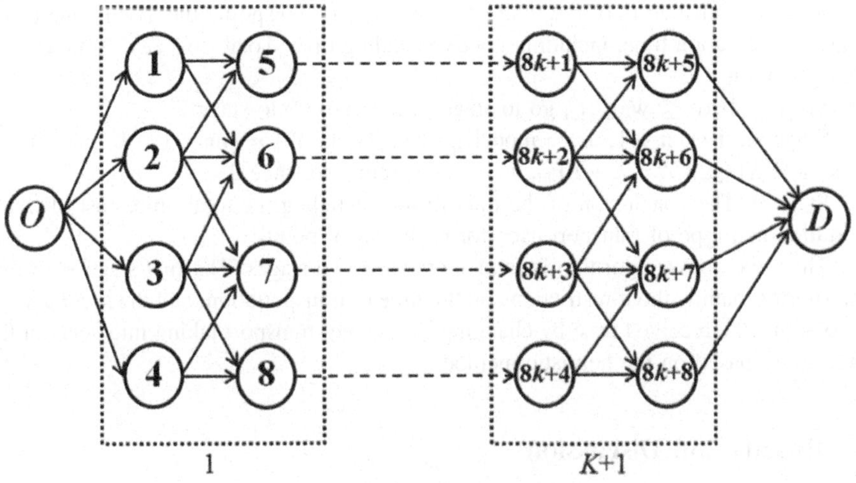

Fig. 2. Virtual multimodal logistics network.

S – many items involved in the implementation of transportation; $\bar{S}$ – many points through which transportation is not carried out from departure point O to destination D: $S, \bar{S} \subset V$ и $S \cup \bar{S} = V$;

Path[i] = k indicates that the previous point i is the node k along the shortest path from point O. Takes the value ∞ if the shortest path from point O to point i does not exist;

Label[i] = {0, 1} – boolean variable indicating whether the shortest path from point O to point i has been found. It takes the value 1 if the path is set; otherwise, 0.

Mode[i] – type of transportation for a couple of items;

ΔT_{kli} – difference between the durations of transportation in case of replacing one mode of transport k with another l in point i;

ΔF_{kli} – additional costs for replacing one mode of transport k with another l in paragraph i;

Max(i, k, l*) indicates that the mode of transport l is the best of all possible transportation options to point i.

The algorithmization process is carried out in stages.

Stage 1. Entering parameters: $S \leftarrow \{O\}$; path[0] ← 0; label[0] = 1; $S \leftarrow V - S$; dist[i] ← C[0] [i]; label[i] = 0; path[i] = ∞; $\forall i \in S$.

Step 2. If label[D] = 1, step 6 is carried out; otherwise – step 3.

Stage 3. Selection min {dist [j]} and dist [k] = min {dist [j]}, where $i \in S$, $j \in \bar{S}$ and $q \leq f_{ij}$. Replacement S, $\bar{S}$ and label[k]: $\bar{S} \leftarrow \bar{S} - \{k\}$, $S \leftarrow S \cup \{k\}$, label [k] = 1.

Stage 4. Dist[j] ← min {dist [i] [j], dist [i] [k] + C[k] [j]}, path [j] ← k, $\in i \in S$, $j \in \bar{S}$; return to stage 2.

Stage 5. Listing the shortest routes from point of departure O to destination D and calculating the total transportation time along the shortest route.

Stage 6. Calculation of the total time and cost of transportation: Total_time (w) – total transportation time, including goods reloading time; Total_cost (Z) – total cost of transportation.

Stage 7. If $t_1 \leq w \leq t_2$, go to stage 9; otherwise – to stage 8.

Stage 8. $i \leftarrow n + 1$; $k \leftarrow$ mode[i]; max $\{i, k, l^*\} = \{\max l \in J\ \Delta T_{kli} \Delta F_{kli}\}$; $w = w - \Delta T_{kli}$; и $Z = Z + \Delta F_{kli}$, $i \leftarrow i + 1$; return to stage 2.

Stage 9. The conclusion of the final result, including the total time, cost of transportation and type of transport used for each pair of points.

Thus, the above algorithm is implemented in two stages. The first stage is to find the shortest path without restrictions on the time of transportation, and the second stage is to adjust the received time by changing the type of transport taking into account the time limit, based on the heuristic method.

3 Results and Discussion

The paper considers a model for optimizing a multimodal logistics system that takes into account the time, cost of transportation, and costs of carbon dioxide emissions. Based on the characteristics of the optimization model, a heuristic algorithm is applied to solve the problem.

As a result of the study, the following conclusions were made: an increase in the duration of the transportation of goods from the point of departure to destination affects the structure of the logistics network and the choice of mode of transport. The introduction of a time interval limitation in the model in order to reduce the congestion time between different modes of transport allows the choice of the optimal route in a multimodal logistics system. Automation of solving the problem of choosing a transportation method in a multimodal logistics system is an urgent study related to improving the efficiency of the transport network as a whole.

References

1. Zhang, D., He, R., Li, S., Wang, Z.: A multimodal logistics service network design with time windows and environmental concerns. PLoS ONE **12**(9), e0185001 (2017). https://doi.org/10.1371/journal.pone.0185001
2. Lebedeva, O., Poltavskaya, J., Gozbenko, V.: Simulation of an integrated public transport system by the example of a compact city. IOP Conf. Ser. Mater. Sci. Eng. **760**(012023), 1–8 (2020). https://doi.org/10.1088/1757-899X/760/1/012023
3. Piecyk, M.I., McKinnon, A.C.: Forecasting the carbon footprint of road freight transport in 2020. Int. J. Prod. Econ. **128**, 31–42 (2010). https://doi.org/10.1016/j.ijpe.2009.08.027
4. Dekker, R., Bloemhof, J., Mallidis, I.: Operations research for green logistics—an overview of aspects, issues, contributions and challenges. Eur. J. Oper. Res. **219**, 671–679 (2012). https://doi.org/10.1016/j.ejor.2011.11.010
5. Lebedeva, O., Kripak, M., Gozbenko, V.: Increasing effectiveness of the transportation network through by using the automation of a Voronoi diagram. Transp. Res. Procedia **36**, 427–433 (2018). https://doi.org/10.1016/j.trpro.2018.12.118

6. Ivanova, S.V., Molchanova, E.D.: Development of a cargo transportation organization system for enterprises of railway transport. Mod. Technol. Syst. Anal. Model. **1**(65), 112–119 (2020). https://doi.org/10.26731/1813-9108.2020.1(65).112-119
7. Tamannaei, M., Rasti-Barzoki, M.: Mathematical programming and solution approaches for minimizing tardiness and transportation costs in the supply chain scheduling problem. Comput. Ind. Eng. **127**, 643–656 (2019). https://doi.org/10.1016/j.cie.2018.11.003
8. Lebedeva, O.A.: Dynamic modeling of the optimal route in the multimodal transport network. Mod. Technol. Syst. Anal. Model. **1**(65), 44–50 (2020). https://doi.org/10.26731/1813-9108.2020.1(65).44-50
9. Huang, Y., Chen, C.-W., Fan, Y.: Multistage optimization of the supply chains of biofuels. Transp. Res. Part E-Logistics Transp. Rev. **46**, 820–830 (2010). https://doi.org/10.1016/j.tre.2010.03.002
10. Levashev, A., Mikhailov, A., Sharov, M.: Special generators in tasks of transportation demand assessment. Transp. Res. Procedia **36**, 434–439 (2018). https://doi.org/10.1016/j.trpro.2018.12.119
11. Ishfaq, R., Sox, C.R.: Hub location-allocation in intermodal logistic networks. Eur. J. Oper. Res. **210**, 213–230 (2011). https://doi.org/10.1016/j.ejor.2010.09.017
12. Shepelev, V., Almetova, Z., Larin, O., Shepelev, S., Issenova, O.: Optimization of the operating parameters of transport and warehouse complexes. Transp. Res. Procedia **30**, 236–244 (2018). https://doi.org/10.1016/j.trpro.2018.09.026
13. Kripak, M.N., Palkina, E.S., Seliverstov, Ya.A: Analytical support for effective functioning of intelligent manufacturing and transport systems. IOP Conf. Seri. Mater. Sci. Eng. **709**(3), 1–8 (2020). https://doi.org/10.1088/1757-899x/709/3/033065
14. Chislov, O.N., Bogachev, V.A., Kravets, A.S., Bogachev, T.V., Filina, E.V.: Multi-agent approach in mathematical modeling of distribution of regional cargo flows. Mod. Technol. Syst. Anal. Model. **64**(4), 87–95 (2019). https://doi.org/10.26731/1813-9108.2019.4(64).87-95
15. Lebedeva, O.A., Gozbenko, V.E., Kargapol'tsev, S.K.: Optimizing urban transportation using entropy model. Mod. Technol. Syst. Anal. Model. **64**(4), 131–137 (2019). https://doi.org/10.26731/1813-9108.2019.4(64).131-137
16. Simangunsong, E., Hendry, L., Stevenson, M.: Supply chain uncertainty: a review and theoretical foundation for future research. Int. J. Prod. Res. **50**(16), 4493–4523 (2012). https://doi.org/10.1080/00207543.2011.613864

Improving the Energy Efficiency of Technological Equipment at Mining Enterprises

Roman Klyuev[1,2(✉)], Igor Bosikov[1], Oksana Gavrina[1], Maret Madaeva[3], and Andrey Sokolov[1]

[1] North-Caucasian Institute of Mining and Metallurgy (State Technological University), 44, Nikolaeva Street, Vladikavkaz 362021, Russia
kluev-roman@rambler.ru

[2] Moscow Polytechnic University, B. Semenovskaya Street, Moscow 107023, Russia

[3] Grozny State Oil Technical University Named After Academician MD Millionshchikov, 100, Isaeva Avenue, Grozny 364051, Russia

Abstract. The article presents the results of the economic effect obtained by increasing the productivity of technological equipment at the mining and processing plant. The purpose of this work is a comprehensive study of power consumption issues for individual technological links and for the processing plant as a whole, as well as issues of increasing energy efficiency by optimizing the operation of equipment during the production and transportation of ore between processing stages. In accordance with the task, the following issues are highlighted: specific power consumption rates for each processing plant are given; the economic effect of implementing calculated scientifically based power consumption rates and optimizing the loading mode of technological equipment is calculated. It is established that the reserve for saving electricity is to increase the productivity of mills to a maximum of 55–60 tonne/h. An increase in the average mill load by 1 tonne/h corresponds to a decrease in the specific consumption of the factory by 0.56 kWh/ton. It is recommended to maintain the productivity of ball mills in the first stage of grinding at the level of 35–40 tonne/h. At the same time, the loading of all technological mechanisms of the processing plant will increase, which will allow obtaining an economic effect of 30.7 million rubles.

Keywords: Technological · Equipment · Mining enterprises

1 Introduction

In modern conditions, the solution of the problem of electrification development is closely related to the tasks of improving the use and saving of electricity in the process of its transportation between separate technological processing units of production. In many enterprises, insufficient attention is paid to the rational use of electricity. This is largely due to the low share of electricity in the cost of production. As a result,

V. Murgul and V. Pukhkal (Eds.): EMMFT 2019, AISC 1258, pp. 262–271, 2021.
https://doi.org/10.1007/978-3-030-57450-5_24

enterprises have poor accounting of electricity consumption, and insufficient technological discipline of energy use.

Scientifically based rationing and forecasting of specific rates of electricity consumption in the conditions of production intensification, reduction of electric capacity and non-productive losses of electricity are important factors in reducing the cost of production and increasing labor productivity. All this fully applies to mining and processing enterprises. With the current global trend of increasing production of copper and barite concentrates, the tasks of rationing and forecasting power losses, determining unproductive power losses in the elements of the network of a mining and processing plant are undoubtedly important and relevant.

2 Characteristics of the Technological Process

In accordance with the planned task, copper-pyrite and barite-containing ores are simultaneously delivered to the processing plant under consideration.

The technological scheme for processing copper-pyrite ores provides for three-stage crushing in an open cycle with preliminary screening, ore grinding to 55–60% of the class −0.074 mm is performed in two stages.

Barite-containing ores are sent to the processing plant, where they are processed alternately on one section. Crushing of all barite-containing ores includes 2 stages. Barite ores make up 54% of the total amount of barite-containing ores; barite-lead ores make up 12%; and borite-polymetallic ores make up 34%.

For the enrichment of all ores, a flotation scheme has been adopted, including 2 stages. Crushing of crushed ore up to 75% of the 0.074 mm class is carried out in two stages.

For crushing, copper-pyrite ore is received in the amount of 6.600 tonne per day. A jaw crusher with a mouth size of 1500 by 2160 mm is installed in the large crushing case. With a discharge slot of 180 mm and a bulk ore weight of 18 tonne/m^3, the productivity of this crusher is 560 tonne/h.

Performance control, as well as uniformity of loading of the jaw crusher is provided by plate feeders.

In the case of medium and small crushing, there is a complex consisting of one medium crushing crusher with a diameter of 2200 mm and two small crushing crushers with a diameter of 2200 mm. All crusher with hydraulic adjustment of the gap. The capacity of the complex of crushers is 600 tonne/h.

Barite-containing ores are received for crushing in the amount of 555 tonne/day. The jaw crusher has a mouth size of 600 by 900 mm. With a 75 mm discharge slot, the capacity of this crusher is 42 m^3/h.

3 Methods and Approaches

Of the existing methods for calculating the power consumption of mining and processing enterprises, three methods are most widely used: statistical, computational and experimental, and computational and statistical, used by many researchers [1–4].

The statistical method for calculating power consumption indicators is based on the use of average operating ratios of the amount of electricity consumed to the amount of produced and processed product. The reporting figure for the specific electricity consumption of the last month (quarter) is used as a base and is approximately extrapolated to the subsequent period or adjusted for measures to save electricity and increase the power capacity of the production and enrichment processes.

Due to the lack of scientifically based calculations and guidelines for determining power consumption in most mining enterprises, power consumption indicators are determined using a reporting and statistical method without technical justification.

Reporting and statistical method should not be confused with mathematical-statistical technique that allows us to scientifically substantiate the reality and the accuracy of the determined energy consumption to give an estimate of possible deviations of the energy consumption when changing the process parameters, to establish the degree of influence of production factors. In this method, experimental and reporting data are used with mandatory preliminary analysis, which allows you to exclude unproductive costs in case of technological process violations and take into account changes in production volumes. Therefore, most researchers refer to the mathematical and statistical method as a computational and experimental method using probability theory and mathematical statistics.

The calculation and analytical method for determining the expected power consumption is based on theoretical calculations that link the installed power of the electric receiver with the indicators of its load and operating mode.

This method of determining power consumption involves determining it depending on the number, purpose and type of electric receivers, as well as the drive system and operating conditions of the mechanism. All these features of the operation of mechanisms in mining enterprises should be expressed in the corresponding values of the coefficients of demand of the k_{demand} (or coefficients of load k_{load}) and the use of the mechanism over time k_{time}. The values of these coefficients vary widely.

4 Determination of the Economic Effect Due to the Introduction of a Scientifically-Based Rate of Electricity Consumption

Based on a large statistical material, the values of the actual specific electricity consumption for the main divisions of the concentrator are established. The values of specific electricity consumption for the processing plant divisions for 2019 are shown in Table 1 [5–8].

Table 1. Values of specific electricity consumption (ω) by concentrator divisions in 2019.

Redevelopment of the processing plant	Copper-pyrite ores, ω, kWh/tons				Barite-containing ores, ω, kWh/tons			
	I quarter	II quarter	III quarter	IV quarter	I quarter	II quarter	III quarter	IV quarter
Ragging (1)	2.23	1.62	1.56	2.57	2.23	1.62	1.56	2.57
Ore reduction (2)	14.22	12.75	12.97	13.13	14.42	13.56	14.75	14.21
Ore transportation, reagent preparation room (3)	0.96	0.61	0.81	1.02	0.96	0.61	0.81	1.02
Flotation (4)	9.94	7.53	8.48	8.03	9.94	7.53	8.48	8.03
Cleaning and baking (5)	3.03	1.7	2.37	2.73	3.03	1.7	2.37	2.73
Lime department (6)	2.22	1.39	1.76	2.12	2.22	1.39	1.76	2.12
Compressor unit (7)	1	0.59	0.62	0.61	1	0.59	0.62	0.61
Tail pumps (8)	2.62	1.96	1.38	2.65	2.62	1.96	1.38	2.65

Compare the calculated specific power consumption with the norms in force at the plant (Table 2).

Table 2. Actual specific power consumption rates for the mining complex in 2019.

Name of the norm	2019 year	Including for individual quarters			
		I quarter	II quarter	III quarter	IV quarter
Norm for processing copper ore, kWh/tonne	31.0	31.0	31.0	31.0	31.0
Norm for processing barite ore, kWh/tonne	42.54	55.19	24.91	34.2	55.87

From the comparison of data in Tables 1 and 2, it can be seen that the recommended rate of electricity consumption for processing copper-pyrite ores differs slightly from the planned rate, while the planned rate for processing barite-containing ores was overstated by more than 25%.

Switching to a scientifically-based rate of electricity consumption will allow you to get an economic effect:

$$E = \left(\omega_{\text{pr}} - \omega_{\text{er}}\right) \cdot Q_{\text{av}} \cdot C_0, \tag{1}$$

where ω_{pr} – planned rate (pr) of electricity consumption for processing barite-containing ores;

ω_{er} – estimated rate (er) of electricity consumption for processing barite-containing ores;

Q_{av} – annual volume (av) of processing of barite-containing ores;

C_0 – cost of electricity.

$$E = (42.54 - 33.76) \cdot 170000 \cdot 5.05 = 7537630 \, \text{rubles}$$

5 Determining the Economic Effect by Optimizing the Loading Mode of Technological Equipment

Analysis of the results of the study of statistical characteristics of electricity consumption and processed ore showed that the specific electricity consumption depends largely on the amount of processed ore per day [9, 10].

The results of the calculation for quarterly arrays for 2016–2019 showed that the correlation coefficient module (r_{QW}), which characterizes the tightness of the relationship between these parameters for the crushing case is in the range of 0.32–0.77; for the main case – in the range of 0.46–0.72 (according to 2019 data), which indicates that the reduction in specific energy consumption can be achieved by increasing the volume of daily ore processing.

Analysis of the source array {Q} for processed ore showed that the array is characterized by a high value of standard deviation, asymmetry and kurtosis.

The high value of the standard deviation is due to the presence of Q_{day} values in the array, which are significantly less than the average m_Q. The total number of such Q_{day} values (emissions) is less than 50%, since the law of distribution of the processed ore mass has a left-sided bevel ($A < 0$).

Thus, a significant factor for increasing the productivity of the concentrator and, thus, for reducing the specific consumption of electricity, is the reduction of operating time in the area of low productivity values, i.e. with an incomplete load [11, 12].

Table 3 shows the results of calculating the characteristics of the source array of processed ore and truncated array (‘), the values of mathematical expectation (m_Q), standard deviation (σ_Q), asymmetry (A_Q) and kurtosis (E_Q), which excludes values Q less than the mathematical expectation of the source array m_Q ($Q_{day} < m_Q$).

The parameters of a truncated array are defined using the moment method. The initial moment of the order s of a random variable $Q \geq m_Q$ is determined by the formula:

$$d_s = \frac{1}{P(Q \geq m_Q)} \cdot \int_{m_Q}^{\infty} Q^s f(Q) dQ, \quad (2)$$

where f(Q) – theoretical differential distribution law of a random variable Q;

Table 3. Determination of probabilistic characteristics of a truncated mass of processed ore.

Year, quarter	Parameters of the source array of processed ore				Parameters of a truncated array of processed ore					Δm_Q, %
	m_Q	σ_Q	A_Q	E_Q	m'_Q	σ'_Q	A'_Q	E'_Q	$P(Q \geq m_Q)$	
1	2	3	4	5	6	7	8	9	10	11
I quarter 2016	2009	626	0.056	2.539	2433	392	1.72	60.43	0.489	21.1
II quarter 2016	2273	756	−0.295	0.838	2831	432	1.191	4.706	0.519	24.55
III quarter 2016	2143	683	−0.535	1.478	2620	379	1.491	6.535	0.535	22.26
IV quarter 2016	2156	637	−0.165	0.305	2647	372	1.059	4.204	0.511	22.77
I quarter 2017	2600	716	1.057	1.965	3210	625	1.446	4.308	0.43	23.46
I quarter 2018	3069	1049	0.259	0.026	3935	684	1.114	4.191	0.483	28.22
II quarter 2018	3014	1189	−0.459	0.073	3906	614	0.52	1.728	0.53	29.6
IV quarter 2018	3388	903	0.538	−0.507	4180	613	0.958	3.572	0.464	23.38
I quarter 2019	3690	1233	−0.649	−0.173	4610	594	0.119	−0.029	0.544	24.93
II quarter 2019	3885	1035	−0.531	−0.323	4668	504	−0.012	−1.075	0.536	20.15
III quarter 2019	3671	980	−0.791	0.254	4371	444	−0.263	−3.099	0.553	19.07
IV quarter 2019	3564	1066	−0.116	−0.679	4426	588	0.454	1.565	0.508	24.19

$P(Q \geq m_Q)$ – the probability that the Q values of the truncated array exceed the expected m_Q values of the original array.

$$P(Q \geq m_Q) = \int_{m_Q}^{\infty} f(Q)dQ. \tag{3}$$

In this paper, the formed arrays are predicted [13–18]. Table 4 shows the forecast values of array parameters for each quarter of 2020. Where r – correlation factor.

Table 4. Forecast values of processed ore array parameters for 2020.

Quarter number	m_Q	σ_Q	r	$m_Q = a_1 \cdot T + b_1$
I quarter	2842	712.5	0.999	$551.2 \cdot T - 1089360.8$
II quarter	3057	806.9	0.972	$513.5 \cdot T - 1014512.6$
III quarter	2907	1080.5	1	$509.3 \cdot T - 1006337$
IV quarter	3036	767.2	0.976	$490.3 \cdot T - 968531.6$

Table 5 shows the results of calculating the parameters of the truncated processed ore array for 2020.

Table 5. Parameters of the truncated processed ore array for 2020.

Quarter number	m'_Q	σ'_Q	r	$m'_Q = a_2 \cdot T + b_2$
I quarter	3547	937.2	0.999	$725.6 \cdot T - 1434229.4$
II quarter	3802	922.9	0.996	$601.6 \cdot T - 1188435.2$
III quarter	3496	1239.1	1	$583.7 \cdot T - 1153040$
IV quarter	3751	964	0.979	$617.8 \cdot T - 1220475.1$

When the mathematical expectation of processed ore increases by an amount $\Delta m = m'_Q - m_Q$, the specific power consumption decreases by an amount $\Delta\omega$:

$$\Delta\omega = \left(a_2 \cdot m'_Q + b_2\right) - \left(a_2 \cdot m_Q + b_2\right) = a_2 \cdot \Delta m_Q, \tag{4}$$

where m_Q и m'_Q – mathematical expectation of the processed ore in the source and truncated arrays, respectively;

a_2, b_2 – coefficients of the regression equation $\omega = a_2 \cdot Q + b_2$.

$$\Delta\omega\% = \frac{\Delta\omega}{m_\omega} \cdot 100\%. \tag{5}$$

Energy savings by eliminating emissions of minimum Q values for the quarter:

$$\Delta W = \Delta\omega \cdot m_Q \cdot n, \tag{6}$$

where n – number of days per quarter.

The results of calculating energy savings are shown in Table 6.

Table 6. Determining energy savings for 2020.

Quarter	m_Q, tonne/day	m'_Q, tonne/day	n	$\Delta\omega$, kWh/tonne	ΔW, %	ΔW, kWh
1	2842	3547	91	4.65	14.2	1500913
2	3057	3802	91	4.92	15	1702231
3	3907	3496	92	3.89	11.9	1251149
4	3036	3751	92	4.72	14.4	1628834
Total for 2020						6083127

The economic effect based on the results of the calculation ΔW for 2020 is:

$$E = \sum_{i=1}^{4} \Delta W_i \cdot C_0 = 6083127 \cdot 5.05 = 30719791 \text{ rubles.}$$

An important reserve for reducing the specific power consumption is to increase the load of ball mills [19, 20]. According to data for 2019, the average mill load is

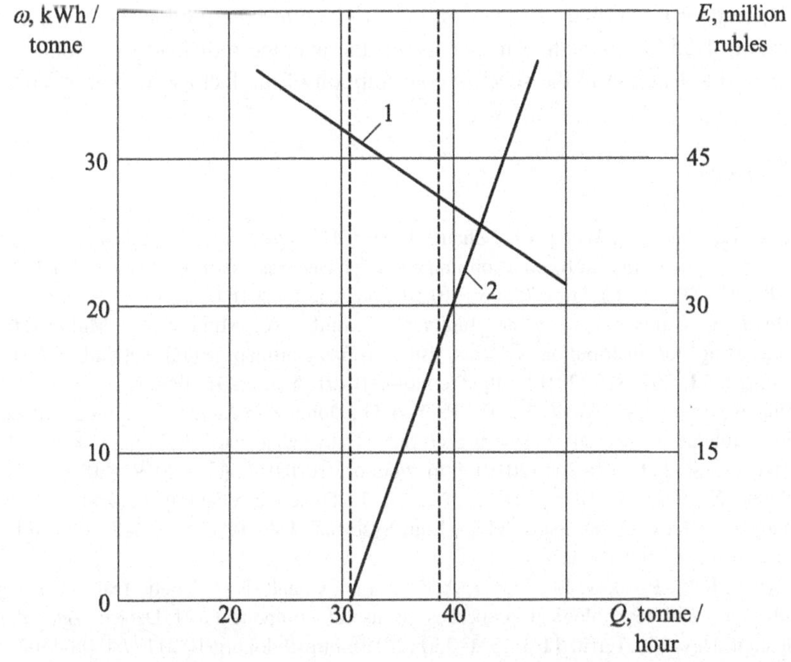

Fig. 1. Dependence of specific power consumption (1) and economic effect (2) on the average productivity of mills.

30.8 tonne/h, while an active experiment has found that without compromising the quality of grinding, the productivity of mills can be increased to 55–60 tonne/h.

An increase in the average mill load by 1 tonne/h corresponds to a decrease in the specific consumption of the factory by 0.56 kWh/tonne.

Figure 1 shows the graphs of the dependence of the specific power consumption (ω) and the economic effect (E) on the productivity of mills (Q).

6 Conclusion

As a result of comparing the calculated specific rates of electricity consumption with the planned ones, it was found that the planned rate of electricity consumption for processing barite-containing ores was overstated by more than 25%. Switching to a scientifically-based rate of electricity consumption will allow you to get an economic effect of 7537630 rubles.

It is recommended to maintain the productivity of ball mills in the first stage of grinding at the level of 35–40 tonne/h. At the same time, the loading of all technological mechanisms of the processing plant will increase, which will allow obtaining an economic effect of 30.7 million rubles.

The reserve for further energy savings is to increase the productivity of mills to a maximum of 55–60 tonne/h. An increase in the average mill load by 1 tonne/h corresponds to a decrease in the specific consumption of the factory by 0.56 kWh/tonne.

References

1. Xiao, L., Shao, W., Wang, Ch., Zhang, K., Lu, H.: Research and application of a hybrid model based on multi-objective optimization for electrical load forecasting. Appl. Energy **180**, 213–233 (2016). https://doi.org/10.1016/j.apenergy.2016.07.113
2. Abreu, T., Amorim, A., Santos-Junior, C., Lotufo, A., Minussi, C.: Multinodal load forecasting for distribution systems using a fuzzy-artmap neural network. Appl. Soft Comput. **71**, 307–316 (2018). https://doi.org/10.1016/j.asoc.2018.06.039
3. Zhang, J., Xiong, G., Meng, K., Yu, P., Yao, G., Dong, Zh.: An improved probabilistic load flow simulation method considering correlated stochastic variables. Int. J. Electr. Power Energy Syst. **111**, 260–268 (2019). https://doi.org/10.1016/j.ijepes.2019.04.007
4. Zhang, X., Gao, H., Huang, H., Li, Y., Mi, J.: Dynamic reliability modeling for system analysis under complex load. Reliab. Eng. Syst. Saf. **180**, 345–351 (2018). https://doi.org/10.1016/j.ress.2018.07.025
5. Klyuev, R.V., Bosikov, I.I., Mayer, A.V.: Complex analysis of genetic features of mineral substance and technological properties of useful components of Dzhezkazgan deposit. Sustain. Dev. Mt. Territ. **11**(3), 321–330 (2019). https://doi.org/10.21177/1998-4502-2019-11-3-321-330
6. Bosikov, I.I., Klyuev, R.V., Egorova, E.V.: Assessment of oil and gas potential prospects of the north eastern unit of the south khulym deposit. Sustain. Dev. Mt. Territ. **11**(1), 7–14 (2019). https://doi.org/10.21177/1998-4502-2019-11-1-7-14
7. Klyuev, R.V., Bosikov, I.I., Gavrina, O.A., Revazov, V.C.: System analysis of power consumption by nonferrous metallurgy enterprises on the basis of rank modeling of individual technocenosis castes In: MATEC Web Conference XIV Int. Scientific-Technical Conf. «Dynamic of Technical Systems» (DTS-2018), vol. 226 (2018). https://doi.org/10.1051/matecconf/201822604018
8. Klyuev, R.V., Bosikov, I.I., Gavrina, O.A.: Development of mathematical model for specific power consumption of resistance furnaces at non-ferrous metallurgy enterprises. In: International Russian Automation Conference (RusAutoCon). Sochi (2018). https://doi.org/10.1109/rusautocon.2018.8501831
9. Golik, V.I., Razorenov, Yu.I., Karginov, K.G.: Mining industry – the basis for sustainable development of North Ossetia-Alania. Sustain. Dev. Mt. Territ. **2**(32), 163–172 (2017). https://doi.org/10.21177/1998-4502-2017-9-2-163-171
10. Zhukovskiy, Y., Batueva, D., Buldysko, A., Shabalov, M.: Motivation towards energy saving by means of IoT personal energy manager platform. J. Phys. Conf. Series **1333**(6) (2019). https://doi.org/10.1088/1742-6596/1333/6/062033
11. Plieva, M.T., Gavrina, O.A., Kabisov, A.A.: Analysis of technological damage at 110 kV substations in JSC IDGC of the North Caucasus- « Sevkavkazenergo » In: International Multi-Conference on Industrial Engineering and Modern Technologies (FarEastCon) (Vladivostok), 19229305. Vladivostok (2019). https://doi.org/10.1109/fareastcon.2019.8934076
12. Buryanina, N.S., Korolyuk, Yu.F., Maleeva, E.I., Lesnykh, E.V.: Power transmission lines with a reduced number of wires in mountain territories. Sustain. Dev. Mt. Territ. **10**(3), 404–410 (2018). https://doi.org/10.21177/1998-4502-2018-10-3-404-410

13. Mcira, E., Oliveira, F., Cyrino, F.: Forecasting mid-long term electric energy consumption through bagging ARIMA and exponential smoothing methods. Energy **144**, 776–788 (2018). https://doi.org/10.1016/j.energy.2017.12.049
14. Kaboli, S., Selvaraj, J., Rahim, N.: Long-term electric energy consumption forecasting via artificial cooperative search algorithm. Energy **115**, 857–871 (2016). https://doi.org/10.1016/j.energy.2016.09.015
15. Wei, N., Li, Ch., Peng, X., Zeng, F., Lu, X.: Conventional models and artificial intelligence-based models for energy consumption forecasting: a review. J. Petrol. Sci. Eng. **181**, 106187 (2019). https://doi.org/10.1016/j.petrol.2019.106187
16. Spiliotis, E., Petropoulos, F., Kourentzes, N., Assimakopoulos, V.: Cross-temporal aggregation: Improving the forecast accuracy of hierarchical electricity consumption. Appl. Energy **261**, 114339 (2020). https://doi.org/10.1016/j.apenergy.2019.114339
17. Xiao, J., Li, Y., Xie, L., Liu, D., Huang, J.: A hybrid model based on selective ensemble for energy consumption forecasting in China. Energy **159**, 534–546 (2018). https://doi.org/10.1016/j.energy.2018.06.161
18. Carvallo, J., Larsen, P., Sanstad, A., Goldman, Ch.: Long term load forecasting accuracy in electric utility integrated resource planning. Energy Policy **119**, 410–422 (2018). https://doi.org/10.1016/j.enpol.2018.04.060
19. Bian, X., Wang, G., Wang, H., Wang, Sh., Lv, W.: Effect of lifters and mill speed on particle behaviour, torque, and power consumption of a tumbling ball mill: Experimental study and DEM simulation. Miner. Eng. **105**, 22–35 (2017). https://doi.org/10.1016/j.mineng.2016.12.014
20. Pawanr, Sh., Garg, G., Routroy, S.: Multi-objective optimization of machining parameters to minimize surface roughness and power consumption using TOPSIS. Procedia CIRP **86**, 116–120 (2019). https://doi.org/10.1016/j.procir.2020.01.036

Energy Indicators of Drilling Machines and Excavators in Mountain Territories

Roman Klyuev[1,2](✉), Olga Fomenko[3], Oksana Gavrina[1], Ramzan Turluev[4], and Soslan Marzoev[1]

[1] North-Caucasian Institute of Mining and Metallurgy (State Technological University), 44, Nikolaeva Street, Vladikavkaz 362021, Russia
kluev-roman@rambler.ru
[2] Moscow Polytechnic University, B. Semenovskaya Street, Moscow 107023, Russia
[3] Southern Federal University, 105/42 Bolshaya Sadovaya Street, Rostov-on-Don 344006, Russia
[4] Grozny State Oil Technical University Named After Academician MD Millionshchikov, 100, Isaeva Avenue, Grozny 364051, Russia

Abstract. The article presents the results of complex studies of calculated electrical loads of drilling machines and excavators. The research was carried out at two open-pit mines. The technical justification of electricity consumption rates and the need to link them with indicators that characterize the influence of the most important production factors on the change in specific electricity consumption is relevant. For the correct and most complete characteristics of power consumption, it is necessary to establish a quantitative assessment of the degree of influence of mining and technological factors and operating modes of mechanisms to identify the most significant factors and establish patterns of power consumption. It is established that the most important factor affecting the power consumption of drilling rigs is the drilling speed for different categories of rocks by thermal conductivity. Individual and group load graphs of electric drives of excavators and drilling machines are studied on the basis of statistical probabilistic calculation methods. Based on the use of these calculation methods, the values of the calculated maximum of electrical loads, the values of mathematical expectation, the average square deviation and the probability of exceeding the loads are determined. For all feeders of quarries, calculated loads of half-hour duration were obtained, varying from 130 kW to 385 kW. The obtained values of load capacities are used in the future to predict the power consumption of drilling machines and excavators in the conditions of changing mountain conditions.

Keywords: Energy indicators · Drilling machines · Excavators · Mountain territories

V. Murgul and V. Pukhkal (Eds.): EMMFT 2019, AISC 1258, pp. 272–281, 2021.
https://doi.org/10.1007/978-3-030-57450-5_25

1 Introduction

Energy performance of drilling machines depends on the type of drilling (type of machine) and physical and mechanical properties of rocks. On open-pit mining operations of the studied quarries No 1 and No 2, roller drilling machines of the SBSH 250 type are used. The technology of drilling with roller machines has not been sufficiently studied, only a few works provide empirical relationships between the power consumption of the rotator engine and the operating parameters of drilling. In the presented dependencies, the energy performance of the rotator drive of roller machines is determined by the rotation frequency of the drilling tool, the axial pressure and the physical and mechanical properties of the drilling operations. The disadvantage of the study is the inability to quantify the impact of these factors on the power consumption of the drilling rig.

It is most likely that the power consumption during drilling is a random value determined by the operating parameters of drilling [1–4]. The selection of the factor that most fully characterizes the power consumption of the drilling machine from a number of parameters allows us to use the position of probability theory for analysis.

2 Methods and Approaches

The paper uses a statistical method to determine the calculated electrical loads. The statistical method makes it possible, using Lyapunov's theorem, to characterize the total impact of all these factors and their variability by two integral indicators: the General average load (P) and the General average square deviation (σ), or in relative units-the General calculated coefficient of use (k_{use}) and the relative deviation ($\sigma_{relative}$).

3 Calculation of Energy Characteristics of Drilling Machines

The drilling process with roller machines is provided by the simultaneous operation of the rotator, compressors, hydraulic pumps, and fans. It is difficult to analytically Express the dependence of the power consumed by mechanisms on the drilling speed (υ) Establishing the desired dependence is possible by conducting experimental studies using the appropriate mathematical apparatus.

Therefore, in production conditions, power consumption and drilling speed were measured when a certain depth of a well was drilled during a specific time.

Experimental data were used to construct the dependence P = f(υ) (Fig. 1, a), which characterizes the power consumption depending on the speed of the rotator. The resulting experimental curve can be approximated by a dependency for ease of analysis:

$$y = a + b \cdot x - c \cdot e^{dx} \tag{1}$$

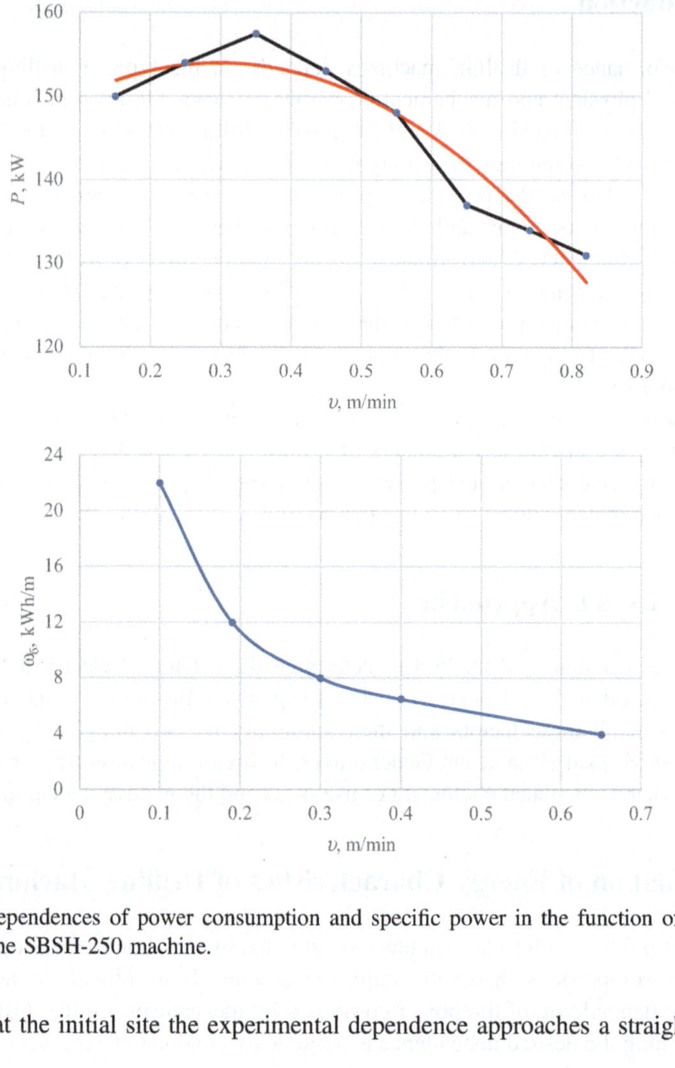

a)

b)

Fig. 1. Dependences of power consumption and specific power in the function of the rotator speed of the SBSH-250 machine.

And at the initial site the experimental dependence approaches a straight line:

$$P = a + b \cdot \upsilon_б \tag{2}$$

Using the least squares method to solve the equation, we find the values of the coefficients in the equation: a = 150 and b = 147. The third coefficient in the equation is found by a well-known method. The experimental and approximated curves are shown in Fig. 1, a. The values of the coefficients in Eq. (1): C = 0.108 and d = 8.2.

Finally, the equation of power consumed by the drilling machine takes the form:

$$P = 150 + 147 \cdot \upsilon - 0.108 \cdot e^{8,2 \cdot \upsilon} \tag{3}$$

The dependence of the specific power consumption on the drilling speed is expressed by the equation:

$$\omega = \frac{P}{60 \cdot \vartheta} \tag{4}$$

Substituting the value P in the equation we get for the technological specific power consumption:

$$\omega = 2.45 + \frac{2.5 - 0.0018 \cdot e^{8.2 \cdot \vartheta}}{\vartheta} \tag{5}$$

The dependence constructed by Eq. (5) is shown in Fig. 1, b.

Power consumption for auxiliary operations does not depend on drilling modes, conditions, and speed. The value of this flow rate is assigned to 1 m of the well. The total power consumption for auxiliary operations when drilling a well with a depth of 8 m is equal to 6.8 kWh, or 0.85 kWh/m for 1 m of the well. Then the equation for the specific power consumption taking into account the expense for auxiliary operations takes the form:

$$\omega = 3.3 + \frac{2.5 - 0.0018 \cdot e^{8.2 \cdot \vartheta}}{\vartheta} \tag{6}$$

Thus, in order to determine the technological specific power consumption, it is necessary to have data on the drilling speed for different categories of rocks by thermal conductivity.

4 Generalized Analysis of Individual and Group Load Schedules and Determination of the Calculated Load

Individual and group load schedules of electric drives of excavators and drilling rigs SBSH-250 performing technological operations at quarries No 1 and No 2 were taken by devices on different feeders of substations.

Analysis of the oscillograms of individual load graphs of excavators and drilling machines allows us to note the different nature and duration of the working sections of the P(t) graphs, despite the repeatability of the technological process operations. The form of individual load schedules for the marked mechanisms is non-cyclical and irregular. The basic time of the completed technological cycle for excavators is T_{basic} = 40–65 s (includes time for scooping, unloading, various turns, etc.). the non-Cyclical nature of individual schedules determines the correlation nature of group load schedules. The formation of group load graphs is influenced by a number of random factors: the total duration of the cycle (t_{cycle}), the duration of its working part, the duration of the pause (t_0), the power consumed in the working part of the cycle (P_{work}), off-load (P_0). The noted indicators of load schedules, along with the coefficient of inclusion ($k_{inclusion}$), vary very widely. Determining the calculated maximum load in a function of these random variables for time θ is a difficult problem to solve.

Using the statistical method does not simplify the task by selecting one or two of the many factors from the entire variety and neglecting the rest. However, no attempt is made to determine the impact of each individual factor or group of factors on the load [5–8].

It is theoretically proved that the normal distribution law is considered to be valid for mains that supply 6–8 electric receivers with a steady technological process. Experimental data confirmed that the normal law is applicable for a smaller number of electric receivers.

According to the statistical method, the group graphs of feeders that characterize the load change over a sufficiently long time are considered, which are divided into m sections with a duration θ equal to the time interval during which the heating of the current-carrying part reaches a steady value. You can take $\theta = T = 3T_0$, where T_0 is the heating time constant. For each section, the average value of the load graph ordinate P_1, P_2, …, P_m is a random value, and it is impossible to predict in advance what value this ordinate will take on a particular section. It is possible, however, with a high probability to specify the limits within which these ordinates will be contained.

The mathematical expectation of the load:

$$\overline{P} = P_{average} = \frac{P_1 + P_2 + \ldots + P_m}{m}, \text{ kW} \tag{7}$$

The corresponding deviation will be equal to:

$$\overline{\sigma} = \sigma_{average} = \sqrt{\frac{\left(P_1 - \overline{P}\right)^2 + \left(P_2 - \overline{P}\right)^2 + \ldots + \left(P_m - \overline{P}\right)^2}{m}}, \text{ kW} \tag{8}$$

The standard deviation can also be determined by the scale of the ordinate load graph. To do this, the entire sample with a volume of m members is divided into several n samples, each with a volume of k members. In each sample, the R_k span is determined, i.e. the difference between the maximum (max) and minimum (min) loads:

$$R_k = P_{max} - P_{min} \tag{9}$$

Then the average span of n samples is determined:

$$\overline{R_k} = \frac{\sum_{K=1}^{n} R_k}{n}, \text{ kW} \tag{10}$$

For a normal distribution law, the value $\overline{\sigma}$ is determined from the expression:

$$\overline{\sigma}_T = \frac{\overline{R_k}}{d_k}, \text{ kW} \tag{11}$$

The average value of the coefficient d_k varies depending on the sample size k and is a random variable with large deviations from the average value. The probabilities of individual d_k values are tabulated and given in mathematical statistics courses.

It is established that the best results are obtained at k = 7–10. According to the law of normal distribution any load and the probability of exceeding it can be determined from the equation:

$$P_k = \overline{P} + \beta \cdot \bar{\sigma}, \tag{12}$$

where β is the accepted multiple of the deviation, and the index T means that the deviation is determined for the duration of T. Then the probability that the average load of any group will exceed $\overline{P} + \beta \cdot \bar{\sigma}$ is equal to the probability (β). This function is also tabulated.

The statistical method for calculating electrical loads allows:

1. Determine the value of the calculated maximum and the probability of its occurrence. For a given feeder mode ($\overline{P}$ and $\overline{\sigma}$), the maximum value and probability of its occurrence are determined by the values β and Probability (β), respectively.
2. By assigning different values to β, we get the range of possible loads and the frequency of their occurrence. Multiplying these frequencies by the amount of working time of consumers, we get the total duration of each load, expressed in hours.

To calculate using Eqs. (7) and (8), a large number of load measurements must be performed with intervals between them equal to $T = 3T_0$. The total duration of the survey, expressed in rotation ($m_{rotation}$), will be significant and not acceptable for practical purposes.

Statistical methods allow us to make a conclusion about the General probabilistic characteristics of the process (mathematical expectation, deviation) based on a small number of observations. In this case, the information obtained is considered as a sample group or random sample taken from the General population, i.e. from the very large number of measurements that would be required to directly determine $\overline{P}$ and $\overline{\sigma}$.

In individual studies, when the deviation is determined from statistical processing of measurements, the number of measurements is selected so that the relative error: $\Delta\overline{\sigma} = \pm\frac{(\sigma-\overline{\sigma})}{\overline{\sigma}}$ with acceptable reliability, it did not go beyond certain limits. For individual studies, you should take $\Delta\overline{\sigma} = \pm 0.15$ a formula that gives, as the calculations show, a tolerance in the value of the calculated load of the order of 3–5%. The number of measurements is m = 60.

Table 1 shows half-hour load measurements of P_{30} feeder No 1 of quarry No 1. Measurements corresponding to the beginning and end of the shift, as well as the break, were excluded from processing. Thus, seven hourly measurements were made in the shift, which are included in the table. The total number of measurements m = 60, based on the tolerance $\Delta\overline{\sigma} = \pm 0.15$.

Table 1. Rated load of feeder No 1 quarry No 1.

No	P_{30}, kW	R, kW	No	P_{30}, kW	R, kW	No	P_{30}, kW	R, kW
1.	287	96	21.	191	48	41.	272	35
2.	296		22.	187		42.	280	
3.	283		23.	192		43.	289	
4.	303		24.	203		44.	285	
5.	272		25.	203		45.	287	
6.	235		26.	210		46.	298	
7.	220		27.	217		47.	305	
8.	225		28.	225		48.	307	
9.	207		29.	230		49.	299	
10.	218		30.	235		50.	297	
11.	224	49	31.	230	50	51.	295	28
12.	208		32.	225		52.	287	
13.	202		33.	218		53.	278	
14.	212		34.	217		54.	277	
15.	210		35.	227		55.	275	
16.	180		36.	232		56.	267	
17.	175		37.	230		57.	275	
18.	193		38.	260		58.	277	
19.	201		39.	265		59.	283	
20.	190		21.	267		41.	285	
$\overline{P} = P_{average} = 245$ kW								

The 60 measurements given in Table 1 for feeder No 1 ($P_{nominal}$ = 2032 kW and n = 4) are divided into 6 samples for k = 10 measurements. In each sample, we define the range $\bar{R} = P_{\max} - P_{\min}$ – the difference between the maximum and minimum members of the sample [9–14].

Determine the average size of all the samples:

$$\overline{R} = \frac{96 + 49 + 48 + 50 + 35 + 28}{6} = 51\,\text{kW} \tag{13}$$

For a normal distribution law in accordance with (11):

$$\overline{\sigma} = \frac{\bar{R}_k}{d_k} = \frac{51}{3.08} = 16.6\,\text{kW}, \tag{14}$$

where dk = d10 = 3.08 for k = 10.

According to the law of normal distribution, the calculated load and the probability of exceeding it are determined from Eq. (3). The value of β in Eq. (13) is assumed to be β = +2.5 and the corresponding Probability (+2.5) = 0.005. The latter eliminates the following of each other load $\overline{P} + 2.5 \cdot \overline{\sigma}$, the duration of the cycle T (i.e., at the maximum load, the temperature of the current-carrying core will not exceed the standard). Therefore, for feeder No1, the estimated load of the half-hour duration will be: $P_{30} = 245 + 2.5 \cdot 16.6 = 286$ kW.

Similarly, we determine the calculated load of the remaining feeders of quarries No 2–8. The results of calculating the average power spans of all samples are shown as graphs in Fig. 2.

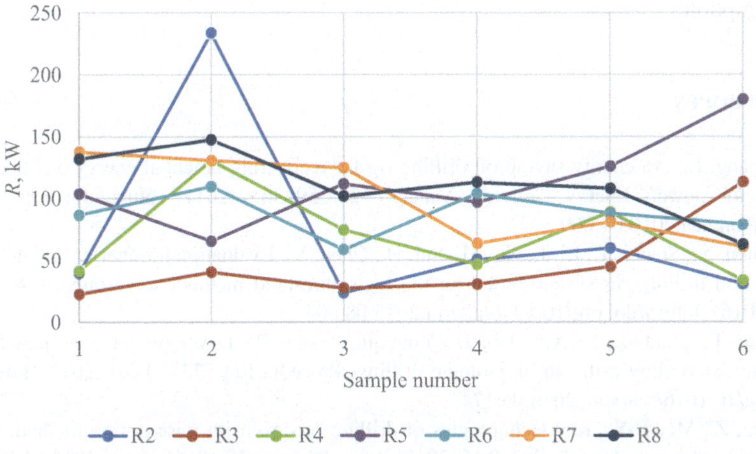

Fig. 2. Results of calculations of average power spans of all samples of feeders of quarries No 2–8.

Values of nominal power ($P_{nominal}$), average power ($P_{average}$) and calculated load of half-hour duration (P_{30}) for 7 feeders of quarries are shown in Table 2.

Table 2. Values of nominal power ($P_{nominal}$), average power ($P_{average}$) and calculated load of half-hour duration (P_{30}) for 7 feeders of quarries.

Number of the quarry feeder	Power values, kW		
	$P_{nominal}$	$P_{average}$	P_{30}
2	2032	166	197
3	1402	130	168
4	2032	312	368
5	2032	266	358
6	2032	274	345
7	2418	304	385
8	2032	260	350

Table 2 shows that the values of calculated loads of half-hour duration for all feeders of quarries vary from 130 kW to 385 kW. The obtained values of load capacities are used in the future to predict the power consumption of drilling machines and excavators under changing mountain conditions [15–20].

5 Conclusion

The nature of power consumption of drilling machines is studied, mining and technological factors that have the greatest impact on their energy performance are determined. Energy characteristics of excavators and drilling machines are constructed in the form of dependencies of total and specific power consumption in the function of productivity and mechanical drilling speed for normalization and planning of power consumption.

References

1. Zhang, L.: An energy-saving oil drilling rig for recovering potential energy and decreasing motor power. Energy Convers. Manag. **52**, 359–365 (2011). https://doi.org/10.1016/j.enconman.2010.07.009
2. Chen, Y., Shang, T., Li, J., Nie, G., Sui, H., Chen, X.: Evaluation for energy-saving effect of hybrid drilling rig system based on the logic threshold method. J. Terramech. **63**, 49–60 (2016). https://doi.org/10.1016/j.jterra.2015.08.004
3. Jian, L., Jianhua, S., Xinmiao, L., Yongqin, Z., Li, P.: Development and application of aluminum alloy drill rod in geologic drilling. Procedia Eng. **73**, 84–90 (2014). https://doi.org/10.1016/j.proeng.2014.06.174
4. Liu, Z., Meng, Y.: Key technologies of drilling process with raise boring method. J. Rock Mech. Geotech. Eng. **7**, 385–394 (2015). https://doi.org/10.1016/j.jrmge.2014.12.006
5. Narciso, D., Martins, F.: Application of machine learning tools for energy efficiency in industry: a review. Energy Rep. **6**, 1181–1199 (2020). https://doi.org/10.1016/j.egyr.2020.04.035
6. Ji, M., Xu, J., Chen, M., Mansori, M.: Effects of different cooling methods on the specific energy consumption when drilling CFRP/Ti6Al4V. Stacks Procedia Manufact. **43**, 95–102 (2020). https://doi.org/10.1016/j.promfg.2020.02.118
7. Zhang, L., Huang, Z., Li, Z., Guo, K.: Research on the correlation of monthly electricity consumption in different industries: a case study of Bazhou county. Procedia Comput. Sci. **139**, 496–503 (2018). https://doi.org/10.1016/j.procs.2018.10.245
8. Javied, T., Rackow, T., Stankalla, R., Sterk, C., Franke, J.: A study on electric energy consumption of manufacturing companies in the German industry with the focus on electric drives. Procedia CIRP **41**, 318–322 (2016). https://doi.org/10.1016/j.procir.2015.10.006
9. Kortiev, L.I., Klyuev, R.V., Kulumbegov, R.P., Kortiev, A.L., Bosikov, I.I., Gavrina, O.A., Madaeva, M.Z.: Technical support of the power lines design - as a linear structure in difficult mountain conditions. In: IOP Conference Series: Materials Science and Engineering, vol. 663 (2019). https://doi.org/10.1088/1757-899x/663/1/012034
10. Klyuev, R.V., Bosikov, I.I., Mayer, A.V.: Complex analysis of genetic features of mineral substance and technological properties of useful components of Dzhezkazgan deposit. Sustain. Dev. Mt. Territ. **11**(3), 321–330 (2019). https://doi.org/10.21177/1998-4502-2019-11-3-321-330
11. Bosikov, I.I., Klyuev, R.V., Egorova, E.V.: Assessment of oil and gas potential prospects of the north eastern unit of the south khulym deposit. Sustain. Dev. Mt. Territ. **11**(1), 7–14 (2019). https://doi.org/10.21177/1998-4502-2019-11-1-7-14

12. Plieva, M.T., Gavrina, O.A., Kabisov, A.A.: Analysis of technological damage at 110 kV substations in JSC IDGC of the North Caucasus- « Sevkavkazenergo » . In: International Multi-Conference on Industrial Engineering and Modern Technologies (FarEastCon) (Vladivostok), 19229305. Vladivostok (2019). https://doi.org/10.1109/fareastcon.2019.8934076
13. Klyuev, R.V., Fomenko, O.A., Gavrina, O.A., Sokolov, A.A., Sokolova, O.A., Plieva, M.T., Kabisov, A.A., Ikoeva, E.Y.: Ensuring the consumer reliability based on retrospective analysis. In: IOP Conference Series: Materials Science and Engineering, vol. 663 (2019). https://doi.org/10.1088/1757-899x/663/1/012033
14. Cadini, F., Agliardi, G., Zio, E.: A modeling and simulation framework for the reliability/availability assessment of a power transmission grid subject to cascading failures under extreme weather conditions. Appl. Energy **185**, 267–279 (2017). https://doi.org/10.1016/j.apenergy.2016.10.086
15. Nepal, R., Sharma, B., Irsyad, M.: Scarce data and energy research: Estimating regional energy consumption in complex economies. Econ. Anal. Policy **65**, 139–152 (2020). https://doi.org/10.1016/j.eap.2019.12.002
16. Meng, M., Wang, L., Shang, W.: Decomposition and forecasting analysis of China's household electricity consumption using three-dimensional decomposition and hybrid trend extrapolation models. Energy **165**, 143–152 (2018). https://doi.org/10.1016/j.energy.2018.09.090
17. Wang, J., Yang, W., Du, P., Li, Y.: Research and application of a hybrid forecasting framework based on multi-objective optimization for electrical power system. Energy **148**, 59–78 (2018). https://doi.org/10.1016/j.energy.2018.01.112
18. Ahmad, T., Chen, H.: Potential of three variant machine-learning models for forecasting district level medium-term and long-term energy demand in smart grid environment. Energy **160**, 1008–1020 (2018). https://doi.org/10.1016/j.energy.2018.07.084
19. Sanstad, A., McMenamin, S., Sukenik, A., Barbose, G., Goldman, C.: Modeling an aggressive energy-efficiency scenario in long-range load forecasting for electric power transmission planning. Appl. Energy **128**, 265–276 (2014). https://doi.org/10.1016/j.apenergy.2014.04.096
20. Luin, B., Petelin, S., Mansour, F.: Modeling the impact of road network configuration on vehicle energy consumption. Energy **137**, 260–271 (2017). https://doi.org/10.1016/j.energy.2017.06.138

Analytical Determination of Fuel Economy Characteristics of Earth-Moving Machines

Vladimir Zhulai, Vitaly Tyunin(✉), Aleksei Shchienko, Nikolay Volkov, and Dmitriy Degtev

Voronezh State Technical University, Moskovsky prospekt, 14, Voronezh, Voronezh Region 394000, Russia
tuninvl@yandex.ru

Abstract. The article considers the problem of the analytical determination of the fuel economy performance of earth-moving machines by the example of the road grader. By fuel economy is understood the vehicle's ability to perform an operation with the minimal fuel consumption per hour or per unit volume of the products being manufactured, which is achieved by the optimization of the operation parameters. The fuel costs constitute a significant part of the net cost of a production unit, and for some earth-moving machines reach up to half of the machine-shifts net cost. Consequently, fuel efficiency is one of the basic operational properties of earth-moving machines. The values of the road grader fuel consumption when performing the technological operations have been obtained and analyzed. The fuel balance of the earth-moving machines in the traction mode is presented. The fuel balance of the motor grader when digging soil has been defined and analyzed.

Keywords: Earth-moving machines · Road grader · Fuel economy · Fuel consumption · Fuel balance

1 Introduction

Improving the fuel economy earth-moving machines (EMM) will allow us to reduce not only the cost of production and to save energy, but also to improve the environment. Therefore, the rational and economical use of the fuel consumed by the EMM is an important task [1, 2].

To solve this problem it is necessary: first, to develop a method for analytical determination of fuel consumption when performing the operations of the EMM working cycle, and second, to establish an analytical relationship between fuel consumption and the design parameters of the EMM [3, 4].

As an example of the EMM, let us consider the grader, which is designed primarily for grading and planning in the construction of earthworks and for road maintenance, as well.

V. Murgul and V. Pukhkal (Eds.): EMMFT 2019, AISC 1258, pp. 282–289, 2021.
https://doi.org/10.1007/978-3-030-57450-5_26

2 Materials and Methods

When constructing a section of the motor road subgrade by the earth cut out of the ditch, the fuel consumed by the motor grader [5–8] is:

$$G_{FL} = G_{FDIG} + G_{FMOV} + G_{FFIN} + G_{FTUR}, \text{kg} \quad (1)$$

where $G_{FDIG}, G_{FMOV}, G_{FFIN}$ *and* G_{FTUR} – the consumption of fuel when digging soil, moving the soil, when finishing the subgrade and when turning around the road grader, respectively, kg.

When digging soil, the fuel consumption is

$$G_{FDIG} = G_{F1} \frac{2L}{\upsilon_{V1AV}} n_D, \text{kg}; \quad (2)$$

when moving the soil, it is

$$G_{FMOV} = G_{F2} \frac{2L}{\upsilon_{V2AV}} n_M, \text{kg}; \quad (3)$$

when finishing the subgrade, it is

$$G_{FFIN} = G_{F3} \frac{2L}{\upsilon_{V3AV}} n_F, \text{kg}; \quad (4)$$

when turning around the road grader, it is

$$G_{FTUR} = G_{F4} 2t_{MOV} (n_D + n_M + n_F), \text{kg}; \quad (5)$$

where $G_{F1}, G_{F2}, G_{F3}, G_{F4}$ – the hourly fuel consumption when digging the soil, moving the soil, when finishing the subgrade and turning around the road grader, respectively, km/h; L – the length of the bank, km; $\upsilon_{V1AV}, \upsilon_{V2AV}, \upsilon_{V3AV}$ – the average actual speeds of the grader motion when digging soil, moving the soil, when finishing the subgrade, respectively, km/h; n_D, n_M, n_F – the number of the operations when digging soil, moving the soil, when finishing the subgrade, respectively; t_{MOV} – the duration of one turn, h.

The hourly fuel consumption G_{Fi} and the actual speeds of movement υ_{ViAV} during passes can be determined analytically [5–8], based on the condition that the traction force T_i must be greater than or equal to the resistance when digging, moving and finishing subgrade.

When digging the soil n_D, the number of the road grader passes is determined by [5–8]:

$$n_D = \frac{k_{OD} F}{2S_{CS}}, \quad (6)$$

where k_{OD} – the coefficient of overlapping passes of the road grader when digging the soil; F – the cross-section area of the bank, m^2; S_{CS} – the projection area of the chips of the soil in the plane perpendicular to the direction of the grader motion, m^2.

When moving soil n_M, the number of passes is determined by [5–8]:

$$n_M = k_{OM} \frac{l_0}{l_M}, \tag{7}$$

where k_{OM} – the coefficient of overlapping passes when moving the soil; l_0 – the average required earth moving (the distance between the centers of the cross section gravity of the reserve and half of the bank), m; l_M – the moving soil in one operation, m.

When finishing the subgrade n_O, the number of passes is determined by [5–8]:

$$n_F = (0,25\ldots0,35)n_D \tag{8}$$

So, having the value of the fuel consumption of G_{FL} one can determine the average hourly fuel consumption of G_{FH} and the average specific fuel consumption per 1 m^3 of the developed and displaced soil g_S:

$$G_{FH} = \frac{G_{FL}}{T_C}, \text{kg/h}, \tag{9}$$

$$g_y = \frac{G_{FL}}{LF}, \text{kg/(h} \cdot \text{m}^3), \tag{10}$$

where T_C – the length of the time of the working cycle, h.

$$T_C = L\left(\frac{n_D}{\upsilon_{V1AV}} + \frac{n_M}{\upsilon_{V2AV}} + \frac{n_F}{\upsilon_{V3AV}}\right) + 2t_{MOV}(n_D + n_M + n_F), \text{h}, \tag{11}$$

In order to establish the analytical relationship between the fuel consumption and the design parameters of the motor grader it is necessary to consider the fuel balance of the road grader. The fuel balance will allow us to estimate the exact distribution of engine energy, obtained during the fuel combustion, to perform the main process, to estimate the losses in the various mechanisms of the machine and the interaction of the wheel mover with the support surface.

3 Results

The fuel balance equation of the motor grader can be written as follows [9–11]:

$$G_F = G_{CL} + G_{ML} + G_{TL} + G_{TR} + G_f + G_{CS} \pm G_h + G_j \tag{12}$$

The left part of this equation shows the hourly engine fuel consumption, and the right side - components of the fuel consumed by the useful traction power for all types of power losses.

The amount of the fuel consumed by calorific losses in the engine is

$$G_{CL} = \left(\frac{G_T h_u}{U} - N_i\right) g_{TSF}, \tag{13}$$

where h_U – the lower fuel consumption heat, kJ/kg; U – the thermal equivalent, i.e. the amount of the heat equivalent to the engine power of 1 kW for 1 h, kJ/(kW h) [12–14]; N_i – the indicated power of the internal combustion engine, kW; g_{TEOP} – the theoretical specific fuel consumption, g/kW h [15].

The amount of the fuel consumed by the mechanical losses in the engine is

$$G_{ML} = \frac{p_{ML} V_h n i}{120} g_{TSF}, \tag{14}$$

where p_{ML} – the pressure of the mechanical losses, MPa [12–14]; V_h – the displacement of the internal combustion engine; n – the frequency of the engine crankshaft rotation; i – the number of cylinders.

The amount of the fuel consumed by the lost power in the transmission is

$$G_{TL} = N_e(1 - \eta_{ET}) g_{TSF}, \tag{15}$$

where η_{ET} – the efficiency of the transmission.

The amount of the fuel consumed by getting the traction power is

$$G_{TR} = T \upsilon_V g_{TSF}. \tag{16}$$

The amount of the fuel consumed by overcoming rolling resistance is

$$G_f = P_f \upsilon_T g_{TSF}, \tag{17}$$

where P_f – the force of the rolling resistance of the wheels; υ_T – the theoretical (circumferential) speed of the wheel mover.

The amount of the fuel consumed by the slipping of the wheel mover is

$$G_{CS} = T \upsilon_{CS} g_{TSF}, \tag{18}$$

where υ_δ – the speed of the wheel mover slipping.

The amount of the fuel consumed by overcoming the land slope ("+" movement on the rise, "–" the motion under the slope) is

$$G_h = \pm G \upsilon_V \sin\alpha g_{TSF}, \tag{19}$$

where G – the force of the machine gravity; α – the inclination angle of the surface motion to the horizon.

The amount of the fuel consumed by overcoming the forces of inertia is

$$G_j = \left(\frac{\chi G \upsilon_Д}{g} \cdot \frac{d\upsilon_Д}{dt} \right) g_{TSF}, \tag{20}$$

where χ – the coefficient considering the rotating masses, g – the acceleration of the gravity, g = 9.81 m/s^2; $d\upsilon_Д/dt$ – the translational machine acceleration, "+" speeding up, "–" braking.

The initial data for the calculation are: the DZ-122 road grader; the A-01MC engine diesel; mechanical transmission; the running equipment: 1 × 2 × 3wheel diagram, 14.00–20 tires; ground surface is horizontal (G_h = 0); ground is cohesive, tight, and dry; the working conditions of the road grader are: the length of the bank section L = 0.5 km, the duration of one turn t_{MOV} = 0.1 h, the coefficient of overlapping the road grader passes when digging soil k_{OD} = 1.5… 1.7, and the coefficient of overlapping the passes when moving soil K_{OM} = 1.15, the average required moving of soil l_0 = 12.5 m, moving soil in one pass l_M = 1.9 m, the steady-state straight moving (G_j = 0).

All the design formulas presented above have been implemented in the program performed in Mathcad, which allows us to determine the fuel consumption of the road grader when changing any parameter included in formulas (2)…(8). For example, Fig. 1 shows the effect of the soil type on the fuel consumption of the road grader. With increase

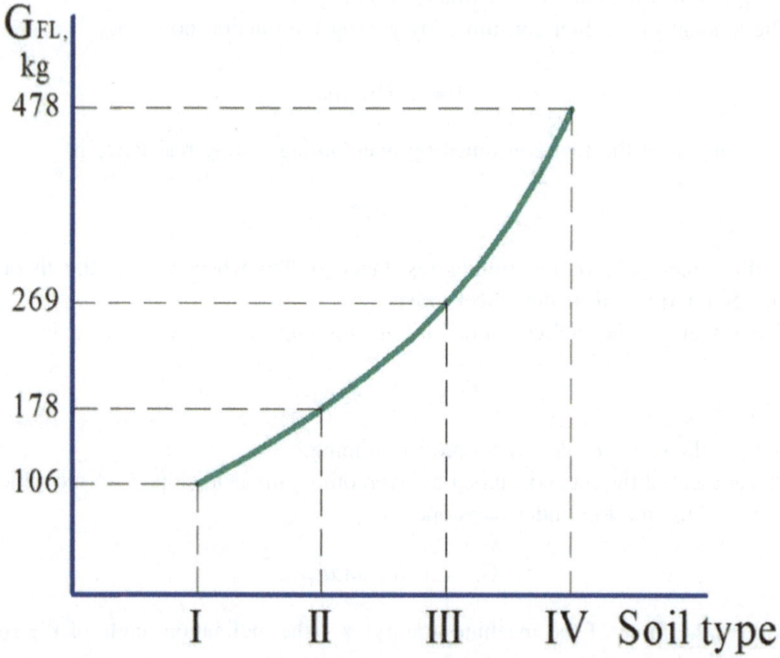

Fig. 1. The effect of soil category on the fuel consumed by the motor grader when constructing the subgrade section.

in the category of soil, the fuel consumed by the motor graders increases, this is due, primarily, to the increase in the coefficient of specific resistance of soil cutting.

The fuel consumption included in formula (1) may also be presented as a percentage of the total fuel consumption (Fig. 2) and then be analyzed.

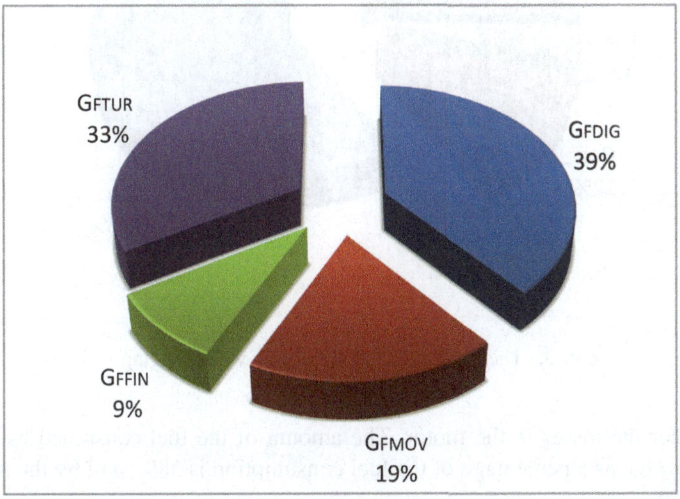

Fig. 2. The components of the road grader fuel consumption from formula (1).

4 Discussion

Figure 2 shows that the fuel consumption when the digging soil G_{FDIG} is 39%, when moving soil G_{FMOV} - 19%, when finishing the subgrade G_{FFIN} is 9% and when turning around the motor grader G_{FTUR} - 33%. In the context of the fuel economy, the most expensive is the process of digging the soil, because of the greatest resistances. Digging soil is performed at the mode of the maximum traction power, when the fuel consumption is also increased, besides, the number of passes when digging and, accordingly, the digging time are also maximized. The part of the fuel required to turns is in the second place. Despite the fact that the turns occur at the minimal resistance, the number of passes is added up, as illustrated in formula (5). When moving soil, the fuel consumption is in the third place, as the resistance when moving the soil is less than that when digging the soil, the grader is operating at a lower traction power, the fuel consumed by the engine is reduced. The number of passes is also less. And the last place is occupied by the consumption of fuel at the finishing operations, as the resistance is not great and the number of passes is less.

According to formulas (13)…(20), it is possible to determine the components of the fuel balance and to analyze the results. For example, Fig. 3 shows the grader fuel balance when digging soil.

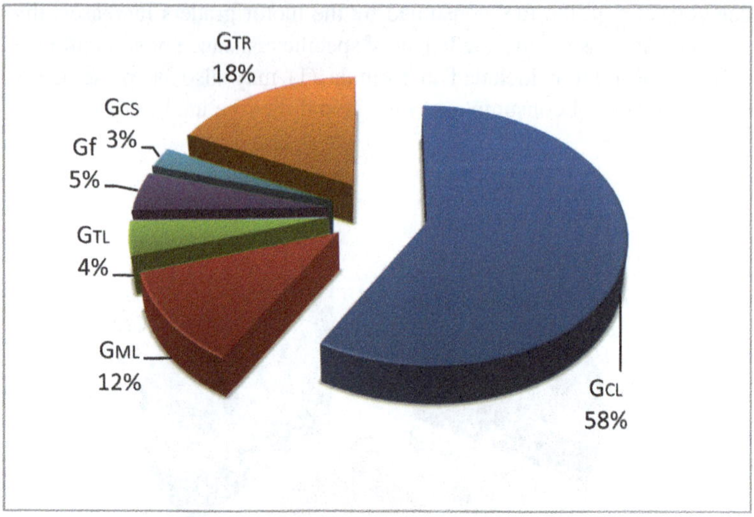

Fig. 3. The road grader fuel balance when digging soil.

Consider the losses in the motor. The amount of the fuel consumed by the motor heat losses G_{CL} as a percentage of the fuel consumption is 58%, and by the mechanical losses - 12%.

The amount of the fuel consumed by the lost power in the transmission G_{TL} is 4%.

Consider the loss in the wheel mover. The amount of the fuel consumed by overcoming rolling resistance G_f is 5%, by the slipping of the wheel mover G_{CS} – 3%.

The amount of the fuel consumed by getting the traction power G_{TR} is 18%.

5 Conclusions

1. By the example of the road grader, the technique of the analytical determination of the EMM fuel consumption when performing the technological operations is presented. Taking into account the characteristics of the work of the other EMM, the use of this method will make it possible to determine the fuel consumption of machines such as bulldozers, scrapers, grader-elevators when performing the technological operations.
2. The example of defining the fuel balance of the DZ-122 motor grader with the analysis of its components when it digs soil has been considered.
3. The results of calculating the fuel consumption for the technological operations of the grader and the components of the fuel balance when digging soil have been analyzed.
4. It is proposed to use the fuel balance to determine the specific design measures aimed at reducing all kinds of the EMM losses and reducing the fuel consumption, respectively.

References

1. Thorpe, S.G.: Fuel economy standards, new vehicle sales, and average fuel efficiency. J. Regul. Econ. **11**(3), 311–326 (1997)
2. Plotkin, S.E., Greene, D.L.: Prospects for improving the fuel economy of light-duty vehicles. Energy Policy **25**(14–15), 1179–1188 (1997)
3. Kryuchin, A.P.: Performance Characteristics and Efficiency of Earth-Moving machines. Transport, Moscow (1975)
4. Govorushenko, N.I.: Fuel Saving and Toxicity Reduction in Road Transport. Transport, Moscow (1990)
5. Ulyanov, N.A., Roninson, E.G., Soloviev, V.G: Self-Propelled Wheeled Earth-Moving Machines. Mashinostroenie, Moscow (1976)
6. Ulyanov, N.A.: The Theory of Self-Propelled Wheeled Earth-Moving Machinery. Mashinostroenie, Moscow (1969)
7. Ulyanov, N.A.: Fundamentals of theory and Calculation of Wheel Mover Machinery. Mashgiz, Moscow (1962)
8. Kholodov, A.M., Nitschke, V.V., Nazarov, L.V.: Earth-Moving Machinery. Vyscha Shkola, Kharkov (1982)
9. Tyunin, V.L.: Methods of calculation of power factors of the wheel mover earth-moving machines. Dis. Cand. Tech. Sci. VGASU, Voronezh (2008)
10. July, V.A., Tyunin, V.L.: Power and fuel balances of wheeled earth-moving machines. Constr. Road Mach. **9**, 42–45 (2014)
11. July, V.A., Tyunin, V.L., Krestnikov, A.V.: Assessment of fuel economy self-propelled wheeled earth-moving machinery. Mech. Constr. **77**(8), 27–31 (2016)
12. Lenin, I.M.: Car and Tractor Engines. Part 1. Higher school, Moscow (1976)
13. Syurkin, V.I.: Fundamentals of Theory and Design of Automotive Engines. LAN, SPb, Moscow, Krasnodar (2013)
14. Kolchin, A.I., Demidov, V.P.: Calculation of Automobile and Tractor Engines. High School, Moscow (2002)
15. Tokarev, A.A.: Theoretical background the design analysis of power, power and fuel balance auto. In: Proceedings of NAMI, the Improvement of the Technical and Economic Performance of Automotive Vehicles, pp. 4–45 (1989)
16. LNCS. http://www.springer.com/lncs. Accessed 21 Nov 2016

Type Analysis of a Multiloop Coulisse Mechanism of a Cotton Harvester

Khabibulla Turanov[1(✉)], Anvar Abdazimov[1], Mukhaya Shaumarova[2], and Shukhrat Siddikov[1]

[1] Tashkent State Technical University named after Islam Karimov, University Street, 2, 100174 Tashkent, Uzbekistan
khturanov@yandex.ru
[2] Tashkent Institute of Irrigation and Agricultural Mechanization Engineers, Qori Niyoziy Street, 39, 100000 Tashkent, Uzbekistan

Abstract. Vertical spindle cotton harvester. Planetary gear. Cotton harvester apparatus. The paper is devoted to the type analysis of the multiloop coulisse mechanism of the spindle drum of a cotton harvester. Provide a description of the design of the multiloop coulisse mechanism of the cotton harvester; identify the cause of breakdowns of individual parts of the specified mechanism; carry out the type analysis of the structure of the studied mechanism; present the optimal type of the studied mechanism. Research methods are based on the type analysis of the kinematic chain of the multiloop coulisse mechanism by increasing the freedom of kinematic pairs. A complete description is made and an analysis of the multiloop coulisse mechanism is performed. The main causes of breakdowns of individual parts of the new design of the spindle drum are clarified. The results of studies on the elimination of redundant constraints in the kinematic chain of the mechanism are presented. The kinematic chain is reduced from a statically indefinable system to a statically definable one. The optimal type of the kinematic chain of the multiloop coulisse mechanism has been developed. Research results will be useful for the wide practical application of spindle drums with multiloop coulisse mechanisms in the construction of vertical spindle cotton harvesters.

Keywords: Vertical spindle cotton harvesters · Cotton harvester apparatus · Planetary mechanism · Type · Kinematic pair · Multiloop coulisse mechanism

1 Introduction

The vertical spindle cotton harvester is mainly designed to collect raw cotton from open cotton bolls [1]. Let us briefly consider the design of the cotton harvester and the technological process of harvesting raw cotton from cotton bushes. It is equipped with a vertical spindle working apparatus (Fig. 1) (see page 13 in [1]).

V. Murgul and V. Pukhkal (Eds.): EMMFT 2019, AISC 1258, pp. 290–305, 2021.
https://doi.org/10.1007/978-3-030-57450-5_27

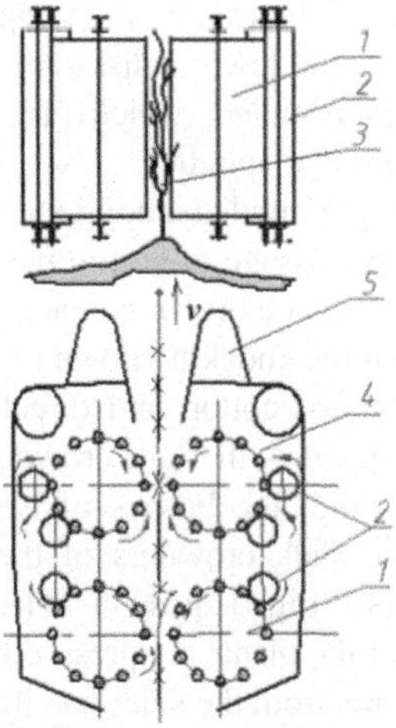

1 - spindle drum; 2 – doffing rollers; 3 - cotton bushes; 4 - spindles; 5 - stalk crowder

Fig. 1. Scheme of the vertical spindle apparatus.

The spindle drum (Fig. 2) is designed for picking raw cotton from open cotton bolls and transporting it to the doffing roller area.

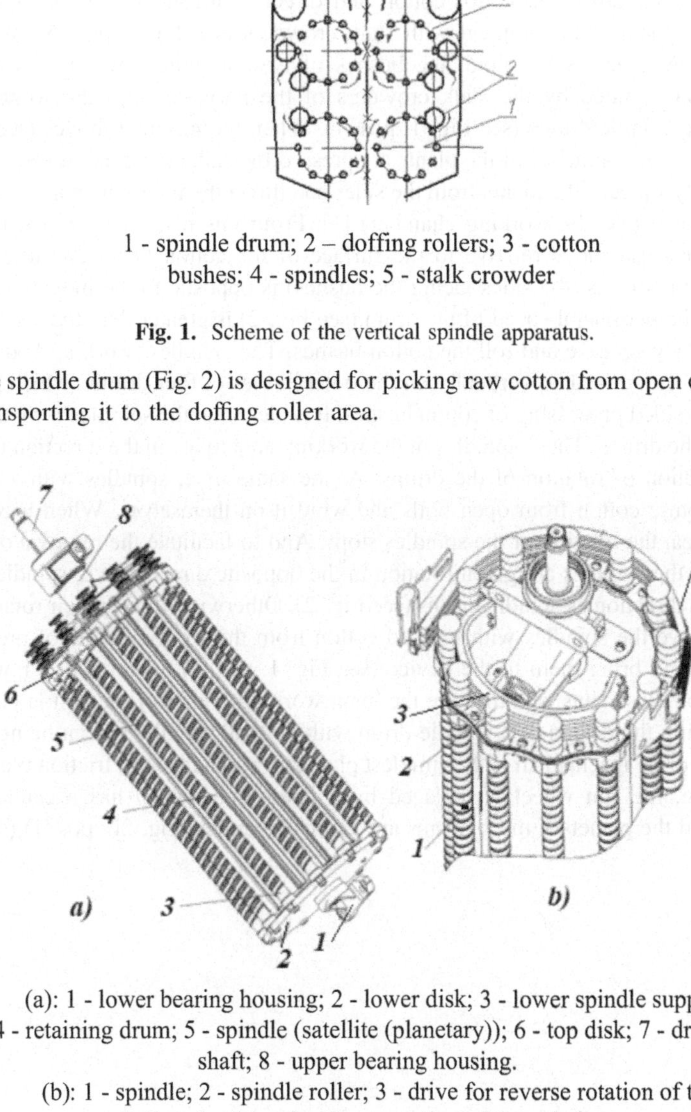

(a): 1 - lower bearing housing; 2 - lower disk; 3 - lower spindle support; 4 - retaining drum; 5 - spindle (satellite (planetary)); 6 - top disk; 7 - drum shaft; 8 - upper bearing housing.
(b): 1 - spindle; 2 - spindle roller; 3 - drive for reverse rotation of the spindles; 4 – drive for direct rotation of the spindles.

Fig. 2. Spindle drum.

Here, the drums are fixed to the frame of the apparatus by the upper 8 and lower 1 bearing housings. The upper 6 and lower 2 disks are supports for spindles. Between the disks, a retaining drum 6 is installed, which prevents the bolly cotton and cotton branches from falling between the spindles.

The working apparatuses are divided among themselves by a working slot (marked with crosses in Fig. 1) for the passage of cotton bushes (see Fig. 1, pos. 5) with open bolls. All parts of the machine that come in contact with the cotton bushes are covered with fairings. Fairings prevent the knocking down of raw cotton from cotton bushes. To raise the beaten down bushes of cotton and direct them into the working slot, the harvesting apparatus is equipped with stalk crowders (see Fig. 1 pos. 5). When the machine moves across the field, the bushes of two or more rows of cotton being processed are guided by the stalk crowders of the harvester into the working slot between the spindle drums (see Fig. 1, pos. 1). When the machine moves (see Fig. 1, pos. 5), the lower branches of the plants are pressed by stalk crowders. Several of them preliminarily squeeze the plants from the sides and direct them into the gap between the spindle drums (into the working chamber) [1]. From this it is clear that parts of the cotton plants also move relative to the surface of the cotton field. The direction of rotation of the drums (the sides facing the bushes) is opposite to the movement of the machine. The peripheral speed of the drum (see Fig. 2) is greater than the speed of the machine. They squeeze and roll the cotton bushes. The spindles (working bodies) (see Fig. 1, pos. 4) are located vertically on the outer circles of four drums, which perform double two-sided processing of cotton bushes. It is clear that the spindles form the outer surface of the drums. These spindles in the working area rotate in the direction opposite to the direction of rotation of the drums. At the same time, spindles with a serrated surface capture cotton from open bolls and wind it on themselves. When leaving the working area, the rotation of the spindles stops. And to facilitate the removal of cotton from them, the spindles are given rotation in the opposite direction, i.e. spindles rotate in the same direction as spindle drums (see Fig. 2). Otherwise, during their rotation, the drums remove the spindles with trapped cotton from the working zone of processing the bushes and bring them to the device (see Fig. 1, pos. 2) for removing raw cotton from the spindles. Thus, spindles are the main working bodies of the cotton harvester.

Analyzing the design of a spindle drum with vertical spindles, it can be noted that the spindle drive mechanism is the simplest planetary mechanism of friction type [2–8]. In this case, the sun wheel is replaced by a fixed tape, which has a central angle $\gamma_1 + \gamma_2$, and the planetary mechanisms are vertical spindles (Fig. 2b, pos. 1) (see page 14 in [1]).

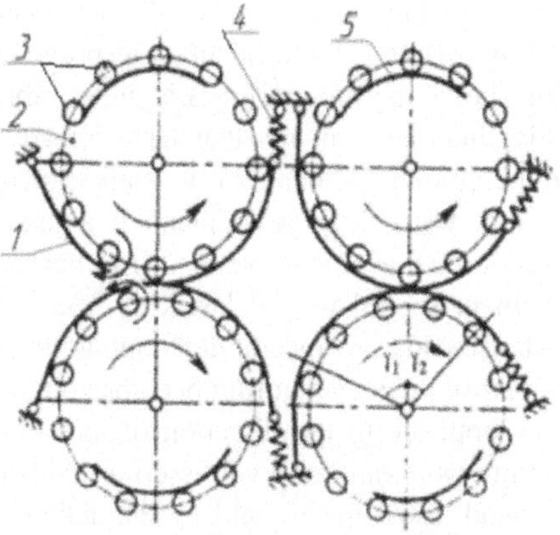

1 - drive for direct rotation of the spindles; 2 - spindle drum;
3 - spindles; 4 - tension spring; 5 - drive for reverse rotation of the spindles.

Fig. 3. Scheme of the spindle drive mechanism.

Here, the V-belts 1 are elastically attached to the frame of the working apparatus, and 2 are elastically connected to the housing of the upper support of the spindle drum (see pos. 7 in Fig. 2). Note that the spindle of the cotton harvester consists of a pipe on which the teeth are cut (or on which serrated bands are attached to its upper part). A metal-ceramic sleeve is pressed into the lower end of the pipe, and there is a three-strand roller in the upper end, on which the roller is mounted. The spindle assembly is inserted into the hole of the upper disk of the drum, and the lower end is put on the pin of the lower disk of the drum.

Note that when the spindle rollers 3 (see Fig. 3) are in contact with the V-belts 1, the spindle drive can be considered as a hypocyclic mechanism, and with V-belts 5 (as with a reverse spindle drive) - as an epicyclical mechanism [2–8]. At that moment when the spindle drum (see Fig. 2) can be considered as a hypocyclic mechanism, the spindles pick up the raw cotton from the opened bolls and wind it onto itself along the path of the elongated hypocycloid, and when it is used as an epicyclic mechanism, self-dropping off and picking raw cotton from the spindles is performed along the path of an elongated epicycloid. Elongated epi- and hypocycloids are formed due to the fact that the average roller radius (6 mm) at the point of contact with the V-belts is less than the radius of the top of the spindle teeth (12 mm).

The study of the technological process of harvesting raw cotton from cotton bushes allows us to note that the process of picking raw cotton depends on the direction of the speed of the top of the spindle tooth relative to the boll. When plants enter the working area (marked with crosses in Fig. 1), the speed of some plants relative to the cotton field is low (see page 19 in [1]). In [1], the calculations proved that moving the spindles in the positive direction (i.e. along the machine) is undesirable, since it can lead to an even greater forward inclination of plants that have already received an inclination when passing through the guiding elements of the stalk crowder (see pos. 3 in Fig. 1). It was noted in [1] that it is desirable to move plants in a negative direction (back) (i.e. opposite to the direction of the machine movement) for their possible straightening to a right angle. This movement should not be large, as it can lead to damage to some plants. Thus, the importance and relevance of developing the design of the mechanisms of the stalk crowder as part of the spindle drum becomes obvious. This would allow for the movement of plants opposite to the direction of movement of the machine. The design of the spindle drum as a multiloop coulisse mechanism of the cotton harvester was developed, created and tested in the field by the authors of the paper 2, 4 in the Tashkent State Technical University named after Islam Karimov.

2 Object

- Provide a description of the design of the multiloop coulisse mechanism of the cotton harvester;
- identify the cause of breakdowns of individual parts of the multiloop coulisse mechanism;
- make a type analysis of the structure of the mechanism under study;
- present the optimal type of the given mechanism (i.e. without redundant loop constraints).

3 Method of Research

The research methods are based on the type analysis of the kinematic chain of the multiloop coulisse mechanism by increasing the freedom of a kinematic pair [2, 9, 12].

4 Research Results

4.1 Description of the Design of the Multiloop Coulisse Mechanism of the Cotton Harvester

We proceed to the description and analysis of the multiloop coulisse mechanism. This mechanism is placed inside the serial spindle drum in two rows along its height (Fig. 4).

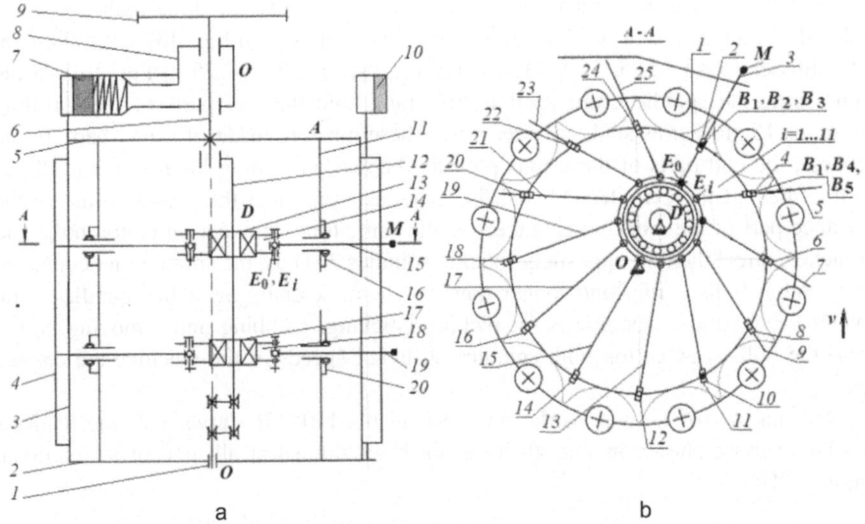

1 - lower drum support; 2 - lower disk of the drum; 3 - spindle; 4 - retaining drum; 5 - upper disk of the drum or crank; 6 - drum shaft; 7 - drive for reverse rotation of the spindles; 8 - upper drum support; 9 – gear wheel; 10 - drive for direct rotation of the spindles; 11 - cylinder of coulisse mechanism; 12 - axis of rotating coulisses; 13 - bearing of the upper rotating coulisse; 14 - upper rotating coulisse; 15 - a rod rigidly connected to the upper rotating coulisse 14 or a rolling coulisse 15_1 (not shown in the figure) that is pivotally connected to the upper rotating coulisse 14; 16 – sliding blocks pivotally connected to the cylinder 11; 17 - bearing of the lower rotating coulisse; 18 - lower rotating coulisse; 19 - a rod rigidly connected to the lower rotating coulisse 18 (or cassette) or a rolling coulisse 18_1 (not shown in the figure) that is pivotally connected to the lower rotating coulisse 18; 20 – sliding blocks pivotally connected to the cylinder 11.

Fig. 4. Principle diagram of the multiloop coulisse mechanism: a) vertical section; b) section along A-A.

It is also indicated in Fig. 4: O - axis of rotation of the drum; D - axis of rotation of the multiloop coulisse mechanism; E_0 - rigid mounting of the rotating coulisse (pin) 15 to the rotating coulisse 14; E_i (where i = 1, …, 11) is the movable connection of the rolling coulisse 151 (not shown in the figure) to the rotating coulisse 14; M - end parts or vertices of the rotating 15 and rolling 151 coulisses (pins); v - direction of movement of the machine.

In addition, in Fig. 4, a section along A-A represents the upper row of the multiloop crank-coulisse mechanism. It is indicated on it: 1 - cylinder of the coulisse mechanism 11, which is intended for fastening sliding blocks 2 to it (position 16 in Fig. 4a); 2, 4, 6, 8, 10, 12, 14, 16, 18, 20, 22, and 24 – sliding blocks pivotally connected to cylinder 11; 3 – rotating coulisses (pins) (position 15 in Fig. 4a) rigidly connected to the rotating coulisse 14; 5, 7, 9, 11, 13, 15, 17, 19, 21, 23, and 25—rolling coulisse (position 151 is not shown in Fig. 4a).

Here, the axis D-D of the rotating coulisses 12 is eccentrical relative to the axis of rotation O-O of the spindle drum 6. The upper rotating coulisse (or cassette) 14 is located at a height of 400… 450 mm, and the lower rotating coulisse (or cassette) 18 - at a height of 200–250 mm from the lower disk 1 of the spindle drum.

As you can see, the crank-coulisse mechanism consists of three main freedoms links: the crank 5 (pos. 1 in Fig. 4b), rotating 13, 17 (pos. 3 in Fig. 4b) and rolling 15, 19 coulisses (positions 5, 7, 9, 11, 13, 15, 17, 19, 21, 23, and 25 in Fig. 4b). At the same time, the spindle drum itself is the crank, and the end parts of these rolling coulisses 15, 19 act as stalk crowder pins. These pins are made of round bars. In the process, the end parts of these pins protrude behind the sliding blocks 16 and 20, as well as behind the cylinders 11 and the retaining drum 4 of the spindle drum in the entrance part of the working chamber. At the same time, they act on cotton bolls and branches, direct them to oppositely rotating spindles 3. Thus, the spindles are contacted with cotton bolls earlier and longer than in a commercially available spindle drum. Therefore, favorable conditions for picking (sticking, grabbing and removing cotton from the bolls) raw cotton with spindles at higher speeds of the machine are created here.

An analysis of known works on mechanisms [10–23] shows that mechanisms similar to those shown in Fig. 4b have not been studied at all, except in rotational engines [4]).

4.2 Cotton Harvester Field Test Results

The results of field tests of a cotton harvester using such spindle drums with mechanisms for directing bushes into the working chamber made it possible to increase the harvest of raw cotton to 85.13% at the first harvest, against the serial design - 79.61%. Otherwise, the new spindle drum design was 5.52% more efficient than the serial design.

At the same time, during the field test, there were failures of spindle drum design with a multiloop coulisse mechanism in the form of breakdowns of individual parts of the mechanism, apparently due to a decrease in structural reliability. So, for example, there was a breakdown and bending of the round rod of the rotating coulisses 15 and 19, disruption of the welds of the sliding blocks 16 and 20 with the cylinder 11, bending of the round rods of the rolling coulisses 15 and 19 (Fig. 5).

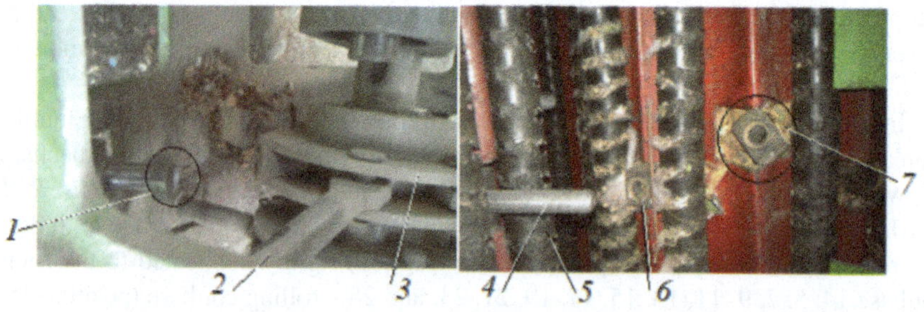

1 - breakdown of the threaded part of the round rod of the rotating coulisse (positions 15 and 19 in Fig. 4a); 2 - rolling coulisse (positions 5, 7, 9, 11, 13, 15, 17, 19, 21, 23, and 25 in Fig. 4*b*); 3 - rotating coulisse (or cassette) (positions 14 and 18 in Fig. 4*a*); 4 - end parts (pins) of rotating (positions 15 and 19 in Fig. 4*a*) or rolling (position 4 in Fig. 4*b*) coulisses; 5 - spindles with a gripping element (pos. 3 Fig. 4*a*); 6 - sliding blocks (positions 16 and 20 in Fig. 4*a*); 7 - failure of welding seams of sliding blocks (positions 16 and 20 in Fig. 4).

Fig. 5. Types of failures of parts of the multiloop coulisse mechanism.

4.3 Identify the Causes of Breakdowns of Individual Parts of the Multiloop Coulisse Mechanism

In our opinion, the main causes of breakdowns of individual parts of the new design of the spindle drum are that all moving joints (i.e. kinematic pairs) of the individual parts of the multiloop mechanism are made with one or two degrees of freedom, as well as their manufacturing and assembly errors. So, for example, rotating coulisse or cassette (pos. 14 and 18 in Fig. 4a) and rolling coulisse (positions 5, 7, 9, 11, 13, 15, 17, 19, 21, 23, and 25 in Fig. 4b) are pivotally connected by kinematic pairs with degrees of freedom (positions 2 and 3 in Fig. 5). To clarify this assumption, we will perform a type analysis of the structure of the studied mechanism below.

4.4 Type Analysis of the Structure of the Studied Mechanism

In the kinematic chain of a spindle drum with a multiloop coulisse mechanism, movement from the gear wheel 9 through the shaft 6 is transmitted to the upper disk 5 (see Fig. 4a). A cylinder 11 is welded to the flanges of the upper 5 and lower 1 disks, which transmits rotational movement to the lower disk 1 and the working links of the multiloop coulisse mechanism (positions 3, 5, 7, …, 25 in Fig. 4b). The crank (cylinder) 11 transfers the rotational movement to the rotating (position 3 in Fig. 4b) and rolling coulisses (positions 5, 7, …, 25 in Fig. 4b), using sliding blocks (positions 2, 4, 6, … 24 in Fig. 4b). At point E_0, the rotating coulisse (pin) (position 3 in Fig. 4b) is rigidly connected to another rotating coulisse (cassette) (pos. 14 and 18 in Fig. 4a). In turn, at the points E_i (where i = 1, …, 11), eleven rolling coulisses are pivotally connected (pos. 5, 7, …, 25 in Fig. 4b) to the rotating coulisse (pos. 14 and 18 in Fig. 4a). These rolling coulisses are driven on one side of the crank (cylinder) 11 by means of sliding blocks (pos. 2, 4, 6, … 24 in Fig. 4b).

Moreover, we emphasize that all movable joints (i.e. kinematic pairs) of individual parts of the upper and lower multiloop mechanism are made with one degree of freedom p1. So, for example, in the hinges O and D, the kinematic pairs are rotational with one degree of freedom $p_1(1r)$, i.e. $p_1(1r) = 1 + 2 = 3$ (since the number of hinges D is 2). At point B1, the sliding blocks are pivotally connected to the cylinder (pos. 1 in Fig. 4b), forming rotational kinematic pairs with one degree of freedom $p_1(1r)$, i.e. $p_1(1r) = 12 + 12 = 24$. At points B_2, B_4, B_6, B_8, B_{10}, B_{12}, B_{14}, B_{16}, B_{18}, B_{20}, B_{22}, B_{24}, the sliding blocks 2 are movably connected to the rolling coulisse, forming one degree of freedom prismatic pair $p_1(1p)$, i.e. $p_1(1p) = 12 + 12 = 24$. At points E_i (where i = 1, …, 11), rolling coulisses (pos. 5, 7, …, 25 in Fig. 4b) are movably connected to the rotating coulisses (pos. 14 and 18 in Fig. 4a). In this case, two degree of freedom prismatic pairs $p_1(1r)$ are formed, i.e. $p_2(2c) = 11 + 11 = 22$.

Thus, the total number of rotational $p_1(1r)$ and translational $p_1(1p)$ kinematic pairs with one degree of freedom in the upper and lower multiloop mechanism is $p_1 = 3 + 24 + 24 = 51$, and $p_2 = 22$, i.e., $p_1 + p_2 = 51 + 22 = 73$. Moreover, in the upper loop, including the kinematic pairs in hinges O and D, $p_1 = 2 + 24 = 26$ and $p_2 = 11$, and in the lower loop, including the kinematic pair in the hinge D, $p_1 = 1 + 24 = 25$ and $p_2 = 11$. The total number of movable links in the upper and lower multiloop coulisse mechanism is n = 49, since the number of crank - 1, rotating coulisse - 2 (pos. 14 and 18 in Fig. 4a), sliding blocks - 24 (pos. 2, 4, 6, … 24 in Fig. 4b), rolling coulisses - 22 (pos. 5, 7, …, 25 in Fig. 4b).

It is well known [8, 9, 23] that, taking into account possible manufacturing and assembly errors, planar mechanisms can be considered as spatial mechanisms. Assuming the degree of freedom of the kinematic chain mechanism (W) to be a known value equal to the number of generalized coordinates, i.e. the number of initial links with a given law of motion, it is possible to determine the degree of redundant constraint q [8, 9, 23]:

$$q = W - 6n + \sum_{i=1}^{5} (6 - i)p_i, \tag{1}$$

where W – the degree of freedom of the kinematic chain, for example, equal to the number of gears (pos. 9 in Fig. 4a), i.e. W = 1; 6 - a number showing that in spatial motion each link has six degrees of freedom; n - the number of degree of freedom of the links; 5, 4, 3, 2, and 1—the number of constraints imposed on the relative motion of the links corresponding to one-, two-, three-, etc., moving kinematic pairs; p_1, p_2, …, p_5 - the number and type of pairs with one, two, three, four and five degrees of freedom, respectively.

Note that the total number of imposed constraints may include a certain degree q of redundant constraints, which duplicate other bonds without decreasing the mobility of the mechanism, only turning it into a statically indefinable system [8, 9, 12, 23]. These constraints can only occur in a closed kinematic chain, and it is impossible to indicate which constraint is redundant, but it is possible to calculate the number of these constraints in the circuit (see page 48 in [1]).

In the studied mechanism (according to Fig. 4b), the initial data are: W = 1, n = 49, $p_1 = 51$, $p_2 = 22$, p_3, …, $p_5 = 0$. Substituting these initial data in formula (1), we obtain:

$$q = W - 6n + 5p_1 + 4p_2 = 1 - 6 \cdot 49 + 5 \cdot 51 + 4 \cdot 22 = 50. \tag{2}$$

As you can see, numerically, the degree of redundant constraints q requiring precise execution turned out to be 50, which is practically not feasible.

The calculation results are summarized in Table 1.

It is well known [8, 9, 12, 23] that numerically the degree of redundant constraints q is equal to the number of dimensions (for example, 50) that require precise execution. Such dimensions are the displacement along the Ox, Oy, and Oz axes (which are not shown in Fig. 4) and the rotations of the closing kinematic pairs around these axes. For the normal functioning of this mechanism, it is necessary that the axes of all rotational kinematic pairs be parallel, without skewing relative to the motion plane of the moving links, which must also not be skewed (i.e. not deformed). As you can see, the studied

Table 1. Calculated results.

Calculation parameters		Upper loop in the coulisse mechanism	Lower loop in the coulisse mechanism	Multiloop coulisse mechanism
The number and type of rotational (**1r** and **2c**) and prismatic (**1p**) kinematic pairs with one or two degrees of freedom (p_1 and p_2)	$\boldsymbol{p_1}$(**1r**) in the hinge $\boldsymbol{O}$	1	–	1
	$\boldsymbol{p_1}$(**1r**) in the hinge $\boldsymbol{D}$	1	1	2
	$\boldsymbol{p_1}$(**1r**) in the hinge $\boldsymbol{B_1}$	12	12	24
	$\boldsymbol{p_1}$(**1p**) in the hinges $\boldsymbol{B_2}$, $\boldsymbol{B_4}$, …, $\boldsymbol{B_{24}}$	12	12	24
	$\boldsymbol{p_2}$(**2c**) in the hinges $\boldsymbol{E_i}$ (**where** $\boldsymbol{i}$ **= 1, …, 11**)	11	11	22
Total number of kinematic pairs $\boldsymbol{p_i}$		37	36	73
Number of cranks (pos. **1** in Fig. 4*a*)		1	–	1
Number of rotating coulisses (cassettes) (pos. **14** and **18** in Fig. 4*a*)		1	1	2
Number of sliding blocks (pos. **2, 4, …, 24** in Fig. 4*b*)		12	12	24
Number of rolling coulisses (pos. **5, 7, …, 25** in Fig. 4*b*)		11	11	22
Total number of moving links $\boldsymbol{n}$		25	24	**49**
Degree of redundant constraints $\boldsymbol{q}$		–	–	**50**

multiloop coulisse mechanism has 50 redundant constraints, which means that it is necessary to precisely perform such a number of dimensions that it is practically unrealizable. So, $q > 0$, which means that this kinematic scheme is statically indefinable.

Thus, the main causes of breakdowns of individual parts of the new spindle drum design recommended by us for practical application are proved.

4.5 Optimal Type of a Multiloop Coulisse Mechanism (I.E. Without Redundant Loop Constraints)

When analyzing a mechanism with an optimal type, it is taken into account that the O and D racks, considered as rigid fixed links, under real conditions are subjected to deformations under the influence of applied loads. In a deformed state, they can affect the relative positions of the moving links not only within the same kinematic pair, but also within the closed kinematic chain of the mechanism. If the type diagram is incorrectly selected, jamming (pinching) of some elements of the kinematic pair is possible during operation, significant additional loads may appear due to skew, bending, stretching of the links, excessive wear of the elements of the kinematic pair, which leads to an increase in the energy consumption of the mechanism, low reliability, and frequent failures.

It is well known [8, 9, 12, 23] that when developing a type diagram of a mechanism without redundant loop constraints, the conditions for assembling closed kinematic chains (loops) of a mechanism should provide the following: the kinematic chain forming a closed loop (or loops of the mechanism) should be assembled without interference even in the presence of deviations in the sizes of the links and deviations in the location of the surface and axes of the kinematic pair elements. For real mechanisms, scientists strive to develop such a type diagram that would eliminate the possibility of additional loads in the kinematic pair due to a change in the configuration of the link loop, regardless of the accuracy of manufacturing parts or the deformability of the rack and other links. If there are no redundant constraints, i.e. $q = 0$, then the mechanism is assembled without deformation of the links. The links are self-adjusted and fully satisfy the requirements of reliability, durability and manufacturability. In practice, mechanisms without redundant constraints work without creaking and noise. A mechanism without redundant constraints has an optimal structure and is called self-adjusted [9, 23]. If there are redundant constraints, i.e. $q > 0$, then the assembly of the mechanism and the movement of its links are possible only with their deformation. Signs of redundant constraints in the mechanism are creaking, screeching and noise during the operation of such mechanisms. Otherwise, the condition $q > 0$ in the structure of the mechanism indicates the non-optimality of its design.

As the results of our research (see p. 3) showed, the kinematic chain has 50 redundant constraints in the main mechanism (see Fig. 4). As is known [9, 23], if $q > 0$, then this kinematic diagram, although it was considered as spatial taking into account manufacturing and assembly errors, is statically indefinable. To bring it into a statistically definable kinematic diagram, the condition $q = 0$ must be met. To achieve this in the multiloop coulisse mechanism, the method of increasing the freedom of kinematic pairs should be applied [8, 9, 12, 23].

To do this, in all kinematic pairs B_1, "crank 1 - sliding block" (pos. 2, 4, …, 24 in Fig. 4b) of the rotating coulisse (pos. 3 in Fig. 4b), we replace lower kinematic pairs $p_1(1r)$ with higher kinematic pairs $p_4(4l)$ (ball in the cylinder). So, for example, we will replace one degree of freedom rotational pairs $p_1(1r)$ of cylinder 1 with sliding blocks 2 in the hinges B_1 of the rotating coulisse (pos. 3 in Fig. 4b) of the upper and lower loops with four degrees of freedom pairs with line contact $p_4(4l)$), i.e. $p_4(4l) = 2 + 2 = 4$. In the hinges E_i (where $i = 1, \ldots, 11$) (see Fig. 4b), two degree of freedom cylindric pairs $p_2(2c)$ of the upper and lower loops is left unchanged, i.e. $p_2(2c) = 11 + 11 = 22$. Further, the one degree of freedom prismatic pairs $p_1(1p)$ with sliding blocks (pos. 2 in Fig. 4b) at points B_2, B_4, B_6, B_8, B_{10}, B_{12}, B_{14}, B_{16}, B_{18}, B_{20}, B_{22}, B_{24} of rolling coulisses (pos. 5, 7, …, 25 in Fig. 4b) of the upper and lower loops are replaced with three degrees of freedom spherical pair $p_3(3s)$, i.e. $p_3(3s) = 11 + 11 = 22$.

In this case, according to Fig. 4b, the initial data are: $W = 1$, $n = 49$, $p_1 = 27$, $p_2 = 22$, $p_3 = 22$, $p_4 = 2$, $p_5 = 0$. Substituting these initial data in formula (1), we obtain:

$$q = W - 6n + 5p_1 + 4p_2 + 3p_3 + 2p_4 = 1 - 6 \cdot 49 + 5 \cdot 27 + 4 \cdot 22 + 3 \cdot 22 + 2 \cdot 2 = 0. \quad (3)$$

The initial data and calculation results are summarized in Table 2.

Table 2. The initial data and calculation results.

Calculation parameters		Upper loop in the coulisse mechanism	Lower loop in the coulisse mechanism	Multiloop coulisse mechanism
The number and type of one, two degree of freedom rotational pairs (1r and 2c) and three degree of freedom spherical pairs (3c), four degree of freedom pairs with linear contact (4l) and prismatic (1p) pairs	p_1(1r) in the hinge *O*	1	–	1
	p_1(1r) in the hinge *D*	1	1	2
	p_4(4l) in the hinge B_1	12	12	24
	p_4(1l) in the hinge B_1	1	1	2
	p_3(3s) in the hinges B_2, B_4, …, B_{24}	11	11	22
	p_2(2c) in the hinge E_i (where $i = 1, \ldots, 11$)	11	11	22
Total number of kinematic pairs p_i		37	36	73
Number of cranks (pos. 1 in Fig. 4*a*)		1	–	1
Number of rotating coulisses (cassettes) (pos. 14 and 18 in Fig. 4*a*)		1	1	2
Number of sliding blocks (pos. 2, 4, …, 24 in Fig. 4*b*)		12	12	24
Number of rolling coulisses (pos. 5, 7, …, 25 in Fig. 4*b*)		11	11	22
Total number of moving links *n*		25	24	49
Degree of redundant constraints *q*		–	–	0

As you can see, q = 0. This means that this kinematic diagram, although it was considered as spatial taking into account manufacturing and assembly errors, is statistically definable. In order to bring it into a statistically definable kinematic diagram, the condition q = 0 must be fulfilled. At the same time, the optimal type of the multiloop coulisse mechanism is achieved (i.e. without redundant loop constraints).

However, it should be borne in mind that the manufacture of three degree of freedom spherical pair $p_3(3c)$ in the hinges B_2, B_4, …, B_{24} is technologically unfeasible. To achieve the manufacturability of manufacturing all kinematic pairs in a multiloop coulisse mechanism of the spindle drum, in all kinematic pairs B_1 "crank 1 – sliding block 2" (pos. 2, 4, …, 24 in Fig. 4b), we replace the lowest kinematic pairs $p_1(1r)$ with higher kinematic pairs $p_4(4l)$ (ball in the cylinder). Otherwise, we replace one degree of freedom rotational pair $p_1(1r)$ with four degree of freedom pair with line contact $p_4(4l)$. In this case, in the multiloop coulisse mechanism, $p_4(4l) = 12 + 12 = 24$.

In this case, according to Fig. 4b, the initial data are: W = 1, n = 49, $p_1 = 27$, $p_2 = 22$, $p_3 = 0$, $p_4 = 24$, $p_5 = 0$. Substituting these initial data in formula (1), we obtain:

$$q = W - 6n + 5p_1 + 4p_2 + 2p_4 = 1 - 6 \cdot 49 + 5 \cdot 27 + 4 \cdot 22 + 2 \cdot 24 = -22. \quad (4)$$

The initial data and calculation results are summarized in Table 3.

Table 3. The initial data and calculation results.

Calculation parameters		Upper loop in the coulisse mechanism	Lower loop in the coulisse mechanism	Multiloop coulisse mechanism
The number and type of one, two (p_1 and p_2) rotational (1r and 2c) and four degree of freedom pairs with linear contact (4l) and prismatic pairs (1p)	$p_1(1r)$ in the hinge O	1	–	1
	$p_1(1r)$ in the hinge D	1	1	2
	$p_1(1r)$ in the hinge B_1	–	–	–
	$p_4(1l)$ in the hinge B_1	12	12	24
	$p_1(1p)$ in the hinges B_2, B_4, …, B_{24}	12	12	24
	$p_2(2c)$ in the hinges E_i (where i = 1, …, 11)	11	11	22

(continued)

Table 3. (*continued*)

Calculation parameters	Upper loop in the coulisse mechanism	Lower loop in the coulisse mechanism	Multiloop coulisse mechanism
Total number of kinematic pairs p_i	37	36	73
Number of cranks (pos. 1 in Fig. 4*a*)	1	–	1
Number of rotating coulisses (cassettes) (pos. 14 and 18 in Fig. 4*a*)	1	1	2
Number of sliding blocks (pos. 2, 4, …, 24 in Fig. 4*b*)	12	12	24
Number of rolling coulisses (pos. 5, 7, …, 25 in Fig. 4*b*)	11	11	22
Total number of moving links *n*	25	24	49
Degree of redundant constraints *q*	–	–	−22

Twenty-two negative redundant constraints were received. A negative sign of redundant constraints means that there are extra links in the structure of the multiloop coulisse mechanism that only increase the role of the structural type optimality (see page 49 in [8] or page 48 in [23]). Extra redundant constraints in the kinematic pairs E_i (where i = 1, …, 11) (see Fig. 4b) in the form of angular rotations are necessary to ensure free rotation around the own axes of the rolling coulisses 5, 7, …, 25 (see Fig. 4b).

Thus, the optimal type of the multiloop coulisse mechanism was achieved (i.e. without redundant loop constraints).

5 Conclusions

Based on the studies, we especially note the following results:

1. A complete description and an analysis of the multiloop coulisse mechanism have been made.
2. The main causes of breakdowns of individual parts of the new design of the spindle drum are clarified. This is due to the fact that all movable joints (i.e. kinematic pairs) of the individual parts of the multiloop mechanism have one or two degrees of freedoms, as well as their manufacturing errors and assembly. So, for example, rotational coulisses or cassette and rolling coulisses are pivotally connected with two degree of freedom pair.
3. A type analysis of the design of a spindle drum with a multiloop coulisse mechanism has been performed. Numerically, the number of redundant constraints q that require precise execution turned out to be 50, which is practically not feasible. Thus, it was found that for q > 0, the kinematic diagram of the multiloop coulisse mechanism of the spindle drum is statically indefinable. This proves the main

causes of breakdowns of individual parts of the new spindle drum design recommended by us for practical application.

4. Summarizing the analysis of the research results, the optimal type of the multiloop coulisse mechanism of a spindle drum of a cotton harvester was established.
5. The obtained research results will undoubtedly be useful for the wide practical application of spindle drums with multiloop coulisse mechanisms in the construction of vertical spindle cotton harvesters.

References

1. Sablikov, M.V.: Cotton Harvesters. Agropromizdat, Moscow (1985)
2. Artobolevsky, I.I.: Theory Mechanisms and Machines. Science, Moscow (1975)
3. Zinovyev, V.A.: Course of the Theory of Mechanisms and Machines. Science, Moscow (1972)
4. Baranov, G.G.: Course of the Theory of Mechanisms and Machines. Engineering, Moscow (1975)
5. Kozhevnikov, S.N., Esipenko, Yu.I., Raskin, Yu.M.: Machinery. Reference Manual. Edited by S.N. Kozhevnikov. Engineering, Moscow (1976)
6. Kraynev, A.F.: Dictionary-Reference Mechanisms. Engineering, Moscow (1987)
7. Turanov, Kh., Shaumarova, M.: Incorrect application of the epicycloid equation to the planetary mechanism of the cotton harvester. E3S Web Conf. **164**, 06034 (2020). https://doi.org/10.1051/e3sconf/202015706034
8. Frolov, K.V.: Theory of Mechanisms and Machines. Higher School, Moscow (2005)
9. Reshetov, L.N.: Self-Adjusting Mechanisms, Reference. Nauka, Moscow (1991)
10. Agrawal, V.P., Rao, J.S.: The mobility properties of kinematic chains. Mech. Mach. Theor. **22**, 497–504 (1987)
11. Jin-Kui, C., Wei-Qing, C.: Identification of isomorphism among kinematic chains and inversions using link's adjacent-chain-table. Mech. Mach. Theor. **29**, 53–58 (1994)
12. Kolovsky, M.Z., Evgrafov, A.N., Semenov, Yu.A., Slousch, A.V.: Advanced Theory of Mechanisms and Machines. Springer, Heidelberg (2000)
13. Turanov, Kh.T., Turanov, Sh.Kh., Tatarintcev, B.E.: Proektirovaniye kulisnykh mekhanizmov v vychislitelynoy srede Matchcad: uchebn. posobye [Design of coulisser mechanisms in Mathcad computing environment: Textbook]. Publishing House SGUPS (NIIZHT), Novosibirsk (2002). (in Russian)
14. Uicker, J.J., Pennock, G.R.: Theory of Mechanisms. Oxford Univ. Press, New York (2003)
15. Litvin, F.L., Fuentes, A.: Gear Geometry and Applied Theory. Cambrige University Press, Cambrige (2004)
16. Gustafsson, L.: Poisson simulation—a method for generating stochastic variations in continuous system simulation. Simulation **74**(5), 264–274 (2000). https://doi.org/10.1177/003754970007400050
17. Rao, A.B.S., Srinath, A., Rao, A.C.: Synthesis of planar kinematic chains. Inst. Eng. **86**, 195–201 (2006)
18. Railway applications. Track. Concrete sleepers and bearers. (n.d.). https://doi.org/10.3403/02717544u
19. Rizvi, S.S.H., Hasan, A., Khan, R.A.: A New for distinct inversions and isomorphism detection in kinematic chains. Int. J. Mech. Robot. Syst. **3**(1), 48–59 (2016)

20. Pozhbelko, V.I., Kuts, E.: Development of the method of structural synthesis of multi-loop lever mechanisms with multi-loop hinges on the basis of basic groups of mechanisms. Theor. Mech. Mach. **4**(16), 139–149 (2018)
21. Pozhbelko, V.I., Kuts, E.: Structural synthesis of planer 10-link–DOF kinematic chains with up to pentagonal links with all possible multiple joint assortments for mechanisms deign. In: New Advances in Mechanism and Machine Science (2018). Mech. Mach. Sci. **57**, 27–35 (2018)
22. Hasan, A.: Study of multiple jointed kinematic chains. Int. J. Comput. Eng. Res. **1**(8), 13–19 (2018)
23. Turanov, Kh.T.: Pricladnaya mekhanika v sfere gruzovykh perevozok: uchebn. posobye [Applied mechanics in the sphere of freight transportation: Textbook]. Publishing House URGUPS, Yekaterinburg (2008). (in Russian)

Mathematical Modeling of a Multiloop Coulisse Mechanism of a Vertical Spindle Cotton Harvester

Khabibulla Turanov[1(✉)], Anvar Abdazimov[1], Mukhaya Shaumarova[2], and Shukhrat Siddikov[1]

[1] Tashkent State Technical University named after Islam Karimov, University Street, 2, 100174 Tashkent, Uzbekistan
khturanov@yandex.ru

[2] Tashkent Institute of Irrigation and Agricultural Mechanization Engineers, Qori Niyoziy Street, 39, 100000 Tashkent, Uzbekistan

Abstract. Vertical spindle cotton harvester. Planetary gear. Cotton harvester apparatus. The paper is devoted to the mathematical modeling of the multiloop coulisse mechanism of the spindle drum of the cotton harvester. Justify the multiloop coulisse mechanism as part of the spindle drum of the cotton harvester using calculated data; present the results of mathematical modeling of the kinematic characteristics of the multiloop coulisse mechanism; present the results of computational experiments on constructing phase trajectories of moving characteristic points of the rotating and rolling coulisses of the multiloop mechanism in the Mathcad system. Research methods are based on the type analysis of the kinematic chain of the multiloop coulisse mechanism and on the methods of vector algebra. The calculated data proved that the spindle drum of the cotton harvester is indeed a multiloop coulisse mechanism and consists of two multiloop coulisse chains, each of which consists of twelve loops. The results of mathematical modeling of the kinematic characteristics of the multiloop coulisse mechanism and the results of computational experiments on constructing the phase trajectories of the image points are presented. Based on the use of the obtained analytical formulas, the problem of mathematical modeling of the kinematic characteristics of the multiloop coulisse mechanism was completely solved when the crank was adopted as the leading link. The research results are of interest for the wide practical application of spindle drums with multiloop coulisse mechanisms in the construction of vertical spindle cotton harvesters.

Keywords: Vertical spindle cotton harvester · Cotton harvester apparatus · Planetary mechanism · Multiloop coulisse mechanism · Mathematical modeling

1 Introduction

Research results [1] established the optimal type of the multiloop coulisse mechanism of the stalk crowder directing bushes to the working area of the spindle drum of the cotton harvester [2]. In [2], the importance of moving cotton bushes with open bolls in

V. Murgul and V. Pukhkal (Eds.): EMMFT 2019, AISC 1258, pp. 306–321, 2021.
https://doi.org/10.1007/978-3-030-57450-5_28

the negative direction (back) (i.e. opposite to the direction of movement of the machine) was noted for their possible straightening to a right angle. From this, the importance and relevance of developing the design of the stalk crowder mechanisms in the spindle drum in order to ensure the movement of plants in opposite direction to the movement of the machine is obvious. It was noted in [1] that such a design of a spindle drum as a multiloop coulisse mechanism of the cotton harvester was developed, created and tested in the field in the Tashkent State Technical University named after Islam Karimov. In particular, in [1], a type analysis of the design of a spindle drum with a multiloop coulisse mechanism was performed, which is designed to direct bushes to the area of collection of raw cotton from cotton bushes. At the same time, the number of redundant constraints q requiring exact execution turned out to be numerically equal to 50, which is practically not feasible. Thus, in [1] it was revealed that for $q > 0$, the kinematic diagram of the multiloop coulisse mechanism of the spindle drum is statically indefinable. This proves the main causes of breakdowns of individual parts of the new spindle drum design recommended by the authors for practical application. In addition, using the method of increasing the mobility of kinematic pairs, the type of the multiloop coulisse mechanism in the spindle drum was optimized.

At the same time, in [1], there is no justification that the coulisse mechanism in the spindle drum of the cotton harvester is of multiloop type; the results of a study on the kinematic characteristics of the coulisse mechanisms giving the possibility of a rational choice of the linear sizes of the links of these mechanisms are not presented. In addition, the study of the kinematic characteristics of the multiloop coulisse mechanism has been overlooked by researchers.

2 Object

- Justify the multiloop coulisse mechanism in the spindle drum of the cotton harvester using calculation data;
- present the results of mathematical modeling of the kinematic characteristics of a multiloop coulisse mechanism;
- present the results of computational experiments on the construction of phase trajectories of the characteristic points of the rotating and rolling coulisses of a multiloop mechanism in the Mathcad system.

3 Method of Research

Research methods are based on the type analysis of the kinematic chain of the multiloop coulisse mechanism and on methods of vector algebra [3–9].

4 Research Results

4.1 Justification of the Multiloop Coulisse Mechanism in the Spindle Drum of a Cotton Harvester

Note that the diagram of the multiloop coulisse mechanism along the A-A cross section is shown in Fig. 4 in [1].

Considering the type diagram of the studied mechanism (see Fig. 4 in [1]) as a spatial kinematic chain, it was established in [1] that this mechanism is statically indefinable. According to Table 1 in [1], in this mechanism, the number of degrees of freedom W = 1, the number of moving links n = 49, the number of one- and two degrees of freedom pairs are $p_1 = 51$ and $p_2 = 22$, respectively, and three-, four- and five degrees of freedom pairs $p_3,\ldots, p_5 = 0$. Using these initial data, the calculations showed that the number of redundant constraints q requiring exact execution turned out to be 50. These redundant constraints can occur only in a closed kinematic chain, and it is impossible to indicate which constraint is redundant, but you can only calculate the number of these constraints in the loop (see page 48 in [7]).

The number k of closed loops of the kinematic chain is calculated by the formula (see formula (3.4) in [8]):

$$k = \sum_{i=1}^{5} p_i - n = p_{\sum} - n. \quad (1)$$

Substituting the initial data in the formula (1), we will have:

$$k = \sum_{i=1}^{5} p_i - n = p_1 + p_2 - n = 51 + 22 - 49 = 24. \quad (2)$$

Thus, the total number of closed loops in the kinematic chain under consideration (see Fig. 4 in [1]) is 24. Moreover, there are upper and lower multiloop coulisse mechanisms inside the spindle drum (Fig. 1) (see Fig. 4a in [1]).

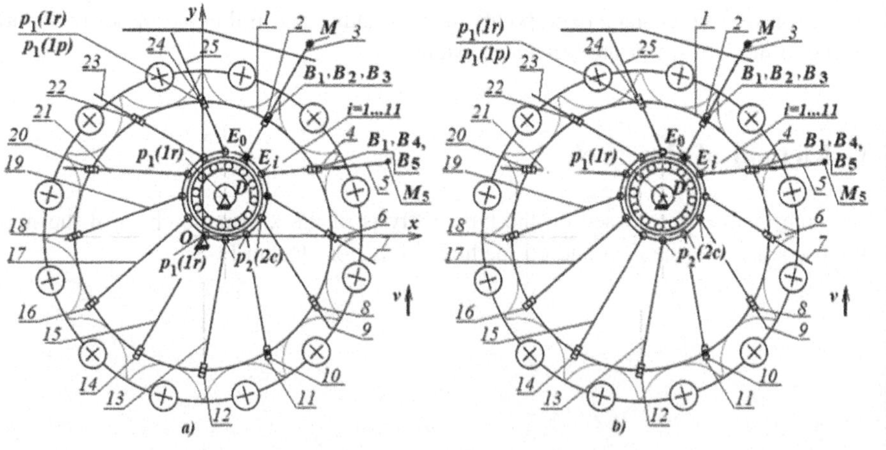

Fig. 1. Diagram of a multiloop coulisse mechanism along cross-section A-A (see Fig. 4 in [1]).

All designations in Fig. 1 correspond to the designation of Fig. 4b in [1]. An exception is that the rotations of the drum O and the axis of rotation of the multiloop coulisse mechanism D are represented by one degree of freedom rotational pair) $p_1(1r)$; the connection of the crank 1 with the sliding blocks 2 in the hinge B_1 is indicated by the one degree of freedom rotational pairs $p_1(1r)$, and the sliding blocks 2 with the rotating and rolling coulisses 3 and 5, 7, …, 25 at point B_3, B_5, B_3, B_7, …, B_{25} are represented by one degree of freedom prismatic pair $p_1(1p)$; the movable connection of the rolling coulisses 5, 7, …, 25 to the rotating coulisses 14 and 18 (see Fig. 4a in [1]) in the hinges E_i (where i = 1, …, 11) are represented by two degree of freedom cylindrical pair $p_2(2c)$.

Figure 4a shows the upper multiloop of the coulisse mechanism, and Fig. 1b - the bottom one. According to the Table 1 in [1], in the upper kinematic chain, the number of moving links is n = 25, the number of one- and two degrees of freedom pairs, is $p_1 = 26$ and $p_2 = 11$, respectively, and in the lower one - n = 24, $p_1 = 25$ and $p_2 = 11$.

Then the number of closed loops in the upper k1 and lower k2 kinematic chains:

$$k_1 = p_1 + p_2 - n = 26 + 11 - 25 = 12; \tag{3}$$

$$k_2 = p_1 + p_2 - n = 25 + 11 - 24 = 12. \tag{4}$$

Thus, the spindle drum of the cotton harvester is indeed a multiloop coulisse mechanism and consists of two multiloop kinematic chains, each of which consists of twelve circuits. Due to such a number of loops, the end parts (point M in Fig. 1) of rotating and rolling coulisses (pins) act on cotton bolls and branches, direct them to oppositely rotating spindles (working bodies) of the cotton harvester. At the same time, the spindles are contacted with cotton bolls earlier and longer than in a standard spindle drum [1].

An analysis of known works on mechanisms [3–20] shows that mechanisms similar to those shown in Fig. 1 have not been studied at all, except in rotational engines (see Fig. 32 in [10]).

5 Presenting the Results of Mathematical Modeling of the Kinematic Characteristics of the Multiloop Coulisse Mechanism, When Crank 1 Is Taken as the Leading Link

Using the closed loop method (or the vector algebra method), known from the course of the theory of mechanisms and machines [3–9], we consider closed loop ODAO or ODB3O (Fig. 2). We will bear in mind that in Fig. 2, point A coincides with point B_3.

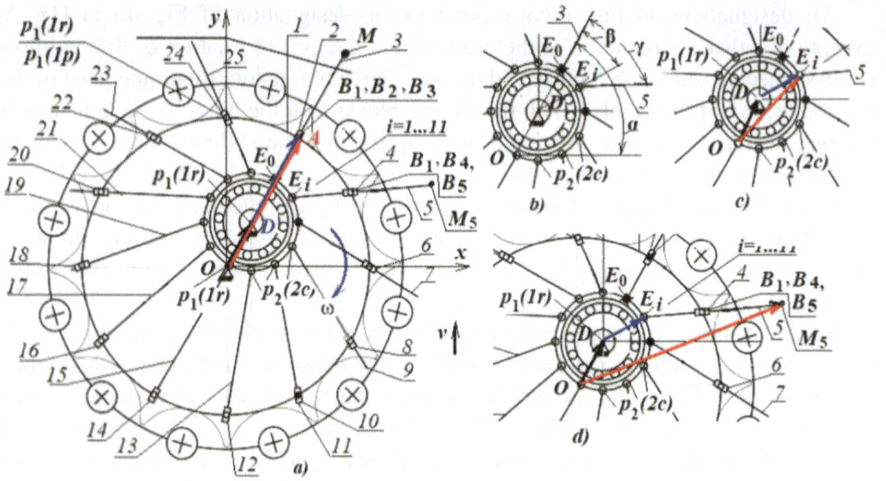

Fig. 2. The multiloop of the coulisse mechanism, when the crank 1 is taken as the leading link.

In Fig. 2, the closed loop ODAO or ODB3O is represented by the line OD, DA and OA, and the directions of rotation of the drum 1 are shown by the arc arrow ω.

The equation of closedness of this loop:

$$\overline{OD} + \overline{DA} - \overline{OA} = 0 \text{ or } \overline{OD} + \overline{DB_3} - \overline{OB_3} = 0. \tag{5}$$

Here the vectors OD and OA (or OB_3) are constant modulo, and the vector DA or DB_3 - variable.

Therefore, projecting Eq. (5) onto the coordinate axes Oxy, we have

$$\left.\begin{aligned} xD + DB_3(\varphi)\cos(\varphi_3(\varphi)) &= R\cos(\varphi); \\ yD + DB_3(\varphi)\sin(\varphi_3(\varphi)) &= R\sin(\varphi), \end{aligned}\right\} \tag{6}$$

or

$$\left.\begin{aligned} DB_3(\varphi)\cos(\varphi_3(\varphi)) &= R\cos(\varphi) - xD; \\ DB_3(\varphi)\sin(\varphi_3(\varphi)) &= R\sin(\varphi) - yD. \end{aligned}\right\} \tag{7}$$

Here R - the radius of the crank OA (or OB1 or OB3) or cylinder (see pos. 11 in Fig. 4a in [1] or link 1 in Fig. 1); φ - the angle of rotation of the crank OA or OB_1 (link 1) relative to the axis Ox in the direction opposite to the reference angles, rad.; xD and yD are the abscissa and the ordinate of the axis of rotation D of the rotating coulisse DM (link 3); DB_3 - the variable distance from the hinge D to the point B_3 of the sliding blocks 2, which form a translational pair with a rolling coulisse 3.

From the first Eqs. (6, 7) we obtain the module of the variable vector DB_3 in the form:

$$DB_3(\varphi) = \left|\frac{R\cos(\varphi) - xD}{\cos(\varphi_3(\varphi))}\right|, \tag{8}$$

and from the second - to control the correctness of constructing the graphical dependence of the variable vector module DB3, the rotation angle φ of the crank OA or OB_1:

$$DB_3(\varphi) = \left|\frac{R\sin(\varphi) - yD}{\sin(\varphi_3(\varphi))}\right|. \tag{9}$$

From Eqs. (6, 7) we find the tangent of the angle $\varphi_3(\varphi)$:

$$\mathrm{tg}(\varphi_3(\varphi)) = \frac{R\sin(\varphi) - yD}{R\cos(\varphi) - xD}. \tag{10}$$

From the last equation we determine the position function of the rolling coulisse DM (link 3):

$$\varphi_3(\varphi) = \arctan\left(\frac{R\sin(\varphi) - yD}{R\cos(\varphi) - xD}\right). \tag{11}$$

It should be borne in mind that the main value of the angle of rotation of the rotating coulisse DM is determined using formula (9), and when choosing the real value, difficulties may arise due to the ambiguity of the inverse trigonometric function (see page 32 in [7]).

Imagine the closed loop ODB_3O (in Fig. 2, point B3 refers to link 3). The equation of closedness of this loop:

$$\overline{OD} + \overline{DB_3} - \overline{OB_3} = 0 \text{ or } \overline{OB_3} = \overline{OD} + \overline{DB_3}. \tag{12}$$

Here, the vectors OD and OB_3 are constant modulo, and the vector DB_3 - variable.

Projecting Eq. (12) on the coordinate axes Oxy, we obtain the position function of the point B_3 of the rotating coulisse (link 3):

$$\left.\begin{aligned} xB_3(\varphi) &= xD + DB_3(\varphi)\cos(\varphi); \\ yB_3(\varphi) &= yD + DB_3(\varphi)\sin(\varphi) \end{aligned}\right\} \tag{13}$$

or using the angle of rotation of the rotating coulisse DM:

$$\left.\begin{aligned} xB_3(\varphi) &= xD + DB_3(\varphi)\cos(\varphi3(\varphi)); \\ yB_3(\varphi) &= yD + DB_3(\varphi)\sin(\varphi3(\varphi)). \end{aligned}\right\} \tag{14}$$

Here DB_3 – the length of the rotating coulisse (link 3 formed by the rigid connection of two parts: cassettes and rod).

Now we consider the closed loop ODE_0O (the line OE_0 is not shown in Fig. 2) or ODE_iO (where i = 1, …, 11) in the rolling coulisse 3. The equation of closedness of this loop:

$$\overline{OD} + \overline{DE_i} - \overline{OE_i} = 0 \text{ or } \overline{OE_i} = \overline{OD} + \overline{DE_i}. \tag{15}$$

Here, all vectors OD and DE_i are constant modulo, and the vector OE_i – variable.

Projecting Eq. (15) on the coordinate axes Oxy, we obtain the position functions of the point E_0 or points E_i (where i = 1, …, 11) of the rotating coulisse (link 3):

$$\left.\begin{aligned} xE_i(\varphi) &= xD + DE_i\cos(\varphi + \beta); \\ yE_i(\varphi) &= yD + DE_i\sin(\varphi + \beta) \end{aligned}\right\} \tag{16}$$

or using the angle of rotation of the rotating coulisse DM:

$$\left.\begin{aligned} xE_i(\varphi) &= xD + DE_i\cos(\varphi3(\varphi) + \beta); \\ yE_i(\varphi) &= yD + DE_i\sin(\varphi3(\varphi) + \beta). \end{aligned}\right\} \tag{17}$$

Here β – the angle formed between the points E_i (where i = 1, …, 11) of the rotating coulisse 3 (see Fig. 2b); $DE_0 = DE_i = r$ - the radius of the rotating coulisse (cassettes).

Let's present a closed loop ODMO (this loop is not shown in Fig. 2). The equation of closedness of this loop:

$$\overline{OD} + \overline{DM} - \overline{OM} = 0 \text{ or } \overline{OM} = \overline{OD} + \overline{DM}. \tag{18}$$

Here, the vectors OD and DM are constant modulo, and the vector OM – variable.

Therefore, projecting Eq. (18) onto the coordinate axes Oxy, we obtain the position function of the point M of the rotating coulisse (link 3):

$$\left.\begin{aligned} xM(\varphi) &= xD + DM\cos(\varphi); \\ yM(\varphi) &= yD + DM\sin(\varphi). \end{aligned}\right\} \tag{19}$$

or using the angle of rotation of the rotating coulisse DM:

$$\left.\begin{aligned} xM(\varphi) &= xD + DM\cos(\varphi3(\varphi) + \pi); \\ yM(\varphi) &= yD + DM\sin(\varphi3(\varphi) + \pi). \end{aligned}\right\} \tag{20}$$

Here, DM = c - is the length of the rotating coulisse DM (link 3, formed by joining a rigid connection of two parts: cassettes and rod).

Next, let's consider the closed loop OE_iM_5O (see Fig. 2c and 2d). The equation of closedness of this loop:

$$\overline{OE_i} + \overline{E_iM_5} - \overline{OM_5} = 0 \text{ or } \overline{OM_5} = \overline{OE_i} + \overline{E_iM_5}. \tag{21}$$

Here, the vectors OE_i and E_iM_5 are constant modulo, and the vector OM_5 □ variable.

Therefore, projecting Eq. (21) onto the coordinate axes Oxy, we obtain the position functions of the points M_5 of the rolling coulisses (links 5, 7, …, 25 in Fig. 2):

$$\left.\begin{aligned} xM_5(\varphi) &= xE_i + E_iM_5 \cos(\varphi + \gamma); \\ yM_5(\varphi) &= yE_i + E_iM_5 \sin(\varphi + \gamma) \end{aligned}\right\} \tag{22}$$

or using the angle of rotation of the rotating coulisse DM:

$$\left.\begin{aligned} xM_5(\varphi) &= xE_i + E_iM_5 \cos(\varphi_3(\varphi) + \gamma); \\ yM_5(\varphi) &= yE_i + E_iM_5 \sin(\varphi_3(\varphi) + \gamma). \end{aligned}\right\} \tag{23}$$

Here γ - the angle formed between the rotating (link 3) and rolling coulisses (links 5, 7, …, 25) (see Fig. 2b); E_iM_5 – the length of the rolling coulisses (links 5, 7, …, 25 in Fig. 2).

We emphasize that the correct presentation of Eqs. (6)–(11), (13) or (14), (16) or (17), (19) or (20), as well as (22) or (23) in the parametric form (a variable parameter is the angle φ of crank rotation of the crank OA or OB_1 (link 1) relative to the axis Ox in the direction opposite to the reference angle) can be checked by the results of computational experiments.

We turn to computational experiments, which allow assessing the correctness and/or incorrectness of writing Eqs. (9) and (10) or (11) in parametric form along the trajectories of the movement of individual points and links of the multiloop coulisse mechanism.

5.1 The Results of Computational Experiments in the Mathcad Syste

The initial calculation data: R = 0.140 - radius of the crank OA or OB_1 or cylinder (see pos. 11 in Fig. 4a in [1] or link 1 in Fig. 1) 1, m; r = 0.039 - radius DE_0 or DE (the first part of link 3 in Fig. 1), m; c = 0.164 - the length of the rotating coulisse DM (link 3, formed by a rigid connection of two parts: cassettes and rod), m; $e_0 = c - r = 0.125$ - the length of the part of circular cross section E_0M (second part of link 3), rigidly connected to the cassette, i.e. to the first part of link 3, m; $e = e_0 = 0.125$ - the length of the part E_1M (link 4), pivotally connected to the cassette, i.e. to the first part of link 3, m; xD = 0.0205 - abscissa of the axis of rotation D of the rotating coulisse DM (link 3), i.e. the projection of the point D on the axis Ox, m; α = 60 • (π/180) = 1.047 - the angle of inclination of the rotating coulisse DM (link 3) relative to the axis Ox (see Fig. 2b), rad.; d = xD/cos(α) = 0.041 - the distance between the axes of rotation O and D (i.e. the base length) of the crank OA or OB_1 (link 1) and the rotating coulisse DM (link 3), m; $DB_3 = R - d = 0.099$ - the distance between the axis of rotation D (i.e. between the rack D) and the point B_3 of the rotating coulisse DM (link 1), m;

$E_iM_5 = e = 0.125$ - the length of the rolling coulisse (links 5, 7, 9, …, 25) pivotally connected to the cassette at points E_1, E_1 and E_2, etc. (rotation centers 5, 7, 9, …, 23), m; $yD = xD/tg(\alpha) = 0.036$ - the ordinate of the axis of rotation D of the rotating coulisse DM (link 3), i.e. the projection of the point D on the axis Oy, m; $DM_0 = c - d = 0.123$ - the distance between the axis of rotation D (i.e. between the rack D) of the rotating coulisse DM (link 3) and its apex (i.e. point M) when this link is rotated by an angle π, m; $\beta = 30 \cdot (\pi/180) = 0.524$ - the angle between the centers of the rolling coulisses E_0 and E_1, E_1 and E_2, etc. (centers of rotation 5, 7, 9, …, 23) (see Fig. 2b), rad.; $\gamma = 30 \cdot (\pi/180) = 0.524$ - the angle between the centers of the rolling coulisses E_0 and $_{E1}$, E_1 and E_2, etc. (centers of rotation 5, 7, 9, …, 23) and points M_5, M_7 and M_9, etc. (the ends of the rolling coulisses 5, 7, 9, …, 25) (see Fig. 2b), rad.; $\varphi_0 = -0 \cdot (\pi/180)$ - the angle of inclination of the rotating wings DM (link 3) relative to the axis Ox in the initial position, rad.; $\varphi_1 = -2 \cdot (\pi/)$ - the angle of inclination of the rotating coulisse DM (link 3) relative to the axis Ox in the first position, rad.; $\Delta_\varphi = \varphi_1 - \varphi_0$ - the step of changing the angle of inclination of the rotating coulisse DM (link 3) (i.e. $\Delta_\varphi \cdot (180/\pi) = -0.035$ rad $\rightarrow -2^o$), rad.; $\varphi_k = -2\pi$ - the angle of inclination of the rotating coulisse DM (link 3) relative to the axis Ox in the final position (i.e. when this link is rotated by an angle of 2π), rad.; $\varphi = \varphi_0, \varphi_1 \ldots \varphi_k$ - variation of the angle of rotation of the crank OA or OB_1 (link 1) relative to the axis Ox in the direction opposite to the angle reference, rad.

First, we present the projections of the characteristic points of the mechanism on the coordinate axes Ox and Oy in the following form.

Let's present the position functions of the crank OA or OB_1 (link 1) (see Fig. 2):

$$\left.\begin{aligned} xA(\varphi) &= OA \cdot \cos(\varphi); \\ yA(\varphi) &= OA \cdot \sin(\varphi). \end{aligned}\right\} \tag{24}$$

where φ – considered position of the initial link (crank) OA or OB_1), which is determined by a linear dependence on the number (index) of the position and which is measured relative to the axis Ox, rad.:

$$\varphi = \varphi_1 + (i - 1) \cdot \Delta\varphi, \tag{25}$$

where Δ_φ – step of changing the angle of rotation of the initial link (crank OA or OB1), for example, $\Delta_{\varphi H} = 0.035$ rad $\rightarrow -2^o$.

Using formula (13), we build the position function of the point B_3 of the rotating coulisse (link 3); according to formula (16) - points E_0 or points E_i (where i = 1,…, 11) of the rotating coulisse (link 3); according to formula (19) - the function of the position of the point M of the rotating coulisse (link 3), and according to formula (22) - functions of the position of the points M_5, M_7, etc. of the rolling coulisses (links 5, 7, …, 25).

Next, we construct the position function $\varphi_3(\varphi)$ of the rotating coulisse DM (link 3) according to formula (11), the position function $DB_3(\varphi)$ of the variable vector DB3 according to formula (8) or (9). Similarly, one can build points B_3 of the rotating coulisse (link 3) according to formula (14); according to formula (17) - points E_0 or points E_i (where i = 1,…, 11) of the rotating coulisse (link 3); according to formula (20) - the function of the position of point M of the rotating coulisse (link 3), according to formula (23) - functions of the position of the points M_5, $_{M7}$, etc. of the rolling coulisses (links 5, 7, …, 25), and according to formula (24) - functions of the position of crank OA or OB_1 (link 1).

It is well known [9, 21, 22] that the law of motion of a point can be illustrated not only by curves in the coordinate system "rotation angle – movement" according to Eqs. (7)–(9) and (10) or (11), as well as (13) and (15) (see Fig. 3). In some cases, it is convenient to use the phase plane to describe the motion. For the problem under consideration, the phase plane is a Cartesian coordinate system in which the xM(φ) movement is plotted along the abscissa, and yM(φ) is plotted along the ordinate axis. In this plane, a phase trajectory can be obtained, i.e. the geometrical location of the image points (for example, the points O, D, E_0 (or Ei (where i = 1, …, 11)) and M) corresponding to successive moments of the angle of rotation φ of crank 1.

Below, we present the results of constructing the laws of motion in the coordinate system "rotation angle—movement" based on formulas (6)–(11), (13) or (14), (16) or (17), (19) or (20), and (22) or (23) in parametric form, as well as the geometric location of the image points (phase trajectory) in the Mathcad environment [23]. These results are presented in Figs. 3, 4, 5 and 6.

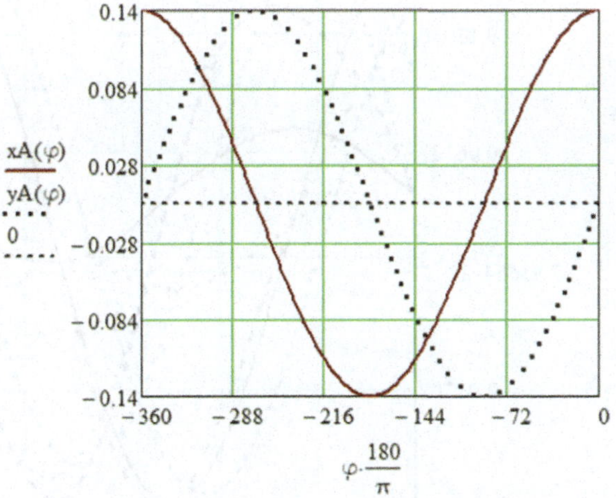

Fig. 3. The law of motion of point A or B_1.

From Fig. 3 it is clear that the law of motion of point A or B_1 in the coordinate system "movement - angle of rotation", according to Eq. (20), is harmonic.

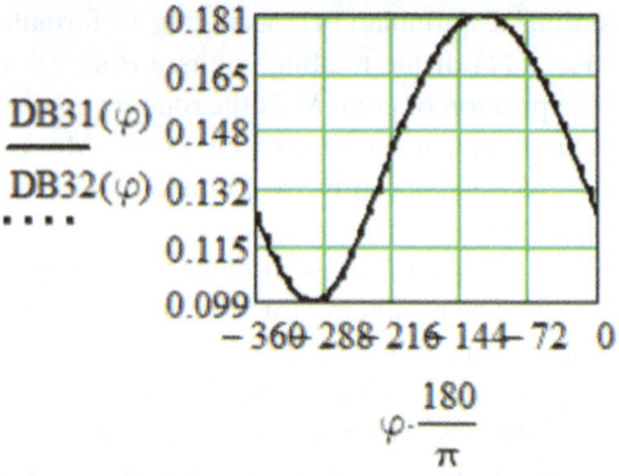

Fig. 4. The law of motion of the module of the variable vector DB_3.

As you can see, the graphs of the module of the vector DB_3 calculated by formulas (6)–(9) coincide, which confirms their correctness.

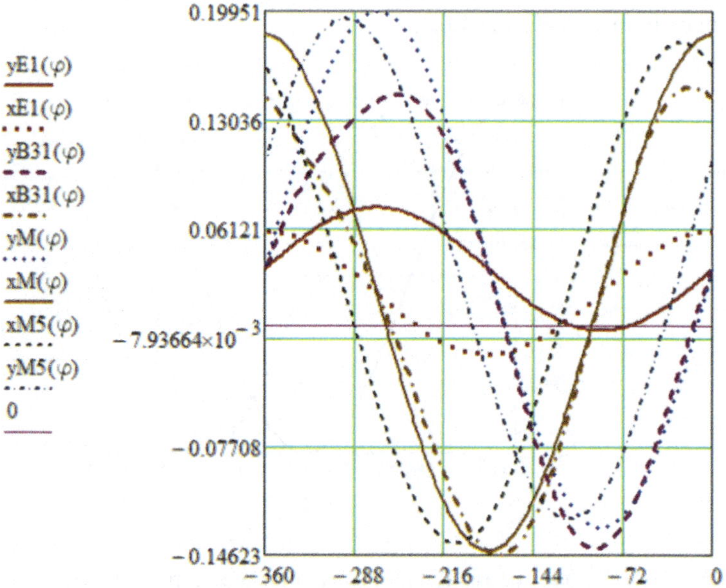

Fig. 5. The law of motion of points E_0 or points E_i (where i = 1, …, 11) and M_5.

Figure 5 confirms that the law of motion of point A or B_1 in the coordinate system “movement - angle of rotation”, according to formulas (13), (16), (19) and (22), is periodic.

For given initial data of the problem, graphic representations of formulas (13), (16), (19), (22) and (24) in the form of phase trajectories of the image points O, D, A or B_1, E_0 (or E_i (where i = 1, …, 11)) and M have the form shown in Fig. 6.

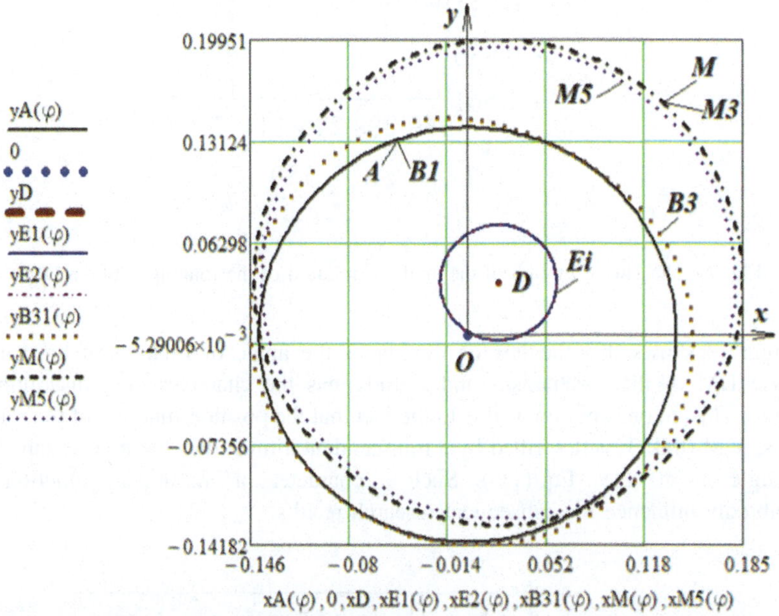

Fig. 6. Phase trajectories of image points O, D, A or B_1, E_0 (or E_i (where i = 1, …, 11)), M and M_5.

As can be seen, the phase trajectories of all the image points O, D, A or B_1, E_0 (or E_i (where i = 1, …, 11)), M and M_5 exactly correspond to the circle, which confirms the correctness of the obtained formulas (13), (16), (19), (22) and (24). The circle described by point B3 of the rotating coulisse 3 relative to the center O lies near the circle described by point A or B_1, which is true. In addition, as it should be, the circle described by point M_5 relative to the center O lies inside the circle described by point M, since the rolling coulisses (links 5, 7, …, 25 in Fig. 1) are deviated from the rotating coulisses (links 3 in Fig. 1) by an angle γ (see Fig. 2b).

Next, we present the results of constructing the laws of motion in the coordinate system "angle of rotation—movement" based on the position function $\varphi_3(\varphi)$ of the rotating coulisse DM by formula (11). Similarly, one can build points B_3 of the rotating coulisse (link 3) according to formula (14); according to formula (17) - points E_0 or points E_i (where i = 1, …, 11) of the rotating coulisse (link 3); according to formula (20) - the function of the position of the point M of the rotating coulisse (link 3), and according to formula (23) - the function of the position of the points M_5, M_7, etc. of the rolling coulisses (links 5, 7, …, 25), as well as the geometric location of the image points (phase trajectory) in the Mathcad environment [23]. These results are presented in Figs. 7, 8 and 9.

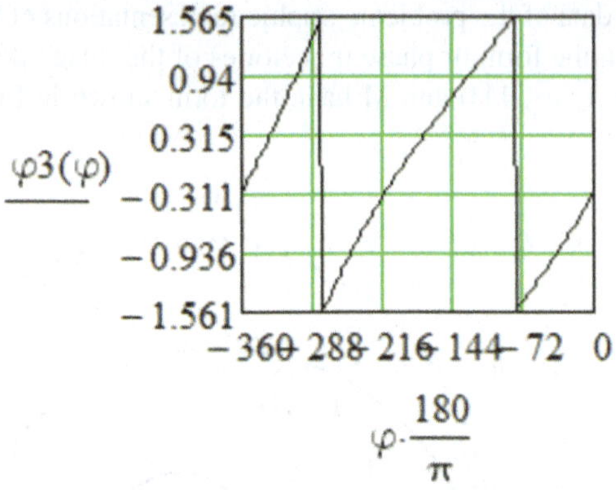

Fig. 7. The law of motion of the angle of rotation of the rotating coulisse DM.

Figure 7 shows that the law of motion of the angle of rotation of the rotating coulisse DM of the mechanism under study has the character of a discontinuous function. This, in our opinion, is due to the fact that the position function of the rotating coulisse DM (link 3) is described by a transcendental function - the inverse tangent of the angle $\varphi_3(\varphi)$ (see Eq. (11)). Such a character of the $\varphi_3(\varphi)$ function will undoubtedly influence the subsequent research results.

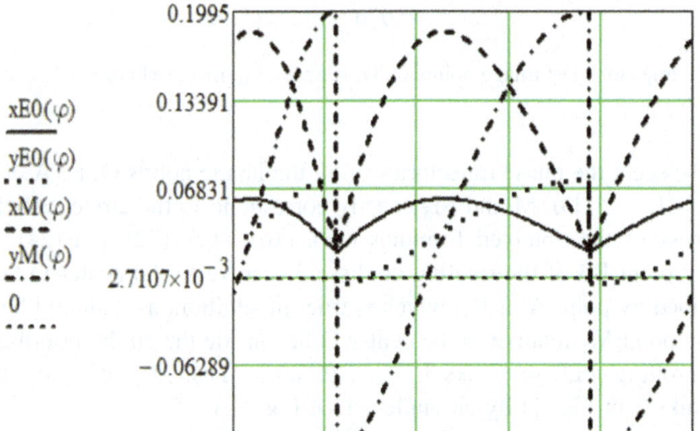

Fig. 8. The laws of motion of points E_0 and M.

Analysis of graphical dependencies in Fig. 8 confirms the effect on their changes of the law of motion of the angle of rotation of the rotating coulisse DM, which has the nature of a discontinuous function. Such a character of the $\varphi_3(\varphi)$ function will

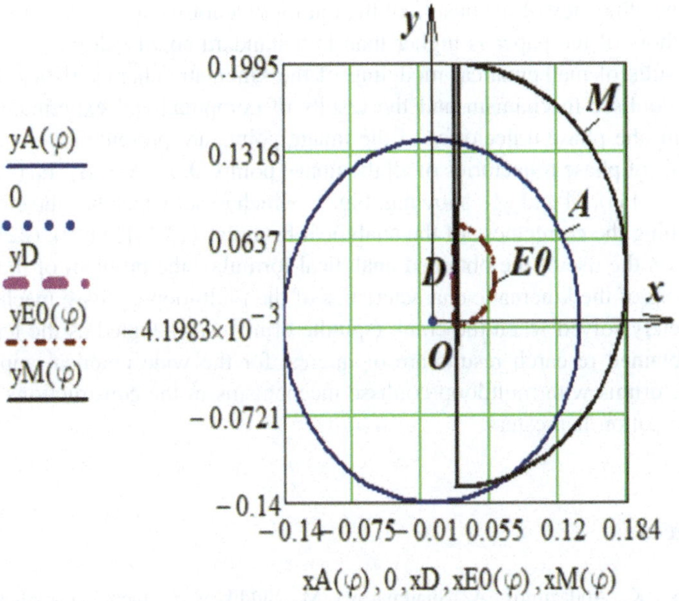

Fig. 9. The phase trajectories of image points O, D, A or B_1, E_0 (or E_i (where i = 1, …, 11)) and M.

undoubtedly influence the subsequent research results. In accordance with this, the characteristic points of the mechanism will also have a form that does not correspond to the movement of these points.

As can be seen, the phase trajectories of all the image points O, D, A or B1exactly correspond to the circle, which confirms the correctness of formula (24). In turn, the nature of the semi-circle described by the points E_0 (or E_i (where i = 1,…, 11)), M or M5 shows the inapplicability of formulas (11), (14), (17), (20), and (23) to plot the phase trajectories of these image points.

Thus, based on the use of analytical formulas (13), (16) (19), (22), and (24), the problem of mathematical modeling of the kinematic characteristics of the multiloop coulisse mechanism was completely solved when the crank (spindle drum) was adopted as the leading link.

6 Conclusions

Based on the studies, we especially note the following results:

1. The calculated data proved that the spindle drum of the cotton harvester is indeed a multiloop coulisse mechanism and consists of two multiloop coulisse chains, each of which consists of twelve loops. Due to such a number of loops, the end parts of the rotating and rolling coulisses (pins) act on cotton bolls and branches, direct them to oppositely rotating spindles (working bodies) of the cotton harvester. At the same

time, the efficiency of the design of the multiloop coulisse mechanism developed by the authors of the paper is higher than in a standard spindle drum.
2. The results of mathematical modeling of the kinematic characteristics of the multiloop coulisse mechanism and the results of computational experiments on constructing the phase trajectories of the image points are presented.
3. Analysis of phase trajectories of all the image points O, D, A or B_1, E_0 (or E_i (where $i = 1, \ldots, 11$)), M and M_5, shown in Fig. 6, which exactly matches the circle, allows confirming the correctness of the analytical formulas (13), (16) (19), (22), and (24). Based on the use of the obtained analytical formulas, the problem of mathematical modeling of the kinematic characteristics of the multiloop coulisse mechanism was completely solved when the crank (spindle drum) was adopted as the leading link.
4. The obtained research results are of interest for the wide practical application of spindle drums with multiloop coulisse mechanisms in the constructions of vertical spindle cotton harvesters.

References

1. Turanov, K., Abdazimov, A., Shaumarova, M., Siddikov, S.: Incorrect application of the epicycloid equation to the planetary mechanism of a vertical spindle of the cotton harvester. E3S Web Conf. **164**, 01008 (2020). https://doi.org/10.1051/e3sconf/202015701008
2. Sablikov, M.V.: Cotton Harvesters. Agropromizdat, Moscow (1985)
3. Zinovyev, V.A.: Course of the Theory of Mechanisms and Machines. Science, Moscow (1972)
4. Artobolevsky, I.I.: Theory Mechanisms and Machines. Science, Moscow (1975)
5. Reshetov, L.N.: Self-Adjusting Mechanisms, Reference. Nauka, Moscow (1991)
6. Kolovsky, M.Z., Evgrafov, A.N., Semenov, Y.A., Slousch, A.V.: Advanced Theory of Mechanisms and Machines. Springer, Heidelberg (2000)
7. Turanov, K.T., Turanov, S.K., Tatarintcev, B.E.: Proektirovaniye kulisnykh mekhanizmov v vychislitelynoy srede Matchcad: uchebn. posobye [Design of coulisser mechanisms in Mathcad computing environment: Textbook]. Publishing House SGUPS (NIIZHT), Novosibirsk (2002). (in Russian)
8. Frolov, K.V.: Theory of Mechanisms and Machines. Higher School, Moscow (2005)
9. Turanov, K., Shaumarova, M.: Incorrect application of the epicycloid equation to the planetary mechanism of the cotton harvester. E3S Web of Conf. **164**, 06034 (2020). https://doi.org/10.1051/e3sconf/202015706034
10. Baranov, G.G.: Course of the Theory of Mechanisms and Machines. Engineering, Moscow (1975)
11. Kozhevnikov, S.N., Esipenko, Y.I., Raskin, Y.M. (eds.): Machinery. Reference Manual. Engineering, Moscow (1976)
12. Kraynev, A.F.: Dictionary-Reference Mechanisms. Engineering, Moscow (1987)
13. Agrawal, V.P., Rao, J.S.: The mobility properties of kinematic chains. Mech. Mach. Theor. **22**, 497–504 (1987)
14. Jin-Kui, C., Wei-Qing, C.: Identification of isomorphism among kinematic chains and inversions using link's adjacent-chain-table. Mech. Mach. Theor. **29**, 53–58 (1994)
15. Uicker, J.J., Pennock, G.R.: Theory of Mechanisms. Oxford University Press, New York (2003)

16. Litvin, F.L., Fuentes, A.: Gear Geometry and Applied Theory. Cambrige University Press, Cambrige (2004)
17. Rizvi, S.S.H., Hasan, A., Khan, R.A.: A New for distinct inversions and isomorphism detection in kinematic chains. Int. J. Mech. Robot. Syst. **3**(1), 48–59 (2016)
18. Pozhbelko, V.I., Kuts, E.: Development of the method of structural synthesis of multi-loop lever mechanisms with multi-loop hinges on the basis of basic groups of mechanisms. Theor. Mech. Mach. **4**(16), 139–149 (2018)
19. Pozhbelko, V.I., Kuts, E.: Structural synthesis of planer 10-link–DOF kinematic chains with up to pentagonal links with all possible multiple joint assortments for mechanisms deign. New Adv. Mech. Mach. Sci. **57**, 27–35 (2018)
20. Hasan, A.: Study of multiple jointed kinematic chains. Int. J. Comput. Eng. Res. **1**(8), 13–19 (2018)
21. Panovko, Y.G.: Fundamentals of Applied Oscillations and Shock. Polytechnics, Leningrad (1990)
22. Bugaenko, G.A.: Fundamentals of Classical Mechanics. Higher school, Moscow (1999)
23. Makarov, E.G.: MathCAD: Training Course (+CD). Piter, St. Peterburg (2009)

Kinematic Characteristics of the Car Movement from the Top to the Calculation Point of the Marshalling Hump

Khabibulla Turanov[1(✉)], Andrey Gordienko[2], Shukhrat Saidivaliev[3], Shukhrat Djabborov[3], and Khasan Djalilov[3]

[1] Tashkent State Technical University Named After Islam Karimov, University Str., 2, 100174 Tashkent, Uzbekistan
khturanov@yandex.ru
[2] Ural State University of Railway Transport, Kolmogorova Str., 66, 620034 Yekaterinburg, Russia
[3] Tashkent Railway Engineering Institute, Temirylchilar Str., 1, 100167 Tashkent, Uzbekistan

Abstract. Purpose: Introduce analytical acceleration formulas that are derived from the classic d'Alembert principle of theoretical mechanics for high-speed sections and for sections of retarder positions; show the possibility of determining the instantaneous car speeds in each section of the marshalling hump according to the formulas of elementary physics both for high-speed sections and for sections of retarder positions; provide formulas for determining the time of movement of a car with uniformly accelerated and/or uniformly retarded motion of the car on the inclined part of the hump, as well as in areas of retarder positions. Research methods: The classic d'Alembert principle of theoretical mechanics is widely used in the paper. Main results: For the first time, the results of constructing a graphical dependence of the estimated height of the marshalling hump over the entire length of its profile are presented in the form of a decrease in the profile height of each section of the inclined part in proportion to the slope of the track. The results of constructing graphical dependences on changes in the speed and time of movement of a car along the entire length of the inclined part of the marshalling hump are fundamentally different from the existing methodology, where, for example, curves of medium (rather than instantaneous) speeds of a car are built. The proposed new methodology for calculating the kinematic characteristics of the car movement along the entire length of the hump allows an analysis of the mode of shunting car at the marshalling humps.

Keywords: Railway · Station · Marshalling hump · Car of the hump · Profile of the marshalling hump · Calculation point of the hump

V. Murgul and V. Pukhkal (Eds.): EMMFT 2019, AISC 1258, pp. 322–338, 2021.
https://doi.org/10.1007/978-3-030-57450-5_29

1 Introduction

A significant number of publications have been devoted to the problem of implementing hump design and technological calculations that simulate the conditions of movement of designed runners (cars) with different running properties [1–22]. Of these, example gratia, in [11], the twelfth counterexamples set forth the content of a critical analysis of the existing methodology for calculating marshalling humps [4, 10], and in [12, 18], an attempt was made to prove the correctness of these methods. Moreover, example gratia, in [18], without substantiated evidence (i.e. analytical proof of correctness supported by calculations), as was done by the authors of [17, 20], it is noted that formulas (1) and (2) in [18] can be applied in any areas with a slope i of the marshalling humps, taking into account the presence of specific values of resistance to movement w and power of braking positions h_b (i.e. height of sections of braking positions) (q.v. first paragraph of the last column on page 36 in [18]). According to the authors of paper [18], it is precisely according to it that hump design and technological calculations are performed, which can be used to simulate the movement conditions of designed runners with different running properties (q.v. the first paragraph of the last column on page 36 in [18]). It was also stated in [18] that "… any new proposed design models for the car movement" should be further compared with formulas (1) and (2) in [18] (q.v. second paragraph of the last column on page 36 in [18]).

However, we believe that the actual operational characteristics of the cars and the variability of the parameters of the railway tracks, as well as the probabilistic nature of many factors affecting the process of movement of cars on the hump, noted in [18], as the main disadvantages of a simplified approximate approach of the authors of the paper [11, 17] to the calculation of the speed of the car along the inclined part of the hump, in our opinion, are unlikely to be taken into account or can be taken into account explicitly or implicitly in the presented formula (2) in [18], which contains the incorrigible gross mistakes listed in [11].

For example, the engineering task of the dynamics of rolling a car along a track profile, taking into account the real rolling friction in bearings, the difference in wheel diameters, deviations between the inner faces of the wheelsets ± 3 mm, rolling (or horizontal cut) along the tape line up to 9 mm, ridge thickness of 33–22 mm, vertical undercut of the ridge up to 18 mm, sliders of 1 … 2 mm, deviations of track gauge from −4 to +10 mm, level difference of rail heads in straight sections up to 6 mm and rail wear, number and the type of sleepers, ballast, etc. (q.v. second and third paragraphs in the first column on page 37 in [18]) is hardly a correctly solvable mathematical problem. So, for example, if there are deviations of the track width from −4 to +10 mm, this means that there is a gap between the wheel flanges and the inner heads of the rail threads. It would seem that taking into account such a simple operational factor of the car movement in the horizontal plane (where the car can undergo lateral movement and wobble within the technological gap) can be elementarily attributed to a solvable engineering problem. However, alas, such an engineering task cannot be analytically solved. Therefore, in order to solve the engineering problem, the presence of a gap in the joints of two parts is not taken into account [5, 24]. It should be noted that it is analytically impossible to solve any engineering problem without their idealization and simplifications of design schemes and mathematical models [21, 22].

It was noted in [18] that "statements of the "unknown" speed of the car in the calculated section v_k are far-fetched" (evidently, in [11]) (p.v. penultimate paragraph of the middle column on page 37 in [18]). It was explained in [18] that v_m is normalized depending on the type of marshalling devices, i.e. $v_m = [v_m]$, the release speed is always known at the beginning of the section v_{init} or it is set at the end of the section according to the permissible impact speed (evidently, in the form of $[v_{imp}]$) or the entrance to the dividing (evidently, in the form of $[v_d]$) or the braking path (evidently, in the form of $[v_{br}]$). It is further noted that, knowing these values, it is possible to determine the medium car speed v_m, from which the specific resistance to the car's movement w_{cr} in the designed area is calculated from the given wind parameters. Also, attention is drawn to the fact that the normalized values $[v_{init}]$ and $[v_K]$ are known in each section, and the medium car movement time tm is determined from them.

In this regard, the arguments of the authors of article [18] that "in the future, it is necessary and justifiable to compare any new design models of car movement with formula (2) in [18] in the method of hump calculations" (q.v. last column on page 36 in [18]) are not supported by any calculated data. Recall that the medium speed v_m characterizes the speed of movement over a given time interval Δt, but does not give an idea of the speed of movement of the body at individual moments t of this time period. For this reason, the instantaneous speed and/or speed of the body v at a given time t should be determined. The time t in [4, 10] is not determined in any way, not mentioning formula (2) in [18].

It is appropriate to note that formulas (1) and (2) in [18] do not appear at all in the normative and technical document [4] and they are not used for any calculations.

In this regard, the authors of paper [18] are strange in their reasoning about the need and legitimacy to compare any new calculation models of car movement in the method of hump calculations with formula (1) and (2) in [18] " in the future (q.v. last column on page 36 in [18]), when these formulas are not used at all in hump design and technological calculations [4], except for textbooks for university students of railway transport (for example, [10]).

Summarizing the results of the discussion of the correctness of the developed universal form of formula (2) in [18], we can conclude that it is inadmissible to perform any hump design and technological calculations using this formula, as having pseudoscientific materials that contradict the principle of theoretical mechanics [5].

In our opinion, the authors of [18] admit absurdity when they argue that it is necessary and legitimate in the future to compare any new proposed design models for the movement of cars with formulas (1) and (2) in [18]. Although, these formulas are not used at all in hump calculations, since they are absent in the normative and technical document [4]. In hump design calculations, the only formula (3) is used in the form $v = \sqrt{2gh_i}$, where h_i is the height of various sections of the hump [4], to determine the free fall rate of the body, taking into account the inertia of the rotating parts, which, unfortunately, was deduced for the ideal constraint connection [4]. For this reason, not only formula (2) in [18], but also formula (3) in [4] cannot be used as a non-ideal surface (constraint), which are rail threads.

Hence the relevance of mathematical modeling of the car movement on the inclined part of the marshalling humps is obvious.

In this paper, similarly to [21], using the analytical formulas given in [22], we describe the kinematic characteristics of the movement of the car along the length of the marshalling hump profile using the developed program to perform design and technological calculations of the marshalling hump when shunting a single car along any of its slopes from the top of the hump (TH) to the calculation point (CP) [17].

2 Objective

- Provide initial data for calculating the kinematic characteristics of the car movement;
- present the analytical acceleration formulas a_i (i - the numbers of the hump sections) obtained on the basis of the classical d'Alembert principle of theoretical mechanics for high-speed sections and for the sections of hump braking positions;
- show the possibilities of determining the instantaneous speeds of the car movement v_i at each section of the marshalling hump according to the formulas of elementary physics for both high-speed sections and sections of hump braking positions;
- present formulas for determining the time of car movement t_i with uniformly accelerated and/or uniformly retarded motion of a car on the inclined part of the hump t_i, as well as on sections of hump braking positions;
- present a formula for calculating the braking distance of the car l_{bi} in the braking zones in the areas of braking positions;
- present the change in the kinematic characteristics of the car movement along the entire length of the inclined part of the marshalling hump in the form of tabular data [21];
- show graphical changes in the estimated height of the studied sections of the hump h_i along the entire length of the track lix, i.e. $h_i = f(l_{ix})$;
- present the change in the kinematic characteristics of the car movement along the entire length of the inclined part of the marshalling hump in the form of graphic dependences [21, 22].

3 Research Methods

Research is based on the classic d'Alembert principle of theoretical mechanics [5, 20–22].

4 Research Results

To perform the calculations, we consider that the marshalling hump consists of the following elements: top of the hump (TH); first and second high-speed section of the hump (HS1 and HS2); first, second and third break positions (1BP, 2BP, and 3BP); intermediate section (IS); switch zone (Sw); first and second sections of the classification track (CT1 and CT2); dividing switch zone (Sw); first, first, second and third switch zones (Sw1, Sw2 and Sw3); section accounting the length of the car's wheelbase (WhB); car breaking zone (SB); the remaining sections of the break release zone (RS). In contrast to [21], we consider the case under the condition of concavity of the profile (for example, in the section HS1 50‰ (per mille), HS2 30‰, RP1 14‰, IS 11‰, RP2 10‰, Sw2‰, CT1 1.6‰, CT2 0.6‰), and the location of the third braking position (3BP) in a straight section of the track with a slope of 0.6‰.

4.1 Example of Calculations

Let us demonstrate the results of calculating the kinematic characteristics of the movement of a car along the entire length of the marshalling hump profile from its top (TH) to the calculation point (CP). For calculation, we accept the following initial data.

1. Initial data for the inclined part of the marshalling hump, except for sections of the brake positions: $G_0 = 650$ is gravity force of the cargo on the car, kN; $G = 908$ is gravity force of a car with cargo, kN; $F_{wx} = 3.2$ - accounting for the projection of the tailwind force of small magnitude, kN; $M_{red} = 9.256 \cdot 104$ - reduced mass of the car with cargo M_{red}, taking into account the moment of inertia of the rotating parts JC, kg (Where $JC = Gr^2/2$ g - the moment of inertia of the wheels of one wheelset relative to the center of inertia C ($r = 0.475$ - radius around the wheel, m)); $g' = 9.635$ - the acceleration of gravity of the body, taking into account the mass of the rotating parts, calculated with a relative calculation error of $\delta g \approx 0{,}184\%$ at $g = 9.81$ m/s^2, n = 4 Ps., $Q = G_0 = 92.56$ tf and/or $G = 908$ kN (according to Table 4.2 in [4], this is a very good runner (VG)), m/s^2; v_{initi} - initial speed and/or speed of the car's entry into the i section of the hump, which is equal to the exit speed of the car $v_{fin(i-1)}$ from the previous section, m/s; ψ_i is the slope angle of the inclined part of the hump (the value taken according to the recommendation, for example, from [4]), degree; lix is the length of the studied inclined part of the hump (the value taken according to the recommendation, for example, from [4]), m; $F_{xi} = f(G, F_{wx}, \psi_i)$ - projection of the gravity of the car G on the direction of movement of the car, taking into account the projection of the tailwind force of small magnitude F_{wx} ($F_{wx} \approx 3.2$ kN) on the i section of the hump); $F_{oi} = k_{oi}G = f(\psi_i)$ - force from the main resistance to the movement of the car (calculated value), kN; $k_{ww} = 5 \cdot 10^{-4}$ - coefficient taking into account the fraction of gravity G when taking into account resistance from the air and wind; $k_{sw} = 2.5 \cdot 10^{-4}$ and $k_{sn.h} = 2.5 \cdot 10^{-4}$ are coefficients that take into account the fraction of gravity G when taking into account the resistance of the switch, snow and hoarfrost; $\alpha_1 = 9.46$, $\alpha_2 = 4.73$, $\alpha_4 = 10.68$, $\alpha_6 = 24.7$, $\alpha_7 = 18.83$ - angles of rotation of the curves in the first and second speed sections, in the intermediate section, in the switch zone, and in the first section of the marshalling hump, respectively, degree; k_{cur} - coefficient taking into account the fraction of gravity G when taking into account the resistance during the transition over the curves, calculated from the dependence $k_{cur} = f(\alpha_i, l_{ix})$ (where lx is the track length along the curve, and i is the number of the curve section of the hump, for example, at

α_4 – 10.68° and l_4 = 41.27 m: k_{cur} = 0.00011); $F_{ww} = k_{ww}G = 5 \cdot 10^{-4} \cdot 908$ – 0.454 - resistance force from air and wind, kN; $F_{sn.h} = k_{sn.h}G = 2.5 \cdot 10^{-4} \cdot 908$ = 0.227 - resistance force from snow and hoarfrost, kN; $F_{sw} = k_{sw}G = 2.5 \cdot 10^{-4}$ • 908 = 0.227 - resistance force when crossing the dividing switch zone, kN; $F_{cur.i} = k_{cur.i}G$ - resistance force during the transition over the curves (and/or resistance from curves) (calculated value), kN; f_{sl} = 0.15, ..., 0.25 - the coefficient of sliding friction of surfaces along the wheel rolling circle (metal on metal); $f_{lol.}$ = 0.001 coefficient of rolling resistance and/or coefficient of friction during rolling [4].

2. The initial data for the braking zone (BZ) of the brake position (BP) sections, at which the car is completely stopped in these zones, are as follows: G_1 = 794 - car gravity together with non-rotating parts, kN; f_{br} = 0.25 - dry friction coefficient of the sliding of the wheel rim on the brake tires of the retarder beams; $F_{br.p}$ = 95 - force pressing brake pads of retarders to the side surfaces of the wheels or the average load on the axis of the car, kN; F_{br} = 23.75 - friction force of the sliding rim of the wheelset on the compressed brake tires, kN; F_{oi} = 198.5 - sliding friction of wheelsets on compressed brake tires, as the main resistance, kN; M_{red0} = 8.869 • 104 - reduced mass of the car with cargo together with non-rotating parts, kg; a_{br} = G1 • 103/M_{red0} = 794 • 103/(8.869 • 104) = 8.953 - conventional designation of the linear acceleration of the car during equidistant movement in the braking zones in the sections of the BP, m/s^2; $g_0 = g' = 9.611$ - according to the methodology [3], the acceleration of gravity due to the mass of rotating parts, m/s$_2$, n = 4 Ps., Q = G_1 = 79.4 tonn power and/or G_1 = 794 kN (according to Table 4.2 in [4], this is a very good runner (VG)).

In [22], it was noted that in order to develop a program for calculating the kinematic parameters of the car along the inclined part of the marshalling hump according to the simplified method proposed by the authors of [17, 20], the formula for each i section of the hump, in accordance with the principles of engineering mechanics [5], i.e. v_{Ki} [v_i], is represented as (2) and (15) in [22]. At the same time, we make a special reservation that, based on the classical d'Alembert principle of theoretical mechanics [5], the acceleration of the car motion with uniformly accelerated motion on the inclined part of the hump ai is calculated by the formula:

$$a_{Ci} = \frac{|\Delta F_{xi}|}{M_{\text{red}}} 10^3, \quad (1)$$

where i - numbers of sections along the entire length of the marshalling hump profile (i = 1, ..., 9); $|a_{Ci}| = |a_i|$ - acceleration of the center of mass C_w of the car to be determined, m/s^2; M_{red} - reduced and/or imaginary mass of the car with cargo, taking into account the moment of inertia of the rotating parts (wheelsets) J_C in all sections of the inclined part of the hump, kg; $|\Delta F_i|$ - the resulting force, under the influence of which the car rolls along the inclined part of the marshalling hump, kN:

$$|\Delta F_{xi}| = F_{xi} - |F_{ci}| \quad (2)$$

taking into account that F_{xi} – the projection of the gravity force of the car G on the C_x axis, taking into account and/or excluding the projection of the tailwind force F_{wx}, under the influence of which the car moves along the slope of the inclined part of the hump, kN:

$$F_{xi} = G \sin \psi_i + F_{Bx} \cos \psi_i.$$

Note that F_{wx} can be neglected due to its smallness: $F_{wx} \ll G$ (example gratia, $3.2 \ll 908$ kN); ψ_i – the slope of the inclined part of the hump, rad.; $|F_{ri}|$ – in the general case, the resistance force of any kind:

$$|F_{ri}| = (F_{oi} + F_{cur.i} + F_{sw} + F_{ww} + F_{sn.hH}) \tag{3}$$

Here, the resistance force of any kind $|F_{ci}|$ taking into account and/or without taking into account the projection of the headwind force of a small value F_{wx}, which can be taken as a fraction of the gravity of a car with cargo G, i.e. $|F_{ci}| = f(G)$, which does not contradict the force relations of the hump calculations. Resistance force $|F_{ri}|$ includes the following forces: sliding friction, taking into account rolling friction forces in axle box bearings, as the force from the main (running) resistance $F_{fri} = F_{oi}$; the resistance appearing during the transition over the curves (and/or resistance from the curves), which depend on the sum of the rotation angles in the curves, including the switch angles in the section under consideration, and the speed of the car, F_{curi}; resistances arising from switch zones (from impacts of wheels against wits, crosses and counter rails) F_{sw}; resistance to air and wind F_{ww}; resistance to overcome additional resistance from snow and hoarfrost within the switch zone and on the marshalling tracks $F_{sn.h}$.

In the braking zones in the areas of brake positions 1BP, 2BP and 3BP, the acceleration abri is calculated by the formula:

$$|a_{kbri}| = \frac{|\Delta F_{Ti}|}{M_{red0}} 10^3, \tag{4}$$

where $|\Delta F_{wi}|$ - the resultant force, under the influence of which the car's wheelsets are forced to slide along the rolling surfaces of the rail threads and the brake tires of the car retarder in the braking zones in the areas of BP, kN:

$$|\Delta F_{Ti}| = F_{xi} + |F_{ci}|; \tag{5}$$

$|a_{kTri}| = a_{kTri} \cdot \mathrm{sgn}\Delta F_{1bri}$ module function, moreover $|a_{kTri}| = -a_{kTri}$, if $|\Delta F_{1bri}| < 0$.

It follows from formula (4) that, subject to the condition $|\Delta F_{1bri}| < 0$ and/or $|F_{ri}| > F_{xi}$, the movement of the car in the braking zone at the brake position at the initial speed $v_{init.bri} > 0$ will be uniformly slowed down until the speed v vanishes.

The formulas of the instantaneous car speeds in each section of the marshalling hump, according to the simplified method adopted in [17, 20], are written in a form convenient for calculation. It should be noted that the rolling speed of a car with uniformly accelerated and/or uniformly retarded motion along the hump profile vi, according to the simplified method of the authors of [17, 20], can also be determined by the formula of elementary physics in all sections i at the accepted length of the studied sections l_i, except for sections of the brake positions. So, for example, the speed of the car on the inclined part of the hump vi is calculated by the formula of elementary physics:

$$v_i = v_{\text{init}.i} + |a_i|t_i \tag{6}$$

where v_{initi} – the initial speed and/or speed of the car's entry into the studied section of the hump profile from the previous section, i.e. the value taken from the results of calculations of the previous sections of the hump; a_i – the acceleration of the car movement (the value calculated by the formula (1)).

Driving time with uniformly accelerated and/or uniformly retarded motion of the car along the inclined part of the hump t_i is calculated by the formula of elementary physics:

$$t_i = \frac{1}{|a_i|}(-v_{\text{init}.i} + \sqrt{v_{\text{init}.i}^2 + 2|a_i|l_i}, \tag{7}$$

and in the areas of braking at the brake positions t_{bri} - according to the formula:

$$t_{\text{bri}} = \frac{v_{\text{init.bri}}}{|a_{k\text{bri}}|}. \tag{8}$$

We also note that the braking distance of the car in the braking zones in the areas of brake positions l_{bri} is calculated by the formula of elementary physics:

$$l_{k\text{bri}} = v_{\text{init.T}} t_{\text{bri}} + \frac{1}{2}|a_{k\text{bri}}|t_{\text{bri}}^2. \tag{9}$$

Formula (9) is valid until the moment $t_{bri} < t$ (t - the current time) of the car in the braking zone.

Each section of the inclined part of the hump is characterized by its own movement conditions [4, 10, 17, 20–22]. Therefore, the power relations that take place in the car-track system at each of the sections of the hump differ from each other.

Because of this, in each section of the marshalling hump, the car rolls with different linear accelerations a_i (i – the numbers of hump sections) and speeds $v_{ei}(t_i)$ for different times t_i, which in this study are determined according to the basic law of dynamics with non-ideal constraint [5, 17, 20] in the Mathcad computing environment [25].

At the same time, the applied problem of studying the movement of the car as it passed through the boundaries between sections of the hump was solved, assuming that the rolling speed of the car at the end of one section v_{ei} corresponds to the initial speed for the next section in the form $v_{init(i+1)}$ [17, 20–22].

4.2 The Results of the Calculation in the Base System Mathcad [25]

The results of the calculated data using the initial data and the final analytical formulas (2)–(15), which were obtained in [20], are summarized in Table 1.

Table 1. The results of the calculations of acceleration, speed and time of car movement along the entire length of the track.

Sections of the inclined hump part	Elements of sections of the inclined hump part	l_{ix}	i_i	h_i	a_i	t_i	v_i	
		Input values		Calculated values				
		m	‰	m	m/s^2	s	m/s	km/h
	TH	–		–	–	–	1.7	6.12
HS1	HS1	39.95	50	1.994	0.51	9.611	6.06	23.78
HS2	Before Sw	15.007	30	0.45	0.314	2.161	7.285	26.23
	After Sw	18.633	18	0.335	0.191	2.477	7.758	27.93
BP1	WhB	8.301	14	0.116	0.157	1.059	7.924	28.53
	SB	13.352		0.187	|2.351|	|3.37|	0	0
	RS	7.348		0.103	0.156	9.672	1.519	5.47
IS	Before Sw	20.001	11	0.22	0.128	9.429	2.723	9.8
	After Sw	21.271		0.234	0.122	6.783	3.549	12.78
BP2	WhB	10.401	10	0.104	0.118	2.8	3.879	14.0
	SB	3.152		0.032	|2.387|	|1.685|	0	0
	RS	17.448		0.174	0.118	17.207	2.028	7.3
Sw	Before Sw1	16.0	2	0.032	0.039	7.369	2.318	8.35
	Sw1	25.69		0.051	0.033	10.334	2.654	9.56
	Sw2	21.0		0.042	0.032	7.567	2.897	10.43
	Sw3	24.0		0.048	0.032	7.934	3.154	11.35
CT1	CT1	59.18	1.6	0.095	0.032	17.241	3.711	13.36
BP3	SB	2.796	1.5	0.043	|2.463|	|1.507|	0	0
	RS	11.704	Straight section	0.018	0.032	27.032	0.866	3.12
CT2	CT2	50.0	0.6	0.03	0.032	38.195	1.752	6.31

Though repeatedly, but we would like to note that in Table 1, in contrast to [19], the third break position (3BP) is located on a straight section of the track.

In Table 1, as and in [21], there are also indicated:

a_i, t_i and v_i - acceleration, travel time and rolling speed of the car under the influence of the projection of the tailwind force of small magnitude F_{wx} and taking into account all kinds of resistance forces (medium, switches, curves, snow and hoarfrost) $|F_{ri}|$.

In this table, $l_{ix} = 0$ corresponds to the position of the car at the top of the hump (TH), and vinit1 = 1.7 m/s or 6.12 km/h - the car's humping speed on the TH (or the initial car speed) in case of designing a hump neck with 24 tracks. Higher and big power humps (HHP and HBP) are considered.

Analysis of the results of studies on finding the acceleration, travel time and rolling speed of a car in various sections of the marshalling hump obtained in Table 1 made it possible to note that when the impact of the projection of the tailwind force of a small value F_{wx}, taking into account all kinds of resistance forces (medium, switches, curves, snow and hoarfrost) on the car with cargo is projected, then |Fri| the collision speed of a car "with a group of standing cars" (6.3 km/h) is within the permissible limits (5 km/h) with a relative error of 26%. It is clear that a "soft" collision of the car "with a group of standing cars" happens in the marshalling yard, which is acceptable. However, if the permissible approach speed of the car to the calculated point is exceeded, even with a "soft" collision of the car "with a group of standing cars", the car and the cargo that are in it can be damaged.

To graphically represent data from Table 1, the length lix and the height hi of each section, and the travel time of the car t_i in these sections should be presented taking into account the length $l_{x(i-1)}$, the height $h_{(i-1)}$, and the travel time of the car $t_{(i-1)}$ in the previous section of the hump (Table 2).

Table 2. The results of the calculations of acceleration, speed and time of car movement along the entire length of the track.

Sections of the inclined hump part	Elements of sections of the inclined hump part	l_{ix}	i_i	h_i	a_i	t_i	v_i	
		Input values		Calculated values				
		m	‰	m	m/s²	s	m/s	km/h
	TH	–		–	–	–	1.7	6.12
HS1	HS1	39.95	50	1.994	0.51	9.611	6.06	23.78
HS2	Before Sw	54.957	30	2.444	0.314	11.772	7.285	26.3
	After Sw	73.59	18	2.779	0.191	14.249	7.758	27.93
BP1	WhB	81.891	14	2.895	0.157	15.308	7.924	28.53
	SB	95.243		3.082	|2.351|	11.938	0	0
	RS	102.591		3.185	0.156	21.61	1.519	5.47
IS	Before Sw	122.592	11	3.405	0.128	31.039	2.723	9.8
	After Sw	143.863		3.639	0.122	37.822	3.549	12.78
BP2	WhB	154.264	10	3.743	0.118	40.622	3.879	14.0
	SB	157.416		3.775	|2.387|	38.937	0	0
	RS	174.864		3.949	0.118	56.144	2.028	7.3
Sw	Before Sw1	190.864	2	3.981	0.039	63.513	2.318	8.35
	Sw1	216.554		4.032	0.033	73.847	2.654	9.56
	Sw2	237.554		4.074	0.032	81.414	2.897	10.43
	Sw3	261.554		4.122	0.032	89.348	3.154	11.35
CT1	CT1	320.734	1.6	4.217	0.032	106.589	3.711	13.36
BP3	SB	323.53	1.5	4.260	|2.463|	105.082	0	0
	RS	335.234	Straight section	4.278	0.032	132.114	0.866	3.12
CT2	CT2	385.234	0.6	4.308	0.032	170.309	1.752	6.31

As you can see, the total estimated (and/or design) length on the slope i of the inclined part of the hump from its top (TH) to the calculation point (CP), in the case when the park brake position (PBP) is located in a straight section of the track, is $L_{com.x} \approx 385.2$ m, the estimated height from the top to the calculated point of the hump is $H_{com.c} \approx 4.3$ m, and the total time of the car's movement throughout the projected length along the slope of the marshalling hump: $t_{com.c} \approx 170.3$ s (or ≈ 2.84 min).

Note that the total estimated (and/or design) slope length i and the height of the inclined part of the hump from its top (TH) to the calculation point (CP) equal to $L_{com.x} \approx 385.2$ and $h_{com.} \approx 4.3$ m correspond to the real geometric parameters of the marshalling yard. For example, we note that the projection of the maximum length $l_{max.h}$ on the horizontal of the inclined part of the odd marshalling hump (the distance from the hump crest to the end closest to the park break position (PBP)) can be equal to 394 m, and the height of the hump (the maximum height difference between the hump crest and the park brake position) Δh is 4.07 m.

Thus, for given initial data of the calculation example (for example, length lix and slope i_i of the profile of each i section of the hump), the calculated height from the top to the calculation point of the hump turned out to be $H_{com.c} \approx 4.3$ m. To reduce $H_{com.c}$, for example, to 3 m, the kinematic parameters of the car movement should be recalculated by varying the lengths lix and slope ii of the profile of each i section of the hump, which is easily carried out by the calculation program [17].

According to the third and fifth columns of Table 2, it is possible to construct graphical dependences of the change in the calculated height hi of the studied hump sections along the entire length of the track l_{ix}, i.e. $h_i = f(l_{ix})$ (Fig. 1).

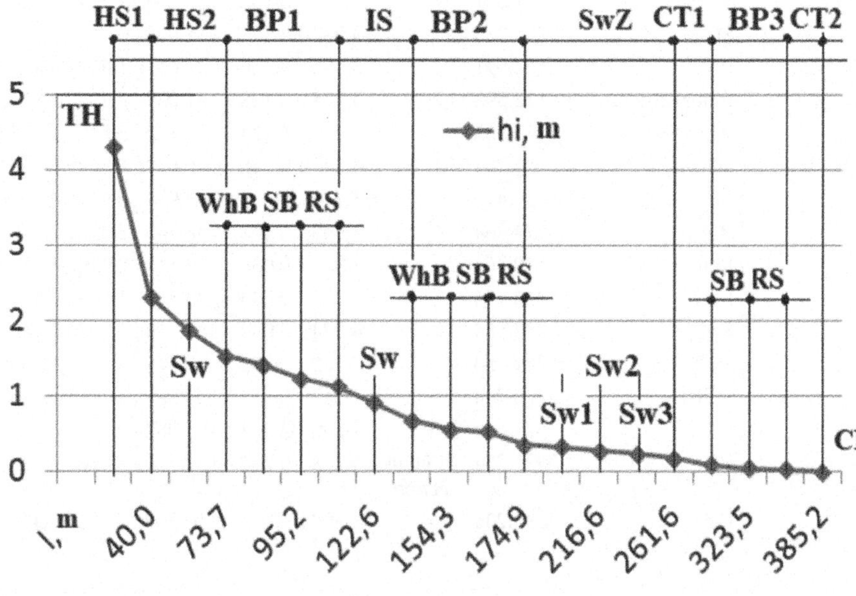

Fig. 1. Graphical changes in the calculated height of the studied hump sections along the entire length of the track – $h_i = f(l_{ix})$.

Designations in Fig. 1 and its explanations are the same as in Table 1 and Table 2.

Analyzing the graphical dependence $h_i = f(l_{ix})$, we note its correspondence to the real profile of the marshalling hump, i.e. reducing the height of the profile of each section of the inclined part of the hump in proportion to the slope of the track i.

We emphasize that the graphical dependence of the structural height on the length of the track profile $h_i = f(l_{ix})$ according to the calculation program [17] was constructed for the first time.

And in [21], according to the third and sixth columns of Table 2, it is possible to construct graphical dependences of the change in the acceleration of the car ai over the entire length of the track lix under the influence of the tailwind force of a small value F_{wx}, taking into account all kinds of resistance forces $F_{ri}|$, i.e. $a_i = f(l_{ix})$ (Fig. 2).

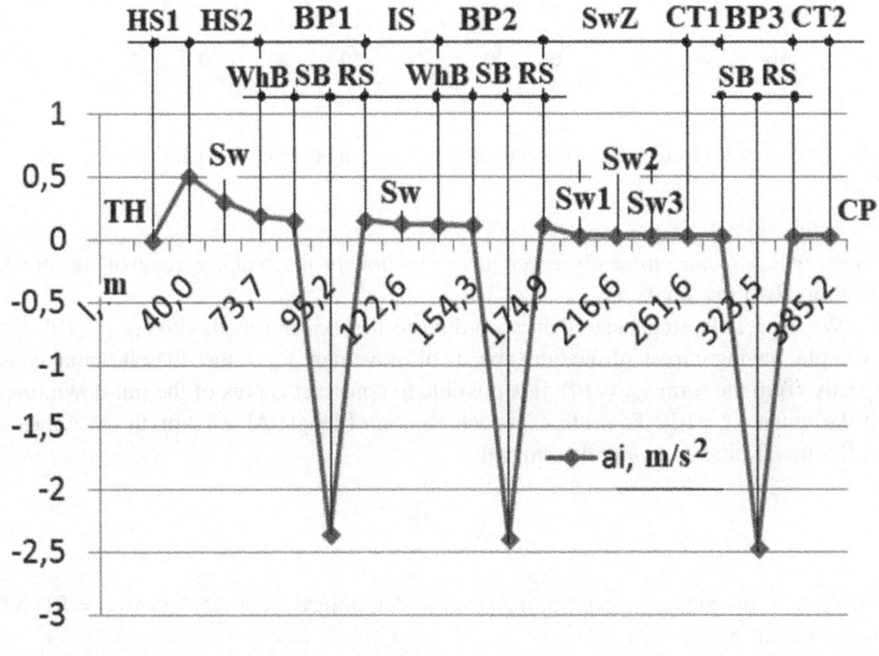

Fig. 2. Graphical changes of the acceleration of the car over the entire length of the track – $a_i = f(l_{ix})$.

It is noticeable in Fig. 2 that in the car retarded sections (RS) in areas of brake positions 1BP, 2BP, and 3BP, the car moves uniformly retardedly with accelerations, which have negative values, i.e. $a_{1br} < 0$, $a_{2br} < 0$ and $a_{3br} < 0$ (where $|a_{1br}| = -a_{1br}$, $|a_{2br}| = -a_{2br}$ and $|a_{3br}| = -a_{3br}$) (q.v. Table 1 and 2).

Similarly, $a_i = f(l_i)$, using the data of the third, seventh and eighth columns of Table 2, graphical dependencies $t_i = f(l_{ix})$ (Fig. 3) and $t_i = f(l_{ix})$ (Fig. 3) are built.

Data analysis of Fig. 3 shows that, over the entire length of the track, the car's travel time t_i and the car's braking sections t_{1br}, t_{2br} and t_{3br} are practically characterized by a change in the slope of the broken lines, which corresponds to negative braking

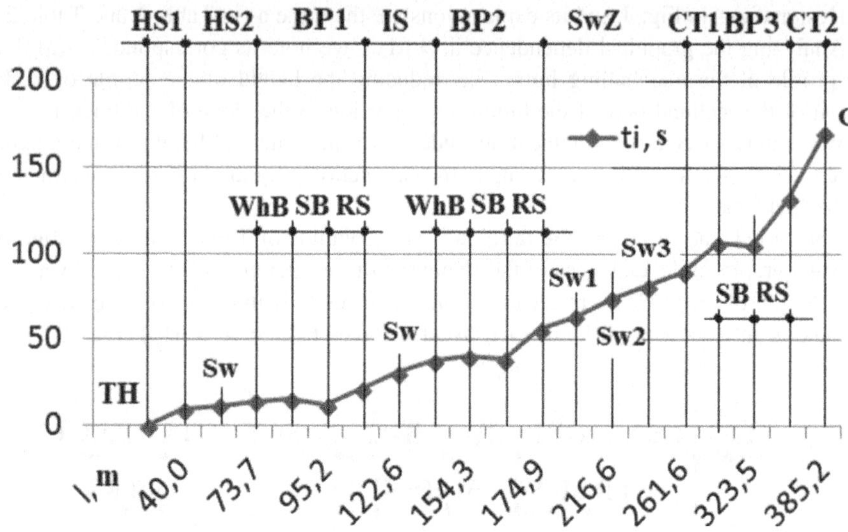

Fig. 3. Graphical changes of the travel time of the car along the entire length of the track – $t_i = f(l_{ix})$.

times, which mean uniformly retarded car motion in the braking zone of the break position (BP) (qv Table 1).

We especially emphasize that, according to the existing methodology [4, 10], for example, having curves of medium speeds of movement v_m (rather than instantaneous speeds vi) in the form $v_m = f(l)$, it is possible to construct curves of the run-down time of the runners $t = f(l)$. To do this, in each section of length $\Delta l = 10$ m, the increments of the travel time Δt_i, s are determined:

$$\Delta t_i = \frac{\Delta l_i}{v_m}, \tag{10}$$

where v_m – the medium speed in the section, determined from the curve $v_m = f(l)$ for each interval Δl_i.

Subsequently, the values Δti in each considered section are summed up and plotted in a selected time scale from the horizontal line at the end of each Δl_i section.

It is well known [10] that, for the convenience of determining the intervals between cars, it is recommended to construct two curves of the running time: one of a very bad runner $t_{iVB} = f(l_i)$ and one of a good runner $_{tiG} = f(l_i)$, or a very good runner $t_{iVG} = f(l_i)$ with braking. The first curve $_{tiVB} = f(l_i)$ is built from the zero point, the curve $t_{iVG} = f(l_i)$ or $t_{iG} = f(l_i)$ is built from the point raised up the time scale by the interval between cars at the top of the hump t_0, the second curve $t_{iVB} = f(l_i)$ - from a point spaced from zero by $2t_0$.

The interval between cars at the top of the hump is found by the formula:

$$t_0 = \frac{l_{iVB} + l_{iG}}{2v_o} \text{ and } t_0 = \frac{l_{iVB} + l_{iVG}}{2v_o}, \tag{11}$$

where l_{VB} - length of a very bed runner, 14.73 m; l_{VG} and l_G - length of a very good and good runner, 13.92 m; $v_o = v_{oc}$ - the estimated shunting speed, which is equal to 1.4 m/s for a hump of a mean power (HMP) and 1.7 m/s for a hump of a big power (HBP).

Typically, the calculations of Δt_i and t_i, although they are calculated using incorrect formulas, are recommended to be tabulated, which subsequently will facilitate their use in determining the shunting speed.

From this it is clear that formulas (1) and (2) in [18] are not really used in the normative and technical document [4].

From Fig. 4 it is clear that in the break sections, where the linear acceleration values are negative (q.v. Fig. 2), as was expected [21], the sliding speed of the car decreases to almost zero.

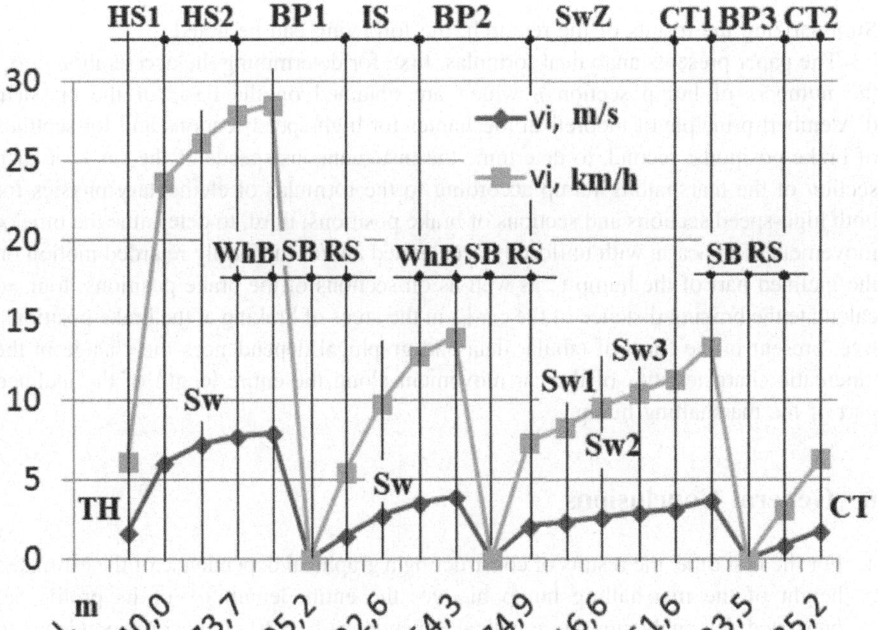

Fig. 4. Graphical changes in the rolling speed of the car along the entire length of the track – vi = f(lix).

We show the mathematical record of the graphical change in the speed of the car when it is retarded (RS) using the example of the first brake position (BP1) in the case of full use of the power of the brake positions in the form (see formula (19) in [22]):

$$v_{1br}(t) = f(t) = \begin{cases} f(\tau_5) = v_{WhB} & \text{under} \quad t < \tau_5, \\ f(t) = f(\tau_5) = -v_{1T} & \text{under} \quad \tau_5 \le t \le \tau_6, \\ f(\tau_6) = v_6 & \text{under} \quad t > \tau_6. \end{cases} \tag{12}$$

Where in the time interval $\tau_5 \leq t \leq \tau_6$: f(t) = f($\tau_5$) – braking zone of the first break position (BP1).

We especially note that the nature of the graphical dependencies of the changes in the rolling speed of the car vi along the entire length of the track l_{ix} are fundamentally different from similar curves that are constructed according to the existing method [4, 10], for example, the curves of medium v_{mi} (and not instantaneous vi) car speeds in the form v_{mi} = f(l).

5 Discussion

Summarizing the results of the research, the following can be noted.

The paper presents analytical formulas, first, for determining the acceleration ai (i - the numbers of hump sections), which are obtained on the basis of the classical d'Alembert principle of theoretical mechanics for high-speed sections and for sections of brake positions; second, to determine the instantaneous speeds of the car v_i at each section of the marshalling hump according to the formulas of elementary physics for both high-speed sections and sections of brake positions; third, to determine the time of movement of the car t_i with uniformly accelerated and/or uniformly retarded motion on the inclined part of the hump t_i, as well as on sections of the brake positions; four, to calculate the braking distance of the car l_{bi} in the areas of braking at the brake positions; five, present in the form of tabular data and graphical dependences the change in the kinematic characteristics of the car movement along the entire length of the inclined part of the marshalling hump.

6 General Conclusions

1. For the first time, the results of constructing a graphical dependence of the estimated height of the marshalling hump hi over the entire length lix of its profile are presented. By analyzing the graphical dependence h_i = f(l_{ix}), its correspondence to the real profile of the marshalling hump is noticed, i.e. reducing the height of the profile of each section of the inclined part of the hump in proportion to the slope i of the track.
2. The results of constructing graphical dependences on changes in the speed and time of movement of a car along the entire length of the inclined part of the marshalling hump fundamentally differ from the existing methodology [4, 10], where, for

example, the curves of medium v_{mi} (and not instantaneous vi) car speeds are constructed in the form $v_{mi} = f(l)$.

3. The proposed new methodology for calculating the kinematic characteristics of the car movement along the entire length of the hump allows analyzing the mode of car shunting at the marshalling humps, combining the power of brake positions and increasing the accuracy of determining the permissible collision speeds of cars in the marshalling yards. This work is the most important step for solving a promising problem: designing an automated system for calculating the dynamic characteristics of a car on a marshalling hump [17, 20–22].

References

1. Prokop, J., Myojin, S.: Design of hump profile in railroad classification yard. Mem. Facul. Eng. Okayama Univ. **27**(2), 41–58 (1993)
2. Zhang, C., Wei, Y., Xiao, G., Wang, Z., Fu, J.: Analysis of hump automation in China. In: Proceedings of Second International Conf. on Transportation and Traffic Studies, pp. 285–290 (2000). https://doi.org/10.1061/40503(277)45
3. Judge, T.: Yard management gets smarter. Railway Age **5**, 33–34 (2007)
4. Design rules and standards of sorting devices on 1520 mm railway gauge. TECHINFORM, Moscow (2003)
5. Komarov, K.L., Yashin, A.F.: Theoretical Mechanics in Railway Transport Problems. Nauka, Novosibirsk (2004)
6. Zářecký, S., Grúň, J., Žilka, J.: The newest trends in marshalling yards automation. Transp. Probl. **3**(4), 87–95 (2008)
7. Ogar, O.M.: East Eur. J. Adv. Technol. **3**(41) (2009)
8. Zhuravel, V.V.: East Eur. J. Adv. Technol. **3**(58) (2012)
9. Dick, C.T., Dirnberger, J.R.: Advancing the science of yard design and operations with the CSX hump yard simulation system. In: 2014 Joint Rail Conference, p. 04022014. American Society of Mechanical Engineers (2014)
10. Apatsev, V.I., Efimenko, Yu.I.: Railway Stations and Junctions: Manual (FSBEI). Railway Transport Educational-Methodical Center, Moscow (2014)
11. Turanov, Kh.T., Gordienko, A.A.: Some problems of theoretical prerequisites for the dynamics of rolling of the car on the slope of the hump yard. Transp. Inf. Bull. **3**(237), 29–36 (2015)
12. Bardossy, M.G.: Analysis of hump operation at a railroad classification yard. In: Proceedings of the 5th International Conference on Simulation and Modeling Methodologies, Technologies and Applications, 21 July 2015–23 July 2015, pp. 493–500. SCITEPRESS - Science and Technology Publications (2015)
13. Bobrovskyi, V., Kozachenko, D., Dorosh, A., Demchenko, E., Bolvanovska, T., Kolesnik, A.: The research of the domain of permissible braking modes of cuts on the gravity humps. In: Transport Problems. IV Symposium of Young Researchers, pp. 632–640 (2015)
14. Rudanovsky, V.M., Starshov, I.P., Kobzev, V.A.: On an attempt to criticize the theoretical positions of the dynamics of rolling the car down the slope of the hump. Transp. Inf. Bull. **6** (252), 19–28 (2016)
15. Boysen, N., Emde, S., Fliedner, M.: The basic train makeup problem in shunting yards. OR Spectr. **38**(1), 207–233 (2015). https://doi.org/10.1007/s00291-015-0412-0

16. Bobrovskyi, V., Kozachenko, D., Dorosh, A., Demchenko, E., Bolvanovska, T., Kolesnik, A.: Probabilistic approach for the determination of cuts permissible braking modes on the gravity humps. Transp. Probl. **11**(1), 147–155 (2016). https://doi.org/10.20858/tp.2016.11.1.14
17. Turanov, K.T., Gordienko. A.A.: The certificate of official registration of computer software programs RU № 2017614017, (2017)
18. Pozoisky, Yu.O., Kobzev, V.A., Starshov, I.P., Rudanovsky, V.M.: On the question of the movement of the car on the slope of the railway track. Transp. Inf. Bull. **2**(272), 35–38 (2018)
19. Kozachenko, D., Bobrovskyi, V., Demchenko, Y.: A method for optimization of time intervals between rolling cuts on sorting humps. J. Mod. Transp. **26**(3), 189–199 (2018). https://doi.org/10.1007/s40534-018-0161-2
20. Turanov, Kh.T., Gordienko, A.A.: Mathematical description of the movement of the car in sections of brake positions of the hump. Transp. Urals **2**(57), 3–8 (2018). https://doi.org/10.20291/1815-9400-2018-2-3-8
21. Turanov, Kh., Gordienko, A.: Movement of a railway car rolling down a classification hump with a tailwind. MATEC Web Conf. **216**, 02027 (2018). https://doi.org/10.1051/matecconf/201821602027
22. Turanov, K., Timukhina, E., Gordienko, A.: Mathematical description of the car's movement on the descent part of the hump. TransSiberia **1115**, 703–716 (2020). https://doi.org/10.1007/978-3-030-37916-2_69
23. Lu, C., Shi, J.: Dynamic response of vehicle and track in long downhill section of high-speed railway under braking condition. Adv. Struct. Eng. **34**(1), 36943321987057 (2019). https://doi.org/10.1177/1369433219870573
24. Ivanov, N.M., Finogenov, V.A.: Machine Detail. Higher School, Moscow (2006)
25. Kiryanov, D.V.: Tutorial MathCAD 13 (Samouchitel MathCAD 13). BHV-Peterburg, Saint-Petersburg (2006)

Analysis of Cross-Distortions in Aircraft Radio Systems with OFDM Signals at Channel Subcarriers Phase Coincidence

Anatoliy Fomin(✉) and Andrey Yalin(✉)

Moscow Aviation Institute (National Research University), Moscow 125993, Russia
fomin1939@yandex.ru, YalinAndrey@hotmail.com

Abstract. The paper considers the effect of cross-distortion resulting from limiting the OFDM signal. It is shown that the OFDM signal has discrete and continuous components. The analysis of the influence of cross-distortions created by both discrete and continuous components on the quality of the OFDM signal reception is carried out. The influence of cross-distortions caused by the restriction of the discrete component and the continuous component is analyzed. The conditions for the occurrence of a discrete component in the OFDM signal are analyzed. Analytical methods have shown that the highest level of cross-distortion occurs due to the restriction of the continuous component. The required energy reserve is calculated analytically in the event of cross-distortions in order to ensure the probability of erroneous reception of 10^{-6}. Cross distortion values are calculated for a signal containing 16 subcarriers and 100 subcarriers.

Keywords: Radiocommunication · Cross-distortion that occurs when the OFDM signal is restricted · Effect of cross-distortion on the quality of OFDM signal reception

1 Introduction

In aircraft high-speed radio systems of transmitting monitoring information, the effect of intersymbol interference formed by the mirror component of the signal reflected by the Earth surface begins to appear at a short duration of the information symbol.

Indeed, at a great distance aircraft from the ground station, the duration delay τ_{dl} of a sufficiently powerful mirror component of the reflected signal can be comparable to the duration of the information symbol $\tau_{dl} \approx \tau$. The addition of these signals at the receiver input is accompanied by inter-character distortions, which leads to a significant decrease the power of the useful signal, to an accidental failure of synchronization and to a restriction of the speed of information transmission. One of the most effective ways to significantly reduce the impact of intersymbol interference is the use signals with orthogonal frequency division multiplexing OFDM [1].

In the OFDM signal, the transmission of N binary symbols of information is carried out simultaneously in N parallel frequency channels by the phase shift keying of the harmonic subcarrier of each channel with symbol of information, and accordingly the

V. Murgul and V. Pukhkal (Eds.): EMMFT 2019, AISC 1258, pp. 339–360, 2021.
https://doi.org/10.1007/978-3-030-57450-5_30

duration of each symbol of information increases by N times compared to serial transmission. Increasing the symbol duration to the value $\tau_1 = N\tau$ allows to practically eliminate the influence of intersymbol interference, which, under the condition of $\tau_{dl} \ll \tau_1$, will occupy a small part of the information symbol, causing minor fluctuations in the amplitude without distorting its phase.

As shown in [2], in order to compensate for the significant Doppler frequency shift in aircraft radio systems in the transmitting part, the formation of channel signals is used by converting a single harmonic signal, which allows us to talk about their common phase at a certain time. In the receiving part, the channels use optimal coherent reception of individual symbols.

In the process of transmitting a OFDM signal through non-linear devices, inter-channel cross-distortion occurs, the structure of which differs from traditional interference in systems with the frequency division channels.

The purpose of this work is to study the characteristics of cross-distortion in the transmission of synchronous OFDM signals.

2 Materials and Methods

2.1 Characteristics of OFDM Signal in the Case of In-phase Channel Subcarriers

As shown in [1], OFDM signal implemented when the signal frequency selected from the condition $f_0 = 0, f_g, 2f_g, 3f_g, \ldots, \ldots$ when the signal frequencies of neighbor channels are shifted at $f_g = 1/\tau_1$. In this case, the maximum value of the signal transmitted at the frequency $f_0 + kf_g$, is generated in the receiver at the output of the k-th integrator with synchronous discharge and zero value of signals transmitted simultaneously in the other channels at frequencies $f_0 + if_g$ *when* $i \neq k$.

OFDM signal at the output of the transmitter, given the PSK of each N harmonic oscillations, for the duration of τ_1 is defined by the expression

$$U_{\sum}(t) = A_i \sum_{k=0}^{N-1} a_k cos\left[(\omega_0 + k\omega_g)t - \varphi_k\right] \quad (1)$$

where $\omega_g = 2\pi f_g$ and $\omega_0 = 2\pi f_0$,

$a_k = (-1)^{\vartheta_k} = \pm 1$ corresponding to the values of harmonic signals phases, determined by the location of binary symbols v_k = 0,1, k = 0, 1, … N − 1 in a code combination of N binary symbols for two-state phase shift keying:

$$\theta_k = \begin{cases} 0 \text{ when transmiting symbol } 1, \\ \pi \text{ when transmiting symbol } 0; \end{cases} \quad (2)$$

ω_0—carrier frequency of first channel;

φ_k—the initial phase of each harmonic oscillation associated with the phase $\Delta\varphi$ of the signal in the first channel by the ratio $\varphi_k = k\Delta\varphi$;

A_i—the amplitude of the harmonic signals.

In-phases of harmonic oscillations at time t_0, which provides coherent addition of signals, achieved when

The condition which provide a in-phase addition of signals is

$$(\omega_0 + k\omega_g)t_0 - \varphi_k = 0. \tag{3}$$

The phase $\varphi_H = \omega_0 t_0$ is common to all summands and does not participate in the formation of coherent addition of signals, so can be ignored.

The second summands $\varphi_k = k\omega_g t_0$ allows to determinate the values of the phases of harmonic oscillations at which the coherent addition of the signal components (1) is performed at time t_0. When k = 1 $\omega_g t_0 = \Delta\varphi$, which allows to determinate the time offset of the maximum $t_0 = \frac{\tau_1}{2\pi}\Delta\varphi$, which is determined only by the phase shift $\Delta\varphi$ of the signal in the first channel.

In the case of transmitting Q follow each other pulses of the same polarity, the sum (1) when $a_k = 1$ $a_k = 1$ can be represented as [2].

$$U_{\sum 1}(t) = A_i \frac{sin\left[\frac{Q}{2}\omega_g(t - t_0)\right]}{sin\left[\frac{1}{2}\omega_g(t - t_0)\right]} \cos\left[\left(\omega_0 + \frac{Q-1}{2}\omega_g\right)(t - t_0)\right] \tag{4}$$

The module of signal envelope

$$|U_0(t)| = A_i \left| \frac{\sin\left[\frac{Q}{2}\omega_g(t - t_0)\right]}{\sin\left[\frac{1}{2}\omega_g(t - t_0)\right]} \right| \tag{5}$$

for Q = 12 is shown in Fig. 1 and almost correspond to the amplitude spectrum of a single pulse. The maximum of the function at $\Delta\varphi = 0$ at the point $t_0 = 0$ after resolving the ambiguity is $U_{max} = A_i Q$.

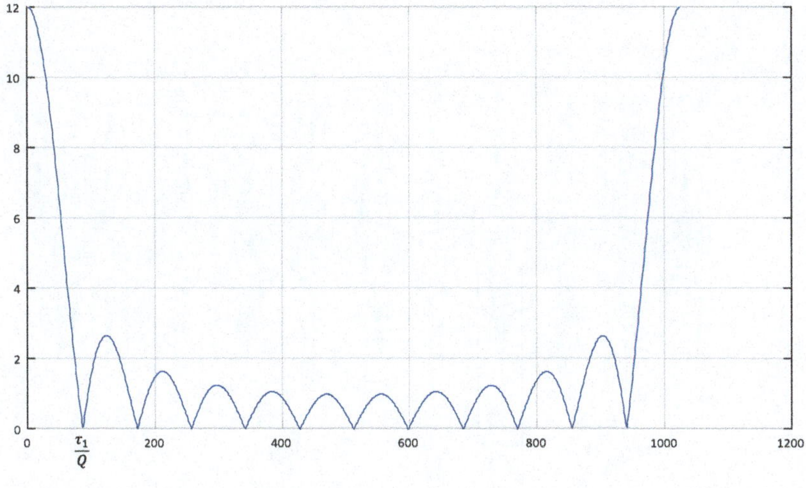

Fig. 1. The module of envelop of OFDM signal at time duration τ_1 and A_i = 1, N = 12, Q = 12.

In general case, Q is the number of pulses of the same polarity, the sum of which according to (5) exceeds the sum of the other N-Q pulses, half of which are positive and the other half are negative.

The implementation of the total signal $U_{\sum 1}(t)$ obtained as result of a computer experiment shown in Fig. 2. The experiment was performed for a sequence of N = 12 binary characters (12 orthogonal frequency channels), allowing to form $L = 2^N =$ 4096 combinations. As follows from the Fig. 2, the signal $U_{\sum 1}(t)$ includes discrete and continuous components, which determined by a random combination of coefficients a_k in the sum of N terms. The value of Q in (4) randomly changes in the range - $-N \ll Q \ll N$, determined by the statistical characteristics of the value Q. As shown in [2], the probability density of Q is described by a binomial distribution

$$W(Q) = \frac{1}{2^N} \frac{N!}{\left(\frac{N+Q}{2}\right)!\left(\frac{N-Q}{2}\right)!} \tag{6}$$

The value Q has the same parity as N, so that $\frac{N+Q}{2}$ and $\frac{N-Q}{2}$ are integers. When N is even, the function (6) is defined for $Q_i = 0, 2, 4\ldots$

The average value Q = 0, and the RMS value $\sigma = \sqrt{N}$. Using the asymptotic Stirling formula, from expression (6), we can proceed to the approximate-Gaussian distribution

$$W(Q) = \frac{1}{\sigma\sqrt{\pi}} exp\left(\frac{-Q^2}{2\sigma^2}\right) \tag{7}$$

Where $N \ll Q \ll N$. For $N \gg 1$, the normalization condition is met and the normalizing coefficient $B \to 1$.

As a rule, when building a radio system, they tend to increase N to values $N \geq 100$, which makes it possible to most effectively use the frequency band occupied by the OFDM signal, which does not have protective intervals between channels.

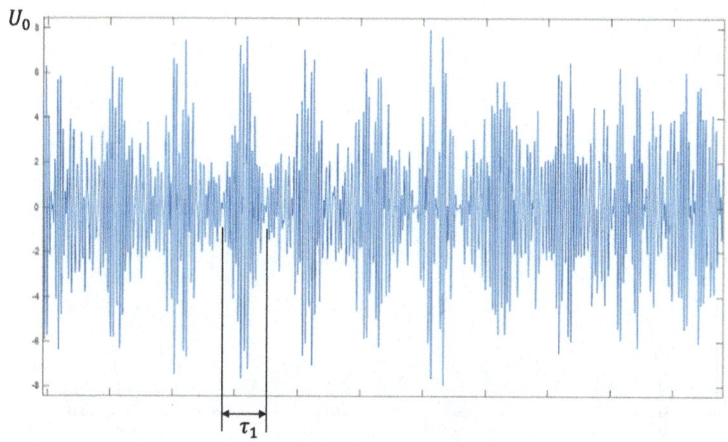

Fig. 2. Implementation of OFDM signal when N = 12.

When the frequency of harmonic oscillations corresponding to pulses of the same polarity are randomly arranged, the sum (1) is converted to the form:

$$U_{\Sigma 1}(t) = A_i \sum\nolimits_{k=q}^{Q_1} \cos\left[(\omega_0 + k\omega_g)t - \varphi_k\right] \qquad (8)$$

where k takes one of Q random values in the range from k = q to k = Q_1.

By analogy with the case of formation of a discrete component in the total OFDM signal considered above, the condition of in-phase $\omega_g t_0 = \varphi_k$ is fulfilled when the equality of $\omega_g t_0 = \Delta\varphi$ determines the moment $t_0 = \frac{\tau_1}{2\pi}\Delta\varphi$ for forming the maximum value of the signal $U_{\Sigma max} = QA_i$. Let's estimate the duration of the main petal.

In the simplest case, the signal $U_\Sigma(t)$ *when* $\varphi_k = 0$ can be presented as

$$U_\Sigma(t) = A_i \sum\nolimits_{k=q}^{Q_1} cos\left[(\omega_0 + k\omega_g)t\right] \qquad (9)$$

The given formula can be considered as the sum of Q discrete values of a harmonic oscillation with frequency $(\omega_0 + \omega_g)$, following randomly in time with an interval repetition of factor $\Delta t_i = \frac{\tau_1}{Q}$ when $q \le k \le Q_1$. Since a sequence of single pulses of the same polarity in a group of N pulses randomly changes its location with equal probability, so the sum $U_\Sigma(t)$ is also random, changing equally likely in the range $-Q \le U_\Sigma \le$ Q,, reaching the maximum value at the moment t = 0. The minimum value at time $t_{min} = \frac{\tau_1}{Q}$ is random and determined by a combination of single pulses of the same polarity in a group of N pulses. Since the combinations are equally probable, the average value of the $U_{\Sigma min}$ signal at the moment t_{min} is equal to zero $U_{\Sigma min} = 0$.

These features do not lead to significant changes in the analysis of interference, so in the future, when describing the signal at the output of the OFDM transmitter, we'll use formula (4).

When digitally generating an OFDM signal the dynamic range of the DAC $U_{out} = f(U_{in})$ selected taking into account the amplitude characteristic of the power amplifier so that the total signal $U_{in} = U_{\sum 1}$ lies on the linear section of the characteristic $U_{out} = f(U_{in})$ in the range $-\mathrm{A}_0 \le U_{in} \le \mathrm{A}_0$. In accordance with the technique common in engineering practice [3], the nonlinear characteristic of the UM is approximated by a linear-polyline dependence, including the linear section $U_{out}(t) = KU_{in}(t)$, where K is the steepness of the characteristic that determines the power amplifier gain, and the restriction section $|U_{out}(t)| = K|\mathrm{A}_0|$ at $|U_{in}(t)| \ge |\mathrm{A}_0|$. The desire to ensure in each channel of the receiver signal/noise ratio $h_i^2 = \frac{P_{ci}\tau_1}{N_0} = \frac{P_{ci}N\tau}{N_0}$, where P_{ci} - power of the input signal of one channel of the receiver is equal to the ratio signal/noise required for the serial transmission of $h_0^2 = \frac{P_0\tau}{N_0}$, where P_0 is the power of the signal at the receiver input when the serial transmission, leads to the choice of amplitude of harmonic subcarriers in each channel is equal $\mathrm{A}_i = \mathrm{A}_0/\sqrt{N}$, where $A_0^2 = 2P_0$. With this choice the average power of the signal in each channel $P_{ci} = \frac{P_0}{\mathrm{N}}$, the value h_i^2 is equal $h_i^2 = h_0^2 = \frac{P_0\tau}{N_0}$. The average power of the total signal $\sigma_\Sigma^2 = NP_i = P_0$, and the amplitude of the total signal equal to the coincidence of the phases of the harmonic components

$A_\Sigma = NA_i = \sqrt{N}A_0$ will exceed the threshold values A_0 in some time intervals, which will lead to the appearance of inter-channel distortion, the spectrum of which covers all the channels of signal processing in the receiver.

As shown in [2], two components can be marked in the total OFDM signal: a discrete $U_D(t)$ and a continuous $U_n(t)$. The discrete component occurs when Q pulses of the same polarity are coherently added at separate time intervals. We assume that the probability of the discrete component of the signal appearing coincides with the probability that the value $|Q|$ exceeds the threshold level $P(|Q| > \sqrt{N})$, determined according to [2] by the value $U_0| = \sqrt{N}$. For the Gaussian probability density (5) $P(|Q| > \sqrt{N}) = 0,32$. The probability of appearing a discrete component $P(|Q| < \sqrt{N}) = 0,32$. In the case of $N \geq 100$, the probability of $Q > 50$ $P(|Q| > 50) \rightarrow 0$ and may not be counted. In the future, the real range of Q values that taken into account in the calculations is $\sqrt{N} \leq |Q| \leq \frac{N}{2}$.

2.2 Evaluation of Cross-Distortions Caused by the Discrete Component

The main peak of the envelope (5) of the signal (4) that exists on the interval $t_0 - \tau_1/Q \leq t \leq t_0 + \tau_1/Q$, can be approximated with a sufficiently high accuracy by a fragment of a harmonic oscillation

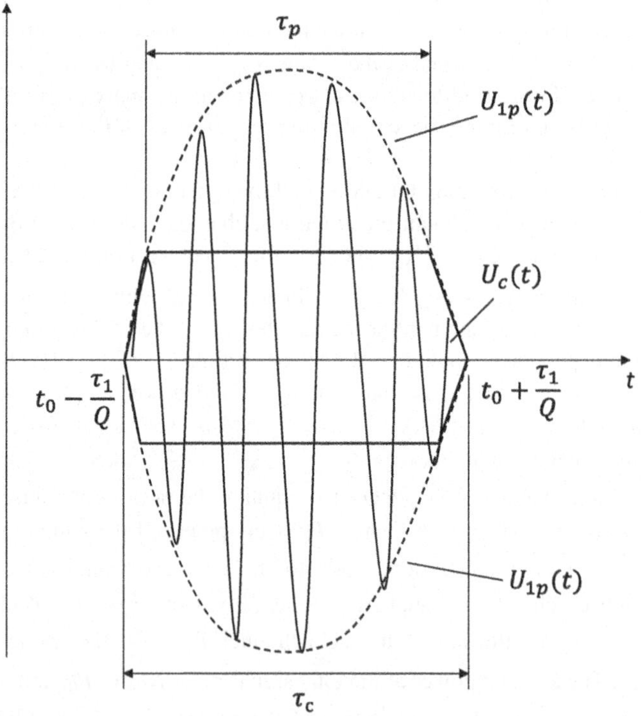

Fig. 3. Signal of the main peak of the discrete component, taking into account the restriction.

$$U_{01}(t) = A_i Q cos \frac{Q\omega_g(t - t_0)}{4} \quad (10)$$

As result of a two-way restriction in the power amplifier of the main peak of the signal (4) at the level $U_{trh} = \pm A_i\sqrt{N}$, a pulse $U_c(t)$ is formed at the output of the limiter (Fig. 3), the envelope of which can be described with sufficient accuracy by a trapezoid. The pulse can be considered as the result of subtracting from the original signal $U_{\sum 1}(t)$ (4) virtual interference pulses $U_{1p}(t)$ and $U_{2p}(t)$, obtained as a result of exceeding the threshold $U_{trh} = \pm A_i\sqrt{N}$.

The interference signal with the amplitude $U_{pA}(t) = A_i(Q - \sqrt{N})$, where $Q = |Q|$ is considered later without special reservation, is shown in Fig. 4, and represents a sequence of positive and negative pulses $U_i(t)$, following with period $T_p = \frac{2\pi}{\omega_p}$, where $\omega_p = \omega_0 + \frac{Q-1}{2}\omega_g$—the frequency of the signal $U_{\sum 1}(t)$, modulated by the function

$$U_{0P}(t) = A_i Q cos \frac{Q\omega_g(t - t_0)}{4} - A_i\sqrt{N} \quad (11)$$

defined on the interval $t_0 - \frac{\tau_p}{2} \leq t \leq t_0 + \frac{\tau_p}{2}$.

The duration of the interference pulse τ_p based on the equality $U_{01} = U_{trh}$ in the formula (10) is determined as a result of the conversion $\frac{Q\omega_\Gamma(t-t_0)}{4} = arccos\frac{U_{trh}}{A_i Q}$ or taking into account the condition, $\omega_g = \frac{2\pi}{\tau_1}$, the value $\Delta\tau = t - t_0 = \frac{2\tau_1}{\pi Q} arccos\frac{U_{trh}}{A_i Q}$.

Finally, for $\tau_P = 2\Delta\tau$, the value of $\tau_P = \frac{4\tau_1}{\pi Q} arccos\frac{U_{trh}}{A_i Q}$, which for $U_{por} = A_i\sqrt{N}$ is defined by the expression

$$\tau_p = \frac{4\tau_1}{\pi Q} arccos\frac{\sqrt{N}}{Q} \quad (12)$$

Each pulse $U_i(t)$ (Fig. 4) is a fragment of the half-period of the harmonic oscillation, obtained as result of this oscillation exceeding the threshold $\pm U_{trh}$. Since a large number of periods of harmonic oscillation with the frequency ω_p are located on the interval τ_p, we can consider the spectrum of the interference signal $U_p(t)$ a discrete. The envelope $G_0(\omega)$ of the spectrum described by a function that coincides in shape with the envelope of the spectrum of the cosine pulse $U_i(t)$. The first zero of the envelope is fixed at the frequency $\omega_1 = 3\pi/\tau_0$, where τ_0 - is the minimum value of the duration τ_i, the harmonic of the carrier frequency of the interference $\omega_p = 2\pi/T_p$, which is the pulse repetition frequency, is located in the range $0 \div \omega_1$.

Frequency band of spectrum $G_0(\omega)$ is associated with a value of ω_1 with ratio $\Delta\omega = 2\Delta\omega_1 = \frac{6\pi}{\tau_0}$ and much superior band of radio system $\Delta\omega_C = N\omega_g$. Linear devices at the output of the transmitting part and at the input of the receiving part of the radio system, tuned to the frequency ω_C, allocate a signal in the frequency band $\Delta\omega_C$ and smooth the function (Fig. 4) at the transition points through zero. Further, we consider the approximation $U_{p1}(t)$ of the cross-distortion signal $U_p(t)$ shown in Fig. 5 as a segment of a harmonic oscillation with a frequency ω_p with the duration τ_p. Formally, the signal $U_{p1}(t)$ combines the positive $U_{1p}(t)$ and the negative pulse $U_{2p}(t)$, shown in Fig. 3, and written as:

$$U_{p1}(t) = U_{0p}(t)cos\left[(\omega_0 + \frac{Q-1}{2}\omega_g)(t - t_0)\right] \quad (13)$$

where $U_{OP}(t)$ defined according to (11).

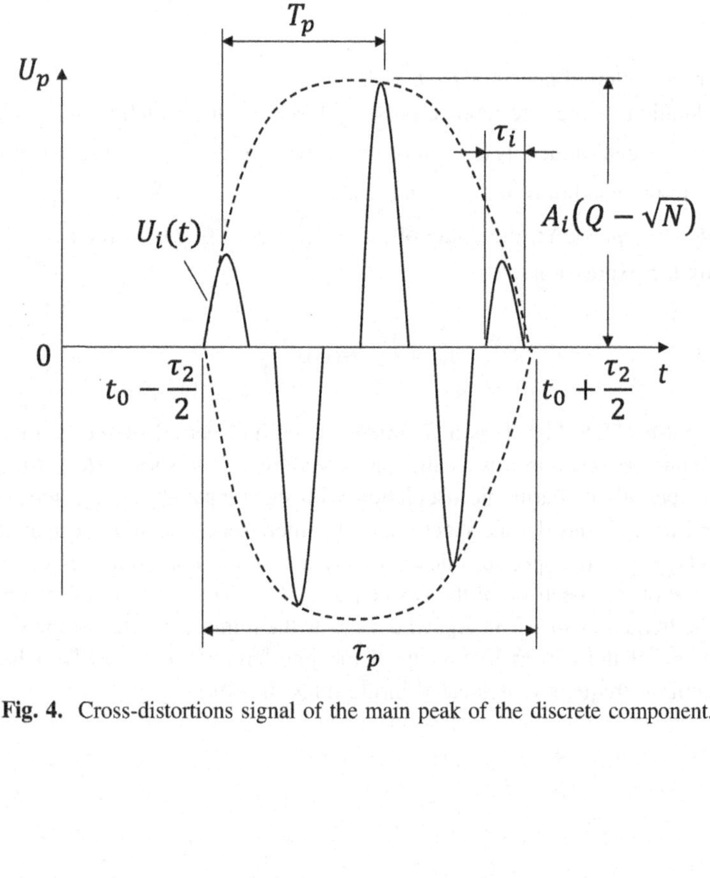

Fig. 4. Cross-distortions signal of the main peak of the discrete component.

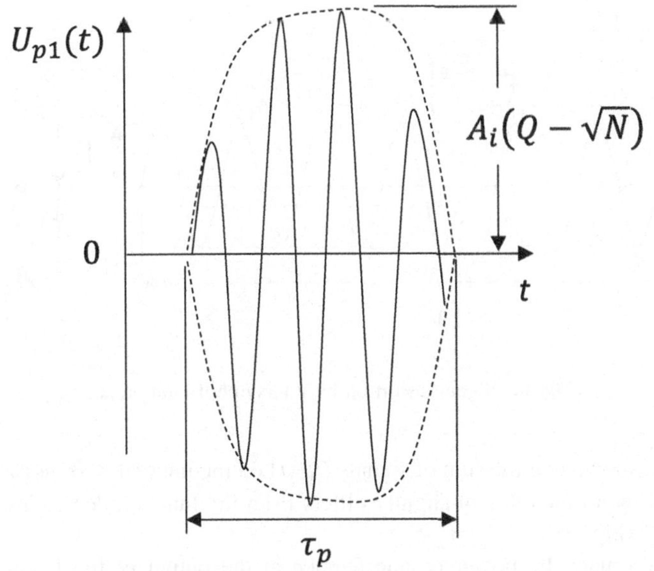

Fig. 5. Approximation of the cross-distortion signal.

Signal formation $U_C(t)$ (Fig. 3) as a result of limitations will be regarded as the result of the subtraction from the main peak of the discrete component of $U_{\sum 1}(t)$ (4) interference signal $U_{p1}(t)$ (13) that will lead to the failure of mainly the peak signal (4), the envelope of which coincides in form with the envelope signal (11).

By analogy with the develops of formulas (4) and (5), we can assume that $U_{p1}(t)$ interference signal, which is a fragment of the signal (1), is formed as a result of the addition of pulses in the interval $t_0 - \frac{\tau_p}{2} \leq t \leq t_0 + \frac{\tau_p}{2}$ in the form of Q segments of harmonic signals that coincide in frequency with the signals that form the main peak of the signal $U_{\sum 1}(t)$.

The result of limiting $U_C(t)$ can be considered as subtracting of the sum Q components of the harmonic signals that form the interference pulse from the corresponding signals that form the main peak of the signal $U_{\sum 1}(t)$. As a result, a dip will appear in each of the q initial signal pulses, the envelope of which coincides in shape with the envelope (11) $U_{Opi}(t)$ with the amplitude

$$U_{pAi}(t) = \frac{U_{pA}}{Q} = A_i \cdot \left(1 - \frac{\sqrt{N}}{Q}\right) \tag{14}$$

with a footing duration of τ_p.

A fragment of one of the Q signals $U_{ci}(t)$ is shown in Fig. 6.

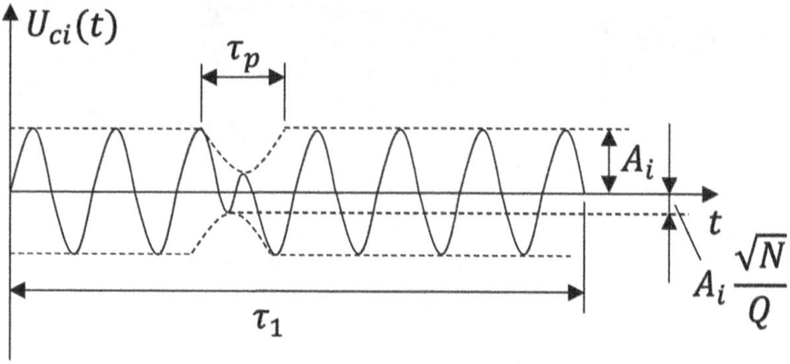

Fig. 6. Signal distortion by intersymbol interference.

It can be shown that the sum of signals $U_{Ci}(t)$ on the interval τ_p of noise subtraction is described by a function that slightly differs from the function describing the limited signal $U_C(t)$ (Fig. 3).

Let us estimate the power of interference at the output of the integrator of one channel for the case under consideration, replacing the interference in each channel with a rectangular radio pulse of duration τ_p with a frequency f_i and an amplitude determined by the expression (14).

The parameters of the interference signal determined by the placement of Q pulses of the discrete component in the band $\Delta f_C = N/\tau_1$.

The probability of occurrence of evenly spaced Q pulses significantly exceeds the probability of occurrence of a block of Q pulses at neighboring frequencies.

If Q pulses are uniformly located at frequencies f_q in the band $\Delta f_C = \frac{N}{\tau_1}$, the signal of the discrete component can be represented as

$$U_{\sum}(t) = A_i \sum\nolimits_{k=0}^{N-1} cos[(\omega_0 + k\omega_r)t - \varphi_k] \tag{15}$$

where $\omega_r = n2\pi/\tau_1$ - the frequency difference of two neighbor signals, n - an integer in the range 2\le n < N/Q2 $\le n < N/Q$. Similar to the previous transformations (2)

$$U_{\sum}(t) = A_i \frac{sin\left[\frac{Q_n}{2}\omega_g(t-t_0)\right]}{sin\left[\frac{n}{2}\omega_g(t-t_0)\right]} \cos\left[\left(\omega_0 + n\frac{Q-1}{2}\omega_g\right)(t-t_0)\right]. \tag{16}$$

As a result of ambiguity resolution, the maximum value of the signal $U_{\sum max} = A_i Q$.

The duration of the interference signal at the limit level according to formula (12) is equal to

$$\tau_p = \frac{4\tau_1}{\pi Q n} arccos \frac{\sqrt{N}}{Q} \tag{17}$$

The value n is a random variable and can change for $Q \leq N/2$ within $2 \leq n \leq N/Q$, where n is an integer. The average value of n is defined by the expression $n_0 = 2 + 0,5\left(\frac{N}{Q} - 2\right)$ and is equal for $n = 100$, $Q = 10$, and $n_0 = 2,65$. The average value of the value τ_p for the boundary values of n is determined by the formula $\tau_{p0} = 0,5(0,5 + Q/N)\frac{4\tau_1}{\pi Q}\arccos\frac{\sqrt{N}}{Q}$ and for $N = 100$, $Q = 50$ is equal to

$$\tau_{p0} = 0,5\frac{4\tau_1}{\pi Q}\arccos\frac{\sqrt{N}}{Q} \tag{18}$$

As shown above, for $N \geq 100$, the probability of $Q \geq \frac{N}{2}$ values appearing is $P(|Q| \geq 50) \to 0$, which limits the scope of Q to the range $\sqrt{N} \leq |Q| \leq N/2$.

When analyzing cross-distortion, you should take into account the features of their formation that arise due to the synchronicity of channel signals. The limited signal (Fig. 3) of the discrete component enters the input of the correlator of each receiver channel and is multiplied with the reference signal having the frequency f_i. At the output of the multiplier of each channel, the video signal $U_i(t)$ of this channel is allocated and the sum $(Q - 1)$ of the signals of the other channels having the difference frequencies $f_K = f_{i+K} - f_i$, where $k = 1, 2...Q$. The video signal $U_i(t)$ is the envelope of the signal $U_{ci}(t)$ (Fig. 6), which includes the dip caused by subtracting the interference pulse with the amplitude $U_{pAi}(t)$ (10). In the absence of interference, the undistorted signals of the remaining $Q - 1$ channels with different frequencies f_k are orthogonal to the undistorted signal $U_i(t)$. The impact of the interference signal leads to a violation of orthogonality, which is manifested in the appearance of cross-distortion. The spectral density of the amplitudes of the interference signal, represented as a rectangular radio pulse $U_{pi}(t)$, is described by the well-known expression

$$G_p(f_k) = A_i\left(1 - \frac{\sqrt{N}}{Q}\right)\frac{1}{2}\ \tau_p\frac{\sin\left[\pi(f - f_k)\tau_p\right]}{(\pi(f - f_k)\tau_p)} \tag{19}$$

The spectrum of each distorted pulse can be considered as the difference between the spectrum $G_C(f)$ of the original undistorted signal (Fig. 7) and the spectrum of the interference pulse $G_p(f)$ with the band $\Delta f_p = \frac{2}{\tau_p}$, which covers part of the channels. After demodulation in the integrator band of each channel, a part of the $G_{pi}(f)$ (shaded) spectrum of $G_p(f)$ will act together with the useful signal, creating inter-channel interference. In the worst case, when summing Q pulses with the same polarity, whose frequencies are located symmetrically relative to the frequency $f_{N/2} = \frac{N}{2\tau_1}$ of the central channel, cross-distortion from Q pulses will act in the integrator band of the central channel.

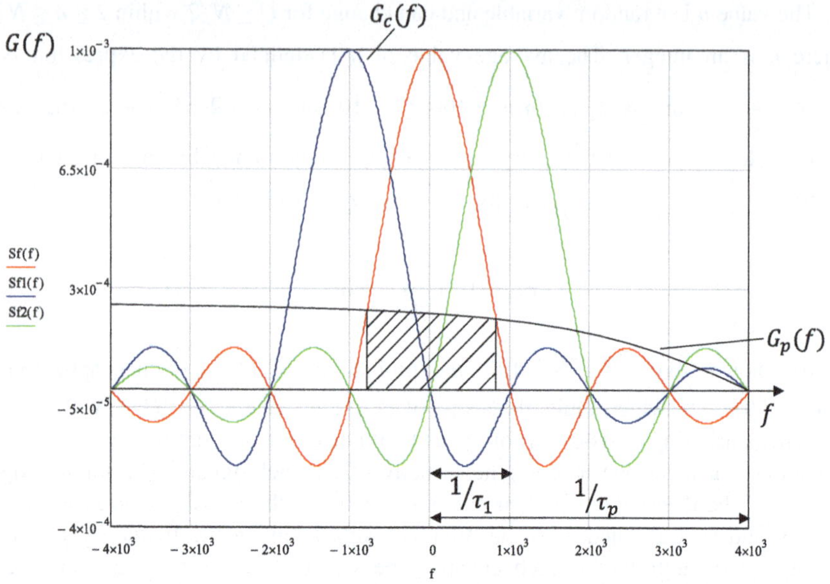

Fig. 7. Specter of signals in the receiver integrator band.

As a result of multiplication with a reference harmonic oscillation with the frequency f_i, individual fragments of the spectra $(Q-1)$ of interference signals (13) located in the frequency range f_i are transferred to the zero frequency in the band of the integrator of the i-th channel. Since all harmonics of the main lobe of the spectrum (13) have the same phase, we can talk about a coherent summation of the interference signal spectra at the integrator input.

We assume that on the left relative to the frequency f_i of the i-th channel there are $Q/2$ channels whose frequencies can take values from f_0 to f_{i-1}, and $Q/2$ channels whose frequencies are located on the right relative to the frequency f_i and can take values from f_{i+1} to f_N. If channels are arranged symmetrically with respect to f_i, which at $N \gg 1$, $Q \gg 1$ is on average fulfilled, then in expression (13) we can consider the one-way spectral density $G_p(t)$ (Fig. 7) in the band $0 \leq f \leq \Delta f_{p1}$, where $\Delta f_{p1} = \frac{1}{\tau_p}$, with doubling the value of the spectral density.

The resulting spectral density in the band of the i-th channel

$$G_p(f_k) = \sum_{k=1}^{Q-1} G_k(f_k) = A_i\left(1 - \frac{\sqrt{N}}{Q}\right) \cdot \tau_p \sum_{k=1}^{Q-1} \frac{\sin\left[\pi(f - f_\kappa)\tau_p\right]}{\left(\pi(f - f_\kappa)\tau_p\right)}. \quad (20)$$

Under the condition $Q \gg 1$, which is fulfilled for the case of exceeding the threshold $Q > \sqrt{N}$, the sum in (20) can be approximated for the value $0 \leq f \leq \Delta f_{p1}$ in the interference band Δf_{p1}.

$$B = \int_0^{\Delta f_{p1}} \frac{\sin(\pi f \tau_p)}{(\pi f \tau_p)} df = \frac{1}{\pi} Si(\pi), \tag{21}$$

which, taking into account the value of the integral sine $Si(\pi) = 1,85$ [3] is equal to $B = 1,85/\pi$.

The transition to the integral leads expression (20) to the form

$$A_{p0} = \sqrt{P_{p0}} = 1,85/\pi \cdot A_i \left(1 - \frac{\sqrt{N}}{Q}\right) \tag{22}$$

a and determines the average value of the spectral density of amplitudes $G_{p0} = G_{pi} \cdot \Delta f_p$ in the band Δf_p if the main lobe of the interference spectrum is represented by an equivalent rectangle with the base Δf_p.

The interference power spectral density is defined by the expression

$$N_{pi} = \frac{P_{p0}}{\Delta f_{p1}} = \frac{2P_{Ci}\left(1 - \frac{\sqrt{N}}{Q}\right)^2 \left(\frac{1,85}{\pi}\right)^2}{\Delta f_{p1}}. \tag{23}$$

The spectral density of the cross-distortion power in the integrator band ΔF_{int}, provided $\Delta f_p \gg \Delta F_{int}$ can be considered uniform and, accordingly, the cross-distortion power in the integrator band of one channel

$$P_{pDi} = \int_0^{\Delta F_{Int}} N_{pi} df = 0,35\left(1 - \frac{\sqrt{N}}{Q}\right)^2 \frac{\tau_p}{\tau_1} \tag{24}$$

or taking into account the average value of τ_{p0} (18)

$$P_{pDi} = \frac{0,35 P_{Ci}}{Q}\left(\frac{4}{\pi} arccos \frac{\sqrt{N}}{Q}\right)\left(1 - \frac{\sqrt{N}}{Q}\right)^2 \tag{25}$$

We determine the power of the useful signal at the output of the integrator of one channel of the receiver, taking into account the distortion caused by subtracting the cross-distortion that occurs due to the restriction of the signal $U_{\sum 1}(t)$. In accordance with the selected distortion model (Fig. 6), we assume that from the information pulse of a single channel $U_{Ci}(t)$ of duration τ_1 transmitted at the frequency $\omega_i = \omega_0 + i\omega_g$, a rectangular interference pulse $U_{pi}(t)$ of duration τ_p is subtracted, representing a segment of harmonic signal that coincides in frequency and phase with the oscillation of the signal $U_{Ci}(t)$. The plot of the signal $U_\Delta(t) = U_{Ci}(t) - U_{pi}(t)$ formed as a result of subtraction is shown in Fig. 6. The signal power at the output of the integrator is determined by the expression

$$P_{CDi} = \frac{1}{\tau_1}\int_0^{\tau_1} U_\Delta^2(t)dt \qquad (26)$$

which, taking into account the dip in the envelope duration τ_p at the interval $t_2 \le t \le t_2 + \tau_p$ can be converted to the form:

$$P_{CDi} = \frac{1}{\tau_1}\left[\int_0^{\tau_1} U_{Ci}^2(t)dt + 2\int_{t_2}^{t_2+\tau_p} U_{Ci}(t)U_p(t)dt + \int_{t_2}^{t_2+\tau_p} U_{pi}^2(t)dt\right] \qquad (27)$$

Since the interference amplitude according to (14) is equal to $U_{pA} = A_i\left(1 - \frac{\sqrt{N}}{Q}\right)$, and the signal and interference fluctuations coincide in frequency and phase, the resulting expression can be converted to the form

$$P_{CDi} = P_{Ci} - 2P_{Ci}\frac{\tau_p}{\tau_1}\left(1 - \frac{\sqrt{N}}{Q}\right) + P_{Ci}\left(1 - \frac{\sqrt{N}}{Q}\right)^2\frac{\tau_p}{\tau_1} = P_{Ci}\left[1 - \frac{\tau_p}{\tau_1}\left(1 - \frac{N}{Q^2}\right)\right] \qquad (28)$$

Taking into account the value of τ_{p0} (18), the signal power at the output of the integrator of one of the Q channels that form the discrete component is determined by the formula

$$P_{CDi} = P_{Ci}\left[1 - \frac{2}{\pi Q}\left(1 - \frac{N}{Q^2}\right)arccos\frac{\sqrt{N}}{Q}\right] \qquad (29)$$

The value of the cross-distortion power and the useful signal of the discrete component for different Q values are shown in Table 1.

Table 1. The value of the cross-distortion power and the useful signal of the discrete component for different Q values.

$N = 16$							
Q	0	6	8	12	16		
P_{PDi}/P_{Ci}	0	$0.62 \cdot 10^{-2}$	$1.4 \cdot 10^{-2}$	$2.2 \cdot 10^{-2}$	$2.6 \cdot 10^{-2}$		
P_{CDi}/P_{Ci}		0.9	0.875	0.88	0.9		
$N = 100$							
Q	12	16	20	40	50	60	80
P_{PDi}/P_{Ci}	$0.63 \cdot 10^{-3}$	$3.5 \cdot 10^{-3}$	$5.8 \cdot 10^{-3}$	$8.3 \cdot 10^{-3}$	$7.8 \cdot 10^{-3}$	$7.13 \cdot 10^{-3}$	$6 \cdot 10^{-3}$
P_{CDi}/P_{Ci}	0.98	0.96	0.95	0.96	0.96	0.97	0.97

As follows from the expression (5), in addition to the main lobe of the discrete component, the side lobes shown in Fig. 1 are formed. As shown in [3], at moments of time $t_k = k\tau_1/2Q$, the maximums of the side lobes are formed and, accordingly, the envelope module (5) takes values at these moments

$$U_{0k} = A_i \left| \frac{\sin \frac{k\pi}{2}}{\sin \frac{k\pi}{2Q}} \right| (k = 3, 5, \ldots) \tag{30}$$

For $k = 2q + 1 (q = 1, 2\ldots)$, the numerator of the given formula is $\left|\sin \frac{k\pi}{2}\right| = 1$. Under the condition $Q \gg 1$, $k \ll Q$, the denominator can be replaced with an argument, and the given expression can be approximated

$$U_{0k} = A_i \frac{2Q}{k\pi} \tag{31}$$

We estimate the maximum value of the side lobes, which is reached (Fig. 1) at $k = 3$ and is fixed in the case of $U_{03} > \sqrt{N}A_i$, when Q reaches the value $Q > 0,5\sqrt{N}k\pi$. As follows from expression (16), the amplitude of the first side lobe at $Q \geq 50$ will exceed the threshold level. However, the probability of $Q \to N$ values appearing is negligible, and the resulting crosstalk can be ignored. For $Q < 50$, the amplitude of the first side lobe satisfies the condition $U_{03} \leq A_i\sqrt{N}$ and may not be taken into account.

2.3 Estimation of the Power of Continuous Component of Cross-Distortion

The continuous component of the signal $U_H(t)$ is formed as a result of addition $N_H = N - Q$ pulses, half of which differ in phase from the pulses of the other half. The frequency values of signals forming a continuous component are randomly located in the band $\Delta f_c = \frac{N+1}{\tau_1} \approx \frac{N}{\tau_1}$ and the spectrum of such a signal is significantly uneven. The probability density of the signal $U_H(t)$ can be considered Gaussian. The value of the dispersion of such a signal $\sigma_H^2 = (N - Q)P_{Ci}$ is determined only by the number of terms N_H and does not depend on the location of channels in the band Δf_c.

To evaluate the cross-distortion of the continuous component of the OFDM signal, we use the technique given in [4] for a bounded random Gaussian process.

Nonlinear signal conversion is performed in a transmitter, the amplitude characteristic of which is approximated by a linear-polyline relationship.$U_{out}(t) = KU_{in}(t)$, where K is the transfer coefficient of a nonlinear device, further satisfies the condition $K = 1$, with a linear section within $U_{in} = U_0$ and a restriction threshold $|U_0| = \mathrm{A}_0 = \mathrm{A}_{Ci}\sqrt{N}$.

Since the phase of the pulses in the channels changes randomly, the energy spectrum of the continuous component is further considered.

The spectral power density of the implementation of $U_{ni}(t)$ at the input of the limiter is described by the expression

$$G_{Hm}(f) = \sum\nolimits_{i=q}^{N} G_i(f_i) \cdot a_i \tag{32}$$

where i - takes one of $N-Q$ random values in the range $1 \le i \le N$; $G_i(f_i) = P_{Ci}/\Delta F$— spectral density of the signal power in the i-th channel; $a_i = \begin{cases} 1 \\ 0 \end{cases}$ - accepts one of two equally probable values.

The energy spectrum of the signal $U_H(t)$ can be defined as the average G_{H0} of the set of spectra of implementations M

$$G_{H0} = \frac{1}{M}\sum\nolimits_{m=1}^{M} G_{Hm}(f) = \frac{1}{M}\sum\nolimits_{m=1}^{M}\sum\nolimits_{i=q}^{N} G_i(f_i)a_i \tag{33}$$

The sum of M component spectra $G_i(f_i)$ at each frequency f_i appears with probability $p_i = \frac{N-Q}{N}$.

The resulting value (33) is the average in the frequency band of the system $\Delta f_c = N\Delta F$ energy spectrum with an average spectral density $G_{H0} = \frac{N-Q}{N} \cdot \frac{P_{Ci}}{\Delta F}$.

To get the average G_H spectrum in the information signal band $\Delta f_n = (N-Q)\Delta F$ the average value of G_{H0} must be multiplied by the band reduction coefficient $K = \frac{N}{N-Q}$, which results in the value $G_H = \frac{P_{Vi}}{\Delta F}$.

Given that the signal strength P_{C_i} and the spectral density value $G_i = \frac{P_{C_i}}{\Delta F}$ in each channel are constant and do not depend on the value Q, the signal dispersion of the continuous component $\sigma^2 = (N-Q)P_{Ci}$ at the limiter input is provided in the signal band $\Delta f_n = (N-Q)\Delta F$.

Under this assumption, the energy spectrum of the continuous component can be approximated by a rectangle with the band Δf_n, the value of the power spectral density $G_H = \frac{P_{Ci}}{\Delta F}$ and the central frequency $f_{cn} = f_0 + \Delta F \cdot N/2$. The correlation coefficient in this case is determined by the expression

$$R_0(\tau) = \frac{sin(\pi\Delta f_n\tau)}{\pi\Delta f_n\tau} cos(2\pi\Delta f_{cn}\tau) \tag{34}$$

According to the method [4], the correlation function of a Gaussian radio signal with a variance $\sigma^2 = (N-Q)P_{ci}$ that passes a limiter with a linear section $\pm U_0$ is defined by the expression

$$K(\tau) = U_0^2\left[\frac{\Phi^2(x)}{x^2}R_0(\tau) + \sum\nolimits_{k=1}^{\infty}\frac{B_{2k+1}(x)}{x^2}R_0^{2k+1}(\tau)\right]\cos(2\pi\Delta f_{cn}\tau) \tag{35}$$

In the given expression the spectrum of the input multi-channel OFDM signal is narrow band in comparison with the value of the central frequency;

$R_0(\tau)$ - correlation coefficient of the original spectrum shifted to zero frequency;

$x = \frac{U_0}{\sigma}$ is determined from a condition $\pm U_0 = A_0 = A_i\sqrt{N}$;
$\sigma^2 = (N-Q)A_i^2/2$;
$\Phi(x) = \frac{2}{\sqrt{2\pi}}\int_0^x exp\left(-\frac{y^2}{2}\right)dy$ - the integral of probability;

$B_{2k+1}(x) = \frac{4\left[F^{(2k)}(x)\right]^2}{(2k!!)^2(k+1)}$, where $F^{(2k)}(x)$ - derivatives of the probability integral presented in the form;

$F(x) = \frac{1}{\sqrt{2\pi}}\int_{-\infty}^{x} exp\left(-\frac{y^2}{2}\right)dy$, tabulated functions whose value tables are given in [5].

In accordance with the transformation $G_0(f) = 4\int_0^{\infty} K(f)\cos(2\pi\Delta f_{ц}\tau)$ power spectral density of the continuous component of the signal at the output of the limiter, in the area of central frequency $f_{cn} = f_0 + \Delta FN/2$, for correlation coefficient (35)

$$G_0(f_{cn}) = \frac{(A_0)^2}{\Delta f_c}\left[\frac{\phi^2(x)}{x^2} + \frac{2}{\pi}\sum_{k=1}^{\infty}\frac{B_{2k+1}(x)}{x^2}\int_0^{\infty}\left(\frac{\sin(z)}{z}\right)^{2k+1}dz\right] \tag{36}$$

The values of a value $C'_{2k+1} = \int_0^{\infty}\left(\frac{sinz}{z}\right)^{2k+1}dz \cong \sqrt{\frac{3\pi}{2(2k+1)}}$ that provides an error of no more than 3% and decreases at $k \geq 2$ are shown in Table 2.

Table 2. The values of a value $C'_{2k+1} = \int_0^{\infty}\left(\frac{sinz}{z}\right)^{2k+1}dz \cong \sqrt{\frac{3\pi}{2(2k+1)}}$ that provides an error of no more than 3% and decreases at $k \geq 2$.

k	1	2	3	4
C_{2k+1}	$\frac{3}{8}\pi = 1.18$	$\frac{115}{384}\pi = 0.94$		
C'_{2k+1}	1.25	0.97	0.82	0.72

We limit the number of summands in (36), taking into account the values of individual multipliers. Taking into account the threshold value $U_0 = \sqrt{N}A_i$ and the variance of the continuous component $\sigma_1^2 = (N - |Q|)A_i^2/2$ division x in (35).

$$x = \frac{\sqrt{2}\sqrt{N}}{\sqrt{N - |Q|}} \tag{37}$$

where $|Q|$ is the modulus of Q that takes into account the probability of both positive and negative Q values appearing. Since the value Q changes randomly in the range $0 \leq |Q| \leq N - 2$, the value x is a random value lying in the range $\sqrt{2} \leq x \leq \sqrt{N}$.

In the future, we will limit the range of values x to the limits of $\sqrt{2} \leq x \leq 2$ for the corresponding values Q within $0 \leq |Q| \leq N/2$, since the probability $P(Q > N/2)$ $|Q| \geq N/2$ is negligible and is $P \approx 10^{-6}$.

Calculation of the amount included in (35)

$$B = \frac{2\left[F^{(2)}(x)\right]^2}{\pi x^2}C_3\sum_{k=1}^{5}\frac{\left[F^{(2k)}(x)\right]^2}{\left[F^{(2)}(x)\right]^2}\frac{8}{(2k!!)^2(k+1)}\frac{C_{2k+1}}{C_3} \tag{38}$$

performed for the number of summands k = 1 … 5 shows that the summands for k = 4 and 5 can be neglected due to their smallness. The amount B can be represented as

$$B = \frac{\left[F^{(2)}(x)\right]^2}{x^2} 0,375M_1 \quad (39)$$

where $M_1 = \sum_{k=1}^{3} \frac{\left[F^{(2k)}(x)\right]^2}{\left[F^{(2)}(x)\right]^2} \frac{8}{(2k!!)^2(k+1)} \frac{C_{2k+1}}{C_3}$.

The value of the sum M_1 for different values of x is shown in Table 3.

Table 3. The value of the sum M_1 for different values of x.

Q	0	10	16	20	40	50	60	80
x	$\sqrt{2}$	1.49	1.54	1.58	1.82	2	2.23	3.15
$[F^2(x)]^2$	0.044	0.0376	0.0344	0.0315	0.02	0.0117	$6{\cdot}10^{-3}$	$6{\cdot}10^{-5}$
M_1	1.04	1.022	1.018	1.0156	1.03	1.08	1.17	3.123

We believe that during the transmission and processing of the signal in the receiving part, the useful signal and the interference signal undergo the same changes.

The first term in the sum (36) determines the spectral power density $G_C(f_{cnt})$ of the useful signal of the continuous component in the band Δf_n. The second term in (36) defines the spectral power density $G_p(f_{cnt})$ of the crosstalk signal in the band Δf_n.

The power of the useful signal of the continuous component at the input of the integrator with a reset in one channel of the receiver with the band $\Delta F = 1/\tau_1$ is determined by the expression

$$P_{CHi} = G_C(f_{cnt})\Delta F_{int} = \frac{(A_0)^2}{\Delta f_H} \frac{\Phi^2(x)}{x^2} \Delta F = P_{ci}\Phi^2\left(\frac{\sqrt{2}}{\sqrt{1-\frac{Q}{N}}}\right), \quad (40)$$

where $(A_0)^2 = 2P_C = 2NP_{Ci}, \Delta f_H = \Delta F(N-Q), x^2 = \frac{2N}{N-Q}$.

The power of the cross-distortion signal of the continuous component at the output of the integrator with the reset of one receiver channel taking into account (35) and (36) is determined by the formula

$$P_{pHi} = G_p(f_{cn})\Delta F_{int} = \frac{(A_0)^2}{\Delta f_H} \Delta F_{int} = P_{Ci}\left[F^{(2)}(x)\right]^2 0,375M_1 \quad (41)$$

The values of the useful signal power and cross-distortion for $N = 100$ are shown in Table 4.

Table 4. The values of the useful signal power and cross-distortion for $N = 100$.

$N = 100$							
Q	0	12	16	20	40	50	60
P_{pDi}/P_{Ci}	0	$0.63 \cdot 10^{-3}$	$3.5 \cdot 10^{-3}$	$5.8 \cdot 10^{-3}$	$8.3 \cdot 10^{-3}$	$7.8 \cdot 10^{-3}$	$7.16 \cdot 10^{-3}$
P_{pHi}/P_{Ci}	$16 \cdot 10^{-3}$	$14.5 \cdot 10^{-3}$	$12.9 \cdot 10^{-3}$	$11.9 \cdot 10^{-3}$	$7.58 \cdot 10^{-3}$	$4.68 \cdot 10^{-3}$	$2.58 \cdot 10^{-3}$
$P_{p\Sigma} = P_{pDi} + P_{pHi}$	$16 \cdot 10^{-3}$	$15.4 \cdot 10^{-3}$	$16.4 \cdot 10^{-3}$	$17.7 \cdot 10^{-3}$	$16 \cdot 10^{-3}$	$12.5 \cdot 10^{-3}$	$10.7 \cdot 10^{-3}$
P_{CHi}/P_{Ci}	0.71	0.74	0.768	0.78	0.86	0.9	0.914
$h_i^2 = P_{CHi}/P_{p\Sigma}$	44.4	48	47	44	53	72	85

2.4 Influence of Interference on Reception Quality

As shown in [1], the power ratio of the signal/noise output of the integrator with the reset of the correlator of each receiver channel is determined by a known expression

$$\left(\frac{P_c}{P_p}\right)_i = \eta \cdot h_{ppi}^2 \tag{42}$$

Where $h_{ppi}^2 = \frac{P_{CHi} \cdot \tau_1}{N_\Sigma}$ - the ratio of the signal energy of the continuous component to the spectral density of interference at the integrator input;

P_{CHi} - power of the useful signal of the continuous component at the integrator input;

$N_\Sigma = N_0 + N_{pHi}$ - the total spectral power density of white noise N_0 and cross-distortion N_{pHi};

η - the coefficient determined by the method of modulation of the harmonic oscillation of the i-th channel by information pulses is equal to $\eta = 2$ for BPSK.

When calculating $h_{ppi}^2 = \frac{P_{CHi} \cdot \tau_1}{N_\Sigma}$, h_{ppi}^2 in the above formula, and $h_i^2 = \frac{P_{CHi}}{P_\Sigma}$ given in Table 4 uses a value of power of the useful signal component of the continuous lower power $P_{CHi} < P_{CДi}$ signal discrete component that corresponds to the worst case minimum values h_i^2.

The total (peak) power of the transmitter P_0 required to generate the sum of the limited useful and cross-distortion signal in the receiver is related to the threshold value A_0 by the ratio [4]

$$P = \left(\frac{4}{\pi}\right)^2 \cdot \frac{A_0^2}{2} = \frac{8}{\pi^2} \cdot A_0^2, A_0^2 = P_0 \frac{\pi^2}{8} \tag{43}$$

The power of the useful signal of the continuous component in the band ΔF of a single channel, taking into account (43), can be represented as

$$P_{CHi} = P_{Ci} \frac{\pi^2}{16} \Phi^2 \left(\frac{\sqrt{2}}{\sqrt{1 - \frac{Q}{N}}} \right) \tag{44}$$

The spectral power density of the cross-distortion at the input of a single channel in accordance with (35) taking into account (36) and the transformation (43) is determined by the formula

$$N_{pH} = \frac{P_{Ci}}{\Delta F} \cdot \frac{\pi^2}{16} \left[F^2(x)\right]^2 0,375 M_1 \tag{45}$$

As a result of multiplication with the reference signal in the correlator, the spectral power density N_p increases by 2 times to the value $N_{p1} = 2N_p$. The reset integrator represents a filter matched to a rectangular video pulse of duration τ_1. The amplitude-frequency response of the integrator $|K(j2\pi f)| = \frac{\sin(\pi f \tau_1)}{\pi f \tau_1}$ and the cross-distortion power at its output

$$P_{pi} = 2N_{pH} \int_0^{\infty} \left|\frac{\sin(\pi f \tau_1)}{\pi f \tau_1}\right|^2 df = N_{pH} \frac{1}{\tau_1} = N_{pH} \Delta F_H \tag{46}$$

Taking into account (46), the power of the cross-distortion at the integrator output in the band is $\Delta F_H = \frac{1}{\tau_1}$:

$$P_{pHi} = P_{Ci} \frac{\pi^2}{16} \left[F^2(x)\right]^2 0,375 M_1 \tag{47}$$

Power ratio signal/noise at the output of the integrator of one channel, taking into account (45), (47)

$$\left(\frac{P_{Ci}}{P_{pH}}\right)_i^2 = \frac{\Phi^2(x)}{\left[F^2(x)\right]^2 0,375 M_1} \tag{48}$$

The values of Table 4 for $Q > \frac{N}{2}$ can be ignored, since the probability of their occurrence is $P\left(Q > \frac{N}{2}\right) \to 0$ for $N \geq 100$. Signal-to-noise ratio at the input of a single receiver channel in the presence of cross-distortion.

$$h_{ppi}^2 = \frac{P_{CHi} \cdot \tau_1}{N_\Sigma + N_0} = \frac{P_{CHi} \cdot \tau_1 \cdot \alpha}{N_0 + N_{pHi} \cdot \alpha_1 + N_{pDi}\beta}, \tag{49}$$

where $\alpha = \frac{\pi^2}{16}\Phi^2(x)$ - a coefficient that takes into account the power loss of the continuous component signal due to its limitation and the power loss caused by the appearance of a cross-distortion signal;

$\alpha_1 = \frac{\pi^2}{16}\left[F^2(x)\right]^2 0,375M$ - coefficient that takes into account the power of the cross-distortion signal of the continuous component;

$\beta = \frac{\pi^2}{16}\frac{0,35}{Q}\left(\frac{4}{\pi} arccos\frac{\sqrt{N}}{Q}\right)\left(1 - \frac{\sqrt{N}}{Q}\right)^2$ – coefficient that takes into account the power of the crosstalk signal of the discrete component;

$N_{pHi} - \frac{P_{Ci}}{\Delta F}; N_{pD} - \frac{P_{CDi}}{\Delta F}$ – spectral power densities of signals in the band integrator ΔF.

The given expression is converted to the form

$$h_{ppi}^2 = \frac{h_{Hi}^2}{1 + h_{Hi}^2 \cdot \gamma}, \tag{50}$$

where $h_{Hi}^2 = \frac{P_{Ci} \cdot \tau_1 \cdot \alpha}{N_0}$ – signal-to-noise ratio of the continuous component;

$$\gamma = \frac{\alpha_1 + \beta}{\alpha} = \frac{1}{h_i^2}.$$

When calculating h_{Hi}^2 and $h_i^2 = \frac{P_{CHi}}{P_\Sigma}$ the worst case is selected, corresponding to the minimum values of the signal power, which correspond to the signal of the continuous component of P_{CHi}. Value h_{Hi}^2 the value $h_{ppi}^2 = 11,3$, required to achieve the specified error probability $p = 10^{-6}$ for optimal reception of BPSK signals can be determined by converting the formula (47) to the form

$$h_{Hi}^2 = \frac{h_{ppi}^2}{1 - h_{ppi}^2 \cdot \gamma} \tag{51}$$

The values h_i^2 are given in Table 4 and for the smallest value $h_i^2 = 44$ ($Q > 20$) the value $\gamma = \frac{1}{h_i^2} = 2,3 \cdot 10^{-2}$ and accordingly, for $h_{ppi}^2 = 11,3$, the value of $h_{Hi}^2 = 15,2$. For $Q = 20$ and $N = 100$, the value $x = 1,58$ and the coefficient $\alpha = \frac{\pi^2}{16}\Phi^2(x) = 0,5$. The original value $h_{0i}^2 = \frac{P_{ci} \cdot \tau_1}{N_0}$ in the formula $h_{Ci}^2 = \frac{P_{Ci} \cdot \tau_1 \cdot \alpha}{N_0} = 15,2$ should be increased to the value $h_{0i}^2 = 30,4$, i.e. approximately 3 times compared to the original value $h_{0i}^2 = 11,3$, necessary to achieve the probability of error in the absence of cross-distortion.

In the absence of a discrete component ($Q = 0$), as follows from Table 4, for $N = 100$, the ratio $h_i^2 = 44$ and the value $\gamma = 2,3 \cdot 10^{-2}$ and, respectively, for $h_{ppi}^2 = 11,3$, the value $h_{Hi}^2 = 15,3$. For $Q = 0$ and $N = 100$, the value $x = 1,41$ and the coefficient $\alpha = 0,44$. The initial value $h_{0i}^2 = \frac{P_{Ci} \cdot \tau_1}{N_0}$ must be increased to the value $h_{0i}^2 = 35$.

Since the probability of occurrence $Q = 0$ is determined by the probability $P(|Q| < \sqrt{N}) = 0,68$ for the Gaussian probability density $W(Q)$, then in the future, when choosing the value h_{0i}^2 the most probable case $h_{0i}^2 = 35$ is taken into account.

3 Conclusions

1. When using OFDM signals, the desire to increase the signal-to-noise ratio in each channel of the receiver to the value required for transmitting a conventional serial

binary signal leads to cross-channel interference, including discrete and continuous components;
2. In the synchronous method of signal generation, the OFDM discrete component is formed with a probability $P(|Q| > \sqrt{N}) = 0,32$ and represents a sequence of outliers with an amplitude equal to QA; The continuous component is formed constantly.
3. The values of the cross-distortion power of the discrete and continuous components for the values $N = 16$ and $N = 100$ are obtained;
4. The values of the signal-to-noise ratio necessary to ensure the probability of error $p = 10^{-6}$ are obtained, provided that the input of the receiver is simultaneously affected by white noise and total cross-distortion.

References

1. van Nee, R., Prasad, R.: OFDM for Wireless Multimedia Communications. Artech House, Norwood (2000)
2. Rabiner, L., Gold, B.: Theory and Application of Digital Signal Processing. Prentice Hall, Englewood Cliffs (1975)
3. Fomin, A.I., Yalin, A.K.: Signal characteristics in radio system with orthogonal frequency division multiplexing in the case of in-phase channel subcarriers. Electrosvyaz **12**, (2017)
4. Spilker, J.J.: Digital Communication by Satellite. Prentice Hall, Englewood Cliffs (1977)
5. Handbook of Mathematical Functions: With Formulas, Graphs, and Mathematical Tables, 0009-Revised edn. Dover Publications, Mineola (1965)
6. Dwight, H.B.: Tables of Integrals and Other Mathematical Data, 4th edn. The Macmillan Company, New York (1961)

Spontaneous Combustion of Pilot Fuel in Dual-Fuel Engine

Vladimir Gavrilov[1(✉)], Valery Medvedev[2], and Dmitry Bogachev[1]

[1] Admiral Makarov State University of Maritime and Inland Shipping, Dvinskaya Str., 5/7, 198035 St. Petersburg, Russia
gavrilov@vg5647.spb.edu

[2] St. Petersburg State Marine Technical University, Lotsmanskaya Str., 3, 190121 St. Petersburg, Russia

Abstract. The present paper offers a kinetic model of pre-flame processes, which represents a theoretical basis for the method of calculating spatiotemporal parameters of working mass of an engine during a deferral of spontaneous combustion. The duration of the local deferral period for combustion of mixture is determined taking into account some factors of both chain and heat acceleration of reactions and a local coefficient of excessive air for combustion. Calculations under certain diesel conditions showed that acceleration of pre-flame reactions depend on the chain mechanism of spontaneous combustion at extent of 85% and on the heat mechanism at extent of 15%. Practical application of this method allow defining coordinates of fuel-air mixture zones and time moments of flame emergence. This gives an opportunity to calculate the subsequent spreading of flame. When calculating the spontaneous combustion process for pilot fuel, one should consider that working mass of a dual-fuel engine contains gas fuel, besides air, liquid and steamy diesel fuel. Therefore, current concentrations of these components should be taken into account. The principal result of the calculation that is expected implies a decrease in the likelihood of detonation during combustion of gas fuel. For application of the proposed method, constants in formulas should be reconfirmed in accordance with the results of indication of the working cycle of engine of a certain type in its different operation modes.

Keywords: Diesel and gas-and-diesel mode · Local mixing process · Kinetics of pre-flame processes · Spontaneous combustion

1 Introduction

The authors of this paper have already proved the relevance of works on creating modern dual-fuel engines that run either on liquid fuel (in diesel mode) or on gas, with the spontaneous combustion of a small dose of pilot fuel (in gas-diesel mode) [1]. Several solutions for improving the process of supplying fuel to a dual-fuel engine [1] and advancing its work cycle were proposed [1]. The efficiency of a dual-fuel engine, its reliability and safety largely depends on the properness of mixture formation and combustion processes in both diesel and gas-diesel modes. Each of these modes implies

V. Murgul and V. Pukhkal (Eds.): EMMFT 2019, AISC 1258, pp. 361–374, 2021.
https://doi.org/10.1007/978-3-030-57450-5_31

significantly different conditions for the run of the processes mentioned. In particular, the differences lie in conditions for development of the main liquid fuel and pilot liquid fuel jets and in conditions for spontaneous combustion.

In a diesel mode, the main liquid fuel spontaneously combusts being a component of fuel-air mixture, while in a gas-diesel mode pilot fuel spontaneously combusts being a part of complex mixture that consists of liquid fuel's vapor, air and gas. The latter case obviously implies more complicated conditions. Therefore, the process of spontaneous combustion of pilot fuel is less studied. Particularly, the issue of localization of a fire source has not been sufficiently studied. At the same time, knowing features of this process will allow controlling it and rationally organizing the following combustion of gas-air mixture. Poorly organized combustion process can entail low efficiency of a work cycle, detonation emergence and disruption of normal operation of an engine at low loads.

From this perspective, the purpose of this study is to develop theoretical foundations for calculating a time point and space position of a spontaneous combustion source when pilot fuel is supplied to a dual-fuel engine.

2 Materials and Methods

Pre-flame processes in a diesel engine that precede the ignition of a combustible mixture are quite complicated. They are even more complicated in a gas-diesel engine. This complexity is explained by the variability of temperature and pressure values in a cylinder over time, the temperature and concentration heterogeneity of combustible mixture, the complex and unstable hydrocarbon and other composition of fuel, the influence of intensity and nature of the turbulence of working mass, the quality of fuel atomization, the multiple variance of pre-flame processes' nature, short duration of processes that obstruct obtaining the actual information on them.

Due to the complexity of theoretical and experimental studies, the process of spontaneous combustion is almost universally evaluated by the duration of the so-called integral deferral period of spontaneous combustion. Researchers have proposed a large number of formulas that are macrokinetic equations containing empirical values. Most formulas are based on the Arrhenius equation [2, 3]. The classical formula is presented, which is an approximate solution of the differential equation of mixture self-heating, which is obtained by O.M. Todes:

$$\tau_i = A(T_0/P_0)^{n-1}\exp(E/R_\mu T_0), \tag{1}$$

where T_0 and p_0 are initial temperature and pressure; n is reaction order; E is perceived activation energy; R_μ is the gas constant; A is an empirical coefficient.

The values of A, n, E and similar values in other formulas, estimated during experiments, are very different [3]. An example of the expression used in the study [4] is given below:

$$\tau_i = a\Phi^{-k} p_{cyl}^{-m} \exp\left(\frac{E}{R_\mu T_{cyl}}\right), \quad (2)$$

Where a, k, m are empirical constants; Φ is equivalence coefficient; p_{cyl}, T_{cyl} are pressure and temperature in the cylinder.

Numerous attempts to describe theoretically the processes occurring during the deferral period of spontaneous combustion usually lead to obtaining a formula that is similar to expression (2) and also contains empirical values [5]. The differences in the formulas discussed are caused by the differences in the types and parameters of the engines under study, the conditions for performing the experiments, and also by the variety of factors taken into account [6]. Researchers are attempting to develop analytical methods for calculating a spontaneous combustion process, taking into account heat and mass transfer processes and pre-flame chemical reactions [7]. Therein, macrokinetic dependencies of a known type are used in a mathematical model. Macrokinetics constants are determined through numerical simulation of fuel combustion process and through comparison of results with experimental data. To verify the adequacy of a simulation, they usually use only the results of measuring the integral duration of a deferral period, without local data on spontaneous combustion. Due to the reasons listed above, the known dependencies are completely not of universal nature. Nevertheless, the published results provide an opportunity to work on the research further. As part of the factors that significantly affect the duration of combustion deferral τ_i, the researchers indicate the coefficient of excess air for burnout α. An example of the indicated dependence obtained for a certain local volume of a diesel combustion chamber of type Ch13/14 is shown in Fig. 1 [7], where τ_i is expressed in degrees of crankshaft's rotation.

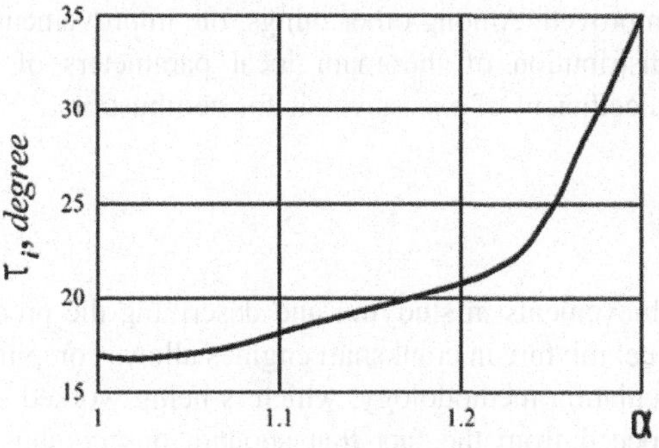

Fig. 1. Dependence of duration of spontaneous combustion deferral in a diesel engine on the coefficient of excess of air for combustion in a Ch13/14 type diesel engine [7].

A much more difficult task is to determine τ_i under conditions of a dual-fuel engine operating in the gas-diesel mode. The difficulty is reasoned by the fact that pilot fuel is spontaneously combusted in the mixture of air and gas fuel instead of almost clean air. Solving the problem of pilot fuel supply, it is necessary to achieve its stable (reliable) spontaneous combustion with the absence of detonation in a wide range of engine loads. In this case, the cyclic dose of pilot fuel should be as small as possible.

Features and variability of the hydrocarbon composition of gas fuels should be taken into account. The main part of natural gas is methane (CH_4) - from 70% to 98%. The composition of natural gas may include heavier hydrocarbons - methane homologues: ethane (C_2H_6), propane (C_3H_8), butane (C_4H_{10}), pentane (C_5H_{12}), etc. Octane numbers of these hydrocarbons differ a lot: from 110 for methane to 70 for pentane. These properties largely determine the difference in the combustion processes for the components of natural gas. In the context of the study topic, bearing in mind the importance of eliminating detonation during combustion in an engine, it is necessary to analyze the features of spontaneous combustion of the considered methane-alkane-air mixture. An admixture of heavier hydrocarbons in methane enlarges a tendency of a mixture to detonate. The reason is that alkanes have significantly lower energy of intermolecular C–H bond. Along with the hydrocarbon composition, a number of factors including temperature, pressure in a reaction zone and concentration of reagents complexly influence the emergence of the detonation-like combustion mode. Mixture's tendency towards detonation can be approximately estimated by the length of a deferral period of spontaneous combustion. The presence of only 1% of C_5H_{12} in stoichiometric mixture of methane and air reduces the spontaneous combustion deferral by two to three times [8, 9]. Moreover, the effect of adding heavy alkanes C3-C5 on the decrease in deferral of spontaneous combustion of methane and, therefore, its detonation resistance, has a strongly pronounced nonlinear nature. Due to high complexity of the described processes, today there is no possibility of proper numerical modeling, which moreover is advisable to be three-dimensional. Experience shows that the error in estimation of a deferral period of spontaneous combustion can exceed 30% [10].

As it can be seen, the method for quantitative description of the air-fuel mixture combustion, which is carried out through spontaneous combustion of pilot fuel in the engine, has to be improved. Among other things, the improvement should involve the consideration of distribution of the main local parameters of working mass, for instance, the local coefficient of excessive air for combustion.

3 Results

The analysis of achievements in studying and describing the process of spontaneous combustion of air-fuel mixture in crankshaft engines allows compiling the requirements for an adjusted calculation methodology, which is being worked out.

One should proceed from the fact that spontaneous combustion processes in a diesel engine represent pre-flame chemical reactions which develop relatively slowly with some acceleration and lead to rapid, explosive oxidation of fuel molecules. Acceleration of pre-flame chemical reactions is of chain-thermal nature. The contribution of chain and thermal mechanisms to the process under discussion should be accessed. In a mixture of heterogeneous composition, spontaneous combustion occurs in certain zones with a certain ratio of fuel vapor and air oxygen that is peculiar to a given fuel. The position of combustion sources should be calculated in order to determine the characteristics of subsequent flame spreading and thereby ensure the required quality of calculation of the combustion process as a whole.

Local deferrals of spontaneous combustion should be calculated with the use of a local approach to describing processes of mixture formation and combustion. Today it is impossible to describe a huge number of various chemical reactions proceeding within a deferral period accurately enough. Therefore, their modeling should be based on common kinetic dependencies with determination of equation constants through experiments. Thereby, it is necessary to solve the problem of taking into account changes in the rate of pre-flame reactions in time and space occupied by a heterogeneous combustible mixture of fuel and oxygen. These are the general principles of the methodology under development for calculating the process of spontaneous combustion of pilot fuel supplied to a dual-fuel engine.

When calculating reaction rates, it is advisable to proceed from the integral period of deferral of spontaneous combustion τ_i, which is quite accurately determined through experiments. To estimate τ_i, one can use a formula of the form (1). The dependence of the deferral period τ_i (s) obtained for a DN 23/30 engine during its operation in a wide range of operating modes:

$$\tau_i = 3.4 \times 10^{-6}(T_0/p_0)^{0.5} e^{\frac{23500}{8.314T_0}}, \tag{3}$$

where T_0, p_0 – temperature (K) and pressure (MPa) in the cylinder at the moment when the fuel injection begins. It should be noted that the following proofs can involve any known formula for calculation of τ_i which contains factor $\exp(E/R_\mu T)$.

The integral period of combustion deferral can be considered as the value that is an inverse of the average rate of pre-flame reactions w_{im}, which means that $w_{im} \sim 1/\tau_i$. The complex $\exp(-E/R_\mu T)$, when T = var, which is the part of expressions similar to formulas (1) and (3), can be interpreted as a factor of thermal acceleration of reactions reasoned by the increase in temperature due to compression of cylinder charge by a piston. Since pre-flame reactions have a chain-thermal nature, their description should contain the chain acceleration factor along with the thermal acceleration one.

When developing the expression of this factor, the dependence of chain reaction rate on time was taken as a basis, which was proposed by academician N.N. Semenov: $w = w_0 \exp(\phi\tau)$ where w_0 is the initial rate; ϕ is a coefficient depending on the ratio of probabilities of branching and breaking of chains. It is not possible to determine the value ϕ precisely. However, when describing pre-flame reactions, it can be assumed that its value is constant and there is an equality $\phi = 1/\tau_i$. This assumption corresponds to the well-known conclusion of D.A. Frank-Kamenetsky on a e-time increase in the rate of these reactions by the time of thermal explosion.

Moving to the relative time $\bar{\tau} = \tau/\tau_i$, the function of rate changing is written:

$$w/w_0 = e^{\bar{\tau}}. \tag{4}$$

It is more convenient to consider the current reaction rate w not in relation to the initial rate w_0, but in the form of its relation to the average rate w_{im}, which, as indicated above, is determined simply by the known integral period of combustion deferral τ_i. In this case, function (4) can be transformed:

$$\bar{w}(\bar{\tau}) = w / w_{im} = e^{\bar{\tau}} / (e - 1) . \quad (5)$$

Despite the negligibility of absolute amount of heat released during pre-flame reactions, their rate can be expressed, in particular, through the rate of heat release. Function (5) is convenient, since taking its integral over a full time gives one. This means that when $\bar{\tau} = 1$ the result of reactions provided by chain acceleration is achieved, which can be interpreted as a certain conditional amount of heat:

$$\bar{Q}_{con} = \int_0^1 \frac{e^{\bar{\tau}}}{e - 1} d\bar{\tau} = 1.0 . \quad (6)$$

Thus, the factor of chain acceleration of reactions can be expressed as a function $e^{\bar{\tau}}/(e - 1)$. The change in relative rate of pre-flame reactions determined only by the chain factor is shown in Fig. 2. The indicated rate increases by 2.7 times over the deferral period. The corresponding change in conditional released heat is presented in Fig. 3.

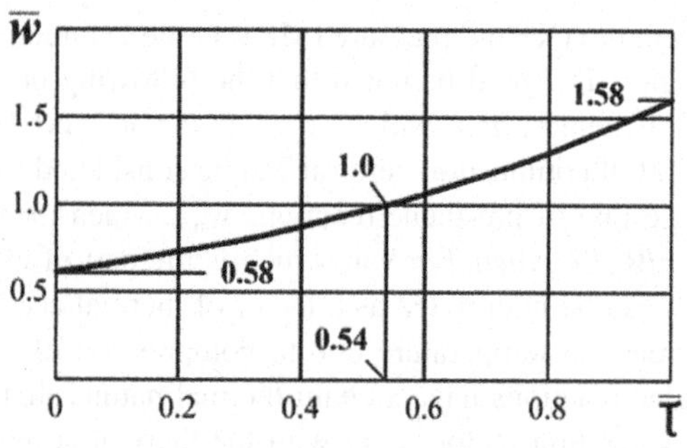

Fig. 2. The dependence of relative rate of pre-flame reactions on relative time.

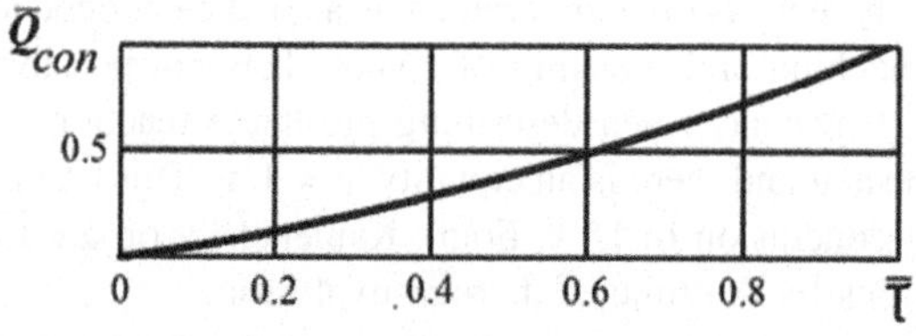

Fig. 3. Change in conditional released heat caused by chain mechanism of pre-flame reactions during spontaneous combustion deferral period.

The concept of current conditional deferral of self-ignition τ_{icon}. This is the deferral determined under the condition of a constant reaction rate that is equal to the current rate. Bearing in mind the aforementioned inversely proportional dependence of combustion deferral on average rate of pre-flame reactions, there can be written:

$$\tau_{icon} / \tau_i = f(\bar{\tau}) = (e - 1) / e^{\bar{\tau}}. \quad (7)$$

If taking into account the obtained dependence and semi-empirical formula (3), then to calculate the absolute value of the current conditional combustion deferral, a function can be proposed that depends on the instantaneous temperature and pressure, as well as on the relative time:

$$\tau_{icon}(T, p, \bar{\tau}) = 3.4 \times 10^{-6} \frac{e - 1}{e^{\bar{\tau}}} \left(\frac{T}{p}\right)^{0.5} e^{\frac{23500}{8.314\,T}}, \quad (8)$$

where T and p are current temperature and pressure in the engine cylinder. It is seen that Eq. (8) contains the values that are inverse to the thermal and chain factors of reactions acceleration.

The calculation made in relation to the nominal mode of a DN 23/30 diesel engine showed that during pre-flame processes, the conditional deferral τ_{icon} reduced, while relative reaction rate $\bar{w}$ 3.2 times increased, respectively. Taking into account quantitative data in Fig. 2, it can be concluded that in this case acceleration of pre-flame reactions depends on the chain mechanism by 85% and on the thermal one only by 15%. In terms of quality, this conclusion is consistent with the well-known statements about the effect of chain reaction mechanism prevailing upon the thermal mechanism during the deferral period of spontaneous combustion.

In order to assess this stage of the study, two notes should be made. First, the temperature and pressure included in Eq. (8) are taken not local, but average in terms of volume of a diesel cylinder. They can be reliably determined by analyzing the indicator diagrams. For switching to the use of local parameters of the working mass, it is advisable to determine the combustion deferral not by the aforementioned traditional method that implies using an indicator diagram, but by the emergence of the first flame sources. Secondly, Eq. (8) can be applied only in order to describe reactions in zones with initial concentration of reagents favorable for the speedy spontaneous combustion.

In order to eliminate the second restriction, which means to enable calculation in all zones of inhomogeneous combustible mixture, it is necessary to have a methodology for accounting local concentrations of reagents when calculating pre-flame reactions. In the future, this will eliminate the need to introduce the conditional combustibility limits into account.

Kinetic models of pre-flame reactions have been developed first of all for homogeneous gas mixtures of simple hydrocarbons and oxygen. For example, the reduced mechanism for methane developed at Chalmers is used, which includes 38 components and 86 reactions [11]. The conditions existing when liquid fuels are injected into a heated air charge of a diesel cylinder are much more complicated. Reactions proceed in a medium with inhomogeneous composition and temperature under the condition of concomitant evaporation of fuel from the surface of droplets in a turbulent flow. If talking about the features of combustion in inhomogeneous fuel stream, mainly qualitative estimates are known. The so-called multi-zone models of combustion mixture formation in piston engines that have become widespread in recent years [12] usually do not have experimental confirmation of distribution of local parameters of working mass, in particular, concentration of mixture components. This circumstance makes the corresponding calculation methods non-universal and limits their scope of application.

In most cases, the authors limit themselves to a shallow description of the influence of temperature and concentration inhomogeneity. It is observed that there is a rich mixture on the axis of a fuel jet, while at the periphery of jet's cross sections there is a diluted mixture. Since the consumption of heat for warming up and evaporating fuel, the temperature of the mixture is significantly lower in areas where fuel concentration is high, while in the diluted zones the temperature is higher. Many authors recall the old facts of increased reaction rates at initial stages of hydrocarbon oxidation in rich mixtures compared to diluted ones. However, in the formation of a general image of pre-flame reactions does not involve the mentioned research results. It is only noted that in enriched zones primary fire sources cannot appear due to reduced temperature. They usually appear in zones corresponding to local coefficients of excessive air for combustion $\alpha_{loc} \approx 1.0$, where the cooling effect of evaporation is less and the specific heat generation per mixture mass unit is maximum.

In the present work, the authors propose a technique for taking into account the heterogeneity of mixture composition when calculating the kinetics of spontaneous combustion. The technique is based on the theory of diffusion combustion of Yu. B. Sviridov. According to this theory, the combustion process is considered as consisting of a number of stages of the conversion of fuel molecules, starting with the formation of heavy radicals and primary oxidation, including the formation of peroxides and aldehydes, as well as the formation of end products CO_2 and H_2O. When changing stages, an increasing amount of oxygen is required to oxidize the intermediate reaction products. Moreover, to ensure the maximum possible reaction rate at each stage, a coefficient of excessive air (oxygen) for combustion that differs from ones of other stages, is required. With some constant α (precisely under such an assumption, pre-flame processes are usually considered), the reaction rate varies from stage to stage. This change for a mixture of hexane and air at various constants α, considered in time, is shown in Fig. 4.

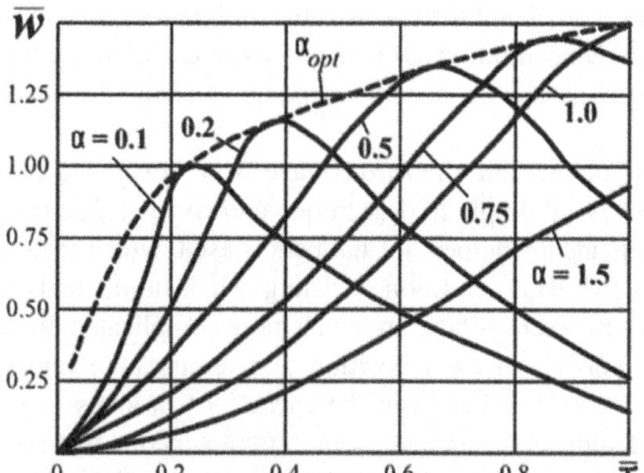

Fig. 4. Changes in relative rate of pre-flame reactions over time for various coefficients of excessive air for combustion.

In the figure, the current velocity w is related to the maximum velocity w_{1max} at the first stage of a reaction. It is worth noting that the early completion of the combustion deferral is ensured when excessive air increases along α_{opt} line.

The rate of reactions is considered the rate of generation of conditional heat, just like in the previous proof. Then

$$\bar{Q}_{con} = \int_0^1 (w/w_{1\,max})\,d\bar{\tau} \tag{9}$$

for some α represents a value that on a certain scale corresponds to the amount of heat in the zone with this α, released at time point $\bar{\tau}_{fin} = 1$. Dividing the calculated integral by time $\bar{\tau}_{fin}$ gives the average relative reaction rate with a constant concentration of reagents in the considered time interval: $w_{m\,(\alpha=const)} / w_{1\,max} = \bar{Q}_{con} / \bar{\tau}_{fin}$. After performing this operation with different values, the dependence of average relative rate on can be plotted as follows:

$$w_{m\,(\alpha=\mathrm{const})} / w_{1\,\mathrm{max}} = f_1(\alpha)\,. \tag{10}$$

The air-fuel mixture is assumed to consist of a number of local zones with $\alpha_{loc} = \mathrm{const}$. In expression (10) the values $w_{m(\alpha=\mathrm{const})}$ are referred not to w_{1max}, but to the average (during the period of combustion deferral τ_i) rate of pre-flame reactions w_{im}, which is considered known taking into account the considerations above. Then the dependence of the average local relative reaction rate on α_{loc} is obtained:

$$\bar{w}_{loc} = w_{m\,(\alpha=\mathrm{const})} / w_{im} = f_2(\alpha_{loc})\,. \tag{11}$$

Dependence (8) for hexane is presented in the Table 1.

Table 1. Values of average local relative rate of pre-flame reactions at different local coefficients of excessive air for combustion.

α_{loc}	0.10	0.20	0.50	0.75	1.00	1.50
$\bar{w}_{loc}$	0.60	0.76	0.99	0.93	0.76	0.44

It turned out that the maximum average rate $\bar{w}_{loc}$ corresponds to $\alpha_{loc} = 0.55$. According to Bon's and Hill's experiments for ethane, a similar point corresponds to a close value of $\alpha_{loc} = 0.6$. Some authors note that in mixtures of air and saturated hydrocarbons of a homologous series, which include ethane C_2H_6 and hexane C_6H_{14}, spontaneous combustion most likely occurs exactly at these values of α. The coefficient of excessive air equaling 0.55 will be considered optimal for these hydrocarbons.

Many authors state that α_{opt} belongs to the value range of 0.8 – 0.9 for diesel fuel. The nature of dependence $\bar{w}_{loc} = f(\alpha_{loc})$ for diesel fuel is assumed to remain the same as for hexane. At this stage of the work, when it is fundamentally important only to determine zones of combustible mixture, in which flame sources appear, and when the accurate estimate of local reaction rates is significant only in close vicinity of point α_{opt}, this assumption can be considered acceptable. In this case, the indicated dependence for diesel fuel can be obtained by multiplying α_{loc} values in the table by a certain coefficient, at which α_{opt} is in the range of 0.8–0.9. The result of the described transformation is shown in Fig. 5 as a row of points. Points are approximated by the dependency

$$\bar{w}_{loc} = \frac{0.005 + 2.73\,\alpha_{loc}}{1 + 0.37\,\alpha_{loc} + 1.4\,\alpha_{loc}^2}. \tag{12}$$

The obtained dependence (12) of the local average reaction rate on the local coefficient of excessive air can be included in formula (8). As a result, the equation of duration of the current local conditional period of combustion deferral is composed

$$\tau_{i\,loc}(T, p, \alpha_{loc}, \bar{\tau}) = A_1 \frac{1}{\bar{w}_{loc}} \left(\frac{e-1}{e^{\bar{\tau}}}\right) \left(\frac{T}{p}\right)^n e^{\frac{E}{R_\mu T}}, \tag{13}$$

where A_1, n are constants from formula (8).

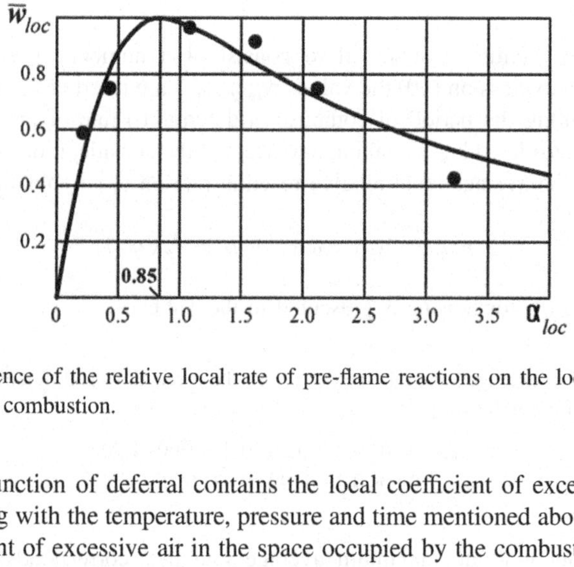

Fig. 5. Dependence of the relative local rate of pre-flame reactions on the local coefficient of excessive air for combustion.

Now the function of deferral contains the local coefficient of excessive air as an argument, along with the temperature, pressure and time mentioned above. The change in the coefficient of excessive air in the space occupied by the combustible mixture is determined by the calculation of the mixture formation.

The equation of current local conditional heat generated can be written in general form as follows:

$$\bar{Q}_{con}(T, p, \alpha_{loc}, \bar{\tau}) = \tau_i \int_0^{\bar{\tau}} \frac{1}{\tau_{i\,con}(T, p, \alpha_{loc}, \bar{\tau})} d\bar{\tau}. \tag{14}$$

During the calculation performed by Eq. (14), the moment when the value $\bar{Q}_{con}$ reaches 1 is coincident with the end of local combustion deferral period.

The proposed kinetic model of pre-flame processes represents the theoretical basis for the calculation method applied for defining spatial and temporal characteristics of the charge of the engine cylinder during the deferral period of spontaneous combustion. Application of this technique in practice with taking into account thermal and chain accelerations of pre-flame reactions, as well as their dependence on the local coefficient of excessive air, enable determining the coordinates of zones of combustible mixture and time points of flame emerged in those zones. This allows calculating the subsequent spreading of flame and, in the long-term perspective, taking into account the influence of a stage of completion of local pre-flame transformations on a fuel combustion process.

4 Discussion

The process of spontaneous combustion of pilot fuel in a dual-fuel engine is much more complicated than in a diesel engine. Extra difficulties can be explained by the fact that working mass contains gas fuel, besides air, liquid and steamy diesel fuel. Moreover, the share of gas fuel varies widely depending on engine load. The most significant difficulty is the risk of detonation during combustion of gas fuels [1].

Both presence and absence of detonation depend on many factors. Dependence of the presence of stable ignition and detonation-free burning on the most important factor – the coefficient of excessive air for combustion α, published by Wärtsilä, is shown in Fig. 6. The figure shows some characteristics of the zones of the combustion process in a gas ICE at various levels of average effective pressure p_{me}, MPa in it.

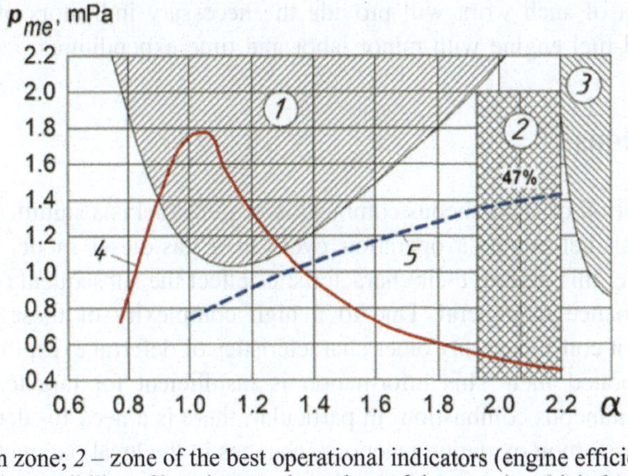

1 – detonation zone; *2* – zone of the best operational indicators (engine efficiency is 47%); *3* – zone of impossibility of burning; *4* – boundary of the zone in which the emission of NO_x does not exceed 1 g/(kW·h); *5* – the nature of the change in engine efficiency

Fig. 6. The dependence of some characteristics and location of the characteristic zones of the combustion process on the coefficient of excessive air for combustion in a gas ICE (according to Wärtsilä: http://mirmarine.net/dvs/toplivnye-sistemy/toplivnaya-apparatura-gazovykh-i-gazodizelnykh-sudovykh-dvigatelej/411-konvertirovanie-dizelej-v-dvigateli-s-vneshnim-smeseobrazovaniem-i-iskrovym-zazhiganiem).

The figure shows that detonation-free, stable and fairly efficient engine operation at a current level of average effective pressure is observed only at a relatively narrow range of $1.9 \leq \alpha \leq 2.2$. Then α is less than 1.9, the engine operates with detonation, and when is more than 2.2, there is a risk of combustion miss. When a dual-fuel engine operates in the diesel mode, the similar situation in terms of quality occurs.

In view of the foregoing, the supply of pilot fuel and its spontaneous combustion should be organized in such a way as on the one hand to ensure the necessary combustion under conditions of a significant change in α over a wide range of engine loads, and, on the other hand, to carry out the working process with the smallest possible cyclic dose of diesel fuel. This may mean that there is a need to have a sufficiently powerful combustion source occupying the maximum possible volume of a combustion chamber of an engine and possessing the necessary quality of fuel atomization. Therefore, the design and operating parameters of fuel equipment designed to supply pilot fuel (the number and diameter of nozzle openings, fuel injection pressure, range of fuel jets, etc.) should be determined from the condition of the maximum possible volume of fuel jets referring to the moment of the end of deferral period for spontaneous combustion of pilot fuel. In this case, sufficient size of a spontaneous combustion zone can be achieved, providing all necessary qualities of the combustion process.

To give a preliminary estimate of spontaneous combustion of pilot fuel, the proposed methodology for calculating pre-flame processes can be used. According to the results of experiments on a certain engine (according to the results of indexing the operation cycle), values of constants can be clarified in the initial formula (3), and a product of concentrations of gas and vapor contained in diesel fuels with the corresponding exponents can be introduced into the pre-exponential factor of this formula. The specified procedure can be carried out for several engine loads, if it is needed. The implementation of such work will provide the necessary indicators of the operation cycle of a dual-fuel engine with minor labor and time expenditures.

5 Conclusions

The deferral period of spontaneous combustion of pilot fuel is a significantly important stage of a dual-fuel engine's operation cycle in a gas-diesel mode. The processes occurring during this period, their characteristics affect the subsequent combustion and engine performance in general. Due to a high complexity of these processes, the authors have not considered any other characteristics of deferral except its duration and mass of evaporated fuel. This information is insufficient for targeted affecting the process of spontaneous combustion. In particular, there is a need for data on pre-flame processes, hydrocarbon oxidation reactions, changes in the local parameters of working mass in time and engine's combustion chamber space.

The present work has theoretically substantiated the possibility of a quantitative assessment of the chain-thermal mechanism of chemical reactions of oxidation of hydrocarbon fuel molecules, acceleration of pre-flame processes that lead to rapid explosive fuel oxidation, which is accompanied by flame emergence. Calculations for specific diesel conditions showed that acceleration of pre-flame reactions depends on the chain mechanism by 85% while on the thermal one – only by 15%. The adopted local approach to description of processes of mixture formation and combustion allows calculating local deferrals of spontaneous combustion. To develop a methodology for such a calculation, the authors have solved the problem of taking into account changes in the rate of pre-flame reactions in time and space of a combustion chamber of a diesel engine, which is occupied by a heterogeneous concentration of a combustible fuel and oxygen mixture. Description of characteristics of pre-flame reactions acceleration may allow assessing, to what extent a mixture is ready for rapid oxidation, which in turn will allow calculating the propagation speed of flame more accurately. All the main factors affecting the spontaneous combustion process are taken into account: local temperature, pressure and local coefficient of excessive air for combustion.

When using the proposed approach to calculation of the process of spontaneous combustion of pilot fuel, it one should keep in mind that in the working mass of a dual-fuel engine contains gas fuel along with air, liquid and steamy diesel fuel. Therefore, there is an additional need to take into account current concentrations of these components. The main expected result of a calculation is the reduction of detonation likelihood when burning gas fuel.

With allowances made for these theoretical basis of modeling the process of spontaneous combustion in a dual-fuel engine, it is necessary to develop a methodology for its calculation. The constants in formulas should be clarified by the results of indexing the working cycle of a particular type of engine in various operation modes. This will provide an opportunity to improve the quality of simulating combustion process and ensure the required engine performance.

References

1. Gavrilov, V., Medvedev, V., Bogachev, D.: Improvement of fuel injection process in dual-fuel marine engine. In: Murgul, V., Pasetti, M. (eds.) EMMFT-2018 2018. AISC, vol. 982, pp. 392–399. Springer, Cham (2020). https://doi.org/10.1007/978-3-030-19756-8_37
2. Kavtaradze, R.Z., Zeilinger, K., Zitzler, G.: Ignition delay in a diesel engine utilizing different fuels. High Temp. **43**(6), 951–960 (2005). https://doi.org/10.1007/s10740-005-0143-z
3. Kuleshov, A.S.: Multi-zone DI diesel spray combustion model and its application for matching the injector design with piston bowl shape. SAE Technical Papers (2007). https://doi.org/10.4271/2007-01-1908
4. Lakshminarayanan, P.A., Aghav, Y.V.: Ignition Delay in a Diesel Engine. Modelling Diesel Combustion, Mechanical Engineering Series, pp. 59–60. Springer (2010). https://doi.org/10.1007/978-90-481-3885-2_5
5. Shankhdhar, V., Kumar, T.: Theoretical study of the effects of ignition delay on the performance of DI diesel engine. Int. J. Res. (IJR) **1**(7), 230–236 (2014)
6. Mikulski, M., Piętak, A.: On the modeling of pilot dose ignition delay in a dual-fuel, selfignition engine. https://www.researchgate.net/publication/270509935. 21 Jan 2020
7. Senachina, A.P., Korzhavinb, A.A., Senachina, P.K.: Simulation of fuel ignition delay in diesel engines with various fuel feeding systems. Proc. Eng. **150**, 190–203 (2016). https://doi.org/10.1016/j.proeng.2016.06.746

8. Troshin, K.Y., Nikitin, A.V., Borisov, A.A., Arutyunov, V.S.: Low-temperature autoignition of binary mixtures of methane with C_3–C_5alkanes. Combust. Explosion Shock Waves **52**(4), 386–393 (2016). https://doi.org/10.1134/S001050821604002X
9. Troshin. K.Ya., Nikitin, A.V., Belyayev, A.A., Arutyunov, A.V., Kiryushin, A.A., Arutyunov, V.S.: Eksperimental'noye opredeleniye zaderzhki samovosplameneniya smesey metana s legkimi alkanami. Fizika goreniya i vzryva **5**, 17–24 (2019)
10. Lata, D.B., Misra, A.: Analysis of ignition delay period of a dual fuel diesel engine with hydrogen and LPG as secondary fuels. Int. J. Hydrogen Energy **36**, 3746–3756 (2011). https://doi.org/10.1016/j.ijhydene.2010.12.075
11. Chomiak, J., Liljenfekit, G.: Performance analysis of a steam injected diesel (STID) engine. In: 23rd CIMAC World Congress of Combustion Engine Technology for Ship Propulsion, Power Generation, Rail Traction, 7–10 May 2001, vol. 2, pp. 372–384. Hamburg (2001)
12. Kuleshov, A., Kozlov, A., Mahkamov, K.: Self-Ignition Delay Prediction in PCCI Direct Injection Diesel Engines Using Multi-Zone Spray Combustion Model and Detailed Chemistry. SAE Technical Paper 2010–01-1960, pp. 1–16 (2010). https://doi.org/10.4271/2010-01-1960

Methods and Algorithms for Controlling Cascade Frequency Converter with High-Quality of Synthesized Voltage

Fedor Gelver, Igor Belousov, and Aleksandr Saushev(✉)

Admiral Makarov State University of Maritime and Inland Shipping, Dvinskaya Str., 5/7, Saint Petersburg 198035, Russia
saushev@bk.ru

Abstract. The paper studies possible circuits and topology of cascade frequency converters with the use of cycloconverters (direct) as well as indirect 2-, 3- and multilevel converters of frequency. There are presented and described ways to improve the quality of the output voltage of a cascade frequency converter. Mathematical description and control algorithm of the cells of a cascade frequency converter, which can produce the required voltage. There were compared values of various methods for quality improvement of the output voltage of a cascade frequency converter. The table of number of possible voltage levels that can be synthesized depending on the topology of the converter's power part, the topology of the location of unit cells and control algorithms is presented. The histogram of the dependence of the number of voltage levels on the construction scheme of the unit cell, the number of cells, and cell control algorithms is presented. The results of mathematical modeling of the output voltage of the converter are given. There are considered various options for the synthesis of the converter output voltage based on the addition and subtraction of voltages of two 3-level cells with differentiated supply voltage. The structure of constructing unit cells of cascade frequency converters with differentiated level of supply that includes reversible electric converters is proposed. There were compared values of various methods for quality improvement of the phase voltage of cascade frequency converters depending on the amount and topology of the cell at every phase of the converter.

Keywords: Cascade frequency converter · Multilevel voltage source inverter · Voltage quality · Electromagnetic compatibility · Single-Phase frequency converter · Voltage level · Pulse-Width modulation

1 Introduction

Nowadays, electric energy converters, that can generate output voltage close to the sine-wave form with the required quality parameters, are in great demand in the world electric power industry [1–5]. Thus, the priority objective in designing electric energy converters and developing algorithms for their control is to ensure electromagnetic compatibility of the load with supply grid as well as fulfilling the required quality parameters of the converted energy. Generally, the quality of the electric energy is

V. Murgul and V. Pukhkal (Eds.): EMMFT 2019, AISC 1258, pp. 375–387, 2021.
https://doi.org/10.1007/978-3-030-57450-5_32

ensured by the supplier, subject to the compliance with the restrictions on the electric load. At the same time, the quality of the voltage, that is synthesized by electrical converters, should be backed with component base, circuitry design and control algorithms of such a converter.

A decrease in the quality of electrical energy after conversion results in a change of the operating mode, degradation of energy characteristics and performance, increase noise and vibration, a shorter service life of the equipment as well as a higher risk of crashes and accidents. As the installed capacity of energy units grows, the relevance of the above issues grows and special attention is required.

A fairly large number of various circuitry solutions is known to construct an electrical converter [6, 7]. Each structure has its own upsides and drawbacks. To synthesize high quality output voltage, either cascade or multilevel voltage source inverter based frequency converters are usually used. It is important to bear in mind when comparing these converters, that multilevel voltage source inverter based one has a more complex topology of the power part and less design workability. Therefore, circuits with multilevel inverters are used in frequency converters of low and medium power. Whereas cascade structures are implemented in high-powered devices [8–12].

The paper discusses the structures and control algorithms of a cascade frequency converter (CFC). The principle of operation of such a converter is based on a series type connection of unit cells (single-phase frequency converters at each of the phases of the CFC) (Fig. 1). Series connection of the unit cells can improve the quality of the synthesized voltage by increasing the number of levels. It also increases the maximum voltage value at the output of the CFC [13]. The operating principle of cascade frequency converters is broken down in lots of sources [6, 7, 14–19] and needs no additional comments.

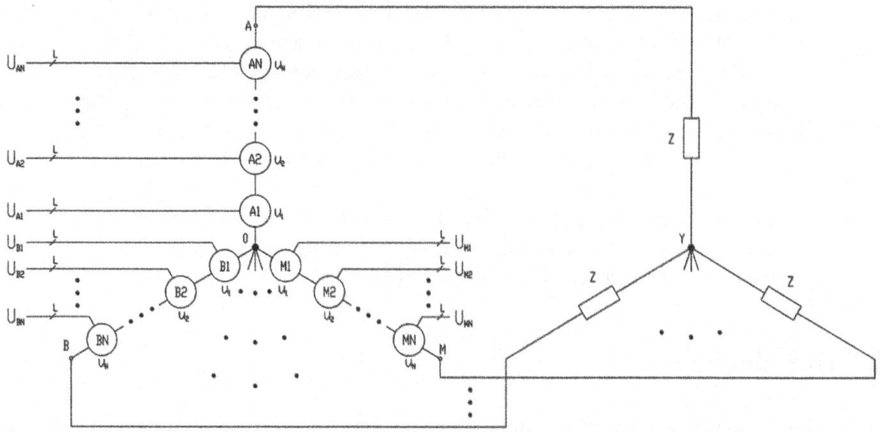

Fig. 1. Connection layout of the unit cells of a cascade frequency converter.

2 Materials and Methods

Unit cells are single-phase frequency converters that can be based on cycloconverters (direct) (Fig. 2, A) [14] and indirect frequency converters with voltage source inverters (Fig. 2, B, C, D) [20]. However, direct frequency converters are not popular due to the constant switching overvoltages. Though, they could potentially reduce weight and size characteristics, improve the energy and electromagnetic compatibility of the converter with the supply mains and the load. Cycloconverters are also able to carry out two-way energy exchange [21].

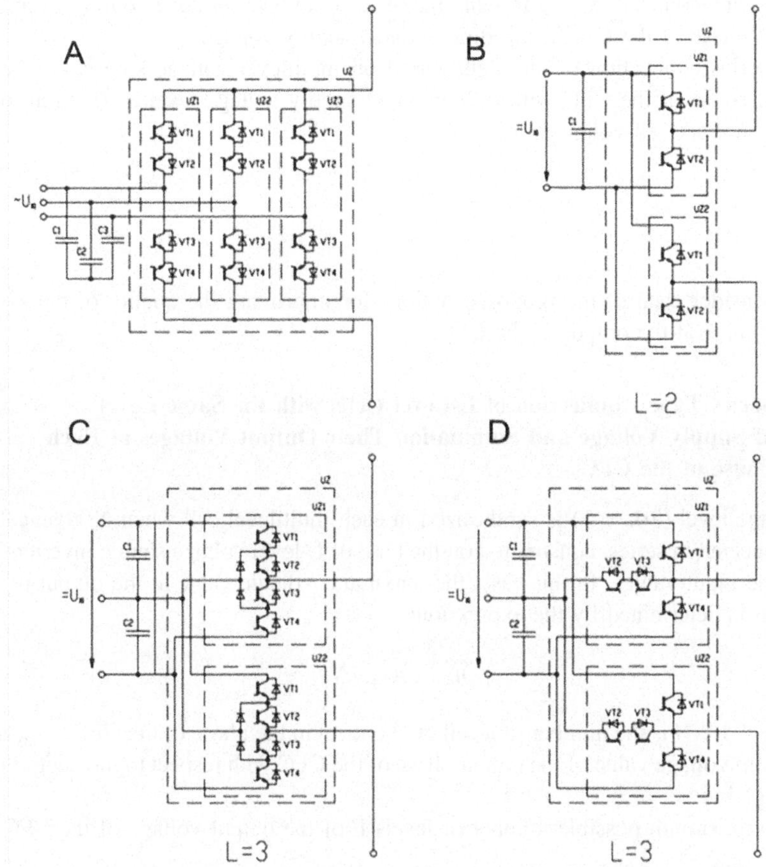

A - a direct frequency converter, cycloconverter; B - a 2-level voltage source inverter; C - a neutral point clamped 3-level voltage source inverter; D - a 3-level voltage source inverter with a T-type transistor connection

Fig. 2. Arrangement of unit cells in a cascade frequency converter based on.

The paper describes the circuitry and control algorithms of the CFC with cells based on voltage source inverters. At the same time, the proposed algorithms can also be used for the CFC with cycloconverter circuitwise cells.

The architecture of CFC has the broadest potential in terms of improving the quality of the synthesized voltage, improving the energy and electromagnetic compatibility of the CFC with the supply mains and load, etc.

The possibility to improve the output voltage will be studied in more detail. This goal can be achieved by the following structural arrangements:

by series connection of unit cells based on multilevel voltage source inverters at each phase of the CFC;
by series connection of unit cells based on multilevel voltage source inverters at each phase of the CFC using differentiated supply voltage;
by series connection of unit cells based on multilevel voltage source inverters at each phase of the CFC with differentiated supply voltage level and option to add and subtract cell voltages.

3 Results

Let's consider each of the proposed options for improving the quality of the synthesized voltage at the output of the CFC.

3.1 Series Type Connection of L-Level Cells with the Same Level of Supply Voltage and Summation Their Output Voltages at Each Phase of the CFC

A voltage level (2•L − 1) is synthesized in each multilevel cell when N identical cell are connected in series, constructed on the base of L-level voltage source inverters with the same input values. In this case, the maximum voltage value at the output of each unit cell is determined by the expression:

$$u_i = u_{\max}/N, \quad (1)$$

where i = 1…N is the number of a cell on the level in the phase of the CFC, u_{max} is the maximum voltage value of the output phase of the CFC with respect to the neutral point "0" (Fig. 1).

The maximum possible number of levels F of the output voltage of the CFC is:

$$F = 2(L - 1)N + 1. \quad (2)$$

According to the obtained dependence, the increase in the number of levels realized by multilevel voltage source inverter of each cell increases the quality of synthesized voltage. This also allows to use low-voltage semiconductor cells. It should be noted that this complicates the circuit implementation of such a unit cell. Moreover, multilevel power supply for the operation of L-level voltage inverter is required.

In order to prevent the maximum permissible value of voltage at the DC link capacitor of the cell from being exceeded, the condition must be fulfilled:

$$C > \frac{i_m \cdot (\sin(\varphi) - \cos(\varphi) \cdot \varphi)}{2 \cdot \omega \cdot \left(u_{\lim} - \frac{u_{\max}}{(L-1) \cdot N}\right)} \tag{3}$$

where C is the capacitance of the capacitor in the cell, u_{lim} is the maximum permissible capacitor voltage, i_m is the peak amplitude of the phase current, φ is the maximum angle between the phase current and voltage, ω is the minimum angular frequency of the phase voltage.

Pulse-width modulation (PWM) is implemented in all cells at each phase of the CFC in sequence and uniformly under the same level of supply voltage of the unit cells. This provides an even distribution of energy losses between the phase cells.

3.2 Series Type Connection of L-Level Cells with Differentiated Level of Supply Voltage and Summation of the Output Voltages at Each Phase of the CFC

Such an architecture of the CFC significantly complicates both the unit cell of the CFC and the power supply. However, this structure and differentiated cell supply can significantly increase the number of levels of instantly synthesized voltage, and therefore, improve its quality.

Let's consider the methodology for selecting cell voltages in the phase of the CFC and the algorithm for the formation of its phase voltage.

Let each phase of the CFC contain N cells of L-level, while a (2•L − 1) voltage level is synthesized in each multilevel cell. Then the minimum level of the voltage to modulus, that is synthesized by i-th cell, should be equal to:

$$u_{\max} \cdot L^{i-1} / (L^N - 1), \tag{4}$$

where u_{max} is the maximum value of the phase voltage of the cascade converter relative to the common point of the cascades ("0" of the converter).

In this case, the cell must generate voltage at its output within the following values:

$$u \in \left[-\frac{u_{\max} \cdot L^{i-1}}{L^N - 1} \cdot (L-1); \frac{u_{\max} \cdot L^{i-1}}{L^N - 1} \cdot (L-1) \right]. \tag{5}$$

To determine the voltage values that need to be generated by each multilevel cell, it is proposed to use the following algorithm.

Let us introduce the switching variable - integer z:

$$z = \left\lfloor \frac{|u|}{u_{\max}} \cdot (L^N - 1) \right\rfloor, \tag{6}$$

where $\lfloor x \rfloor$ is the integer part of the number, u is the required instantaneous value of the phase voltage of the cascade converter relative to the common point of the cascades ("0" of the converter).

Let us assume z in an L-base positional number system with the number of digits equal to N: $z = \sum_{i=0}^{i=N-1} b_i \cdot L^i$, where b_i are non-negative integers less than L.

To find the b_i coefficients, the following recurrence formulas are used:

$$b_i = d_i - d_{i+1} \cdot L;\ d_0 = z;\ d_{i+1} = \lfloor d_i/L \rfloor, i = \overline{0, (N-1)}. \tag{7}$$

If $z = L^N - 1$, then we take $c_i = L - 1$, where i varies over the range from 0 to $(N - 1)$. Otherwise, assume the value z + 1 in a L-base positional number system with the number of digits equal to N:

$$z + 1 = \sum_{i=0}^{i=N-1} c_i \cdot L^i, \tag{8}$$

where c_i are non-negative integers less than L.

Then, in order to form the instantaneous value of phase voltage of the cascade converter relative to the common point of the cascades ("0" of the converter), the i-th cells should simultaneously synthesize voltage $uh_i = u_{\max} \cdot c_{i-1} \cdot L^{i-1} \cdot \mathrm{sign}(u)/(L^N - 1)$ within t_i and $uh_i = u_{\max} \cdot c_{i-1} \cdot L^{i-1} \cdot \mathrm{sign}(u)/(L^N - 1)$ within t_h during PWM period T_{PWM}, where

$$t_l = \left(1 - \left\{\frac{|u|}{u_{\max}} \cdot (L^N - 1)\right\}\right) \cdot T_{PWM};\ t_h = \left\{\frac{|u|}{u_{\max}} \cdot (L^N - 1)\right\} \cdot T_{PWM} \tag{9}$$

T_{PWM} – pulse-width modulation period; {x} is fractional part of x; sign(x) is sign function.

Consequently, voltage u will be generated at the phase of the CFC, that equals to:

$$\begin{aligned} u &= \sum_{i=1}^{N} \left(ul_i \cdot \frac{t_l}{T_{PWM}} + uh_i \cdot \frac{t_h}{T_{PWM}}\right) \\ &= \frac{u_{\max}}{L^N - 1} \cdot \left(\left(1 - \left\{\frac{|u|}{u_{\max}} \cdot (L^N - 1)\right\}\right) \cdot \sum_{i=1}^{N} (b_{i-1} \cdot L^{i-1}) + \left\{\frac{|u|}{u_{\max}} \cdot (L^N - 1)\right\} \cdot \sum_{i=1}^{N} (c_{i-1} \cdot L^{i-1})\right) \cdot \mathrm{sign}(u) \\ &= \frac{u_{\max}}{L^N - 1} \cdot \left(\left\lfloor \frac{|u|}{u_{\max}} \cdot (L^N - 1) \right\rfloor + \left\{\frac{|u|}{u_{\max}} \cdot (L^N - 1)\right\}\right) \cdot \mathrm{sign}(u). \end{aligned} \tag{10}$$

The number of levels F, that are formed at the phase of CFC, consisting of N cells of L-level, equals to:

$$F = 2 \cdot L^N - 1. \tag{11}$$

If the generated phase voltage and load current of the CFC have different signs, then the energy will be transferred from the load to the constant voltage source of phase cells of the CFC. If these sources cannot accept the energy, then it is directed to the capacitor. In order to prevent exceeding the maximum permissible voltage value on the capacitor of the DC link of the unit cell, for all cells that satisfy: $\sin(\varphi) > (L^{i-1} - 1)/(L^N - 1)$, the following approximate condition should be right:

$$C_i > \frac{i_m \cdot (1 - \cos(\varphi))}{2.\omega \cdot \left(u_{\lim} - \frac{u_{\max} \cdot L^{i-1}}{L^N - 1}\right)}, \tag{12}$$

where C_i is the capacitance of the capacitor in the i-th cell, u_{lim} is the maximum permissible capacitor voltage, i_m is the peak amplitude of the phase current, φ is the maximum angle between the phase current and voltage, ω is the minimum angular frequency of the phase voltage.

The following could be done in case this condition is not met. An N + 1-th cell should be added for every phase of the CFC, that is capable to receive the recuperative energy from the load and generate either of the set voltage values: $-u_{max}$; 0; u_{max}.

The following formula determines the voltage generated by the cell:

$$u_{N+1} = \begin{cases} 0, \text{ if } u \cdot i \geq 0 \\ u_{\max} \cdot \operatorname{sign}(u), \text{ if } u \cdot i < 0 \end{cases}. \tag{13}$$

The total voltage value generated by cells with numbers from 1 to N equals to: $u - u_{N+1}$.

This voltage is generated according to the above algorithm.

The use of differentiated supply voltages of the unit cells of the CFC allows to significantly increase the number of levels of the output voltage of the CFC. Therefore, it improves the quality of the voltage and the electromagnetic compatibility of the CFC with the load.

3.3 Series Type Connection of L-Level Cells with Differentiated Supply Voltage Based on Addition and Subtraction of the Output Voltage of the Cells at Each Phase of the CFC

These topology and arrangement of the power supply for the CFC are the most difficult to design both the unit cells and the entire CFC as a whole. The architecture of the power part of the cell implies the use of a reversible electrical converter capable to regulate the energy flow in both directions.

Let us consider the method of selecting cell voltages at the phase of the CFC and the algorithm for generating the phase voltage of the CFC. In this case, it is assumed that each phase of the CFC contains N cells of L-level, while (2·L − 1) voltage levels are implemented in every multilevel cell. Then the minimum level of the voltage to modulus, that is synthesized by i-th cell, should be equal to:

$$\frac{2 \cdot u_{\max} \cdot (2 \cdot L - 1)^{i-1}}{(2 \cdot L - 1)^N - 1}, \tag{14}$$

where u_{max} is the maximum value of the phase voltage of the cascade converter relative to the common point of the cascades ("0" of the converter).

The cell must generate output voltage u within the following values:

$$u \in \left[-\frac{u_{\max} \cdot 2 \cdot (2 \cdot L - 1)^{i-1} \cdot (L - 1)}{(2 \cdot L - 1)^N - 1}; \frac{u_{\max} \cdot 2 \cdot (2 \cdot L - 1)^{i-1} \cdot (L - 1)}{(2 \cdot L - 1)^N - 1} \right] \quad (15)$$

To determine the voltage values that need to be generated by each multilevel cell, it is proposed to use the following algorithm.

Let us introduce the switching variable - integer z:

$$z = \left\lfloor \left(\frac{u}{u_{\max}} + 1 \right) \cdot \frac{(2 \cdot L - 1)^N - 1}{2} \right\rfloor, \quad (16)$$

where $\lfloor x \rfloor$ is the integer part of the number, u is the required instantaneous value of the phase voltage of the cascade converter relative to the common point of the cascades ("0" of the converter).

Let us assume z in an (2 L − 1)-base positional number system with the number of digits equal to N: $z = \sum_{i=0}^{i=N-1} b_i \cdot L^i$, where b_i are non-negative integers less than (2 L − 1).

To find the bi coefficients, the following recurrence formulas are used:

$$b_i = d_i - d_{i+1} \cdot (2 \cdot L - 1);\ d_0 = z;\ d_{i+1} = \lfloor d_i / (2 \cdot L - 1) \rfloor, i = \overline{0, (N - 1)}. \quad (17)$$

If z = $(2 L - 1)^N - 1$, then we take c_i = 2 L − 1, where i varies over the range from 0 to (N-1). Otherwise, assume the value (z + 1) in a (2 L − 1)-base positional number system with the number of digits equal to N:

$$z + 1 = \sum_{i=0}^{i=N-1} c_i \cdot (2.L - 1)^i, \quad (18)$$

Then, in order to form the instantaneous value of phase voltage of the cascade converter relative to the common point of the cascades ("0" of the converter), the i-th cells during PWM period T_{PWM} should simultaneously synthesize voltage

$$ul_i = u_{\max} \cdot \left(\frac{2 \cdot b_{i-1} \cdot (2 \cdot L - 1)^{i-1}}{(2 \cdot L - 1)^N - 1} - 1 \right) \text{ and}$$
$$uh_i = u_{\max} \cdot \left(\frac{2 \cdot c_{i-1} \cdot (2 \cdot L - 1)^{i-1}}{(2 \cdot L - 1)^N - 1} - 1 \right) \quad (19)$$

within:

$$\begin{aligned} t_l &= \left(1 - \left\{\left(\frac{u}{u_{\max}} + 1\right) \cdot \frac{(2 \cdot L - 1)^N - 1}{2}\right\}\right) \cdot T_{PWM}, \\ t_h &= \left\{\left(\frac{u}{u_{\max}} + 1\right) \cdot \frac{(2 \cdot L - 1)^N - 1}{2}\right\} \cdot T_{PWM}, \end{aligned} \tag{20}$$

where T_{PWM} – pulse-width modulation period, {x} is fractional part of x.

Consequently, voltage u will be generated at the phase of the CFC, that equals to:

$$\begin{aligned} u &= \sum_{i=1}^{N} \left(ul_i \cdot \frac{t_l}{T_{PWM}} + uh_i \cdot \frac{t_h}{T_{PWM}}\right) \\ &= u_{\max} \cdot \left(\frac{2}{(2 \cdot L - 1)^N - 1} \cdot \left(\begin{array}{l} \left(1 - \left\{\left(\frac{u}{u_{\max}} + 1\right) \cdot \frac{(2 \cdot L - 1)^N - 1}{2}\right\}\right) \cdot \sum_{i=1}^{N}\left(b_{i-1} \cdot (2 \cdot L - 1)^{i-1}\right) \\ + \left\{\left(\frac{u}{u_{\max}} + 1\right) \cdot \frac{(2 \cdot L - 1)^N - 1}{2}\right\} \cdot \sum_{i=1}^{N}\left(c_{i-1} \cdot (2 \cdot L - 1)^{i-1}\right) \end{array}\right) - 1\right) \\ &= u_{\max} \cdot \left(\frac{2}{(2 \cdot L - 1)^N - 1} \cdot \left(\left\lfloor\left(\frac{u}{u_{\max}} + 1\right) \cdot \frac{(2 \cdot L - 1)^N - 1}{2}\right\rfloor + \left\{\left(\frac{u}{u_{\max}} + 1\right) \cdot \frac{(2 \cdot L - 1)^N - 1}{2}\right\}\right) - 1\right). \end{aligned} \tag{21}$$

The number of levels F, that are formed at the phase of CFC, consisting of N cells of L-level, equals to: $F = (2\,L - 1)^N$.

4 Discussion

The proposed topology of unit cells of the CFC with differentiated supply levels when using reversible electrical converters, allows both the summation of the output voltages of the unit cells and their subtraction. As a result, the quality of the synthesized voltage is significantly improved. The form of the voltage can be maximum close to the set one. The proposed options for arrangement of unit cells and their differentiated power supply allow to get the maximum possible number of voltage levels, and accordingly, with optimal control algorithms and other things being equal, achieve the extremum of the quality index of the synthesized voltage.

Figure 3 shows a quantitative comparison of various options for improving the quality of the synthesized phase voltage of the CFC depending on the number (N = 1…3) and topology (L = 2 or 3) of the cells at each phase of the CFC.

Table 1 presents the number of levels of voltage at the output of the phase of the CFC.

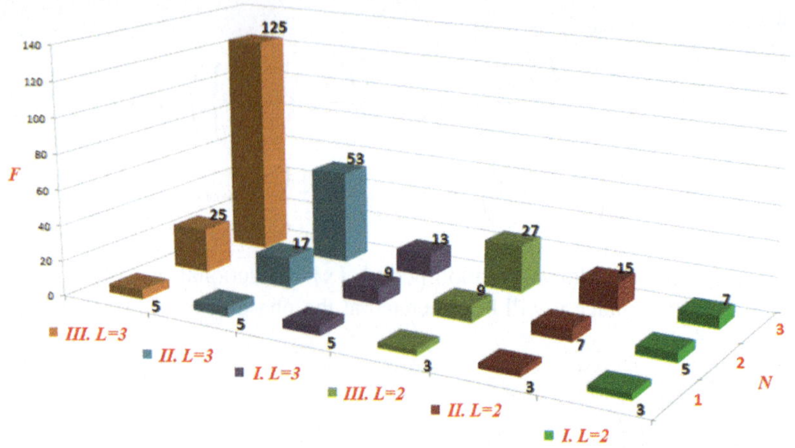

Fig. 3. Quantitative comparison of proposed options for the quality improvement of the synthesized phase voltage of the CFC depending on the number of cells at each phase of the CFC.

Table 1. Number of levels of voltage at the output of the phase of the CFC.

Number of voltage levels generated by a unit cell	Way to improve the quality of the output voltage of CFC	Number of cells at a phase of CFC				
		1	2	3	...	N
2	3.1	3	5	7	...	$2 \cdot N + 1$
	3.2	3	7	15	...	$2^{N+1} - 1$
	3.3	3	9	27	...	3^N
3	3.1	5	9	13	...	$4 \cdot N + 1$
	3.2	5	17	53	...	$2 \cdot 3^N - 1$
	3.3	5	25	125	...	5^N

Table 2 presents the synthesis data on the phase voltage at the output of the CFC, consisting of two 3-level cells. Cells have differentiated supply voltages. In this case, only summation of cell voltages is allowed.

Table 2. Synthesis data on the output voltage of the phase of CFC by summation of two 3-level cells with differentiated power supply.

Level number of the output voltage of the phase of CFC	Voltage		
	Output voltage of the phase of CFC	Output voltage of the first cell	Output voltage of the second cell
−8	$-U_{max}$	$-U_{max}/4$	$-3/4 \cdot U_{max}$
−7	$-7/8 \cdot U_{max}$	$-U_{max}/8$	$-3/4 \cdot U_{max}$
−6	$-3/4 \cdot U_{max}$	0	$-3/4 \cdot U_{max}$
−5	$-5/8 \cdot U_{max}$	$-U_{max}/4$	$-3/8 \cdot U_{max}$
−4	$-U_{max}/2$	$-U_{max}/8$	$-3/8 \cdot U_{max}$
−3	$-3/8 \cdot U_{max}$	0	$-3/8 \cdot U_{max}$
−2	$-U_{max}/4$	$-U_{max}/4$	0
−1	$-U_{max}/8$	$-U_{max}/8$	0
0	0	0	0
1	$U_{max}/8$	$U_{max}/8$	0
2	$U_{max}/4$	$U_{max}/4$	0
3	$3/8 \cdot U_{max}$	0	$3/8 \cdot U_{max}$
4	$U_{max}/2$	$U_{max}/8$	$3/8 \cdot U_{max}$
5	$5/8 \cdot U_{max}$	$U_{max}/4$	$3/8 \cdot U_{max}$
6	$3/4 \cdot U_{max}$	0	$3/4 \cdot U_{max}$
7	$7/8 \cdot U_{max}$	$U_{max}/8$	$3/4 \cdot U_{max}$
8	U_{max}	$U_{max}/4$	$3/4 \cdot U_{max}$

Figure 4 shows the result of the synthesis of the output sinusoidal voltage of the phase of the CFC by summing the voltages of two 3-level cells with differentiated supply voltages.

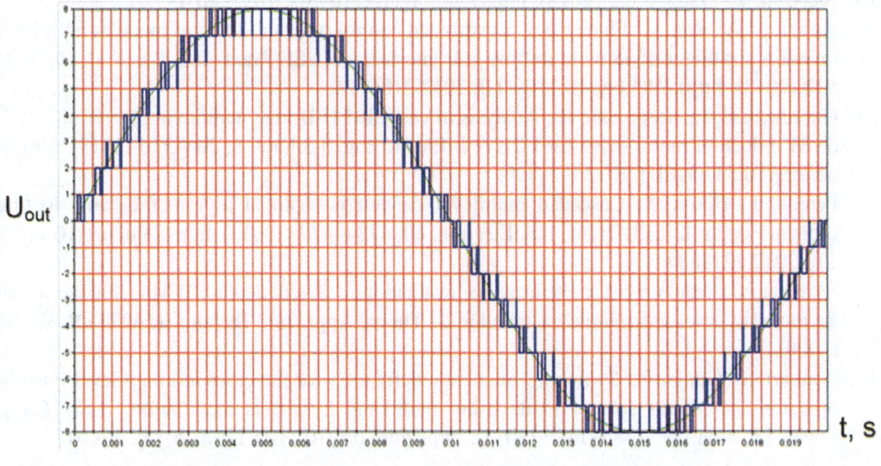

Fig. 4. Result of synthesis of the output sinusoidal voltage of the phase of CFC by summation of the voltages of two 3-level cells with differentiated supply voltages.

Similar estimates were obtained in the synthesis of the phase voltage at the output of the CFC, consisting of two 3-level cells with differentiated supply voltages. In this case, both summation and subtraction of cell voltages were allowed.

5 Conclusion

The proposed topology solutions and control algorithms of CFC offer leading edge opportunities to synthesize high-quality out voltage. At the same time, there is no need to install additional higher harmonic filters in order to obtain electrical energy of the required parameters. The use of the proposed arrangement of unit cells of the CFC as well as options for their power supply will result in quality supply for the critical loads, that require voltage of high-quality.

References

1. Mekhilef, S., Kadir, M.N.A., Salam, Z.: Digital control of three phase three-stage hybrid multilevel inverter. IEEE Trans. Ind. Inform. **9**, 719–727 (2013). https://doi.org/10.1109/TII.2012.2223669
2. Dixon, J., Pereda, J., Castillo, C., Bosch, S.: Asymmetrical multilevel inverter for traction drives using only one DC supply. IEEE Trans. Veh. Technol. **59**, 3736–3743 (2010). https://doi.org/10.1109/TVT.2010.2057268
3. Young, C., Chu, N.: A single-phase multilevel inverter with battery balancing. IEEE Trans. Ind. Electron. **60**, 1972–1978 (2013). https://doi.org/10.1109/TIE.2012.2207656
4. Thitichaiworakorn, N., Hagiwara, M., Akagi, H.: Experimental verification of a modular multilevel cascade inverter based on double star bridge cell. IEEE Trans. Ind. Appl. **50**, 509–519 (2014). https://doi.org/10.1109/TIA.2013.2269896
5. Rajeevan, P., Gopakumar, K.: A hybrid five-level inverter with common-mode voltage elimination having single voltage source for IM drive applications. IEEE Trans. Ind. Electron. **27**, 3505–3512 (2012). https://doi.org/10.1109/tia.2012.2226197
6. Parker, M.A., Ran, L., Finney, S.J.: Distributed control of a fault-tolerant modular multilevel inverters for direct drive wind turbine grid interfacing. IEEE Trans. Ind. Electron. **60**, 509–522 (2013). https://doi.org/10.1109/TIE.2012.2186774
7. Gholinezhad, J., Noroozian, R.: Analysis of cascaded H-Bridge multilevel inverter in DTC-SVM induction motor drive for FCEV. J. Electr. Eng. Technol. **8**, 304–315 (2013). https://doi.org/10.5370/JEET.2013.8.2.304
8. Pereda, J., Dixon, J.: Cascaded multilevel converters: optimal asymmetries and floating capacitor control. IEEE Trans. Ind. Electron. **60**, 4784–4793 (2013). https://doi.org/10.1109/TIE.2012.2219834
9. Seyezhai, R., Mathur, B.L.: Performance evaluation of inverted sine PWM technique for an asymmetric cascaded multilevel inverter. J. Theor. Appl. Inf. Technol. (JATIT) **9**, 91–98 (2009)
10. Gautam, S., Gupta, R.: Switching frequency derivation for the cascaded multilevel inverter operating in current control mode using multiband hysteresis modulation. IEEE Trans. Power Electron. **29**, 1480–1489 (2014). https://doi.org/10.1109/TPEL.2013.2262807

11. Cho, Y., Labella, T., Lai, J., Senesky, M.K.: A carrier-based neutral voltage modulation strategy for multilevel cascaded inverters under unbalanced DC source. IEEE Trans. Ind. Electron. **61**, 625–636 (2014). https://doi.org/10.1109/TIE.2013.2254091
12. Li, Z., Wang, P., Zhu, H., Chu, Z., Li, Y.: An improved pulse width modulation method for chopper-cell-based modular multilevel converters. IEEE Trans. Power Electron. **27**, 3472–3481 (2012)
13. Liu, L.M., Li, H., Wu, Z.C., Zhou, Y.: A cascaded photovoltaic system integrating segmented energy storages with self-regulating power allocation control and wide range reactive power compensation. IEEE Trans. Power Electron. **26**, 3545–3559 (2011). https://doi.org/10.1109/TPEL.2011.2168544
14. Irusapparajan, G., Periyaazhagar, D.: Asymmetric three-phase cascading trinary-DC source multilevel inverter topologies for variable frequency PWM. Circ. Syst. **7**, 506–519 (2016). https://doi.org/10.4236/cs.2016.74043
15. Corzine, K.A., Wielebski, M.W., Peng, F.Z.: Control of cascaded multilevel inverters. IEEE Trans. Power Electron. **19**, 732–738 (2004). https://doi.org/10.1109/TPEL.2004.826495
16. Rotella, M., Peñailillo, G., Pereda, J., Dixon, J.: PWM method to eliminate power sources in a nonredundant 27-level inverter for machine drive applications. IEEE Trans. Ind. Electron. **56**(1), 194–201 (2009)
17. Ramani, K., Krishnan, A.: New hybrid 27 level multilevel inverter fed induction motor drive. Int. J. Recent Trends Eng. **2**, 38–42 (2009)
18. Mahato, B., Mittal, S., Nayak, P.: N-level cascade multilevel converter with optimum number of switches. In: 2018 International Conference on Recent Trends in Electrical, Control and Communication (RTECC), pp. 228–233 (2018)
19. Ajami, A., Reza, M., Oskuee, J., Mokhberdoran, A.: Advanced cascade multilevel converter with reduction in number of components. J. Electr. Eng. Technol. **9**, 127–135 (2014). https://doi.org/10.5370/JEET.2014.9.1.127
20. Rotella, M., Peñailillo, G., Pereda, J., Dixon, J., Abstract, A.: PWM method to eliminate power source in a nonredundant 27-level inverter for machine drive applications. IEEE Trans. Ind. Electron. **56**, 194–201 (2009). https://doi.org/10.1109/TIE.2008.927233
21. Filho, F., Maia, H.Z., Mateus, T.H.A., Ozpineci, B., Tolbert, L.M., Pinto, J.O.P.: Adaptive selective harmonic minimization based on anns for cascade multilevel inverters with varying DC source. IEEE Trans. Ind. Electron. **60**, 1955–1962 (2013). https://doi.org/10.1109/TIE.2012.2224072

Preventive Protection of Ship's Electric Power System from Reverse Power

Alecsandr Saushev(✉), Nikolai Shirokov, and Sergey Kuznetsov

Admiral Makarov State University of Maritime and Inland Shipping, Dvinskaya Str., 5/7, Saint Petersburg 198035, Russia
saushev@bk.ru

Abstract. The present paper considers the task of developing an algorithm of preventive protection of a ship's electric power system (SEPS) from reverse power through implementing a method of preventive control. The paper gives a list of drawbacks of the existing method for organizing protection of electric generators from operation in motoring mode. A new approach is proposed that is based on consistent consideration of the problem taking into account information on operational limits of an object. Methods for partitioning a working range of ship's electric power systems were elaborated in order to obtain reduced sections of operability for different operation conditions of a system. From these methods, some logical expressions were derived that during operation enable timely forecasting the emergence of reverse power and providing preventive protection of ship's power mains from overloading. Several ways to complete the task for systems of different structural adaptation degree were considered. A method of early identification of inoperative generator set is proposed, which allowed elaborating the algorithm of preventive protection of an SEPS under failure of its elements. Therewith the forecasting of operation modes of the system in case of generator set's failure is fulfilled, as well as its structural adaptation to a malfunction emerged. Unlike the existing approaches, implementation of preventive control system will enable a failure-free transit of an SEPS to a partly operable state excluding accidental situations, which will positively affect the safety of the whole ship operation.

Keywords: Ship's electric power system · Generator set · Reverse power · Protection of ship's power mains · Functional diagnostics

1 Introduction

Ship's electric power systems (SEPS) manage electric power flows required to ensure the basic functions of a vessel. The key tasks of operation of an SEPS are: to ensure the operability of a system and its elements [1–3], to compensate reactive powers [4, 5], to provide high-quality electrical energy to all consumers of a vessel [6–9]. Technical diagnostics takes a significant part in the process of solving these problems [10–12]. The article discusses the issues of protection of SEPS from reverse power [13]. Reverse power (N_{rev}) emerges in a network of a ship's electric power system (SEPS) when at

V. Murgul and V. Pukhkal (Eds.): EMMFT 2019, AISC 1258, pp. 388–398, 2021.
https://doi.org/10.1007/978-3-030-57450-5_33

least one of simultaneously running generator sets (GS) shifts to the motoring mode. This mode of SEPS operation is pre-emergency; the load is redistributed among remaining generators, while a GS switched to the motoring mode loads the network additionally with an amount of its losses. This can lead to overloading the system, triggering protection and shutting down working GSs and, as a result, powering off the vessel. The delay in the decommissioning of a failed GS in most cases leads to the further development of a malfunction. This determines the need for the application of protective shutdown of the GS when it enters the motor mode of operation, which should occur automatically using a special control system (CS).

A generator usually shifts to the motoring mode because of a malfunction of a GS's primary motor. Most often, this is associated with the failure of fuel equipment or air supply system of this motor. The signal of the reverse power appearance is considered a generalized diagnostic sign of this type of failure, so all the known control systems of SEPSs use this exact signal to perform control actions in this pre-emergency situation. However, this process may involve the appearance of false control signals and actuation of protectors (type 1 error) reasoned by the presence of such operating modes of the system, in which GSs can consume electrical energy from the network and remain operable. These cases must exclude the triggering of a protection system.

In order to eliminate a chance of this error, the developers of control systems for SEPS introduce additional restrictions when diagnosing a GS. A GS is identified as inoperative only when the reverse power exceeds the permitted value (N_{lim}) which is has been observed for longer than set time (T_{lim}). For transport ships, the value N_{lim} is usually chosen to be 10% of nominal power of a generator (N_{nom}) with a delay time of 5–6 s when triggered. For vessels of technical fleet equipped with hoisting machines, for example, floating cranes, these values are chosen with advanced significance. For GSs using diesel as the primary motor, the Rules of the Russian Maritime Register of Shipping assigned the maximum permitted value of reverse power equaled to 15% of N_{nom}, and of T_{lim} – to 10 s. Such an approach, on the one hand, eliminates false shutdowns of operable generator sets, but, on the other hand, reduces the efficiency of protection in case of a GS failure, which often leads to an SEPS emergency situation related to de-energization and the loss of control of a vessel. In this regard, elaboration of new approaches, which would ensure, first, well-timed identification of inoperative state of a GS and, second, its shutdown before the emergence of reverse power, appears to be an urgent scientific and technical task.

2 Materials and Methods

Serious shortcomings of the existing method of protecting a GS from operation in motoring mode after the its primary motor's operability loss are reasoned by the fact that this GS is considered during diagnostics as a separate, independent object. Therefore, parameters of other GSs working simultaneously are not taken into account, which is fundamentally wrong. This methodological mistake in the approach ultimately leads to a significant methodological margin of error when implementing the protection algorithms and to serious shortcomings in operation of a control system of a SEPS.

In this regard, it is proposed to consider GSs that work simultaneously as an autonomous power generation system (APGS) consisting of elements – GSs, which are interconnected and meant to supply a vessel with electric energy of required quality. The technical state of an APGS should be assessed based on information about operational limits G, which are the set of admissible values of primary parameters at which all the requirements for output and internal parameters of a system are fulfilled, as it is discussed in the work [14].

To the present day, a number of methods have been developed for defining an operability area for various objects [15]. According to papers [15, 16], the set G can be represented in general terms as an intersection of sets D_x, M_y and M_z, which is written as follows:

$$G = D_x \bigcap M_y \bigcap M_z, \tag{1}$$

where D_x is the tolerance area of the primary parameters, which has a brick shape and in Euclidean space can be described as $D_x = \bigcap_{k=1}^{n} D_k,\ D_k = D_{k\,\min} \bigcap D_{k\,\max}$, it corresponds to the internal condition of operability.

M_y is the area described by the depiction in space of primary parameters $\Phi_{yx} : D_y \to M_y,\ M_y = \bigcap_{i=1}^{m} M_i$ tolerance area $D_y = \bigcap_{i=1}^{m} D_i$, corresponding to the external condition of operability.

M_z is the area described by the depiction in space of primary parameters $\Phi_{zx} : D_z \to M_z,\ M_z = \bigcap_{r=1}^{p} M_r$ tolerance area of the parameters of system's functional blocks $D_z = \bigcap_{r=1}^{p} D_r$, corresponding to the internal condition of operability.

Since the problem being solved implies that diagnostics depth is determined by parameters of GSs, which are the elements of a system, so there are no larger functional blocks, for this particular case the following can be written:

$$G = D_x \bigcap M_y \tag{2}$$

The work [10] showed that operability area can be represented as the sum of reduced areas of proper functioning w_j^q:

$$\forall w_j^q \in G,\ G = \bigcup_{j=1}^{q} w_j^q,\ j = \overline{1, q}. \tag{3}$$

Segmentation of the operability area is to be carried out in such a way as to separate the reduced areas of proper functioning $H_x = \bigcup_{j=1}^{g} w_j^g$, characterized by APGS operating modes that imply at least one of GSs switching to the motoring mode.

It follows from expression (3) that the operability area G belongs to the tolerance range D_x, which means that $G \in D_x$, therefore, the following expression can be written:

$$D_x = B_x + H_x = \bigcup_{i=1}^{v} w_i^v + \bigcup_{j=1}^{g} w_j^g i = \overline{1,v};\ j = \overline{1,g}, \tag{4}$$

where $B_x = \bigcup_{i=1}^{v} w_i^v,\ i = \overline{1,v};$ represents an area where unloading and reverse power emergence allow a GS to be identified as inoperative and turn it off immediately.

3 Results

The point characterizing the parameters of the APGS at a given point in time as S, then the disconnection condition of the k-th GS can be written as the following condition:

$$(N_k < 0) \wedge (S \notin H_x), k = \overline{1,n}. \tag{5}$$

To develop an operation algorithm for a control system, the functional-and-logical method [17] will be used and operating modes of an APGS will be determined, in which GSs can switch to the motoring mode, while remaining operable; so the area $H_x = \bigcup_{j=1}^{g} w_j^g$ is set. In this case, the following circumstances should be considered.

Firstly, a generator can shift to the motoring mode at the moment of paralleling. One of the conditions for synchronizing GSs is the coincidence of frequencies of operating sets, maintained with a given accuracy. In this case, the allowance for difference between the frequencies of power mains and a generator (Δf) is usually set within 0.05 Hz ≤ Δf ≤ 0.5 Hz. At the moment of connecting a generator, frequencies are equalized due to the phenomenon of synchronism, therefore, the rotation frequency of one GS increases by means of energy received from another GS. When synchronization is carried out from below, the voltage frequency of the connected GS is less than the frequency of power mains voltage, so this generator rapidly switches to a motoring mode and creates an additional load. When an operating set is working at low power and the connection of GS occurs at higher speeds, the synchronized generator may undertake the entire load and shift the operating set into a motoring mode. Reduced section of proper functioning of the APGS corresponding to the mode of paralleling of GSs is indicated as w_1^g.

Secondly, many SEPSs include power consuming units, for example, thrusters, power of which is comparable to power of one generator. These units' shutdown leads to a sharp decrease in power mains' load. At the same time, power developed by each set is sharply reduced. Due to differences in constant-error behavior of speed characteristics of the primary motors and to inertia of fuel injection equipment, one of the GSs being operable can switch to a motoring mode. The reduced area of proper functioning corresponding to the operation mode of an SEPS at the moment of a sharp decrease in load is indicated as w_2^g.

Thirdly, the operation mode for an SEPS is possible, if the load of the GSs working simultaneously is very small and close to off-load running. In this case, one of the sets can shift to motoring mode, while the other set will work to compensate for losses of APGS. The reduced area of proper functioning corresponding to this mode of operation is indicated as w_3^g.

Fourth, technical fleet vessels, special-purpose vessels and floating cranes are usually equipped with high-capacity hoisting mechanisms. Therefore, a situation is possible when GS's load is shed and the set shifts to a motoring mode because of recuperation of energy into the ship's mains. The reduced area of proper functioning characterized by the appearance of a powerful source of recuperation energy is indicated as w_4^g.

Considering the above, the condition for disconnection of the k-th GS, k = 1, 2, …, n can be expressed as follows:

$$(N_k < 0) \wedge \{(S \notin w_1^g) \vee (S \notin w_2^g) \vee (S \notin w_3^g) \vee (S \notin w_4^g)\}. \tag{6}$$

Each reduced area of proper functioning $w_1^g, w_2^g, w_3^g, w_4^g$ corresponds to a certain SEPS operating mode, which can be identified by the distinctive markers x_1, x_2, x_3, x_4.

Moreover, expression (2) can be written as follows:

$$(N_k < 0) \wedge \{(S \notin x_1) \vee (S \notin x_2) \vee (S \notin x_3) \vee (S \notin x_4)\}, \tag{7}$$

Where x_1, x_2, x_3, x_4 – are the distinctive markers of operating modes of the system.

A SEPS may have the following markers:

x_1 is a logical signal that characterizes a time interval from the moment of closing the circuit breaker of the synchronized generator to the moment of undertaking a full load in accordance with work [12];

x_2 is a potential-free contact of controls, characterizing the disconnection of an electricity consuming unit, the power of which is comparable to power of a generator;

x_3 is a logical signal generated at a time when the SEPS is working under a load close to off-load running;

x_4 is a potential-free contact of controls that characterizes operation of a recuperative braking contactor.

The logical conditions presented in expression (7) can be considered as a warning signal, on the basis of which the SEPS structure is controlled by shutting down an inoperative GS for the purpose of adapting the system to a malfunction. In accordance with work [17], such a control is called preventive. In this case, the controlling impact is formed until the moment when a failed unit shifts to a motoring mode. At the same time, the preventive protection of an SEPS from reverse power can be discussed.

The simplicity of technical implementation is an ultimate advantage of the considered approach. It gives opportunities for using special approach-based devices instead of reverse power relays, which are widely used on ships, yet not effective.

It should be mentioned that while successfully solving the problem of preventing ship mains from loading with reverse power, the considered method does not provide a solution to the problem of SEPS overload in the event of failure of primary motors of a GS. Therefore, devices that implement control algorithms based on logical expression (7) should be used in fairly simple electric power systems, which are adapted to possible defects in a low extent.

A significant drawback of the considered approach is the identification of the GS's inoperative state at the time of its complete unloading, when the entire load of the set is redistributed among the remaining electrical machines. In this case, when, for instance, two generators are operating simultaneously, the load on a GS that has remained operable doubles, which in most cases leads to a shutdown of a primary motor working and to blackout of a whole vessel.

In this regard, the energy state of an APGS is to be considered in case of a failure of fuel equipment of a primary motor. In this case, power generated by the failed generator set will decrease, while the load initially accepted by it will be redistributed between GSs working simultaneously with it. The unloading time of an inoperative set usually ranges from 4 to 20 s and depends on a number of factors. The main conditions affecting the speed of load redistribution include the following: location and nature of a defect; power and load of a machine at the time of defect occurrence; availability and type of fuel filters; type and quality of fuel. Usually the reason for the GS's switch to the motoring mode is a sudden failure of its functional units (fuel pump, fuel filter, loss of integrity of pipe junctions and pipes for fuel transfer, fuel control equipment), and in terms of energy processes occurring in an SEPS, malfunction of a whole set has signs of gradual failure.

A condition that implies the reduction of load on one GS while the load on the remained GSs working simultaneously is increasing is indicated as L_1. This condition can serve as a diagnostic sign of a failure of the unloaded generator, but only if the sets are already operating in automatic mode. In actual practice, there is a possibility of a situation when machines operate in a manual mode and are loaded differently. When automatic control is turned on, the load on one of the machines will be increasing, while on another it will be decreasing. This can entail the appearance of type 1 error during diagnostics and consequently lead to disconnection of an unloaded generator's circuit breaker. This can be avoided by excluding this mode of diagnostic process, as it was done in the previous case.

The reduced area of proper functioning corresponding to this mode of SEPS operation is indicated as w_5^g. For the same reason, one need to exclude the process, which was previously indicated as w_1^g, consisting in determination of SEPS's technical state during a synchronization mode until a generator, first, is paralleled, and second, undertakes the load.

On the other hand, it is necessary to take into account the fact that fuel equipment even of one-type primary motors can have different inertia. Moreover, in dynamic operating modes, when load of mains increases and then rapidly drops, a fuel regulator of one of generator sets may respond to changes in external impacts and will start reducing fuel supply, while other regulators will remain in a position of ensuring the increase in supply. To exclude this type of errors, an additional diagnostic parameter is introduced: a value of load difference between simultaneously operating GSs (L_2). At the moment of identification of the inoperative state of a set, this difference must be greater than the specified value (L_{lim}), which is determined based on the permissible accuracy of load distribution between generators, which is given in the technical regulatory documentation for control systems of SEPS.

A diagnostic marker that allows identifying GSs with decreasing load is indicated as an inoperative set through F_i = 1, 2, …, n, where n is the number of currently working GSs.

Then the following expression can be written:

$$F_i = L_1 \wedge (L_2 > L_{\lim}) \wedge (\overline{w_1^g} \vee \overline{w_5^g}). \tag{8}$$

Parameter F_i characterizes the inoperative state of a set before it shifts to a motoring mode; it can be used for functional diagnostics of SEPS.

According to codes and standards as well as generally accepted definitions, the operable state of a technical system determines only its ability to perform its functions. For failure-free operation of equipment, a system must work in this mode under specified merit index, or in other words, must be in operable condition. For this purpose, certain requirements must be fulfilled. Regarding SEPS, such conditions are specified by load vector $\overline{P}$ of the controlled parameters and by vector of external conditions $\overline{V}$ characterizing environment conditions.

Operating conditions of ship equipment shortly before a GS failure and immediately after a set goes into an inoperative state usually do not change significantly and remain within the specified limits established by the technical documentation. Thus, excluding special situations, in order to make a diagnostic model of an SEPS, one can abstract from the influence of vector $\overline{V}$ on the processes of control and assessment of its technical state. In this case, the condition for an SEPS to be in operable and operable state H is written as the intersection of areas G and M_p, where M_p is the tolerance area described by the following reflection:

$$\Phi_{xp} : D_p \rightarrow M_p, M_p = \bigcap_{c=1}^{e} M_c. \tag{9}$$

Area M_p characterizes the correspondence of a tolerance area determined by the limitations of a load value developed by a GS in each power mode. Then, taking into account (1), the following can be written:

$$H = G \bigcap M_p = D_x \bigcap M_y \bigcap M_p. \tag{10}$$

The segmentation of area H is performed in accordance with SEPS's functioning modes determined by a number of operating GSs and a final value of reduced areas of proper functioning w_j^p is obtained. In order to preserve a system in operable condition and being able to fulfill its functions, its technical condition must belong to the reduced area of proper functioning at the given moment. In this case, $H = \bigcup_{j=1}^{e} w_j$, where e is the number of operating modes of SEPS. Most often, APGSs of modern ships are equipped with GSs of the same power. The number of operating modes is equal to the number of

GSs, which means that e = n. In this case, it is easy to recognize the reduced area of proper functioning, which must correspond to the point S_i corresponding to the current technical state of an SEPS in case of failure of any number of operating GSs. The condition for safe shutdown of generators can be written as follows:

$$S_i \in w^p_{i-q}, \tag{11}$$

where q is the number of disabled GSs. If requirement (6) is not fulfilled, the network load should be reduced, for example, by disconnecting a group of any power consuming units. Under condition (11), GSs remained operation will undertake the load of inoperative units, so their shutdown will not entail an overload or a break in the power supply of a vessel.

Using the obtained expressions (8) and (11), precautionary control algorithms can be developed that provide preventive protection of an SEPS from GS switching to the motoring mode, and, therefore, from reverse power emergence.

Figure 1 shows a flowchart of the algorithm for preventive protection of an SEPS from reverse power, using two GSs as an example. By the starting command, the following data is entered: on the load of the first and second generator sets (P_1, P_2); on the fulfillment of logical conditions x^g_1, x^g_5; on limitation $P_{lim\ off}$; on information defining the reduced area of the proper functioning of one GS w^p_1. Next, for each GS, the logical condition (8) is checked. The fulfillment of this condition is denoted by a logical unit signal $F_1 = 1$, $F_2 = 1$, respectively. Changing values of the generators' load taken from load sensors are indicated as P_{sen1}, P_{sen2}.

When condition (8) is satisfied, the possibility of failure-free shutdown of an inoperative set is checked in accordance with expression (11). In this case, the total power in mains is calculated at the moment of occurrence of malfunction P_s, and then the obtained value is compared with load P_{lim1} which can be undertaken, for example, by the first GS. If when two generators are working, the current state of the SEPS S_2 characterized by load P_s does not fulfill requirement (11) and $P_s > P_{lim1}$ then a signal is made to turn off some of power consuming units. When $S_2 \in w^p_1$ and $P_s \leq P_{lim1}$, failure-free shutdown of an inoperative set is possible. However, the immediate shutdown is usually undesirable, since large inrush of load current may cause unwanted transitional processes in an SEPS. In order to prevent them, it will be sufficient to wait until the inoperative set is unloaded to the permissible load $P_{lim\ off}$ for example, by 20% of its power consumption N_{con}, after which the inoperative set should be turned off.

The proposed algorithm allows the shutting down an inoperative set much earlier than reverse power appears. This circumstance provides effective protection of an SEPS from overload. Moreover, in some cases, the operating time of a failed GS is reduced by more than 90%, which reduces a chance for further growth of the defect and its possible consequences.

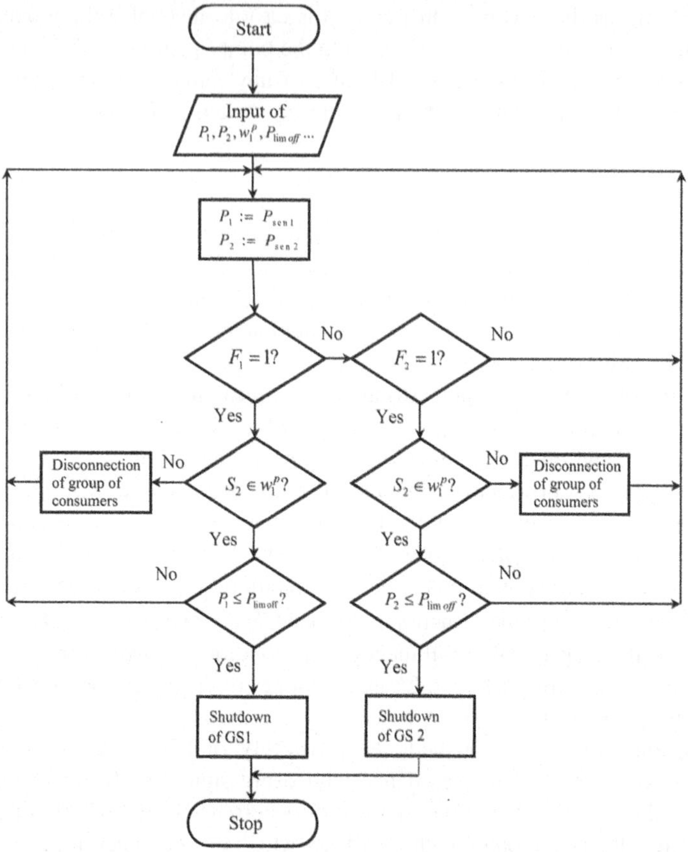

Fig. 1. Flowchart of the algorithm for preventive protection of the SEPS from reverse power.

4 Discussion

Modern methods for the functional diagnosis of ship's electrical equipment and automation systems do not fully meet the modern requirements for the operation of SEPSs. In most cases, triggering of protective devices leads to a blackout of mains and to possible development of emergencies on a vessel. This fact is the basis for recommendations of the IEC and owners of shipping companies on the increased reliability regime usage when a vessel is used in narrow places or under stormy conditions. In these cases, operation of an SEPS is ensured by a large number of generator sets, which leads to an increase in the consumption of combustive and lubricating materials, deterioration and reduction in the residual operation time of sets.

The approaches to fulfilling the task considered in the article partially solve this problem. They can be applied when developing control systems of the highest level of the hierarchy. At the same time, a significant improvement is possible in the work of the algorithm proposed, for instance, through identifying options for optimal disconnection of power consuming units under various SEPS operating conditions. The problem of load value of an inoperative set, that should imply the formation of signal to turn it off, remains unresolved. In terms of protecting an SEPS from overload and smoothing transition processes, it is desirable to turn off a generator's circuit breaker at the moment of complete unloading. However, the moment of detecting a malfunction formalized in expression (8) should be considered the most preferable for the safety of the most inoperative set. The improvement in quality indicators of one process leads to the decline of quality indicators of another process. The problem of choosing the only optimal solution from the set of Pareto-optimal solutions remains the subject of further research.

5 Conclusion

The proposed approach enables a swift identification of an inoperative generator set during SEPS operation by means of technical diagnostics. If necessary, it allows unloading the mains and ensuring the shutdown of an inoperative GS under the most favorable conditions for an SEPS. This helps timely forecast the process of generator's shifting to the motoring mode and preventively protect an SEPS from reverse power due to the structural adaptation of a system to a malfunction occurred. Unlike the existing control systems, implementation of the preventive control system will allow a failure-free transition of an SEPS to a partially operable state without any accidental situations, which will ensure a safer and more cost-effective operation of a whole vessel.

References

1. Abbasian, N.S., Salajegheh, A., Gaspar, H., Brett, P.O.: Improving early OSV design robustness by applying 'Multivariate Big Data Analytics' on a ship's life cycle. J. Ind. Inf. Integr. **10**, 29–38 (2018)
2. Alizadeh, S., Sriramula, S.: Reliability modelling of redundant safety systems without automatic diagnostics incorporating common cause failures and process demand. ISA Trans. **71**(2), 599–614 (2017)
3. Saushev, A.V., Kuznetsov, S.E., Karakayev, A.B.: System approach to ensure performance of marine and coastal electrical systems during operation. In: IOP Conference Series: Earth and Environmental Science, vol. 194, no. 8, p. 082037. https://doi.org/10.1088/1755-1315/194/8/082037
4. Dudko, S.: Digital control by the discrete systems of reactive power compensation in ship electrical power plants. In: IOP Conference Series: Earth and Environmental Science 4. Ser. "4th International Scientific Conference SEA-CONF 2018", p. 012011. Constanta (2018)
5. Kornev, A.S., Kuznetsov, V.I., Pan, H., Senkov, A.P.: Sposoby kompensacii vysshih garmonik napryazheniya v sudovyh elektroenergeti-cheskih sistemah. Morskie intellektual'nye tehnologii **1**(47) (2020). https://doi.org/10.37220/mit.2020.47.1.035
6. Kalinin, I.M.: Kontseptsiya sozdaniya otechestvennoy sistemy skvoznogo proyektirovaniya sudovykh elektroenergeticheskikh system. Trudy Krylovskogo gosudarstvennogo nauchnogo tsentra **1**(387), 61–72 (2019). https://doi.org/10.24937/2542-2324-2019-1-387-61-72

7. Serebryakov, A., Steklov, A., Titov, V.: New method of technical condition diagnostics of ship electric power plants. In: 2016 IX International Conference on Power Drives Systems (ICPDS) (2016). https://doi.org/10.1109/icpds.2016.7756713
8. Blanke, M., Lunau, C.P., Lyngsø, S.T., Hornsby, C.: KBSSHIP - communicating expert systems for ship-wide decision support. IFAC Proc. **25**(3), 73–86 (1992)
9. Integrated electric ship power systems (n.d.). AccessScience. https://doi.org/10.1036/1097-8542.yb051930
10. Doerry, N., Amy, J.: Electric ship power and energy system architectures. In: 2017 IEEE Electric Ship Technologies Symposium (ESTS) (2017). https://doi.org/10.1109/ests.2017.8069355
11. Tymkiv, O.: Ways to improve ship power plants. Aust. J. Tech. Nat. Sci. 49–51 (2019). https://doi.org/10.29013/ajt-19-1.2-49-51
12. Campora, U., Cravero, C., Zaccone, R.: Marine gas turbine monitoring and diagnostics by simulation and pattern recognition. Int. J. Naval Archit. Ocean Eng. **10**(5), 617–628 (2018)
13. Dale, S.J.: Ship power system testing and simulation. In: IEEE Electric Ship Technologies Symposium (2005). https://doi.org/10.1109/ests.2005.1524675
14. Saushev, A.V., Shirokov, N.V.: Diagnostirovaniye sostoyaniya elektrotekhnicheskikh sistem v prostranstve parametrov ikh elementov. Vestnik gosudarstvennogo universiteta morskogo i rechnogo flota imeni admirala S.O. Makarova **2**(36), 143–156 (2016). https://doi.org/10.21821/2309-5180-2016-8-2-143-156
15. Saushev, A.V.: Oblasti rabotosposobnosti elektrotekhnicheskikh sistem. Politekhnika, Saint Petersburg (2013)
16. Saushev, A.V., Bova, E.V.: Solution of problems of parametric optimization and control of electric drives state based on information about operability area boundary. In: IOP Conference Series: Materials Science and Engineering, vol. 327, no. 5, p. 052029 (2018)
17. Shirokov, N.V.: Predupreditel'noye upravleniye sudovoy elektroenergeticheskoy sistemoy pri otkaze istochnikov elektroenergii. Vestnik gosudarstvennogo universiteta morskogo i rechnogo flota imeni admirala S.O. Makarova **11**(2), 396–405 (2019). https://doi.org/10.21821/2309-5180-2019-11-2-396-405

The Role of Water Transport in the Formation of the Brand of the Coastal Regions: The Example of St. Petersburg

Anton Smirnov and Mikhail Zenkin(✉)

Admiral Makarov State University of Maritime and Inland Shipping,
Dvinskaya Str., 5/7, Saint Petersburg 198035, Russia
zenkinma@gumrf.ru

Abstract. Over the past ten years, conferences and events of the all-Russian and international level have discussed the development of infrastructure for the passenger and small fleet in Russia and, in particular, in St. Petersburg - the sea capital of Russia. The areas that form St. Petersburg as the Sea Capital are sea cruises, river cruises, intercity and suburban passenger transportation, recreational shipping, yachting. Development of cruise tourism in Saint-Petersburg was boosted by opening of the new "Passenger Port of Saint Petersburg "Marine Façade". Full scale operation opened sea gates for leaders of cruise business. Hundreds of thousands international tourists, coming by cruise ships and ferry vessels start to perceive Saint-Petersburg as a marine city. River cruises have a strong potential for territory branding. There are 60–80 cruise motor ships operating on cruise lines during navigation period. Every cruise motor ship enters Saint-Petersburg from 1 to 5–8 times during navigation period. Considerable segment of water tourism is occupied by intercity and suburban passenger transportation; rivers and canals cruises and boat tours have become a business card of the port city. Yacht tourism makes a considerable contribution into development of the "Sea capital" brand. It is one of the most progressing types of water tourism. According to forecasts of Committee on transport-transit policy of Saint-Petersburg number of small size vessels (yachts and motor boats) is going to increase by 37% and will make up approximately 60 000 vessels in the nearest 5 years.

Keywords: Water transport · Water tourism · Branding of coastal regions

1 Introduction

Branding of territories is a modern effective tool for attracting tourists to the region. The formation of the image of the destination is carried out by means of media, a significant segment of which, as a rule, is occupied by the Internet. As noted by Xiang, Z., Wang, D., O'Rielly, J.T., Fesenmaier, D.R., "In the era of the information society, international communication channels are much more accessible to all kinds of destinations, and also far more important" [1]. When planning a visit to a particular region, a modern tourist carefully studies the reviews and photos of travelers on social networks, web-sites

V. Murgul and V. Pukhkal (Eds.): EMMFT 2019, AISC 1258, pp. 399–408, 2021.
https://doi.org/10.1007/978-3-030-57450-5_34

dedicated to the review of tourism in this region. Information from social media, as noted by Avraham, E., Ketter, E., he perceives "as the "reality" of the place" [2].

In order to express themselves, to positively position themselves in the minds of potential tourists, the destination, according to Qu, H., Kim, L. H., Im, H. H., should "use online media to 'talk' about themselves" [3]. The effect of Internet communications on the formation of a destination brand is described in an article by Vinyals-Mirabent, S., Kavaratzi, M., Fernández-Cavia J. "The role of functional associations in building destination brand personality: When official websites do the talking" [4]. The team of authors analyzed the twelve official sites of European cities with the greatest tourist attractiveness and came to the conclusion that it is necessary to create more balanced content: between official information and various entertainment attractions of the destination, which will increase the awareness of tourists and strengthen the brand of the territory.

Destination brand promotion, as a rule, is carried out around the semantic core, reflecting the specifics of the territory. Reliance of the brand on history, cultural heritage and existing infrastructure is the key to its viability.

2 Hypothesis Development and Methodology

In St. Petersburg, due to its rich history, diversity of architectural and cultural heritage, and geographical features of the region, several tourist brands have developed at once: "the cultural capital of Russia", "White Nights", "Sea Capital of Russia", "Northern Venice", etc. Throughout the history, the brand "Sea Capital of Russia" is actively formed and supported in St. Petersburg. The brand is driven by historical, cultural and infrastructural factors. In the framework of this paper, we focused on the physical, geographical and infrastructural factors that influence the formation of the brand "Sea Capital of Russia".

St. Petersburg is located on the shores of the Gulf of Finland of the Baltic Sea and has an extensive network of rivers and canals. The main waterway of the city is the Neva River, which, when it flows into the Gulf of Finland, is divided into several branches, forming an extensive delta. The total length of all water bodies in St. Petersburg reaches 282 km, and their surface is about 7% of the total area of the city.

This separation of St. Petersburg by the Neva River Delta, rivers and canals into separate parts leaves an imprint on the development of its transport complex: on the one hand, rivers and canals break the logistic unity of the land transport system, and on the other, they themselves act as communication routes, which creates the prerequisites for the development of water passenger transport and excursion and recreational shipping in St. Petersburg. Water excursions increase interest in the objects of tourist attractiveness of St. Petersburg and act as a tool for branding destination as "Northern Venice".

The following areas of water transport development can be distinguished in St. Petersburg, which form the brand "Sea Capital of Russia":

- sea cruises;
- river cruises;

- intracity transportation, excursions on water transport;
- suburban passenger transportation by water;
- yacht tourism and yachting.

3 Results

The driver of the development of cruise tourism in St. Petersburg was the opening of the new passenger port "St. Petersburg "Marine Facade" in 2008. The port is a modern complex with seven berths, including for the reception of ocean liners up to 340 m long. In 2019, the passenger turnover amounted to 1 104 479 passengers. The dynamics of passenger turnover over 10 years has increased by almost 2.5 times.

The 72-h visa-free regime for tourists arriving by cruise lines has significantly increased the flow of foreign tourists to St. Petersburg. Among passengers arriving in St. Petersburg by cruise lines, according to the JSC Passenger Port "St. Petersburg "Marine Façade", representatives of Germany (29%), USA (20%) and the UK (12%) prevail.

The high-capacity passenger port opened the sea gate for the leaders of the cruise business: Carnival Corporation & plc., Royal Caribbean International and Celebrity Cruises, Norwegian Cruise Line, MSC Cruises S.A. Thanks to this, in the minds of hundreds of thousands of foreign tourists arriving on cruise ships and ferries, the perception of St. Petersburg as a sea city, as a port city is being formed.

River cruises also have significant potential in terms of branding territories. Cruises with an average duration of 4–8 days are being formed in St. Petersburg and Moscow. The "two capitals" route, including calls at the ports of Moscow and St. Petersburg, as well as circular cruise routes to the islands of Valaam, Konevets, Kizhi from St. Petersburg, are popular among foreign tourists.

The number of cruise ships serving cruise lines during the navigation period ranges from 60 to 80 units or more in different years. Each of the ships performs from 1 to 5–8 calls to St. Petersburg during navigation.

At the same time, the capacity of the berths of the River Station of St. Petersburg has exhausted itself in 2000. And since 2012, the River Station ceased to function. Part of the fleet was relocated to Utkina Zavod berths constructed by Passenger Port OJSC. Meanwhile, tourists' interest in river cruises has steadily increased over the past 10 years. The lack of berthing infrastructure and the aging of the fleet are a serious deterrent to the development of river cruise tourism in the region.

A significant segment of water tourism is occupied by intra-city and suburban passenger transportation, recreational navigation on the rivers and canals of St. Petersburg. Excursions on ships, rising bridges during the white nights - a hallmark of marine St. Petersburg.

The largest player in the market for the provision of tourist services using water transport in the inner-city water area is the Association of Passenger Ship Owners of St. Petersburg, established in 2003 and uniting more than 85% of the operating fleet. The purpose of the association is the expansion of excursion services using water transport, the development of small business, as well as protecting the interests of shipowners of passenger ships of St. Petersburg. The Association unites 16 shipping companies owning more than 150 passenger ships of various types.

In order to stimulate intercity passenger transportation, excursion and recreational navigation, the External Transport Agency of Saint-Petersburg (subordinate to the Committee for Transport of St. Petersburg) developed the project "City berths of St. Petersburg". The aim of the project is to ensure the accessibility of the berth in the city center for the development of recreational navigation. From 2014 to 2018, 18 city public berths were opened in St. Petersburg. They are located in the historical center of the city, on Elagin Island, on Krestovsky Island, on Vasilyevsky Island from the side of Makarova embankment, as well as on the right bank of the Neva (Sverdlovskaya embankment), on the Petrograd embankment, and in other places.

As of 2018, 13,866 moorings were carried out under the project, and 4,838,400 rubles were received in the city budget.

The main results of the implementation of the project "City berths of St. Petersburg":

- an increase in the volume of an equally accessible berth in the central part of the city;
- launching regular tourist routes using city berths;
- increasing the level of control over the safety of inland water transport during boarding and disembarking of passengers;
- systematization of the inland water transport market.

The project "City berths of St. Petersburg" makes a significant contribution to the formation of the brand of the sea capital, to the formation of the appearance of a European city, equally accessible both from the coast and from the water.

Yacht tourism, one of the most developing and at the same time debatable types of water tourism, makes a significant contribution to the formation of the brand "sea capital". In 2019, the Committee for Tourism Development of St. Petersburg developed the "Concept for the Development of Yacht Tourism in St. Petersburg", which involves the comprehensive development of infrastructure for the small fleet.

According to the forecasts of the Committee for Transport and Transit Policy of St. Petersburg, in the next five years, the number of small vessels (yachts and boats) in St. Petersburg will grow by 37% and amount to about 60,000 vessels. Today, there are 48 basing and maintenance facilities for the small fleet with a total area of 131 hectares in the city. At the same time, the overwhelming majority of ship mooring and storage facilities do not comply with the Preliminary national standard PNST 153-2016/ISO 13687: 2014 "Services to the population. Yacht ports. Minimum requirements". Lack of developed infrastructure in marinas: poorly equipped berths, lack of opportunities for bunkering with water and fuel, as well as minor repairs, remoteness from shops, etc. restrains the development of inbound yacht tourism and does not allow fully using the coastal areas of St. Petersburg in the interests of tourism.

The most important factor restraining the development of yacht tourism in St. Petersburg, in addition to infrastructure, are non-normative restrictions: natural and geographical, technical, social.

Natural and geographical factors are determined by seasonality (short navigation) and climatic conditions (rainy climate, windy weather), as well as geographical distance from the main locations of foreign yachts, remoteness of the sea checkpoint and customs control "Fort Konstantin" from the sea state border of the Russian Federation.

Technical limitations are associated with the unsatisfactory condition of the berths for yacht mooring, the insecurity of the berths with water and electricity columns, the lack of adapters for connecting the on-board power supply, the lack of life ladders and lifebuoys on the berths. Development of coastal infrastructure - the presence of a gas station, shower/sauna, toilets, washing machines, ATMs, the ability to pay by credit card, the presence of a nearby first-aid post, pharmacy, etc. significantly increases the attractiveness of the marina in the eyes of a tourist arriving in the city on a yacht.

Social factors are related to the information support of marina facilities in English and other major foreign languages (signposts, information sheets, travel directions to the city, etc.), including knowledge of the information and foreign languages of the main marina specialists and staff. The remoteness of the berth from the city center and tourist attractions, the availability of municipal transport stops, especially metro stations, transport accessibility are also an important factor determining the choice of a tourist.

The presence/absence of information tourist centers on the territory of the marina or nearby, ensuring security in the marina (security, proximity to the police station, the ability to call an ambulance with staff speaking a foreign language) affect the feeling of comfort and safety of a tourist.

These main factors determining the quality of the yacht mooring infrastructure play an important role for the formation of an attractive brand of the coastal region.

Of the 48 berths in St. Petersburg, the seven most promising and to a greater or lesser extent meeting the requirements of the Preliminary National Standard are:

1. St. Petersburg River Yacht Club of Trade Unions (Petrovskaya Kosa, 9);
2. Krestovsky Yacht Club (South Road, 4, building 1);
3. Imperial Marine Yacht Club of St. Petersburg (Martynova emb., 92);
4. Yacht Club of St. Petersburg (Lakhta village, Beregovaya St., 19, letter A);
5. Baltiets Yacht Club (Peterhof highway, 75, building 2);
6. Yacht port "Tirijoki" (Zelenogorsk, Gavannaya St., 1, letter A);
7. Yacht Club "Fort Konstantin" (Kronstadt, Fort Konstantin, lit. A).

An analysis of the data of yacht ports showed that all of them are located in a fenced or partially fenced (impeding the free passage of vehicles) territory with security posts, equipped with a video surveillance system (overview of mooring places and territory), guarded berth.

An express analysis of the official websites of the marinas under consideration revealed a description of the security service with the specification (24/7, guarded berth, etc.) in more than half of the cases. Thus, the management of the port, focused on receiving guests and residents, understands the importance of property and personal security for the yachtsman.

It is worth noting that in more than half of the marinas under consideration there is an opportunity for unhindered entry of unauthorized persons into the territory. This is primarily due to the presence of entertainment venues, hotels, recreation centers on the territory of the yacht port.

This reduces, firstly, the level of property safety of a yachting tourist, and secondly, the level of his comfortable stay in a marina - a yachtsman may be disturbed not only by noise in the evening and at night, but also by excessive interest in his life and yacht from visitors to entertaining institutions. The combination of these factors can become a significant limitation for the development of inbound yacht tourism.

The remoteness of the police stations from the yacht port also affects the feeling of tourist safety. The remoteness of the nearest police station for half of the marinas under consideration does not exceed 2.2 km:

- Krestovsky Yacht Club - 1.7 km;
- Imperial Marine Yacht Club of St. Petersburg - 1.7 km;
- Yacht Club of St. Petersburg - 2.2 km;
- Yacht port "Tirijoki" - 0.8 km.
- For the other half - more than 3 km:
- Baltiets Yacht Club - 3.1 km;
- St. Petersburg River Yacht Club of the Trade Unions - 4.4 km;
- Yacht Club "Fort Konstantin" - 5.8 km.

The presence of a local police station within walking distance (up to 2.2 km) seems to be a significant factor increasing the level of comfort of a yachtsman.

It should be noted that for foreign tourists arriving in St. Petersburg on yachts, the possibility of obtaining medical care, including emergency care, from English-speaking staff can play an important role. This service is provided by private medical clinics in St. Petersburg, which have private emergency departments.

Measures to improve property, personal security, comfort:

1. Establishment of restrictions for unimpeded access to the berth and direct access to yachts in the form of checkpoints equipped with electronic passes or other means.
2. If possible, allocation of guest mooring spaces in the most remote part of the marina from entertainment establishments. Or providing discounts for guests arriving on a yacht for accommodation in a hotel or guest house located on the territory of the yacht port (if possible). In the absence of such capabilities, this non-normative restriction seems unavoidable.
3. Improving information services - posting in public places information about the nearest police stations, indicating hours of work and contact numbers, as well as information about the possibility of calling private ambulance with English-speaking medical staff.

In addition to the above infrastructural, geographical, and social factors that create difficulties for the development of inbound yacht tourism in St. Petersburg, information technology and positioning in the media environment play a significant role in forming the brand of the yacht capital of Russia. As noted by Ruixia, Ch., Zhou Zh., Zhan, G., Zhou, N., "User-generated content has a major influence on destination branding and destination image" [5].

Let us conduct a comparative analysis of the most attractive marinas meeting PNST requirements (in accordance with the list above).

When comparing the data of yacht ports, we proceed from the following provisions:

1. 40% of European tourists make decisions based on information available on the Internet (according to PriceWaterhouseCoopers research);
2. 83% of the traveling population of Russia aged 18 to 45 years use the Internet when organizing trip (according to expert estimates).

The availability and quality of information services has a significant impact on the tourist flow, including in the field of yacht tourism (Table 1).

Table 1. Comparative analysis of information services of the yacht ports most developed in terms of inbound and domestic tourism.

Name of the yacht port	Site content analysis				Position of the site in the Yandex search system for queries: "yacht port of St. Petersburg"/"yacht mooring in St. Petersburg"/"yacht marina in St. Petersburg" and similar (included/not included in TOP-30)	Representation of the yacht port in Internet catalogs	Rating of marina on Internet resources (Yandex, 5-point scale)	Availability of the site in English/site content
	Description of location, approach routes	Information on the dimensions of the vessels accepted, berth scheme	Information about services with telephone numbers	Possibility to reserve a service on-line				
St. Petersburg River Yacht Club of Trade Unions	Yes	Yes	Yes	No	Not included in TOP-3/Not included	Yes	4.2	No
Krestovsky Yacht Club	Yes	Yes	No	No	No/no/no	Yes	4.1	No
Imperial Marine Yacht Club of St. Petersburg	Yes	Yes	Yes	Yes (partly)	No/TOP-5/TOP-10	Yes	–	No
Yacht Club of St. Petersburg	Yes	No	Yes	No	TOP-3/TOP-30/no	Yes	4.0	Yes (contact information, how to get there)
Baltiets Yacht Club	Yes	No	No	No	No/no/no	Yes	4.0	No
Terijoki Yacht Port	Yes	Yes (partly)	Yes	No	TOP-5/TOP-10/no	Yes	–	Yes (1 page - contacts)
Fort "Constantine" Yacht Club	Yes	Yes	Yes	Yes	No/TOP-3/no	Yes	–	No

A comparative analysis of the most developed and adapted for the reception of tourists yacht ports of St. Petersburg demonstrates a generally satisfactory level of filling the sites with significant tourist information - detailed information on the access routes, dimensions of the vessels received, the berth scheme is presented on almost all the sites under consideration. Information about the service provided is present in 70% of cases, and contact information in one form or another - in 100%

A vulnerability for most of the sites in question can be considered insufficient attention to search engine optimization (SEO). A quick search analysis of the three low-frequency queries revealed that the semantic core for which sites are optimized is often narrowed. This does not allow the user to easily and freely find yacht ports suitable for him.

In modern times, when, according to Marine-Roig, E., & Clavé, S.A., "The volume of data generated in social media has grown from terabytes to petabytes, and data stored and analysed by big companies are set to move from the petabyte to exabyte magnitude soon" [6], and all the major world tourism players use huge amounts of tourist content and pay considerable attention not only to SEO (Search Engine Optimization) and SMM (Social Media Marketing) tools, but also SERM (Search Engine Reputation Management), such an insufficient attention of yacht ports of St. Petersburg to the issues of integrated promotion on the Internet seems to be a significant limiting factor for the development of tourism.

This remark is smoothed by the good work of aggregator sites - RusYachting and YachtInform, which contain basic information about all the marines under consideration.

Quite high ratings of the service of considered marinas on Yandex indicate a well-developed infrastructure. But it should be noted a relatively small number of reviews, which indirectly indicates a low tourist traffic.

Perhaps the most serious drawback of the sites of the considered yacht ports can be considered the absence of the English version of the site in 70% of cases. At the same time, the English version of the site of the Terijoki Yacht Port can hardly be considered full-fledged, since it amounts to no more than half the web page of the translated text. Thus, it is completely legitimate to speak of the absence of an English version of the site in 85% of cases.

Particularly noticeable is the absence of the English version on the site of the yacht club "Fort Konstantin", since it is located in the immediate vicinity of the customs checkpoint and inspection. At the same time, Konstantin presents one of the most progressive sites - the only one on which online services are fully implemented.

Thus, we can conclude that 85% of the yacht ports of St. Petersburg are focused on domestic yachting tourism and on intracity yachting.

Recommendations:

1. Develop recommendations for the yacht ports of St. Petersburg on positioning in the Internet telecommunication network, including for the English-speaking segment of the network;
2. Develop recommendations for the yacht ports of St. Petersburg on the development of online services;
3. Establish interaction with sites-aggregators of yacht information (RusYachting, YachtingInfo, etc.) with the aim of regularly updating data on yacht ports in St. Petersburg;
4. Develop a section on yacht tourism in St. Petersburg (including the English version of this section) on the basis of the official city tourism portal of St. Petersburg "Visit Petersburg";

5. Carry out search engine optimization of the developed section of the "Visit Petersburg" website, ensure its promotion in Yandex and Google search engines, including in the English-speaking segment of the network, ensure regular updating of content.

The development of yacht tourism, despite a number of significant constraints, is promising and important for maintaining the brand of the sea capital. Development of a draft federal law "On Yachting Tourism", planned for 2020 (according to the head of the Crimean State Council Committee on Tourism, Resorts and Sports Alexei Chernyak), implementation of the "Concept for the Development of Yacht Tourism in St. Petersburg", developed in 2019 by the Committee for Tourism Development of St. Petersburg, the consistent elimination of non-regulatory restrictions, the development of infrastructure (large-scale reconstruction of the yacht port "Hercules") open up new prospects for positioning St. Petersburg as a sea - yacht - capital of Russia.

4 Discussion

Prospects for the development of water tourism in St. Petersburg are presented in the Table 2:

Table 2. SWOT analysis of the prospects for the development of water tourism in St. Petersburg.

Strengths	Weaknesses
1. Favorable physical and geographical conditions 2. Favorable climatic conditions (navigation from May to October) 3. Rich historical and cultural heritage. 4. Relatively inexpensive offers of goods and services	1. Poorly developed infrastructure for receiving river cruise ships 2. Lack of targeted advertising and marketing of water tourism 3. A set of problems related to environmental protection
Opportunities	**Threats**
1. Constantly growing interest in the region from European and Russian tourists 2. Deepening international cooperation with the countries of the Baltic region, implementing international projects aimed at creating a single tourist and recreational space of St. Petersburg 3. Organization of thematic water tourism events (river cruises, boating and excursions along rivers and canals, yacht festival)	1. The aging of the fleet 2. Lack of a river port and related infrastructure 3. Deterioration of existing berthing infrastructure

5 Conclusions

Thus, we see that the entire complex of water transport affects the formation and maintenance of the brand of coastal territories and St. Petersburg in particular: the "front sea gates" and the sea facade of St. Petersburg open for passengers arriving by cruise lines; tourists who choose river cruises can see picturesque landscapes of the inland waterways of Russia; on the inner-city excursion routes tourists get acquainted with the unique architecture and transport infrastructure of the "Northern Venice" from the water; islands of the Gulf of Finland and Gulf of Vyborg open their waters for yachtsmen. This whole complex is impossible without the development of water transport and related infrastructure.

References

1. Xiang, Z., Wang, D., O'Rielly, J.T., Fesenmaier, D.R.: Adapting to the Internet: trends in travellers' use of the web for trip planning. J. Travel Res. **54**(4), 1–17 (2014)
2. Avraham, E., Ketter, E.: Media Strategies for Marketing Places in Crisis: Improving the Image of Cities, Countries and Tourist Destinations. Routledge, London (2008)
3. Qu, H., Kim, L.H., Im, H.H.: A model of destination branding: integrating the concepts of the branding and destination image. Tourism Manage. **32**(3), 465–476 (2011)
4. Vinyals-Mirabent, S., Kavaratzi, M., Fernández-Cavia, J.: The role of functional associations in building destination brand personality: when official websites do the talking. Tourism Manage. **75**(12), 148–155 (2019)
5. Ruixia, Ch., Zhou, Z., Zhan, G., Zhou, N.: The impact of destination brand authenticity and destination brand selfcongruence on tourist loyalty: the mediating role of destination brand engagement. J. Destination Marketing & Management **15**, 100402 (2020). https://doi.org/10.1016/j.jdmm.2019.100402
6. Marine-Roig, E., Clavé, S.A.: Tourism analytics with massive user-generated. content: a case study of Barcelona. J. Destination Market. Manage. **4**(3), 162–172 (2015)

Hardening Peculiarities of Metallic Materials During Wear Under Ultrasonic Cavitation

Yuriy Tsvetkov(✉), Evgeniy Gorbachenko, and Yaroslav Fiaktistov

Admiral Makarov State University of Maritime and Inland Shipping, 5/7, Dvinskaya str, Saint-Petersburg 198035, Russia
yuritsvet@mail.ru

Abstract. An attempt was made to determine the mechanism of energy transfer (by microjets or shock waves) from the cavitation zone to the metal surface, which is based on the analysis of metal hardening during cavitation wear. The research was carried out using technical copper and silumine AK12pch. Two batches of cylindrical samples were made from each material. Frontal surfaces of the samples from the first batch were ground, polished, and subjected to cavitation exposure. It was carried out on an ultrasonic magnetostrictive vibrator in fresh water at a frequency and amplitude of oscillations of the concentrator face equal to 22 kHz and 28 μm, respectively. During the incubation wear period, the microhardness of copper and silumin was measured and the maximum achievable microhardness, corresponding to the end of the hardening period, was determined. Samples of the second batch were plastically deformed under uniaxial compression to various degrees of deformation. Then, the deformed samples were cut into two halves along their axis, the plane of the obtained section was ground and polished, and after that the microhardness was measured in the center of the section. According to the measurement results, the correlation between the microhardness and the intensity of plastic deformation was determined for each material. Using the dependence of microhardness on the deformation intensity, an assessment was made for the strain, which corresponds to the onset of fracture of a metal surface under cavitation. It turned out that the plastic deformation value obtained in this way corresponds to the stress state of the surface layers.

Keywords: Hardening · Cavitation wear · Ultrasonic cavitation

1 Introduction

Cavitational wear is the destruction of the surface in a fluid stream due to jets and shock waves generated by the collapse of cavitations. It is a common phenomenon occurring on the surfaces washed by water, for example on bushings and cylinder blocks of high-speed diesel engines [1], turbine blades [2], blades and guides propeller nozzles [3], impellers of pumps for pumping coolants in nuclear reactors [4], etc. Centers of cavitation wear can affect the strength of wear parts and equipment efficiency, therefore special attention is paid to preventing the cavitation damage, to using the wear-resistant

V. Murgul and V. Pukhkal (Eds.): EMMFT 2019, AISC 1258, pp. 409–420, 2021.
https://doi.org/10.1007/978-3-030-57450-5_35

alloys and special surface treatment technologies in particular [5–7], which implies testing of materials for cavitation wear. Currently, such tests are carried out mostly in conditions of vibrational cavitation on ultrasonic magnetostrictive vibrators (MSV) [8]. However, all the features of tests on MSVs have not yet been determined. For example, there is an open question regarding how mechanical action on the surface is carried out when tested on MSV, by shock waves from the cooperative collapse of cavitation bubbles or mainly by micro-jets from the collapse of individual bubbles. For example, according to [9] it was concluded that cavitation destruction is caused by shock waves from the collective collapse of bubbles. Such conclusion was based on analyzing the nature of the pulses measured using a miniature pressure sensor installed under the end face of an MSV concentrator vibrating in a liquid. At the same time according to [10], the researchers showed that the destruction is carried out by shock microjets from separately collapsing bubbles. The high-speed visualization of the cavitation region in combination with measuring pressure using piezoelectric sensors was used.

It is well known that equipment subjected to cavitation wear is often used in corrosive liquids. Therefore, materials are also tested in sea water or other aggressive solutions, and various electrochemical methods are used for analyzing the synergistic effect and the relative contribution of corrosion to the total wear [11]. Answering the question about the mechanism of action on the surface when tested on ultrasonic MSVs will help explaining some seemingly illogical phenomena that occur when tested on ultrasonic MSVs in corrosive environments. For example, reduced wear when switching from testing in fresh water to the tests in sea water in some operational modes. Answering the mechanism question also allows assessing the reliability of the results obtained when testing metal materials on the MSV, with regard to the operating conditions of parts of various hydraulic and marketing equipment.

This question could be answered not by examining the cavitation area itself, but by analyzing the metal reaction to mechanical impact from the side of the cavitation area. This approach has not been applied so far.

A distinctive feature of cavitation wear for metallic materials is the presence of an initial (incubation) period, during which there are practically no mass losses. During the incubation period, metal hardens until its plasticity is completely exhausted, after which the separation of wear particles from the material surface begins.

It is known that the superposition of ultrasonic vibrations significantly changes the nature of the plastic flow of the metal under the influence of static stresses on the material. The strain force decreases and the hardening degree of the metal increases [12, 13]. The observed effects are associated with an increase in the mobility of dislocations under the action of ultrasonic stresses. It occurs in case the opportunity of "crawling" of dislocations from blocked slip planes into free ones is converted. This is due to an increase in the concentration of vacancies in the material, which results in a significant increase in the density of dislocations [13]. Correspondingly, plastic deformation of a metal under the a load with an ultrasonic frequency should be accompanied by a more substantial increase in hardness than in the case of its static application. This is due to the fact that the plastic deformation resistance is proportional to the dislocation density to the degree of about 0.5 [14]. Indeed, as noted in numerous researches, the impact of ultrasonic vibrations on the tool under surface plastic deformation causes an increase in surface hardness. Such increase is significant compared to the mode when only the static indenter is pressed against the surface, while the hardening is more uniform and stable. Similar effects also occur during surface plastic deformation in the tear-off mode, i.e., without the use of clamping force, by unsecured (free) balls deforming the surface under impacts produced with ultrasonic frequency.

According to the research [15], the destruction of metals under cavitation impact has a character similar to the quasistatic. Thus, during the incubation period, one-sided accumulation of plastic deformations from cycle to cycle occurs until a critical degree of deformation is achieved. Thus, by comparing the maximum degrees of surface hardening under cavitation exposure on MSV with static loading (for example, under uniaxial compression) it can be concluded whether the frequency affects metal hardening when tested on an ultrasonic MSV. As a consequence, the conclusion can be made about the mechanism of cavitation impact on the surface.

The goal of the research is to evaluate the mechanism of energy transfer from the cavitation zone to the metal surface, which undergoes wear under conditions of ultrasonic cavitation on the MSW, by the reaction of the metal to plastic deformation under cavitation.

Hardness is a sensitive characteristic of metal hardening during plastic deformation. Taking into account that a very thin surface layer is subjected to plastic deformation when tested on an ultrasonic MSV, it is necessary to use the microhardness method. Therefore, in order to achieve the goal set, the following issues should be resolved: 1) to obtain the graph for change in the microhardness of the metal during the incubation period of the cavitation effect on the ultrasonic MSV; 2) to determine the correlation between the microhardness and the degree of deformation during static compression of metal samples; 3) to compare the value of the maximum metal microhardness obtained by cavitation exposure using MSV with the value obtained by plastic deformation of samples of this metal under static loading.

2 Materials and Methods

Two materials were selected for the experiments, which are very producible and have appreciable hardening during cold plastic deformation, technical copper M3 and silumin AK12pch. Moreover, these materials sharply differ in structure, one has a homogeneous structure, the other (AK12pch) has a pronounced heterogeneous structure.

Copper samples were cut from a hot-rolled bar with a diameter of 16 mm, then they were annealed at 700 °C. Silumin samples were cut from the casting and no additional heat treatment was performed.

All samples had a cylindrical shape. Two batches of samples were made from each material. In the first batch, the samples had a diameter of 16 mm and a height of about 10 mm, they were intended for cavitation wear testing. In the second batch, the diameter and height of the silumin samples were 12 and 18 mm, respectively, and that for copper samples that was 16 and 24 mm, these samples were intended for uniaxial compression tests.

The end surfaces of the samples intended for testing for cavitation wear were ground before testing with abrasive cloths of different grain sizes and polished. The experiments were carried out on a UZDN-2T magnetostrictive vibrator. Sample 1 was mounted in a special mandrel 2 and installed in a transparent container 3 filled with soft fresh water 4 (Fig. 1). The distance Z between the flat surface of the sample and the frontal surface of the concentrator 5 of the vibrator was set equal to 0.5 mm. The water temperature was maintained in the range of 17…23 °C by using a coil 6, along which cooling water 7 was pumped. The oscillation frequency of the concentrator was about 22 kHz, and the oscillation amplitude of the end face of the concentrator, which was measured with an eddy current sensor, was maintained equal to 28 μm.

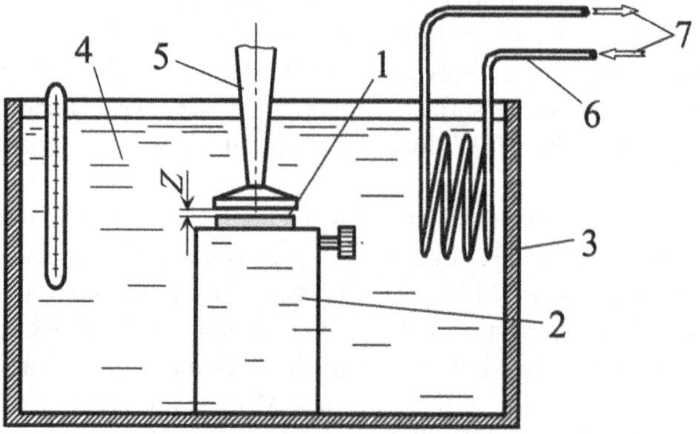

Fig. 1. Model for cavitation wear resistance testing.

During the tests, the samples were periodically weighed on an VL–224V–S analytical balance with a readability of 0.1 mg. After that the surface microhardness was measured within the area of cavitation wear. The mass loss was used to plot the correlation between the wear and time.

The microhardness was measured by PMT-3 microhardness tester within the incubation period at certain time intervals (not exceeding 1 min) of cavitation exposure. In order to avoid the effect of the uneven deformation distribution throughout the thickness of the surface layers on the measurement result, the microhardness values were determined at three loads on the Vickers indenter. Load values of 0.196; 0.49; 0.98 N were used for silumin; 0.098; 0.196 and 0.49 N were used for copper. Six prints were applied at each load value, three prints on each of the two tested samples, which is a total of 18 prints, and the arithmetic mean was taken as the result.

Samples intended for deformation under uniaxial compression conditions were set on the press to various deformation degrees. The deformation intensity ei was calculated according the following formula:

$$e_i = -ln\frac{h_k}{h_0}, \quad (1)$$

where h_0 and h_k—sample height before and after compression.

In order to reduce the friction on the contact surfaces and, as a consequence, to deviate from the uniaxial compression scheme, the frontal surfaces of the cylinder samples were lubricated with Litol-24 non-fluid oil. After compression, the samples were cut along a plane passing through the axis of the cylinders. The surface of the sections was ground using abrasive cloth of different grain sizes, and then polished on wet cloth with the addition of chromium oxide paste. Microhardness measurements were carried out in the area adjacent to the intersection point of the cross section diagonals. This area was selected in order to exclude the impact of deviations on the microhardness. The deviations are caused by the uniaxial stress state taking place in the areas of the sample adjacent to the frontal and cylindrical surfaces of the cylinders. The loads applied to the indenter and the number of prints were the same as for measuring the surface microhardness after cavitation. According to the results of measurements, the correaltaion between the microhardness and the strain intensity was plotted.

3 Outcome of Experiments

The correlations between the wear ΔM, which is expressed by mass loss units of the samples, and the time t of the cavitation effect are presented in Fig. 2 and 3 for copper and silumin, respectively. Since there is no unified methodology for determining the incubation period, it was estimated conditionally using the ΔM(t) correlations. Namely, intersection point of time axis with the tangent 1 drawn to the section with the highest wear rate [8] was used. Also, a graph $H_\mu(t)$ is shown under each correlation ΔM(t), which indicates the change in the microhardness of the material during the incubation period. Microhardness measurements were carried out until (this moment in the graphs is shown by a vertical dashed line) the relief of the wearing surface made it possible to produce clear prints. As can be seen, with the onset of cavitation, the microhardness increases and then decreases, i.e., after the hardening stage, the softening stage begins. This process looks very clear in Fig. 3. After the softening stage of copper (Fig. 2), the surface relief made it possible to register the beginning of the next hardening stage. However, according to the ΔM(t) curve, noticeable mass losses have not yet begun. That is, the beginning of the second hardening stage cannot be attributed to hardening of underlying layers, which should have occurred after the removal of the upper extremely riveted layer. The reason for this behavior of copper remains to be determined in the future. According to the present study, the maximum microhardness obtained at the end of the hardening stage was determined using the $H_\mu(t)$ dependences. It is this value that was used below when conducting a comparative analysis of hardening under cavitation and during static deformation.

Figure 4 and 5 show the correlations $H_\mu(e_i)$ between the microhardness and the strain intensity obtained from uniaxial compression tests of copper and silumin, respectively. At the same time, the horizontal lines 1 in Fig. 4 and 5 correspond to the maximum microhardness of copper and silumin, which were detected on their surface during the incubation period of cavitation wear (Fig. 2 and 3). For copper, the maximum value is H_μ = 388 MPa, 665 MPa for silumin.

Let us assume that the dependence of hardness on the strain intensity is uniform, that is, independent of the stress state diagram [16], and the ultrasonic frequency does not affect the kinetics of hardening of the alloy. In such case, at the intersection point of the graph $H_\mu(e_i)$ with line 1, the value of the deformation corresponding to the destruction beginning of the surface during cavitation exposure can be determined. The correlation $H_\mu(e_i)$ for silumin is limited by a strain value of approximately 0.8 (Fig. 5) due to the formation of cracks at higher specimen strains under compression. Thus, the ultimate strain of silumin under cavitation was determined by extrapolating the obtained plot of the dependence $H_\mu(e_i)$ to the intersection with line 1.

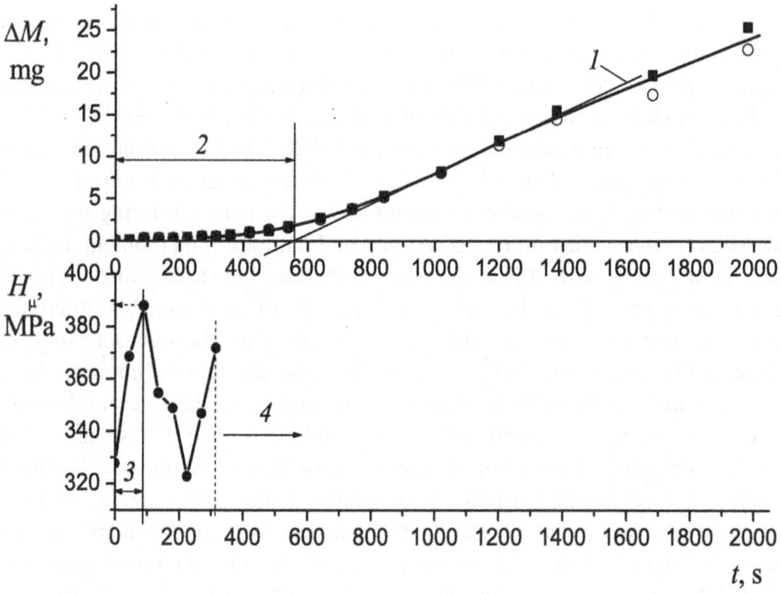

■ — the first sample; ○ — second sample; *1* — tangent (see text); *2* — conditional incubation period; *3* — stage of hardening; *4* — zone of developed relief (clear prints are not obtained)

Fig. 2. Kinetics of mass loss changes (above) and surface hardening (below) during cavitation wear of copper.

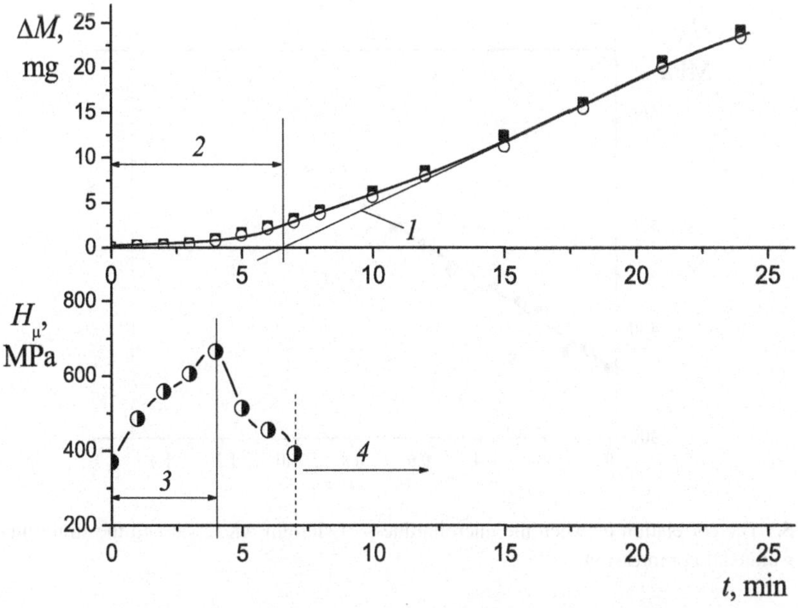

■ — the first sample; ○ — second sample; *1* — tangent (see text); *2* — conditional incubation period; *3* — stage of hardening; *4* — zone of developed relief (clear prints are not obtained)

Fig. 3. Kinetics of mass loss changes (above) and surface hardening (below) during cavitation wear of silumin AK12pch.

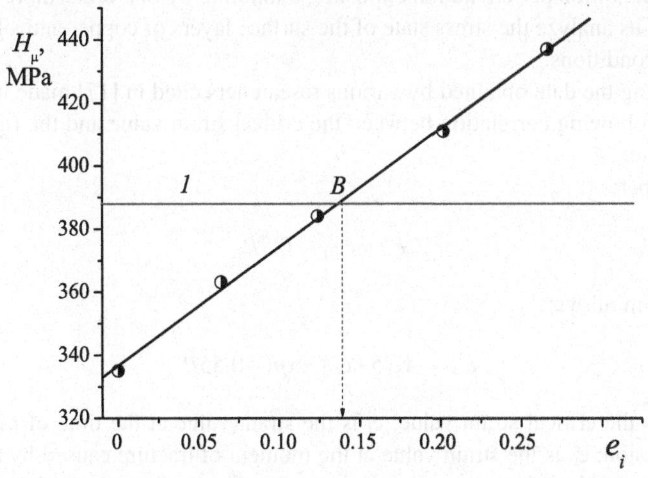

Fig. 4. Correlation between the microhardness of copper and the strain intensity under uniaxial compression.

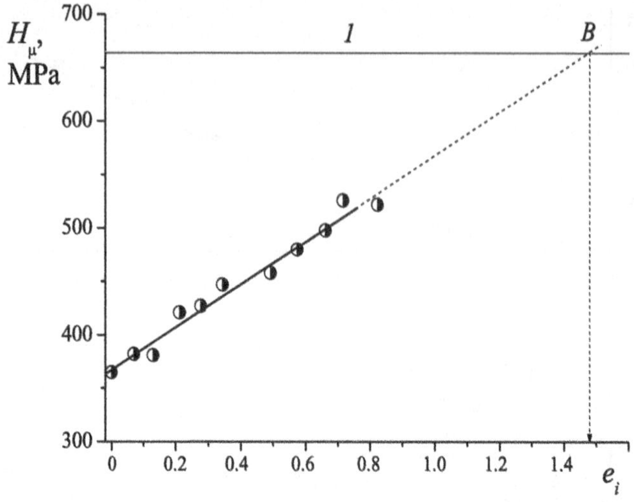

Fig. 5. The correlation between the microhardness of silumin AK12pch and the strain intensity under uniaxial compression.

Having lowered the perpendicular from point B to the abscissa axis, the following values of critical deformation (i.e., corresponding to the destruction attainable by cavitation) were obtained: for copper (Fig. 4) $e_{cr} = 0.14$, and for silumin (Fig. 5), $e_{cr} = 1.47$.

A large difference in the obtained values of the critical deformation is noticeable, i.e., under conditions of cavitation exposure, silumin is by one order more plastic than copper. Let us analyze the stress state of the surface layers of copper and silumin under cavitation conditions.

Analyzing the data obtained by various researchers cited in [17] made it possible to obtain the following correlation between the critical strain value and the rigidity of the stressed state:

for copper:

$$e_{cr} = e_0 - 0.7P, \tag{2}$$

for aluminum alloys:

$$e_{cr} = 1{,}75 \cdot e_t \cdot exp(-0{,}55P), \tag{3}$$

where e_{cr} is the critical strain value; e_t is the strain value at the time of rupture under uniaxial tension; e_0 is the strain value at the moment of fracture caused by torsion. P is the stiffness coefficient of the stress state model proposed by G. A. Smirnov-Alyaev [18] in the following form:

$$P = \frac{\sigma_1 + \sigma_2 + \sigma_3}{\sigma_i} \tag{4}$$

where σ_1, σ_2, σ_3 are the main stresses, σ_i is the stress intensity.

Unfortunately, torsion experiments couldn't help obtaining e_0 for copper. Therefore, the e_0 value was determined as follows. Uniaxial rupture of copper samples was tested, the shape and size of the samples were consistent with the recommendations of GOST 1497-84. The samples were machined from the same rod as the samples for wear and uniaxial compression tests, and then subjected to the same heat treatment. It was determined that $e_t = 2.02$. Since $e_{cr} = e_t = 2.02$ for the uniaxial tension, and the coefficient P = +1 according to (4), then by substituting the indicated values in formula (1), we obtained $e_0 = 2.72$. After substituting the values $e_0 = 2.72$ and $e_{cr} = 0.14$ in the formula (1), the following was calculated for the conditions of cavitation exposure:

$$P = \frac{e_0 - e_{cr}}{0.7} = \frac{2.72 - 0.14}{0.7} = 3.68. \tag{5}$$

In order to assess the stress state on the surface of silumin according to formula (3), it is necessary to have the e_t value. However, there were a large number of pores in the AK12pch alloy, which was due to the used casting technology. This was the reason for the extreme sensitivity of the uniaxial tensile test results to the size of the samples. Therefore, according to the results of microhardness measurements on oblique thin sections prepared from samples after cavitation exposure of different durations during the incubation period, the maximum achievable riveted layer thickness was determined. It turned out to be approximately 0.055 mm. Uniaxial tensile specimens, the shape and dimensions of which corresponded to GOST 1497-84, were machined from the same casting, from which the samples were made for wear and uniaxial compression tests. The diameter of the working part of the samples was d = 10, 6, 5, and 3 mm. 5 samples of each diameter were produced. According to the results of tensile tests, the correlation $e_t(d)$ was plotted. By extrapolating this correlation to a value of d = 0.055 mm, it was determined that $e_t = 1.0$. After substituting the values of $e_t = 1.0$ and $e_{cr} = 1.47$ in the formula (3), the following expression was obtained for the conditions of cavitation exposure:

$$P = -\frac{1}{0.55} ln\left(\frac{e_{cr}}{1.75 \cdot e_t}\right) = -\frac{1}{0.55} ln\left(\frac{1.47}{1.75 \cdot 1}\right) = 0.32. \tag{6}$$

4 Results and Discussion

The stiffness coefficient value for the stress state model of copper P = 3.68, obtained for cavitation conditions on an ultrasonic MSV, indicates that a very stiff stress state appears on the surface of copper, stiffer than under uniaxial tension. The value P = 3.68 corresponds to a state close to biaxial tension (according to (4) with equal biaxial tension, P = +2, and with equal triaxial tension P = +∞).

When testing silumin using the ultrasonic MSV, the stiffness coefficient value for the stress state model is P = 0.32, which corresponds to a state that occupies an intermediate position between torsion and uniaxial tension.

The fact that a stiffer stress state arises on the surface of copper is due to the higher hardness of silumin (83 HV) compared to copper (57 HV). The surface stress state under wear is sensitive to surface hardness, the higher the hardness, the softer the stress state is in the surface layers under constant conditions of cavitation [15].

At first glance, the obtained values of the stiffness coefficient seem to be overestimated, and they should be expected to be negative rather than positive. However, it is known that with cavitation exposure, a wavy relief is first formed on the surface. The depth of the troughs gradually grows, craters are formed with the material displaced along the periphery of the craters in the form of ridges, for example, A1 and A2 (Fig. 6). The formation of wear particles occurs, most likely, as a result of material over-deformation of the ridges. This is also confirmed by the results of experiments on indentation of cones into the surface, the highest values of the coefficient P corresponding to the shear (P = 0) were found for the metal on the contour of the contact spot [19]. Obviously, the material at the surface points at the top of the ridge has a more stiff stress state than at the bottom of the dent. Obviously, it also differs from the state corresponding to compression, since tensile stresses appear on the surface of the ridge. An analysis of the literature data carried out in [15] allows concluding that the lower the hardness of the metal is, the greater is the height h of the ridge and smaller is the radius of its rounding. This means that the ratio of tensile stresses is greater in the material adjacent to the top of the ridges, t. e. in the area where the cracks appear (Fig. 6), and destruction occurs.

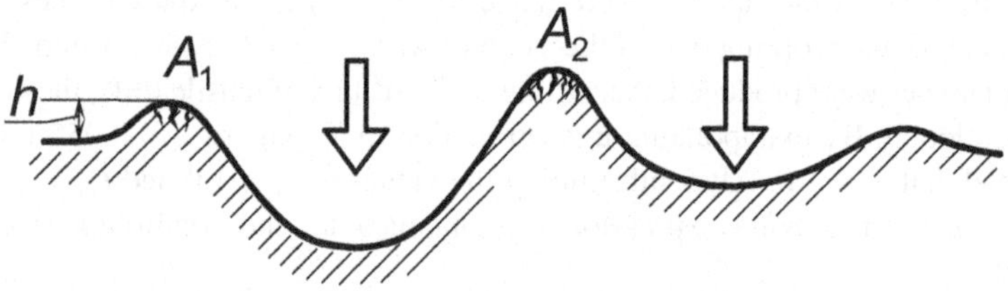

Fig. 6. Pattern model of the tensile stresses occurring in the surface upon shock impact of micro-jets (the direction of impact is shown by arrows).

Thus, it can be argued that such a hard stress state on the surface of the tested alloys can be caused only by the influence of micro-jets on the surface. If the main mechanism of energy transfer to the surface were shock waves from the cooperative collapse of bubbles in a cavitation area, then a much higher surface hardening would be recorded. The mechanical action on the surface would occur with the oscillation frequency of the MSV concentrator, i.e., about 22 kHz. Using the $H_\mu(e_i)$ correlation in this case (which is obtained according to the plastic deformation results of samples under static loading) to estimate the critical deformation degree under cavitation would lead to significantly higher values of the critical deformation degree.

When energy is transferred from a cavitation cloud to the surface by shock jets, it is unlikely that the micro-jets will hit the same micro-section of the surface. That means that the loading frequency of a specific micro-volume of the surface will be significantly lower than the oscillation frequency of the MSV concentrator, and the ultrasonic frequency effect disappears. At the same time, when the jets simultaneously hit in the areas surrounding a specific micro-section of the surface, the occurrence of an even more severe stress state than at the tops of the ridge surrounding an isolated dent is likely. In Fig. 6 the ridge A_2 corresponds to this region.

5 Conclusion

As a result of analyzing the stress state of the metal surface under the cavitation effect on ultrasonic MSVs, positive stiffness coefficient values of the stress state pattern are obtained. Thus, the plastic deformation of the surface of the tested alloys occurs under the prevalence of tensile stresses in the pattern.

The ultrasonic frequency effect on the hardening kinetics of a metal when tested on an ultrasonic MSV can occur only if the surface is affected by shock waves. They are generated as a result of collective collapsing of bubbles in each oscillation cycle of the concentrator end. In this case, using the dependence of the microhardness on the intensity of plastic deformations (plotted according to the results of uniaxial compression of the samples), significantly higher values of the critical degree of deformation of the surface layers under cavitation would be obtained than those recorded in this work. Therefore, the values of the stiffness coefficient of the stress state model would be negative.

The main energy transfer mechanism from a cavitation cloud to the surface of a metal tested on an ultrasonic MSV is shock micro-jets. Due to the random distribution of bubbles in the cavitation cloud under the concentrator, it seems unlikely that the impact of micro-jets in each vibration cycle of the frontal surface of the concentrator will fall on the same micro-volume of the surface. Thus, the influence of ultrasonic frequency is excluded. At the same time, the simultaneous impact of several micro-jets into the areas surrounding a specific micro-area of the surface can lead to the appearance of significant tensile stresses in the latter.

In order to confirm the conclusions obtained in this work, further studies should be aimed at conducting similar experiments on a wider range of alloys, the hardness and plasticity of which would vary over a wide range.

References

1. Gravalos, I., Kateris, D., Xyradakis, P., Gialamas, Th.: Cavitation erosion of wet-sleeve liners: case study. J. Middle Eur. Constr. Des. Cars (MECCA) **4**(3), 10–16 (2006)
2. Kumar, P., Sain, R.P.: Study of cavitation in hydro turbines—a review. Renew. Sustain. Energy Rev. **14**, 374–383 (2010). https://doi.org/10.1016/j.rser.2009.07.024

3. Boorsma, A., Whitworth, S.: Understanding the details of cavitation. In: Proceedings of the Second International Symposium on Marine Propulsors, Hamburg, Germany, vol. 11, pp. 319–327 (2011)
4. Sreedhar, B.K., Albert, S.K., Pandit, A.B.: Cavitation damage: theory and measurements – a review. Wear **372–373**, 177–196 (2017). https://doi.org/10.1016/j.wear.2016.12.009
5. Kwok, C.T., Man, H.C., Cheng, F.T., Lo, K.H.: Developments in laser-based surface engineering processes: with particular reference to protection against cavitation erosion. Surf. Coat. Technol. **291**, 189–204 (2016). https://doi.org/10.1016/j.surfcoat.2016.02.019
6. Qiao, Y., Cai, X., Chen, Y., Cui, J., Tang, Y., Li, H., Jiang, Z.: Cavitation erosion properties of a nickel-free high-nitrogen Fe-Cr-Mn-N stainless steel. Mater. Technol. **51**(6), 933–938 (2017). https://doi.org/10.17222/mit.2017.034
7. Momeni, S., Tillmann, W., Pohl, M.: Composite cavitation resistant PVD coatings based on NiTi thin films. Mater. Des. **110**, 830–838 (2016). https://doi.org/10.1016/j.matdes.2016.08.054
8. ASTM G32-10 Standard test method for cavitation erosion using vibratory device. ASTM International (2010)
9. Brujan, E.-A.: Cavitation Erosion. Cavitation in Non-Newtonian Fluids 155–174 (2010). https://doi.org/10.1007/978-3-642-15343-3_5
10. Zubrilov, S.P., Rastrygin, N.V.: Issledovanie protsessa kavitatsii i vozmozhnosti snizhe-niia erozionnogo iznosa. Vestnik Gosudarstvennogo universiteta morskogo i rechnogo flota im. admirala S. O. Makarova **11**(4(56)), 705–717 (2019). https://doi.org/10.21821/2309-5180-2019-11-4-705-717
11. Amann, T., Waidele, M., Kailer, A.: Analysis of mechanical and chemical mechanisms on cavitation erosion-corrosion of steels in salt water using electrochemical methods. Tribol. Int. **124**, 238–246 (2018). https://doi.org/10.1016/j.triboint.2018.04.012
12. Han, Q.: Ultrasonic processing of materials. Metall. Mater. Trans. B **46**(4), 1603–1614 (2015). https://doi.org/10.2172/859314
13. Tsuboi, R., Kakinuma, Y., Aoyama, T., Ogawa, H., Hamada, S.: Ultrasonic vibration and cavitation-aided micromachining of hard and brittle materials. In: Procedia CIRP 1 (2012), 5th CIRP Conference on High Performance Cutting 2012, pp. 342–346 (2012). https://doi.org/10.1016/j.procir.2012.04.061
14. Belyakov, A.: Microstructure and mechanical properties of structural metals and alloys. Metals **8**(9), 676 (2018). https://doi.org/10.3390/met8090676
15. Tsvetkov, Y.: Behavior of surface layers of material in cavitational wear. Int. J. Multiph. Flow **22**, 145 (1996). https://doi.org/10.1016/s0301-9322(97)88555-2
16. Larsson, P.-L.: Determination of residual stresses utilizing the variation of hardness at elastic-plastic indentation. J. Test. Eval. **47**(4), 20170525 (2018). https://doi.org/10.1520/jte20170525
17. Ma, F., Kuang, Z.-B.: Stresses, deformations and porosities in standard fracture specimens. Acta Metall. Mater. **42**(2), 497–507 (1994). https://doi.org/10.1016/0956-7151(94)90504-5
18. Segal, V.: Review: modes and processes of severe plastic deformation (SPD). Materials **11**(7), 1175 (2018). https://doi.org/10.3390/ma11071175
19. Stefaniv, B.V.: Investigation of wear resistance of protective coatings under conditions of hydroabrasive wear. Paton Weld. J. **2016**(9), 26–29 (2016). https://doi.org/10.15407/tpwj2016.09.05

Technology Level and Development Trends of Autonomous Shipping Means

Vladimir Karetnikov, Evgeniy Ol'Khovik, Aleksandra Ivanova, and Artem Butsanets(✉)

Admiral Makarov State University of Maritime and Inland Shipping, 5/7, Dvinskaya Str., Saint Petersburg 198035, Russia
butsanetsaa@gumrf.ru

Abstract. The modern technical level already allows us to build small crewless ships. Today, more than 60 models of such vessels are known to be developed. However, usually the scientific literature describes the development of elements, mechanisms and intelligent control systems, but studies on the assessment of trends and development tendencies of autonomous shipping means by analyzing patent literature have not been identified. This paper presents the results of a thematic search for titles of protection (applications, patents for inventions and utility models, certificates for computer programs) in relation to autonomous shipping means. As a result, applications and patents for inventions/utility models, as well as registered certificates for computer programs, were identified. The analysis of existing solutions on the sources of patent information for the further study of the technical level of autonomous navigation means is carried out. Over 70% of applications have been submitted in the last 2 years. It was confirmed that the market of autonomous small vessels and vessels under the operator's control is actively developing.

Keywords: Autonomous shipping facilities · Crewless ships · State of the art · Development trends · Patent search

1 Introduction

A promising way to increase the efficiency of the functioning and development of water transport is the implementation of crewless shipping. Such shipping implies a complete absence of crew on vessel board, while the movement of the vessel is carried out under the remote control of the operator or using software and hardware in an autonomous mode.

The modern technical level makes it possible to build crewless ships, on board of which a complex of software and hardware is functioning, including information and communication systems, telematic technologies, navigation and communication equipment, and automated remote control systems [1, 2]. This will make it possible in the near future to switch to crewless navigation technologies. According to experts [3, 4], commercial operation of crewless vessels is possible in a few years. There are known developments of analytical models for assessing the risk of introducing a crewless vessel [5].

V. Murgul and V. Pukhkal (Eds.): EMMFT 2019, AISC 1258, pp. 421–432, 2021.
https://doi.org/10.1007/978-3-030-57450-5_36

To date, it is known about the development of more than 60 variants of crewless robotic vehicles [6], however, the literature describes particular cases of the development of elements and mechanisms for autonomous shipping, the design of intelligent ship control systems in general [7, 8], including using IOT technology [9], in particular using the MQTT protocol [10], as well as exhaust gas control [11, 12], automated mooring [13] and authentication tools [14] but studies to assess trends and development tendencies of autonomous navigation tools through the analysis of patent literature is not revealed.

Objects of research are means of autonomous (crewless, unmanned) shipping, including the control system of such means. In the framework of this study, a thematic search for titles of protection (applications, patents for inventions and utility models, certificates for computer programs) was performed. As a result, applications and patents for inventions/utility models, as well as registered certificates for computer programs, were identified. The analysis of existing solutions on the sources of patent information for the further study of the technical level of autonomous navigation means is carried out.

Means of autonomous shipping, as a rule, include traffic systems, power systems, a communications complex, a navigation complex, a coast station and additional equipment. A patent search was conducted to evaluate the trends and development tendencies of autonomous shipping facilities.

2 Materials and Methods

In the process of conducting patent research, the main area of International Patent Classification (IPC) search was determined: B63 – vessels and other floating equipment; equipment for them, – in the classification headings of which a substantive search was conducted among Russian, Eurasian, European and international patents, as well as among applications for inventions and utility models filed from 01/01/2014 to 09/31/2019. Search was carried out by electronic databases of abstracts and full descriptions of patents for inventions and utility models of the Russian, Eurasian, European and international patent offices, namely through the database of the Federal Institute of Industrial Property (FIIP) and the database of European Patent Office, containing national patents of European countries, Japan, the US, the world's patents, published by the World Intellectual Property Organization (WIPO), as well as other patents of the national patent offices.

The methodology of the search was to study abstracts of patent documents located in the indicated electronic databases using the keywords corresponding to the research topic: unmanned vessel, unmanned ship, unmanned boat, autonomous ship, autonomous vessel, autonomous boat, unpiloted boat.

To analyze the level of technology and development trends of autonomous shipping means, 436 patent documents were selected, which were then classified according to common criteria into groups and subgroups.

3 Results

The selected 436 documents include 58 patents for inventions and 148 patents for utility models, 227 applications for inventions, as well as three certificates for computer programs. Figure 1(a, b) displays information on the number of published patent documents by year (from 2014 to 2019) and their types, and in Fig. 1(c) presents the distribution of patent documents by applicant countries.

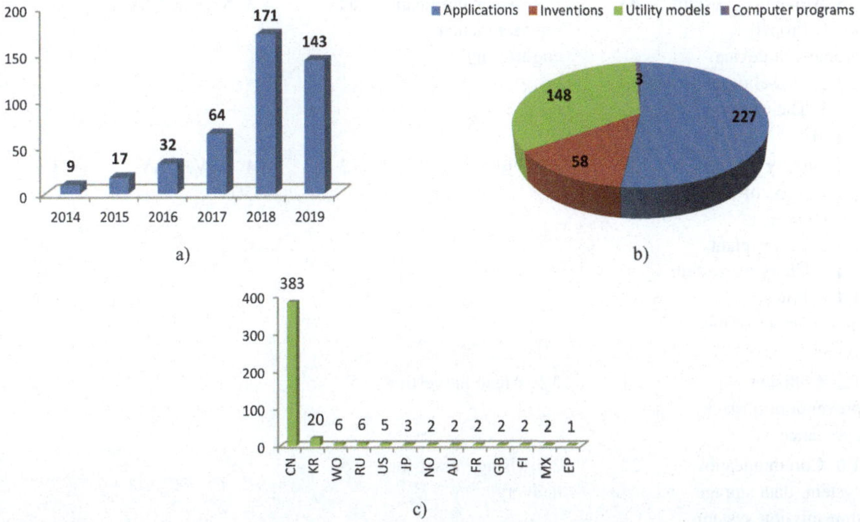

Fig. 1. The number of published patent documents by year (a), by type (b) and by applicant country (c).

All selected patents can be divided into three groups. The first group of patents relates to the technical means of autonomous shipping (MAS), the second group of patents relates to the field of application of MAS, the third group of patents is a macro system of several MAS. The most numerous is the first group of patents (Table 1).

Table 1. Distribution of patent documents into groups and subgroups.

First group of patents	Amount of patents	Second group of patents	Amount of patents	Third group of patents	Amount of patents
1. Technical field of autonomous navigation	246	2. Scope of unmanned vehicle (UV)	164	3. Systems of several UV	26
1.1. Mooring, docking, anchoring of the vessel, parking lock	12	2.1. Monitoring of the marine environment, water areas, observation	22	3.1. Unmanned surface vehicle (USV) +Unmanned aerial vehicle (UAV)	**12**

(continued)

Table 1. (*continued*)

First group of patents	Amount of patents	Second group of patents	Amount of patents	Third group of patents	Amount of patents
1.2. Re-equipment of the vessel	2	2.2. Water quality control (sampling, control of water pollution, measurement)	**48**	3.2. Unmanned surface vehicle (USV) +Unmanned underwater vehicle (UUV)	6
1.3. Movement system 1.3.1. Propeller, propulsion device 1.3.2. Vessel drive 1.3.3. The engine of the vessel	**43**	2.3. Water purification (environmental engineering)	**32**	3.3. Several USV	5
1.4. Supply system 1.4.1. Type of energy source used 1.4.2. Power plant 1.4.3. Charging system 1.4.4. Power management of the vessel	**38**	2.4. Fishing	12	3.4. USV+UAV +UUV	3
1.5. Collision prevention, obstacle avoidance	7	2.5. Algae harvesting	5		
1.6. Communication system, data storage, transmission system, navigation system	22	2.6. Using vessel for delivery	5		
1.7. Monitoring the position of the vessel, its stability	31	2.7. The study of geographical elements of the area	10		
1.8. Control and monitoring system of the vessel	**42**	2.8. Rescue vessel (rescue and/or fire fighting)	24		
1.9. Cooling system	3	2.9. Icebreaking vessel	2		
1.10. Atypical UV hull	5	2.10. Shoreline survey	2		
1.11. Deployment of the vessel (removal/launch of the vessel	22	2.11. Laying of pipes	1		
1.12. Auxiliary equipment/system	7	2.12. Search of the mines	1		
1.13. Protection system, protection device, method	11				
1.14. Installation of equipment on a vessel	1				

The first group consists of 14 subgroups and includes 246 patent documents, the second group consists of 12 subgroups and 164 patent documents, the third group is divided into 4 subgroups and contains 26 patent documents.

In the first group, three areas of autonomous shipping can be distinguished, according to which patent applications are filed more often than the rest: the vessel movement system – 43 documents, the vessel control system – 42 documents and the vessel power system – 38 documents. From 20 to 30 objects are patented in such technical areas of autonomous shipping as vessel stability, communication system and vessel deployment (vessel removal and launch). Then there are areas related to the mooring (docking) of the vessel, protection of the vessel (hull or devices on the vessel), collision prevention (obstacle avoidance) and additional auxiliary equipment. The smallest number of objects are patented in the following areas: atypical hull of the vessel, cooling system, re-equipment of the vessel and installation of equipment on the vessel.

All patents in the subgroup "Vessel movement system" can be divided into three sections: propulsion, drive and engine. The largest number of patent objects relates to propulsion devices: a shaftless propeller (CN208931621 U), a grease seal and leak prevention device for propelling a small unmanned vessel (CN109538764 A), a ship propulsion device with a cutting function (CN108263587 A), waterproof fittings of ship propeller (CN208264538 U), safety device of marine propeller (CN208248479 U), wind power propulsion (CN108557047 A), integrated electric screw of unmanned ship (CN107554738 A), monitoring of the cycloid propeller control system (CN205186500 U).

Patent documents of the "Vessel management system" subgroup disclose either a vessel management system located on board and monitoring the work of the vessel independently according to a predetermined program, or developers patent remote methods for controlling the vessel. It can be a remote control (CN207208405 (U), CN107402568 (A), CN104192261 (B)), cloud technology for coast communications (CN108200175 (A)), communication via Bluetooth and Google glasses (CN107264731 (A)), as well as using cameras (KR20180046803 (A)) or an entire control center for monitoring vessels (FI20175133 (A)). Patents GB2511731 (B), CN204642100 (U), CN106444776 (B) represent fully autonomous vessels, and patent CN107145145 (B) discloses dual-vessel control: autonomous navigation mode and remote management mode with a remote control. There are also applications for inventions in which an unmanned vessel drives other vehicles. These are applications filed by the Norwegian companies ROLLS ROYCE MARINE (NO20171498 (A1)) and KONGSBERG MARITIME CM AS (EP3448748 (A1) - European patent).

All patent documents in the subgroup "Vessel power system" can be divided into the following categories: type of energy used, power plant, charging system and power management of the vessel. The most common patenting area in this subgroup is the type of energy used.

According to the energy used, developments are patented in which the vessel moves due to the wave, the sun, fuel, gas, and the battery. Most of the developments use a hybrid power system, i.e. based on several energy sources (mainly two, three).

The developments mentioned in patent documents: US2019118920 A, CN109334935 A, CN208198727 U, CN106394824 B use wave energy to move an unmanned vessel. A patent for the invention was obtained by the PRC in December 2018, a utility model was obtained also in 2018, and two applications for the invention were registered in 2019. It turns out that the use of this type of energy to propel a vessel is a fairly new and promising development.

The energy of the sun is used in patent documents: CN108357642 (B), CN208021679 (U), CN205837137 (U), CN205661628 (U), wind energy is used in CN106741782 (A); jointly the energy of the sun and wind is used in the developments indicated in CN108438138 (A), CN205738030 (U), CN105644752 (A), and the hybrid power system, which provides for the use of fuel in conjunction with some environmental energy source, is used in CN109774909 (A), CN109436272 (A), CN203946267 (U). For example, patent application CN106114802 (A) is a self-powered unmanned surface vessel provided by wind and gas turbines and solar panels, and the converted electrical energy is stored in a lithium battery. Patent application CN108583810 (A) discloses a portable dock for an unmanned vessel that provides automatic charging and data exchange. Utility model CN207060369 (U) discloses a battery and supercapacitor in the form of a composite power source for an unmanned vessel.

The subgroup "Stability of the vessel, control of its position, stabilization of the vessel" is small compared to the sections already considered (31 patent documents). The following patents belong to this subgroup: patent application CN109911114 A is an unmanned vessel with a three-stage damping self-stabilizing system; patent application CN109278945 A discloses a stable binary micro-light unmanned boat comprising two hulls with FRP fiberglass material; the purpose of the application of the invention CN109501971 A is to create a kind of unmanned boat system, which is not easily susceptible to the shadow of ring waves, can efficiently and stably move in order to effectively carry out accurately check of a large-scale water field; utility model CN208915391 U is a carbon-fiber unmanned hull of the vessel that can reduce vibration; utility model CN208789876 U reveals a carbon fiber protective hull that improves device stability; application for the invention CN109367718 A contains an unmanned vessel with a sliding path type device for adjusting, due to which the effect of better reduction of the transverse angle of inclination and radius of rotation is achieved; patent application CN108945282 A presents an unmanned vessel carbon fiber protective device for resistance to wind and waves, in which a streamlined plate allows for unimpeded navigation of an unmanned vessel, avoids the problem of rocking left and right during a voyage, the vessel becomes more steady by extending the telescopic end of the electric pusher, that is why the problems of turning over of the vessel caused by the wind can be avoided; patent application CN108016576 A discloses a self-healing unmanned vessel, the first propeller of which is used to rotate in the opposite direction when the unmanned vessel is in an inverted state.

The "Communication systems" subgroup includes methods for detecting underwater (reefs, water grasses, etc.) and surface obstacles (vessels of various types), image transmission systems, sending/receiving data using satellite communication systems (Beidou, GPS) or a smartphone, the use of special technologies and systems (eNodeB and Hadoop), as well as base stations of the coastal control center (CN206932215 U, FI20175127 A).

Basically, the patent documents of the subgroup "Deployment of the vessel (removal/launch of the vessel)" disclose a lifting mechanism (CN208882044 U, CN108945300 A, CN108298026 A, CN108945316 A) or a lifting and launching device (CN109591963 A, CN208979067 U, CN108248765 A, US 9199699 B, CN203996800 U); either it can be a system and method for automatically uncoupling and engaging when lifting and placing an unmanned vessel (CN109353456 A) or an automatic system for distributing and restoring an unmanned boat, which includes a lifting mechanism, a propulsion and a mechanism for opening and closing the door (CN109229284 A).

The trend in the area of "Mooring, docking of the vessel" is magnetic mooring equipment (CN108820134 A, CN106965907 A) and methods/systems for controlling the discharge of the anchor (CN109733537 A, CN109720523 A, CN206797651 U).

Patent documents in the subgroup "Vessel cooling system" are represented by the following developments: patent application WO2019163397 A1, which is an unmanned boat equipped with a CS cooling structure for refrigeration the central processing unit CPU1 for image recognition and the central processing unit CPU2 for control; patent application CN109011264 A discloses a fire protection and cooling method for a small unmanned cockpit hull; utility model CN208089378 U discloses a cooling pipe system for power plant that introduces cooling water and dissipates heat. Thus, the cooling system can be provided for the entire hull of the vessel, as well as for individual devices (processor, power plant).

As an auxiliary equipment, developments related mainly to winding/unwinding devices (CN208291433 U, CN207809690 U, CN207658003 U, CN207658002 U), as well as photo and/or video fixing equipment (CN207926709 U, CN108111733 A), ventilation systems (CN208053608 U) are patented.

In the "Vessel protection" subgroup, developments are patented that imply protection of both the entire hull of the vessel (CN108438151 A) and individual devices on it. In the first case, the vessel can have a corrosion-resistant hull (CN107697234 A), have a shock-absorbing bumper to increase the impact resistance of the hull (CN207141340 U), have a special Foretell hull protection mechanism based on the elastic effect (CN206691334 U, CN108438151 A), and have a device that prevents impact to an unmanned vessel of various objects, such as fishing nets, aquatic plants, etc. (CN104960628 B). In the second case, the developers propose a camera protection system for an unmanned vessel (CN208079231 U), an overvoltage protection system (CN109606580 A). Also, an invention for saving a vessel while stranded (CN106741790 B) is patented.

The atypical shape of the vessel is contained in the application for invention CN108995775 A, filed and published in 2018, where the hull of the vessel is in the form of a rugby ball, i.e. it is a kind of enclosed unmanned boat that resists storm waves, has excellent carrying capacity, the precision equipment placed on the vessel will not vibrate, and the measurement will stabilize. Another form of a vessel in the form of a sailboat (CN208278282 U, CN109606579 A) is already quite common for unmanned vehicles compared to the previous patent document.

The field of collision prevention, obstacle avoidance is not a common area of patenting. The following developments are presented here: CN206704473 U - an unmanned vehicle with an accurate collision avoidance function; KR101823029 B1 - a system and method for controlling obstacles on a vessel (avoiding obstacles), KR20170058719 A - a method for controlling the tracking of an unmanned vehicle and preventing collisions with an obstacle, CN205801453 U, CN106043617 B - an unmanned vessel that dodges obstacles and prevents capsizing.

In the second group, two areas of autonomous shipping can be distinguished, according to which patent applications are filed more often than others: water quality control – 48 documents and environmental engineering – 32 documents. From 10 to 25 objects are patented in such fields of application of autonomous shipping as rescue, water area monitoring, fishing and the study of geographical elements of the area. Further, there are areas in which up to five developments are patented: algae collection, use of a vessel for delivery, icebreaking vessel, shoreline survey, pipe laying and mine search.

Developments in the field of water quality control are devices/systems for sampling water/oil stains, or elements associated with a water sampler (butt connection of a sampler (CN108502101 A), liquid supply device (CN109437344 A)), or methods/mechanisms for determining quality of water (for example, spectral absorption method, as in CN208125606 U), or fully automatic methods/devices for controlling water pollution (CN108760388 A, CN207964362 U, CN207241970 U, CN107585266 A).

Development in the field of environmental engineering are unmanned vessels for cleaning the water surface. Most of them are able to automatically search and dispose of debris on the surface of the water (CN108425351 A, CN108275246 A, CN108058790 A, etc.). Some of them are designed for the treatment of oil pollution on water (CN109024522 A) or for clearing the river bed and removing sludge (CN205530419 U and CN107757835 A).

Development in the field of rescue is mainly a rescue vessel that can move autonomously or be controlled remotely. Some patented objects differ from others in the use of binocular vision (CN208931612 U, CN108995782 A), the presence of additional medical functions on board (CN109050801 A), or the possession of special means on board for automatic ejection of a lifebuoy (CN207191347 U). A small part of the development is designed to extinguish a fire at offshore facilities (CN109292048 A, CN108714279 A, CN105235824 B).

Developments in the field of water area monitoring relate to vessels intended for monitoring the aquatic environment: above-water, underwater (CN108516058 A), or both. These can be devices for measuring the spectral characteristics of the environment (CN109520938 A), temperature, salinity and the speed of sound (CN109425328 A). Some patent objects are capable of observing in any weather (CN109436217 A, CN103803045 A) and using special technologies, for example, based on big data (CN109336198 A).

Patenting in the field of fisheries is mainly related to areas such as search (CN109644956 A, CN109501972 A), fishing (CN208963283 U, CN109430184 A, CN109577295 A) and feeding fish (CN108812479 A, CN107439425 A). Moreover, the latter direction is the most relevant, which is associated with an increase in the number of fish farms in open water, where they are cultivated.

The most common area in the study of geographical elements of the area is underwater topography (CN208683068 U, KR101946542 B, CN108955653 A, CN206171741 U, CN205396467 U). Also patented are objects in the direction of mapping (CN109606040 A, CN105937899 A) and related to unmanned seismic vessels (MX2018008831 A, MX2018004667 A).

In the field of delivery, unmanned vessels for the transport of equipment and special devices (CN208576697 U), underwater robots (CN206437172 U), various goods (CN107791761 A), liquid cargo (WO2017030446 A), as well as fast unloading systems on coast (CN207274931 U) are patented.

The least common patent areas related to the second group of patents are "ice-breaking vessel", "shoreline survey", "pipe laying" and "mine search", where a total of 1-2 patents were registered.

The third group of patents, which is a system of several unmanned vehicles, consists of four subgroups differing in the types of unmanned vehicles used (surface, underwater and air): the combined use of the vessel and the aircraft, the vessel and the underwater vehicle, the vessel together with the aircraft and underwater vehicles and several vessels. The use of a vessel with other autonomous devices is the most common patenting area in this group – 21 pieces and exceeds the number of developments using only autonomous surface vehicles by four times.

The most relevant developments are those involving the joint use of an unmanned surface vessel and an unmanned aerial vehicle (UAV). Here, the vessel and the aircraft can work together, for example, in search and rescue operations (CN108945342 A), when monitoring the water area (CN107727081 A), when mapping (KR101863123 B), when cleaning the water area (CN107097910 B) or the vessel can only be a platform for transportation and/or maintenance of UAVs (CN108820130 A, CN108466703 A, CN105292398 A).

Developments related to the joint operation of unmanned surface and underwater vehicles (CN206307246 U, CN106394815 A) or when an unmanned surface vessel is used only for servicing an unmanned underwater vehicle (CN109367706 B, CN109367707 B, US10363996 B) are patented two times less. Developments related to the joint use of several unmanned vessels (CN108725704 A, CN206520723 U, CN107097908 B) are recorded with the same intensity.

Objects associated with the use of all three modes of transport are rarely patented: WO2019113137 A, KR101913391 B and KR20170043035 A. Such developments are published twice less than objects of the second (USV + UUV) and third (several USV) subgroups and four times less than the objects of the first subgroup (USV + UAV).

4 Discussion

The field of technical means of autonomous shipping is the most numerous and quite diverse field of patenting. The most actively developments are patented which related to the vessel movement system (43 pieces), the vessel control system (42 pieces) and the vessel power supply system (38 pieces). Among the developments, preference is given to vehicles based on environmentally friendly energy sources (wind, sun, waves), or, at least, based on a hybrid power system. Moreover, the use of wave energy to advance the vessel began to appear in the developments only in 2018-2019, which indicates the emergence of a new trend in the development of power sources for means of autonomous shipping.

A little less often, but also in large numbers, patent documents are registered in such areas as vessel stabilization (31 pieces), communication system (22 pieces), vessel deployment (launch/retrieval) – 22 pieces. Among the identified patent documents, the most widely used direction of stabilization of the vessel is the use of special hull material (usually carbon). Often, the stability of a vessel is ensured by designing a two- or three-hull boat, either by installing a special device on the vessel that controls its position, or by using propulsions.

All patents in the field of deployment of a vessel in one way or another relate to the mechanism for extracting/launching the vessel: either directly represent the raising/lowering mechanism itself, or relate to parts (connecting devices, for example) of the lifting and launching device.

The area related to the mooring (docking) of the vessel, despite the fact that it is quite young (the first patent document was registered in 2017) and few in terms of identified patents (12 pieces), is developing dynamically and the number of patented developments in this area increases many times from every year. Here, the trend is magnetic mooring equipment, as well as methods/systems for controlling the discharge of the anchor.

Among 12 areas of application of autonomous shipping means, there are such areas as water quality control (48 pieces) and environmental engineering (32 pieces). Objects in the field of water area monitoring (22 pieces) and rescue operations (24 pieces) are patented a little less often. The new and least common patent areas related to the second group of patents are "icebreaking vessel", "shoreline survey", "pipe laying" and "mine search", where a total of 1–2 patents were registered.

The use of a vessel with other autonomous devices is the most common area of patenting in the third group of patents – 21 pieces and exceeds the number of developments using only autonomous surface vehicles by four times. The most relevant here are developments that include the combined use of an unmanned surface vessel and an unmanned aerial vehicle (UAV). Despite the fact that the combined use of several unmanned vehicles is a less patentable area of autonomous shipping (26 pieces) compared with the above areas, nevertheless, it is actively developing and improving.

5 Conclusion

An analysis of the technology level and trends in the development of autonomous shipping means using selected patent documents showed that with the development of unmanned technologies, the market for fully autonomous small vessels or vessels under the operator's control is actively developing, new technologies and new areas of application for vessels are being introduced. Given the large number of applications for registration of patents for inventions (52% of the selected patent documents), this industry is rather young and actively developing (of the analyzed patent documents 72% of the applications were filed over the past two years).

The vast majority of patent documents belong to the People's Republic of China (PRC) – 88% of the total volume of the analyzed documents, and are represented by organizations such as Suzhou Vocational University, Shandong Haoce New Material Technology Co., Ltd., Systems Engineering Research Institute Of CSSC, Dalian

University of Technology, Beijing Institute of Technology, China Ship Design & Research Center Co. Ltd., Zhongshan Institute of Technology, Wuhan University of Technology, Hainan University, Beijing Yudi Technology Co., Ltd., etc.

Among other countries, the Republic of Korea stands out, which owns 5% of all registered documents and which is represented by organizations: Korea Institute of Ocean Science and Technology, Daewoo Shipbuilding & Marine Engineering Co., Ltd., Marine Research Co Ltd., Marine Spatial Information Technology, etc.

The European part of the continent is represented by such companies as the US-British L3 Harris Technologies, the Finnish subsidiary of the British multinational corporation Rolls-Royce Holdings plc – Rolls-Royce Oy Ab, the Norwegian companies Rolls Royce Marine AS and Kongsberg Maritime. American companies such as Liquid Robotics Inc, Ion Geophysical Corp, Maritime Robotics AS, Sea Machines Robotics, Inc., are also actively developing autonomous shipping facilities.

56% of registered patents relates to technical MAS, where the leading role belongs to such areas as the vessel movement system (17.5%), the vessel control system (17%) and the vessel power system (15%). Although the application scope of MAS is inferior in the number of patentable documents of the previous group of patents (38% versus 56%), it also has its own development trends: water quality control and environmental engineering.

Thus, over the past six years, the most relevant and actively patented areas related to means of autonomous shipping are areas bound up with the motion, control and power systems of the vessel, as well as using an unmanned surface vehicle to control water quality and clean the water area. New trends in the development of autonomous shipping means are the directions of the vessel mooring (docking) and the use of an unmanned vessel for the delivery of goods/equipment, as well as the use of an unmanned vessel in conjunction with other types of unmanned vehicles, especially with unmanned aerial vehicles.

References

1. Im, I., Shin, D., Jeong, J.: Components for smart autonomous ship architecture based on intelligent information technology. Proc. Comput. Sci. **134**, 91–98 (2018). https://doi.org/10.1016/j.procs.2018.07.148
2. Kwon, Y.: Korean technical innovation: toward autonomous ship and smart shipbuilding to ensure safety. In: Proceedings of the International Seminar on Safety and Security of Autonomous Vessels (ISSAV) and European STAMP Workshop and Conference (ESWC) 2019, pp. 83–94. Sciendo (2020)
3. Wróbel, K., Montewka, J., Kujala, P.: System-theoretic approach to safety of remotely-controlled merchant vessel. Ocean Eng. **152**, 334–345 (2018). https://doi.org/10.1016/j.oceaneng.2018.01.020
4. Hogg, T., Ghosh, S.: Autonomous merchant vessels: examination of factors that impact the effective implementation of unmanned ships. Aust. J. Marit. Ocean Affairs **8**(3), 206–222 (2016)

5. Karetnikov, V.V., Kozik, S.V., Butsanets, A.A.: Risks assessment of applying unmanned means of water transport in the water area. Vestnik Gosudarstvennogo universiteta morskogo i rechnogo flota imeni admirala S. O. Makarova **11**(6), 987–1002 (2019). https://doi.org/10.21821/2309-5180-2019-11-6-987-1002
6. Liu, Z., Zhang, Y., Yu, X., Yuan, C.: Unmanned surface vehicles: an overview of developments and challenges. Ann. Rev. Control **41**, 71–93 (2016). https://doi.org/10.1016/j.arcontrol.2016.04.018
7. Geertsma, R.D.: Design and control of hybrid power and propulsion systems for smart ships: a review of developments. Appl. Energy **194**, 30–54 (2017). https://doi.org/10.1016/j.apenergy.2017.02.060
8. Qiu, B.: Hybrid cloud based cyber-enabled ship control and management system. In: 2018 IEEE International Conference on Prognostics and Health Management (ICPHM). IEEE (2018). https://doi.org/10.1109/icphm.2018.8448547
9. Yang, S.: Development of ship structure health monitoring system based on IOT technology. In: IOP Conference Series: Earth and Environmental Science, vol. 69, no. 1. IOP Publishing (2017). https://doi.org/10.1088/1755-1315/69/1/012178
10. Atmoko, R.A., Riantini, R., Hasin, M.K.: IoT real time data acquisition using MQTT protocol. J. Phys.: Conf. Ser. **853**(1) (2017). https://doi.org/10.1088/1742-6596/853/1/012003
11. Raptotasios, S.I., Sakellaridis, N.F., Papagiannakis, R.G., Hountalas, D.T.: Application of a multi-zone combustion model to investigate the NOx reduction potential of two-stroke marine diesel engines using EGR. Appl. Energy **157**, 814–823 (2015). https://doi.org/10.1016/j.apenergy.2014.12.041
12. Jenaru, A., Arsenie, P., Hanzu-Pazara, R.: The estimation of the pollutant emissions on-board vessels by means of numerical methods. In: IOP Conference Series: Materials Science and Engineering, vol. 145, no. 8. IOP Publishing (2016). https://doi.org/10.1088/1757-899x/145/8/082016
13. Butsanets, A., Ol'khovik, E.: Development of technical means for mooring the unmanned vessels. In: 2019 International Multi-Conference on Industrial Engineering and Modern Technologies (FarEastCon), pp. 1–3. IEEE (2019). https://doi.org/10.1109/fareastcon.2019.8934708
14. Karetnikov, V.V., Butsanets, A.A., Mitrofanova, A.V. Selection of a rational authentication method for the remote systems of interaction with the vessels. In: IOP Conference Series: Earth and Environmental Science, vol. 378, no. 1, pp. 012090. IOP Publishing (2019). https://doi.org/10.1088/1755-1315/378/1/012090

Quality Assessment of the System of Filling a Shipping Lock Chamber from Under the Segmental Guillotine Gate

Anatolii Gapeev[1], Konstantin Morgunov[1(✉)], and Mariya Karacheva[2]

[1] Admiral Makarov State University of Maritime and Inland Shipping, St. Petersburg, Russia
morgunovkp@gumrf.ru
[2] FSBI «Kanal imeni Moskvy», Petrovka, d. 3/6, Moscow, Russia

Abstract. The present paper provides the results of studies aimed at determining the main parameters of a system of filling a chamber from under a segmental guillotine gate with cylindrical culverts, using the example of the low-pressure shipping lock №2 of the Moscow Canal. The main parameters of the chamber filling system of a shipping lock include its hydraulic, energetic and kinematic parameters, functional design and physical dimensions estimated through the composition of elements and their arraignment. The quality of a filling system is estimated through the complex of certain parameters, which ensure safe conditions of ship lockage, required pass-through capacity and fail-safe operation of equipment of the facility. The navigation lock under consideration is located in Tempy village, Dmitrovsky district of the Moscow region. This single-chamber single-line lock was commissioned in 1937. It has a chamber filling system standardized for all locks of the Moscow Canal: a head system consisting of segmental guillotine gate, four trench culverts and screen-beam baffles behind them. A type of a chamber filling system for shipping locks used in Russia is mainly chosen proceeding from the results of laboratory hydraulic studies aimed at substantiation of safe modes of filling and draining chambers for different groups of ships and vessels and determination of hydraulic parameters. The influence of other parameters on the quality of the chamber filling system was sometimes ignored.

Keywords: Shipping lock · Chamber filling system · Segmental gate · Screen-Beam baffle

1 Introduction

According to building codes and regulations (Construction Rules SP101.13330.2012 Retaining walls, shipping locks, fish passing and fish protecting facilities. Updated version of SNiP 2.06.07 – 87. M: Ministry of Regional Development of Russia, 2012. – 70 p.), the type of a filling system is chosen in accordance with the size of both a lock chamber and a design ship (vessel). Codes define the main hydraulic requirements to chamber filling and draining systems for time and conditions for vessels mooring in a

V. Murgul and V. Pukhkal (Eds.): EMMFT 2019, AISC 1258, pp. 433–441, 2021.
https://doi.org/10.1007/978-3-030-57450-5_37

lock chamber. Speed of flow in a chamber and approach canals is also prescribed in codes. Efficient time modes of filling and draining chambers, which provide safe conditions for locked vessels, should be chosen based on the results of experimental and theoretical researches. Recommendations on other parameters of chamber filling systems are not given.

Experience of designing and operating shipping locks shows [1, 2] that hydraulic parameters themselves do not enable effectively assessing hydraulic-and-mechanical processes running during ship lockage in a chamber. Hydraulic parameters help understand how the discharge, water levels and other parameters depending on accepted lockage modes are changing in time. They specify the time of filling and draining required for assessing the lock's pass-through capacity. Safe conditions for vessel mooring in a lock chamber are provided at the stage of their design, usually through the application of modes of gate lifting with constant speeds, substantiated by the results of laboratory hydraulic tests.

When assessing the quality of a chamber filling system, one should take into account its structural properties, as well as changes in hydraulic, energetic and kinematic parameters. It was proved that composition of components of a chamber filling system, their outline and arrangement significantly influence the system's basic parameters [3]. The sufficient number of components of a chamber filling system and their sound arrangement provide the effective damping of flow speeds at a chamber entrance, and their uniform depth-distribution at a negligible distance from the flow inlet [4, 5]. In this case, safety of the ship lockage can be provided by using non-linear modes of gate lifting (with variable speeds), in which the time of filling and draining the chamber is significantly reduced [6]. Energetic parameters are used to calculate the volume of a stilling basin and the length of a damping section, within which flow speeds are finally depth-smoothed. Kinematic parameters allow estimating flow speeds in culverts formed by the components of the chamber filling system.

A great number of works related mainly to the functional design and studies on mooring conditions for different types of vessels are devoted to chamber filling systems of shipping locks with head filling chambers. Some works provide the results of laboratory and field studies aimed at the estimation of operation of chamber filling systems with short cills, performing the charge from under flat guillotine gates. Those estimations are made based on kinematic parameters of the flow. The results of these studies are specified in works [7, 8]. A.V. Mikhailov [4] was the first who suggested assessing the operation of lock chamber filling systems on the basis of flow energy parameters. He elaborated a method for determining the energy capacity of the flow within the volume of a stilling basin and along the length of a damping section, where flow speeds are depth-smoothed. However, this method does not take into account the composition of components of a stilling basin and does not allow determining the size of culverts, placed on the upper head of the shipping lock. Work [4] does not give any recommendations on determining vortex shedding zones in flow stilling basins and does not provide the assessment of influence of those zones on kinematic parameters of a chamber filling system.

Theoretical definition of the structure and location of vortex flows appearing when a non-steady flow slips around a significant number of different-shaped and different-sized components is still considered a challenging problem when studying chamber filling systems. For today, there are some works describing vortex flows appearing when slipping around simple-shaped components of a chamber filling system. For instance, works of A. V. Vasin and co-authors discuss the flow in a draining system of a circular-shaped chamber [9, 10]; works of I. V. Lipatov [11, 12] describe a filling system with rectangular components, which was adopted only for the purpose of complete damping of the flow energy, without taking into account kinematic parameters of the flow. There are also works of some other authors [13–19].

The quality of the system of filling the chamber of the Moscow Canal's shipping lock № 2 has not been assessed in terms of considered performance parameters before.

2 Materials and Methods

Shipping lock №2 of the Moscow Canal is located in Tempy village, Dmitrovsky district of the Moscow region. This single-chamber and single-line navigation lock was put into operation in 1937. It has a chamber filling system which is typical for locks of the canals: a head system consisting of segmental guillotine gates, four trench culverts and screen-beam baffles behind them [20]. It differs from chamber filling systems of other canal locks only by the size of certain components and change of a bottom of a segmental gate.

When analyzing changes of hydraulic, energetic and kinematic parameters of the system of filling the chamber of lock №2, the existing methods, known principles of hydraulics and results of experimental studies were used.

The existing method for calculating volume of a stilling basin and the length of a damping section, which is based on the use of maximum and specific flow capacity, is valid for any type of head system of filling, yet it does not allow determining a componential composition or a size of culverts formed by those components. On the basis of this method, when analyzing changes in flow energy within the limits of a stilling basin, there was developed a method for determining a size and mutual location of components of a system of filling a shipping lock's chamber through a flat guillotine gate [21]. This method was applied in the design of lock №8 of the Sheksna hydro-engineering complex and of the second line of the Nizhnesvirsky shipping lock.

Vortex zones of stilling basins are visually detected when performing research via models of shipping locks or are approximately estimated by using the fundamental principles of the theory of turbulent jets motion. To assess hydraulic parameters of the filling system, the results of theoretical and experimental research are used, methods to carry out which are widely known [22].

3 Results

Figure 1 shows the kinematic scheme of flow motion in the upper head's stilling basin of the Moscow Canal's lock №2 from under the guillotine gate.

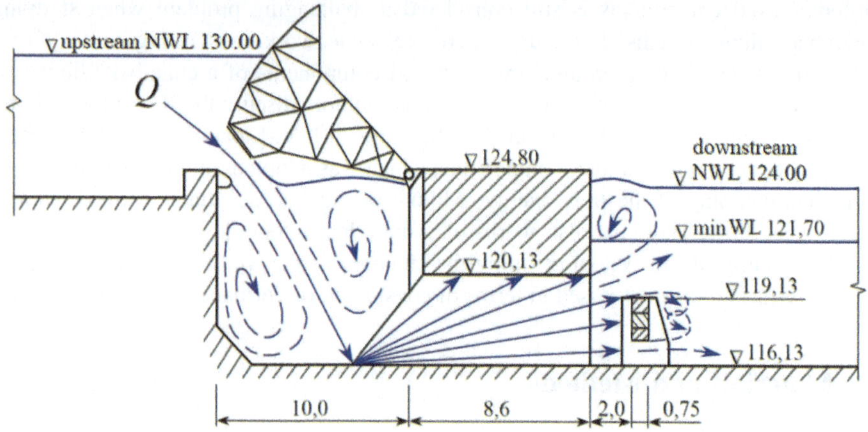

Fig. 1. Scheme of the upper head of lock №2 of the Moscow Canal.

Vortex swirls appear in a gate recess when flow jets are detached from a spillway lip, the edge of a spillway crest, as well as from the upper edges of trench culverts and from behind the screen-beam baffles. The flow loses a great part of energy there. Hydraulic characteristics defined during 1961 field studies are presented in Fig. 2. They are used to determine energy flow parameters.

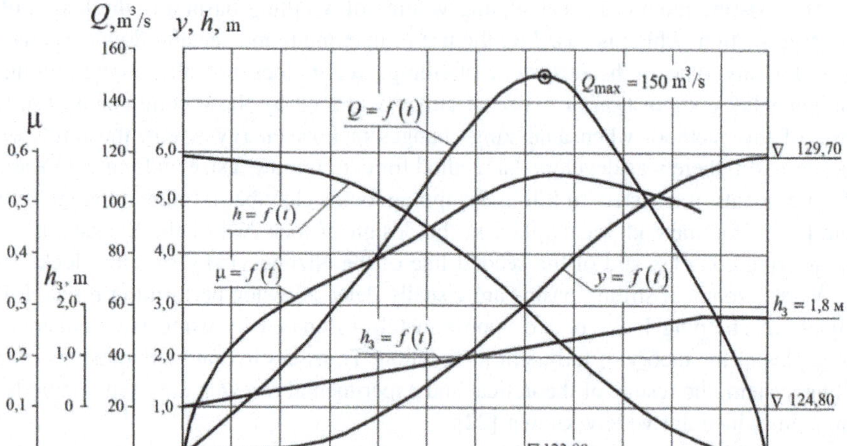

Fig. 2. Hydraulic characteristics of the process of filling the chamber of Moscow Canal's lock №2.

Hydraulic research on the shipping lock №2 was carried out by scientific research sector Hydroproject named after S. Y. Zhuk in order to improve conditions of mooring in the chamber for large range vessels. When the chamber with such vessels as Volgoneft and Volgo-Don, displacing up to 70 000 kN of water weight, was filled according to the project schedule of gate lifting up to the altitude of 1.6–1.8 m with constant speed of 0.2 m/min, breaks of mooring ropes were observed and, consequently, ships reclined against the lock structure. During the surveys, the operational speed of gate lifting was slightly lower than the design speed: 0.17 m/min.

As a result of researches it was proposed to fill the chamber at the altitude of gate lifted of 1.22 m, and place large range vessels at the lower head of the lock in order to reduce the hydrodynamic impact of the flow.

Later, some other measures for ensuring safety of large range vessels passing through the locks of the Moscow Canal were considered, including the use of automooring devices, but they were not practically implemented.

According to the presented values of changes in water discharge and water levels in the chamber, a curve of changes in the discharge coefficient $\mu = f(t)$ were plotted (see Fig. 2), as well as curves of changes in the total energy of the flow $E = f(t)$, entering the chamber, of specific energy $E_{sp} = f(t)$, referred to the area of the clear opening of the chamber (Fig. 3).

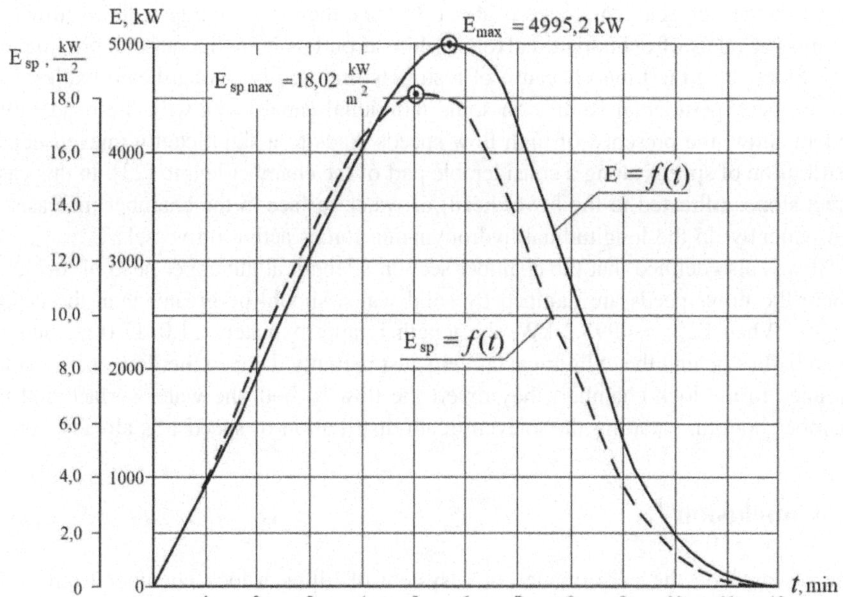

Fig. 3. Change of total energy $E = f(t)$ and specific energy $E_{sp} = f(t)$ of the flow.

The highest value of discharge coefficient (0.56) is observed at $Q_{max} = 150.0\ m^3/s$, while at the altitude of gate lifted of 1.8 m, the discharge coefficient $\mu = 0.48$. The time of filling the chamber is from 12.0 to 12.3 min. The highest energy values are $E_{max} =$ 4 995.2 kW and $E_{sp\ max} = 18.02\ kW/m^2$ respectively. They are observed 1.6–2.3 min earlier than Q_{max}.

4 Discussion

The kinematic scheme of the flow motion in the stilling basin of the Moscow Canal's lock №2 slightly differs from the schemes under consideration with filling from under the guillotine gate [5, 23]. The difference is only in the absence of a vertical girder grid at the flow entrance into the lock chamber: instead of that grid there are screen-beam baffles at the distance of 2.0 m behind the trench culverts. The chamber filling system under consideration basically ensures the suppression of flow energy, but when passing through cylindrical culverts the flow faces considerable resistances. This fact is associated with the insufficient accuracy of sizes of inlets and outlets of cylindrical culverts, as well as the lack of devices providing the even flow distribution at the entrance to the lock chamber. When the maximum discharge rate is passing, the average flow speed in the gap under the segmental gate is about 3.8 m/s, then it is slightly reduced (down to 3.0 m/s) at inlets of cylindrical culverts, while at outlets it can increase approximately by 2 times (up to 6.0 m/s) because of restraints imposed by screen-beam baffles.

As the experimental studies on some individual canal locks with the head filling system show, the presence of high flow speeds leads to a significantly uneven depth-distribution of speeds along a considerable part of the chamber length [23]. In this case, direct slopes (directed to the lower head) of water surface in the chamber increase, as consequently do the longitudinal hydrodynamic forces acting on vessels.

It was also defined that the chamber section's length at the upper head of lock №2, where the flow speeds are damped by solid walls, should be greater than the design length. When $E_{max} = 4995.2$ kW, the length is approximately 11.0–13.0 m. Screen-beam baffles behind the cylindrical culverts just partially dampen the flow energy at the entrance to the lock chamber; they divert the flow to both the water surface and the chamber bottom, entailing the uneven depth-distribution of speeds, as already noted.

5 Conclusion

The research on the performance of a system of filling a lock chamber from under segmental guillotine gates with cylindrical culverts allowed evaluating the changes in its basic parameters during the process of filling the lock chamber. They do not meet the requirements to the quality of chamber filling systems in terms of design and kinematic flow parameters.

The existing structure of the upper head of the Moscow Canal's lock №2 enables suppressing considerable part of flow energy before its entering the lock chamber, however due to large resistances in culverts of the system, high flow speeds appear leading to the increase in direct slopes of water surface in the lock chamber. Reduction of gate lifting altitude down to 1.22 m, which is advisable according to the results of field studies, only entails the increase in time of filling a chamber, while hydrodynamic forces acting on large-range vessels decrease negligibly. Therefore, it is the complete reconstruction of the chamber filling system, what will help achieve the required pass-through capacity and satisfactory conditions for large range vessel mooring in the considered lock chamber.

Figure 4 presents the concept of reconstructing the upper head of lock №2, which was considered via the existing method for determining the componential composition of dampening devices, their outlines and sizes of culvert openings formed by them, regarding the system of filling a chamber from under the flat guillotine gate [21].

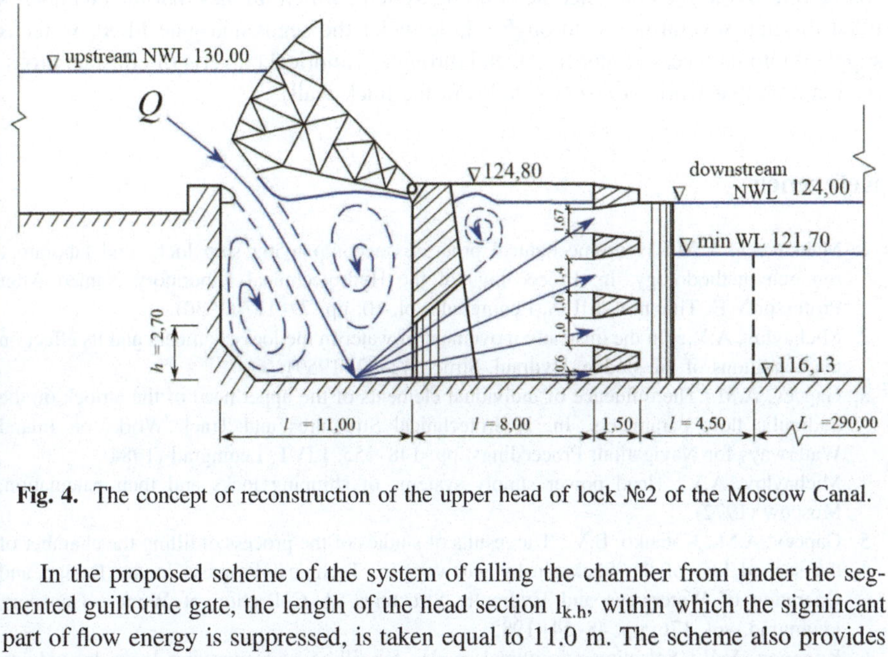

Fig. 4. The concept of reconstruction of the upper head of lock №2 of the Moscow Canal.

In the proposed scheme of the system of filling the chamber from under the segmented guillotine gate, the length of the head section $l_{k.h}$, within which the significant part of flow energy is suppressed, is taken equal to 11.0 m. The scheme also provides for a section of flow distribution within the distance of 8.0 m, with of four 1.5-meter-long and 1.0-meter-high trapeze-shaped beams installed at the entrance. Beams of vertical distribution grid are installed with irregular interval between holes, which increases from bottom to top: 0.60; 1.00; 1.40 and 1.67 m.

All elements of the reconstructed filling system are located within the existing abutment with cylindrical culverts and a screen-beam baffle behind them; they do not make the actual head length bigger than the design head's length.

Under considered change in element composition of the upper head of the Moscow Canal's lock №2, the area of culverts' openings from inlet to outlet into the lock chamber increase approximately by 1.5–2.0 times, which essentially lowers the flow speed. For instance, when segmental gate is lifted to the height of 1.55 m, which is permissible under the mooring of a design ship with a displacement of 64,000 kN and under the 150.0 m^3/s of maximum discharge passing, flow speeds averagely vary from 3.20 m/s to 2.15 m/s in the opening under the screen wall and drop down to 1.20 m/s when exiting gaps of the girder grid. The length of the lock chamber's stilling basin designed for the final depth-smoothing of flow speeds is approximately 4.5 m. When flow speeds are evenly distributed in the lock chamber, safe mooring conditions are provided for vessels.

The proposed concept of reconstruction of the Moscow Canal's lock №2 will further require theoretical and field studies aimed at choosing favorable modes of gate lifting, which will provide safe conditions for mooring of large-range vessels and the effective time of filling the chamber.

It is also necessary to specify the existing method of hydraulic calculation of the chamber filling system of the Moscow Canal lock chambers. In terms of structure, it can be referred to the combined head filling system, which implies that the chamber is filled through several holes: through a hole under the segmental gate lifted, water is supplied into its recess (water inlet), and through cylindrical culverts of constant cross-section area that work as nozzles (holes in the thick wall).

References

1. Makkaveev, V.M.: Hydromechanical processes accompanying ship locks and laboratory research methodology. In: Proceedings of the Hydrotechnical Laboratory Named After Professor V.E. Timonov, LIIPS, Leningrad, vol. 10, pp. 79–117 (1930)
2. Michaylov, A.V.: On the unsteady movement of water in the lock chambers and its effect on the conditions of the sludge. Hydraul. Struct. 42–62 (1959)
3. Gapeev, A.M.: The influence of individual elements of the upper head of the airlock on the hydraulic flow parameters. In: Hydrotechnical Structures and Track Works on Inland Waterways for Navigation: Proceedings, pp. 148–155. LIVT, Leningrad (1984)
4. Michaylov, A.V.: Head power supply systems of shipping locks and their calculation, Moscow (1972)
5. Gapeev, A.M., Kolenko, B.V.: The results of studies of the process of filling the chamber of the second line of the Sheksninsky gateway. In: Technical Progress in the Design and Operation of Waterways and Hydraulic Structures: A Collection of Works, Transport, Leningrad, vol. 176, pp. 48–54 (1983)
6. Kononov, V.V.: Selection of optimal modes for filling and emptying locks by solving equations of transient motion. In: Waterways and Waterworks: Proceedings, vol. 158, pp. 86–95. Transport, Leningrad (1977)
7. Gapeev, A.M., Kononov, V.V.: Water systems for shipping locks. GUMRF, Saint Petersburg (2013)
8. Gapeev, A.M.: Improving the performance of shipping locks with a head feed system: thesis for the degree of Doctor of Technical Sciences. SPGUVK, Saint Petersburg (1999)
9. Vasin, A.V., Timofeeva, O.A.: Modeling of vortex flow zones in drainage galleries of lock chambers. Hydraul. Eng. **2**, 17–21 (2013)
10. Timofeeva, O.A.: Analysis of numerical methods for calculating fluid flow in culverts. Vestnik Gosudarstvennogo universiteta morskogo i rechnogo flota imeni admirala S.O. Makarova **4**, 104–108 (2015)

11. Lipatov, I., Sitnov, A., Chichkin, I.: Analytical approach to the assessment of operation efficiency of elements of energy dissipation in transport hydrotechnical facilities. Mar. Intellect. Technol. **2**(4), 170–176 (2018)
12. Chichkin, O.I.: Numerical modeling of hydrodynamic processes in a quench chamber of a typical Volga-Don lock. In: Design, Construction, and Operation of Waterway Structures, Conference Proceedings, vol. 2, pp. 73–80. GUMRF, Saint Petersburg (2018)
13. Levachev, S.N., Gorgin, A.G., Shaytanov, A.M.: Comparison of gateway camera calculation methods. Hydraul. Eng. **6**, 32–39 (2018)
14. Emelyanov, A.V.: The results of full-scale studies of the process of passing large-tonnage vessels VDSK and the choice of a method for justifying the safe dimensions of shipping connecting channels. Vestnik Volzhskoy akademii vodnogo transporta **42**, 255–260 (2015)
15. Guselnikova, E.N., Gimgin, P.A.: Hydraulic processes of filling (emptying) the lock chamber during normal operation. Trudy Novosibirskogo gosudarstvennogo arkhitekturno-stroitel'nogo universiteta (Sibstrin) **3**(69), 70–80 (2018)
16. Verelst, K., Vercruysse, J., De Mulder, T.: Hydraulic design of a filling emptying system for the new Royers lock in the port of Antwerp (Belgium). In: Proceedings of the 36th IAHR World Congress: Deltas of the Future and What Happened Upstream, Delft, Netherlands, pp. 4563–4574 (2015)
17. Yang, Q., Zhai, J., Jiang, Z.: Numerical simulation of saltwater intrusion in a lock with long canal adopting coupled box model and three-dimensional model. In: Proceedings of the 36th IAHR World Congress: Deltas of the Future and What Happened Upstream, Delft, Netherlands, pp. 181–190 (2015)
18. Zubkova, E., Klementyev, A., Lobanov, V., Khvostov, R.: Features of modeling the process of entering ship into the lock-chamber at great restriction of the flow area section. Mar. Intellect. Technol. **2**(4), 166–169 (2018)
19. Ven, V.D., Loon, V.: The interaction of a lock's filling jet and the ship in the lock chamber, using scale model measurements. In: 7th IAHR International Symposium on Hydraulic Structures. ISHS, pp. 402–410 (2018)
20. Gapeev, A.M., Morgunov, K.P.: Analysis of changes in the main parameters of the filling system of the chamber of the shipping lock no. 7 of the Moscow Canal. Vestnik Gosudarstvennogo universiteta morskogo i rechnogo flota imeni admirala S.O. Makarova **10**(6), 78–86 (2018)
21. Gapeev, A.M.: Determining the size and relative position of the elements of the filling system of the chambers of shipping locks from under the flat lifting and lowering gates. Jurnal gosudarstvennogo universiteta vodnich communicaciy **II**(XIV), 34–40 (2012)
22. Gapeev, A.M., Kononov, V.V., Morgunov, K.P.: Hydraulic Calculations of Shipping Locks: A Monograph, in 2 Parts. GUMRF, Saint Peterburg (2019)
23. Gapeev, A.M., Morgunov, K.P., Podreshetnikova, A.V.: Improving the filling system of the navigation lock chambers of the Saratov hydroelectric complex. Vestnik Gosudarstvennogo universiteta morskogo i rechnogo flota imeni admirala S.O. Makarova **9**(5), 984–992 (2017)

Principles of Interaction of Agents During Cooperative Maneuvering of Unmanned Vessels

Sergey Smolentsev(✉)

Admiral Makarov State University of Maritime and Inland Shipping, 5/7 Dvinskaya Str., St. Petersburg 198035, Russia
sswasily@mail.ru

Abstract. The article investigates the problem of collision avoidance for a group of ships at sea. To solve this problem, it is necessary to ensure the cooperative maneuvering of several vessels belonging to the group. The article looks into the case of unmanned autonomous ships, when each vessel is operated automatically. The formalization of the process of the occurrence of a hazardous approach situation and its resolution by the collective efforts of agents is carried out. The state graph of set of vessels and the transition conditions between these states are described. A list of agent conditions is determined depending on the status of the vessel that it controls and the current navigation situation. The article shows that for cooperative maneuvering it is necessary to ensure the exchange of information between agents. A protocol for the interaction of agents in solving the problem of cooperative maneuvering is proposed. Addressing system and message types are defined. The messages that agents must exchange during the process of detecting and resolving a situation of hazardous approaching are listed.

Keywords: Protocol · Unmanned ship · Cooperative maneuvering · Collision avoidance · COLREG

1 Introduction

Today designing an unmanned vessel is a high-potential direction of development of the maritime industry. In the nearest future, numerous unmanned vessels will appear in the vast of the World's ocean. This will significantly influence many aspects of shipping, primarily the safety of navigation [1].

The problem of safe ship separation is an essential component of navigation safety. When ships are moving with a risk of approaching to a distance, which is less than prescribed, one or more vessels must execute maneuvers to exclude this approach. The ship separation principles are specified in The International Regulations for Preventing Collisions at Sea (COLREGs). Unmanned vessels must comply with these regulations in order to separate from each other safely. The papers [2–4] have formulated the issues of taking into account COLREG by control systems of unmanned vessels.

Currently, active research is being carried out in the field of developing systems for the safe separation of unmanned vessels. Work [5] provides a review of modern

V. Murgul and V. Pukhkal (Eds.): EMMFT 2019, AISC 1258, pp. 442–452, 2021.
https://doi.org/10.1007/978-3-030-57450-5_38

research in this area. The work [6] considers the use of neural networks in safe separation algorithms, the use of the VO (Velocity obstacle) algorithm is considered in works [7, 8], the OZT method is described in work [9].

In case of dangerous closure of several unmanned vessels, the task of cooperative maneuvering arises, which implies mutual actions of several ships with the aim of safe separation. Papers [10, 11] focus on this issue. Work [12] discusses the need for coordination distributed between all unmanned vessels; works [13, 14] consider distributed algorithms for solving the problem of safe separation.

In work [15], the authors have proposed an algorithm to solve the problem of safe ship separation at sea. When the solution of this problem concerns unmanned vessels, it is reasonable to speak of a group of automation devices operating those vessels, which develop and implement a collective solution. Therefore, the problem may be solved by using the theory of multi-agent systems (MAS). A similar approach to the problem of safe ship separation was proposed in work [16]. This paper suggests using agents to control each unmanned vessel, so a cooperative solution on safe separation is elaborated by a group of agents that are able to exchange information. This article discusses the basic principles of organizing the interaction of agents controlling unmanned vessels.

2 Materials and Methods

The present paper considers the problem of cooperative maneuvering of a group of ships. This problem arises when two or more ships of the group dangerously approach to each other. Therewith, safe separation cannot be ensured by actions of only one vessel, or actions aimed to separate from a dangerous vessel may lead to hazardous approaching to other vessels. In this case, several vessels from the group must perform coherent maneuvers for the safe separation of all vessels in the group. This means, the vessels should jointly (cooperatively) solve the problem of safe separation.

In the present work, unmanned (autonomous) vessels operated by agents are considered as objects. All the agents are universal and equal, which means that the agent group does not have a selected central (or priority) agent that coordinates actions of other agents. Depending on the current situation of approaching and the status of the vessel, each of the agents can perform a different function in solving the particular problem of ship separation.

3 Problem Statement

A set (group) of vessels U in a given water area is to be considered. In the set U, a subset of vessels G can be defined, which includes objects belonging to the following three classes:

- Class A – vessels that navigate dangerously for other ships, yet cannot maneuver (because of their privileges in accordance with the IRPCS or due to other reasons);
- Class B – vessels that navigate dangerously for other ships and are required to maneuver;

- Class C – vessels that do not pose danger to other vessels, but can maneuver;
- Class F – vessels that maneuver to separate safely.

Therefore, $G = A \cup B \cup C \cup F$.

Vessels belonging to class D may also navigate in the given water area: D = U/G. None of D-class vessels is sailing dangerously towards any vessels from the set U. Otherwise, those vessels are included in the set G. Vessels, that are not included in the set G, act as limits imposed on solutions made for vessels from the set G.

Each vessel is connected to an agent managing this particular vessel. Each agent has its own interest (objective function) – a prescribed ship route in a given water area.

The problem of cooperative maneuvering consists in the fact that agents must act to resolve the conflict of dangerous approach of vessels within the group G. This requires such a change in routes for the vessels within the group, which eliminates the risk of dangerous approaching of the vessels, meanwhile minimizing the total deviation of all ship routes from the initial ones. The principles of building sets A, B, C, D, as well as algorithms for finding solutions for safe maneuvering are proposed in work [16].

The same work proves that the information exchange between the agents controlling the vessels must be ensured in order to solve the problem of safe separation of those vessels at sea in accordance with the proposed algorithm. First of all, information to exchange includes data on planned routes and parameters of ship dynamics. This data is used to forecast travel paths of the ships for fulfilling tasks of safety assessment and looking for ship separation solutions. In addition, there is a need for exchange of the obtained solutions between agents for executing coordinated separation maneuvers.

The present paper proposes a protocol for agents' interaction during cooperative maneuvering.

4 Results

First of all, possible states of each vessel of the group are to be determined. In accordance with the current state of a vessel, an agent controlling it transmits certain messages. To determine possible conditions, the processes of both arising of the situation of dangerous approaching and its resolving are to be considered as the sequence of states of set U. These states differ in the composition of its subsets:

$$U = A \cup B \cup C \cup D \cup F; \tag{1}$$

$$G = A \cup B \cup C \cup F; \tag{2}$$

$$A \cap B = A \cap C = A \cap D = A \cap F = B \cap C = B \cap D = B \cap F = C \cap D = C \cap F = D \cap F = \varnothing. \tag{3}$$

The states of set U are listed below:

U1 – All vessels in the water area move safely:

$$A = \varnothing, B = \varnothing, C = \varnothing, D = U, F = \varnothing, G = \varnothing. \tag{4}$$

U2 – Situation of hazardous approaching:

$$A \neq \varnothing, B \neq \varnothing, C = \varnothing, D = U/G, F = \varnothing, G = AB. \tag{5}$$

U3 – To solve the problem of safe separation, additional vessels are involved:

$$A \neq \varnothing, B \neq \varnothing, C \neq \varnothing, D = U/G, F = \varnothing, G = ABC. \tag{6}$$

U4 – Maneuvering for safe separation:

$$A = \varnothing, B = \varnothing, C = \varnothing, D = U/F, F \neq \varnothing, G = F. \tag{7}$$

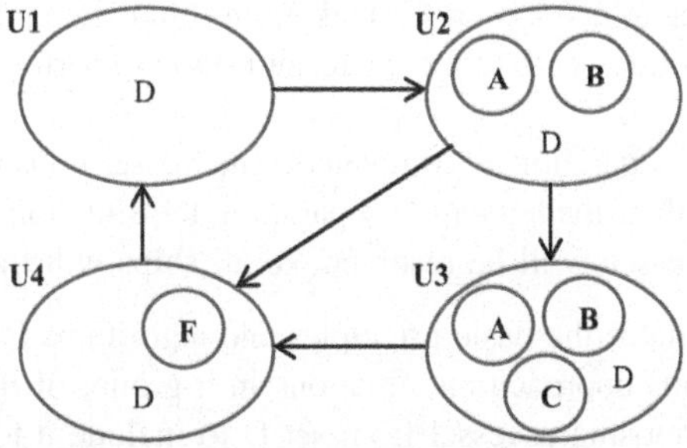

Fig. 1. State graph of set U.

Figure 1 shows the state graph of set U of all vessels in the water area. Arrows indicate transitions between the states. Each vessel can belong to any subset of set U depending on current navigational situation at different moments of time. The vessel's state is denoted in accordance with the set, which the vessel currently belongs to. The state graph of the vessel can be represented as follows (Fig. 2).

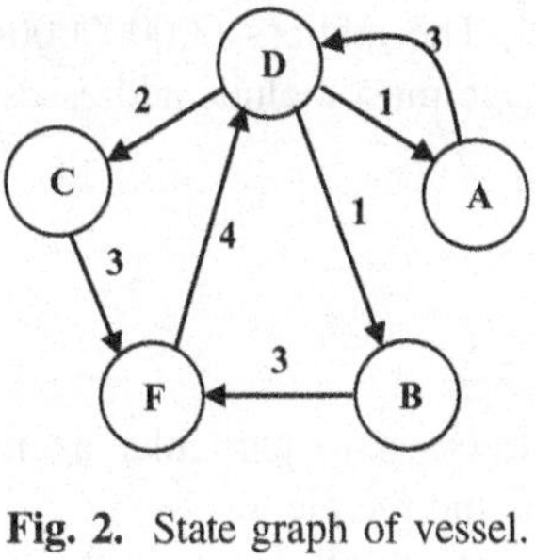

Fig. 2. State graph of vessel.

Transitions between the states correspond to the following events:

1. A situation of hazardous approaching occurs, which be detected by one or more agents. Depending on the status and position relative to opposing vessels, the vessel must either take actions to safely separate (transit to state B) or keep its traffic elements (transit to state A). In addition, the situation of hazardous approaching may not affect the vessel, then it remains in state D.
2. A vessel that goes safely towards other vessels undertakes a maneuver to ensure the separation of other vessels. This situation arises when actions of vessels belonging to set B (obliged to maneuver for safe separation) cannot resolve the dangerous approach situation. This situation may occur due to vessels belonging to set D, which serve as limits for the safe maneuver taken by the vessels belonging to set B. In this case, the vessel belonging to set D is invited to undertake a maneuver to loosen those limits. If the ship undertakes such a maneuver, it transits to state C.
3. A decision on maneuvering that ensures safe separation is made. The agent controlling the vessel in state B or C, makes the final decision on a maneuver after coordinating its actions with other agents and starts to perform it. The vessel transits to state F then.
4. The maneuver of separation is completed. The vessel transits to state D. When all vessels complete their maneuvers of separation, the situation of dangerous approach in the group of vessels will be resolved, so all ships will transit to state D.

The work [15] formulates the basic principles and algorithms for solving the problems of detecting dangerous approaching situations in a group, of distributing responsibilities among ships, choosing a vessel from set D to include it to set C and finding the optimal solution for separation of vessels. However, the solution of all these problems is carried out jointly by all agents of the ships belonging to set U, so it is necessary to ensure the exchange of information between them. For this purposes, the protocol described below is proposed.

4.1 Addressing

Each agent must have a unique address. It is proposed to use the Maritime Mobile Service Identity (MMSI) as the address. This identifier consists of nine digits and has the form MIDXXXXXX (MID is a country code, XXXXXX is a ship code). It is a unique identifier for a vessel. The address 000000000 is proposed to be used as a broadcast address. Each message must include addresses of a sender and a recipient of this message.

4.2 Message Types

There are two types of messages:

Addressable messages addressed to particular agents. The address of the corresponding agent is indicated as the recipient.

Broadcast messages addressed to all agents in the water area. The recipient is the broadcast address.

4.3 Messages

Messages Not Related to a Specific Situation of Hazardous Approach. READY is a broadcast message periodically transmitted by an agent in state D. This message confirms by the protocol for coordinated maneuvering that this agent is ready to interact with other agents. In fact, the transmission of such a message signals that the vessel is unmanned and automatically controlled. This message conveys information on the vessel, which is necessary to give a general context of the current navigational situation. The content of the context and, accordingly, the content of the information transmitted by the agent are subjects for another study. However, it is obvious that the following data must be conveyed: safety criterion used by the agent, parameters of a planned route (coordinates and speeds at route points for a given period of time, for example, for an hour, forward + next point), and vessel dynamics parameters. Using this information will enable predicting the movement of all vessels in the water area more accurately and thereby correctly assessing the navigational situation.

Messages Related to a Specific Situation of Hazardous Approach. These messages must contain the tag ID_SIT, which is a unique identifier for a hazardous approach situation. The size of this identifier is 2 bytes.

COLLISION_DETECTED is a broadcast message reporting of detection of a hazardous approach situation. This message is sent by any of the agents, which first detected the situation. Then the agent generates a unique identifier ID_SIT for this situation, which is put in the message. All further messages, that all agents will exchange during resolving the situation, should contain this identifier. Moreover, the message should contain the list of MMSIs of hazardously approaching ships.

COLLISION_NEGOTIATION is a broadcast message which means negotiation and approval of the hazardous situation. The agents of dangerously approaching ships exchange these messages in order to decide whether the situation is considered hazardous by all agents and which agent must undertake maneuvering in this situation. Each agent puts in this message a list of MMSIs of vessels considered dangerously approaching by a particular agent, indicating a class, which those vessels should belong to (A or B). All agents can make different suggestions (especially regarding a particular ship's controlled by a particular agent capability to maneuver). Agents exchange these messages until they reach consensus.

COLLISION_AGREEMENT is a broadcast message that implies the agreement on a situation. This message can be sent by any agent involved in negotiating on a hazardous situation. It signals to all other agents that the agreement was reached and the one of dangerously approaching ships to maneuver was clearly determined as well as those to follow their previous routes. The message contains a list of MMSIs of the ships with indicating of their classes (A or B). Thus, the transmission of this message divides all vessels belonging to set U into sets A, B and D.

DESITION_OFFER is a broadcast message, a proposal for maneuvering. These messages are transmitted by agents of the ships belonging to sets B and C, which have undertaken responsibility to maneuver for resolution of the situation of hazardous approaching. In this message, each agent suggests its own maneuvering option: parameters of its own modified route (coordinates and speeds at route points). Each of the ship's agents from sets B and C analyzes suggestions of other agents and contextually adjusts its own suggestion. As a result of the exchange of these messages, modified ship routes are determined, which resolve the situation of hazardous approach.

DECISION_AGREED is a broadcast message, an agreed decision on maneuvering. These messages are transmitted by ship agents belonging to sets B and C after coordinating their further maneuvers. In this message, each agent transmits its modified route (coordinates and speeds at route points). By transmitting this message, the agent undertakes the responsibility to follow this route in order to resolve the situation of hazardous approaching. The vessel controlled by this agent transits to set F.

DECISION_INVITE is an addressable message that implies an offer to participate in resolving the situation of hazardous approach. This message is transmitted by one of the agents of ships that belong to set B when the attempt to resolve the hazardous approach situation by maneuvers performed only by B set ships did not succeed. As it is said above, vessels belonging to set D can serve as limits when solving the problem of safe separation. Then maneuvering of one or several vessels belonging to set D can help solve the problem of safe separation. This message is transmitted to an agent of such a vessel. It contains invitation to take part in fulfilling the task of safe separation, which means the transition of this vessel to set C.

DECISION_INVITE_ACCEPTED is a broadcast message, acceptance of invitation to participate in resolving the situation of hazardous approach. This message is a positive response from the agent which received the DECISION_INVITE message. Having transmitted this message, the agent undertakes the obligation to maneuver in order to resolve the situation of hazardous approaching of other vessels belonging to the group. In this case, the vessel controlled by this agent transits to set C.

DECISION_INVITE_DENIED is a broadcast message implying the refusal to invitation to participate in resolving the situation of hazardous approach. This message is a negative response from the agent which received the DECISION_INVITE message. Having transmitted this message, the agent signals that for some reasons it cannot maneuver in order to resolve the situation of hazardous approaching of other vessels belonging to the group. In this case, the vessel controlled by this agent stays in set D.

DECISION_COMPLETED is a broadcast message on maneuver completion. This message is transmitted by the agent of the vessel maneuvering to resolve the situation of dangerous approach. The message is transmitted at the moment of completion of the separation maneuver or when the dangerous vessel passed safely and was left behind the crossbeam. Thus, the agent signals that it has fulfilled the obligations undertaken in order to resolve this situation. The vessel controlled by this agent transits to set D.

4.4 Protocol Diagram

The protocol diagram defines the connection of messages transmitted by an agent to states of this agent and events belonging to the process of arising and resolution of a hazardous approach situation.

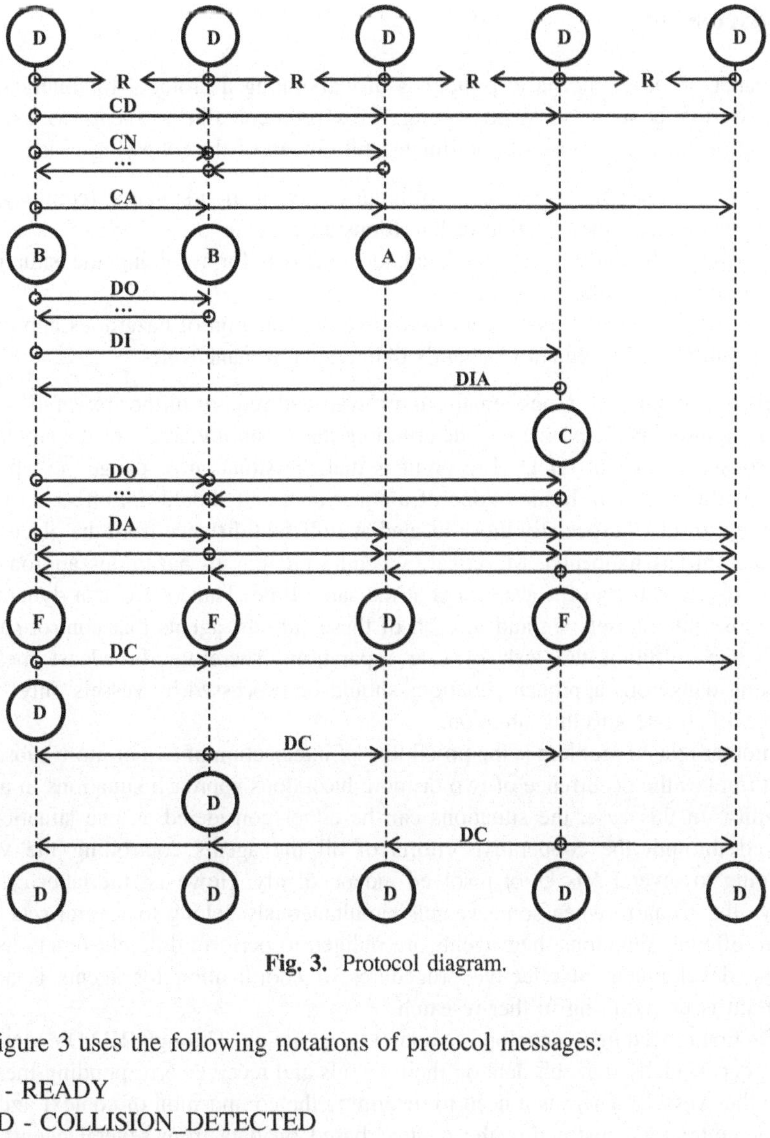

Fig. 3. Protocol diagram.

Figure 3 uses the following notations of protocol messages:

R - READY
CD - COLLISION_DETECTED
CN - COLLISION_NEGOTIATION
CA - COLLISION_AGREEMENT
DO - DESITION_OFFER
DA - DECISION_AGREED
DI - DECISION_INVITE
DIA - DECISION_INVITE_ACCEPTED
DC - DECISION_COMPLETED

5 Discussion

This paper proposes the basic principles of elaborating a protocol for interaction of agents during cooperative maneuvering. This protocol allows agents to exchange information for the purpose of coordinated fulfillment of the following tasks:

- to give a consistent interpretation of a current navigation situation (context);
- to detect exactly the situation of hazardous approaching;
- to assign coherently vessels obliged to maneuver for resolving the situation of hazardous approach;
- to define coordinated maneuvers to resolve the situation of hazardous approach;
- to monitor the fulfillment of agent's obligations to maneuver.

However, a number of issues remain unresolved and require further research.

The first one is the problem of determining the "visibility area" for the situation of dangerous approach in set U. The point is that the situation of dangerous approach occurs within a certain limited zone of a water area and should not affect all vessels belonging to set U, especially those located at sufficient distance from the place where the situation has happened. Moreover, several situations of hazardous approach can arise in different parts of a water area at the same time. Thus, set U can dynamically decompose into subsets U_i, and in each of these subsets, agents that control subset's vessels are fulfilling the task of safe separation. Therefore, broadcast messages regarding dangerous approach situations should be processed by vessels only from a subset Ui that refers to this situation.

Another related problem is the possibility of intersection of two or more subsets U_i, which implies the occurrence of two or more hazardous approach situations in a close proximity. In this case, the situations can be either considered as one situation and resolved through the cooperative efforts of all the agents controlling the vessels belonging to several sets U_i or resolved independently. However, the latter case can involve the scenario when some vessels simultaneously belong to several sets U_i, so that in different situations their agents are obliged to perform different functions.

The development of effective procedures of coordination for agents is another important issue requiring further research.

The first procedure to develop is a context negotiation. Through READY messages, agents occasionally transmit data on their vessels and receive corresponding messages from other vessels. There is a need to determine the composition of context and form the procedure for constructing the context based on data from several agents. This procedure should be identical for all agents. In this case, agents will construct the same context from the same data and will understand current conditions the same way.

It is also important to develop effective procedures for coordinating a situation and maneuvers required. At the initial stage, when transmitting COLLISION_NEGOTIATION messages, different agents present their viewpoints on the emerging navigational situation. This message provides a list of hazardously approaching ships and division of vessels into certain roles: which vessels should perform maneuvers to separate. Although in the abstract situation mutual obligations of vessels during hazardous approach are regulated by COLREG, in real life different agents may have different viewpoints on the situation. Therefore, a mechanism is needed to negotiation and approve various proposals for assessing the situation.

A similar problem arises at the stage of coordinating the maneuvers of agents that are in states B and C. In DESITION_OFFER message, each agent suggests a maneuver, which it agrees to perform for solving the problem of safe separation. That means, the agent offers its contribution to the collective solution. At the same time each agents tries to minimize its vessel's deviation from the planned route. Moreover, each agent makes its suggestion with taking into account of already received suggestions from other agents. Therefore, it is necessary to develop an effective iterative algorithm for coordinating the proposals of all agents in order to find a solution to the problem of safe ship separation while minimizing route deviations of all maneuvering vessels.

6 Conclusion

The article considers the issue of interaction of agents controlling autonomous unmanned vessels when solving the task of cooperative maneuvering.

It was proved that all agents are equal, but under different circumstances they can be in different states and perform different functions. The protocol for interaction of agents during solving the problem of cooperative maneuvering was proposed. The addressing system and message types were defined. The paper provides the list of messages for agents to exchange during the process of detecting and resolving a situation of hazardous approaching. Implementation of the proposed protocol will ensure effective interaction of agents during the process of cooperative maneuvering when fulfilling the task of safe separation.

References

1. Wróbel, K., Montewka, J., Kujala, P.: Towards the assessment of potential impact of unmanned vessels on maritime transportation safety. Reliab. Eng. Syst. Saf. **165**, 155–169 (2017). https://doi.org/10.1016/j.ress.2017.03.029
2. Naeem, W., Irwin, G.W., Yang, A.L.: COLREGs-based collision avoidance strategies for unmanned surface vehicles. Mechatronics **22**(6), 669–678 (2012). https://doi.org/10.1016/j.mechatronics.2011.09.012
3. He, Y.X., Jin, Y., Huang, L.W., Xiong, Y., Chen, P.F., Mou, J.M.: Quantitative analysis of COLREG rules and seamanship for autonomous collision avoidance at open sea. Ocean Eng. **140**, 281–291 (2017). https://doi.org/10.1016/j.oceaneng.2017.05.029
4. Johansen, T.A., Perez, T., Cristofaro, A.: Ship collision avoidance and COLREGS compliance using simulation-based control behavior selection with predictive hazard assessment. IEEE Trans. Intell. Transp. Syst. **17**(12), 3407–3422 (2016). https://doi.org/10.1109/Tits.2016.2551780
5. Huang, Y., Chen, L., Chen, P., Negenborn, R.R., van Gelder, P.H.A.J.M.: Ship collision avoidance methods: State-of-the-art. Saf. Sci. **121**, 451–473 (2020). https://doi.org/10.1016/j.ssci.2019.09.018
6. Praczyk, T.: Neural anti-collision system for autonomous surface vehicle. Neurocomputing **149**, 559–572 (2015). https://doi.org/10.1016/j.neucom.2014.08.018

7. Huang, Y., van Gelder, P.H.A.J.M., Wen, Y.: Velocity obstacle algorithms for collision prevention at sea. Ocean Eng. **151**, 308–321 (2018). https://doi.org/10.1016/j.oceaneng.2018.01.001
8. Huang, Y., Chen, L., van Gelder, P.H.A.J.M.: Generalized velocity obstacle algorithm for preventing ship collisions at sea. Ocean Eng. **173**, 142–156 (2019). https://doi.org/10.1016/j.oceaneng.2018.12.053
9. Fukuto, J., Imazu, H.: New collision alarm algorithm using obstacle zone by target (OZT). IFAC Proc. **46**(33), 91–96 (2013). https://doi.org/10.3182/20130918-4-jp-3022.00044
10. Alonso-Mora, J., Beardsley, P., Siegwart, R.: Cooperative collision avoidance for nonholonomic robots. IEEE Trans. Rob. **34**(2), 404–420 (2018). https://doi.org/10.1109/Tro.2018.2793890
11. Chen, L., Huang, Y., Zheng, H., Hopman, J.J., Negenborn, R.R.: Cooperative multi-vessel systems in urban waterway networks. IEEE Trans. Intell. Transp. Syst., 1–14 (2019). https://doi.org/10.1109/tits.2019.2925536
12. Li, S.J., Liu, J.L., Negenborn, R.R.: Distributed coordination for collision avoidance of multiple ships considering ship maneuverability. Ocean Eng. **181**, 212–226 (2019). https://doi.org/10.1016/j.oceaneng.2019.03.054
13. Kim, D., Hirayama, K., Okimoto, T.: Distributed stochastic search algorithm for multi-ship encounter situations. J. Navig. **70**(4), 699–718 (2017). https://doi.org/10.1017/s037346331700008x
14. Kim, D.G., Hirayama, K., Park, G.K.: Collision avoidance in multiple-ship situations by distributed local search. J. Adv. Comput. Intell. Intell. Inform. **18**(5), 839–848 (2014). https://doi.org/10.20965/jaciii.2014.p0839
15. Smolentsev, S.V.: Automatic synthesis of decisions on vessels collision avoidance at sea. Vestnik Gosudarstvennogo universiteta morskogo i rechnogo flota imeni admirala S.O. Makarova **2**(36), 7–16 (2016). https://doi.org/10.21821/2309-5180-2016-8-2-7-16
16. Smolentsev, S.V., Sazonov, A.E., Iskanderov, Y.M.: Cooperative maneuvering of unmanned ships for collision avoidance at sea. Vestnik Gosudarstvennogo universiteta morskogo i rechnogo flota imeni admirala S. O. Makarova **10**(4), 687–695 (2018). https://doi.org/10.21821/2309-5180-2018-10-4-687-695

Methodological Approaches to Setting the Goal of Multimodal Transportation Management

Elena Karavaeva(✉) and Elena Lavrenteva(✉)

Admiral Makarov State University of Maritime and Inland Shipping,
5/7, Dvinskaya Str., Saint Petersburg 198035, Russia
{kaf_oeodto,LavrentievaEA}@gumrf.ru

Abstract. The goal of the research is to formulate the general problem of multimodal transportation management. The used empirical research methods include studying the literary sources on the planning and organization of multimodal transportation, as well as analyzing the obtained information. The implemented theoretical research methods include a systematic approach to describing multimodal transportation as a system, a mathematical modeling method for formulating the problem of managing multimodal transportation. The following research results were obtained: the definition of multimodal transportation is proposed, the peculiarities of this cargo transportation type are presented, as well as the advantages for customers. The main risks for the operator of multimodal transportation, such as delays, losses, defects and damages to cargo, are determined. It is indicated that the timely processing of the order by the multimodal transportation operator is hindered by the need for information exchange with all participants in the transportation process. Moreover, the processing is affected by the complexity of the calculations associated with possible delivery options, the lack of an effective mechanism for tracking and further accounting for failures, failures at the implementation stage. It is concluded that the competitive advantage of the transport company is the use of software and hardware that provide planning, optimization and control of the transportation process. The goal was set for the multimodal transportation management; methodological approaches to managing such transportations are presented.

Keywords: Mixed traffic · Multimodal transportation · Multimodal transportation operator

1 Introduction

The analysis of scientific and practical literature on the research subject showed there is no uniform definition of the multimodal transportations term at the present development stage of domestic transport science, which would reveal the understanding and essence of the research subject. The vast majority of scientists and practitioners behind them give only contextual definitions of this concept. In our opinion, the wording of the multimodal transportation term, which is cited by the vast majority of scientists and practitioners, contains a mistake of too broad definition.

V. Murgul and V. Pukhkal (Eds.): EMMFT 2019, AISC 1258, pp. 453–462, 2021.
https://doi.org/10.1007/978-3-030-57450-5_39

The first and only legal source at the international level containing the multimodal transportation term, and revealing its concept is the 1980 United Nations Convention On International Multimodal Transport of Goods [1], which, unfortunately, has not yet gone into effect.

According to the main provisions of this document, the main features characterizing transportation as multimodal, i.e. organized on the principles developed by the Western Doctrine of the Law of Transport (formulated as international multimodal transportation of goods), include the following:

- transportation of goods should be international;
- no less than two types of transport should be involved in transportation;
- transportation is organized by the transportation operator or by a person acting on his behalf;
- the transportation operator in the contract of carriage between him and the person ordering the carriage (client) acts as an equal party to the contract, and not as an agent, or on behalf of the consignor or carriers involved in the carriage operations;
- the transportation operator assumes responsibility for performing the carriage contract;
- the document issued by the transportation operator covers the entire route of the goods from the consignor to the consignee;
- the operator's liability for the goods covers the period from the moment he takes over the goods to his charge until the goods are transferred to the recipient.

Thus, multimodal transportation can be defined as international transportation carried out by two or more modes of transport, organized by the operator assuming responsibility for the carriage of goods in general, which issues a document to the sender for multimodal transportation covering the entire route of the goods. It is the presence of the multimodal transportation operator assuming responsibility for all the risks [2] associated with the carriage of goods sent under a single shipping document covering the entire route of the goods, that is the fundamental difference between multimodal transportation and all other related to it.

Thus, the multimodal transportation term characterizes the transportation process from the standpoint of its legal and organizational support, in contrast to the terms "combined" and "intermodal transportation". They, in turn, characterize the transportation process depending on the technological operations carried out with cargo or the vehicle transporting it [3].

Table 1 shows the main existing approaches to studying the multimodal transportation. A comparative analysis of the represented studies showed that the multimodal transportation system can include and consider various factors, which significantly affect the transportation result, when planning cargo transportation. Moreover, the result itself can be evaluated according to different criteria.

Table 1. Existing approaches to the studying the multimodal transport.

№.	Authors	Approaches to the studying the multimodal transportation system	Criteria used in the system/proposed criteria for the system assessment
1.	Narayasnawami and Rangaraj [4]	A multimodal transportation system is affected by a violation, which is defined as an incident related to the unavailability of the path between some stations in the work area. Such incident may lead to a change in certain critical and/or preferred restrictions of the system	Time delays, specific route break point/delay
2.	Zilko, Kurowicka and Goverde [5]	A system violation is defined as an unforeseen accident affecting the railway schedule	Delay time, repair time/estimated average delay time
3.	Huang, Hu and Zhang [6]	Taking into account force majeure circumstances. The system is impacted by an unforeseen event (hurricane, snow accident, traffic accident, etc.) occurring in one link of a multimodal chain	Duration of force majeure circumstances/use of statistics on similar events
4.	Burgholzer et al. [7]	Operational disruption of individual sections (links) is possible in the multimodal transportation system, which can lead to a decrease in the level of chain performance	The chain performance level drops in case of connection failure. The model uses a chain performance indicator/transport time for measuring it
5.	Ishfaq [8]	The multimodal transport system takes into account external interference in the supply chain	Delay time/transportation cost
6.	Uddin and Hyunh [9]	The multimodal transportation system includes considering the capacity of the node (termanal)	Capacity, transporting time/cost of transportation
7.	Miller-Hooks, Zhang and Faturechi [10]	The multimodal transportation system takes into account disasters of different nature	Five disaster scenarios (bombing, terrorist attack, flood, earthquake and other attacks on an intermodal terminal)/stability level
8.	Pant, Barker and Landers [11]	First of all, various operational violations of inland waterways entailing economic losses are considered	Considering falling supply and demand/transportation costs

(*continued*)

Table 1. (*continued*)

№.	Authors	Approaches to the studying the multimodal transportation system	Criteria used in the system/proposed criteria for the system assessment
9.	Liu, Zheng and Zhang [12]	The effect of equipment breakdown during unloading and loading of containers on ships during multimodal transportation is investigated	Delay time, repair time/negative deviation from the originally planned schedule
10.	Fialkoff et al. [13] Daskin and Snyder [14] Starita et al. [15]	The impact of natural disasters in multimodal transport is investigated	Delay degree/Shipping Costs
11.	Chen and MillerHooks [16]	It is necessary to take into account not only natural disasters, but also the human factor when organizing multimodal transportation	Transportation costs
12.	Udenta et al. [17]	The multimodal transport system takes into account various modes of transport and possible failures associated with them	β-distribution (either 0 or 1)/ number of vehicles used, transportation time

The importance of the transport industry is difficult to overestimate. For any economy in the world, transport is an analogue of the human circulatory system. The balanced and timely development of transport infrastructure is the foundation for a confident long-term socio-economic development, increasing the volume of foreign and domestic trade. The total cargo turnover of Russia is gradually increasing, continuing the trend that began in 2014. In 2017, freight turnover grew by 5.5%, and the volume of cargo transportation grew by 1.6%. Key modes of transport are rail, road and pipeline. Today, their total share in the cargo turnover of the Russian Federation is about 98%. An important role in the transport system of Russia is played by port infrastructure, the railway, road and pipeline routes are locked on it. Today, seaports provide transshipment of about 60% of Russia's foreign trade cargo. From 2000 to 2017, the volume of cargo transshipment in seaports increased almost four times and reached 786 million tons.

Transportation is an integral part of logistics, including the movement and storage of raw materials, stocks, work in progress and finished goods from the place of origin to the place of consumption. The most promising development direction for technology of goods transportation is multimodal transportation. The containerization of international cargo flows has significantly impacted not only the material and technical basis of transport, but also the technology of the international transportation process. Customers and transportation operators abandoned the traditional cargo delivery system by each transport mode in isolation from each other, switching to integrated multimodal transportation. The multimodal transportation term is more properly implemented not as a simple translation of the mixed transportation term, but as representing a new step in the interaction of transport modes on the principles of logistics.

The following peculiarities of multimodal transportation can be distinguished:

- the coordinated use of more than one mode of transport in transportation;
- the transportation is organized and carried out by one person – the multimodal transportation operator;
- the relationships between the customer and the provider of integrated transport services are regulated on the basis of one contract;
- multimodal transportation may have international status.

Multimodal transportation can be defined as international transportation carried out by two or more modes of transport, organized by the operator, who assumes responsibility for the carriage of goods in general. The operator gives sender a single transportation document covering the entire route of goods. It is the presence of a multimodal transportation operator, who assumes responsibility for all the risks associated with the carriage of goods, that is the fundamental difference between multimodal transportation and all other related ones.

Multimodal transport operators face a serious problem of planning cost-effective routes in the face of uncertainty and risks of high financial losses while strictly observing the final consumer's requirements for the quality of order execution [18]. An effective tool is required for solving this problem, the development of which this work is devoted to.

From the customer's point of view, the main advantages of multimodal transportation are as follows:

- ensuring the safety of the cargo;
- the possibility of minimal participation in the process of organizing transportation;
- the possibility of customs clearance and the provision of other related logistics services;
- the possibility of issuing a multimodal transportation document accepted by domestic and foreign banks;
- financial transparency of customer relationships.

Planning multimodal transportation can be represented as a series of aggregates consisting of elementary work, which must be sequentially performed. Each mode of transport has its own advantages and disadvantages that must be taken into account when choosing a transportation method, type of transport, type of vehicles and a particular carrier. When organizing multimodal transportation, a high degree of consistency is necessary regarding the actions of owners of various transport modes [19].

The multimodal transportation includes:

- trucking enterprises;
- all railway stations open for freight operations;
- river ports and hithes;
- seaports;
- civil aviation airports.

The following carriages are not accepted for multimodal transportation:

- goods carried by filling;
- wood, which is carried by waterways in rafts;
- explosive and potent toxic substances.

The list of goods accepted for transportation is established by the relevant transport legislation.

The transport process in multimodal transportation consists of the sequential delivery of goods by various modes of transport and intermediate transshipment. Thus, the planning of multimodal transportation can be represented as a series of aggregates consisting of elementary operations that must be sequentially performed.

The technological stage of multimodal transportation is characterized by the type of transport, type of vehicle, specific carrier.

2 Materials and Methods

Multimodal transportation requires a specific approach regarding criteria selection for evaluating transportation. For a long time prime cost was considered the main selection criterion. However, this criterion does not represent numerous significant factors. Today, the problem of selecting the transportation type must be solved on the basis of a multi-criteria approach.

Three following groups of criteria for selecting the transportation method and vehicle type are distinguished: quantitative qualitative and technical criteria for the carriage of goods. The main quantitative criterion today is the reliability of cargo transportation. The reliability is provided not only by the internal factors of individual vehicles (for example, wear), but also by external factors, such as natural phenomena, situations on the routes, interaction with other vehicles and objects. Thus, the transportation reliability is a complex parameter. It is characterized, in particular, by the transportation terms, the preservation of the batch and of the cargo usability during transportation. Therefore, it seems appropriate to include the consideration of the reliability criterion for cargo transportation when organizing multimodal transportation.

Multimodal transportation is considered from the perspective of the logistics operator, who is the organizer of this transportation. Such perspective involves maximizing profits when all the conditions of the transportation agreement are met. This is ensured by minimizing the costs associated with the transportation process and possible losses in case of violating the agreement terms (penalties for delays, damage to goods and etc.). The single, atypical transportation is considered.

Let us analyze the process of multimodal transportation as an object of management. The quality of multimodal transportation management can be assessed using the following criteria:

- total costs;
- profit;
- transportation profitability;
- total delivery duration;

- the value of cargo losses, cargo synchronization rate, compliance with the complete preservation of the cargo as a possible option;
- adherence to delivery dates;
- change in the actual cost of delivery compared to the agreed.

3 Results

Let us consider the factors affecting the quality indicators of transportation. First of all, this is the selection of route and transport modes for each step. The total costs depend on the cost of performing a single stage, determined by a specific transport unit, technical means, tariff of a specific subcontractor, direction, distance, speed of delivery, etc. The duration of delivery is determined by the duration of the individual operations (it depends on the technical means used, and on the execution speed as a result) Various delays may occur, which can be associated, for example, with a shortage of individual technical means, their work schedule, failures, weather conditions, and with force majeure circumstances. The amount of losses depends on the characteristics of the cargo, delivery type, packaging, distance, duration of delivery, speed, weather conditions, features of the equipment, failures and force majeure circumstances. In general, all factors can be attributed to one of the following types: technological, natural, political, economic, legal, social.

The control parameters of multimodal transportation can be a route (a set of intermediate nodes), modes of transport, specific transport subcontracting companies, specific vehicles, technical means, packaging methods.

Input parameters are as follows: information about the cargo (origin, quantity), place of departure, destination, restrictions on the delivery terms and cost, time of year, weather conditions, availability of communication lines, data on carriers and other subcontractors (tariffs, business geography, technical means, work schedules, etc.).

Status coordinates: current total costs, current duration, current losses, average speed.

The problem of managing multimodal transportation can be generally represented as follows. It is necessary to select a type of control actions for the system, so that the cargo is delivered with the best values of quality indicators under given restrictions and initial conditions. Intermediate nodes, transport modes, specific transport companies, specific vehicles, packing methods are the examples of the control actions. The restrictions are technological peculiarities of transporting the specific cargo, as well as the availability of certain technical means for one or another step performer at the time of the service. Storage conditions and conditions of transportation for the territory of a particular country, region, city, speed regime, axial load, climatic conditions, environmental restrictions, etc. are these technological peculiarities. In addition, certain conditions set by the customer act as limitations, for example, the need to use a particular type of transport, to integrate a certain step in the sequence, etc.

4 Discussion

In general, the algorithm of the multimodal transportation operator is presented below.

The operator accepts an order from the consignor. The order contains information about the cargo, an indication of the place of departure and destination, restrictions on the delivery terms and cost. The preliminary problem of optimal management is solved. Possible options are determined, including the route, transport modes, subcontractors, etc., the values of management criteria are calculated. It should be noted that the implementation of a specific step is possible both by the multimodal transportation operator and by external subcontractors. After that, using the methods of decision-making theory, operator makes a choice in favor of the option meeting all the conditions best and concludes contracts with the client and subcontractors. At the implementation stage, the transportation process is directly organized by using the sequential implementation of the transport and technological protocol developed at the initial stage. Operational control is carried out over the transportation process and the activities of subcontractors involved in it. During the transportation course, the external conditions can change, for example, a failure can occur at any step, so adjustments are made to the delivery conditions. These can be changes in the route of cargo movement, changing delivery terms and requests for additional services, such as custom clearance, provision of storage facilities, etc. In this case, the control actions are recalculated, and then the cargo is delivered in another optimal way. The process of multimodal transportation ends with the final settlements between all participants, giving the cargo to the consignee and analyzing the effectiveness of the whole complex of management decisions accompanying to this process.

In order to implement the management, it is proposed to develop an automated multimodal transportation management system, which can be achieved by solving the following problems:

- development of analytical models of multimodal transport linking management criteria with input and control parameters;
- analytical formulation of problems regarding optimal control of multimodal transportation, selecting the methods for solving them;
- development of software and hardware for a management system that implements the optimal management of multimodal transportation.

5 Conclusion

Thus, when developing the management model, multimodal transportation is represented as a multi-graph (directed graph that has multiple arcs). Multiple arcs are alternative operational options. Initially, a multi-graph is converted into a directed graph, which does not contain multiple arcs, by the means of a specially designed algorithm. After that, according to the directed graph and taking into account the limitations of the problem associated with the characteristics of the cargo and with other restrictions, a set of delivery options is constructed.

The management task can be represented as follows. The delivery option must be selected, for which the criteria vector adopts the best value from the perspective of the decision maker (DM).

The problem is multi-criteria. During the conversion of the designed analytical model of multimodal transportation, the total duration of the transportation, total delay and the total amount of losses is reduced to costs. The criteria for profitability and efficiency also depend on costs. As a result, the multi-criteria problem is reduced to a single criterion by the criterion of total costs. However, such a reduction is not always possible, for example, in the case when time losses or quantitative losses cannot be adequately estimated in terms of value (transportation of goods in emergency situations, transportation of especially dangerous substances, cultural property, etc.). In addition, if the problem is considered from the customer's perspective, the resulting penal reimbursements received from the multimodal transportation operator may not cover the real losses caused by delay or by cargo damage.

If the parameters of the model are considered determinate, then decision methods for the multi-criteria problem under certainty are used in order to solve the problem. If any parameters of the model are considered as random variables, then the decision-making methods are implemented in conditions of uncertainty and risk.

References

1. United Nations Convention of International Multimodal Transport of Goods (1980). https://treaties.un.org/pages/ViewDetails.aspx?src=TREATY&mtdsg_no=XI-E-1&chapter=11&clang=_en. Accessed 15 Apr 2020
2. Seo, Y.J., Chen, F., Roh, S.Y.: Multimodal transportation: the case of laptop from Chongqing in China to Rotterdam in Europe. Asian J. Shipping Logist. **33**(3), 155–165 (2017). https://doi.org/10.1016/j.ajsl.2017.09.005
3. Glossary for Transport Statistic/Document prepared by the Intersecretariat Working Group on Transport Statistics EUROSTAT, ECMT, UN/ECE Second edition (1997). https://www.tsi.lv/sites/default/files/editor/science/Research_journals/Tr_Tel/2001/V1/art04_glossary.pdf. Accessed 09 Apr 2020
4. Narayanaswami, S., Rangaraj, N.: Modelling disruptions and resolving conflicts optimally in a railway schedule. Comput. Ind. Eng. **64**(1), 469–481 (2013)
5. Zilko, A.A., Kurowicka, D., Goverde, R.M.P.: Modeling railway disruption lengths with Copula Bayesian networks. Transp. Res. C **68**, 350–368 (2016)
6. Huang, M., Hu, X., Zhang, L.A.: Decision method for disruption management problems in intermodal freight. Transport 13–21 (2011)
7. Burgholzer, W., Bauer, G., Posset, M., Jammernegg, W.: Analysing the impact of disruptions in intermodal transport networks: a micro simulation-based model Decis. Support Syst. **54**(4), 1580–1586 (2013). https://doi.org/10.1016/j.dss.2012.05.060
8. Ishfaq, R.: Resilience through flexibility in transportation operations. Int. J. Logist. Res. Appl. **15**, 215–229 (2012). https://doi.org/10.1080/13675567.2012.709835
9. Uddin, M.M., Huynh, N.: Routing model for multicommodity freight in an intermodal network under disruptions. Transp. Res. Rec. J. Transp. Res. Board **2548**, 71–80 (2016). https://doi.org/10.3141/2548-09
10. Miller-hooks, E., Zhang, X., Faturechi, R.: Measuring and maximizing resilience of freight transportation networks Comput. Oper. Res. **39**(7), 1633–1643 (2012)

11. Pant, R., Barker, K., Landers, T.: Dynamic impacts of commodity flow disruptions in inland waterway networks. Comput. Ind. Eng. **89**, 137–149 (2015)
12. Liu, C., Zheng, L., Zhang, C.: Behavior perception-based disruption models for berth allocation and quay crane assignment problems. Comput. Ind. Eng. **97**, 258–275 (2016)
13. Fialkoff, M.R., Omitaomu, O.A., Peterson, S.K., Tuttle, M.A.: Using geographic information science to evaluate legal restrictions on freight transportation routing in disruptive scenarios. Int. J. Crit. Infrastruct. Prot. **17**, 60–74 (2017)
14. Snyder, L.V., Daskin, M.S.: Reliability models for facility location : the expected failure cost case reliability models for facility location : the expected failure cost case. Transp. Sci. **39**(3) 400–416 (2005). https://doi.org/10.1287/trsc.1040.0107
15. Starita, S., Scaparra, M.P., Hanley, J.R.O.: A dynamic model for road protection against flooding. J. Oper. Res. Soc. **68**(1), 74–88 (2016)
16. Chen, L., Miller-hooks, E.: Resilience : an indicator of recovery capability in intermodal freight transport. Transp. sci. **46**(1), 109–123 (2012). https://doi.org/10.1287/trsc.1110.0376
17. Udenta, F.C., Jha, M.K., Mishra, S., Maji, A.: Srategies to improve the efficiency of a multimodal interdependent transportation system in disasters. Proc.- Soc. Behav. Sci. **104**, 805–814 (2013). https://doi.org/10.1016/j.sbspro.2013.11.175
18. Yatskiv (Jackiva), I., Budilovich, E.: A comprehensive analysis of the planned multimodal public transportation HUB. Transp. Res. Proc. **24**, 50–57 (2017). https://doi.org/10.1016/j.trpro.2017.05.067
19. Jarašūnienė, A., Batarlienė, N., Vaičiūtė, K.: Application and management of information technologies in multimodal transportation. Proc. Eng. **134**, 309–315 (2016). https://doi.org/10.1016/j.proeng.2016.01.012

Factors Determining Thermohydraulic Efficiency of Liquid Cooling Systems for Internal Combustion Engines

Vladimir Zhukov(✉), Valentin Erofeev, and Olesya Melnik

Admiral Makarov State University of Maritime and Inland Shipping, Dvinskaya Street 5/7, Saint Petersburg 198035, Russia
va_zhukov@rambler.ru

Abstract. The cooling system is one of the most important systems of internal combustion engines and has a significant impact on the resource, economic and ecological performance of the engine. In order to assess the performance of the cooling system, it is proposed to use the energy coefficient of thermohydraulic efficiency. The main factors affecting the thermohydraulic efficiency of the cooling system are determined, suggestions for ensuring its improvement are made. The most significant factor affecting the thermohydraulic efficiency of the cooling system is the specific heat capacity of the coolant. An increase in the thermohydraulic efficiency of the ICE cooling system due to an increase in the temperature difference of the coolant at the engine outlet and its entrance is possible only with intensive heat removal from the coolant in the cooler. This requires further research and development on the modernization of heat exchangers for ICE cooling systems and optimization of their operation parameters.

Keywords: Cooling system · Heat extraction · Power loss · Cooling liquids · Thermal capacity · Liquid temperature · Pressure in the cooling system

1 Introduction

Conventional engines are currently widely used as the main elements of the power plants for various vehicles and stationary units [1, 2]. The main directions for improving internal combustion engines are increasing their fuel efficiency and ensuring environmental safety. Environmental safety is characterized primarily by the concentration of harmful substances in the exhaust gases of engines and is regulated by international and national standards for engines of various purposes. In order to meet the requirements for modern internal combustion engines (ICE), the engine's operating processes are continuously studied and improved, the main systems and mechanisms are upgrading [3–5]. The cooling system, along with fuel supply, lubrication and turbocharging systems, is the system determining the economic and ecological performance of internal combustion engines. However, much less attention is paid to its improvement compared to other mentioned systems [6]. In accordance with the functional purpose, the cooling system has to provide the optimum thermal state of the engine for all operating conditions. The cooling system performs its functional purpose

V. Murgul and V. Pukhkal (Eds.): EMMFT 2019, AISC 1258, pp. 463–472, 2021.
https://doi.org/10.1007/978-3-030-57450-5_40

due to the heat removal from parts, which are heated due to the contact with combustion products having a high temperature and as a result of friction. The temperature state of the engine has a major impact on the reliability of its main parts [7]. A heat carrier is used to remove heat, air and liquid cooling systems are distinguished depending on its type. The optimal temperature state is understood as one, in which the unproductive heat loss released during the combustion of fuel is minimal, but at the same time, the engine's resource indicators remain at the established level. The impact of cooling parameters on the economic and resource indicators of the engine is analyzed in the research [8]. Maintaining the optimum temperature level of the engine at various operating conditions is ensured by improving the control [9, 10] and automatic adjustment in cooling systems [11, 12]. Studying the processes occurring in ICE cooling systems and developing the ideas for their modernization are carried out using numerical simulation methods [13, 14].

Based on the results of such studies, the most promising scientifically based technical solutions are proposed. In order to reduce unproductive heat losses in ICE, it is proposed to use heat-resistant materials for manufacturing the parts of the cylinder-piston group [15]. This will increase the maximum temperature of the engine's duty cycle and its efficiency. Moreover, applying heat-insulating coatings to the cooled cylinder-piston group of the engine parts [16–18] is proposed, which can also reduce heat loss along with the coolant. A promising way for improving cooling systems is to reduce the cost of providing the circulation of coolant along the internal circuit of the cooling system. In order to solve this problem, more efficient methods are proposed for cooling the air for inflation [19–21], upgrading such elements of the cooling system as the circulation pump [22–24] and heat exchangers [25, 26]. Increasing the efficiency of using secondary energy resources of engines remains the promising research area, primarily using the heat removed from the exhaust gases and coolant [27–29]. An important area of research is the optimization of cooling modes and the constructive improvement of cooling systems for engines using alternative fuels. The results of such studies are presented in [30–32].

In order to further improve ICE cooling systems regarding the above mentioned aspects, it is necessary to assess the impact of various factors on the thermohydraulic efficiency of ICE cooling systems. The conducted research is devoted to solving this problem.

2 Materials and Methods

From a thermohydraulic point of view, the internal circuit of the ICE's liquid cooling system can be considered as a closed circuit, which includes a heater 1, a cooler 2 and a circulation pump 3 (Fig. 1). In the presented model, heater 1 combines all the following elements, in which heat is supplied to the coolant: the cooling cavities in the engine itself, the engine oil cooler, the inflation air cooler. In cooler 2, heat from the coolant is discharged into the environment through sea water in marine diesels or through atmospheric air in automobile and diesel engines. The power loss for the drive of the circulation pump 3 is determined by the hydraulic resistance of the circulation circuit. The hydraulic resistance of the ICE cooling circuit depends mainly on the structural characteristics of the elements included in the circuit and on the length of the channels and pipelines forming the circuit.

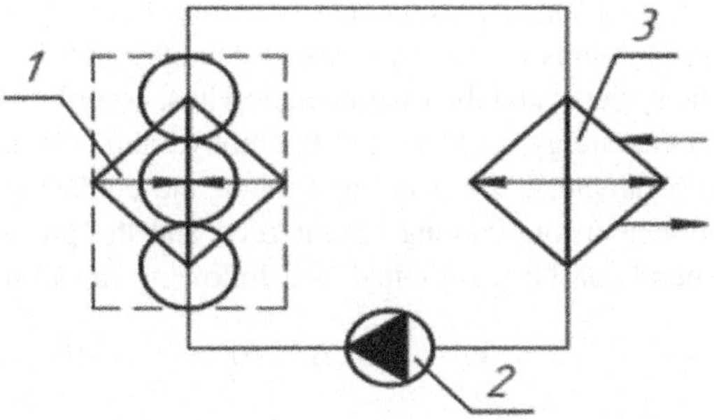

1– heater (piston part, oil cooler, inflation air cooler); 2 - circulation pump, 3 - cooler (water-to-water cooler, radiator)

Fig. 1. Internal circuit diagram of ICE's cooling system.

The energy coefficient, which characterizes the thermohydraulic efficiency of the cooling system, can be used as a criterion for the fluid system performance. This coefficient is the ratio of the amount of heat Q_c dissipated through the cooling system to the power spent on pumping the heat carrier along the coolant circuit N_c

$$E_c = \frac{Q_c}{N_c} \tag{1}$$

The amount of heat removed through the cooling system is determined by the expression:

$$Q_c = G_c c_p \Delta T_c \tag{2}$$

Where G_c is the mass flow rate of the coolant, kg/h; c_p is average isobar heat capacity of the coolant, kJ/(kg·K); ΔT_c is the difference between the coolant temperature at the outlet of the engine and the at the input, K.

The power consumed by the circulation pump drive of the internal circuit in the cooling system can be calculated according to the following formula:

$$N_c = \frac{p\,G_c}{\eta_v \eta_m} \tag{3}$$

where p is the pressure in the cooling system, MPa; $\eta_v = 0.6\ldots0.7$ is hydraulic efficiency of the pump, $\eta_m = 0.7\ldots0.9$ is the mechanical efficiency of the pump.

Using expressions (1) – (3), the energy coefficient equation of the thermohydraulic efficiency of the cooling system can be represented as follows:

$$E_c = \frac{c_p \Delta T_c}{p} \eta_v \eta_m \tag{4}$$

The numerical values of the hydraulic and mechanical efficiency of the circulation pumps in the cooling systems are in rather narrow ranges and do not depend on the design features of the systems and their parameters. Thus, according to the equality (4), it can be argued that the energy coefficient of thermohydraulic efficiency is determined by three following parameters: the heat capacity of the coolant c_p, the temperature difference of the coolant in the cooling circuit ΔT_c, and the pressure in the cooling system p and in general can be represented as a following function:

$$E_c = f(c_p, \Delta T_c, p) \tag{5}$$

The goal of the research was to determine the impact of these factors on the thermohydraulic efficiency of ICE cooling systems and to distinguish the most significant of them. It was taken into account while analyzing that the currently used ICE liquid cooling systems, are divided into two groups according to the temperature level. These are low-temperature systems, in which the temperature of the coolant at the engine outlet does not exceed the boiling point and is about 90…95 °C and high-temperature ones, in which the temperature of the coolant exceeds the boiling point and is 120…130 °C or more. Studying the high-temperature cooling systems is of interest, because such systems are used in the construction of modern forced engines from leading manufacturers such as MAN B&W Diesel Ltd, Caterpillar, General Motors, Wartsila/Sulzer, Deutz AG, OAO Barnaultransmash. The main advantages of high-temperature ICE cooling systems compared to low-temperature ones are as follows [33]:

- reduction of thermal stresses in the cylinder liner due to a decrease in the temperature difference on its surfaces and stabilization of temperature fields in the parts of the piston-cylinder group;
- reduction of the heat part removed through the cooling system due to redistribution of the heat balance components;
- reduction of mechanical losses and wear of parts in a "piston – cylinder liner" pair;
- more possibilities for recycling secondary energy resources discharged with exhaust gases and coolant;
- improvement of the ignition conditions and fuel combustion in the combustion chamber by increasing the average temperature of the cycle;
- reduction of weight and size indicators of heat exchangers in the cooling system.

Raising the temperature of the coolant in high-temperature cooling systems is achieved by increasing the boiling point T_s of the fluid. The boiling point of water, depending on the pressure, was calculated using the following empirical formula obtained on the basis of the data given in [34]:

$$T_s = 273 + 174.47p^{0.2391} \tag{6}$$

where p is the pressure, MPa (ranging from 0.1 MPa to 22 MPa).

The formula is valid for pressures ranging from 0.1 to 22 MPa.

3 Results

During the research, the variation ranges of the parameters included in Eq. (4) were determined.

The most common coolants for liquid cooling systems are as follows:

- water, which has a specific isobaric heat capacity of 4.19 to 4.25 kJ/(kg·K) in the operating temperature range (80…120 °C);
- aqueous solutions of ethylene glycol (antifreezes) having a specific isobar heat capacity of 3.05 to 3.46 kJ/(kg·K) in the operating temperature range (80…120 °C).

Motor oils that have a specific isobar heat capacity in the same temperature range from 2.05 to 2.12 kJ/(kg·K) are used as a coolant in the cooling systems of some engines.

Liquids modified with nanoparticles of graphene and multigraphene oxide [35–37], which are capable of impacting the heat capacity of the base fluid, are considered promising coolants for cooling systems. Based on the data presented, the variation range of the specific isobaric heat capacity of the coolant was set from 2.0 to 4.5 kJ/(kg·K).

The correlation between the energy coefficient of the thermohydraulic efficiency of the cooling system and the specific isobaric heat capacity of the coolant is presented in Fig. 2.

According to the data presented in studies the temperature difference between the coolant at the engine outlet and inlet for low-temperature cooling systems

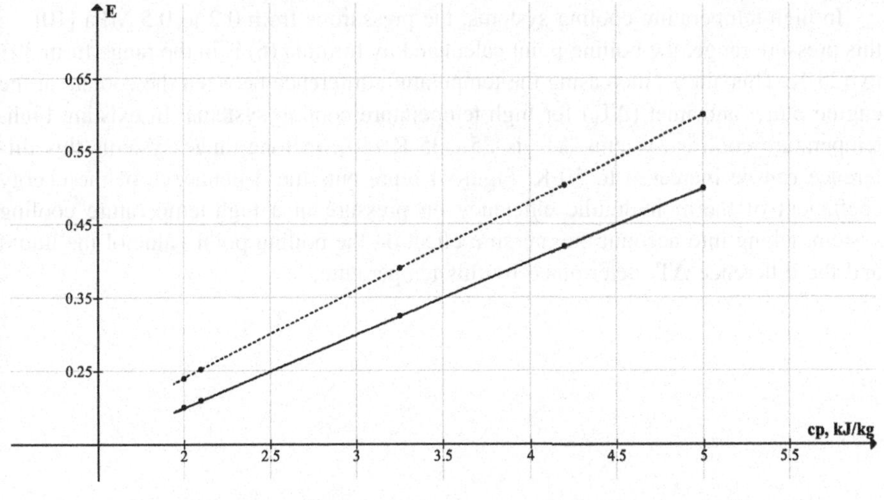

——— – low-temperature cooling system; - - - - - – high-temperature cooling system.

Fig. 2. The influence of the coolant's (c_p) heat capacity on the coefficient of thermohydraulic efficiency (E).

ΔT_c is 5…15 K for low-temperature cooling systems. The pressure in such systems is in the range from 0.11 to 0.15 MPa, depending on their design features. The effect of pressure in the cooling system on the energy coefficient of thermohydraulic efficiency of a low-temperature cooling system with a temperature difference of coolant ΔT_c equal to 10 K is presented graphically in Fig. 3

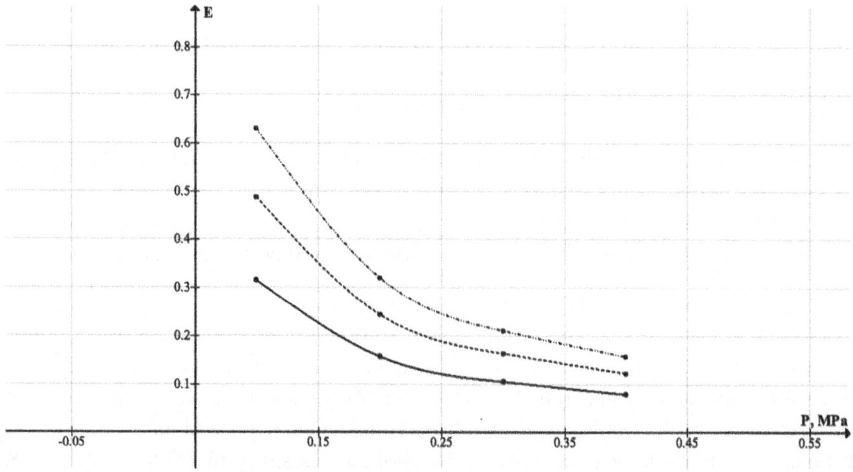

——— – c_p = 2.1 kJ/(kg·K) (motor oil); —— —— – c_p = 3.2 kJ/(kg·K) (aqueous ethylene glycol solution); - - - - - c_p = 4.2 kJ/(kg·K) (water).

Fig. 3. The effect of pressure in a low-temperature cooling system on the thermo-hydraulic efficiency coefficient of at various heat capacities of the coolant.

In high-temperature cooling systems, the pressure is from 0.2 to 0.5 MPa [10]. In this pressure range, the boiling point calculated by formula (6) is in the range from 393 to 425 K. This allows increasing the temperature difference between the coolant at the engine outlet and inlet (ΔT_c) for high-temperature cooling systems. In existing high-temperature cooling systems ΔT_c is 25…35 K [10], in long-range systems this difference can be increased to 50 K. Figure 4 represents the dependence of the energy coefficient of thermohydraulic efficiency on pressure in a high-temperature cooling system, taking into account the pressure effect on the boiling point value of the liquid and the difference ΔT_c determined by this temperature.

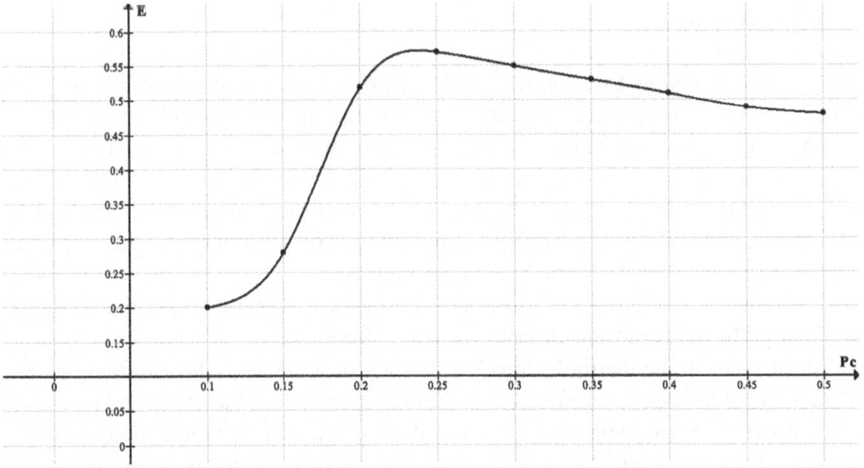

Fig. 4. The pressure effect in a high-temperature cooling system on the thermo-hydraulic efficiency coefficient (coolant – water).

4 Discussion

Analyzing the results allows drawing the following conclusions.

Thermohydraulic efficiency of ICE cooling systems is proportional to the specific heat capacity of the heat carrier. When switching to high-temperature cooling, the effect of the coolant's heat capacity slightly increases. Among the widespread heat carriers in cooling systems, the use of water is most preferable from the thermohydraulic efficiency point of view. Due to the fact that previous studies showed low efficiency and high complexity of using metals with low melting temperatures as cooling liquids for ICE, using water or aqueous solutions of ethylene glycol modified with nanoparticles that can increase the heat capacity of a liquid has more potential.

In low-temperature cooling systems operating at coolant temperatures below the boiling point, the complication of the circulation circuit leads to a pressure increase in it and a to a decrease in the thermohydraulic efficiency of the system.

Due to the higher temperature difference between the coolant at the engine outlet and its inlet (ΔT_c), high-temperature cooling systems have greater thermohydraulic efficiency regardless of the coolant type used.

An increase in pressure in high-temperature cooling systems to 0.25 MPa leads to a significant increase in the thermohydraulic efficiency of cooling systems. At pressures in the system of more than 0.35 MPa, the thermo-hydraulic efficiency begins to gradually decrease, which is explained by a significant increase in the power loss necessary for pumping the coolant along the cooling circuit.

5 Conclusion

Studies have shown that an additional advantage of high-temperature cooling systems compared to low-temperature systems is their higher thermohydraulic efficiency.

The most significant factor affecting the thermohydraulic efficiency of the cooling system is the specific heat capacity of the coolant. Therefore, studies aimed at developing modifiers, which increase the heat capacity of the main coolants in the ICE's cooling systems (water and aqueous ethylene glycol solutions) without significant increase in their viscosity, seem to be long-ranged.

The pressure range from 0.3 to 0.35 MPa in high-temperature cooling systems is the most preferred regarding thermohydraulic efficiency.

An increase in the thermohydraulic efficiency of the ICE cooling system due to an increase in the temperature difference of the coolant at the engine outlet and its entrance is possible only with intensive heat removal from the coolant in the cooler (cooler or radiator). This requires further research and development on the modernization of heat exchangers for ICE cooling systems and optimization of their operation parameters.

References

1. Internal Combustion Engines: Performance, Fuel Economy and Emissions. Editor-in-Chief IMechE. Woodhead (2013)
2. Woodyard, D.: Pounder's Marine Diesel Engines and Gas Turbines. Butterworth-Heinemann, Oxford (2014)
3. Advanced Direct Injection Combustion Engine Technologies and Development Diesel Engines. Editor-in-Chief Hua Zhao. Woodhead Publishing (2010)
4. Xin, Q.: Diesel Engine System Design. Woodhead Publishing, Cambridge (2011)
5. Okubo, M., Kuwahara, T.: New Technologies for Emission Control in Marine Diesel Engines. Butterworth-Heinemann, Oxford (2019)
6. Zhukov, V.: Outlook of improving of ship diesel engine's cooling system. Vestnik Gosudarstvennogo universiteta morskogo i rechnogo flota imeni admirala SO Makarova **4**(32), 131–137 (2015). https://doi.org/10.21821/2309-5180-2015-7-4-131-137
7. Zhukov, V.A., Pulyaev, A.A., Melnik, O.V., Nyrkov, A.P.: Ensuring the permissible temperature state of parts of the cylinder-piston group of forced diesels. In: 2019 IEEE Conference of Russian Young Researchers in Electrical and Electronic Engineering, pp. 385–388. https://doi.org/10.1109/eiconrus.2019.8657247
8. Bezjukov, O., Zhukov, V., Pulyaev, A.: The Choice Of Cooling Parameters For Marine Diesel Engines. Vestnik Gosudarstvennogo Universiteta Morskogo I Rechnogo Flota Imeni Admirala S.O. Makarova **2**(48), 379–389 (2018). https://doi.org/10.21821/2309-5180-2018-10-2-379-389
9. Zhukov, V.A., Zhelezniak, A.A., Bezmennikova, L.N., Erofeev, V.L.: Diagnosis of thermal processes in motors of the electrical objects. J. Phys. Conf. Ser. **803**(1), 012184. https://doi.org/10.1088/1742-6596/803/1/012184
10. Zhukov, V.A., Sherban, S.A., Melnik, O.V., Sokolov, S.S., Kolesnichenko, S.V.: Improvement of methods and means of thermal control of ship power plants. In: 2019 IEEE Conference of Russian Young Researchers in Electrical and Electronic Engineering, pp. 389–392. https://doi.org/10.1109/eiconrus.2019.8656799

11. Bezyukov, O.K., Zhukov, V.A., Timofeev, V.N.: Sovremennaya koncepciya regulirovaniya oxlazhdeniya sudovy`x dizelej. Vestnik GUMRF imeni admirala S.O. Makarova. **3**(31), 93–103 (2015). https://doi.org/10.21821/2309-5180-2015-7-3-93-103
12. Zhukov, V., Melnik, O., Logunov, N., Chernyi, S.: Regulation and control in cooling systems of internal combustion engines. E3S Web Conf. **135**, 02015. https://doi.org/10.1051/e3sconf/201913502015
13. Nahim, H.M., Younes, R., Shraim, H., Ouladsine, M.: Modeling with fault integration of the cooling and the lubricating systems in marine diesel engine: experimental validation. IFAC-PapersOnLine **49**(11), 570–575 (2016). https://doi.org/10.1016/j.ifacol.2016.08.083
14. Gholinia, M., Pourfallah, M., Chamani, H.R.: Numerical investigation of heat transfers in the water jacket of heavy duty diesel engine by considering boiling phenomenon. Case Stud. Therm. Eng. **12**, 497–509 (2018). https://doi.org/10.1016/j.csite.2018.07.003
15. Pierce, D., Haynes, A., Hughes, J., Graves, R., Daniel, C.: High temperature materials for heavy duty diesel engines: historical and future trends. Prog. Mater Sci. **103**, 109–179 (2019). https://doi.org/10.1016/j.pmatsci.2018.10.004
16. Taymaz, I., Çakır, K., Mimaroglu, A.: Experimental study of effective efficiency in a ceramic coated diesel engine. Surf. Coat. Technol. **200**(1–4), 1182–1185 (2005). https://doi.org/10.1016/j.surfcoat.2005.02.049
17. Taymaz, I.: The effect of thermal barrier coatings on diesel engine performance. Surf. Coat. Technol. **201**(9–11), 5249–5252 (2017). https://doi.org/10.1016/j.surfcoat.2006.07.123
18. Taymaz, I.: An experimental study of energy balance in low heat rejection diesel engine. Energy **31**(2–3), 364–371 (2006). https://doi.org/10.1016/j.energy.2005.02.004
19. Zegenhagen, M.T., Ziegler, F.: Feasibility analysis of an exhaust gas waste heat driven jet-ejector cooling system for charge air cooling of turbocharged gasoline engines. Energy **160**, 221–230 (2015). https://doi.org/10.1016/j.apenergy.2015.09.057
20. Momeni, S.M., Salehi, G., Nimvari, M.E.: Modeling and thermoeconomic optimization of marine diesel charge air cooler. Energy **162**, 753–763 (2018). https://doi.org/10.1016/j.energy.2018.08.092
21. Mezher, H., Chalet, D., Migaud, J., Chesse, P.: The application of a wave action design technique with minimal cost on a turbocharged engine equipped with water cooled charge air cooler aimed for energy management. Energy Procedia **36**, 948–957 (2013). https://doi.org/10.1016/j.egypro.2013.07.108
22. Dere, C., Deniz, C.: Load optimization of central cooling system pumps of a container ship for the slow steaming conditions to enhance the energy efficiency. J. Cleaner Prod. **222**, 206–217 (2019). https://doi.org/10.1016/j.jclepro.2019.03.030
23. Cipollone, R., Battista, D.D.: Sliding vane rotary pump in engine cooling system for automotive sector. Appl. Therm. Eng. **76**, 157–166 (2015). https://doi.org/10.1016/j.applthermaleng.2014.11.001
24. Cipollone, R., Bianchi, C., Battista, D.D., Fatigati, F.: Fuel economy benefits of a new engine cooling pump based on sliding vane technology with variable eccentricity. Energy Procedia **82**, 265–272 (2015). https://doi.org/10.1016/j.egypro.2015.12.032
25. Yih, J., Wang, H.: Experimental characterization of thermal-hydraulic performance of a microchannel heat exchanger for waste heat recovery. Energy Convers. Manag. **204**, 112309 (2020). https://doi.org/10.1016/j.enconman.2019.112309
26. Wang, H., Peterson, R.B.: Performance enhancement of a thermally activated cooling system using microchannel heat exchangers. Appl. Therm. Eng. **31**(14–15), 2951–2962 (2011). https://doi.org/10.1016/j.applthermaleng.2011.05.026
27. Zhu, Y., Li, W., Sun, G., Li, H.: Thermo-economic analysis based on objective functions of an organic Rankine cycle for waste heat recovery from marine diesel engine. Energy **158**, 343–356 (2018). https://doi.org/10.1016/j.energy.2018.06.047

28. Hoseini, S.S., Najafi, G., Ghobadian, B.: Experimental and numerical investigation of heat transfer and turbulent characteristics of a novel EGR cooler in diesel engine. Appl. Therm. Eng. **108**, 1344–1356 (2016). https://doi.org/10.1016/j.applthermaleng.2016.08.018
29. Aly, W.I.A., Abdo, M., Bedair, G., Hassaneen, A.E.: Thermal performance of a diffusion absorption refrigeration system driven by waste heat from diesel engine exhaust gases. Appl. Therm. Eng. **114**, 621–630 (2017). https://doi.org/10.1016/j.applthermaleng.2016.12.019
30. Aydin, H.: Combined effects of thermal barrier coating and blending with diesel fuel on usability of vegetable oils in diesel engines. Appl. Therm. Eng. **51**(1–2), 623–629 (2013). https://doi.org/10.1016/j.applthermaleng.2012.10.030
31. Aydin, S., Sayin, C., Aydin, H.: Investigation of the usability of biodiesel obtained from residual frying oil in a diesel engine with thermal barrier coating. Appl. Therm. Eng. **80**, 212–219 (2015). https://doi.org/10.1016/j.applthermaleng.2015.01.061
32. Han, G., Yao, A., Yao, C., Wang, Q., Liu, M.: Theoretical and experimental investigations on energy balance on DMDF engine. Fuel **164**, 393–402 (2016)
33. Koch. F.W., Haubner, F.G.: Cooling SYSTEM DEVELOPMENT AND OPTIMIZATION. SAE Technical Paper Series, 15 (2000). https://doi.org/10.1016/j.fuel.2015.10.024
34. Akhiani, H., Metselaar, S.C.: A comprehensive review on graphene nanofluids: recent research, development and applications. Energy Convers. Manag. **111**, 466–487 (2016)
35. Goharshadi, E., Ahmadzadeh, H., Samiee, S., Hadadian, M.: Nanofluids for heat transfer enhancement. Phys. Chem. Res. **1**(1), 1–33 (2013). https://doi.org/10.1016/j.enconman.2016.01.004
36. Shanthi, R., Shanmuga Sundaram, A., Velraj, R.: Heat transfer enhancement using nanofluids. Therm. Sci. **16**(2), 423–444 (2012). https://doi.org/10.2298/TSCI110201003S
37. Lam, P.A.K., Prakash, K.A.: Thermodynamic investigation and multi-objective optimization for jet impingement cooling system with Al2O3/water nanofluid. Energy Convers. Manag. **111**, 38–56 (2016). https://doi.org/10.1016/j.enconman.2015.12.018

Impact Study of Basalt and Polyacrylonitrile Fibers on Performance Characteristics of Asphalt Concrete

Sergey Andronov[1(✉)], Yuri Vasiliev[2], Eduard Kotlyarsky[2], Natalia Kokodeeva[1], and Andrey Kochetkov[3]

[1] Saratov State Technical University Named After Gagarin Yu.A., 77 Polytechnic Street, Saratov 410054, Russia
atomic08@yandex.ru

[2] Moscow State Automobile and Road Engineering University, 64 Leningradsky Prospect, Moscow 125319, Russia

[3] Perm National Research Polytechnic University, 29 Komsomolsky Prospect, Perm 614990, Russia

Abstract. The adhesion interaction at the bitumen-fiber interface was examined in test ranges for temperatures from 0 °C to 50 °C. Mechanisms of adhesion interaction are defined, as well as values of adhesion strength of compounds of asphaltic cement with basalt and polyacrylonitrile fibers, determining values of strength properties of fiber-containing road asphalt concrete mixtures. Qualitative effect of fiber surface relief on adhesion strength of bitumen-fiber joint is determined. The paper has been studied the effect of mixed fiber reinforcement of basalt and polyacrylonitrile fiber on physical and mechanical properties of asphalt concrete. Mixed introduction of basalt and polyacrylonitrile fibers reduces the speed of rutting on road asphalt concrete surfaces by 2.3 times higher as compared to physical and mechanical indices of asphalt concrete without addition of fiber. The average rate of rutting has a maximum value when the ratio of polyacrylonitrile fiber to 35% and basalt fiber to 65% in the combined sample is reduced by an average of 33 to 43%.

Keywords: Asphaltic cement · Basalt fiber · Polyacrylonitrile fiber · Asphalt concrete · Mixed fiber reinforcement

1 Introduction

Nowadays. the development of new cost-effective asphalt mixes and asphalt concrete for the construction of pavements is an urgent task. During the operation of the road structure or other objects of the road infrastructure from mechanical impact on the object, weather, climate and soil-hydrological factors, a gradual decrease in their strength and uniformity occurs, associated with internal irreversible changes in each of the structural elements and their relationship between themselves and external influences. Asphalt concrete pavement is primarily exposed to static and dynamic traffic loads, on the one hand, and to adverse environmental and climatic factors, on the other [1–3].

V. Murgul and V. Pukhkal (Eds.): EMMFT 2019, AISC 1258, pp. 473–485, 2021.
https://doi.org/10.1007/978-3-030-57450-5_41

One of the ways to increase the resistance of asphalt concrete to external loads is the use of reinforcing elements in its composition, which are used fibers and threads [4]. Basalt fiber and polyacrylonitrile fiber, which is a precursor for making carbon thread [2], are promising on the basis of technical and economic considerations as reinforcing fibers in asphalt concrete mixture [2]. A typical list of literature on the topic of research is characterized by works [5–7]. An analysis of the normative, technical, and educational literature showed, on the whole, limited information on the possibility of uniform introduction of fiber into hot asphalt mixes, as well as the lack of standardized testing methods for the interaction of fiber and bitumen of asphalt concrete. The introduction of small (discrete) elements into the mixture makes it possible to achieve their uniform distribution (dispersion) in the mixture and to obtain a "composite" material with higher physical and mechanical properties in the finished structural element.

One of the key roles in ensuring the strength of fiber-reinforced asphalt is the adhesive strength of the fiber-bitumen joint. The functional characteristics of the operation of road asphalt concrete pavement in actual temperature operating conditions substantially depend on the adhesion contact mechanism [8–10].

2 Patent Search on Research Topic

The present studies included the search for accessible scientific and technical information that combines normative and technical documentation, scientific articles, valid patents for inventions and utility models, information about advanced production technologies, close technical analogues, and provisions of normative and design documentation. The depth (retrospective) of information retrieval was not limited. Patent studies were conducted on the full electronic Databases of the Russian Federation, including information until 1993 and since 1993 (FIPS website www.fips.ru).

The website of the European Patent Organization (http://ep.espacenet.com) has been searched for patent documents of the European Patent Organization (EPO), World Intellectual Property Organization (WIPO), Japan, Austria, Belgium, Cyprus, Denmark, Finland, France, Germany, Greece, Ireland, Italy, Liechtenstein, Luxembourg, Monaco, the Netherlands, Portugal, Spain, Sweden, Switzerland, England and other countries. Based on the results of the search through the website http://ep.espacenet.com there has been searched for patent information in the respective national patent databases with open access. The authenticity of the results is guaranteed with respect to patent documentation funds available in the electronic databases of the listed countries. The search for scientific and technical documentation has been carried out using the Internet. The array of scientific and technical information includes state standards (GOSTs), quality specifications (TUs), reference books, catalogs, books, articles, reviews, promotional materials.

Patent research on this development has been carried out in the following areas:

Use of mineral fibers (basalt, mineral and fiberglass) for reinforcing the asphalt concrete mixture, their properties and processing; reinforcement of asphalt concrete mixtures with mineral fibers; interaction of silanes with mineral fibers; interaction of silanes with bitumen in an asphalt concrete mixture; interaction of bitumen with mineral fibers; interaction in the system of bitumen-mineral fibers-silane.

Indices of the International Patent Classification (IPC) are determined based on these areas of the technical field of the topic under study:

When searching for information, the following keywords were used: silane; concrete reinforcement; basalt fiber; mineral fibers.

Search results for patent information.
Database of the Russian Federation since 1993
Request "C03C25/40" - 6 patents found.
Request "C03C25 *" - 152 patents found.
Request "silane and IPC: C03C25 *" - 64 patents were found.
Request "silane and IPC: C04B16 *" - 2 patents found.
Request "silane and IPC: E01C7 *" - 2 patents found.
Request "silane and IPC: C10C3 *" - 2 patents found.
Request "silane and IPC: C08G77 *" - 108 patents found.
Request "silane and IPC: C08K5 *" - 189 patents found.
Request "mineral fibers and IPC: C08G77/44" - 1 patent found.
Request "mineral fibers and IPC: C08K5 *" - 112 patents found.
Request "asphalt mix" - 329 patents found.
Request "bitumen and IPC: C04B16 *" - 11 patents found.
Request "bitumen and IPC: C08G77 *" - 2 patents found.
Request "bitumen and IPC: C08K5 *" - 75 patents found.
Request "bitumen and IPC: E01C7 *" - 95 patents found.
Request "bitumen and IPC: E01C7 *" - 95 patents found.
Patents RU No. 2102353, 2129103, 2135426, 2190576, 2351561, 2389698 and 2583387 were selected from the database of the Russian Federation for the analysis of the patent situation.
European Patent Organization Database (espacenet.com).
Request "silane and C03C25" - 500 patents found.
Request "silane and C04B16" - 71 patents found.
Request "silane and E01C 7" - 55 patents found.
Request "silane and E01C 3" - 12 patents found.
Request "concrete reinforcement silane" - 24 patents found.
Request "basalt fiber silane" - 50 patents found.
Patents selected from the European Patent Organization database for analysis of the patent situation are: CN No. 102515570, CN No. 102898044, CN No. 103936302, CN No. 105271830, JP No. H07150050, JP No. H08311349, US No. 4243426, US No. 4273588, US No. 4278470, US No. 4292371.

In accordance with the results of the information search, this study's objective is determined - to assess the effectiveness of the combined impact of basalt and polyacrylonitrile fibers on the performance characteristics of asphalt concrete. Such a study is being carried out for the first time.

Initial data. For research, an asphaltic cement grade BND 90/130 was used, which meets the requirements of GOST 22245-90 [4]. This asphaltic cement is recommended by GOST 9128-2013 [5] for the production of road asphalt concrete mixtures in the climatic conditions of the Saratov region. The basalt and polyacrylonitrile fibers used met the requirements [6, 7].

3 Research Methods

The physicochemical mechanism of the interaction of polyacrylonitrile fibers with bitumen and basalt fibers with bitumen was studied. As a basic criterion for the bond quality of the components of the bitumen-fiber composite cell, the strength of breaking the bond between the reinforcing fiber and the bitumen matrix was considered.

As the main criterion for the bond quality of unit cell's components of the composite material "bitumen-fiber", we considered the fracture strength of the compound between the reinforcing fiber and the bitumen matrix.

The value of the fracture force was determined by the intensity of the required force for pulling the fiber from the cured matrix at a constant depth of embedding (immersion of the fiber in bitumen). The pulling force was measured using a dynamometer by pulling the fiber in the perpendicular direction to the surface of the bitumen. It should be noted that the measurement of the pulling force of individual fibers with a diameter of 8 to 12 μm from the matrix material is associated with difficulties in fixing the mechanical clamp on the fiber; therefore, 10 bundles with a diameter of 1.2 to 1.5 mm were formed from the fibers.

After that, each bundle was immersed by approximately 5 mm in the bitumen melt at a temperature of 120 °C, which then solidified at indoor temperature (Fig. 1, 2, 3 and 4).

Fig. 1. Polyacrylonitrile fiber.

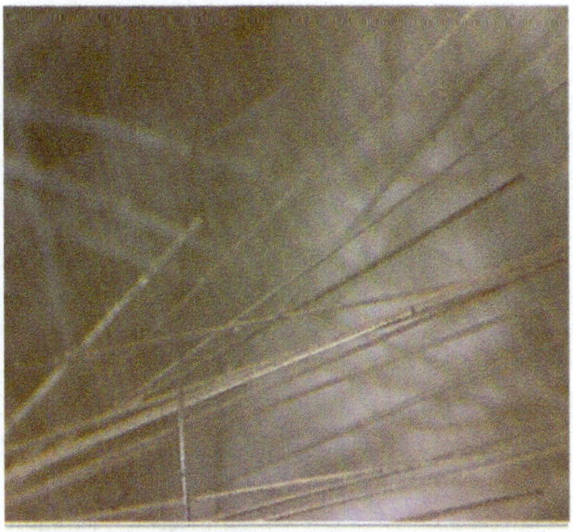

Fig. 2. Basalt fiber.

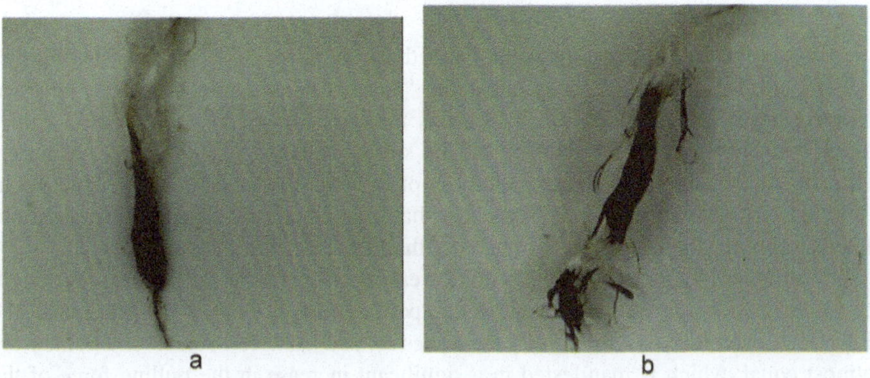

Fig. 3. Polyacrylonitrile threads drawn from bitumen: a - at T = 50 °C, b - at T = 25 °C.

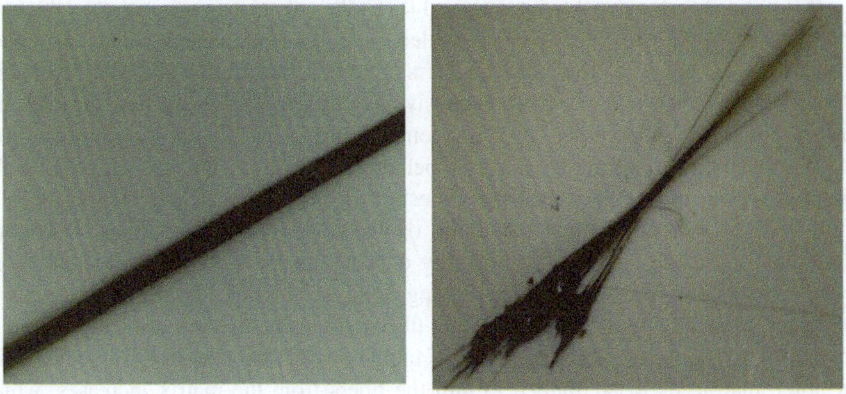

Fig. 4. Basalt threads drawn from bitumen: a - at T = 50 °C, b - at T = 25 °C.

4 Results

Table 1 shows the results of the bitumen-fiber model composite tests carried out at temperatures of 0 °C, 25 °C and 50 °C.

Table 1. Dependence of pullout force value on test temperature and fiber type.

Fiber	Temperature, °C	F, N	Note
Polyacrylonite thread, 56 tex	50	0.5 to 1	Thread covered with bitumen
	25	25 to 35	Thread covered with bitumen
	0	130 to 145	Thread broken
Basalt thread, 50 tex	50	0.5 to 1	Thread covered with bitumen
	25	18 to 25	Thread covered with bitumen
	0	270 to 330	Thread broken

Note. Tex is an off-system unit of linear fiber density (fiber mass in gg of 1 km long).

Based on the analysis of the data obtained, it can be concluded that for the systems under consideration, the interaction at the "bitumen-fiber" interface in the selected temperature range is carried out by various mechanisms, which is associated with a change in the viscosity of bitumen and its surface energy. At T = 50 °C, polyacrylonitrile and basalt filaments are pulled out of bitumen due to cohesive destruction of bitumen. This is a consequence of the fact that it is in a plastic state and the energy of interaction with the fiber surface is greater than the cohesion of bitumen.

A different picture is observed with decreasing temperature - the cohesive strength of bitumen increases, which begins to compete more and more with adhesion at the interface. Therefore, at T = 25 °C, the adhesion and cohesion forces of bitumen are almost equal, which is manifested in a significant increase in the pulling force of the threads. At T = 0 °C, the energy of cohesion of bitumen and adhesion at the border exceed the energy of cohesive interaction of fibers and bitumen, and therefore the threads are destroyed when stretched.

It is noticed that at a test temperature of 0 °C, the adhesion forces between the fiber and bitumen do not allow the tow to be pulled out of the composite, which is associated with the "swelling" of the tow onto the fibers due to the penetration and wetting of each fiber with bitumen. As a result, the adhesive strength of the connection between the matrix and the filler increases. At a test temperature of 50 °C, polyacrylonitrile fiber and basalt fiber, as reinforcing elements, behave identically, while the adhesion to the fibers is significantly higher than the cohesion of bitumen, which is manifested in a significant amount of bitumen remaining on the fibers during stretching. Therefore, hardening of asphalt concrete significantly depends on the quality of bitumen - softening temperature. The higher the softening point of bitumen, the more noticeable is the effect of bitumen reinforcement during fiber reinforcement. At a test temperature of 25 °C, the cohesion and adhesion forces of bitumen with fibers begin to compete, while the force that needs to be applied to pull the fibers from the matrix increases, which indicates the possibility of real hardening of asphalt concrete with fibers. At zero temperature and below, cohesion of bitumen and interfacial interaction between bitumen and fiber significantly increase. In this case, the decisive role is played by the strength of the fibers. As can be seen from Table 1, in this temperature range, reinforcement with basalt fibers is preferable, since their strength characteristics are much higher than that of polyacrylonitrile fibers.

Assuming the cross-section of the filaments to be round and knowing the depth of their immersion in bitumen, the values of adhesive strength (τ_i) at the "bitumen-fiber" interface have been estimated using the formula:

$$\tau_i = F_i / S_i \tag{1}$$

where: F_i is the pulling force of the threads; S_i is the area of the adhesive joint — the area of individual circular sections of each strand immersed in bitumen by 5 mm, multiplied by their number.

Table 2 shows the measurement results of series of 10 experiments on determining the adhesive strength of the bitumen-fiber joint at a temperature of 25 °C.

Table 2. Dependence of adhesive strength of bitumen-fiber joint at temperature of 25 °C on pulling force, type of fiber, diameter of bundle, area of adhesive joint.

Type of fiber	Strand diameter, mm	S_i, mm^2	F_i, N	τ_i, N/mm^2
Polyacrylonitrile	1.2	18.84	25	1.33
	1.25	19.62	26	1.32
	1.22	19.15	24	1.25
	1.38	21.67	27	1.25
	1.32	20.72	27	1.3
	1.4	22.0	28.5	1.3
	1.45	22.76	32	1.4
	1.5	23.55	35	1.49
	1.28	20.1	24	1.19
	1.36	21.35	25	1.17
Basalt	1.4	22	22	1.0
	1.3	20.4	22	1.08
	1.2	18.8	18	0.96
	1.5	23.55	24	1.02
	1.5	23.55	22	0.93
	1.4	22	25	1.14
	1.2	18.8	19	1.01
	1.4	22	24	1.09
	1.3	20.4	21	1.03
	1.3	20.4	19	0.93

The surfaces of the polyacrylonitrile and basalt fiber were examined by scanning electron microscopy. It has been found that on the surface of the polyacrylonitrile fiber, defects in the form of growth are visible, which contribute to the creation of a more developed morphology.

Figure 5 shows images of the surface of the polyacrylonitrile fiber. Figure 6 shows the images, the analysis of which indicates the morphology of the basalt fiber surface. It can be seen from Fig. 6 that the surface of the basalt fiber is smooth and has no advanced morphology.

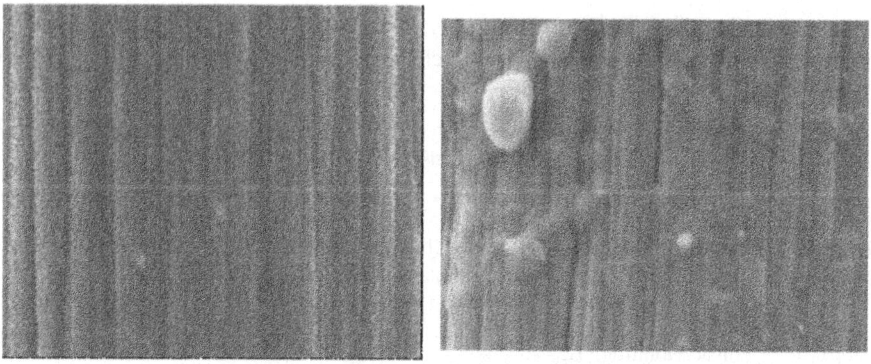

Fig. 5. Polyacrylonitrile fiber surface.

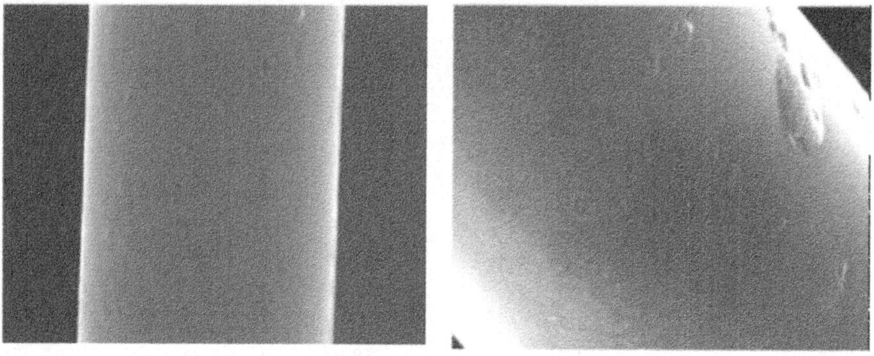

Fig. 6. Basalt fiber surface.

An analysis of the data shows that the adhesion energy of polyacrylonitrile fibers to bitumen is on average 25 to 28% higher than that of basalt fibers. This can be explained by the more developed surface relief of polyacrylonitrile fibers, which is formed by fibrillar supramolecular structures that form the corrugated outer fiber border.

5 Description of Test Procedure

The physical and mechanical properties of fiber-containing asphalt concrete prepared by the method of joint reinforcement with basalt and polyacrylonitrile fibers were examined. In order to compare the results, the physical and mechanical properties of asphalt concrete without fiber fiber additives were also examined, and the physical and mechanical properties of asphalt concrete with polyacrylonitrile fiber additives were examined. The tests were carried out on mixtures of type B as per GOST 9128-2013. Basalt fiber with a cut length of 15 mm was used. Polyacrylonitrile fiber was used with a cut length of 12 mm. Optimal fiber dosages were pre-established. In asphalt concrete with addition of basalt fiber, optimal dosage of fibers is 0.4% of mixture weight. In asphalt concrete with addition of polyacrylonitrile fiber, optimal dosage of polyacrylonitrile fiber is 0.2% of mixture weight.

A 10-liter laboratory mixer has been developed for research into the production technology of fiber-containing asphalt concrete mixtures (see Fig. 7). The fiber was injected by the method of blowing a pre-fluff hinge into a laboratory mixer. Mixer design and operation mode are developed based on the principle of simulating operation of asphalt concrete plant mixer with capacity of 100 t/h. On the basis of the Saratov State Technical University named after Gagarin Yu. A., an experimental sample of a laboratory mixer was assembled. Counter rotation of agitator shafts provides transfer of material from action range of one shaft blades to the area of other shaft blades.

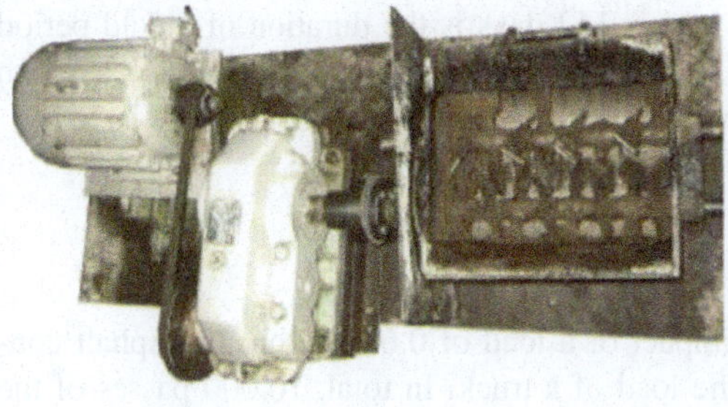
a

b

a - general view; b - working chamber

Fig. 7. Photo of laboratory mixer (with horizontally located shafts with blades).

Production and determination of physical and mechanical properties of samples of fiber-containing asphalt concrete samples have been carried out in accordance with the procedure of GOST 12801-98. Since a distinctive feature of fiber-containing asphalt concrete is a lower rutting rate under the influence of traffic and weather and climate factors, an additional determination was made of the resistance index of fiber-containing asphalt concrete to rutting, namely, the rate of rut formation on an asphalt concrete pavement. To test the resistance to rutting of road fiber-containing asphalt concrete, a device for testing rutting of road composite mixtures was developed and patented [3]. At the Saratov State Technical University named after Gagarin Yu. A., a prototype device was assembled. Figure 9 schematically depicts the inventive device.

The stand includes a chamber 1, with a coated layer of heat-insulating material, two side, rear and front walls. The front wall is made with a transparent window in it. Inside the chamber 1, a Table 3 is mounted on the rods 2, which is driven by an actuator 4 with an electric motor 5. The Table 3 on top has a place for fixing an asphalt concrete sample 6. The stand has a force generating unit 7 on a sample 6, which includes a rubberized wheel 8 fixed in a metal holder 9. The stand is equipped with a temperature controller 16 and a meter 17 of rut depth located inside chamber 1 and displayed on the control panel 15. In the chamber 1 on the Table 3 is a sample. According to the signals received from the control panel 15 to the actuator 4 with the electric motor 5, the latter provides reciprocating motion of the Table 3 with the sample 6 fixed on it. Wheel 8, rotating around the axis and being in contact with the surface of the sample 6, causes its deformation in the form of rut, the depth of which is proportional to the vertical force generated by the node 7, and is determined by the meter 17. Simulation of the inclined section of the roadway is carried out by lifting from the loose side of the working plate 10, carried out by moving the rail 12 and the inclined section thereon towards the central part of the plate 13. Motion of the rail 12 is provided by rotating the wheel 13 mounted on the shaft of the motor 14, which is meshed with the rail 12. The edge of the plate 10 is lowered by moving the rail 12 in the direction from the center of the plate 10 and rotating the wheel 13. Figure 10 shows a photo of a rubberized wheel on the surface of a tested sample made of fiber-containing asphalt concrete and a rut formed by moving the rubberized wheel.

In accordance with Russian State Standard SNiP 23-01-99 Construction climatology, in the conditions of the Saratov region 142 days is the duration of a cold period with an average temperature of 7.5 °C. The remaining 223 days comprise a warm period with an average temperature of 25 °C.

6 Study Results

The rubberized sector simulated the impact of a load of 0.6 MPa on the asphalt concrete coating, which corresponds to the load of a truck. In total, 10,000 passes of the rubber sector on asphalt concrete were made. At the same time, to simulate the cold and warm period of the year, 3,890 passes of the rubberized sector were made at a temperature of −7.5 °C, and the remaining 6,110 passes at a temperature of 25 °C. The average rut formation rate was estimated at a load of 0.6 MPa for 10,000 passes of the rubberized sector. Table 3 shows the physical and mechanical properties of asphalt concrete with the addition of fiber and original asphalt concrete without the addition of fiber those obtained by the methods of Russian State Standard GOST 12801-98.

To study the effectiveness of the method of joint dispersed reinforcement of asphalt concrete with basalt and polyacrylonitrile fiber, polyacrylonitrile and basalt fibers in the ratios 20:80, 35:65, 50:50, 65:35 and 80:20 were jointly introduced into the composition of type B asphalt mixtures by the method of blowing previously fluffed hinges respectively. Physical and mechanical characteristics of asphalt concrete with an additive consisting simultaneously of basalt and polyacrylonitrile fibers are shown in Table 4.

Table 3. Physical and mechanical characteristics of the original asphalt concrete, fiber-containing asphalt concrete with the addition of polyacrylonitrile fiber and fiber-containing asphalt concrete with the addition of basalt fiber.

Name of indicator	GOST 9128-2013 requirements for grade I type B	Actual physical and mechanical properties		
		Original asphalt concrete without addition of fiber	Asphalt concrete with addition of polyacrylonitrile fiber	Asphalt concrete with addition of basalt fiber
Compressive strength at 50 °C, MPa	Not less 1.30	1.7	2.9	2.5
Shear resistance by internal friction coefficient	Not less 0.83	0.85	0.89	0.91
Shear resistance by grip at shear at 50 °C	Not less 0.38	0.6	0.79	0.7
Crack resistance by ultimate strength tensile strength at split at 0 °C and strain rate 50 mm/min, MPa	4.0 to 6.5	3.6	5.0	6.0
Average rutting speed, mm/10,000 passes	n/a	3.5	1.4	1.5

Table 4. Physical and mechanical characteristics of asphalt concrete.

Name of indicator	Actual physical and mechanical properties of fiber-containing asphalt concrete				
	Polyacrylonitrile fiber 20%, basalt fiber 80%	Polyacrylonitrile fiber 35%, basalt fiber 65%	Polyacrylonitrile fiber 50%, basalt fiber 50%	Polyacrylonitrile fiber 65%, basalt fiber 35%	Polyacrylonitrile fiber 80%, basalt fiber 20%
Compressive strength limit at 50 °C, MPa	2.45	2.5	2.57	2.64	2.8
Shear resistance by internal friction coefficient	0.9	0.89	0.9	0.89	0.89
Shear resistance to adhesion at shear at 50 °C	0.7	0.72	0.74	0.77	0.78
Crack resistance by ultimate strength tensile strength at split at 0 °C and strain rate 50 mm/min, MPa	5.8	5.5	5.2	5.0	4.8
Average rutting speed, mm/10,000 passes	1.5	1.05	1.2	1.35	1.4

7 Discussion

The analysis of the test results presented in Table 3 showed that the physical and mechanical characteristics of fiber-containing asphalt concrete are higher in comparison with the physical and mechanical properties of asphalt concrete without the addition of fiber. The use of fiber in the composition of asphalt concrete helps to reduce the rate of rutting by about 2.3 times. As a result of the analysis of the data presented in Table 4, it was found that with a decrease in the dosage of basalt fiber and, accordingly, with an increase in the dosage of polyacrylonitrile fiber, the index of compressive strength at 50 °C and the index of shear resistance to adhesion at shear at a temperature of 50 °C increase.

It was found that the adhesion of polyacrylonitrile fibers to bitumen is on average 25 to 28% higher than that of basalt fibers. This can be explained by the more developed surface relief of polyacrylonitrile fibers, which is formed by fibrillar supramolecular structures that form the corrugated outer fiber border. On the surface of the fibers there are growths that contribute to the creation of a more developed morphology. At a temperature of 50 °C, polyacrylonitrile fibers and basalt fiber, as reinforcing elements, behave identically, while the adhesion to the fibers is significantly higher than the cohesion of bitumen, which is manifested in a significant amount of bitumen remaining on the fibers during stretching. Therefore, hardening of asphalt concrete significantly depends on the quality of bitumen - softening temperature. The higher the softening point of bitumen, the more effective is fiber reinforcement.

At a test temperature of 25 °C, the cohesion and adhesion forces of bitumen with fibers begin to compete, while the pulling forces of the fibers from the matrix increase, which indicates the possibility of real hardening of asphalt concrete with fibers. At zero temperature and below, cohesion of bitumen and interfacial interaction between bitumen and fiber significantly increase. In this case, the decisive role is played by the strength of the fibers. At this temperature, reinforcement with basalt fibers is preferable, since their strength characteristics are much higher than that of polyacrylonitrile fibers.

With an increase in the dosage of basalt fiber and, correspondingly, with a decrease in the dosage of polyacrylonitrile fiber, the fracture toughness index with respect to the tensile strength at break at a temperature of 0 °C and the strain rate of 50 mm/min increases.

The index of shear stability by the coefficient of internal friction with a change in the dosages of polyacrylonitrile and basalt fiber in the combined sample as a whole remains unchanged. The average rate of rutting has a maximum value when the ratio of polyacrylonitrile fiber to 35% and basalt fiber to 65% in the combined sample is reduced by an average of 33 to 43%. The decrease in the rutting rate is explained by different adhesive interactions of viscous oil road bitumen and fiber, taking into account the properties of the fibers and temperature [11].

8 Conclusions

The scientific novelty of the performed work reflects the possibilities to improve the operational properties of composite asphalt concrete by modifying the binder with prestressed (basalt) and elastic (polyacrylonitrile) fibers.

The studied mechanism affects improvement of properties of the end material due to increase of uniformity of strength indices of interaction between bitumen and mineral fillers and timely redistribution of changes in stress-strain state of fiber-containing asphalt concrete (reducing the number and volume of weak points and the risk of damage accumulation).

Features of combination of polyacrylonitrile and basalt fibers in proportion providing optimal index of rut formation rate are determined.

Based on the studies with the participation of the authors, the Federal Road Agency ODM 218.3.054–2015 developed Guidelines for surface treatment and thin wear layers using various types of fiber, which is used at the road design stage when assigning the parameters of the surface roughness depending on traffic conditions, when choosing the appropriate types of coatings and methods of distribution of materials, types and quality of materials used, technologies and organization of work arrangement of rough surface layers with application of different types of fibers, as well as for quality control.

References

1. Rudensky, A.V.: Asphalt concrete pavement. Transport, Moscow (1992)
2. Berkovich, A.K.: Synthesis of Acrylonitrile-Based Polymers. Technology for producing PAN and carbon fibers. Moscow State University named after M.V. Lomonosov, Moscow (2010)
3. Kotlyarsky, E.V., Voyeyko, O.A.: Durability of road asphalt pavements and factors contributing to destruction of asphalt structure during operation. Technical Polygraph Center, Moscow (2007)
4. Andronov, S.Yu., Artemenko, A.A.: Patent RU 173579 U1 Device for testing the rutting of road composite mixtures. IPC G01N 3/56. Bull. Number 25. Accessed 31 Aug 2017
5. Yu, D., Wang, W., Cheng, Y., Gong, Y.: Laboratory investigation on the properties of asphalt mixtures modified with double-adding admixtures and sensitivity analysis. J. Traffic Transp. Eng. (2016). https://doi.org/10.1016/j.jtte.2016.09.002
6. Cheng, Y., Yu, D., Tan, G., Zhu, C.: Low-temperature performance and damage constitutive model of eco-friendly Basalt fiber–diatomite-modified asphalt mixture under Freeze-Thaw cycles. Materials **11**(11), 2148 (2018). https://doi.org/10.3390/ma11112148
7. Celauro, C., Praticò, F.: Asphalt mixtures modified with Basalt fibres for surface courses. Constr. Build. Mater. **170**, 245–253 (2018). https://doi.org/10.1016/j.conbuildmat.2018.03.058
8. Gong, Y., Bi, H., Liang. C., Wang, S.: Microstructure analysis of modified asphalt mixtures under Freeze-Thaw cycles based on CT scanning technology. Sciences **8**(11), 2191 (2018). https://doi.org/10.3390/app8112191
9. Qin, X., Shen, A., Guo, Y., Li, Z.: Characterization of asphalt mastics reinforced with Basalt fibers. Constr. Build. Mater. **159**, 508–516·(2018). https://doi.org/10.1016/j.conbuildmat.2017.11.012
10. Gong, Y., Bi, H., Tian, Z., Tan, G.: Pavement performance investigation of nano-TiO2/CaCO3 and Basalt fiber composite modified asphalt mixture under Freeze-Thaw cycles. Sciences **8**(12), 2581 (2018). https://doi.org/10.3390/app8122581
11. Andronov, S.Y., Trofimenko, Y.A.: Influence of temperature regime for preparation of composite dispersion-reinforced asphalt mixtures on quality indicators. J. Basic Res. **3**, 451–455 (2016)

Using the Response Surface to Assess the Reliability of the Russian Cryolithozone Road Network in a Warming Climate

Anatolii Yakubovich(✉) and Irina Yakubovich(✉)

Moscow Automobile and Road Construction State Technical University (MADI), 64, Leningradsky Prospect, Moscow, Russia
54081@mail.ru, yakubovich_irina@mail.ru

Abstract. The paper predicts the functional state of the main part of the road network of the Magadan region with a total length of 1219.3 km in the conditions of climate warming up to 2°. It is shown that in this scenario of climate changes, the road network as a whole will retain its functionality, with an expected decrease in throughput on individual network fragments. To assess the risks of reduced functionality, we performed simulations of the temperature regime of permafrost at the base of the road, determining the maximum depth of its thawing and the amount of sediment of thawing soil, which manifests as numerous randomly located defects on the surface of the road profile. Since modeling of the soil temperature regime is a computationally time-consuming procedure, it was performed only in individual reference points – points of systematic observations of the climate, for which there were arrays of long-term values of climate parameters. Based on the modeling results, the value of the climate risk for reference points was determined, reduced to a range from 0 to 1000 points (the maximum risk value corresponded to the complete impossibility of continuing operation of the road), and the response surface for the risk was constructed as a function of the coordinates of an arbitrary point. The reference road network was divided into 1602 sections, for each of which the response surface was used to determine the climate risk. There were considered 4 types of soil possible for the foundation of the highway, for each of them a separate response surface was built.

Keywords: Road network · Cryolithozone · Climate change · Climate risks

1 Introduction

One of the main manifestations of global climate change is a systematic increase in near-surface air temperature. Overall, 2018 was the fourth warmest year for the entire period of instrumental observations; the period from 2015 to 2018 is the warmest four-year period both on land and in the World's oceans. With an undoubted warming trend, the temperature regime in almost all territories of the Northern hemisphere is characterized by significant anomalies and contrasts, almost in every month of 2018, the presence of large and intense anomalies of both signs was noted; the total number of negative anomalies (local cooling) increased by 2.5-3 times compared to 2017 [1].

V. Murgul and V. Pukhkal (Eds.): EMMFT 2019, AISC 1258, pp. 486–495, 2021.
https://doi.org/10.1007/978-3-030-57450-5_42

Accordingly, for the temperature dynamics of near-surface air, it is impossible to build a reliable forecast model even in a fairly short-term perspective, and scenario assumptions are the main means of predicting the future state of the climate. The factual basis for constructing the most probable scenarios should be constantly updated information about the current values of climate parameters in a certain territory; a specialized information system is a means for recording these parameters [2].

Global warming is most intense in the Arctic zone and is accompanied by the degradation of permafrost soils: their thawing and significant reduction in load-bearing capacity. As a result, engineering structures located on permafrost are subject to significant deformations that are not considered during the design and construction; when the maximum possible deformations are reached, the structure goes into the stage of destruction, with the complete loss of its operational suitability. As global climate changes increase, the prospect of a massive decrease in the functionality of infrastructure facilities in the cryolithozone becomes more and more real [3].

In relation to the road network, deformations (subsidence) of the thawing soil at the base of the road lead to the appearance of numerous sinkholes located randomly on its surface. As a result, the capacity of the road segment decreases; the amount of this decrease is determined by the depth of defects in the upper part of the road profile, which, in turn, depend on the intensity of climate change and the characteristics of thawing permafrost soil. In relation to the climate conditions of Yakutsk, the projected capacity of the road decreases by 7% with an increase in the average annual temperature by 1 °C and dry sandy soil; warming by 3 °C for wet clay soils at the base of the road causes a 43% decrease of the capacity of the segment [4]. Thus, the possible negative consequences for the road network of the cryolithozone are significant, which makes it necessary to forecast them for the most likely scenarios of climate changes.

Determining the forecast sediment for thawing permafrost soil (in accordance with the climate and soil characteristics of the territory, as well as the accepted scenario of climate change) is a key point in the procedure for predicting the operational suitability of the road network. To determine sediment, it is necessary to perform numerical modeling of heat transfer in the ground mass, including the road profile and the base under it, for at least one full year; when taking into account measures to stabilize the soil temperature regime (installation of thermal insulation coatings or seasonal cooling devices), the required modeling period may be 5–7 years [5]. This leads to significant costs of computational resources and to the inability to determine forecast sediments for a significant number of road sections that are located in different climatic conditions and form a long road network together.

A significant reduction in computational procedures is possible when using a cellular model, which forms a set of cells on the territory of the road network; within each cell, the climatic and ground conditions are considered unchanged. The forecast sediments of thawing soil is determined for a limited number of cells (10–15 reference cells), for which the above computationally expensive procedure is fully implemented. For the remaining cells (and sections of the road network located within each of them), forecast sediments is determined using the response surface, - a function whose parameters depend on the precipitation values in the reference cells.

Cellular (cell) models are often and with high efficiency used in the study of the road network, in particular, when predicting the parameters of traffic flows formed on this network. In [6], a cellular model is proposed, which is used to study the traffic flow in various conditions, such as the bend radius, the length of the bend arc, and the coefficient of friction of the road. The cellular automaton as a one-dimensional model of transport flow is considered in [7]. A broader generalization of the basic model of the cellular automaton describing traffic flow along the highway was made in [8], where the influence of acceleration, confusion, and slow start on the length of the queue at the entrance to the highway was studied.

The feasibility of using the response surface is also confirmed by numerous studies. The response surface based on high-order polynomials with mixed terms, approximating the true limit function of the state of a complex engineering structure, was used in [9], the surface parameters were determined on the basis of step regression. As a result, it became practically possible to calculate the probability of failure of a structure using the Monte Carlo method with a large number of implementations. The response surface, called adapters, was proposed in [10], when forming it, weighted regression was used instead of normal regression, and experimental (reference) points were selected from the area where the existence of the calculated point was most likely. It was shown that the use of the response surface significantly facilitates the computational load during engineering analyses. The use of the response surface, which was constructed using an adaptive improved method based on a combination of the least squares method and the original weight function, is shown in [11]; the effectiveness of this approach to the construction of the response surface is confirmed by numerous calculations of real structures.

2 Characteristics of the Research Object

The Magadan region is a major mining region of Russia. The arrival of industrial cargo necessary for the development of mineral deposits occurs through the seaport in the city of Magadan; the deposits themselves are concentrated at a considerable distance from the coast, at a distance of 300 to 500 km. The movement of goods to mining territories is possible only by road, which makes the road network of the region an important element of the regional logistics system and gives special importance to ensuring its smooth functioning under any operating conditions.

The road network diagram is shown in Fig. 1, the main part of this network is marked out - as a year-round road with an increased quality of road surface (compared to other road segments). The rest of the road network (marked with a dotted line in Fig. 1a) is either temporary roads (used only during the summer or winter period), or has a low capacity. The total length of the reference network is 1219.3 km, and the geometric distance between its most remote points (the cities of Magadan and Susuman) is 394 km.

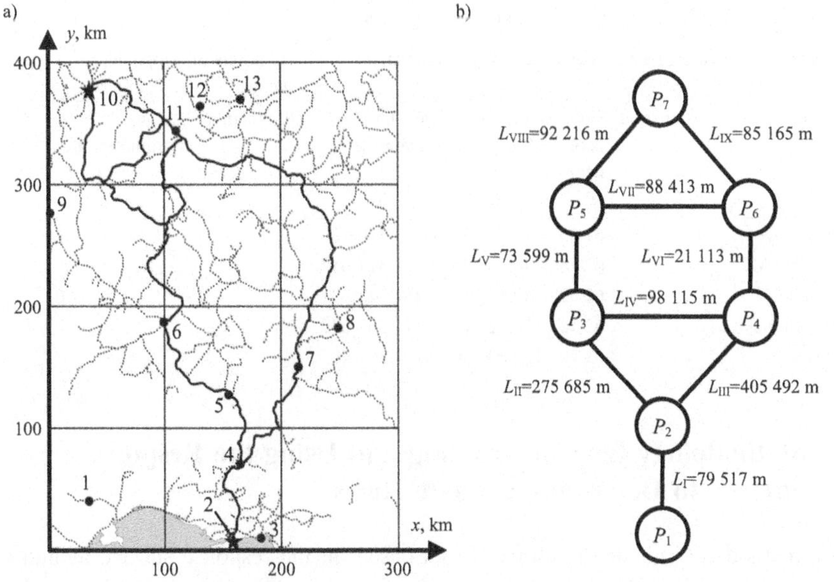

Fig. 1. Geographical (a) and topological (b) scheme of the reference road network of the Magadan region; the names and coordinates of points 1–13 are given in Table 1.

The region has 13 points of systematic instrumental observations of climate, the results of these observations can be used for quantify characteristic of climatic parameters and for further building of the response surface of climate risks. The coordinates of these points are shown in Table 1; the coordinates of the key points of the reference road network are also shown there.

Table 1. Coordinates of reference points of the road network of the Magadan region.

Points for systematic measurements of climate parameters						Key points of a topological scheme		
Name	Designation	Geographic coordinates, degrees		Rectangular coordinates, m		Designation	Rectangular coordinates, m	
		B	*L*	*x*	*y*		*x*	*y*
Talon	1	59.77	148.65	34 658	40 189	P_1	159 995	2 238
MAGADAN	2	59.57	150.80	158 662	7 541	P_2	165 158	71 810
Ola	3	59.58	151.30	182 332	9 631	P_3	102 467	267 900
Palatka	4	60.10	150.94	164 777	70 630	P_4	115 388	337 500
Madaun	5	60.61	150.70	154 701	127 044	P_5	48 565	302 249
Ust-Omchug	6	61.15	149.63	99 608	186 847	P_6	99 869	352 929
Atka	7	60.84	151.79	214 620	150 107	P_7	35 601	376 435
Talaja	8	61.13	152.39	248 715	182 214	–	–	–

(continued)

Table 1. (*continued*)

Points for systematic measurements of climate parameters						Key points of a topological scheme		
Name	Designation	Geographic coordinates, degrees		Rectangular coordinates, m		Designation	Rectangular coordinates, m	
		B	*L*	*x*	*y*		*x*	*y*
Stokovoye	9	61.85	147.66	1 766	276 126	–	–	–
SUSUMAN	10	62.78	148.15	35 788	377 657	–	–	–
Yagodniy	11	62.52	149.62	110 235	343 337	–	–	–
Khatyngnakh	12	62.72	150.02	131 011	363 637	–	–	–
Elgen	13	62.79	150.69	165 510	367 467	–	–	–

3 Methodology for Constructing and Using the Response Surface to Determine Climate Risks

The method for calculating climate risks based on the response surface includes six stages of the computational process. At the first stage, the predicted climate risks are determined at the reference points of the response surface – points of systematic climate observations (the total number of such points is equal to nb). Based on the results of long-term instrumental observations, the average monthly temperature values for the selected period are revealed; the average monthly wind speed values are determined in the same way. Each value of the climate parameter (temperature or wind speed) refers to the middle of the corresponding month, after which two polynomials are constructed that approximate the values of air temperature t and wind speed v for one year (365 days). Thus, for each of the nb points, a formalized description of the average year p is created as a set of two interrelated climate parameters:

$$p(\tau) = \{t(\tau), v(\tau)\}, \tag{1}$$

where τ – is the number of seconds since the beginning of the year.

The average year is divided into consecutive intervals of length Δτ; for the middle of each interval, the values of climate parameters are determined using (1). The result is a mass P:

$$\mathbf{P} = \{(t_1, v_1), \ldots (t_k, v_k)\}, \tag{2}$$

where k – is the number of intervals.

At the second stage, the G mass is formed as a structured set of parameters that characterize the ground mass, including the road profile and the base under it. These include the physical, mechanical and thermal parameters of the soil, the geometric dimensions of the modeled soil mass, and so on. Next, heat transfer in the soil mass is modeled (this procedure is described in more detail in [4, 5]), and for each interval, temperature values are determined for all points in this mass, which together characterize the temperature dynamics of the soil D throughout the average year.

The third stage consists of analyzing the temperature dynamics D and determining the depth of thawing soil h, the maximum throughout the year, after which the calculation of the amount of sediment of thawed soil m is performed. Thus, the ground sediment at a certain reference point can be represented as:

$$m = F(\mathbf{P}, \mathbf{G}), \tag{3}$$

where F – is a generalized transformation operator based on numerical modelling of heat transfer in a soil mass.

The fourth stage is to define a climate change scenario and create the P_c mass as a formalized description of this scenario. In the simplest case, the warming scenario is described by an increase in the values of all temperatures in P_c by the same amount; there are also more complex patterns that reflect the predicted temperature dynamics of near-surface air. Then the second and third stages of computational procedures are repeated, in which instead of the mass P (which characterizes the base climate), the mass P_c corresponding to the changed climate is used. In the course of calculations, for each reference point, the temperature dynamics of the soil in the changed climatic conditions D_c is formed, the increased depth of thawing of the soil h_c and its sediment mc are determined (its calculation can be represented as an expression (3), in which P_c is used instead of P). The stage ends with the determination of additional soil sediment Δm, predicted as a result of climate change for all reference points:

$$\Delta m = m_c - m = F(\mathbf{P_c}, \mathbf{G}) - F(\mathbf{P}, \mathbf{G}). \tag{4}$$

In the fifth stage, the climate risk is assessed for each of the n_b reference points. The extreme points of dependence R(Δm) are fixed: it is assumed that there is no risk in the base climate conditions (at Δm = 0) (R = 0), and the limit value of additional sediment Δm_{max} is set, corresponding to the maximum risk R_{max} = 1000 points. The general form of the dependence R(Δm) is set a priori, taking into account the requirements for its monotonous increase over the interval [0; Δm_{max}]; in the simplest case, this dependence can be assumed to be linear. Based on R(Δm), n_b values of climate risk R are determined for individual reference points.

The sixth stage is to construct the response surface as a function of the coordinates of point a, for which the climate risk R is determined:

$$R(x,y) = \sum_{i=1}^{n_b} \frac{R_{b,i} d(b_i, a)^{-2}}{\sum_{j=1}^{n_b} d(b_j, a)^{-2}}, \tag{5}$$

where $R_{b,i}$ – the risk value determined at the previous stage at the i-th reference point; b_i – the i-th reference point, characterized by coordinates (x_i; y_i); a – the point for which the risk is determined, characterized by coordinates (x; y); d(b,a) – the geometric distance between points b and a.

Based on (5), it becomes possible to quantify the climate risk R for an arbitrary point a(x,y) without performing computationally time-consuming procedures related to modeling the temperature dynamics of the road profile soil for at least one average year (as it was done for reference points when constructing the response surface).

4 The Results of Numerical Modeling

The temperature regime of the near-surface air temperature which is typical for the base climate, in which the risks of loss of functionality are considered to be absent, was determined on the basis of averaged results of instrumental observations for the period 1961–1990. During this period, the main part of the existing road network of the Magadan region was designed and built (based on the actual values of climate parameters at that time). The climate change scenario adopted for numerical modeling assumed an increase in air temperature by +2 °C, the same for all days of the average year. Other climate parameters, including wind speed, were assumed to be unchanged.

Since it is not possible to accurately determine the type of permafrost soil at the base of the road network throughout its entire length, the climate risk was quantified for 4 possible types of soil: sandy (hereinafter type A), sandy- loam (type B), loam (type C) and clay (type D); the humidity of all types of soil was assumed to be average.

The values of the predicted risks for the points of systematic observations of climate (reference points) are shown in Table 2; these results were obtained directly by modeling the soil temperature regime with the subsequent determination of excess thawing and risk. Table 2 also shows the results of climate risk assessment for road network keys P1–P7, obtained using response surfaces (5); a separate response surface was constructed for each type of soil.

Table 2. Predicted risks of reducing the functionality of the road network for reference points.

Points for systematic measurements of climate parameters					Key points of a topological scheme				
Item	Predicted risk *R* for soils				Point	Predicted risk *R* for soils			
	A	B	C	D		A	B	C	D
1	103	147	202	267	P_1	160	217	279	360
2	161	218	281	362	P_2	114	163	222	278
3	145	204	264	336	P_3	89	134	183	229
4	114	163	222	278	P_4	82	131	177	217
5	90	141	186	235	P_5	86	132	179	225
6	85	121	172	222	P_6	83	131	177	218
8	83	131	178	216	P_7	78	121	165	212
10	78	121	165	212	–	–	–	–	–
11	82	130	176	216	–	–	–	–	–
12	87	137	184	226	–	–	–	–	–

Using the response surface, forecast climate risks R (corresponding to the accepted climate change scenario) were determined for each type of soil at the base of the road for the entire road network; the results are shown in Fig. 2.

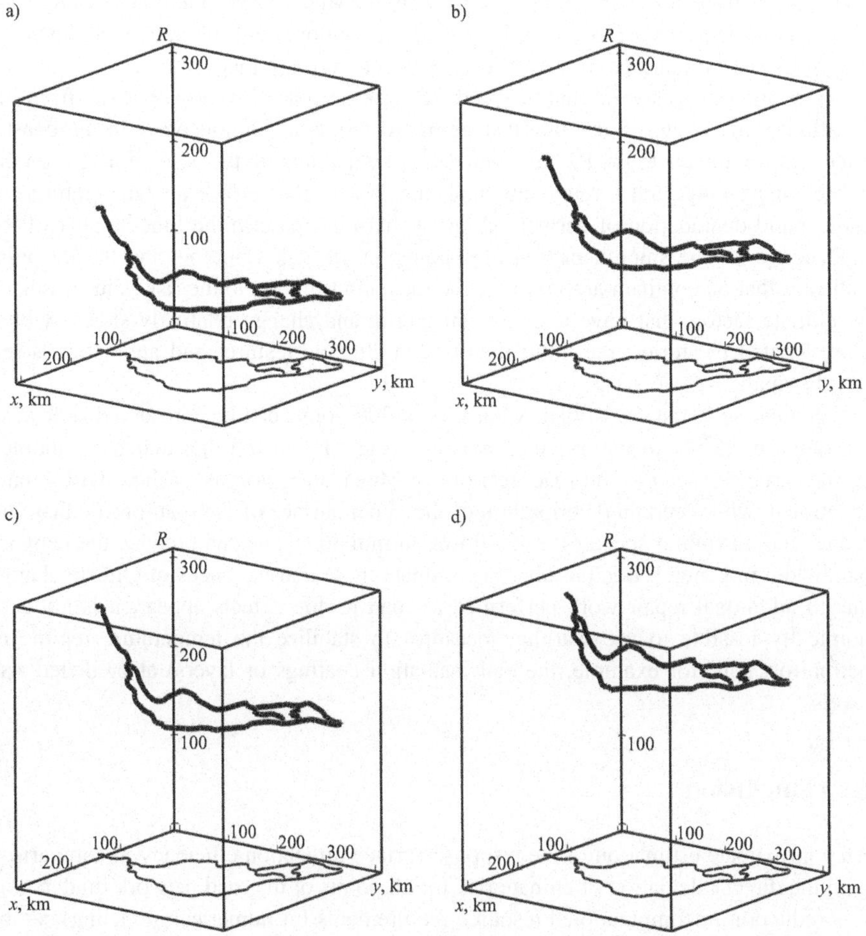

Fig. 2. Risks of reducing the functionality of the reference road network of the Magadan region with a warming climate by 2 °C, depending on the soil base: a) – sandy, b) – sandy-loam, c) – loam, d) – clay.

When performing numerical modeling, a square grid was used, the cell size of which was equal to 1 km. A fragment of the road network that fell within the boundaries of each of the cells was characterized by a risk value that was constant throughout this fragment; the risk value was estimated from the response surface for the center of the cell. The total number of cells that a fragment fell within boundaries (non-empty cells) was 1602.

5 Discussions

Figure 2 shows a significant increase of risk as the soil changes from sandy to clay. The spread of risk values for the same point, depending on the type of soil at the base of the road, ranges from 134 to 201 points. In general, the values corresponding to sandy soils can be considered as a risk assessment from the bottom, and the values obtained for clay soils can be considered as a risk assessment from the top.

It is also well observed that the predicted risk decreases as moving away from the coastline. This is due to the fact that permafrost soils are fragmentary in the coastal zone (approximately to the P2 key point), their temperature in the base climate is on the border of thawing, and a relatively small increase in the surface air temperature can cause rapid degradation of permafrost and a sharp decrease in the functionality of the road network. The smooth pattern of changes in the risk value beyond the P2 point indicates that on continental territories, the main contribution to the risk values is made by climate factors that have a significant inertia and change relatively slowly within close-located territories; the influence of local climatic features and anomalies is relatively small.

Qualitatively, risks with their values up to 300 points can be considered as low, in the range from 301 to 600 points – average. It can be argued that under the adopted climate change scenario the road network in Magadan region as a whole will remain functional, with a certain deterioration of the upper surface of the road profile that will reduce the maximum speed of traffic flows on individual sites and reduce the network bandwidth in general. It is possible to eliminate these consequences of climate change due to additional repair work performed as road profile defects appear; it is not economically feasible to use warning measures to stabilize the temperature regime of permafrost soil (for example, thermal insulation coatings or layers) at predicted risk levels.

6 Conclusion

An approximate estimation of the response surface utilization efficiency in comparison with the direct calculation of climate risk for all points of the road network (understood as a reduction in computational resource requirements for numerical modeling) can be performed as follows. The duration of direct risk calculation for a single reference point, including modeling of the temperature regime of the soil mass over the course of an average year for two climate variants (basic and modified) and four types of soil, on the used computational system was about 8200 s. The procedure for calculating the risk based on the response surface for 1602 square grid cells on the same computational system took less than 4 s of machine time. The calculation acceleration thus exceeded 3 million times.

References

1. Gruza, G.V.: Features of the temperature regime at the surface of the globe in 2018. Fundam. Appl. Climatol. **1**, 97–127 (2019). https://doi.org/10.21513/2410-8758-2019-1-97-127. (in Russian)
2. Yakubovich, A.N., Yakubovich, I.A., Trofimenko, Yu.V., Shashina, E.V.: Intelligent management system of the automobile road's technical and operational condition in the cryolithozone. In: 2019 Systems of Signals Generating and Processing in the Field of on Board Communications, SOSG 2019. Institute of Electrical and Electronics Engineers Inc. (2019). https://doi.org/10.1109/sosg.2019.8706742
3. Shiklomanov, N.I., Streletskiy, D.A., Swales, T.B., Kokorev, V.A.: Climate change and stability of urban infrastructure in Russian Permafrost regions: prognostic assessment based on GCM climate projections. Geograph. Rev. **107**(1), 125–143 (2017). https://doi.org/10.1111/gere.12214
4. Yakubovich, A.N., Trofimenko, Yu.V., Yakubovich, I.A., Shashina, E.V.: A forecast model for a road network's section traffic capacity assessment on a territory of the cryolithozone in conditions of the climate change. Periodicals Eng. Nat. Sci. **7**(1), 275–280 (2019). https://doi.org/10.21533/pen.v7i1.380
5. Yakubovich, A., Trofimenko, Yu., Pospelov, P.: Principles of developing a procedure to assess consequences of natural and climatic changes for transport infrastructure facilities in permafrost regions. Transp. Res. Procedia **36**, 810–816 (2018). https://doi.org/10.1016/j.trpro.2018.12.076
6. Junwei, Z., Yongsheng, Q., Dejie, X., Zhilong, J., Zhidan, H.: Impact of road bends on traffic flow in a single-lane traffic system. Math. Prob. Eng. **2014**, 6 (2014). https://doi.org/10.1155/2014/218465
7. Liu, Y.: A cellular automaton traffic flow model with advanced decelerations. Math. Prob. Eng. **14** (2012). https://doi.org/10.1155/2012/717085
8. Benjamin, S.C., Johnson, N.F., Hui, P.M.: Cellular automata models of traffic flow along a highway containing a junction. J. Phys. A **29**(12), 3119–3127 (1996). https://doi.org/10.1088/0305-4470/29/12/018
9. He, J., Zhang, P., Li, X.: Reliability analysis for bypass seepage stability of complex reinforced earth-Rockfill dam with high-order practical stochastic response surface method. Math. Prob. Eng. **15** (2019). https://doi.org/10.1155/2019/8261961
10. Kaymaz, I., McMahon, C.A.: A response surface method based on weighted regression for structural reliability analysis. Probabilistic Eng. Mech. **20**(1), 11–17 (2005). https://doi.org/10.1016/j.probengmech.2004.05.005
11. Eshghi, A.T., Lee, S.: Adaptive improved response surface method for reliability-based design optimization. Eng. Optim. **51**(12), 2011–2029 (2019). https://doi.org/10.1080/0305215X.2018.1561885

Needed Additions to the Diagnostic System of High-Speed Lines

Viktor Pevzner, Kirill Shapetko(✉), and Alexander Slastenin

JSC "VNIIZhT" (Railway Research Institute), 10 3rd Mytischinskaya Street, Moscow 129626, Russia
Shapetko.kirill@vniizht.ru

Abstract. The paper describes topical issues related to the movement of speed and high-speed rolling stock operated on the railway network of JSC "RZD". The organization of speed and high-speed traffic requires stricter requirements for assessing the state of the track. This indicates a lack of existing regulatory documentation. The questions on the sinuous movement of the train on the rail track, the influence of irregularities of the longitudinal profile on the dynamic performance of the rolling stock are considered. The paper discusses the possible options for the diagnosis of these parameters in conditions of high-speed traffic on the Russian Railways. The technique of mathematical modeling on the basis of the method of sinuous movement on the track sections with long irregularities is applied. To solve these problems, the theory of oscillatory motion of the train on the basis of a physical pendulum on the sections of the sinuous movement path is used.

Keywords: Rail cant · Irregularities of the longitudinal profile · Sinuous movement · Increase of movement speeds · Traffic safety

1 Introduction

Organization of speed traffic and high-speed traffic operations on the railway network of JSC "RZD" revealed several systemic disparities between the parameters of infrastructure setup and maintenance, and the requirements for railroad track and speed rolling stock interaction.

There are two main reasons for arising discrepancies. Globally, speed and high-speed railway traffic operate on specially designed or radically modified tracks [1, 2]. "RZD" operates speed and high-speed traffic on the tracks built in the 19th and beginning of the 20th century (Moscow - St. Petersburg, Moscow - Smolensk, Moscow - Nizhny Novgorod). Current regulatory standards for railroad track maintenance, developed in the first half of 20th century, significantly differ from modern European and Asian standards for speed railroad traffic [3–5].

There are two problems of current interest:

- a control of factors which lead to increased intensity of snaking motion of speed and high-speed train cars;

V. Murgul and V. Pukhkal (Eds.): EMMFT 2019, AISC 1258, pp. 496–505, 2021.
https://doi.org/10.1007/978-3-030-57450-5_43

- a control of irregularities of a length up to 150 m of longitudinal profile, which cause low frequency train car oscillations in vertical and horizontal planes.

In Europe, the "effective taper" ("effective conicity") indicator is used to assess the intensity of the snaking (winding, sinuous) motion, according to [1, 3].

2 Methods

Analysis of possibility of applying this norm on Russian railways showed that:

Procedures to calculate effective taper in Europe is defined in standard [1] developed by International Union of Railways (UIC), where the following assumptions are made:

1. Wheels are rails are considered rigid bodies (non-deformable);
2. Rails are considered straight-line, with a constant cross-section;
3. Elastic deformations at the contact of wheel with the rail are ignored.

"Universal Mechanism" software is used for modeling an interaction between the wheelset and the rail track. The wheelset is moving along the rail track of a constant gauge of 1520 mm, wheel profile "VNIIZhT-RM-70", rail profile "P65" [5]. Distance between rotation disks of a wheelset is 1580 mm (constant), wheel rolling radius is 460 mm, rail canting is constant 1/20.

This set of simplified assumptions, applicable to the European speed railways, does not reflect the actual conditions on the speed railways of JSC "RZD". The reasons are:

- Constant adjustments of the rail gauge lead to rails not being straight;
- Insufficient stiffness of the rail lining does not rule out the occurrence of elastic deformations;
- Rail canting is not a constant; main problem – rail canting is variable along the rail with gradient $\frac{\partial \varphi}{\partial l}$, which "Universal Mechanism" software does not take into account.

The method was developed for modeling of a wheelset/rail track interaction which closer reflects the reality of field conditions based on the study [4–6].

Using Russian research studies [7] related to snaking motion of a locomotive or car underframe, a theory of physical (compound) pendulum can be applied to the sinusoidal motion of an underframe [5, 8]. This approach allows evaluation of the effect of rail canting on the lateral forces which determine critical train speeds. Figure 1 illustrates the trajectory of an underframe motion along the rails, including degree of freedom of a conical-shaped wheelset.

Period of oscillation of the physical pendulum is defined by Eq. (1) [8]:

$$T = 2\pi\sqrt{\frac{J}{mgl}} \tag{1}$$

where: T – period of oscillation of a physical pendulum, s;

m – mass of pendulum, kg;

l – pendulum suspension height, m;
J – moment of inertia, kg m^2;
g – gravitational acceleration, m/s^2.

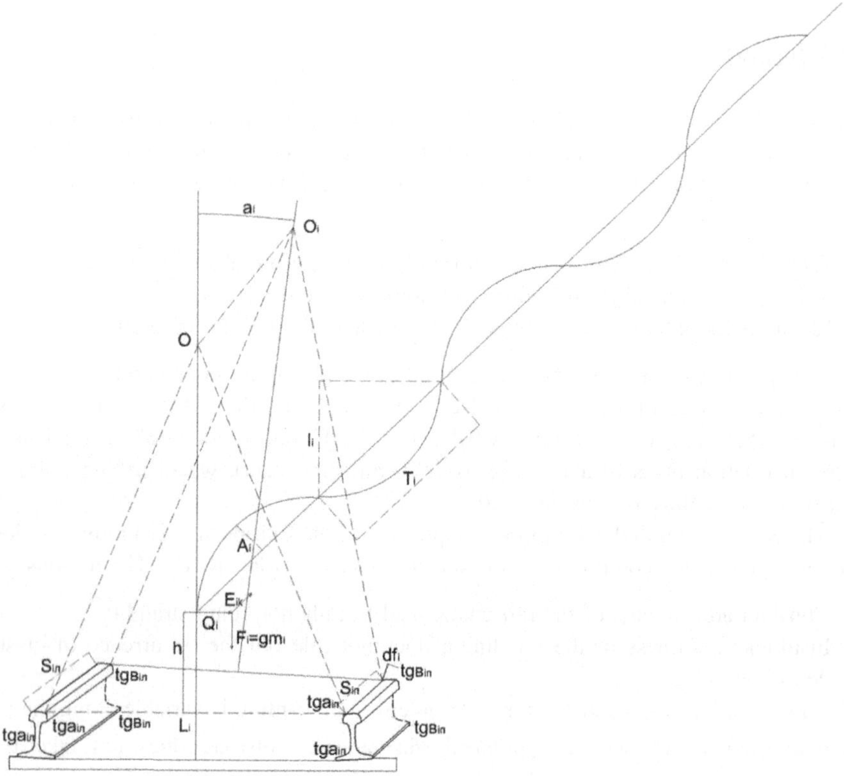

Fig. 1. The scheme of movement of rolling stock.

Period of oscillation of a physical pendulum is proportional to the square root of its moment of inertia and inversely proportional to the square root of a product of a pendulum mass, gravitational acceleration, and an arm of acting forces. Knowing the oscillation parameters, one can find the forces acting at the end points as moment of inertia of a period of small-amplitude oscillations of a physical pendulum using Eq. (2) [8].

$$T = 2\pi\sqrt{\frac{l}{g}}\left(1 + \frac{1}{4}sin^2\left(\frac{a}{2}\right)\right) = \frac{\pi}{4}\sqrt{\frac{l}{g}}(9 - cosa) \quad (2)$$

For a cross-section, pendulum suspension length at time i can be calculated using Eq. (3) [8].

$$L_i = \frac{h_i}{2} tg\alpha \tag{3}$$

where: h_i – track gauge at time i in a cross-section;
tga – rail cant.

Normal force at the contact point between the rail and the wheel is determined for left and right points using Eqs. (4) and (5) [9].

$$N_L = -N_L \sin(\partial_L + \varphi) j_T + N_L \cos(\delta_L + \varphi) k_T \tag{4}$$

$$N_R = N_R \sin(\partial_R - \varphi) j_T + N_R \cos(\delta_R - \varphi) k_T \tag{5}$$

where: N_L, N_R – normal forces at point of contact between the rail and the left and right wheel respectively;
φ – turning angle along the center line of a track;
j_T – axis tangent to the center line of a track;
k_T – axis normal to the plane of a track;
∂_L, ∂_R – tilt angles of the contact (bearing) surface of the rail at the point of contact between the rail and the left and right wheel respectively.

Equation of oscillation is defined by the following expression (6) [9]:

$$\frac{d}{dt}\left[\frac{\partial E_K}{\partial q_i}\right] - \frac{\partial E_K}{\partial q_i} + \frac{\partial E_p}{\partial q_i} + \frac{\partial E_D}{\partial q_i} = Q_i \tag{6}$$

where: E_K – kinetic energy of the system;
E_p – potential energy of the system;
E_D – Rayleigh function, describing dissipating energy of the system;
q_i – generalized coordinates by time.

Due to the mathematical modeling, this expression evaluates isolated force Q_i acting on the track during estimated time interval at a given speed in a system of "physical pendulum" of snaking motion.

Data calculated using this method can be used to analyze parameters of the snaking motion of the underframe, which is crucial for evaluation of traffic safety and smoothness of train motion.

These calculations should be conducted within the framework of continuing monitoring of the track condition [10].

Based on the foregoing, it can be noted that the use of mobile diagnostic tools provides a large amount of data on the state of the upper structure, but at the moment, there are no standards for using the entire output information potential to monitor the position of the upper structure of the track and its infrastructure. According to the instructions [11], to improve the safety of mixed traffic tracks and the predictability of the train's behavior when the wheel and rail interact, in order to timely predict the state of the track's upper structure, the mathematical pendulum principle was applied with the possibility of using data obtained from mobile diagnostic tools.

Based on studies [12], it is possible to note a number of interrelations between the state of the track's upper structure and the state of the wheelset with the wheel surface rolling directly along the rail head with a changing profile from natural or uneven wear. Taking into account the uncertainty of the outline of the wheel and rail surfaces at a particular moment in time, the trajectory of movement is formed depending on:

- the load (train weight);
- the duration of the dynamic load (train length);
- the intervals of movement of trains.

It is possible to note the possibility of accumulation of fatigue of the upper structure of the track and the roadbed, as a result of a change in the position of the rail with a change in the rolling surface.

As a result of this, on the mixed traffic tracks, the impact through the wheelset on the upper track structure is not linear, as a result of which the problems of stability of motion, smoothness of ride, comfort and ensuring traffic safety remain not completely resolved. A change in the position of the rail in space and, consequently, changes in the incremental forces as a result of the ratio of two indicators $\frac{\partial\varphi}{\partial l}$, as a result, can lead to emergency situations. Figure 2 shows an example of track sections at wheel and rail contact points that can flow one into another due to accumulation of fatigue of the track's upper structure due to increased axial load, reduced intervals between trains, increased train length and weight.

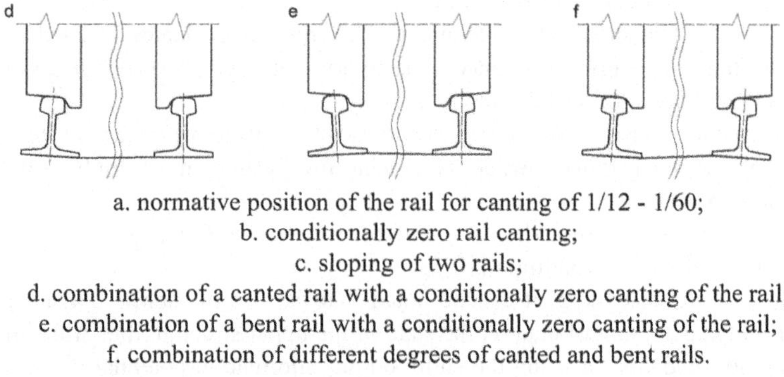

a. normative position of the rail for canting of 1/12 - 1/60;
b. conditionally zero rail canting;
c. sloping of two rails;
d. combination of a canted rail with a conditionally zero canting of the rail;
e. combination of a bent rail with a conditionally zero canting of the rail;
f. combination of different degrees of canted and bent rails.

Fig. 2. The position of the rail at the point of contact with the wheel.

It is necessary to use the physical property of the mathematical pendulum, since there is no clear connection in what position the rail is, namely, its "cant" or "bend", as a result of which the mechanics of the work of the upper track structure can be radically different. This phenomenon is especially true on sections of the track with mixed traffic, taking into account the interchangeability of the operating seasons.

Considering the movement of a train with an indefinite riding surface, this process can be represented in the mechanics of movement. Based on a mathematical pendulum, it allows taking into account the change of the parameter $\frac{\partial \varphi}{\partial l}$, which in turn will make it possible to consider the upper structure of the track as not a static system under a passing train (Fig. 3), but a dynamic one with forces transmitted to the body.

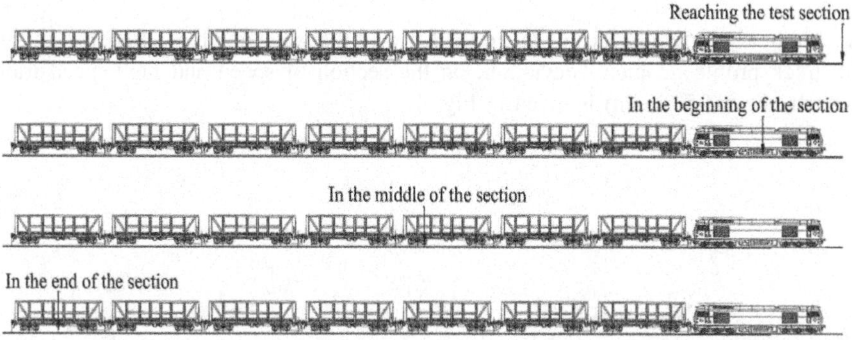

Fig. 3. The position of the train on the track.

Topics of low-frequency oscillations of a train car body of a high-speed rolling stock were emphasized in European and Japanese studies [3, 10, 13–16].

Measurement methods which allow control of actual track positions are used to obtain parameters of the irregularities in the longitudinal profile of a rail track. It should be noted that standard track gauge equipment, which uses chordal method, do not output these characteristics. Experience of application of field irregularities method [17] is summarized in European standards documents [1].

It should be borne in mind that during operation on the track with mixed traffic, there is a transition from elastic deformations of the upper structure of the track to residual deformations [18]. This process is not systematic in nature and may arise as a result of a number of operational factors, and the use of standard methods for diagnosing the state of the upper structure of the track may be insufficient. As was noted in the studies [7, 12, 19], it is possible to combine the operating conditions of the upper track structure with the determination of the possible resonance arising under operating conditions on mixed traffic tracks. It must be remembered that the process of interaction between the track and the rolling stock is affected by deformations that form irregularities of more than 25 m, information about which is not fully taken into account.

European standards [1, 2] account for a significantly wider range of irregularities (Table 1), which, at the high speeds, have a significant impact on the dynamics characteristics of the train.

Table 1. Ranges of the irregularities in the physical profile of a track line (settlement).

Range	Range of irregularity, m
D1	$3 < \lambda < 25$
D2	$25 < \lambda < 70$
D3	$70 < \lambda < 150$

3 Results and Discussions

An example of irregularities of range D3 [2] with the wavelength of up to 150 m, of a rail track profile geometry deviation, on the section of speed and high-speed traffic (speeds above 250 km/h) is given in Fig. 4.

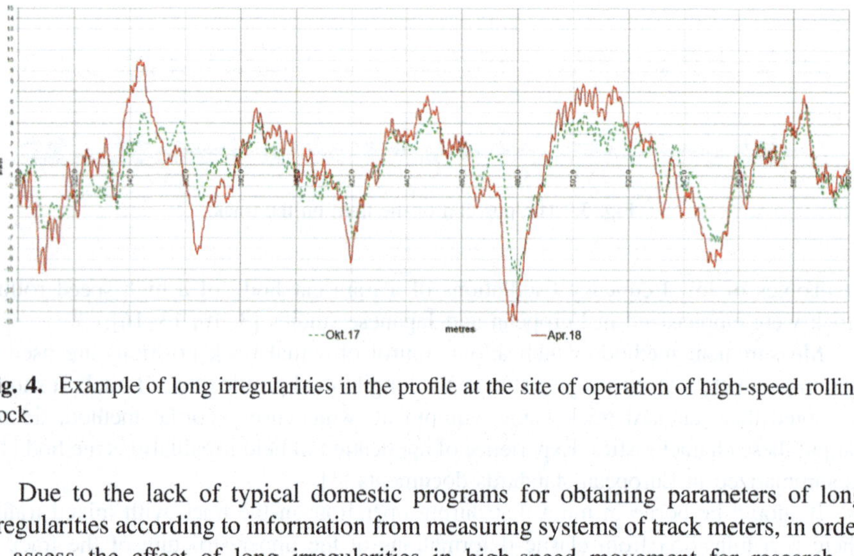

Fig. 4. Example of long irregularities in the profile at the site of operation of high-speed rolling stock.

Due to the lack of typical domestic programs for obtaining parameters of long irregularities according to information from measuring systems of track meters, in order to assess the effect of long irregularities in high-speed movement for research, a generally known method for obtaining parameters of full-scale irregularities was practically implemented.

Taking into account domestic and foreign experience, the problem can be solved using the Kotelnikov theorem and the Fourier transform. According to the Kotelnikov theorem, it is possible to restore the data of an analog signal to obtain a trajectory describing a railway track.

Since the railway track has different stiffness along the entire length, it is necessary to use the Fourier transform formulas having the form (7) and (8):

$$X(k) = \sum_{j=1}^{N} x(j) e^{2x/N(j-1)(k-1)} \quad (7)$$

$$X(k) = \frac{1}{N}\sum\nolimits_{k=1}^{N} X(k)e^{-2x/N(j-1)(k-1)} \quad (8)$$

Based on the research of domestic and foreign experts, it is necessary to apply the data conversion scheme shown in Fig. 5.

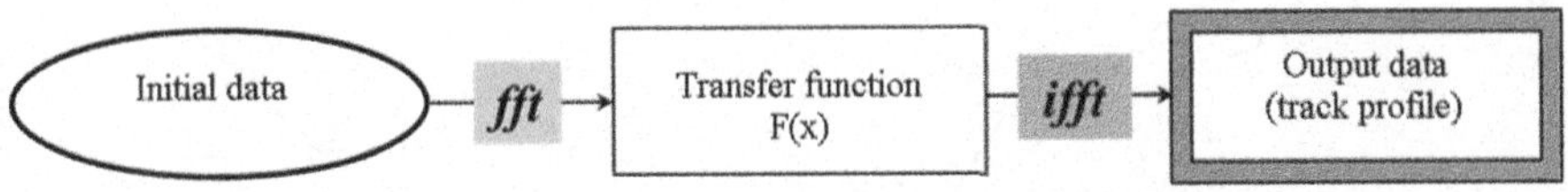

Fig. 5. Scheme for converting data received from the track recording car.

The used transfer function indicated in the scheme (Fig. 5) F(x) has the form (9):

$$F(x) = 1 - \frac{b}{c} \cdot e(-1i \cdot w \cdot a) - \frac{a}{c} \cdot e(1i \cdot w \cdot b) \quad (9)$$

where: a, b, c – the dimensions of the chords used on the track recording car (Fig. 5), which may vary for different car manufacturers (Fig. 6);
e – the exponent;
w – the cyclic frequency.

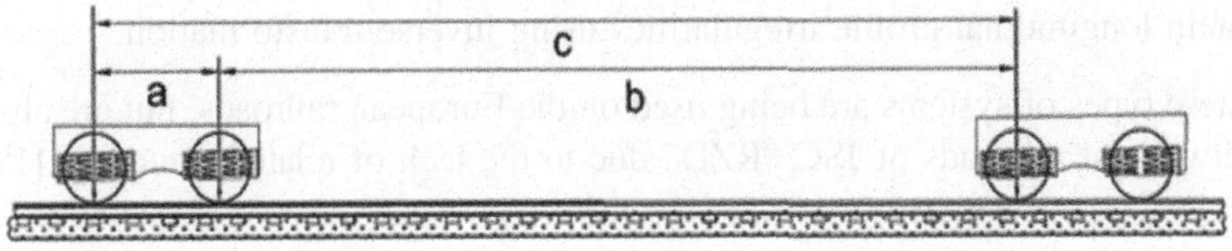

Fig. 6. Example of a chordal measuring system.

Applying the above algorithm shown in Fig. 5, we obtain data describing the trajectory of the railway track profile, which can be used to obtain full-scale irregularities shown in Fig. 4.

After all the actions have been done, we can get full-size irregularities shown in Fig. 7.

As can be seen from the data presented in Fig. 5, the irregularities obtained during geodetic surveys have convergence with the data obtained by converting the data of the measuring systems of the track recording car. Thus, the use of these field irregularities in the diagnosis of the track in the areas of operation of speed and high-speed rolling stock should significantly improve the quality of track maintenance. This requires expanding the content standards and taking into account the parameters of the irregularities of the longitudinal profile.

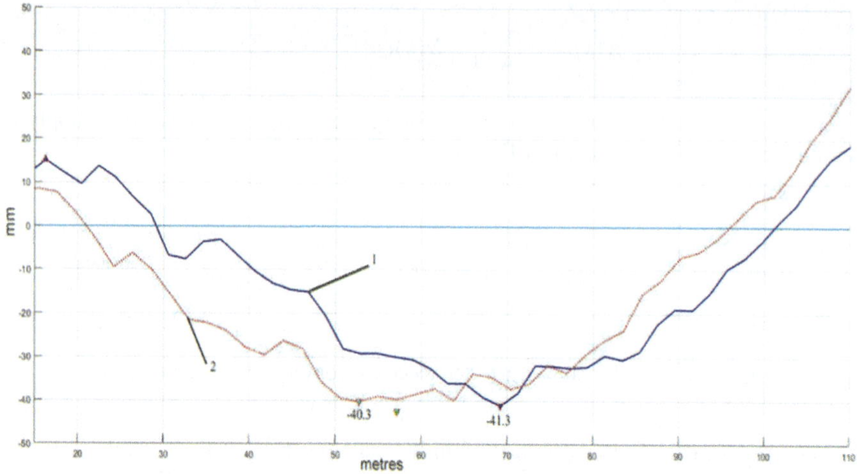

Fig. 7. Parameters of long irregularities: (1 - according to the data of the track recording car using the considered method; 2 - according to geodetic survey).

The method of study provides for the following stages:

- obtain initial data from the track gauge car (settlements) using chordal measurement system;
- convert obtained data to frequency domain, using Fourier transform and transfer function;
- obtain longitudinal profile irregularities using inverse transformation.

These types of systems are being used on the European railroads, but are absent on the network of railroads of JSC "RZD" due to the lack of related standards [19].

4 Conclusion

The analysis that was conducted concluded that existing track diagnostic system does not fully meet the conditions of the speed traffic and must be supplemented by the developed system of evaluation of the underframe snaking motion. For that purpose, a model defining the period of oscillation of small-amplitude physical pendulum can be used. The need to control irregularities in longitudinal profile on the sections of freight and speed passenger traffic is demonstrated.

References

1. EN 15302:2008 + A1:2010: Railway applications - Method for determining the equivalent conicity. Required by Directive 2008/57/EC
2. EN 13848-5:2008 + A1:2010: Railway applications - Track - Track geometry quality - Part 5: Geometric quality levels - Plain line. Required by Directive2008/57/EC

3. Kerchof, B., Wu, H.: Causes of rail cant and controlling cant through wheel/rail interface management. In: Proceedings of the 2012 Annual AREMA Conference, p. 19 (2012)
4. Potapov, A.V.: System for evaluating, monitoring and regulating the canting of the rails lying on the track. Sci. Eng. Transp. **1**, 23–25 (2019)
5. Potapov, A.V.: Rail canting and ensuring the stability of the train movement. Transp. Constr. **8**, 23–25 (2018)
6. Pevzner, V.O., Belotsvetova, O.Y., Potapov, A.V.: Results of observations to evaluate the impact of operational factors on the side rail wear. Vestn. Railway Res. Inst. **75**(4), 242–247 (2016). https://doi.org/10.21780/2223-9731-2016-75-4-242-247. (in Russian)
7. Pevzner, V.O., Potapov, A.V., Belotsvetova, O.Y.: Canting of rails and its relationship with traffic safety and lateral wear. Railway Track Facil. **2**, 30–33 (2018)
8. Bishop, R.E.: Oscillations. In: Marcuson, I.A. (ed.) Nauka, Moscow (1968)
9. Garg, V.K., Dukkipati, R.V.: Dynamics of rolling stock: Transl. from English. In: Pankina, N.A. (ed.) Transport, Moscow (1988)
10. Kampczyk, A.: Überwachung der Gleisgeometrie bei den polnischen Bahnen (PKP). Eisenbahningenieur **4**, 37–41 (2011)
11. Instructions for assessing the state of the rail gauge with measuring instruments and measures to ensure the safety of train traffic: approved by the order of JSC Russian railways No. 436r. Accessed 28 Feb 2020
12. Pevzner, V.O.: Influence of rail canting on the work of the track's upper structure. Railway Transp. **7**, 41–45 (2010)
13. Gerber, U., Fengler, W.: Setzungsverhalten des Schotters. ETR: Eisenbahntechn. Rdsch **59** (4), 170–175 (2010)
14. Takahara, K., Miyamoto, K.: Survey automation and management improvement of profile and alignment for the Shinkansen track. Jap. Railway Eng. **24**(1), 13–17 (1984)
15. Köhler, J.: Zur plastischen Verformung des Gleises-Signal und Schiene. Zeitschrift Für Angewandte Mathematik Und Mechanik **28**(4), 130–132 (1984)
16. Wolter, K.U., Erhard, F.: Beurteilung von Gleislage-abweichungen mit Hilfe von Fehlerreferenzfunktionen. Esenbahntechnische Rundschau **9**, 48–51 (2013)
17. Shapet'Ko, K.V.: Research of the accumulation of railway track deformations in the section of testing of cars with an axial load of 27 tons. Vestn. Railway Res. Inst. **76**(4), 238–242 (2017). https://doi.org/10.21780/2223-9731-2017-76-4-238-242. (in Russian)
18. Pevzner, V.O., Romen, Yu.S.: The basics of developing standards for maintaining the track and establishing speeds (2013)
19. Pevzner, V.O.: Improvement of normative documents for assessing the state of the track in high-speed traffic. Railway Track Facil. **12**, 2–6 (2014)

Planning and Modeling of Urban Transport Infrastructure

Angela Mottaeva[1,2(✉)] and Asiiat Mottaeva[3]

[1] Moscow Region State University, Radio str., 10A, 105005 Moscow, Russia
doptaganka@yandex.ru
[2] Moscow State University of Civil Engineering, 26 Yaroslavskoye Shosse, 109377 Moscow, Russia
[3] State Corporation of Rostechnology, Usacheva st., 24, 119048 Moscow, Russia

Abstract. The relevant issue of urban transport infrastructure development is studied. The growth of social-and-economic processes, the rise in the standard of the population living, has led to the rapid increase in the number of passenger transport. The development of construction in cities, the consolidation of residential areas involves the improvement and development of transport and social infrastructure as well as the transport conditions. According to the domestic and foreign statistical data, the analysis of the state of transport infrastructure development is carried out and the structure of the forecast transport model is proposed. The impact of modern information technologies on improving the quality and safety of transport is also assessed. The developed forecast model for optimal development of transport infrastructure and the use of territories will allow to simplify the process of building a road network at the micro- and macro- levels; to display all the specifications of public and private transport in a single model; to solve operational and strategic tasks of transport planning for the preparation of projects for the organization and analysis of traffic flows at intersections and junctions. The practical significance is that the planning of transport flows of infrastructure will improve the efficiency and safety of urban transport through the use of modern communication technologies, computer systems and specialized software.

Keywords: Urban transport · Infrastructure · Passenger transport · Road network

1 Introduction

The development of the city's road transport network has been going on for decades. The significant amounts of time and investment are being spent on its upgrading. The increase in the speed of social-and-economic processes has led to the increase in the standard of the population living, which has led to the rapid increase in the number of passenger transport. These processes have affected the state of the commercial,

V. Murgul and V. Pukhkal (Eds.): EMMFT 2019, AISC 1258, pp. 506–517, 2021.
https://doi.org/10.1007/978-3-030-57450-5_44

housing, construction and other sectors of the urban structure. The high rate of development of the city's infrastructure has worsened the capacity of the road transport network and the environmental component, and excessive use of certain types of transport has led to significant over-runs of passenger transport as well as the increase in the required travel time among the districts of the locality [1].

The availability of a functioning transport complex affects the speed and scale of development of the urban structure from the economic and technological aspect. New types of transport are being introduced in order to change the transport situation for the better and existing types are being modernized.

For a long time, public passenger transport has been the priority for the development of the road transport network in the Russian Federation. Its standard value was 60 cars per 1000 people. It was on the basis of the established standards that the entire road transport network of Russia was structured [2, 3].

Every year, developing countries increasingly use intelligent transport systems (ITS) when planning their infrastructure transport flows. ITS allows to improve the performance and safety of urban transport through the use of modern communication technologies, computer systems and specialized software. At last the positive effect of ITS application can still not reach more than 20%. Therefore, it is necessary to adhere not only to traffic control measures, but also search for other ways to solve the problems [4].

One of the most radical and expensive measures is the development of the city's road network. Increasing the share of the city's territory involved can be achieved by adjusting existing documentation and creating new guidelines for the development of city districts.

According to the foreign experience, the most effective way to improve the efficiency of urban roads is to introduce some measures to reduce the use of personal transport while improving the public transport system. These issues are also studied by some domestic scientists [5–8].

Another direction of urban transport infrastructure development is the development of public transport, in which the use of public passenger transport becomes more profitable and priority for the population.

The most important directions of the transport flow planning are [9]:

1. Reducing traffic intensity in the Central part of the city.
2. Focus on the use of passenger public transport.
3. Special policy regarding the placement of Parking spaces.

During the past decades, the understanding of the functional component of the road system has expanded. The main functions are: transit; accessibility of movement; formation of urban infrastructure; social; economic; environmental, etc.

These functions make it necessary to apply some special approaches to planning movements, which are presented in the form of mathematical models that help to identify the human transport behavior.

2 Methods

Nowadays the use of mathematical models is considered to be the most effective way of modeling. To get the most reliable information, several models are used; and they consider the aforesaid problem from different points of view [10].

There are several groups of models among the many mathematical models, which are designed for analyzing the traffic flows [11]:

- forecast models;
- simulation models;
- optimization models.

We consider it appropriate to use the predictive models for this study, because they are focused on the smallest characteristics of the traffic flow and allow to assess the actual transport situation accurately.

3 Results

The transport model consists of the information block and calculation block. The information blocks include the necessary set of data for further forecasting of traffic flows. The calculation blocks include data on the need for movement and the calculation of traffic flows that meet this need.

To correctly compile any mathematical model, we need to collect a huge block of data that allows to see the whole situation. Therefore, the stage of collecting information when building a forecast model requires the greatest amount of time and labour resources.

The next two stages are inseparable from each other: they are creating supply and calculating the transport demand.

The final stage involves improving the way transport demand according to supply and calibrating the final model.

The structure of creating a forecast model of traffic flows is presented in Fig. 1.

The transport supply includes the indicators which help the transport system (city, region) to meet the demand. The transport supply determines the quality and volume of the transport system that will have to meet the demand.

The transport demand demonstrates what qualitative and quantitative characteristics residents expect from the transport system.

The entire set of input data for creating a forecast transport model is presented in Fig. 2.

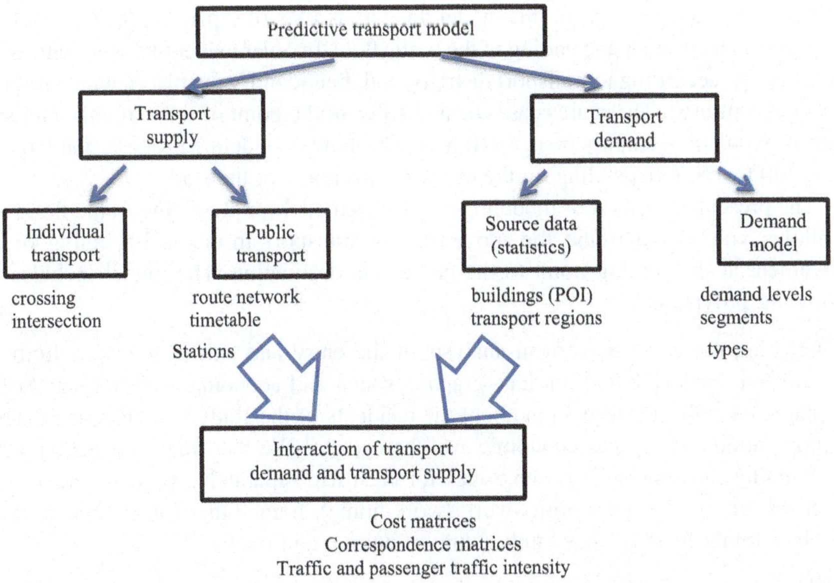

Fig. 1. Structure of the forecast transport model [3].

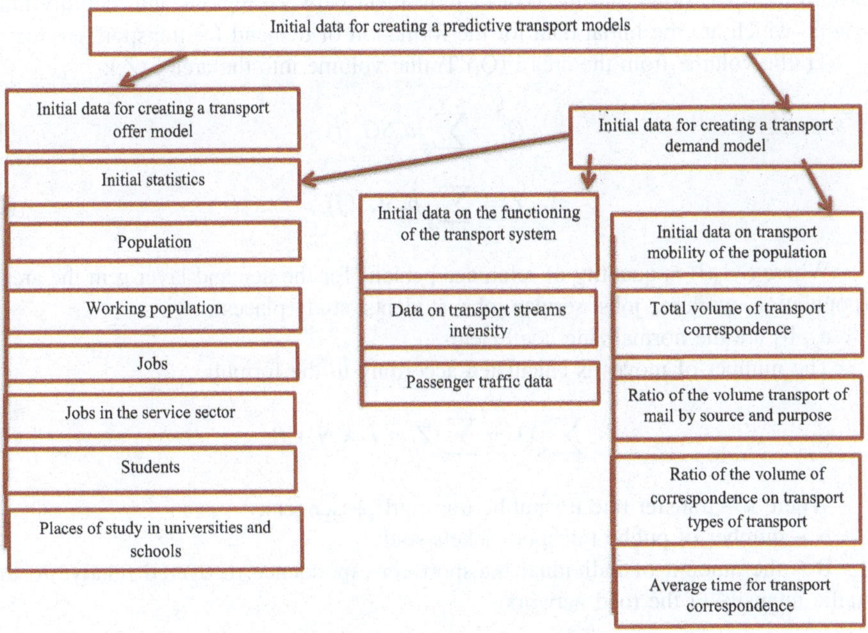

Fig. 2. Initial data for creating predictive transport models [3]

From the point of view of this model, the city is a set of separate centers, which are the points of beginning and ending of the traffic flow. In order to identify such centers, we zone the city according to transport districts, and then identify the places where traffic is most concentrated. These areas are considered from the point of view of characteristics such as population, employment, and others. The number of defined zones usually ranges from 500 to 1000, depending on the estimated accuracy of the study.

The presented method of modeling is a 4-step model for predicting traffic flows and it allows you to determine the movement of transport from the beginning of the movement at the starting point to the end at the destination. The model includes the following provisions:

1. Creating routes of movement-analysis of the entry and exit of transport from the selected territory based on demographic, social and economic information. At this stage, we collect information about the residents of the study area: labour capacity, population, social and economic well-being, etc. We use this information when planning movements between zones for each trip separately. Trips for travel purposes are divided into: home-work (work-home), home-other (other-home), work-other (other-work), home-study (study – home), and others.

The concept of a trip goal is identical to the concept of demand. The normalizing coefficients are calculated for each demand layer in accordance with Table 1.

Transport demand is calculated based on the number of indicators that generate and absorb transport flows and the cost of movement between regions and mobility indicators, which are the initial data for the formation of demand for transport services.

Traffic volume from the area i (Q_i) Traffic volume into the area j (Z_j):

$$Q_i = \sum\nolimits_g a_g SG_g(i) \tag{1}$$

$$Z_j = \sum\nolimits_g b_g SG_g(j) \tag{2}$$

Where $SG_g(i)$ is quantity of reference persons for the demand layer g in the area i (population, workers, jobs, service jobs, students, study places);

a_g, b_g are the normalizing coefficients.

The number of moves is calculated according to the formula:

$$\sum\nolimits_i Q_i = \sum\nolimits_j Z_j = k \times N + B \tag{3}$$

Where k – transfer rate for public transport passengers;

N – number of public transport tickets sold;

B – the amount of individual transport correspondence recovered relative to the traffic intensity of the road network.

2. Redistribution of traffic flows among zones is the estimation of inter-district movements. The number of movements detected in the previous stage between zones should be redistributed among the other zones. That allows you to build a two-dimensional matrix of relationships relative to each movement goal.

Table 1. Goals for moving around the network according to demand layers.

Demand layers	Source	Goal
Home-Work	Working population	Places of employment
Work-Home	Place of employment	Working population
House-Other	Working population	Places of employment in the service sector
Other-House	Places of employment in the service sector	Working population
Work-Other	Places of employment	Places of employment in the service sector
Other-Work	Places of employment in the service sector	Places of employment
Work – Work	Places of employment	Places of employment
Other – Other	Places of employment in the service sector	Places of employment in the service sector
Home-Study	Students	Training place
The School – House	Training place	Students
Work-Study	Places of employment	Training place
Study-Work	Training place	Places of employment
Study-Other	Training place	Places of employment in the service sector
Other – Study	Source	Training place
Study – Study	Working population	Training place

When distributing movements, they are based on two models:

The first is the growth factor model (arithmetic mean growth factor, General growth factor). Based on the forecast of existing movements, it allows you to identify the number of movements expected in future periods.

Training place
Training place
Places of employment in the service sector
Training place
Other – Study
Study – Study

$$A_{ijF} = F \times A_{ij} \tag{4}$$

$$A_{ijF} = 0,5 \times (F_i + F_j) \times A_{ij} \tag{5}$$

$$A_{ijF} = (F_i + F_j)/F \times A_{ij} \tag{6}$$

Where A_{ijF}, A_{ij} - future and current number of movements between zones i and j; F, F_i, F_j - growth coefficient of zones I and j.

The second model is theoretical.

The gravitational model is preferred:

$$A_{ij} = P_i \times \frac{A_j \times F_{ij} \times K_{ij}}{\sum_j (A_j \times F_{ij} \times K_{ij})} \quad (7)$$

Where A_{ij} is the number of movements between zones i and j; P_i is the number of incipient movements in zone i; A_j is the number of ending movements in zone j; F_{ij} - the resistance factor between zones i and j; K_{ij} - is the social-and-economic factor of zones i and j.

When using the model, the balance of departures and arrivals is calculated for two districts. It is assumed that the number of movements between two areas is proportional to the total volume of departure, i.e. arrival and the function, which depends on the distance of the trip. When using the model, only a couple of districts are considered separately from the total set of zones, which is the main disadvantage of using this method of distribution.

3. Choice of transport type is the load distribution on the road network according to the types of transport used. This is the most important stage of movement modeling. Movement can be carried out by public transport, private car or on foot. According to the obtained calculations, we determine which type of movement is the most preferable for the use. It is important to determine what influences the selection process. All proposals to change or improve the transport issue should be taken into account during selecting priority transport.

The most popular model when choosing a mode of transport is the logit model. It allows to predict, which transport consumers will prefer, according to the cost of movement. The concept of cost includes not only the actual price for the transportation service, but also the time, convenience of the selected type of transport, etc. Sometimes this characteristic is referred to as the "level of service".

This stage is sometimes called splitting, because the movement matrices, as a result of its implementation, obtained at the previous stage are divided into traffic modes (public and individual transport).

The steps for selecting a mode are presented in Fig. 3.

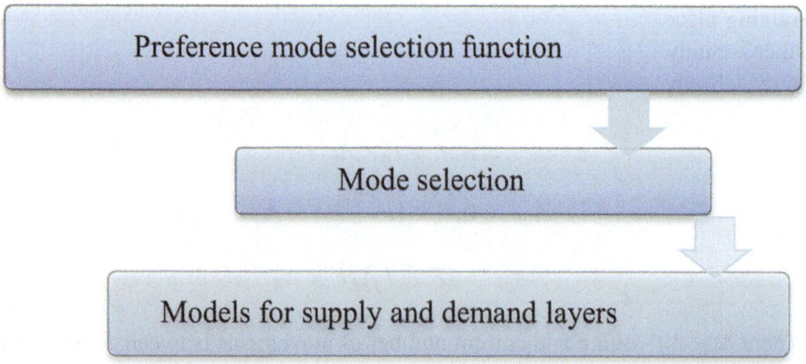

Fig. 3. Stages of selecting the driving mode.

4. Distribution of movements within the network. After redistributing traffic flows by transport type, we select the transport network from the initial to the final destination.

Reallocation is the main method for identifying the transport supply. It is based on the principle of equilibrium flows, according to which drivers are guided by the time spent on the trip when choosing the optimal route for their movement.

The distribution of traffic on the network allows you to determine the share of load on transport network objects and parameters for evaluating the quality of connection between transport areas.

During distribution, we model passenger trips. The network user does not choose the route, but chooses the departure time, because they are primarily interested in achieving the goal of the trip.

When distributing movements, we use several models that allow us to use information effectively. As a result of using several approaches, the average values of traffic congestion at the entrance to the network and the average number of cars turning at intersections are detected.

The first applied model is "all or nothing". It does not take into account the bandwidth of zones, but focuses on the shortest path between two sections. When using this model, there is a high probability of getting an unrealistic picture of movements.

The second model is stochastic. As the previous model does not take into account the capacity of zones, the distribution of movements between sections is inversely proportional to the length of the path.

The next applied model is the equilibrium one. When it is used, the traffic capacity of the road is compared with the total traffic flows in the zone. When the traffic limit is exceeded, the speed decreases, which leads to the increase in travel time between sections and the search for alternative routes.

The aforesaid models are related to the redistribution of transport demand.

The main principle of the procedure for this redistribution is that when determining the load, public transport is guided by all individual routes and timetables. It also takes into account not only the time spent for each route option, but also the time of transfers, as well as the time on foot from the center of gravity of the source district to the stop and from the stop to the center of gravity of the target district.

The probability of choosing a path is estimated using the formula:

$$P(R) = \exp[-16(\frac{R}{R_{min}} - 1)^2] \tag{8}$$

Where R is resistance of a particular route,

R_{min} is minimum resistance of all connections.

Assessment of the adequacy of the created model of transport flows is carried out at the final stage of modeling, i.e. during the calibration of the resulting model.

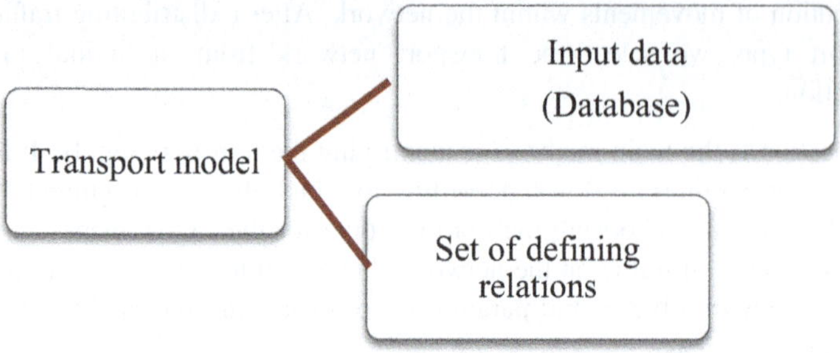

Fig. 4. Components of the transport model.

The transport model consists of the database and relationships that connect these data (Fig. 4). Therefore, it is necessary to determine which structural parts of the model are subject to modernization.

Continuous improvement of the set of defining ratios determines the quality of the models obtained at the output, so this indicator, first of all, affects the quality of the initial information laid down at the initial stage of creating the model.

The basis for calibration of the transport model is data on the intensity of traffic and pedestrian flows, collected on the basis of an already functioning network (Fig. 5).

Calibration is performed based on the correlation of the calculated values of intensity parameters with the actual data obtained during the study of the traffic flow.

The calibration stage also captures the process of refining a set of parameters embedded in the defining relationships of the model based on full-scale and verified input data.

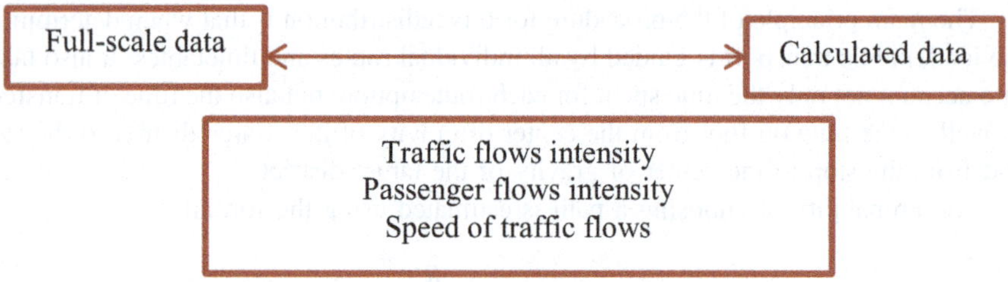

Fig. 5. Information base for calibration.

During the entire process of creating transport models, the following errors may be detected, leading to the need to calibrate the network:

1. Model specification errors (As a result of incorrect interpretation of incoming data from network users, because of the lack of understanding of their behavior and reaction to changes in the transport situation).
2. Detail errors (errors of the quality of abstraction and aggregation models.)

3. Field data errors in collecting and processing data is suitable for calibration.
4. Input data errors (errors in the network parameter, calculation procedures, statistics, etc.)

Thus, the consistent process of modeling traffic flows was presented, which allows to independently and with minimal costs form a transport model for a modern city.

However, we should keep in mind that all of the models mentioned above have some disadvantages and limitations. When the model is improved, the complexity of its application increases accordingly and the required amount of source data increases.

4 Discussion

It should be noted that all the discovered patterns of action of traffic flows are algorithmized and programmed as special software products. These programs make it possible to simplify the process of building a road network at micro and macro levels [12].

The most detailed transport model in the world is the German model created using the PTV Vision VISUM.

One of the most well-known programs is EMME/2, which first appeared in Canada and Finland.

There are also many software products on the Russian market, the most famous of them are the products of PTV AG company. These programs allow to display all the specifications of public and private transport in the single model.

For example, PTV VISION is successfully used for preparing projects for organizing and analyzing traffic flows at intersections and crossings, and simultaneously solves operational and strategic tasks of transport planning. Due to the versatility of this program, it has been distributed among a wide range of people: transport departments, engineering companies, railway traffic management, and others [12].

Nowadays, there are many software products for micro-modeling of the transport system in the world, such as VISSIM, TRANSIMS, PARAMICS, EMME/2, and SATURN.

Automated programs allow to take into account the transport costs of network passengers, process and output huge amounts of data, create 3-D models of the road sections under consideration, and so on.

All the programs use indicators to analyze the transport situation. The indicators can include: the number of road accidents, the number of cars in parking spaces, weather conditions, etc.

The main advantage of using automated software products is the ability to visualize the predicted information, speed up the process of data collection, analysis and output.

Developers use the classification to understand the purposes for which certain software features are created. According to the classification, software products can offer a model with different levels of detail and geometric parameters of the network. The single software system includes those products, which solve the matters of the organization of planning of the road.

All the existing programs can be divided by the type of tasks and purpose, but to maintain a proper level of competitiveness, manufacturers strive to create universal products that cover many models of different classes.

5 Conclusion

The consistent process of modeling traffic flows was presented as a result of the research, which will allow to independently and with minimal costs form the transport model for a modern city.

The developed forecast model for the optimal development of transport infrastructure and the use of territories will allow:

- to simplify the process of building the road network at the micro- and macro-levels;
- to display all public and private transport specifications in the single model;
- to solve operational and strategic tasks of transport planning for the preparation of projects for the organization and analysis of traffic flows at intersections and junctions.

The practical significance is that the planning of transport flows of infrastructure will improve the efficiency and safety of urban transport through the use of modern communication technologies, computer systems and specialized software.

References

1. Kapsky, D.V., Kasyanik, V.V., Evtukh, A.V., Kaptsevich, O.A.: Assessment of the efficiency of traffic flows based on the processing of the navigation data on the movement of vehicles. Sci. Technol. **5** (2017)
2. Klimovich, A.N., Shut, V.N.: System analysis of the main trends in the development of adaptive methods of transport flow management. Syst. Anal. Appl. Inform. **3** (2017)
3. Moroz, D.G., Titova, S.S., Korotaev, A.S.: Features of planning and organization of transport and transfer hubs. Sci. Technol. Educ. **2**(32) (2017)
4. Nikolaev, N.N.: Modeling of transport processes Tutorial. Azov-black sea engineering Institute-branch of Donskoy GAU, Zernograd (2016)
5. Poltavskaya, Yu.O., Kripak, M.N., Gozbenko, V.E.: Estimation of traffic flow conditions using geoinformation technologies, modern technology. Syst. Anal. Model. **1**(49) (2016)
6. Popova, E.E.: Organizational bases of functioning of regional transport and logistics systems in passenger transport. In: Proceedings of BSU, vol. 1 (2015)
7. Miloslavskaya, S.V., Pochaev, Yu.A.: Transport systems and technologies of transportation. INFRA-M, Moscow (2015)
8. Turpishcheva, M.S., Nurgaliev, E.R., Dzhakhyaeva, S.B.: Research of passenger transportation processes by automobile transport. Vestnik of ASTU **1**(63) (2017)
9. Seliverstov, S.A., Seliverstov, Ya.A.: Modeling of megapolis transport flows with the introduction of new types of water intra-city passenger transport. Bull. Admiral Makarov State Univ. Sea River Fleet **2**(30) (2015)
10. Shamlitsky, Ya.I., Okhota, A.S., Mironenko, S.N.: Modeling of transport flows in the AnyLogic environment. Softw. Prod. Syst. **3** (2018)

11. Skalozub, V.V., Panik, L.A.: Modeling of heterogeneous transport flows with variable tariffs. Transp. Syst. Transp. Technol. **10** (2015)
12. Feyzullaev, A.R., Kazakov, L.V.: Forecasting the distribution of transport flows in PTV vision VisSim. Young Don Res. **3**(6) (2017)

Energy Management and Economics

Management of Innovations in the Field of Energy-Efficient Technologies

Evgeniya Sizova(✉), Evgeniya Zhutaeva, Olga Volokitina, and Vladimir Eremin

Voronezh State Technical University, Moscow Avenue, 14, Voronezh 394026, Russia
sizova@vgasu.vrn.ru

Abstract. The article is devoted to innovation management in the field of energy-efficient technologies. The key factors that determine the problem field for improving energy efficiency of enterprises are identified. The feasibility of enhancing innovation to characterize the tasks of improving energy efficiency and its main provisions from the perspective of organizational and economic features is described. The specifics of large business as an entity generating innovations is presented. This allowed us to characterize the strategic orientation of innovation and the relationship of factors in the formation of an innovation management system in the field of energy-efficient technologies. Peculiarities of carrying out activities under the conditions of multi-design as a factor complicating the management of innovations at large enterprises are considered. Identified areas of interdependence of projects within the innovation portfolio. The requirements for monitoring indicators are formulated based on the results of management procedures, and a system of estimated indicators for monitoring the effectiveness of innovative activities of large enterprises is proposed. Problem places in the implementation of innovative activities in the field of energy-efficient technologies are identified.

Keywords: Energy efficient technology · Large enterprises · Innovations · Innovation management · Innovation performance

1 Introduction

The search for ways to improve energy efficiency is one of the most important tasks facing business entities at the present stage of development. Those questions were considered in [1–15] works in terms of strategic enerprise management. On the basis of the analysis of existing approaches to assessing energy efficiency of an enterprise we conclude that it depends on many factors which cannot be we considered separately because it will falsify information about whole system state. The most important factors that determine level of enercy efficiency from the point of ensuring its sustainable development and the possibility of exerting a controlling influence on processes and sub processes are:

V. Murgul and V. Pukhkal (Eds.): EMMFT 2019, AISC 1258, pp. 521–531, 2021.
https://doi.org/10.1007/978-3-030-57450-5_45

1. The degree of progressiveness of the used equipment and technologies, its compliance with the world and domestic level of development.
2. The degree of introducing into the management practive the scientific substantiation of the adopted production and management decisions.
3. The qualification of employees and degree of the focus of the introduction of advanced energy saving solutions.
4. The presence and level of development of the system of strategic planning, accounting and control in the field of energy efficiency.
5. The potential of energy efficiency of enterprice in infrastructure, buildings and other objects [5–7].

One of the measures to improve energy efficiency, allowing to take into account the above aspects and achieve significant results for the enterprise in this area, is the intensification of innovation. Innovations are power tools to influence on the efficiency of organization activity. It influences on organizational status, scale and relations with other actors. This is the reason why innovational way of development should follow only after the analysis of organizations functioning aims, its tasks and priorities, current organizations position, its strong and week sides as well as material and technical base. It is also necessary to try to predict a potential change in the market, including changing needs, and also take into account the difference in the timing of achieving goals, since most of the goals of innovation can be achieved only over a significant period of time. The goals of innovation are correlated with the goals of the business strategy of the business entity, as a rule.

The implementation of innovative activities in the field of energy-efficient technologies for a number of reasons is the most affordable specifically for large enterprises, which are traditionally characterized by number of advantages that facilitate innovation. This is confirmed by a much higher innovative activity compared to other market participants, as well as a larger share of funds allocated for innovation, in their total amount. The high business activity of large enterprises and significant volumes of turnover increase the rate of return on investment, which makes it possible to divert funds for research and development. Unlike small businesses, large enterprises are more willing to carry out innovative transformations, since it is less critically affected by the unsuccessful results of introducing innovations, failure to achieve the planned result, increasing the capital intensity of innovative products and other unforeseen negative consequences [8].

At the same time, the introduction of innovations in large enterprises is significantly complicated by the cumbersomeness and complexity of management structures both in the field of innovation and enterprise management as a whole, as well as multi-personality and multi-project orientation. As a result, huge number of innovational management moments should be considered for innovational activity of large enterprises, because this will increase the efficiency of its functioning.

2 Materials and Methods

Coordination of quantitative and qualitative elements and processes that make innovative activity unique, complex and high-risk should be ensured through the implementation of a combination of organizational, economic and other methods and methods of innovation policy, formed for the needs of a particular large enterprise. In this case, conditions are created to ensure the possibility of its development on its own basis in the future, to achieve a high level of competitiveness and a number of indirect strategic advantages.

The strategic aspect of the innovation activity of a large enterprise is very important for the following reasons:

1. Expanding spheres of influence, increasing competitiveness, ensuring leadership in the market and other advantages that are of value to the enterprise itself, which are ensured by innovational activity, could be achieved only through the gradual and systematic implementation of a number of consecutive steps according to a pre-worked out strategy.
2. The instability of the external environment, which generates specific types of risks, imposes flexibility requirements on innovation management structures along with ensuring sufficient stability under the influence of destructive factors. Therefore, the company needs to develop several alternative development scenarios for the long term.
3. Economical innovational results which are planned to be achieved (not only income, but tax reduction, cost reduction, productivity increase, increase in investments efficiency, etc.) should be provided by following the investment strategy, which will increase the potential of the company.
4. Innovation strategy must be developed with connection to all other functional strategies of the enterprise. This will provide consistency and coordination in time, as well as available and predicted labor, material, financial and other resources.
5. The development of an innovative strategy based on the specifics of a particular enterprise in this case allows us to identify key areas for further development and critical areas, as well as to maximize the positive result from the innovative synergy of large enterprises, manifested under the influence of a number of specific effects.
6. The form of industrial and economic integration of a large enterprise is important in developing an innovative strategy and forming a portfolio of innovative projects, since the scale and direction of the innovative advantages of a large enterprise will depend on it, which can be the basis for making managerial decisions in the field of energy efficiency.

From the foregoing, we can conclude that the strategic management of innovation implies the systematic development of measures to organize innovation and its continuous improvement throughout the entire life cycle of innovations, and also covers all the basic subsystems of the enterprise [9]. That is why one of the most important steps to improve the efficiency of innovation is to identify the individual characteristics of a large enterprise. Such features can develop under the influence of various events, managerial decisions, factors that take place during the development of a large enterprise, manifest themselves in the process of interaction with other entities, stem from the specifics of the structure of innovation, previous experience in introducing innovations, the presence of partnerships with external innovative enterprises, as well as the existing corporate innovation culture. A full and comprehensive consideration of such features affecting innovation, coupled with key aspects that determine the level of energy efficiency of an enterprise, will contribute to the construction of an innovation management system that will ensure an adequate level of efficiency and effectiveness of activities (Fig. 1).

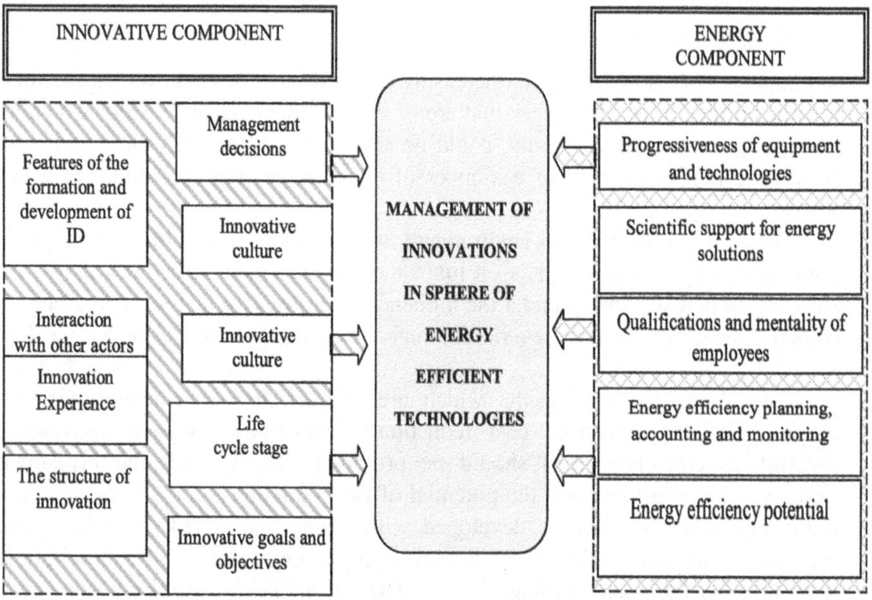

Fig. 1. The relationship of factors in the formation of an innovation management system in the field of energy-efficient technologies.

3 Results

To achieve more pronounced orientation to increase energy efficiency, it is necessary to give this direction a higher priority status. This is particularly important in cases where a large enterprise to innovate in a multi-project, which is a constraint to innovation management process. The consequence of multi-project is the necessity to make priorities that influence on realization queue, which stempts due to possible interconnection of projects on different directions. This makes it impossible and unefficient to implement them simultaneously.

Determining the priority of innovative projects is a difficult task [10]. The process of prioritizing projects in large enterprises is exacerbated by the multidimensionality and diversification of their activities. Since innovative projects of any company are characterized by a variety of tasks, directions, characteristics, parameters, it is difficult to give an exhaustive description of the significance parameter outside the context that determines the specific stage of development of the company. At its turn, the definition of projects priority can be based on the completness of their compliance with the list of priority areas, which are identified by the company in accordance with their development strategy.

Inconsistency and contradictions in the companys innovative projects can lead to non-compliance with deadlines, costs overruns, not meeting expectations results, as well as the forced completion of projects. To avoid such consequences, it is necessary to evaluate the level and nature of dependencies, which are considered, implemented and deferred for the implementation of innovative projects both in the field of energy-efficient technologies and in the whole projects portfolio. This leads to the use of additional tools, which implies a deeper level of goal setting and planning. Tools allow us not only to manage the implemetation of each individual project rationally, but also monitoring the timely identification of dependencies between project and their operational coordination. We proposed to analyze dependencies between projects in three areas - the result, structure, resources.

The dependence on the result when projects, in accordance with their internal logic and characteristics, must be implemented sequentially (or on its specific stage), it in not possible to begin the implementation of the next. The main impact of this dependence can be in two types: firstly, start and end dates of implemetation of the first priority project affect the start of the subsequent one. Secondly, the qualitative and quantitative results obtained at the output at the form of concrete characteristics of a product or a technology can affect the requirements for a subsequent project or affect its pricipal components. Since innovative activity is characterized by a high degree of uncertainty of the result due to the uniqueness of each project, the volume of potential impact on the subsequent project is most difficult to predict..

Dependence on the structure stems from the requirements for the structure of the innovation network and its subjects. One and the same subject can be part of an innovation network simultaneously for several ongoing projects. This may concern both internal and external entities of a large enterprise engaged in innovative activities [11]. This will be favored by the presence of partnerships and cooperation agreements with other enterprises, research institutes, universities and other entities that provide not one-time, but regular interaction in the field of innovation.

Since the same subject can be simultaneously involved in several innovative projects, it is necessary to analyze the level of workload of participants in advance. Moreover, entities external to a large enterprise are responsible for the performance of work in accordance with contracts or other relevant documents, and, therefore, independently evaluate their ability and willingness to participate in a specific project. The composition of internal actors is determined based on the needs of the project. However, it must be considered that involving the same subject in the process of implementing a significant number of projects can lead to a decrease in the efficiency of its work, deterioration in the quality or delay of work due to excessive workload.

Resource dependence of project (material, technical, labor, etc., which are required in different projects) is coordinated through the implementation of a number of activities traditional for project management (determining needs, drawing up calendar plans, assessing labor reserves, etc.) using tools production, distribution and financial logistics, as well as human capital management methods.

An indicator of the quality of innovation management procedures is its effectiveness as an integral characteristic of the effectiveness of the implementation of innovations in a large enterprise. The result of the implementation of innovative projects included in the portfolio according to the results of the coordination and prioritization procedures is an increase in the effectiveness of innovative activities, which should be regularly monitored in order to track the impact of the results of managerial decisions on the achievement of planned indicators in the field of innovation. We formulated the requirements for a system of performance indicators in order to carry out monitoring, taking into account the implementation by a large enterprise of innovative projects in the field of energy efficiency (Fig. 2).

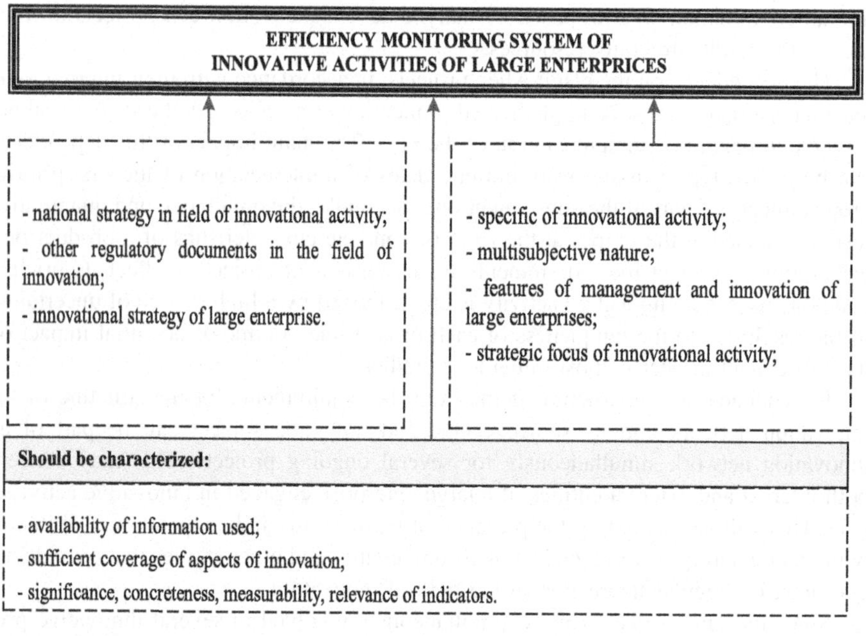

Fig. 2. Requirements for the system of performance indicators for the effectiveness of innovation activities of large enterprises.

Monitoring involves the systematic view on the results of innovation, assessing the dynamics, forecasting and identifying trends in the innovative development of an enterprise in order to develop and evaluate the effectiveness of management decisions to optimize the subject structure and form a balanced portfolio of projects. Based on the stated requirements for the system of indicators, taking into account the complexity and diversity of innovative activity and based on the need to ensure the convenience of monitoring its effectiveness, we have formed a system of estimated indicators of the effectiveness of innovative activity of large enterprises (Table 1).

Table 1. The system of performance indicators for the effectiveness of innovation activities of a large enterprise.

Category	Indicator	Calculation formula
Structure	The share of joint research and development projects in which the organization participates	$I_z = \frac{Z_s}{Z}$, (1) I_z - the proportion of joint research and development projects in which the organization participates; Z_s – the number of joint research and development projects implemented in the reporting period, units.; Z - the number of research and development projects implemented in the reporting period, units.;
	The involvement of divisions in innovation	$I_u = \frac{U_i}{U}$, (2) I_u – indicator of the involvement of units in innovation; U – the number of business units of the holding at the end of the reporting period, units.; U_i – the number of divisions, subsidiaries and dependent structures of the holding engaged in innovative activities for the period under review, units.
Resources	The share of innovation costs in revenuesdf	$I_c = \frac{C_i}{R}$, (3) I_c – share of innovation costs in revenue; C_i – expenses for all types of innovations for the considered period, million rubles; R – revenue for the period under review, million rubles.
	Proportion of procurement of innovative and high-tech products in total procurement	$I_{ht} = \frac{C_{ht}}{C}$, (4) I_{ht} - the proportion of purchases of innovative and high-tech products in total purchases; C_{ht} - expenses for the acquisition of innovative and high-tech products for the period under review, million rubles. C – expenses for the purchase of goods, works, services for the period under review, million rubles.
Market	Share of projects completed on time	$I_t = \frac{K_t}{K}$, (5) I_t – the proportion of projects for which during the reporting period all the stated deadlines for the implementation of the stages of the innovation process were observed, in the total number of innovative projects implemented in the reporting year; K_t - the number of innovative projects implemented in the reporting period for which the stated deadlines for the implementation of the stages of the innovation process were observed during the period, units K - the number of innovative projects implemented in the reporting period, units
	The share of innovative products in total production	$I_{vi} = \frac{V_i}{R}$, (6) I_{vi} – the proportion of innovative products in the total volume of products, work, services produced during the period, million rubles; V_i – the cost of innovative products, works, services produced during the period, million rubles. R – revenue for the period under review, million rubles

The presented system of indicators is based on the application for evaluation purposes of a system in general (quantitative) and specific (qualitative and quantitative) indicators and involves the implementation of a set of structural and design activities. The indicators are also differentiated in three key areas of manifestation of the advantages of large enterprises (structure, resources, market). This allows us to determine the advantages of which of the enterprise areas are used more or less efficiently, which are the value for strategic innovation management, building a corporate innovation management system and interaction with other participants in the innovation environment.

In many respects, the effectiveness of the activity will depend on the chosen form of organization and control of the interaction of science and production, as well as other units and entities whose participation in innovation is provided at any stages. Their functionality, size and status are characterized by significant differences, which leads to a variety of organizational structures for managing innovative activities [12, 13]. The choice of the optimal management structure will not only reduce the costs of organizing and managing the innovation process, but will also have a significant impact on the effectiveness of innovation in general. Therefore, when evaluating effectiveness, the following fact should be taken into account. Since innovative resources are distributed asymmetrically within a large enterprise, it is necessary to carry out activities to implement an innovative project, achieving such a subject structure and building the structure of relations in such a way that, on the one hand, it fully meets the requirements of the project for its successful implementation, and on the other hand, it ensures satisfaction of the requirements of the subjects to achieve a sufficient level of profitability from their participation in the innovation network. Consequently, efficiency as an integral characteristic of the activities of a large enterprise as a whole will not be identical to the effectiveness of its individual structural elements, divisions and subsidiaries.

4 Discussions

The implementation of innovative projects is not a guarantee of an immediate increase in the energy efficiency of the enterprise. As practice shows, innovation in the field of energy-efficient technologies is currently characterized by a number of problem areas:

- a significant level of risk as a consequence of the totality of the features of innovative projects in a given area;
- the unacceptability of hard goal setting, incorrect goal setting in the early stages of the project and the mismatch of the result with the expected;
- predominantly individual nature of energy surveys;
- the need to cover a significant number of aspects when building a system for managing energy-saving innovations (including motivation, logistics, marketing, etc.)
- significant diversion of funds from turnover due to the cost of energy projects and a long payback period;
- applying standard leverage to innovations similar to other activities;

- high energy consumption and high energy saving potential;
- lack of understanding of innovative activity as an activity susceptible to factors of the external and internal environment;
- internal corporate discrepancy of the declared organization and the structure of innovation activity management [14, 15].

All this factors can not only make the ongoing project ineffective, but also lead to more significant negative consequences, both financial and otherwise, since innovative activity affects all aspects of the economic activity of enterprises.

One of the ways to eliminate such problems can be innovative outsourcing as engaging environmental actors outside the large enterprise to carry out individual stages or the entire innovation process. Despite the combination of advantages of holdings, outsourcing is an integral component of such structures, providing increased efficiency in the functioning of the complex of interconnected enterprises.

We can distinguish a number of the most common reasons why large enterprises resort to this method of implementing innovative activities. Firstly, since innovation is complex and requires a significant amount of diverse resources, the holding may not have the necessary equipment, raw materials, materials and similar resources for the project. In this case, the transfer of innovation to outsourcing solves the problems associated with the material and technical basis for the implementation of innovations. Secondly, innovative activity imposes special requirements on the employees involved (both those who generate innovations and those engaged in the management of innovative activities) who must have sufficient qualifications, the required knowledge and, preferably, previous experience working with innovations at one stage or another. If the holding's human potential is insufficient, innovation outsourcing can also be effective. In addition, even with the availability of resources and capacities, outsourcing can be used if its application is economically and organizationally more efficient than the performance of work on its own. In this case another productivity should be renown. ndirect advantages can be achieved not only financial benefits, but also the accumulation of knowledge, the strengthening of innovative corporate culture, the development of the ecosystem, increasing the resilience of the organization and other sustainable competitive advantages, manifested due to internal factors.

In the case of a positive decision on outsourcing, it is necessary to identify alternative options for the types, volumes and forms of transfer of functions for the implementation of innovative processes. In addition, it will be necessary to analyze the outsourcers themselves and select the outsourcer on the conditions that are most appropriate for the situation and according to the criteria that are most priority in the opinion of management (terms, cost, reliability, partnerships, etc.).

5 Conclusion

Innovative activity implies the progressiveness and intensity of initiated changes, which is ensured through the transition of enterprises to the strategic path of innovative development. Along with the non-linearity of innovation processes, as well as their ability to initiate internal (in relation to the enterprise) structural changes, innovations become one of the aspects of achieving sustainable competitive advantages of the enterprise. The implementation of innovative projects in the field of energy-efficient technologies requires an integrated approach to management, since innovative activities are subject to a significant number of factors. Identification and systematization of such factors, as well as an assessment of the directions of their impact, will help to eliminate the negative impact and increase the effectiveness of innovations by forming the most effective innovative strategy. In addition, the company needs to regularly analyze the level of project compatibility within the portfolio and timely identify fundamental obstacles to the implementation of the project in the field of energy-efficient technologies throughout the entire planned period. In this case, it is necessary to take into account the degree of significance of the projects selected as part of the portfolio on the basis of the list of priority areas for the enterprise, as well as the level of coordination of projects according to results, structure and resources.

In order to make timely adjustments, develop measures and make operational management decisions at large enterprises, the effectiveness of innovative activities should be regularly monitored. This will avoid fragmentation and lack of system in the implementation of innovative activities, which will contribute not only to a phased increase in energy efficiency throughout the entire planning horizon, but also provide an opportunity for further sustainable development of the enterprise.

References

1. Melnik, A., Lukishina, L.: Methodological foundations of the formation of the energy strategy of an enterprise. World Appl. Sci. J. **23**, 1085–1089 (2013). https://doi.org/10.5829/idosi.wasj.2013.23.08.13131
2. Vatin, N., Nemova, D., Khazieva, L., Chernik, D.: Distant learning course "energy efficient refurbishment management". Appl. Mech. Mater. **635–637**, 2057–2062 (2014). https://doi.org/10.4028/www.scientific.net/AMM.635-637.2057
3. Saik, P., Petlovanyi, M., Lozynskyi, V., Sai, K., Merzlikin, A.: Innovative approach to the integrated use of energy resources of underground coal gasification. Solid State Phenom. **277**, 221–231 (2018). https://doi.org/10.4028/www.scientific.net/SSP.277.221
4. Gitelman, L., Kozhevnikov, M.: Energy strategies of industrial enterprises. In: Garcia, J.L. M.I., Brebbia, C.A. (eds.) Ecosystems and Sustainable Development X. WIT Transactions on Ecology and the Environment, vol. 192, pp. 297–307. WIT Press (2015). https://doi.org/10.2495/ECO150271
5. Vatin, N., Gorshkov, A., Nemova, D., Tarasova, D.: Energy efficiency of facades at major repairs of buildings. Appl. Mech. Mater. **633–634**, 991–996 (2014). https://doi.org/10.4028/www.scientific.net/AMM.633-634.991
6. Harmati, N., Jakšić, Z., Vatin, N.: Energy consumption modelling via heat balance method for energy performance of a building. Procedia Eng. **117**(1), 786–794 (2015). https://doi.org/10.1016/j.proeng.2015.08.238
7. Ursul, A., Ursul, T.: From planetary to space mining: prospects for sustainable development. MATEC Web Conf. **265**, 06015 (2019). https://doi.org/10.1051/matecconf/201926506015
8. Ilin, I., Kalinina, O., Makarov, V., et al.: To the problem of effective organization of procurement activities of construction companies. MATEC Web Conf. **170**, 01008 (2018). https://doi.org/10.1051/matecconf/201817001008

9. Ilin, I., Kalinina, O., Petrenko, L.: The far seeing planning systems and models for the construction management. MATEC Web Conf. **170**, 01007 (2018). https://doi.org/10.1051/matecconf/201817001007
10. Kulakov, K., Belyaeva, S., et al.: Sustainable economic development at the meso level: factors and ratings. MATEC Web Conf. **170**, 01118 (2018). https://doi.org/10.1051/matecconf/201817001118
11. Uvarova, S., Bukreev, A., et al.: Conceptual approach to the formation of innovation strategies of economic systems of the micro- and meso-level based on self-organization. MATEC Web Conf. **251**, 05044 (2018). https://doi.org/10.1051/matecconf/201825105044
12. Lukmanova, I., Sizova, E., et al.: Justification of the methodological approach to formation of the parties of innovation activities in holdings. MATEC Web Conf. **239**, 04014 (2018). https://doi.org/10.1051/matecconf/201823904014
13. Minnullina, A., Astafieva, N.: Ranking of transport companies innovative activity. MATEC Web Conf. **239**, 04015 (2018). https://doi.org/10.1051/matecconf/201823904015
14. Ershova, N., Yutkina, O., Pashkov, A., et al.: Influence of human capital on the level of innovation activity of an enterprise. MATEC Web Conf. **239**, 04011 (2018). https://doi.org/10.1051/matecconf/201823904011
15. Makovetskaya, E., Sumachev, A.: Evaluation of personnel development system based on the example of a transport company. MATEC Web Conf. **239**, 04005 (2018). https://doi.org/10.1051/matecconf/201823904005

Barriers and Limitations of Innovative Road Projects Aimed at Improving Energy Efficiency

Ivan Provotorov(✉), Valentin Gasilov, Alshammari Haidar Fazel Mohammed, and Alexander Fedotov

Voronezh State Technical University, Moscow Prospect, 14, Voronezh 394026, Russia
ivanprovotorov@yandex.ru

Abstract. The subject of barriers and restrictions for innovative projects of the road sector, the purpose of which is to increase energy efficiency, is considered. A methodological rationale for the implementation of adaptive traffic control systems is proposed. An assessment of the economic efficiency of the project in one of the cities of Russia was carried out. A conclusion was made about high rates of commercial, social, and regional effectiveness. In order to assess the feasibility of an innovative project, a number of limitations must be taken into account that may affect its effectiveness and achievement of the goals set. In accordance with the research concept, barriers and limitations for the project are considered by the following groups: economic; financial; HR, technical and technological; forecasting innovation; resistance of the existing system; specific; barriers to networks and interactions; socio-psychological; organizational and management. To interpret the level of impact of each barrier, it was assessed by a developed scale. Based on the proposed approach, an expert assessment of barriers and limitations was carried out for the project of implementing an adaptive automated traffic control system. It is concluded that the project does not have critical barriers. Its implementation is advisable, but it is necessary to implement a set of measures to smooth out the possible negative impact. As a result of this, a high socio-economic effect will be achieved due to a significant improvement in the operating conditions, which will contribute to increasing the energy efficiency of the transport complex.

Keywords: Barriers · Limitations · Innovation · Road economy · Energy efficiency

1 Introduction

Road economy is one of the backbone parts of the infrastructure support of the economy of any state. Of particular importance is the road economy for such a huge country as Russia. The dynamics of the development of the road economy and the significant problems that it faces are the subject of many publications of Russian scientists.

One of the main methods for overcoming the problems characteristic for this national economic complex is the development of the road economy on the basis of

V. Murgul and V. Pukhkal (Eds.): EMMFT 2019, AISC 1258, pp. 532–542, 2021.
https://doi.org/10.1007/978-3-030-57450-5_46

innovative technologies. And if in the 90s of the twentieth century, innovative development in Russia is not at the forefront, in the twentieth century, there was some intensification of innovative activity in the road sector through the implementation of targeted public policies. The emphasis was not only on the use of innovative materials, technologies, and the use of modern road-building equipment, but also on the financial, economic and managerial innovations. In Russia, the strategy for the development of innovative activities of the Federal Road Agency for 2011–2015 [1] has been replaced by a road development strategy for 2016–2020 [2]. At the same time, the dynamics of innovation indicators show a decrease in indicators.

A hasty implementation of innovative activities in the road sector could have a significant positive impact on various aspects: increasing throughput and optimizing the level of traffic load on roads; increasing the service life of pavement and road surfaces; increasing the service life of artificial structures on roads; road safety [3]. Of particular importance in the road sector are innovations aimed at reducing transport energy costs [4].

In the field of road facilities in Russia, there is sufficient innovative potential, which is determined by the large number of research organizations and the large resources that the industry possesses [5–7]. However, the active use of innovative solutions is hindered by a number of barriers and limitations. Based on the methodology proposed by the authors in scientific articles, it is advisable to identify and systematize the main types of barriers and limitations in the road sector according to the following groups:

- economic barriers;
- financial barriers;
- personnel barriers;
- technical and technological barriers;
- forecasting innovation;
- resistance of the existing system;
- specific barriers;
- barriers to networks and interactions;
- socio-psychological barriers;
- organizational and managerial barriers.

Identification and overcoming of these barriers is a prerequisite for ensuring the feasibility of innovative projects in the road sector, aimed at improving energy efficiency.

2 Materials and Methods

For the feasibility study of innovations in the road sector, aimed at improving energy efficiency in relation to traffic flow, the development of appropriate methodological support is necessary. Consider the features of the justification of such projects on the basis of a variety of automated traffic control systems (ATCS), an adaptive traffic light control system ("smart" traffic lights), which is actively introduced in Russia in recent years. The use of this technology is one of the current methods of traffic management, solving one of the most important tasks - increasing the capacity of the existing street-road network without significant investments in creating new infrastructure or developing new modes of transport. The result is a reduction in congestion, travel time, an increase in speed, and a significant increase in energy efficiency of the traffic flow is achieved [8–10].

To ensure the effectiveness of the project of implementing "smart" traffic lights, the development of appropriate methodological support is necessary. In accordance with the legislation in force in Russia, ODM 218.9.011-2016 [10] is used to justify the effectiveness of the implementation of adaptive ATCS [11], on the basis of which the architecture of performance indicators is developed. The structure of target indicators consists of the following categories:

- road safety;
- ensuring environmental safety;
- increase in cargo turnover;
- increase in passenger turnover;
- increase in the financial attractiveness of the project;
- increase in the user comfort.

However, the economic assessment of all these parameters is rather laborious and often requires expert estimates that do not have a solid foundation. The most understandable and appropriate indicators are the assessment of reduced travel time and reduced fuel consumption.

Assessing the effects of energy conservation by reducing fuel consumption involves the use of a number of standards and approaches:

- average fuel consumption is taken equal to the average basic fuel consumption of cars in accordance with current regulatory documents [12];
- the consumption coefficient is calculated based on the dependence of the fuel consumption coefficient on the speed of vehicles;
- the economic effect of reducing fuel consumption (potential total annual energy savings) is determined taking into account the reduction in fuel consumption per year and the cost of gasoline.

The economic effect of reducing travel time for the i-th group of vehicles (Evp_i) calculated by the following formula:

$$Evp_i = \sum\nolimits_{i=1}^{n} (N_i \times \Delta T_i \times \mathrm{VoT}) \times 247, \tag{1}$$

where

N_i - peak intensity in the section of the road network (cars/hour);

ΔT_i - reduction of travel time on the road network section, hours;

VoT - the cost of time for the population (rubles/hour);

247 - the average annual number of working days in a year.

For the purpose of economic justification, in order to evaluate the effects of saving time, it is necessary to take into account the number of passengers (including the driver) in the vehicle. For this, it is necessary to take into account the presence in the traffic stream of not only cars, but also buses.

In accordance with the "Guidelines for predicting traffic intensity on highways" [12], we introduce the following notation:

N_l - the average number of passengers in cars, taking into account drivers, passenger/car;

W - average bus capacity, passenger/bus;

N_a - the filling ratio of buses.

Accordingly, formula 1 is converted for cars into the formula:

$$Evp_i = \sum_i (N_i \times \Delta T_i \times \mathrm{VoT} \times \mathrm{H}_л) \times 247. \quad (2)$$

And for buses in the formula:

$$Evp_i = \sum\nolimits_i (N_i \times \Delta T_i \times \mathrm{VoT} \times \mathrm{H_a} \times W) \times 247. \quad (3)$$

As part of the study, traffic intensities at the experimental intersections in the morning rush hour were collected, and the reduction in travel time was analyzed. As a result, the following values of reduction of travel time were obtained: when moving east - average decrease by 0.12 h; when moving west - average decrease by 0.008 h.

The cost of time is calculated on the basis of GDP per capita of the able-bodied population of the Voronezh region:

$$\mathrm{VoT} = \mathrm{GRP}/(\mathrm{NP} \times 12 \times 168), \quad (4)$$

Where:

VoT - the cost of one h of time;

GRP - gross regional product;

NP - the number of employed people.

3 Results

The development of transport communications is one of the main engines for the development of the socio-economic sphere of a modern large city. The reverse is also true: the unsatisfactory state of the transport sector can be a serious brake on the development of most sectors of the economy. Using the example of the city of Voronezh (Russia, population 1,054 million people), a number of negative trends can be noted: a significant increase in travel time; excessive consumption of fuel when driving in traffic congestion; environmental degradation caused by transport; decreased public transport performance; increased downtime of freight vehicles.

To solve the transport problem in the city of Voronezh, many measures are applied: paid parking, regulation of the sphere of passenger traffic, development of infrastructure, etc. In recent years, the direction of introducing adaptive traffic control at the busiest intersections of the city has begun to actively develop. The use of "smart" traffic lights is one of the relevant methods of traffic control, solving one of the most important tasks - increasing the capacity of the existing street-road network without significant investment in creating new infrastructure or developing new modes of transport.

On the basis of the above-proposed method, the overall economic effect (ΣE) for the entire network of 50 intersections equipped with ATCS, 1-year operation of the project will amount to 756,466,000 RUB. In the structure of the total effect, the effects of energy saving due to lower fuel consumption are slightly more than 19% (19.03%). The forecast of effects from the implementation of projects for the implementation of the adaptive automated traffic control system in the city of Voronezh by years of project implementation is presented in Fig. 1.

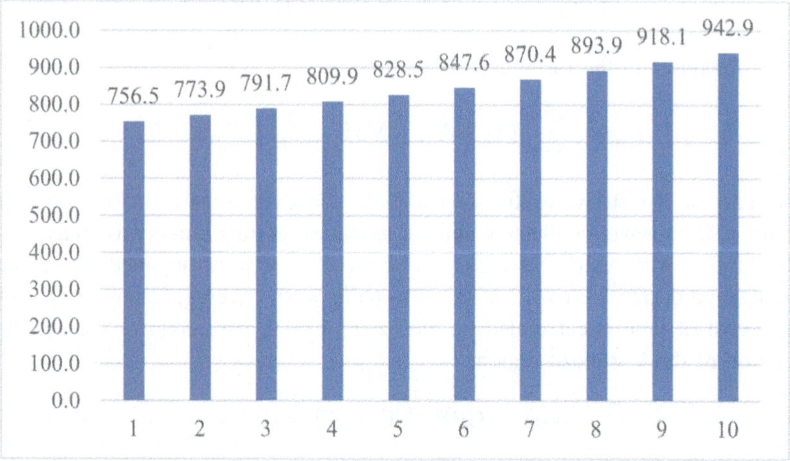

Fig. 1. Effects of the project per year, million rubles.

The presented results show a rather high efficiency of the implementation of the ATCS project. Net present value amounted to 2955.719 million rubles. Here are other performance indicators:

- the total socio-economic effect of the project will amount to 8433 million rubles;
- return on investment index is 14.23.

Calculations show that on the calculation horizon of 10 years, the implementation of projects for the introduction of an adaptive automated system of traffic control gives a significant cumulative socio-economic effect - 8.433 billion rubles.

The effectiveness function of the smart traffic lights (F_{US}) implementation project will depend on many variables:

$$\text{Fuc.} = f(Z_{vn};\ Z_{ek}.;\ N_i.;\ \text{VoT};\ \Delta T_i.;\ \text{GRP.};\ \text{K.};\ \text{TI.};\ \text{KS.}), \quad (5)$$

where

Z_{vn} - the costs of introducing ATCS;
Zek - annual costs associated with the operation of ATCS;
K - the amount of capital investment in the implementation of ATCS;
TI. - inflation rate;
KS - the key rate of the Bank of Russia.

However, to assess the feasibility of an innovative project, it is necessary, in addition to the objective function, to take into account a number of limitations that may affect its effectiveness and achievement of the goals set.

In accordance with the concept of the study, we consider the barriers and limitations for the implementation of the project for the creation of an ATCS in the city of Voronezh - Table 1.

Table 1. Barriers and limitations for projects implementing an adaptive automated traffic control system.

Barrier categories	Barriers	Degree of exposure	Barrier measures
Economic barriers	Disadvantages of existing methodologies for assessing the effectiveness of ATCS implementation projects	Average	In-depth study of existing domestic and foreign methods, adaptation to project conditions
	Lack of understanding of multiplier effects	High	Development of a methodology for assessing multiplicative effects taking into account regional characteristics
Financial barriers	The high cost of implementing ATCS	High	Careful development of design estimates
	The high cost of maintaining ATCS	Average	Organizational structure optimization and approval of content service regulations
	Financial constraints on road development	Average	Attracting loans and joint Russian-Japanese financing
Personnel barriers	The need to attract foreign developers	Average	The conclusion of long-term cooperation agreements with Japanese specialists, the formation of a system for the exchange of experience
	The lack of domestic personnel to support the system in working condition	Average	
Technical and technological barriers	Problems of introducing foreign technologies in domestic conditions	High	Adapting the system to domestic conditions, taking into account regional characteristics
	Errors in the operation of ATCS	High	Using a trial period, system setup

(continued)

Table 1. (*continued*)

Barrier categories	Barriers	Degree of exposure	Barrier measures
Forecasting innovation	Inconsistency between real performance indicators and planned	High	Careful development of the rationale, the use of conservative approaches to assessing effects
	Project survey complexity	Low	User behavior modeling, public awareness work
	The difficulty of predicting user behavior	Average	
	The difficulty of predicting the economic effects of project implementation	High	Integration of ATCS in the development plans of the transport complex of the region, assessment of the consequences of measures taken
	The difficulty of predicting the future development of city transport in the context of the impact on the project	High	
Resistance of the existing system	Public resistance due to a lack of understanding of the essence of the work of ATCS	Low	Public education, publication of data in open sources
	Insufficient understanding of the obtained effects of ATCS by the decision-maker from the authorities	Average	Feasibility study of the project, negotiations, integration of experience of various countries and regions
Specific barriers	The difficulty of interacting with Japanese experts regarding the implementation of the ATCS project in the conditions of an unstable geopolitical situation	Average	The conclusion of long-term agreements
Network barriers and interactions	Problems of interaction between Russian, Japanese developers, authorities of the constituent entities of the Russian Federation and the public	Average	Educational work with the public, publication of data in open sources, holding public hearings, conferences, etc.

(*continued*)

Table 1. (*continued*)

Barrier categories	Barriers	Degree of exposure	Barrier measures
Socio – psychological barriers	The difficulty of understanding the individual and collective (public) effect of the project	High	Development of a sound methodology for assessing the transport and other effects of the project
Management barriers	Lack of experience in managing the various stages of implementation of ATCS implementation projects	Low	Adaptation of existing domestic and foreign experience in managing various stages of implementation of ATCS projects
	The complexity of the formation of the management structure for the content of ATCS	Low	Development of regulations for the content of ATCS, staff structure, cost standards

Some of the above barriers and limitations have already been manifested, despite a short period of operation. The ATCS in Voronezh had to be reconfigured, adapted to the behavior of Voronezh drivers, which was not taken into account in the software development process. Also, a resident's lack of understanding of the principles of the system, especially in conditions of insufficient efficiency during the setup of the system, should be called as psychological barriers to the introduction of ATCS.

To determine the feasibility of an innovative project, it is proposed to assess the barriers and limitations for innovative road projects by an expert method based on Table 2.

Table 2. Gradation of assessments of barriers and limitations for innovation in road projects.

Value, points	Description of the state of scientific and methodological support
0	Barrier or limitation does not occur
0–0.2	Insignificant level of impact on the feasibility and achievement of the goals of the innovation project
0.2–0.4	Insignificant level of impact on the feasibility and achievement of the goals of the innovation project
0.4–0.6	The average level of impact on the feasibility and achievement of the goals of the innovation project
0.6–0.8	Serious impact on the feasibility and achievement of the goals of the innovation project
0.8–0.99	Critically strong impact on the feasibility and achievement of the goals of the innovation project
1	The barrier makes the project completely unrealizable

To interpret the level of impact of each barrier, it should be assessed by the proposed scale, taking into account the possible positive impact of the proposed measures to overcome barriers and limitations. Based on the proposed approach, an expert assessment of barriers and limitations was carried out for the project of implementing an adaptive automated traffic control system. The results are presented in Table 3.

To determine the combined level of barriers and limitations for the considered project, it is necessary to take into account that the significance of the given barriers can be different. In this case, we proceed from the assumption that barriers and limitations are equivalent. Table 4 shows the approach to assessing the feasibility of a project in terms of the value of the composite indicator.

Table 3. Assessment of barriers and limitations for innovation for a project of implementing an adaptive automated traffic control system in Voronezh.

Area	Estimation of the limit value
Economic barriers	0.5
Financial barriers	0.6
Personnel barriers	0.4
Technical and technological barriers	0.7
Forecasting innovation	0.6
Resistance of the existing system	0.3
Specific barriers	0.3
Network barriers and interactions	0.4
Socio-psychological barriers	0.5
Management barriers	0.2
Summary indicator	**4.5**

Table 4. Assessment of the summary level of barriers and limitations for an innovative project.

Value, points	Characterization of the summary level of barriers and limitations
0–2	An innovative project does not have significant barriers and limitations. Its implementation is advisable
2–4	An innovative project has certain barriers and limitations that do not have a decisive influence. Its implementation is advisable
4–6	The average level of impact on the feasibility and achievement of the goals of the innovation project. Its implementation is advisable if there are effective measures to mitigate the negative impact
6–8	Serious impact on the feasibility and achievement of the goals of the innovation project. The implementation of the project is extremely risky, its feasibility is unlikely
8–10	Critically high number of barriers and limitations. The implementation of the project is impractical

4 Conclusions

Based on the above methodology, we can conclude that the project for the implementation of an adaptive automated traffic control system does not have barriers that would make it completely unrealizable. Moreover, there are no barriers and limitations that have a critically strong impact on the feasibility and achievement of the goals of an innovative project. The value of the composite indicator (4.5) is in the range of 4–6, which indicates the average level of impact of barriers and limitations on the feasibility and achievement of the goals of the innovation project. Its implementation is advisable, but a set of measures is required to smooth out the negative impact of barriers and limitations. Possible activities are shown in Table 2, and they can significantly increase the likelihood of a project being implemented and achieving its goals. As a result of this, a high socio-economic effect will be achieved due to a significant improvement in the operating conditions of the transport complex.

References

1. Gavrilov, T., Kolesnikov, G.: Cracks in upper road layer at negative temperature change: modelling and forecasting. MATEC Web of Conf. **239**, 05013 (2018). https://doi.org/10.1051/matecconf/201823905013
2. Gasilov, V., Provotorov, I., et al.: Assessment of the impact of transport energy costs on the efficiency of public-private partnership projects. E3S Web Conf. Electron. Ed. 02128 (2019). 10.1051 / e3sconf / 201911002128 /
3. Obradović, N.: Evaluation of the benefits of utilization of fly ash as a material for road subgrade. MATEC Web Conf. **239**, 05016 (2018). https://doi.org/10.1051/matecconf/201823905016
4. Jocković, S., Pujević, V., Marjanović, M.: Utilization of ash from power plants for high embankments on soft soil. MATEC Web Conf. **239**, 05017 (2018). https://doi.org/10.1051/matecconf/201823905017
5. Arata, M., Petrangeli, M., Longo, F.: Innovative approaches to implement road infrastructure concession through public-private partnership (PPP) initiatives: a case study. Transp. Res. Procedia **14**, 343–352 (2016). https://doi.org/10.1016/j.trpro.2016.05.086
6. Gasilov, V., Anisimova, N., Provotorov, I.: Structure of the scientific-methodical support of realization of infrastructure concession projects. MATEC Web Conf. **106**, 48–52 (2017). https://doi.org/10.1051/matecconf/201710608035
7. Aleksandrov, A., Aleksandrova, N., Chusov, V., Riabov, A.: Ways of application of the provisions of mechanics of bodies with cracks to the calculation of asphalt concrete on strength and plasticity. MATEC Web Conf. **239**, 05018 (2018). https://doi.org/10.1051/matecconf/201823905018
8. Aleksandrov, A., Dolgih, G., Ignatov, V., Kalinin, A.: The application of the principles of the theory of shakedown to the calculation of pavement layers of granular materials in shear. MATEC Web Conf. **239**, 05019 (2018). https://doi.org/10.1051/matecconf/201823905019
9. Ognjenovic, S., Donceva, R., Vatin, N.: Dimensioning of the speed-transition lanes at the entering ramps on the motorway and urban road intersections. Procedia Eng. **117**(1), 544–550 (2015). https://doi.org/10.1016/j.proeng.2015.08.210
10. Ognjenovic, S., Ristov, R., Vatin, N.: Designing of rehabilitations of urban and non-urban roads. Procedia Eng. **117**(1), 568–573 (2015). https://doi.org/10.1016/j.proeng.2015.08.215

11. Isakov, A., Razuvaev, D., Gudkova, I., Chakhlov, M.: Modeling the operation of road pavement during the thawing of soil in the subgrade of highways. MATEC Web Conf. **239**, 05001 (2018). https://doi.org/10.1051/matecconf/201823905001
12. Usanova, K., Rybakov, V., Udalova, V., Kovylkov, A.: Examination of quality and operational properties of vibropressed paving elements. MATEC Web Conf. **73** (2016). https://doi.org/10.1051/matecconf/20167304013

Organization of Combined Heat Energy Generation for Municipal Facilities

Andrey Ovsiannikov(✉), Vladimir Bolgov, Anna Vorotyntseva, and Alexey Efimiev

Voronezh State Technical University, Moscow Avenue, 14, Voronezh 394026, Russia
ovsyannikovas@yandex.ru

Abstract. Modern society consumes more and more energy. In accordance with the geographical location and climatic conditions, fuel costs both for providing the population with heat and for production in Russia are very high. Russia is considered the coldest country in the world, both in terms of the duration of the heating season, and in the proportion of the population living in regions with negative average annual temperature. Increasing the amount of municipal facilities requires an increase in the production of thermal and electric energy to ensure their functioning. Considering on current economic situation there is a necessity of huge investments to create generation capacity for energy production. Also there is a necessity to maintain these capacities during the off-season period, when the generation of thermal energy is at minimum level. When switching to combined energy generation for municipal infrastructure, it will become possible to use a more optimal scheme for converting thermal energy of gaseous fuels into thermal energy directed to heating water with the associated generation of electric energy. Thus the efficiency of system is increased and cost to provide facilities with thermal and electric energy is decreased.

Keywords: Heat supply · Municipal services · Cogeneration · Combined energy production

1 Introduction

Heating of buildings provides thermal comfort for people or the fulfillment of technological requirements for indoor air parameters, depending on the purpose of the room and installed equipment.

In harsh climatic conditions of cold and long winters in a significant territory of Russia, it is impossible to live on premises without a heating system, which compensates for heat loss through the external fencing and heats up to the sanitary norm of the external supply air.

According to researches the following optimal standards were applied in the habitable zone of residential public and administrative buildings (those standards were established for people, located in rooms more than two hours continuously):

V. Murgul and V. Pukhkal (Eds.): EMMFT 2019, AISC 1258, pp. 543–552, 2021.
https://doi.org/10.1007/978-3-030-57450-5_47

- air temperature 20–22 °C;
- relative humidity 30–45%;
- air speed no more than 0,2 m/s [1].

In the Labor Code of the Russian Federation, the length of the working day is interconnected with the air temperature in the working area of the room - the lower it is, the shorter the working day. Increase in temperature in working zone will lead to increase the length of working day, which will increase the efficiency of working day.

The constructive solution of heating primarily depends on the type of thermal energy used to increase the air temperature in the room, including the type of fuel used to generate energy.

Russia is characterized by the presence of large forests and therefore, from ancient times, wood has been the main type of fuel, from the combustion of which thermal energy was generated for heating the premises. In 1900–1920 wood accounted for 50–60% of the country's fuel balance. At the beginning of 20 centuries the necessity to generate not only thermal but electric energy arose. The use of wood became inefficient and its waste is minimalized nowadays (Fig. 1).

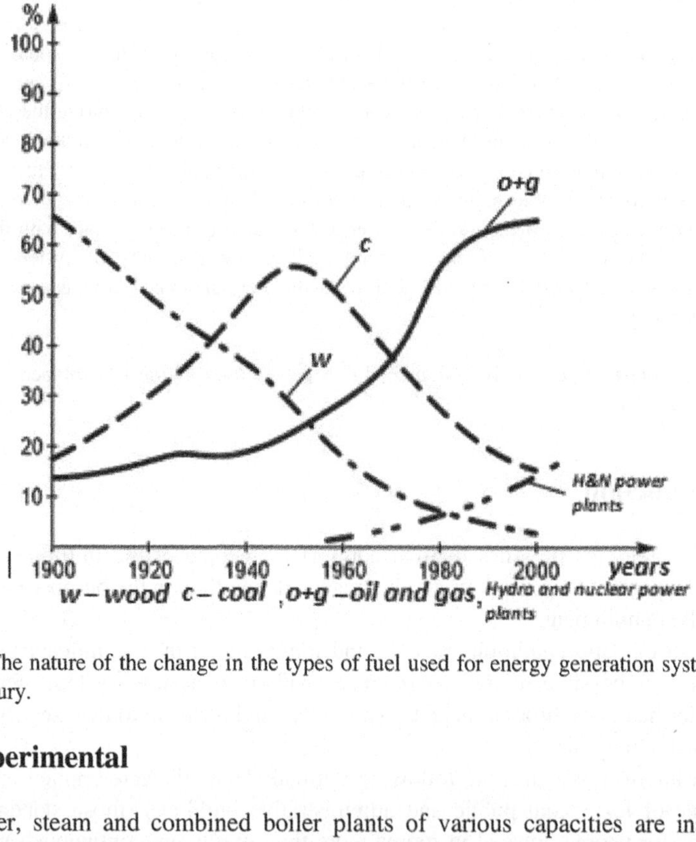

Fig. 1. The nature of the change in the types of fuel used for energy generation systems in the 20th century.

2 Experimental

Hot water, steam and combined boiler plants of various capacities are installed in district heating cities and towns for centralized heating purposes. Various types of fuel are used to provide heat for the communal infrastructure in the Russian Federation, among them are: solid (coal), liquid (fuel oil), gaseous (natural gas). The share of electricity in thermal production is negligible. Individual boiler units have this possibility, as a rule.

Table 1 shows data on the total number of boiler houses, as well as their division by types of fuel use and the amount of energy generated.

Table 1. The structure of fuel consumption for the needs of the housing system.

Type of fuel	The number of boiler rooms, thousand units	Heat produced, million HJ, (million Hcal)	Share in total production, %
Solid	22,4	709 (169,2)	48
Liquid	4,4	147,5 (35,2)	10
Gaseous	19,8	620 (148)	42
Total	46,6	1477 (352,4)	100

In addition to the consumption of various types of fuel in the production of thermal energy for housing and civilian purposes, the structure of enterprises for the generation of this energy is quite various (Table 2).

Table 2. Volumes of thermal energy generation.

Power source, MW (Gcal/h)	Thermal energy production	
	Amount of thermal energy generated, million GJ (million Gcal)	Share in total production, %
House boiler rooms - up to 3.5 (3)	302 (72)	9
Group boiler houses (GBH) - from 3.5 to 23.3 (3–20)	557 (133)	17
Quarterly boiler rooms (QBR) - from 23.3 to 116 (20–100)	754 (180)	23
District boiler rooms (RTS) - more than 116 (more than 100)	587 (140)	18
CHP	1027 (245)	33
Total	3227 (770)	100

It should be noted that the main producers and suppliers of thermal energy in the housing and communal services system are specialized municipal energy enterprises under the jurisdiction of municipalities and executive bodies of the constituent entities of the Russian Federation. In 2019, public utilities provided about 2220 million GJ (530 million Gcal) per year, which accounted for 64% of the total demand for housing and communal and social spheres. The rest of the thermal energy is supplied by regional joint-stock companies of energy and electrification, as well as other enterprises and organizations of ministries, departments, concerns and associations [2, 3].

About 1477 million GJ (352.4 million Gcal) per year are produced by utilities in their own thermal sources (boiler houses), and about 964 million GJ (230 million Gcal) are purchased from other manufacturers with the subsequent transfer to customers - consumers on communal thermal distribution networks [4–7].

Table 1 and 2 shows that half of boiler houses – 22,4 thousand units are working with solid fuel and generate about 35% of all thermal energy, needed to housing stocks. This causes significant pressure on people natural habitat. By replacing many small boiler houses with centralized thermal sources or transferring them to ecologically cleaner fuels: gas, liquid boiler-furnace, as well as non-traditional renewable energy resources (for example, the energy of the sun, waves, wind, geothermal sources, etc.). an ecological improvement of residential neighborhoods could be reached.

By a decision of the Government of the Russian Federation, rural thermal supply systems should be transferred to the balance sheet and commissioned to municipal formations of local administrations. This work is ongoing, and the number of utilities is increasing.

In the production of thermal energy, it is necessary to pay attention to the cost of this energy (Table 3). When producing it from different types of fuel, its cost varies quite widely.

Table 3. The cost of thermal energy.

Type of fuel	Cost for one unit, rub	Typical efficiency %	Cost of 1 kV/h heat taking into account the efficiency, rub.
Main gas, m^3	7.40	90%	0.88
Coal, kg	5.80	80%	0.93
Dry coniferous firewood (20%), kg	4.45	70%	1.61
Pellets, kg	7.80	80%	2.04
Firewood of natural moisture, coniferous (40%), kg	3.85	70%	2.44
Electricity	3.74	95%	3.74
Liquefied gas, l	22.60	90%	4.45
Diesel fuel, l	44.00	85%	5.11

Table shows that the most economically efficient is the production of thermal energy with the use of main gas.

In addition to the cost of thermal energy, it is necessary to take into account the convenience of using this type of fuel. The most optimal by performance is electricity and main gas, the operation of boiler houses on them is possible fully automatically. When using other types of fuel, it is almost impossible to do without periodic human influence (liquefied gas or diesel fuel) or the constant presence of personnel providing fuel to the boilers (firewood, coal, pellets, etc.).

Although the use of natural gas is economically feasible and makes it possible to operate the boiler house in automatic mode, its widespread use has a significant drawback - this is a high explosive hazard and the devastating consequences of incidents. Therefore, the question of creating a system for providing thermal energy based on its centralization, which will make it possible to maximize the benefits of natural gas as energy fuel and reduce the possibility of negative consequences arises.

Also the possibility of using systems with different types of fuel should be considered. By reserve types of power supply, it will be possible to increase the reliability level of the heat supply system of housing and civil facilities and to ensure uninterrupted heat supply in a wide variety of unforeseen situations.

3 Evaluation

When considering boiler houses of different capacities, a pattern is revealed: the higher the power of the boiler room, the higher the efficiency of the boiler room itself, the higher the performance and reliability of the heat supply system. But in this case, the question of heat loss during transportation to consumers arises. With increasing transportation distance, losses increase. For the conditions of the European part of the Russian Federation, the effective radius of heat supply does not exceed 9 km (along the length of the networks). At present, the main heat production in the Russian Federation for residential and administrative-civil facilities is accounted for by thermal power plants (33%) and quarterly boiler houses (23%) [8, 9].

Figure 2 shows a comparison of the efficiency of using the combined system for the production of thermal and electric energy at combined heat and power plants (CHP) with the separate production system at power plants and boiler houses based on the processing of the same type of energy fuel.

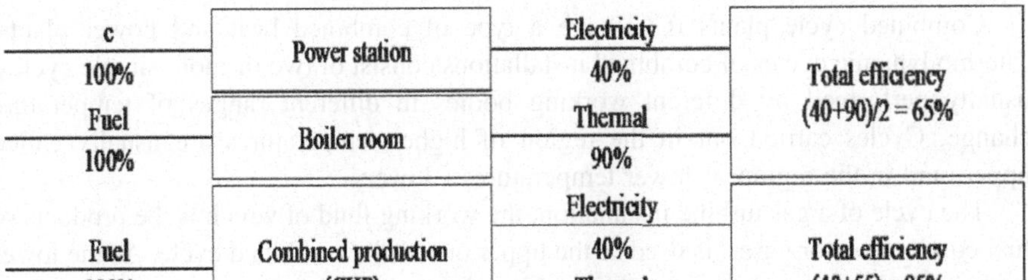

Fig. 2. Methods of energy production for housing facilities.

The combined heat and power plant (CHP) is one of the types of thermal power plants that use the combined production system of two main types of energy: electric and thermal (in the form of steam and hot water, including for providing hot water and heating for residential and industrial facilities).

Fig. 3. Dzerzhinskaya CHP (CHP-22 Mosenergo).

Figure 3 show Dzerzhinskaya CHP as an example. Figure gives us a chance to imagine the view of CHP.

The widespread use of the combined energy production system with the use of thermal power plants has its advantages and disadvantages.

The advantage of thermal power plants is the combined generation of heat and electricity, which can significantly reduce the total cost of energy production. With joint production, we get a reduction in energy costs for the facility.

The main advantage of the CHP is the use for the production of heat energy of the exhaust secondary gases obtained for the production of electric energy through the use of combined-cycle plants.

Combined cycle plants (CCP) are a type of combined heat and power plants. Thermodynamic cycles of combined installations consist of two or more simple cycles, usually performed by different working bodies in different ranges of temperature change. Cycles carried out in the region of higher temperatures are usually called upper, and in the region of lower temperatures - lower.

The cycle of a gas turbine installation, the working fluid of which is the products of fuel combustion, or gases, is used as the upper one in the combined cycle. As the lower cycle is used steam turbine unit, the working fluid of which is water vapor. Therefore, the names of the cycle and installations are combined-cycle ones.

CCP is a relatively new type of generating stations operating on gas or liquid fuel. The principle of operation of the most economical and widespread classical scheme is based on the simultaneous use of gas turbine and steam power plants. In a gas turbine installation, the turbine is rotated by gaseous products of fuel combustion. Passing through a gas turbine, the combustion products give it only part of their energy and still have a high temperature at the exit of the gas turbine. From the exit of the gas turbine, the combustion products enter the steam power plant, into the recovery boiler, where the water and the resulting steam are heated.

In the first gas turbine cycle, the efficiency rarely exceeds 38%. Spent in a gas turbine installation, but still maintaining at high temperature, the combustion products enter the so-called waste heat boiler. They heat steam to a temperature up to 500 °C and a pressure of 80 atm. Suck indicators are sufficient for steam turbine to work, which is connected to another generator. In the second - steam-power cycle, about 20% of the energy of the burned fuel is used. In total, the efficiency of the entire installation is approximately 58%. Steam power units are well mastered, reliable and durable. Their unit capacity reaches 800–1200 MW, and the coefficient of performance, which is the ratio of the generated electricity to the heat production of the used fuel, is up to 40–41%, and at the most advanced power plants abroad—45–48% [9].

The increase in efficiency when combining steam turbine and gas turbine plants is obtained due to two factors:

- the implementation of the superstructure of the gas cycle over the steam;
- reduced total flue gas consumption.

In most schemes, both factors are used at the same time, giving an increase in efficiency. However, there are schemes in which only one of them is used.

The main advantages of combined-cycle plants with a conventional steam generator are:

- the ability to operate the steam generator of a gas turbine installation on any fuel (70–85% of the total fuel is burned in the steam generator);
- the ability to use conventional steam generators, which facilitates the creation of combined-cycle plants on the basis of serial equipment and allows for gas superstructure of existing power plants while maintaining all installed main equipment.

Researches and international experience show that the development and widespread use of combined-cycle plants of various types are the main direction for improving the efficiency of thermal power plants, which recently accounted for up to 70% of all electricity generation. The best indicators of efficiency among all types of combined cycle plants have installations with a waste heat boiler. When operating on natural gas with a rated load, they ensure the production of electricity with a net efficiency of up to 60% [10].

The disadvantage of using CHPs is due, firstly, to their high cost. Tens of millions of rubles and the high minimum power of the installation impose a limitation on the use of this type of heat generation method: their construction and use are effective only in large compact settlements.

This problem can be solved through the use of mini-CHP based on CCP, the main advantages of which are:

1. Low losses during the transportation of thermal and electric energy in comparison with centralized heat and power supply systems.
2. Autonomous functioning of the mini-CHP (independence from an external energy system) and the possibility of selling surplus generated electricity or heat to the energy system.
3. Low cost of generated heat and electricity (2–2.3 times less than in centralized heat and power supply systems).

4. mproving the reliability of energy supply, since possible interruptions in the supply of electricity from an external energy system do not lead to the cessation of the mini-CHP.
5. Low fuel consumption for electricity and thermal production.
6. Large motor resource and durability of the equipment of mini-CHP (Table 4 and Fig. 4).

Table 4. Comparative characteristics of various CCGT.

Indicator	CCP-40	CCP-20
Gas turbine		
Electric power, MW	32	16
Flue gas temperature, °C	510	510
Waste heat boiler		
Productivity, t/h	43	22,6
Vapor pressure, ata	60	55
Vapor temperature, °C	500	500
Steam turbine		
Electric power, MW	8,5	5,2
Vapor pressure, ata	60	55
Steam consumption, t/h	38,2	19,4
Heating capacity, Gcal/h	28	13,5
Back pressure, ata	0,8	0,8
Fuel utilization rate,%	89	88
Size	18 × 12 × 16	15 × 9 × 14

Fig. 4. Appearance of a mini-CHP (CCP).

4 Conclusions

The investment attractiveness of the mini-CHP construction projects is beyond doubt. It is characterized by relatively high performance with small dimensions and weight. This makes it possible to install it in almost any boiler house building without significant volumes of construction work. The competitiveness of power plants with combined cycle power plants is ensured by a relatively low unit cost per MW of installed capacity compared to other types of equipment.

The choice of one or another mini-CHP scheme requires solving a complex technical and economic problem, taking into account factors such as the efficiency and reliability of the installation, the cost of development, manufacture and operation, etc.

References

1. Korsun, V., Vatin, N., Korsun, A., Nemova, D.: Physical-mechanical properties of the modified fine-grained concrete subjected to thermal effects up to 200 °C. In: Applied Mechanics and Materials, vol. 633–634, pp. 1013–1017 (2014). https://doi.org/10.4028/www.scientific.net/AMM.633-634.1013
2. Korsun, V., Korsun, A., Volkov, A.: Characteristics of mechanical and rheological properties of concrete under heating conditions up to 200 °C. In: MATEC Web of Conferences, vol. 6, p. 07002 (2013). https://doi.org/10.1051/matecconf/20130607002
3. Korsun, V., Vatin, N., Franchi, A., Korsun, A., Crespi, P., Mashtaler, S.: The strength and strain of high-strength concrete elements with confinement and steel fiber reinforcement including the conditions of the effect of elevated temperatures. Procedia Eng. **117**(1), 970–979 (2015). https://doi.org/10.1016/j.proeng.2015.08.192
4. Korniyenko, S.V., Vatin, N.I., Gorshkov, A.S.: Thermophysical field testing of residential buildings made of autoclaved aerated concrete blocks. Mag. Civil Eng. **64**(4), 10–25 (2016). https://doi.org/10.5862/MCE.64.2
5. Harmati, N., Jakšić, Z., Vatin, N.: Energy consumption modelling via heat balance method for energy performance of a building. Procedia Eng. **117**(1), 786–794 (2015). https://doi.org/10.1016/j.proeng.2015.08.238
6. Grinfeldi, G.I., Gorshkov, A.S., Vatin, N.I.: Tests results strength and thermophysical properties of aerated concrete block wall samples with the use of polyurethane adhesive. In: Advanced Materials Research, vol. 941–944, pp. 786–799 (2014). https://doi.org/10.4028/www.scientific.net/AMR.941-944
7. Baiburin, A.K., Rybakov, M.M., Vatin, N.I.: Heat loss through the window frames of buildings. Mag. Civil Eng. **85**(1), 3–14 (2019). https://doi.org/10.18720/MCE.85.1
8. Barabanshchikov, Y., Fedorenko, I., Kostyrya, S., Usanova, K.: Cold-bonded fly ash lightweight aggregate concretes with low thermal transmittance: review. In: Advances in Intelligent Systems and Computing, vol. 983, pp. 858–866 (2019). https://doi.org/10.1007/978-3-030-19868-8_84

9. Dudin, M.O., Vatin, N.I., Barabanshchikov, Y.G.: Modeling a set of concrete strength in the program ELCUT at warming of monolithic structures by wire. Mag. Civil Eng. **54**(2), 33–45 (2015). https://doi.org/10.5862/MCE.54.4. https://www.scopus.com/inward/record.uri?eid=2-s2.0-84991407896&doi=10.5862%2FMCE.54.4&partnerID=40&md5=95bfd294704c3e47ee913d7dfff20f3c
10. Gorshkov, A.S., Rymkevich, P.P., Vatin, N.I.: Simulation of non-stationary heat transfer processes in autoclaved aerated concrete-walls. Mag. Civil Eng. **52**(8), 38–66 (2014). https://doi.org/10.5862/MCE.52.5. https://www.scopus.com/inward/record.uri?eid=2-s2.0-84991380377&doi=10.5862%2FMCE.52.5&partnerID=40&md5=50a3105694e0ec8f0799098244a315d0

Cost Management for Fuel and Energy Resources in the Creation and Operation of Urban Infrastructure

Olga Kutsygina(✉), Margarita Agafonova, Andrei Chugunov, and Irina Serebryakova

Voronezh State Technical University, Moscow Avenue, 14, Voronezh 394026, Russia
olga.kutsigina@rambler.ru

Abstract. The object of the study is the fuel and energy resources used as part of the costs in the creation and operation of construction and investment facilities. The classification of factors determining the quality of buildings and the urban environment, consumer properties of buildings at the stages of the life cycle is given. In order to realistically achieve the target cost indicators of fuel and energy resources established at the design stage, a descriptive model for managing the costs of construction facilities as a single system is proposed, which shows a consistent scheme for the formation of standard costs for subsequent periods depending on the results of decisions made at the previous stages of the life cycle of buildings and structures. To prevent the risk of negative deviations from the target indicators of the effective consumption of fuel and energy resources, the organization of cost management in the creation and operation of urban infrastructure facilities is proposed, which is based on the use of a multi-level analysis of deviations over the life cycle of buildings.

Keywords: Fuel and energy resources · Cost management · Building life cycle · Multi-level deviation analysis

1 Introduction

An integral part of urban real estate is a combination of housing, social and engineering infrastructure. Unlike industrial products, construction projects are individual, require significant amounts of one-time investments and ongoing operating costs, are created in places of subsequent use, which complicates the processes of organization and management of construction production. The intended useful life is dozens of years and is accompanied by regularly increasing annual operating costs, including the cost of fuel and energy resources. Their value is determined by many factors formed during the design, construction and operation, which multidirectionally affect the resource and cost indicators of objects [1].

As part of the annual operating costs for the maintenance of building utility systems (heating, hot water, ventilation and air conditioning), up to 80% is the share of the cost of heat generated by heat supply systems [2]. Therefore, in this study, the main

V. Murgul and V. Pukhkal (Eds.): EMMFT 2019, AISC 1258, pp. 553–565, 2021.
https://doi.org/10.1007/978-3-030-57450-5_48

indicator characterizing the consumer properties of buildings is the heat costs necessary to compensate for heat losses through external building envelopes and hot water supply, air treatment in ventilation and air conditioning systems.

An analysis of information on forecasts of the total amount of energy in the earth's interior shows their inexhaustibility in the foreseeable future. Russia ranks third in the world in total consumption of primary energy resources (1.2 trillion tons). However, the energy intensity of Russian GDP at the beginning of the 21st century was 2.5 times higher than in the USA and 3.5 times higher than in Europe. The energy intensity of utilities in Russia is 3–4 times higher than similar indicators of countries with similar climates [3]. Despite the measures taken to reduce the consumption of fuel and energy resources during the operation of buildings [4–7], the further development of energy-efficient technologies used by housing and utilities infrastructure are defined by the "Energy Strategy of Russia for the period until 2035" as first-priority technologies designed to ensure the independence and sustainability of the development of the power system as the basis for economic growth. The authors consider that one of the reasons for the lack of effectiveness in the practical implementation of energy-efficient technologies is their local application at the stages of the life cycle, lack of managerial control over the results of their implementation, insufficient study of the possibilities of organizational measures as a means of energy conservation of fuel and energy resources [8].

2 Materials and Methods

The time period including the duration of the design, construction, intended use and demolition of buildings and structures forms a life cycle, which requires a one-time investment in the period of construction of facilities and annual current costs in the period of their intended use. About 2% of the energy consumed during the life cycle of buildings is consumed in the construction process, in the construction industry - about 8%, and 90% falls on the operational period. In modern practice of designing buildings and structures, the introduction of energy-saving technologies and ensuring compliance with a certain class of energy efficiency has become an integral part of construction projects in order to reduce annual operating costs. However, energy-saving design solutions are not always fully implemented in practice, are too costly and economically inefficient.

The annual operating costs reflect the consumer properties of buildings and structures, which are characterized by quality factors of the surrounding building of the urban environment and design decisions of buildings. Factors are represented by many indicators that determine the technical level, socio-economic, climatic and other conditions of functioning of capital construction objects. Some of them are formed at the design and construction stages, others, in turn, establish target operating requirements that determine possible solutions at the design and construction stages of the building life cycle (BLC) in accordance with the diagram in Fig. 1.

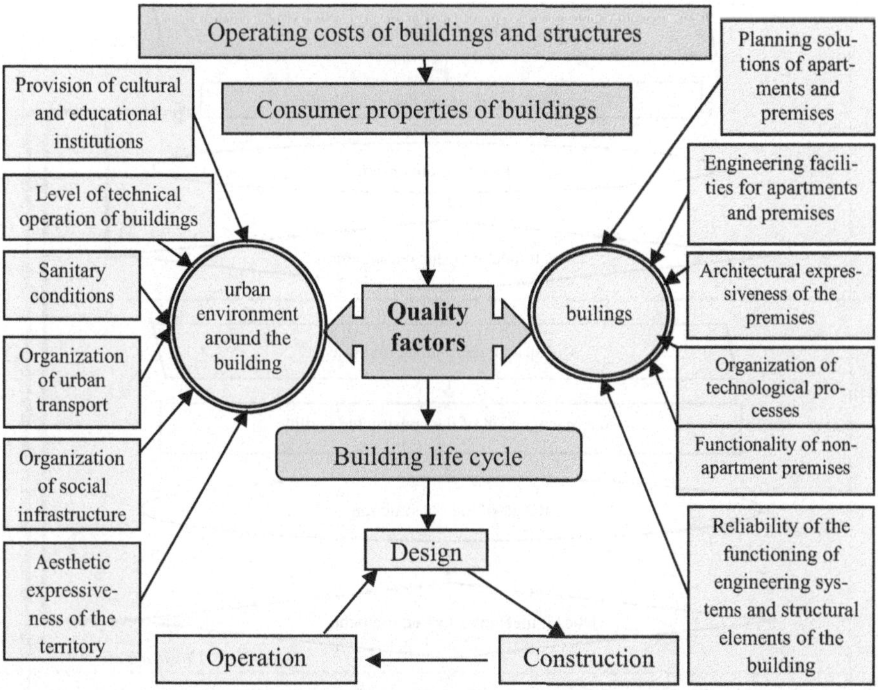

Fig. 1. The scheme of formation of quality factors that determine the consumer properties of buildings at the stages of the life cycle.

The implementation of investments at the stages of the BLC is carried out by independent business entities exposed to risks of wasting capital, including fuel and energy resources (FER). To prevent them, the cost management methodology (CM) is used as an effective tool of modern management aimed at obtaining positive results from capital investment. Cost management methods make it possible to monitor actual indicators, quickly monitor quantitative and qualitative changes in comparison with the planned standards, timely identify the causes that caused them and make management decisions, reducing or completely preventing the negative consequences of forecasted situations. The CM system is based on the principles of management accounting and allows for "transparency" of accounting systems.

Since the construction design, erection and maintenance of the operational needs of real estate is carried out by economically independent organizations, the relationships between these successive stages of the life cycle are not well coordinated, which negatively affects the indicators of investment and construction projects and the consumption of fuel and energy resources during the operation of buildings. In order to ensure the interdependence of the technical and economic characteristics and parameters of construction and investment projects formed at each stage of the BLC, the authors propose a descriptive model (algorithm) for the organization of cost management during the life cycle of urban infrastructure facilities as a single system (Fig. 2).

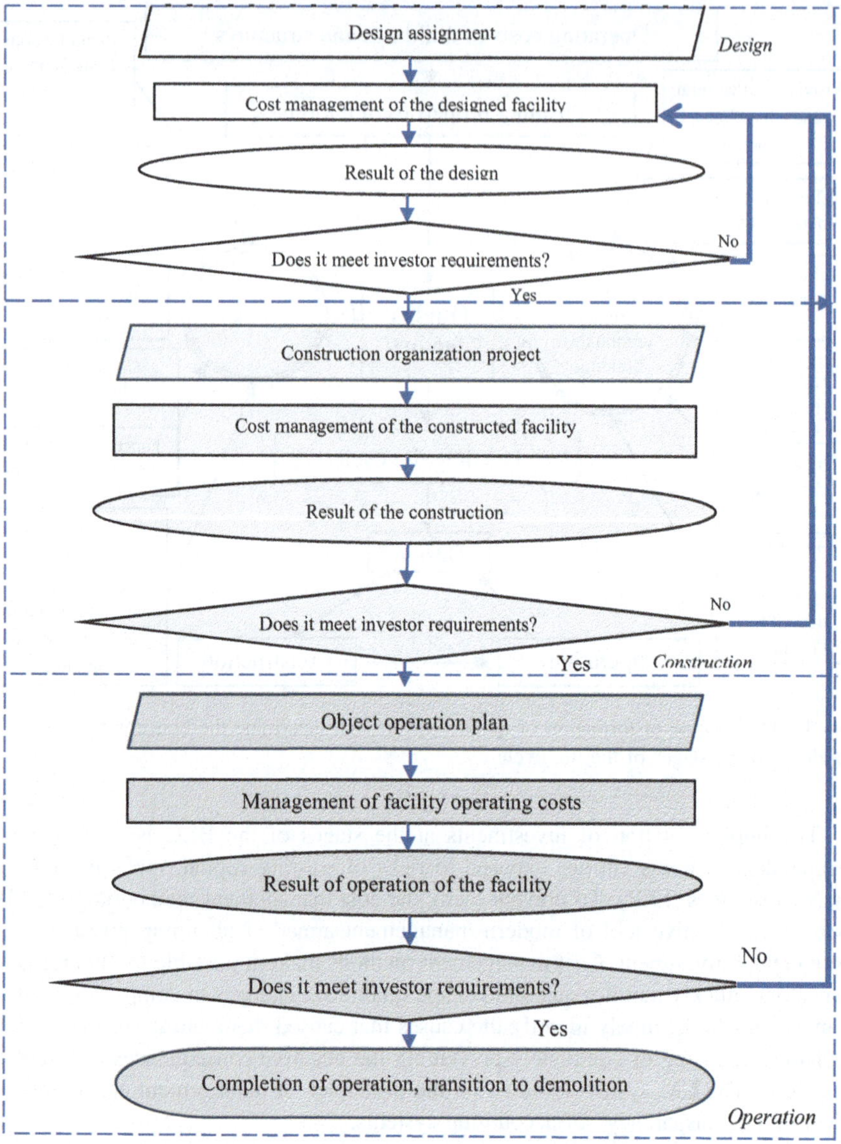

Fig. 2. Descriptive model of cost management during the life cycle of construction facilities as a single system.

The model characterizes the interconnection of the stages of the life cycle (without the demolition stage) and shows a consistent scheme for the formation of regulatory costs of subsequent periods depending on the results of decisions made at the previous stages of the life cycle of buildings and structures.

In the process of managing the implementation of construction and investment projects, it is proposed to apply management control as a method of cost management at each stage of the life cycle of facilities. At the design stage, special attention should be paid not only to the budget indicators of the design organization, but also to the indicators of the project that is being developed by the designers. If in the process of managing costs of a project organization, actual and planned indicators of its activity as an economic entity are subject to comparison, then in the process of managing costs of a developed project, the calculated (design) indicators and their target values are subject to comparison. In variant design, the most effective option is determined [8].

In relation to construction orders, calendar schedules are used to fulfill the terms of construction and installation works, costs and revenues. Compliance with schedules is crucial. Their indicators are monitored during the course of the work by comparing the planned indicators determined on the basis of the project and the actual ones with the frequency necessary for making timely decisions (once a week, ten days or another period). As a result of monitoring the implementation of the schedule, deviations, their causes and quality are revealed, their influence on the financial result is determined, and measures are taken to eliminate possible negative effects. A new forecast of planned costs is made. If the additional costs of cost control are less than the achieved result in the form of prevented losses, then further detailing of the control system is considered inappropriate. The use of calendar schedules for construction and installation works in combination with financial statements and estimated cost is an instrument of professional management of a construction and installation company. And the indicators of the constructed facility are recorded in the passport of the facility and serve as the basis for planning annual operating costs. These are cost of consumed resources necessary for the production of products, the performance of work or the provision of services. The amount of costs is determined by the amount of resources used.

The used resources presented in monetary terms have two sides - quality and price. The scheme of a multilevel analysis of cost deviations is based on identifying the difference between the planned and actual cost indicators, determining the factors under the influence of which the deviation occurred; identification and timely elimination of the causes of negative deviations; analysis and decision making to reduce the negative impact of negative deviations on the financial result or expenditure of fuel and energy resources as a target result. The scheme for conducting a multi-level cost analysis is shown in Fig. 3. Multilevel cost analysis is a tool for analyzing budget execution in the practice of management accounting and controlling. For its implementation, indicators of planned, adjusted planned (flexible) and actual budgets of the enterprise are considered and compared. The planned (static) budget is a calculation of the financial result based on standard indicators of the plan (main budget). A flexible budget is compiled to determine the financial result with planned indicators of both unit and fixed costs, prices of selling and purchasing materials, but for actual indicators of sales.

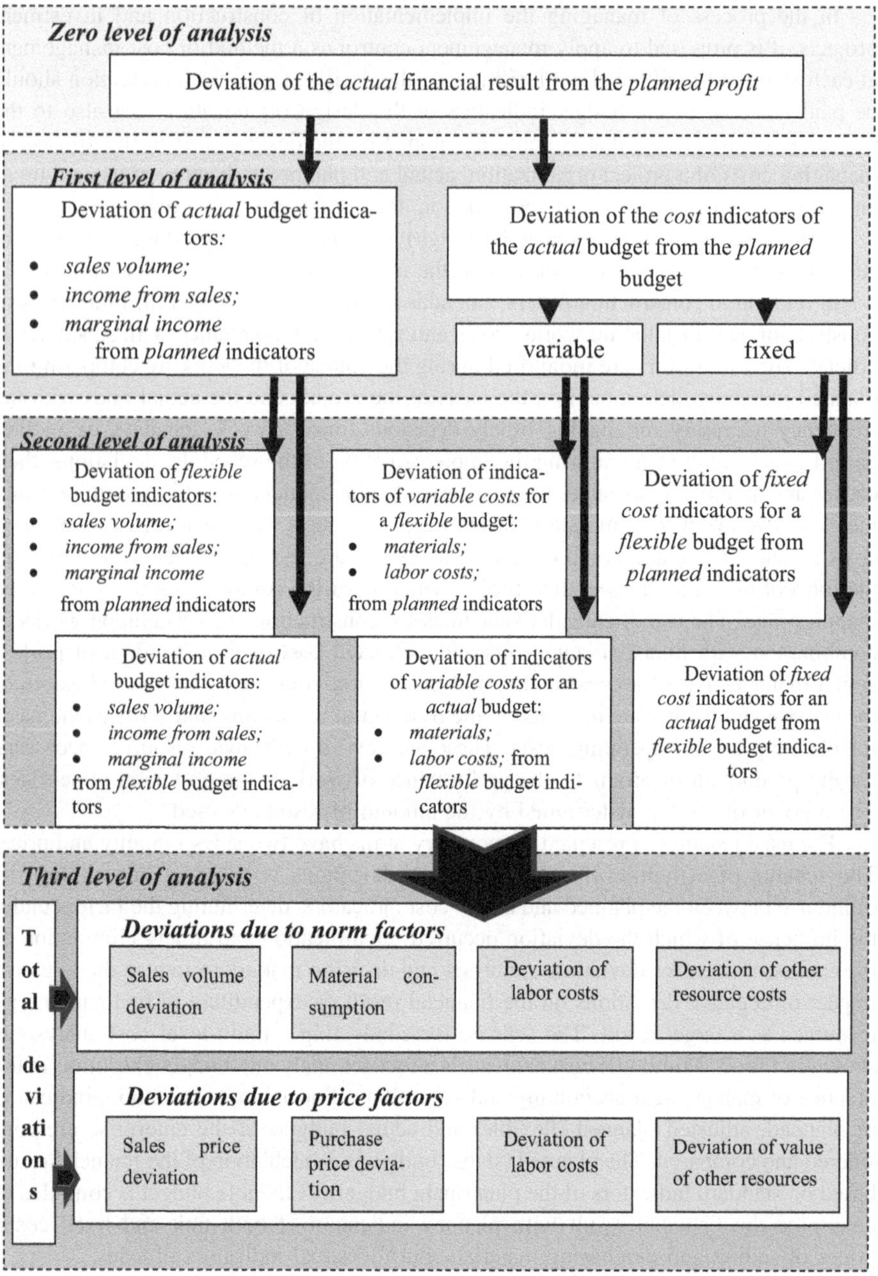

Fig. 3. The scheme for conducting a multilevel analysis of deviations of actual indicators from planned standards.

Zero (initial) level of analysis shows the magnitude and quality of deviation of the actual financial result from implementation of its planned value. However, deviation from profit, both positive and undesirable, can occur under the influence of many factors. To identify them, the analysis process is further detailed.

The next or first level of analysis allows comparing the actual and planned indicators, which does not provide deviations from the volume of sales, cost components and other conditions in the planning period.

The second level of analysis compares the indicators of the plan, flexible budget and actual data. The analysis of the second level involves a pairwise comparison of the indicators of the plan and flexible budget, flexible budget and actual data. The share of deviations in profit, which is revealed between the indicators of a static and flexible budget, is explained by the deviation of sales volumes. The share of deviation in profit, which is revealed between the indicators of the flexible budget and actual indicators, is explained by the deviation of the factors of the norms of resource consumption and/or labor costs and prices and/or tariffs.

Since costs are a product of the norms of the resource's consumption by the acquisition price, a further analysis will make it possible to identify due to what factors —norms or prices—a deviation in costs occurred. To do this, the third level of cost analysis is used.

A multilevel cost analysis allows detailing the composition of costs and identifying specific places where deviations of actual indicators from plan indicators occur, which makes reporting "transparent" and makes it possible to timely identify adverse deviations and make decisions to prevent negative consequences.

3 Results

A multilevel analysis of deviations of actual indicators from planned standards is given using the example of heat supplying organizations of a district utility company, including 14 boiler houses with a total planned capacity of 110,000 Gcal per year for servicing the district center. Boiler houses operate on gas, backup fuel is not provided.

Multilevel cost analysis showed the following results.

1. At the zero level, the excess of actual profit over the planned one by 213 thousand rubles was revealed, which is 0.63% (last row of Table 1). The quality of the deviation (positive or negative) is marked by signs (+, −) after the quantitative indicator.
2. At the first level of analysis, deviations of the actual indicators of heat generation from the planned ones were revealed (Table 1), showing the multidirectional effect on the financial result.

An analysis of the data in Table 1 showed that the volume of heat sold for the year actually turned out to be less than according to the plan (by 6160 Gcal), and the actual income from sales was lower than planned (by 2992 thousand rubles). The analysis of cost indicators allows concluding about the reduction of variable fuel costs (by 3553 thousand rubles) and the excessive consumption of electricity (by 624 thousand rubles). Despite the decrease in income and the excess of planned targets for electricity and fixed costs, as a result of lower fuel consumption and other fixed costs, the financial result exceeded the planned figure by 213 thousand rubles [9–11].

Table 1. Input data for the first level of multi-level cost analysis.

Indicators	Actual indicators, thousand rubles	Share, %	Planned indicators, thousand rubles	Share, %	Deviations of actual indicators from planned ones, thousand rubles
Sales volume, Gcal	103840	–	110000	–	6160 (−)
Income	124608	100	127600	100	2992 (−)
Variable expenses, including: -Fuel (gas) - Electricity	75647 6729	60.71 5.4	79200 6105	62.07 4.8	3553 (+) 624 (−)
Marginal income	42232	33.9	42295	33.1	63 (−)
Fixed costs, incl. -Water -Wage, including UST -Other fixed costs	8207 915 4118 3174	6.6 0.7 3.3 2.5	8483 880 3900 3703	6.6 0.7 3.1 2.9	276,2 (+) 34,8 (−) 218 (−) 529 (+)
Financial result	34025	27	33812	26	213 (+)

3. At the second level of analysis, the indicators of the plan and the adjusted plan for the actual volume of production and sales are compared, as well as the actual indicators with the indicators of the plan adjusted for the actual volume of production and sales. In the first case, deviations are determined by the sales volume factor. And in the second - by other factors, including factors of consumption rates and resource acquisition prices (Table 2).

Due to a decrease in sales by 6160 Gcal, 2369 thousand rubles of profit were lost, which is determined by the formula

$$\Delta P = \left(V_p - V_p^{cor}\right) \cdot MD_p, \tag{1}$$

where V_p – planned sales volume, thousand rubles; V_p^{cor}-volume of sales according to the adjusted plan, thousand rubles; MD_p - planned specific marginal income, thousand rubles/unit.

Deviations of actual indicators from indicators of the adjusted plan are caused by factors of sales prices (heat tariff), resource consumption rates and acquisition prices. A positive deviation of the financial result from the indicator for the adjusted plan is 2582 thousand rubles.

Deviations caused by the selling price factor can be defined as the product of the difference between the actual and planned price per unit of heat and the actual sales volume. So, a positive deviation in income of 4154 thousand rubles due to an increase in the selling price by 0.04 thousand rubles/unit is determined by the formula

$$\Delta D = \left(T_f - T_p^{cor}\right) \cdot V_f, \tag{2}$$

Table 2. Input data for the second level of multi-level cost analysis.

Indicators	Actual indicators, thousand rubles	Deviations of actual indicators from planned ones (column 2–column 4)	Planned indicators adjusted for the actual volume of sales, thousand rubles	Deviations of the adjusted plan from the base one (column 4–column 6)	Planned indicators, thousand rubles
1	2	3	4	5	6
Sales volume, Gcal	103840	0	103840	6160 (−)	110000
Income	124608	4154 (+)	120454	7146 (−)	127600
Variable costs, including: -Fuel -Electric power	82376 75647 6729	1848 (−) 882 (−) 966 (−)	80528 74765 5763	4777 (+) 4435 (+) 342 (+)	85305 79200 6105
Marginal income	42232	2306 (+)	39926	2369 (−)	42295
Fixed costs, incl. -Water - Wage, including UST	8207 915 4118	276 (+) 35 (−) 218 (−)	8483 880 3900	0 0 0	8483 880 3900
Other fixed costs	3174	529 (+)	3703	0	3703
Financial results	34025	2582 (+)	31443	2369(−)	33812

where ΔD - deviation of actual income from income by the adjusted plan, caused by the selling price factor, thousand rubles, T_f, T_p^{cor}- actual price per unit of sales and according to the adjusted plan, thousand rubles/unit; V_f - the volume of actual sales, units.

A favorable deviation of actual income indicators from indicators of the adjusted plan is neutralized by the excess of actual variable costs compared with the data of the adjusted plan by 1878 thousand rubles (882 + 966) due to the increase in specific variable costs and is determined by the formula

$$\Delta PZ = \left(UPZ_p^{cor} - UPZ_f\right) \cdot V_f, \quad (3)$$

where ΔPZ - deviation of variable costs between actual indicators and indicators of the adjusted plan, thousand rubles; UPZ_p^{cor}, UPZ_f- unit variable costs according to the adjusted and actual plan, respectively, thousand rubles/unit.

Fixed costs for water and labor actually exceed the figures of the adjusted plan by 253 thousand rubles, and other expenses of 529 thousand rubles less than planned. Therefore, the total positive deviation of fixed costs amounted to 276 thousand rubles, and the positive financial result exceeds the indicator of the adjusted plan by 2582 thousand rubles, which is in total with 2369 thousand rubles of loss due to lower sales will be only 213 thousand rubles of the additionally earned profit [12–14].

4. At the third level of analysis, it is possible to identify the effect of the factor of norms and the price factor on deviations. The initial data for the third level of a multilevel cost analysis are presented in Table 3.

Based on the analysis, it follows that the negative deviation of variable fuel costs (882 thousand rubles) was formed due to a negative deviation in the price factor (3218 thousand rubles) in the direction of their increase and due to a decrease in the consumption rate factor (2336 thousand rubles). In this case, the negative deviation of the actual fuel costs from the planned value is due to an increase in the fuel tariff from 4.5 to 4.7 rubles/m^3 of gas, and positive - by reducing the rate of flow from 160 m^3/Gcal up to 155 m^3 Gcal. Thus, the use of energy-saving measures in sources of heat generation can lead to cost reduction by the factor of norms, which can become a reserve for lowering tariffs. Negative deviation of electricity costs in the amount of 966 thousand rubles is caused by a positive deviation in the price factor (187 thousand rubles) and a negative deviation in the factor of norms (1153 thousand rubles). A positive deviation of the actual energy costs from the planned value occurred due to a reduction in the fuel tariff from 3.7 to 3.6 rubles/kW·h, and a negative deviation in the rate factor due to overruns from 15 to 18 kW·h/Gcal

In general, the negative deviation in variable and fixed costs is 2101 thousand rubles, including 2963 thousand rubles - an overspending due to the price factor. This deviation is caused by overspending on fuel (3218 thousand rubles) and water (88 thousand rubles), reduced by savings in electricity (187 thousand rubles) and labor costs (156 thousand rubles).

Table 3. Initial data for the third level of multilevel cost analysis.

Type of costs	Unit price of the resource (planned), rubles/unit of measure	Specific rate of resource consumption (planned), unit of measure/Gcal	Unit costs of resources, (planned), rub/Gcal	Actual cost of resources, thousand rubles	Adjusted plan costs, thousand rubles	Deviations, thousand rubles		
						Total	Due to price factor	Due to norm factor
1	2	3	4	5	6	7	8	9
Fuel (gas)	4.5 rub/m^3	160 m^3/Gcal	720 rub/Gcal	75647	74765	882 (−)	3218 (−)	2336 (+)
Electric power	3.7 rub/kW·h.	15 kW·h./Gcal	55.5 rub/Gcal	6729	5763	966 (−)	187 (+)	1153 (−)
Total variable costs			775.5 rub/Gcal	82376	80528	1848 (−)	3031 (−)	1183 (+)
Water	–	–	–	915	880	35 (−)	88 (−)	53 (+)
Wage	–	–	–	4118	3900	218 (−)	156 (+)	374 (−)
Total fixed costs (excluding other costs)				5033	4780	253 (−)	68 (+)	321 (−)
Total other fixed costs				–	–	529 (+)	–	–
Total fixed costs (including other costs)					–	276 (+)	–	–
Total costs (excluding other costs)					85308	2101 (−)	2963 (−)	862 (+)
Total costs (including other costs)					–	1572 (−)	–	–

Deviation due to the factor of norms has a positive value of 9862 thousand rubles, which is for 2336 thousand rubles caused by the fuel component of costs, 53 thousand rubles - for water, and there was overspending for electricity and wage in the amounts of 1153 and 374 thousand rubles, respectively.

4 Conclusions

1. To increase the efficiency of the practical implementation of energy-efficient technologies, it is advisable to develop organizational measures for cost management in the creation and operation of urban infrastructure facilities as a full means of energy conservation of fuel and energy resources.
2. In order to ensure the interdependence of the technical and economic characteristics and parameters of construction and investment projects formed at each stage of the building life cycle, including the consumption of fuel and energy resources, a descriptive model for the organization of cost management during the life cycle of the infrastructure as a single system is proposed. The model shows a consistent scheme for the formation of regulatory costs of subsequent periods depending on the results of decisions made at the previous stages of the life cycle of buildings and structures, which contributes to the organization of management control throughout the life cycle and the prevention of unproductive losses of fuel and energy resources.

3. To organize management cost control over the life cycle of buildings, it is proposed to apply the methodology of multi-level analysis of costs and revenues, which makes reporting transparent and allows not only timely identification of adverse deviations, making decisions to prevent them, but also an economic justification for the formation of prices and tariffs for fuel and energy resources.

References

1. Kutsygina, O.A., Shakir, YA.A.: Konceptual'nyj podhod k upravleniyu zatratami po rezul'tatam na etapah zhiznennogo cikla stroitel'noj nedvizhimosti. Sovremennye problemy nauki i obrazovaniya, №1, pp. 287–295 (2015). https://science-education.ru/ru/article/view?id=18853
2. Nezhnikova, E., Santos, S., Egorycheva, E.: Management of the investment design process at the enterprises of the energy sector. In: Murgul, V., Pasetti, M. (eds.) EMMFT-2018 2018. AISC, vol. 983, pp. 127–137. Springer, Cham (2019). https://doi.org/10.1007/978-3-030-19868-8_12
3. Tinkov, S.A., Tinkova, E.V.: Indicator to assess the level of development of productive capacity and quality of life. In: Solovev, D.B. (ed.) FarEastCon 2018. SIST, vol. 139, pp. 567–576. Springer, Cham (2019). https://doi.org/10.1007/978-3-030-18553-4_69
4. Kutsygina, O., Uvarova, S., Belyaeva, S., Chugunov, A.: Technical and economic aspects of energy saving at the stages of the building life cycle. In: Murgul, V., Pasetti, M. (eds.) EMMFT-2018 2018. AISC, vol. 983, pp. 36–44. Springer, Cham (2019). https://doi.org/10.1007/978-3-030-19868-8_4
5. Kankhva, V.: Using the entropy of cover method in the analysis of investment risks. In: MATEC Web of Conferences, vol. 212, p. 08003 (2018). https://doi.org/10.1051/matecconf/201821208003
6. Lukmanova, I., Golov, R.: Modern energy efficient technologies of high rise construction. In: E3S Web of Conferences, vol. 33, p. 02047 (2018). https://doi.org/10.1051/e3sconf/20183302047
7. Butturi, M.A., Lolli, F., et al.: Renewable energy in eco-industrial parks and urban-industrial symbiosis: a literature review and a conceptual synthesis. Appl. Energy **2551**, 113825 (2019). https://doi.org/10.1016/j.apenergy.2019.113825
8. Hawkey, D., Webb, J., Winskel, M.: Organisation and governance of urban energy systems: district heating and cooling in the UK. J. Clean. Prod. **501**, 22–31 (2013). https://doi.org/10.1016/j.jclepro.2012.11.018
9. Bugge, M.M., Fevolden, A.M., Klitkou, A.: Governance for system optimization and system change: The case of urban waste. Res. Pol. **48**(4), 1076–1090 (2019). https://doi.org/10.1016/j.respol.2018.10.013
10. Kuznetsova, E., Cardin, M.-A., et al.: Integrated decision-support methodology for combined centralized-decentralized waste-to-energy management systems design. Renew. Sustain. Energy Rev. **103**, 477–500 (2019). https://doi.org/10.1016/j.rser.2018.12.020
11. Uyarra, E., Gee, S.: Transforming urban waste into sustainable material and energy usage: the case of greater Manchester (UK). J. Clean. Prod. **501**, 101–110 (2013). https://doi.org/10.1016/j.jclepro.2012.11.046

12. Mesjasz-Lech, A.: Reverse logistics of municipal solid waste – towards zero waste cities. Transp. Res. Proc. **39**, 320–332 (2019). https://doi.org/10.1016/j.trpro.2019.06.034
13. Blumberg, R., Mincyte, D.: Infrastructures of taste: rethinking local food histories in Lithuania. Appetite **1381**(July), 252–259 (2019). https://doi.org/10.1016/j.appet.2019.02.016
14. Łapniewska, Z.: Cooperatives governing energy infrastructure: a case study of Berlin's grid. J. Co-op. Organ. Manage. **7**(2), 100094 (2019). https://doi.org/10.1016/j.jcom.2019.100094

Model for the Development of an Energy Enterprise

Yulia Bondarenko(✉), Tatiana Azarnova, Irina Kashirina, and Ekaterina Vasilchikova

Voronezh State Technical University, Moskovskiy Prospekt, 14, Voronezh 394026, Russia
bond.julia@mail.ru

Abstract. The article is devoted to the development of a model for the formation of the trajectory of development of the energy complex enterprise in the context of the transition to a digital economy. A feature of the proposed approach is the emphasis on quality management of production factors, including labor resources (human capital) and fixed assets, which is relevant for modern electric power enterprises. Dynamic changes in the quality of resources are accepted as the most important component of the development path. The paper presents mathematical methods for obtaining an integral quantitative assessment of the quality of each resource. To take into account the influence of the quality and quantity of production factors on the volume of products and the profit of the energy complex enterprise, the concept of a production function is introduced taking into account the quality of resources, its properties are investigated and construction methods are given. This function expands the concept of the production function of an enterprise in the direction of adaptation to the conditions of the digital economy. To build the optimal trajectory of the enterprise, a mathematical model is proposed, which is an optimization problem. The variables of the model are the amounts of available financial resources of the enterprise invested in each of three areas: the expansion of fixed assets through modern equipment, improving the quality of labor resources (through continuing education programs, etc.), creating new jobs.

Keywords: Energy complex · Digital economy · Development path · Quality · Mathematical model

1 Introduction

In modern conditions of intensive technological transformations, digitalization is a driver of accelerated development of the national economy and a key factor in increasing its competitiveness. The introduction of digital technologies is already changing the current processes in all sectors of the economy in the direction of increasing efficiency and reducing costs [1].

A special role in the digital transformation of the economy belongs to the enterprises of the fuel and energy complex (FEC). The transition to new energy production technologies should not only meet the growing needs of enterprises and end

V. Murgul and V. Pukhkal (Eds.): EMMFT 2019, AISC 1258, pp. 566–577, 2021.
https://doi.org/10.1007/978-3-030-57450-5_49

consumers, but also lead to a significant reduction in energy intensity, environmental friendliness, accessibility of energy resources and energy infrastructure [1–3].

At the same time, it should be noted that the digital economy in the energy sector (digital energy) requires significant changes not only in the production and distribution of energy technologies, but also in the technologies for managing enterprises in the FEC. The emergence of a digital platform, the displacement by machines of inefficient, requiring the routine participation of human resources transactions, necessitates the management of each company to review and radically change production factors (labor and capital). Such transformations should affect not only proportions and quantities, but also the qualitative characteristics of production factors. In this case, the central direction of the development strategy of the enterprise of the FEC should be a coordinated change in the quantity and quality of resources, ensuring effective development and integration into a single digital space. The development of theoretical, applied, and software tools that provide the management of the FEC with intellectual decision-making support to formulate an enterprise development strategy taking into account the modern needs of the digital economy is an urgent task.

The issues of transition to the digital economy are actively discussed and investigated in modern foreign and domestic literature and open sources [1–4]. Various aspects of the transition to digital energy, its differences from classical energy, especially the organization of a digital platform, pricing, smart grids, etc. are discussed in [5–7]. Research is devoted to new technologies in the energy industry and methodological issues of personnel training [8]. In [9], a formalized model of specialist knowledge in the digital economy is presented.

In [10], such problems of the transition to digital energy as the lack of scientific research in the field of intellectual support for strategic decision-making on the development of technological infrastructure are discussed. The author notes that in order to eliminate these shortcomings, it is necessary to develop tools based on the integration of existing backlogs in the field of mathematical modeling, situational management, intelligent computing, etc. Agreeing with the opinion of the author, it should be noted that the existing, proven models and methods of the traditional economy should be modified, supplemented and brought into line with the requirements and features of the modern digital economy. Mathematical research tools are based on scientific achievements in the development of models and decision-making mechanisms and management of socio-economic systems [10].

The aim of this work is to develop mathematical methods and models for forming the trajectory of an enterprise's development that provide decision-makers at the FEC with analysis of the state and prospects of development, as well as the reasonable distribution of financial resources for the expansion and development of production, which ensures maximum profit in digital transformation. The specification of the proposed models is based on the use of information available for the management of the enterprise, and the developed software product makes it convenient to use them in the practice of managing the enterprise and the region.

2 Materials and Methods

Consider the enterprise of the FEC, operating in the territory of a certain region. We will represent each enterprise as a complex, dynamic system. The state of such a system is usually described by a set of values of the most significant quantitative and qualitative indicators (state parameters). Researchers associate the change in time of quantitative indicators with the growth of the system, and the change in qualitative indicators with its development.

We believe that the planning period of the enterprise is T-years. We consider the time interval $\{1, 2, \ldots, T\}$ to be discrete, and we choose the time cycle equal to one year (which is associated with the period of statistical reporting).

Due to the special importance of the products of the fuel and energy complex for life support and socio-economic development of each region, we will consider the most significant indicators of the activity of the enterprise of the FEC:

- y_t – volume of products (energy) of good quality at time t;
- Z_t – enterprise profit at a time t.

Essential indicators of the development of the enterprise in the digital economy is the use of advanced technologies, modern equipment and skilled labor. Therefore, we will consider the quality of production factors, their compliance with the development of the digital economy, to be no less significant indicators of the state of the enterprise:

- $\bar{K}_t$ – quantitative indicator of the quality of capital (fixed assets) at time t;
- $\bar{L}_t$ – quantitative indicator of the quality of labor resources at time t.

The development path of the FEC in this case is the following sequence:

$$\{(y_t, Z_t, \bar{K}_t, \bar{L}_t)\}_{t=1}^{T} = \{(y_1, Z_1, \bar{K}_1, \bar{L}_1), \ldots, (y_T, Z_T, \bar{K}_T, \bar{L}_T)\}. \tag{1}$$

The initial state of the FEC in the selected indicators is an ordered three $(y_0, Z_0, \bar{K}_0, \bar{L}_0)$.

We formulate the problem of forming the optimal trajectory of the development of the fuel and energy complex.

It is required at the moment t = 0, taking into account the initial resources and capabilities of the enterprise, to form a development trajectory that meets the region's needs for high-quality energy and obtains the largest total profit for the period:

$$\sum\nolimits_{t=1}^{T} Z_t = \sum\nolimits_{t=1}^{T} (\sigma^t \cdot p \cdot y_t - c_t(y_t)) \rightarrow \max, \tag{2}$$

where p – unit cost of energy, $\sigma \geq 1$ – energy tariff growth rate, $c_t(y_t)$ – energy production costs at a point in time t.

It is proposed to carry out the search for the optimal development path for the FEC using methods of economic and mathematical modeling based on the solution of the optimization problem. The limitations of the task describe the main dependencies between resources and indicators of the functioning and development of the enterprise, and the goal function has the form (3).

One of these dependencies, widely used in modeling practice, is the production function that describes the technological relationship between the maximum volume of output and the volume of expended production factors (capital and labor) [10]:

$$y = F(K, L), \tag{3}$$

where y – output volume, K – the amount of capital of the enterprise, L – labor force, F – production function.

Based on studies related to the construction of the production function (3), and taking into account modern features of the influence on the volume of output not only the number of factors of production, but also their quality, in this paper we introduce the concept of production function taking into account the quality of resources.

Under the production function, taking into account the quality of resources, we understand the quantitative dependence of the maximum volume of products on the quantity and quality of production factors:

$$y = f(K, L, \bar{K}, \bar{L}), \tag{4}$$

where f – production function taking into account the quality of resources, $\bar{K}$ – quantitative measure of capital quality, $\bar{L}$ – quantitative indicator of the quality of labor.

We believe that the production function, taking into account the quality of resources, satisfies the following axioms.

1. For any quality of resources $\bar{K}^0, \bar{L}^0$ acceptable for the production technology, the function of two variables $F(K, L) = f(K, L, \bar{K}^0, \bar{L}^0)$ satisfies the axioms of the production function.
2. The mapping (4) does not decrease in each of the variables $\bar{K}$ and $\bar{L}$.
3. If the quality value of at least one of the resources is below the acceptable level, then the volume of the enterprise's output will be 0. In other words, production with unacceptably low quality of resources is impossible:
4. $f(K, L, \bar{K}, \bar{L}) = 0$, if $\bar{K} < \varepsilon_K$ or $\bar{K} < \varepsilon_L$, where $\varepsilon_K, \varepsilon_L$ – minimum quality requirements for resources.
5. If the value of the quality indicator of at least one of the resources used in the production process is close to the minimum acceptable value, then the output of the enterprise is close to zero, i.e. at $\bar{K} \to \varepsilon_K$ or $\bar{L} \to \varepsilon_L$ performed: $y = f(K, L, \bar{K}, \bar{L}) \to 0$.

The above axioms are satisfied, for example, by functions of the following form:

1. $y = f(K, L, \bar{K}, \bar{L}) = A \cdot K^{\alpha_1} \cdot L^{\alpha_2} \cdot (\bar{K} - \varepsilon_K)^{\alpha_3} \cdot (\bar{L} - \varepsilon_L)^{\alpha_4}$, where $A > 0, \sum_{i=1}^{4} \alpha_i = 1, \alpha_i > 0, \quad i = 1, \ldots, 4$;
2. $y = f(K, L, \bar{K}, \bar{L}) = (\alpha_1 K + \alpha_2 L) \cdot (\bar{K} - \varepsilon_K)^{\alpha_3} \cdot (\bar{L} - \varepsilon_L)^{\alpha_4}$, where $\alpha_i > 0, \quad i = 1, \ldots, 4$.

The parameters of the production function, taking into account the quality of resources, are determined on the basis of retrospective statistical information by econometric methods.

The solution to the problem of forming the optimal trajectory for the development of the fuel and energy complex is proposed to be divided into the following stages.

Stage 1. Building quality indicators of resources.
Stage 2. The formation of the production function of the FEC, taking into account the quality of resources.
Stage 3. Building a model for the formation of an optimal trajectory for the development of the FEC.
Stage 4. Settlement.

Consider the practical implementation of each stage.

3 Results

At the first stage, we will consider the quality indicators of labor resources as integral indicators, taking values in the interval [0, 1], where a value of 0 corresponds to an extremely low quality of the resource, and 1 to high.

Here is the rationale for calculating the quality indicator of labor resources.

We believe that the management of the enterprise can choose M particular characteristics whose values can be measured. For an indicator of each particular characteristic with index m:

- $P_m^{\min}, P_m^{\max}$ – respectively, the minimum and maximum possible values;
- P_m^* – normative value (requirement) in accordance with production technology;
- P_m – the value of the performance indicator for this enterprise, $P_m^{\min} \leq P_m \leq P_m^{\max}$.
- As such characteristics can be selected:
- the proportion of employees of individual units in the total number of employees of the enterprise;
- average category of workers;
- the proportion of employees with higher or secondary specialized education;
- average work experience in the specialty of managers and certain specialists of the enterprise;
- staff turnover in hiring and firing workers; Percentage of staff retrained in the digital economy;
- the number of employees using digital technologies in the process of their activities, etc.

For each characteristic, its normalized value μ_m and the normalized requirement value λ_m are calculated in accordance with the following rule:

- if an increase in the value of the indicator leads to an increase in the quality of labor resources, then

$$\mu_m = \frac{P_m - P_m^{\min}}{P_m^{\max} - P_m^{\min}}, \lambda_m = \frac{P_m^* - P_m^{\min}}{P_m^{\max} - P_m^{\min}}; \quad (5)$$

- if an increase in the value of the indicator leads to a decrease in the quality of labor resources, then

$$\mu_m = \frac{P_m^{\max} - P_m}{P_m^{\max} - P_m^{\min}},\ \lambda_m = \frac{P_m^{\max} - P_m^*}{P_m^{\max} - P_m^{\min}}. \tag{6}$$

Based on the obtained values, M "private" quality indicators are calculated:

$$d_m = 1 - \frac{\lambda_m(1 - \mu_m)}{\mu_m(1 - \lambda_m)}. \tag{7}$$

Moreover, we assume $d_m = 0$ for $\mu_m = \lambda_m = 0$ and $d_m = 1$ for $\mu_m = \lambda_m = 1$.

The integral indicator of the quality of labor resources is proposed to be calculated according to the following formula:

$$\bar{L} = \prod_{m=1}^{M} d_m. \tag{8}$$

A similar approach can be used to obtain an indicator of the quality of capital of an enterprise. The differences are in the number of private quality characteristics and the characteristics themselves. Among them, such characteristics as physical depreciation of fixed assets can be taken; obsolescence of fixed assets; degree of capacity utilization, etc.

Quality indicators of labor resources and capital (fixed assets) of the FEC, calculated on the basis of retrospective data, are used as independent factors in the formation of the production function taking into account the quality of the resources of the enterprise. Based on the data of one of FEC's companies, a function was obtained by regression analysis:

$$y_t = \left(0{,}04 \cdot \left[\bar{K}_t^{0{,}24} \cdot K_t\right]^{-0{,}127} + 0{,}356 \cdot \left[\bar{L}_t^{0{,}078} \cdot L_t\right]^{-0{,}127}\right)^{-0{,}634}. \tag{9}$$

To build a model for the formation of an optimal trajectory of enterprise development, we believe that at the initial time t = 0, the enterprise management knows the following information:

- B_0 – the amount of available financial resources that the enterprise management can invest in the development and expansion of its own production;
- K_0 – volume of capital (fixed assets); $\bar{K}_0$ – the value of the quality indicator of fixed assets;
- L_0 – labor force; $\bar{L}_0$ – the value of the indicator of the quality of labor;
- β – the average cost of creating one job;
- γ – fixed assets liquidation ratio; λ – fixed assets renewal rate;
- ρ – share of deductions from the wage fund;
- η – current income tax rate;
- Q_t – energy needs of the region at the time t, where $t = 1, \ldots, T$;

- $\tilde{B}_t$ – financial resources received by the enterprise from external sources (for example, state support funds);
- $Z_t^{\min}$ – minimum profit of an enterprise at a time t;
- ω_t – company average nominal wage;
- b_t – material costs per unit of production.

Model variables are: B_t^L – amount of financial resources spent on creating new jobs in the year t; $\bar{B}_t^L$ – financial resources intended to increase the quality of labor resources; B_t^K – financial assets invested in the expansion and updating of fixed assets.

Regarding the features of accounting in the model for changing the quality of factors of production, we make the following assumptions.

1. The change in the quality of fixed assets occurs due to the involvement of new funds, the quality of which is not lower than those already involved in the production process:

$$\bar{K}_t = \frac{(1-\gamma)K_{t-1}}{K_t} \cdot \bar{K}_{t-1} + \frac{B_t^K}{K_t} \cdot \beta_t \cdot \bar{K}_{t-1},$$

where $\beta_t \geq 1$ – indicator characterizing the growth rate of the quality of new funds.
2. The change in the quality of labor resources is affected by the amount of funding in this area:

$$\bar{L}_t = \varphi\left(\bar{B}_t^L, \bar{L}_{t-1}\right),$$

where φ – continuous function that describes the dependence of the quality of labor resources on the amount of funding.

The model for the formation of the optimal trajectory for the development of the FEC enterprise is:

$$\sum\nolimits_{t=1}^{T} Z_t = \sum\nolimits_{t=1}^{T} \left(\sigma^t \cdot p \cdot y_t - c_t(y_t)\right) \to \max, \tag{10}$$

subject to restrictions:

- restrictions on the volume of products:

$$y_t = f(K_t, L_t, \bar{K}_t, \bar{L}_t), \tag{11}$$

$$y_t \geq Q_t, t = 1, \ldots, T; \tag{12}$$

- change in the quantity and quality of fixed assets:

$$K_t = (1-\gamma+\lambda)K_{t-1} + B_t^K, \tag{13}$$

$$\bar{K}_t = \frac{(1-\gamma+\lambda)K_{t-1}}{K_t} \cdot \bar{K}_{t-1} + \frac{B_t^K}{K_t} \cdot \beta_t \cdot \bar{K}_{t-1}, t = 1, \ldots, T; \tag{14}$$

- dynamics of the quantity and quality of labor resources:

$$L_t = L_{t-1} + \frac{B_t^L}{\beta}, \tag{15}$$

$$\bar{L}_t = \varphi\left(\bar{B}_t^L, \bar{L}_{t-1}\right), t = 1, \ldots, T; \tag{16}$$

- profit generation:

$$Z_t = \sigma^t \cdot p \cdot y_t - c_t(y_t), \tag{17}$$

$$c_t(y_t) = \left(b_t \cdot y_t + \frac{1}{1-\rho}\omega_t \cdot L_t + \lambda K_t\right), t = 1, \ldots, T; \tag{18}$$

- restrictions on the formation and distribution of financial resources

$$B_t = B_{t-1} + (1-\eta)Z_{t-1} + \tilde{B}_t, \tag{19}$$

$$B_t = B_t^L + \bar{B}_t^L + B_t^K, t = 1, \ldots, T; \tag{20}$$

- variable restrictions:

$$B_t^L \geq 0, \bar{B}_t^L \geq 0, B_t^K \geq 0, t = 1, \ldots, T. \tag{21}$$

Model (10)–(21) is an optimization problem with one criterion and can be reduced to an equivalent problem with variables $B_t^L, \bar{B}_t^L, B_t^K$. To solve it, it is recommended to use nonlinear optimization methods.

Denote the optimal solution to the problem (10)–(21):

$$\left(B_t^L\right)^*, \left(\bar{B}_t^L\right)^*, \left(B_t^K\right)^*, t = 1, \ldots, T. \tag{22}$$

Based on the obtained solutions, as well as relations (11), (13), (16) and (17), the optimal trajectory of the development of the FEC is calculated:

$$\left\{\left((y_t)^*, (Z_t)^*, (\bar{K}_t)^*, (\bar{L}_t)^*\right)\right\}_{t=1}^T.$$

For the practical implementation of the model, a software package has been developed in the language C#. Consider the results of calculations based on data provided by one of the enterprises of the FEC of the Voronezh region.

Table 1 shows the data of the enterprise, on the basis of which the calculation of the qualities of production factors (resources) in a year t = 0 was carried out.

Labor force and capital quality indicators:

Table 1. Data for calculating quality factors of production factors.

Name of indicator	P_m	P_m^{min}	P_m^{max}	P_m^*	d_m
The proportion of employees of individual units in the total number of employees, un.	0,34	0,1	0,8	0,3	0,21
The average rank of workers, un.	4	1	6	3	0,30
The proportion of employees with higher or secondary special education, un.	0,8	0,2	0,94	0,6	0,73
Average work experience in the specialty of managers and certain specialists, years	10	3	25	5	0,79
Staff turnover rate, un.	0,02	0	0,98	0,03	0,34
The coefficient of physical depreciation of fixed assets, un.	0,3	0	0,98	0,4	0,27
Equipment obsolescence ratio, un.	0,2	0	0,98	0,5	0,61
Capacity utilization rate, un.	0,8	0,2	0,98	0,6	0,69

$$\bar{L} = 0{,}21 \cdot 0{,}3 \cdot 0{,}73 \cdot 0{,}79 \cdot 0{,}34 = 0{,}015;$$

$$\bar{K} = 0{,}34 \cdot 0{,}27 \cdot 0{,}61 \cdot 0{,}69 = 0{,}039.$$

The constructed production function, taking into account the quality of resources for this enterprise, has the form (9).

Table 2 shows a fragment of the main data for the model for the formation of the optimal development path for the FEC.

Table 2. Basic data for the model.

Indicator name	Units	Value
Number of employees	people	1863
Fixed assets (end of year)	thous. roubles	1 992 221
Fixed assets renewal rate (at the end of the year)	in % of the total value of fixed assets	60,2
The liquidation ratio of fixed assets (end of year)	in % of the total value of fixed assets	0,2
The share of deductions from the wage fund	%	30
Average monthly nominal wage	thous. roubles	32,46
Profit before tax (retained earnings)	thous. roubles	165 589
The volume of net supply of thermal energy	thous. Gcal	1 908,4
Revenues from sales	thous. roubles	3 990 193

For the calculations, a period of 3 years is selected. Expected receipt of funds from external sources: 1 year - 147.1 mil. rubl., 2 year - 245, 27 mil. rubl., 3 year - 189.67 mil. rubl.

The results of calculations of indicators of the optimal development path are presented in Table 3.

Table 3. Calculation results.

Name of indicator	t = 1	t = 2	t = 3
Volume of heat supply, thous. Gcal	2 110,1	2 379,8	2 502, 7
Profit before tax, thous. rubles	165 001	190 456	278 139
Labor force quality indicator	0,017	0,022	0, 034
Fixed assets quality indicator	0,056	0, 096	0,124

The calculations showed that the bulk of the available financial resources should be invested in improving the quality of fixed assets. However, in order to obtain the greatest profit, an increase in the quality of fixed assets should be consistent with an increase in the quality of the workforce of the enterprise.

The calculated values for the optimal distribution of financial resources were taken into consideration by the management of the enterprise in 2018 and were taken into account in the formation of the strategy, development and modernization program.

4 Discussions

The study showed the possibility of the effective use of mathematical modeling and mathematical methods to solve tasks relevant to the management of the FEC related to the development of a development strategy in the context of the transition to a digital economy. Based on advanced research in the field of digital economy and digital energy, this article proposes an approach to the formation of a development path based on a quantitative consideration of the quality of production factors as a driver for the effective development of an enterprise.

The model of formation of the optimal trajectory of the FEC proposed in the study was repeatedly discussed with the leadership of one of the enterprises of the Voronezh region, its limitations were clarified, and the dependencies used were corrected. The results of the calculation do not affect the implementation of specific activities to increase the efficiency of the enterprise, but allow us to identify general trends and prospects for its development, mathematically substantiate the direction of distribution of financial resources.

From a scientific point of view, the article proposes a new approach to modeling enterprise development processes, based on taking into account the quality of resources and building the dependence of the volume of products not only on the number of factors of production, but also on their quality. The constructed production function, taking into account the quality of resources, has good statistical properties and confirms the assumptions made.

A discussion of the results of model calculations with the management of the FEC allowed us to receive positive feedback on the practical value of the study and outlined ways to develop options for the proposed model for situational enterprise management. One of these options is to expand the model with the possibilities of reasonably forming a package of measures for the enterprise development program.

5 Conclusion

The approach proposed in this paper to solving the urgent problem of managing a FEC - forming a development path reflects the main trends in the digital economy - emphasis on the high importance of the quality of resources. To take into account the quality of production factors in the strategic planning process, the paper considers the issues of calculating quantitative indicators of labor resources and capital. To take into account the influence of the quality of factors of production on the volume of output, the article introduces the concept of a production function taking into account the quality of resources, describes its properties, examines the issues of practical construction. The dynamics of the quality of production factors is carefully taken into account in the model for the formation of the optimal development trajectory of the energy complex enterprise. The model is an optimization problem with a criterion for maximizing total profit. The limitations of the model reflect the main dependencies between the factors of production, their quality and indicators of the economic activity of the enterprise - profit and volume of output (electricity). A distinctive feature is the account of the dynamics of the enterprise. The result of the calculations is the distribution of financial resources to increase the quantity and quality of production factors. Based on the implementation of the model, an optimal development path is formed, the components of which are indicators of the number of products, profit and quality of production factors. The developed software package allows us to analyze the state of the FEC, assess development prospects, identify the main channels for distributing financial resources for development along the optimal path, and calculate their optimal distribution. The introduction of the results of the work into the management practice of one of the FEC enterprises made it possible to draw a conclusion about the appropriateness of applying the proposed mathematical methods and models for making managerial decisions and determined the ways for further research development.

Acknowledgment. The reported study was funded by RFBR, project number 19-29-07400 mk.

References

1. Vatin, N.I., Gorshkov, A.S., Nemova, D.V., Staritcyna, A.A., Tarasova, D.S.: The energy-efficient heat insulation thickness for systems of hinged ventilated facades. Adv. Mater. Res. **941–944**, 905–920 (2014). https://doi.org/10.4028/www.scientific.net/AMR.941-944.905
2. Vatin, N., Nemova, D., Khazieva, L., Chernik, D.: Distant learning course "energy efficient refurbishment management". Appl. Mech. Mater. **635–637**, 2057–2062 (2014). https://doi.org/10.4028/www.scientific.net/AMM.635-637.2057

3. Pukhkal, V., Vatin, N., Murgul, V.: Central ventilation system with heat recovery as one of the measures to upgrade energy efficiency of historic buildings. Appl. Mech. Mater. **633–634**, 1077–1081 (2014). https://doi.org/10.4028/www.scientific.net/AMM.633-634.1077
4. Harmati, N., Jakšić, Z., Vatin, N.: Energy consumption modelling via heat balance method for energy performance of a building. Procedia Eng. **117**(1), 786–794 (2015). https://doi.org/10.1016/j.proeng.2015.08.238
5. Ananin, M., Perfilyeva, N., Vedishcheva, I., Vatin, N.: Investigation of different materials usage expediency for a low-rise public building from the energy efficiency standpoint. In: IOP Conference Series: Materials Science and Engineering, vol. 365, no. 2 (2018). https://doi.org/10.1088/1757-899x/365/2/022014
6. Bondarenko, Yu.V., Sviridova, T.A., Averina, T.A.: IOP Conference Series: Materials Science and Engineering, vol. 537, p. 042045 (2019). https://doi.org/10.1088/1757-899X/537/4/042045
7. Bondarenko, Yu.V., Goroshko, I.V., Kashirina, I.L.: The task of coordinating social and economic indicators of the development of the region and the mathematical approach to its solution. J. Phys: Conf. Ser. **1203**, 012037 (2019). https://doi.org/10.1088/1742-6596/1203/1/012037
8. Gorshkov, A.S., Vatin, N.I., Rymkevich, P.P., Kydrevich, O.O.: Payback period of investments in energy saving. Mag. Civil Eng. **78**(2), 65–75 (2018). https://doi.org/10.18720/MCE.78.5
9. Gorshkov, A., Vatin, N., Nemova, D., Shabaldin, A., Melnikova, L., Kirill, P.: Using life-cycle analysis to assess energy savings delivered by building insulation. Procedia Eng. **117**(1), 1080–1089 (2015). https://doi.org/10.1016/j.proeng.2015.08.240
10. Barkalov, S.A., Averina, T.A.: Constructing a model for managing the trajectories of innovative development based on their integral characteristics. **16**(2), 82–90 (2016). https://doi.org/10.14529/ctcr160209

Integrated Assessment System Based on Dichotomous Tree

Vladimir Burkov[1,2], Irina Burkova[2,3](✉), Alla Polovinkina[2], and Lyudmila Shevchenko[2]

[1] V.A. Trapeznikov Institute of Control Sciences of Russian Academy of Sciences, Profsoyuznaya Street, 65, Moscow 117997, Russia
[2] Voronezh State Technical University, Moskovskiy Prospekt, 14, Voronezh 394026, Russia
irbur27@gmail.com
[3] Academy of Management of the Ministry of Internal Affairs of Russia, Moscow, Russia

Abstract. To assess the state of complex objects (regions, enterprises, programs, etc.), integrated assessment systems (IA) based on dichotomous trees and matrix convolution of assessment criteria are currently widely used. The dichotomous tree determines the structure of the IA-system (the order of convolution of criteria), and the matrix convolutions at the vertices of the tree determine the generalized estimates obtained as a result of the convolution of two estimates. The task of synthesizing an IA-system includes determining the structure of the system (dichotomous tree) and convolution matrices at each vertex of the tree, with the exception of the initial. The second problem is considered in the article, i.e. convolution matrix determination problems based on training options set by experts. The task is to determine (m − 1) matrices, where m – is the number of criteria so that for any training option, the score obtained on the basis of the IA-system coincides with the expert one. The problem is considered for the case of a consistent tree structure, when the criteria are added one at a time (criteria 1 and 2 are aggregated, criterion 3 is added to the resulting generalized estimate, etc.). The idea of the proposed approach consists in the sequential construction of convolution matrices, starting from the top level, the next matrix is built on the basis of the previous one. At the same time, the conditions for the existence of a matrix in the form of inequalities for generalized estimates of learning sub-options are obtained. These inequalities are associated with a graph whose vertices denote subvariants, and arcs reflect inequalities connecting generalized estimates.

Keywords: Integrated assessment system · Expert options · Dichotomous tree · Matrix convolutions · Graph

1 Introduction

The effective solution of many control problems requires the assessment of complex multi-criteria objects. It can be obtained using the integrated assessment method [1, 2]. It allows using aggregation or convolution estimates of a large number of private

V. Murgul and V. Pukhkal (Eds.): EMMFT 2019, AISC 1258, pp. 578–587, 2021.
https://doi.org/10.1007/978-3-030-57450-5_50

indicators, taking into account the degree of their influence, to obtain an integrated assessment (IA) of the object.

To use the method, it is necessary to select a set of indicators, bring them to a single point scale for evaluating, create a tree of pairwise convolutions of these indicators and develop convolution algorithms for two indicators for all vertices of this tree, except for the initial ones (which contain initial estimates of all directions). The dichotomous structure of the tree allows using for the aggregation of two vertices at the top of the top level of the logical convolution matrix.

The implementation of the integrated assessment (IA) method consists in sequentially performing certain steps. Below we give their enlarged sequence.

At the first stage, a set of indicators is formed that adequately and adequately characterizes the subject.

Further, for all indicators, a qualitative scale is formed and all indicators are evaluated according to it. The most widespread 4-point scale with ratings of 1 (poor), 2 (satisfactory), 3 (good) and 4 (excellent). The translation of quantitative assessments into qualitative ones is done by dividing the scale of possible quantitative assessments into intervals by the number of qualitative assessments and setting an assessment depending on what interval the quantitative value of the indicator falls into.

At the next step, a pairwise convolution tree of initial or already aggregated indicators (criteria) is formed, and at the upper level, an IA is obtained. Different trees can be formed for the same set of indicators - indicators can be combined in different ways in pairs, either a second aggregated estimate, or an entry-level indicator, etc. can be added to the aggregated assessment. Moreover, the appearance of the tree may affect the final IA [3–5].

For each vertex of the tree where the aggregation of criteria takes place, logical convolution matrices are formed that reflect which assessment of the aggregated criterion will correspond to each pair of estimates of the criteria of the lower level that are included in it. Moreover, after the formation of all matrices, we get not just an IA for a specific combination of estimates of the indicators of the selected object, but a tool to obtain IA for any combination of ratings. Obviously, the value of the IA depends on the formed convolution matrices (which can be formed in different ways).

Let us pay attention to the fact that each of the four stages can be implemented ambiguously - different sets of criteria, rating scales can be selected, and ratings of indicators are set differently. Features of stages 3 and 4 are discussed above, and it is these stages that affect the final result especially strongly. Therefore, for the most adequate reflection of the opinions of the person conducting the comprehensive assessment, one should be prepared to solve two important problems:

The choice of the most effective structure of the dichotomous tree IA.

Formation of adequate logical convolution matrices at the vertices of the IA-tree.

The paper considers task 2 for a given structure of the IA-tree.

2 Materials and Methods

2.1 Formulation of the Problem

So, we assume that the structure of the dichotomous tree of the integrated assessment system (IAS) is given.

We consider the case of the sequential structure shown in Fig. 1. This structure can be called a "tree branch".

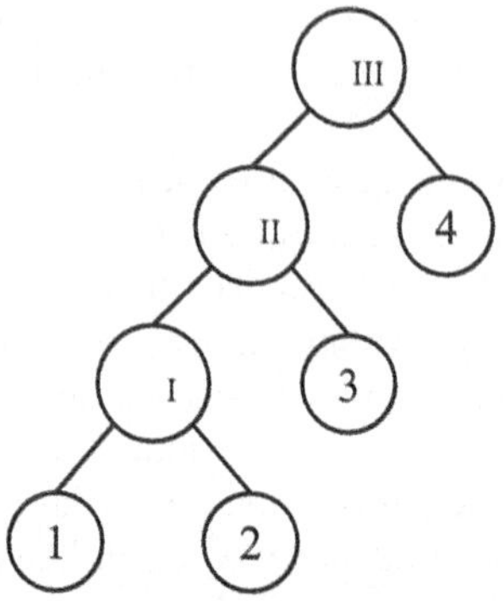

Fig. 1. «Tree branch» type structure.

The set of indicator values $\pi = (x_1, x_2, \cdots, x_m)$, will be called the option, and part of this set $\pi = (x_1, x_2, \cdots, x_q)$, where is the $q < m$ – q - subvariant.

It is required to define convolution matrices at vertices I, II, and III. We assume that learning options are given, i.e. for a certain set of options π values of indicators, experts determined complex estimates $K(\pi)$. The task is to determine $(m - 1)$ the matrices (m – s the number of evaluation criteria) such that the IA of any option π is equal to $K(\pi)$.

2.2 Solution Method

First, consider the case of a two-point rating scale and three criteria (m = 3). In this case, we need to define two matrices. The dichotomous tree for m = 3 is shown in Fig. 2.

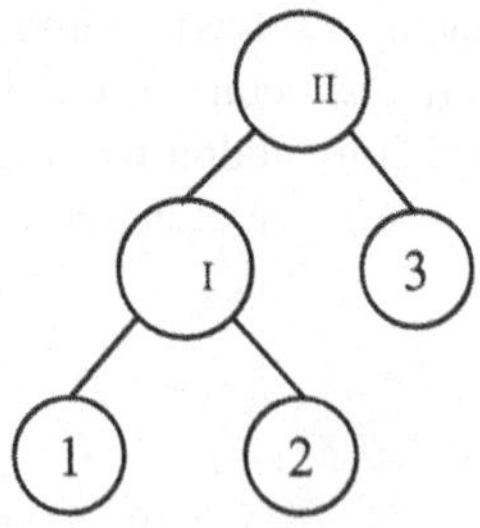

Fig. 2. Dichotomous tree for m = 3.

Note that for m = 3 and a two-point scale, there are $q = 2^3 = 8$ various options. Table 1 shows an example of expert integrated assessments of these options.

Table 1. .

IA	x	y	z
0	0	0	0
0	0	0	1
1	0	1	0
1	0	1	1
0	1	0	0
0	1	0	1
1	1	1	0
1	1	1	1

Based on Table 1, Table 2.

Table 2. .

(x, y)	$z = 0$	$z = 1$
(0, 0)	0	0
(0, 1)	1	1
(1, 0)	0	0
(1, 1)	1	1

We must assign to all possible combinations of estimates x and y generalized estimates $K(x, y)$ equal to 0 or 1. We obtain the necessary condition for the existence of a two-point IA-system. Suppose that for z = 0 there are exist pairs (x, y) such that IA = 1. In this case, the generalized estimate of these pairs $K(x, y)$ should be equal to 1. Indeed, if $K(x, y) = 0$, then IA at z = 0 cannot be 1. Moreover, for pairs (x, y), having a complex estimate of 0 (either for z = 0, either for z = 1), the generalized estimate should be equal to 0. And, finally, for all pairs having for z = 1 a complex estimate 0, the generalized estimate should be 0. It is convenient to present the obtained conditions in the form of a graph whose vertices correspond to pairs (x, y), and the arcs reflect the difference in the generalized estimates (Fig. 3).

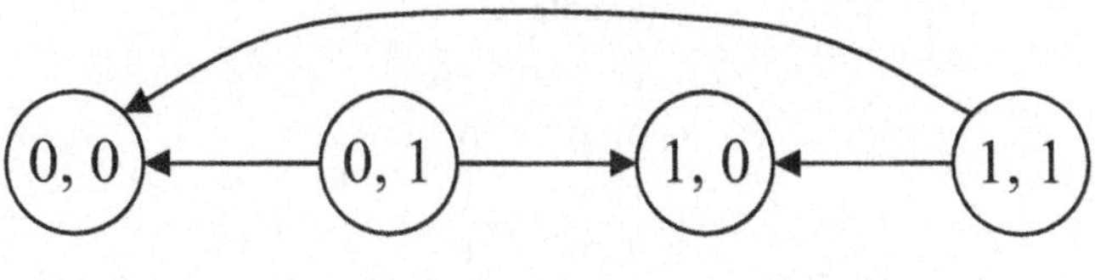

Fig. 3. Graph representation of a condition.

It follows from the graph that the generalized estimates of the pairs (0, 0) and (1, 0) are 0, and the pairs (0, 1) and (1, 1) are equal to 1. We obtain the convolution matrix I of the exponents (x, y) (Table 3).

Table 3. .

1	1	1
0	0	0
y/x	0	1

From Table 2 we obtain the convolution matrix II of the generalized indicator (x, y) and indicator z (Table 4).

Table 4. .

1	0	1
0	0	1
$z/(x, y)$	0	1

We give an example of a violation of the necessary conditions, Table 5.

Table 5. .

IA	x	y	z
0	0	0	0
0	0	0	1
0	0	1	0
0	1	0	0
1	1	1	0
1	1	0	1
1	0	1	1
1	1	1	1

Build Table 6.

Table 6. .

(x, y)	$z = 0$	$z = 1$
(0, 0)	0	0
(0, 1)	0	1
(1, 0)	0	1
(1, 1)	1	1

The corresponding graph is shown in Fig. 4.

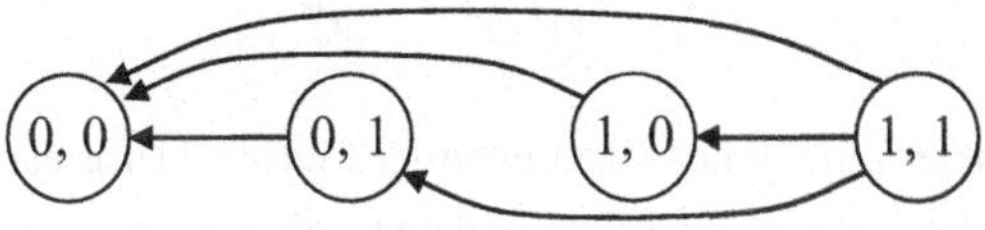

Fig. 4. Presentation of the condition in the form of a graph.

From this graph it follows that a two-point IA-system cannot be built. Indeed, the pair (0, 1) must have a generalized estimate of 1, since there is an arc [01, 00]. On the other hand, it should have a generalized estimate of 0, since there is an arc [11, 01]. A similar situation with the pair (1, 0). However, it is possible to construct an IA-system if we allow a three-point scale for the generalized estimate (x, y). It is given below. Table 7 defines matrix I, and Table 8 defines matrix II.

Table 7. .

1	1	2
0	0	1
y/x	0	1

Table 8. .

1	0	1	1
0	0	0	1
$z/(x,y)$	0	1	2

3 Results

3.1 General Case

In the general case of successive structures, there are m-indicators (criteria) and a rating scale with ni-gradations for the i-th indicator. The algorithm for solving the problem consists in sequentially obtaining generalized estimates, starting with the matrix (m − 1) of the upper level. Consider the algorithm for the matrix (m − 1).

1 step. Imagine any option

$$u_i = \left(x_i^1, x_i^2, \cdots, x_i^m\right)$$

as

$$u_i = \left(u_i^{m-1}, x_i^m\right),$$

where

$$u_i^{m-1} = \left(x_i^1, x_i^2, \cdots, x_i^{m-1}\right).$$

Our task is to determine generalized estimates $w\left(u_i^{m-1}\right)$ for each u_i^{m-1}. To do this, we build a table whose rows are u_i^{m-1}, and the columns are – x_i^m. Accordingly, the number of rows N is equal to the number u_i^{m-1}, i.e.

$$N = \prod_i n_i,$$

and the number of columns is nm. In the cells of the table we write a comprehensive assessment of the corresponding option (examples are Tables 1 and 4).

2 step. Check the necessary conditions for the existence of a solution. To do this, we consider all pairs of variants ui and uk such that $K(u_i) > K(u_k)$ and at the same time $x_i^m \leq x_i^m$. In this case, the generalized estimate $w\left(u_i^{m-1}\right)$ must necessarily be larger than the generalized estimate $w\left(u_k^{m-1}\right)$.

By analogy with the graphs in Figs. 3 and 4, we construct a graph whose vertices correspond to the subvariants ui, nd the arcs connect the two subvariants u_i^{m-1} and u_k^{m-1}, if the generalized estimate of one of them is greater than the generalized estimate of the other. The following statement holds:

Statement. If the graph has no contours, then there exists a convolution matrix A(m − 1). In this case, the minimum number of gradations of the scale is equal to the maximum length of the paths of the graph (the length of the path is equal to the number of its arcs).

The proof is obvious.

3.2 Arbitrary Structures

In the case of arbitrary structures of the IA-tree, the problem becomes significantly more complicated. We show this by the example of four indicators and the structure shown in Fig. 5.

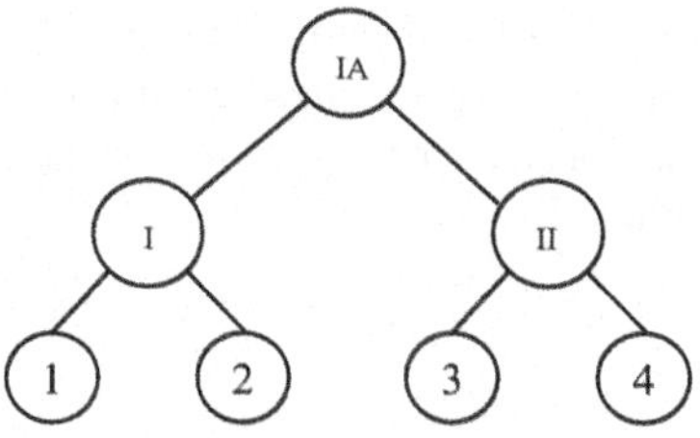

Fig. 5. Dichotomous tree option for m = 4.

The following is a fragment of a table (Table 9) similar to Tables 2 and 6. Table 9 indicates:

$$x_1 = (0,0),\ x_2 = (0,1),\ x_3 = (1,0),\ x_4 = (1,1),$$
$$y_1 = (0,0),\ y_2 = (0,1),\ y_3 = (1,0),\ y_4 = (1,1).$$

Consider a couple of options, for example, (x_1, y_4) and (x_2, y_2), with different complex estimates. Since $K(x_1, y_4) > K(x_2, y_2)$, the following conditions must be met:

$$\text{either } w(x_1) > w(x_2),\ \text{either } w(y_4) > w(y_2). \quad (1)$$

Table 9.

x_4			1	
x_3		1	0	
x_2		0		1
x_1	0			1
x/y	y_1	y_2	y_3	y_4

This type of condition is written out for any pairs of options that have different complex estimates. The resulting system of logical inequalities is a complex combinatorial problem, which currently does not have effective methods of solution. With small dimensions (the number of indicators), it can be solved by simple exhaustive search. So for the data in Table 9 we have:

$$w(x_4) > w(x_3) > w(x_2).$$

Therefore, the dimension of the rating scale should be at least 3 (a two-point scale is not enough for this task). Similarly, we have

$$w(y_4) > w(y_2) > w(y_3),$$

therefore, the IA-system can be implemented only on the basis of a three-point scale. The corresponding matrices A (I) and A (II) are given below (Tables 10 and 11).

Table 10. Matrix A(I).

1	1	2
0	0	0
1/2	0	1

Table 11. Matrix A(II).

1	0	2
0	0	1
$^{3}/_{4}$	0	1

The resulting IA-matrix has the form (Table 12):

Table 12. .

2	1	1	1
1	0	1	1
0	0	0	1
$^{12}/_{34}$	0	1	2

4 Conclusion

The method of forming IA-systems based on expert options proposed in the article is a rather effective means of forming convolution matrices for sequential structures. It is of interest to generalize this approach to arbitrary structures of a dichotomous tree. However, a number of problems arise here [6, 7]. The fact is that if in sequential structures the columns of convolution matrices are ordered by increasing ratings, then in other structures this is not always the case. This problem requires further research.

Another problem is the assessment of the minimum number of expert options sufficient to construct a full scale of generalized estimates of sub-options.

It also makes sense to conduct further research to assess the computational complexity of the proposed algorithms with the development of appropriate software.

Acknowledgement. This work was supported in part by the Russian Foundation for Basic Research (project 18-07-01258 A).

References

1. Masuin, R., Latief, Y., Zagloel, T.Y.: Development of information systems in integrated management systems in order to increase organisational performance in a construction company. IOP Conf. Ser.: Earth Environ. Sci. **258**, 012012 (2019). https://doi.org/10.1088/1755-1315/258/1/012012
2. Niu, H., Yu, F., Li, B., Chen, J.: Research on operation optimization of integrated energy system. IOP Conf. Ser.: Earth Environ. Sci. **267**, 032094 (2019). https://doi.org/10.1088/1755-1315/267/3/032094

3. Barkalov, S.A., Perevalova, O.S., Averina, T.A.: Development of the algorithm for maximizing the financial results of the investment program. In: 2018 Eleventh International Conference "Management of Large-Scale System Development, Art. no. 8551863 (2018). https://doi.org/10.1109/MLSD.2018.8551863
4. Abramov, I.L.: Systemic integrated method as the basis for high-quality planning of construction production. IOP Conf. Ser.: Earth Environ. Sci. **603**, 052077 (2019). https://doi.org/10.1088/1757-899X/603/5/052077
5. Palacios-Quiñonero, F., Rubió-Massegú, J., Rossell, J.M., Karimi, H.R.: Advanced design of integrated vibration control systems for adjacent buildings under seismic excitations. IOP Conf. Ser.: Earth Environ. Sci. **744**, 12163 (2016). https://doi.org/10.1088/1742-6596/744/1/012163
6. Li, B., Du, Z., Xie, D., Huang, L.: Research on engineering construction safety integration based on BIM and RFID. MATEC Web Conf. **246**, 03035 (2018). https://doi.org/10.1051/matecconf/201824603035
7. Wang, Y., Zhang, P.Q.: Research on the integrated design strategy of green building. MATEC Web Conf. **63**, 02039 (2016). https://doi.org/10.1051/matecconf/20166302039

Integrated Technology for Creating a Development Management Systems in the Field of Energy Saving

Vladimir Burkov[1,2], Irina Burkova[1,2,3], Tatiana Averina[2](✉), and Olga Perevalova[2]

[1] V. A. Trapeznikov Institute of Control Sciences of Russian Academy of Sciences, Profsoyuznaya Street, 65, Moscow 117997, Russia

[2] Voronezh State Technical University, Moskovskiy Prospekt, 14, Voronezh 394026, Russia

ta_averina@mail.ru

[3] Academy of Management of the Ministry of Internal Affairs of Russia, Moscow, Russia

Abstract. Energy saving and energy efficiency are one of the most important factors determining the effectiveness of activities in any industry. Raising their level requires the introduction of innovative projects in these areas. The article gives a brief description of the technology for creating a development management system created at ICS of RAS together with leading specialists in strategic management and considers the task of adapting this technology to a system for managing innovative activities in the field of energy efficiency and energy conservation. Six mathematical models are proposed that increase the efficiency of its application. The first three models describe methods for comprehensive assessment of the state of the program for the development of energy conservation and energy efficiency, taking into account the presence of multi-purpose and interdependent projects. The last three describe the methods of risk management, adjustments to the program and the formation of calendar plans for its implementation.

Keywords: Energy conservation · Energy efficiency · Development management system · Innovation

1 Introduction

Energy saving and energy efficiency are one of the most important factors determining the effectiveness of activities in any industry. Raising their level requires the introduction of innovative projects in these areas [1–5].

In particular, a number of standards have been issued by JSCo «RZD» that govern innovation. Note the standards that determine the principles for the implementation of innovative activities, the innovation management system, the rules for selecting innovative projects, as well as the basics of implementing an energy management system.

This work was partially supported by the Russian Science Foundation. Grant No. 16-19-10609.

V. Murgul and V. Pukhkal (Eds.): EMMFT 2019, AISC 1258, pp. 588–600, 2021.
https://doi.org/10.1007/978-3-030-57450-5_51

The article gives a brief description of the technology for developing a development management system (DMS), created at ICS of RAS together with leading specialists in strategic management [6–10] and discusses the task of adapting this technology to a system for managing innovative activities in the field of energy efficiency and energy conservation.

2 Materials and Methods

DMS technology description.

The DMS technology is based on three fundamental principles. The first is program-targeted management, the essence of which is the clear setting of goals and the development and implementation of a program to achieve them. The second is the project approach, the essence of which is to reliably bring the program to the final result. And finally, the third is the effective implementation of the program based on a set of smart mechanisms.

A holistic DMS, like any management system, includes the set of management mechanisms necessary for obtaining high results and efficiency, covering the full management cycle, and the organizational structure that implements these mechanisms both at each level of the control object and during interaction between them.

The full development management cycle includes:

1. Forecasting.
2. Goal setting.
3. Planning.
4. Organization of the implementation of the plan.
5. Control of the actual implementation of the plan and obtaining real results.
6. Feedback:
   - analysis of deviations "plan-fact" and forecast their consequences,
   - adjustment, if necessary, of control actions (resource allocation work plans, etc.) and, possibly, goals.
7. Reporting.

In relation to DMS, the full control cycle (ideally) should include the solution of the following tasks:

1. Preliminary goal setting (desired state in the future).
2. Development of a long-term (for 20 years) and medium-term (for 3 years) forecast, including scenario analysis and identification of a working innovative development scenario.
3. Development of a strategy for innovative development consistent with the development strategy, including the following main results of its development:
   - refinement taking into account the forecast of the ultimate development goals (changes in the values of the criteria in relation to the inertial development scenario);
   - setting a multi-level system of development goals, assessing the innovative potential of their achievement;

- determination of priorities and corresponding target programs, the required development budget;
- organizational structure of change management (development).

4. Development of a medium-term development program coordinated with a comprehensive development program (mainly in terms of priorities) for a 5-year period, consisting of a set of targeted programs for the implementation of priority areas and development projects defined in the strategy (iteratively, returning to paragraph 3).
5. Development of a comprehensive indicative plan (priority projects with approved project teams and budgets and "other") development for 3 years, including the development budget, with a return, if necessary, to paragraph 4.
6. Development and approval of the annual comprehensive development plan (quarterly).
7. Current planning (a comprehensive calendar of specific works and their performers per quarter for a month) and operational monthly management;
8. Management accounting, monthly monitoring of actually achieved results and costs.
9. The formation of a quarterly report, including an analysis of the reasons for the deviations of the "plan-fact" by the timing, results and costs and proposals (if necessary) to adjust subsequent quarterly plans.
10. Adjustment (if necessary) of the complex calendar plan-assignments for the next quarter and three quarters forward, return to paragraph 5.
    Comprehensive plan includes:
    - plan of results, i.e. changes in the values of criteria (target indicators) in dynamics;
    - a schedule for the implementation of projects and development programs;
    - development budget in the format of cash flows, including identification of sources.
11. Annual report generation.

Accordingly, when adjusting (the current re-planning from today to the future) at every moment of time, management at all levels should have a complete picture of the development process, expected results and costs for all planning periods and for the future.

The approved comprehensive calendar development work plan for the next (first) year is a quarterly task plan, binding.

Moreover, the local quarterly plan has sufficient detail for the implementation and initial accounting of planned and actual terms, costs and results.

For another two years, information is provided in the format of a less detailed comprehensive three-year plan (accurate to the stages of the targeted programs) as part of a medium-term development program. Further (for a period of up to 20 years) - in the format of a long-term strategy and/or target forecast.

At the end of the next year, the picture will continue, but with a shift of one year, i.e. continuous rolling planning.

DMS technology contains a complex of 6 mathematical models:

1. Model for a comprehensive assessment of the state in the field of energy efficiency and energy saving and a method for developing a development program that maximizes the effect for a given amount of funding.
2. The model of accounting for multi-purpose measures, that is, measures that have an effect in several areas of energy efficiency and energy conservation.
3. A model for accounting for interdependent events, that is, events whose joint inclusion in the program provides an additional (synergetic) effect (either an increase in energy efficiency and energy saving, or cost savings for the implementation of measures).
4. A risk management model in which a program is formed that provides the maximum effect at a given risk level.
5. The program adjustment model, which takes into account the additional costs that arise when using projects from the program.
6. A model for creating a schedule for a given financing schedule by the criterion of minimizing lost profits.

3 Results

3.1 Formation of a Program Based on an Integrated Assessment System

The development program usually consists of several areas.

We assume that there is a mechanism (technique) for assessing the state in each direction. Consider possible ways to assess the condition.

1. The easiest method when there is a clear quantitative state meter.
2. The expert method, when assessing the state in the direction is carried out by a group of experts (for example, experts evaluate the state in a one-hundred-scale scale).
3. The method of integral assessment when a hierarchical structure of indicators is formed.

For example, the level of energy saving and energy efficiency are divided into several indicators. This structure can be extended both down and in breadth.

At each level, an integral assessment is formed by aggregating indicators of the lower level. Aggregation methods may vary.

The simplest is a weighted linear convolution of indicators

$$y = \sum_i \lambda_i x_i,$$

where x_i – values of indicators, λ_i – indicator weight reflecting its priority.

More complex is additive convolution

$$y = \sum_i \lambda_i f_i(x_i),$$

where $f_i(x_i)$ nonlinear (usually concave) function. Recently, matrix convolutions have become popular, which will be discussed below. So, we assume that there is a mechanism for assessing the state in each of the areas.

Define the boundary levels in each direction A1, A2, A3, A4. If the state estimate is Y < A1, then this is a catastrophic (emergency) state (0). If A1 $\leq$ Y < A2, then this is an unsatisfactory condition (1), if A2 $\leq$ Y < A3, then this is a satisfactory condition (3), if Y $\geq$ A4, then this is an excellent condition (4). Boundary levels are determined expertly. Consider a system of integrated state assessment based on matrix convolutions. The structure of the system is a dichotomous tree, an example of which is shown in Fig. 1.

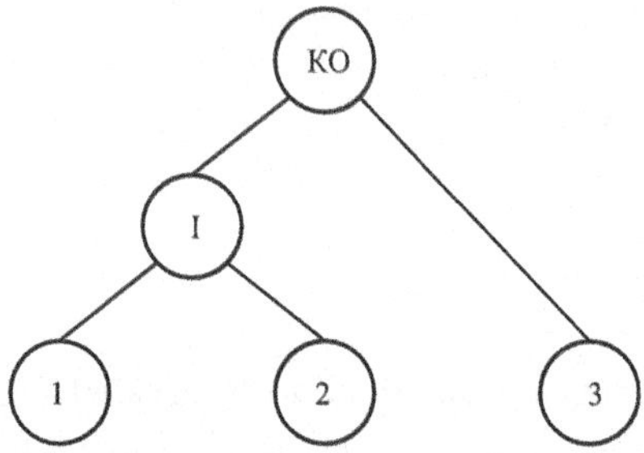

Fig. 1. An example of dichotomous tree.

First, the estimates of directions 1 and 2 are aggregated into an integral estimate, which is then aggregated with a level 3 estimate into a complex state estimate.

Aggregation is based on matrix convolutions at the vertices (I) and (KO). If the number of directions is m, then (m − 1) matrix convolutions are required. Convolution matrices reflect leadership policy, that is, priorities in the development of various areas. An example of an integrated assessment system is shown in Fig. 2. Let, for example, the estimates in the directions have the form (1, 2, 3).

4	2	3	3	4
3	2	2	3	4
2	1	2	3	3
1	1	2	2	3
1 / 2	1	2	3	4

⇨

4	2	3	4	4
3	2	2	3	4
2	1	2	2	3
1	1	1	2	3
I / Б	1	2	3	4

Fig. 2. An example of the integrated assessment system.

From the left matrix we obtain that the integral estimate is 1 (bad), and from the right matrix we obtain a complex estimate of the state - 2 (satisfactory). Having a mechanism for assessing the state, it is possible to set and solve problems of increasing a comprehensive state assessment based on the development and implementation of development programs in the field of energy conservation and energy efficiency. In accordance with the DMS development methodology, a development potential is formed for each direction, that is, a lot of projects that produce an effect (increasing the assessment of the state in the direction). Let aij be the effect and cij be the costs of implementing project i in the direction j, i = $\overline{1, n_j}$, j = 1, 2, 3, where nj is the number of projects in the j-th direction. We define for each direction the minimum costs, sij required to achieve estimates of j. To do this, we calculate the necessary increase in state estimates in the directions

$$\Delta_{\mathrm{kj}} = A_{kj} - Y_k, j = 2, 3, 4$$

where Akj – boundary estimates in the k-th direction, Yk - existing state in the k-th direction.

To determine the set of projects, the implementation of which allows us to achieve a state with an estimate of j we use the "cost-effect" information technology.

3.2 Accounting for Multi-purpose Projects

Multipurpose projects are called that give an effect in several directions (goals) at once. If the number of multi-purpose projects is not large, then you can consider all the options for joining the program of multi-purpose projects (the number of options is 2q, where q – is the number of multi-purpose projects). For each option, the task described in paragraph 2 is solved. Of all the options, the best one is selected.

If the number of multi-purpose projects is large, then the method of enumerating all the options leads to a large amount of computation. In this case, the network programming method becomes more efficient. The idea is that the costs of multi-purpose projects are divided arbitrarily into several parts according to the number of directions in which this project gives an effect. Further, the problem is solved as for the case of single-purpose projects. Here are the main theorems from the theory of network programming [3].

Theorem 1. Solving a problem with single-purpose projects gives a lower estimate of the costs for the original problem. The definition of the division of costs of multi-purpose projects, so that the lower estimate is maximum, is called the generalized dual task (GDT).

Theorem 2. GDT is a convex programming task. The lower bound obtained can be used in the branch and bound method.

3.3 Accounting for Interdependent Projects

Interdependent projects are those whose inclusion in the program gives an effect greater than the sum of the effects of individual projects (the so-called synergistic effect). In this case, the "cost-effect" method is not applicable. We define a graph of interdependencies. The vertices of the graph correspond to the designs. Two vertices i, j are connected by an edge if the corresponding projects are interdependent. The length of the edge is equal to the additional effect b_{ij}. In Fig. 3 is shows an example of a dependency graph for five projects.

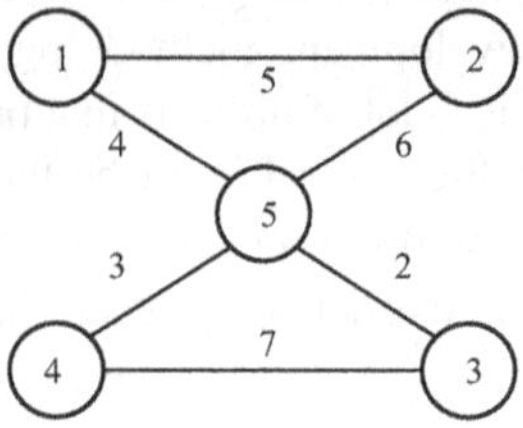

Fig. 3. An example of interdependence graph for five projects.

We first consider a special case when the interdependence graph is a pair combination (a graph in which no edges have a common vertex). In this case, the dichotomous programming method can be applied to solve the problem. The structure of the dichotomous representation of the problem is chosen so that the pairs of interdependent projects are considered first.

Consider the general case when the graph is not a vapor combination.

Definition. A subgraph - matching (P-subgraph) is a subgraph of a graph that is a matching.

Note that the concept of a P-subgraph differs from the well-known concept of matching of a graph, which is understood as a subset of the edges of the graph, none of which have two common vertices. In Fig. 4 shows the graph (Fig. 4a), matching of the graph (Fig. 4b) and the P-subgraph (Fig. 4c).

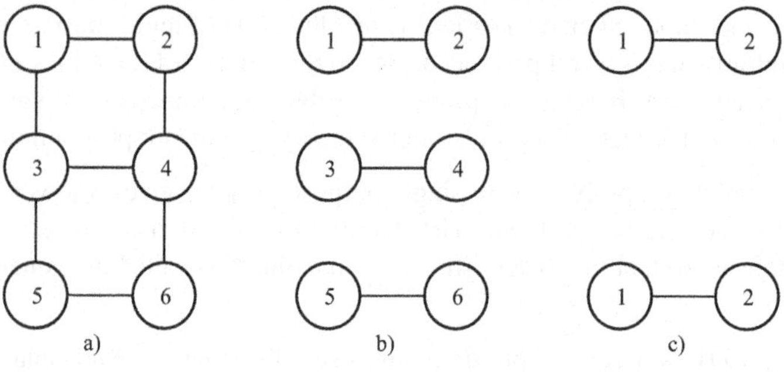

Fig. 4. The graph, matching of the graph and the P-subgraph.

The idea of solving the problem in the general case is to remove a set of vertices from the graph and obtain a P-subgraph. Next, we consider all the options for entering remote projects into the program and solve for each option the problem for the matching case discussed above. Of all the options, we determine the best. If the number q of deleted vertices is small (the number of options is 2q), then the method has a fairly effective. Naturally, it is desirable to minimize the number of deleted vertices and, therefore, maximize the number of vertices of the P-subgraph.

Task. Determine the maximum P-subgraph, that is, the P-subgraph with the maximum number of vertices (arcs).

The task is a complex task of discrete optimization. Consider a simple heuristic algorithm for solving it.

Algorithm description

1. step. For each edge (i, j) of the graph, we determine the number of vertices nij, with which the boundary vertices of the edge are adjacent.
2. step. We include an edge with minimal nij in the P-subgraph.
3. step. We remove the edge included in the P-subgraph along with adjacent vertices.
4. step. Repeat step 1 for the remaining graph..

Example. Consider the graph in Fig. 5.

It is easy to see that for the edge (1, 2) n12 = 3 (similarly for the edges (6, 7) and (8, 9). After including the edges (1, 2) in the P-subgraph and excluding the vertices 3, 4 and 5 we get the graph of Fig. 6.

Now the minimal nij has an edge (6, 7). Excluding the vertices 6, 7, 10, we get the edge (8, 9). We get a P-subgraph consisting of vertices 1, 2, 6, 7, 8, 9. The vertices 3, 4, 5, 10 are excluded. Thus, it is necessary to consider 24 = 16 options.

3.4 Management of Risks

Risk management includes such basic processes as risk identification, determination of their main characteristics, choice of ways to respond to risks (reduction, transfer, avoidance, acceptance). The risk is described by two indicators - the probability of a risky event and the damage upon its occurrence.

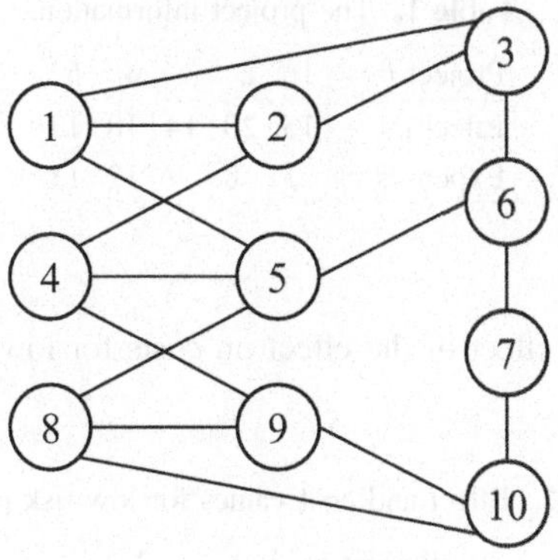

Fig. 5. An original graph.

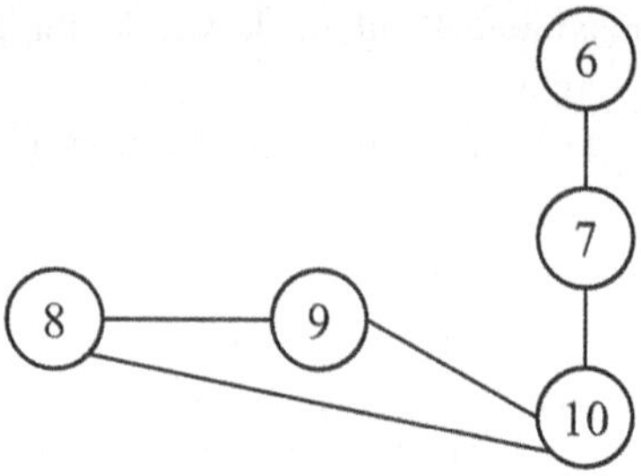

Fig. 6. The resulting graph.

A generalizing characteristic is the degree of influence (or rank) of risk, defined as the expected damage. Recently, a great deal of attention has been attracted to the study of risk management problems based on qualitative estimates of their characteristics. The fact is that in practice, as a rule, it is precisely qualitative risk assessments that are used. The simplest is a two-grade scale (low probability, high probability; small damage, large damage; low degree of influence, high degree of influence). This is understandable, since the project is, by definition, a unique event, which does not allow fully relying on statistical data. One of the methods used to reduce the risk of the program is to limit the financing of high-risk projects. Let us describe a modification of the "cost-effect" method, taking into account the presence of high-risk projects. We will describe the method using the data below. Let projects 1, 2 and 3 are high-risk. Let us accept the restriction on financing high-risk projects Rв = 12. Table 1 shows the project information.

Table 1. The project information.

Project i	1	2	3	4	5
Effect a_i	15	20	14	18	13
Expenses c_i	5	8	7	12	13

1 step. We build the effect of the effect on costs for low-risk projects (Table 2).

Table 2. Effect and cost values for low-risk projects.

Project i	0	1	2
Effect a_i	0	12	25
Expenses c_i	0	18	31

2 step. Build cost-benefit effects for high-risk projects.

Table 3. Effect and costs values for high-risk projects.

Project i	0	1	2	3
Effect a_i	0	5	8	12
Expenses c_i	0	15	20	29

3 step. We form the final table.

Table 4. Results.

2	25;31	30;46	33;51	37;60
1	12;18	17;33	20;38	24;47
0	0	5;15	8;20	12;29
H	0	1	2	3
B				

We calculate at target settings Δ12 = 18, Δ13 = 37, Δ14 = 53 (Table 3).

$$s_{12} = 8, s_{13} = 20, s_{14} = 37.$$

To achieve score 2, a high-risk project 2 is included in the program, to achieve score 3 also a high-risk project 2, to achieve score 4 a high-risk project 2 and project 4 (Table 4).

3.5 Program Composition Adjustment

The operational management tasks associated with adjusting the composition of the program arise for many reasons. Firstly, in the case of a reduction in the amount of funding for the program. Secondly, in the case of the emergence of new highly effective projects, a decrease in the effects of projects included in the program, an increase in project rice, etc. A feature of the tasks of adjusting the composition of programs is the fact that when a project is excluded from the program, additional costs arise associated with the termination of contracts, payment of compensation to contractors, etc. With a large value of these costs, it is more profitable to leave the project in the program than to exclude it. Denote l_i – additional costs when project i is excluded from the program, Q – many new projects, P – many old projects.

Denote $x_i = 1$, if project i is included in the program, $x_i = 0$, otherwise. The cost limit will look like

$$\sum_{i \in Q} c_i x_i + \sum_{i \in P} (c_i x_i + l_i(1 - x_i)) \leq \mathrm{R}$$

or

$$\sum_{i \in Q} c_i x_i + \sum_{i \in P} (c_i - l_i) x_i \leq \mathrm{R} - \sum_{i \in P} l_i$$

Example 9. Take the project data from model 4. Let projects 1, 2, 3 be new, and projects 4, 5 old. We take $l_4 = 7$, $l_5 = 6$. The ordering of projects by efficiency has the form

$$4 \rightarrow 1 \rightarrow 2 \rightarrow 3 \rightarrow 5$$

Let R = 31. Using the "cost-effect" method, we get that projects 4, 1 and 2 are included in the program, that is, project 4 remains in the program, although there is a more effective project 3.

3.6 The Formation of Calendar Plans

When the composition of the projects of the program is formed, it is necessary to develop a schedule for the implementation of the program with a given funding schedule. The criterion is the amount of lost profit $\mathrm{F} = \sum_i a_i\, t_i$, where t_i – is the moment of completion of the i-th project. The task is a difficult optimization problem that does not have effective exact algorithms for solving.

Consider a heuristic algorithm based on project priority rules. There are two priority rules. The first rule is $q_i = \frac{a_i}{c_i}$, that is, the cost-effectiveness of the project. The second rule $p_i = \frac{a_i}{\tau_i}$, τ_i - the duration of the i-th project, characterizes the effectiveness of the project in time. The first rule gives a solution that is close to optimal for the case when the program is implemented and funded by periods, and each project can be fully implemented during this period. The second rule gives an optimal solution when projects are carried out sequentially. In the general case, we take a convex linear combination of these rules

$$r_i(\alpha) = \alpha\, q_i + (1 - \alpha) p_i, \text{ where } 0 \leq \alpha \leq 1$$

Solving the problem for various values of α, we choose the best solution.

4 Discussions

The integrated development management technology considered in the article can certainly be adapted to the development of development management systems for organizations of any type (enterprises, regions, etc.). Of course, the proposed set of six optimization models does not cover the entire development management cycle. It is desirable to supplement it with models and mechanisms of forecasting, control, stimulation, planning and others. Moreover, planning models should take into account the duration of the planning period (long-term, medium-term, annual plan).

We also note that large development programs affect the interests of many participants, and these interests must be taken into account. In other words, we are talking about developing models of optimal coordinated planning, i.e. models for the formation of plans taking into account the interests of participants. Otherwise, it is possible to receive false information, as well as non-fulfillment or poor-quality implementation of plans.

5 Conclusion

The article discusses the integrated technology for the development of development management systems in the field of energy conservation and energy efficiency. Its feature is a complex of six optimization models, the use of which allows for the effective selection of projects, taking into account their multi-purpose nature and the presence of a synergistic effect. As noted above, further research can be carried out in the direction of expanding the composition of models to cover the full development management cycle, as well as improving the proposed models (taking into account stakeholders, the duration of planning periods, etc.). In the future, we are talking about creating a software product that allows you to get options for development programs when applying the required information to the input.

References

1. Gorshkov, A., Vatin, N., Nemova, D., Shabaldin, A., Melnikova, L., Kirill, P.: Using life-cycle analysis to assess energy savings delivered by building insulation. Proc. Eng. **117**(1), 1080–1089 (2015). https://doi.org/10.1016/j.proeng.2015.08.240
2. Gorshkov, A.S., Vatin, N.I., Rymkevich, P.P., Kydrevich, O.O.: Payback period of investments in energy saving. Mag. Civil Eng. **78**(2), 65–75 (2018). https://doi.org/10.18720/MCE.78.5
3. Harmati, N., Jakšić, Z., Vatin, N.: Energy consumption modelling via heat balance method for energy performance of a building. Proc. Eng. **117**(1), 786–794 (2015). https://doi.org/10.1016/j.proeng.2015.08.238
4. Barkalov, S.A., Perevalova, O.S., Averina, T.A.: Development of the Algorithm for Maximizing the Financial Results of the Investment Program. Art. no. 8551863 (2018). https://doi.org/10.1109/mlsd.2018.8551863
5. Bondarenko, Yu.V., Sviridova, A., Averina, T.A.: IOP Conference Series: Materials Science and Engineering, vol. 537, p. 042045 (2019). https://doi.org/10.1088/1757-899X/537/4/042045
6. Bondarenko, YuV, Goroshko, I.V., Kashirina, I.L.: The task of coordinating social and economic indicators of the development of the region and the mathematical approach to its solution. J. Phys: Conf. Ser. **1203**, 012037 (2019). https://doi.org/10.1088/1742-6596/1203/1/012037
7. Novikov, D.A.: Complex models of system optimization for the production and economic activity of an enterprise. Autom. Remote Control **80**(11), 2068–2089 (2019). https://doi.org/10.1134/S0005117919110109

8. Mak, S.T.: Improvement of service reliability and optimization of energy delivery (asset management). In: New Technologies for Smart Grid Operation, Chapter 5, pp. 5–1 to 5–22 (2015). https://doi.org/10.1088/978-0-7503-1158-8ch5
9. Kravets, O.J., Podvalny, E.S., Barkalov, S.A.: Quality assessment of a multistage process in the case of continuous response functions from resource influences. Autom. Remote Control **76**(3), 500–506 (2015). https://doi.org/10.1134/S0005117915030145
10. Avdeeva, E., Davydova, T., Skripnikova, N., Kochetova, L.: Human resource development in the implementation of the concept of "Smart Cities". In: E3S Web of Conferences, vol. 110, p. 02139 (2019). https://doi.org/10.1051/e3sconf/201911002139

Development of Engineering Services in the Implementation of Investment-and-Construction Projects

Irina Vladimirova(✉), Kseniia Bareshenkova, Galina Kallaur, and Anna Tsygankova

Plekhanov Russian University of Economics, Stremyanny lane, 36, Moscow 117997, Russia
kaf-stroy@yandex.ru

Abstract. The advanced market of engineering services is essentially important for a shift to the age of industry 4.0, since it determines the potential capacity of a country for creating science-driven productions, improving technologies under conditions of digital transformation, increasing the existing capacities through implementation of large high-tech investment-and-construction projects. Within the study, the stage of development of engineering in Russia was assessed as a method of promoting innovations applied within the process of elaboration and implementation of investment-and-construction projects under digital transformation of the economy. The study involved analysis, comparison, generalization, grouping and modeling methods. Legislative acts of the Russian federation were used, as well as regulatory documents on activities in construction engineering, statistical and analytical proceedings and documentation, materials from scientific conferences, official publications on problems of engineering development when implementing investment-and-construction projects. During the study, some external, legal and technological factors that restrain the development of the engineering services market were revealed and some ways to neutralize them were found. The following factors of technological revolution were identified: transition to digital economy and thereby transition from investment-and-construction engineering to digital engineering in the framework of investment-and-construction activities. Processes of digital engineering were investigated in terms of projects implemented by ViPS Group of Companies and their general contractor ZAO Strabag. The main advantages of digital engineering were analyzed when implementing investment-and-construction projects in Russia.

Keywords: Engineering · Investment-and-construction project · Digitalization of economy · Information modeling · Engineering service market · Digital engineering

V. Murgul and V. Pukhkal (Eds.): EMMFT 2019, AISC 1258, pp. 601–615, 2021.
https://doi.org/10.1007/978-3-030-57450-5_52

1 Introduction

Challenging tasks on modernization of production plants that Russian economy has to deal with and, particularly, the focus on digital transformation and transition to the age of industry 4.0, explain the demand for designing and creating new innovative products, improving technologies, expanding the existing capacities. This, in turn, entails the implementation of large high-tech investment-and-construction projects (ICP) of production facilities or projects of their modernization [1]. Currently the following trends are observed, when implementing large ICPs in Russia: an increase in the technological complexity of design solutions and practices for implementing ICP; increase in the information asymmetry between a purchaser and contractors at both design and construction stages; rise in a number of deadline failures and budget overruns when implementing a large ICP; the growing shortage of project management experts in most industries. Therefore, a professional skill of managing large ICPs is becoming more and more necessary for a customer-developer. These conditions reveal the demand in elaborating certain measures on state support and incenting of the engineering services market.

According to the State Standard GOST R 57306–2016, engineering is defined as the engineering and consulting activity, which implies the solution of engineering problems associated with producing or advancing products, systems or processes.

According to the State Standard GOST R 58179–2018, engineering in the framework of implementing investment-and-construction projects implies engineering and consulting services provided by consulting engineers in the construction sector or by engineering companies according to a contract with a purchaser. The ultimate goal of these services is to gain the best possible outcome of capital investments or other expenditures in construction. Throughout the whole life cycle of a capital object, engineering companies work out new processes and management systems, amend existing business processes in order to improve management of investment-and-construction projects, monitor the implementation of organizational-and-technical, managerial, financial-and-economic, information models of systems, objects and processes in accordance with goals set by a purchaser.

The present study is aimed to analyze and assess the level of engineering development in Russia as of a method of promoting innovations when developing and implementing an ICP in the context of the digital transformation of the economy.

2 Materials and Methods

Analysis of the global market for engineering services predicts its growth by an average of 3.95% in the period 2018–2022. The marketing research showed that in 2017 the global market of engineering services was valued at 1,022.9 billion dollars. Western Europe has become the largest geographical area, accounting for 312.9 billion dollars, or 30.5% of the global market. The United States accounted for 218.6 billion dollars, or 21.3% of the global market of engineering services. According to the Federal State Statistics Service "Rosstat", the share of Russia amounted to 2.5% in the same period.

Today, Russia observes a tendency for the increase in exporting engineering services. However, this market segment is considered small and insignificant, since there are few such engineering companies. Therefore, statistical indicators are very sensitive to large projects. The current growth of export is associated with the implementation of several large industry projects, mainly foreign ones that involve design and construction stages in the mining industry, nuclear energy, etc. When the projects are completed, a sharp decrease in the volume of exports may be observed if no new projects are started. Comparative analysis of import and export of engineering services shown in Table 1 indicated that import is also increasing during recent years, but the rate is lower than export's.

Table 1. Trade in engineering services with foreign countries, according to Rosstat.

Reporting period	Export		Import	
	Amount of agreements, items	Cost of agreement subject, mln $	Amount of agreements, items	Cost of agreement subject, mln $
2016 г.	771	26 452,4	1667	10 671,9
2017 г.	1036	25 068	2133	14 475
2018 г.	1030	30 932	2351	12 941

According to the Institute of Statistical Studies and Economics of Knowledge (ISSEK) of the Higher School of Economics, in 2016 the Russian market for engineering services valued 97.8 billion rubles. Analysis of amount of engineering services provided by types of economic activities performed by purchasing organizations in Russia allowed compiling a rating chart for these activities, which is shown in Fig. 1.

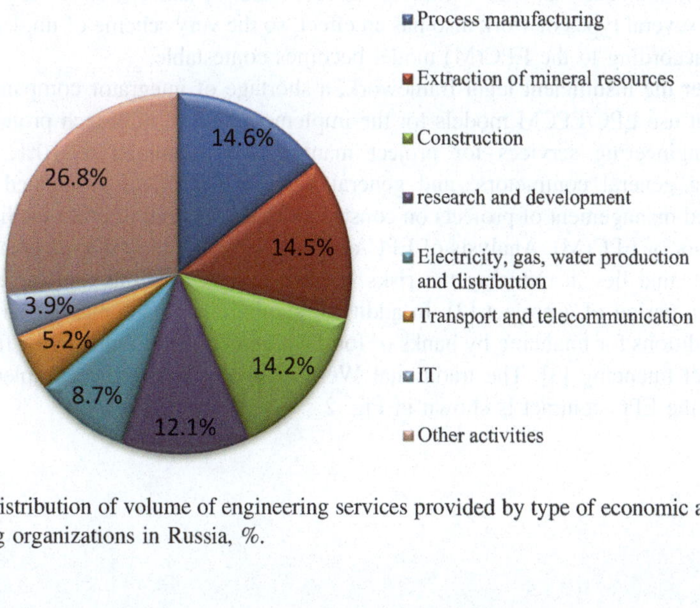

Fig. 1. Distribution of volume of engineering services provided by type of economic activity of purchasing organizations in Russia, %.

It is worth mentioning that the development level of engineering within an industry accords with a state of this industry, which proves that the role of engineering is to conduct innovations. Thus, process manufacturing sector and extractive industries provide the largest amount of engineering services: 14.6% and 14.5% respectively; 14.2% is for construction sector. Research and development sector rests on engineering services in the amount of 12.1%. Meanwhile, such high-tech activities as production and distribution of electricity, gas and water (8.7%), transport and telecommunications (5.2%), IT activities (3.9%) turned out to be quite humble consumers of engineering services.

Some theoretical and field studies allowed defining and systematizing the factors restraining the development of the Russian market of engineering services.

The low pace of development of the engineering services market in Russia is reasoned by the fact that its formation and development started only in the 2000s. At the same time, experience of the Soviet Union has been still greatly influencing the economy entailing contradictions global trends both in terms of organizing business processes and in scientific activities.

External factors include poor investments made in fixed assets due to the economy stagnation. According to the Federal State Statistics Service Rosstat, in 2018, investments in fixed capital at constant prices were estimated to be 73.6% of the level of the Russian SFSR in 1990. The share of complex infrastructure projects is declining; in 2017, state investments in infrastructure projects fell by almost a quarter compared to 2014, which does not help attract engineering companies to the investment-and-construction process.

Legal factors were identified. The lack in a legal framework obstructs using EPC-contracts when implementing complex investment-and-construction projects in the Russian Federation: these contracts are not enshrined by law so do not have legal force. In fact, there is a complex contract agreement that is adapted to EPC(M)-contract. However, its preparation and maintenance is often associated with additional costs. Moreover, the fact that one contractor is restricted by antimonopoly legislation to perform several types of work also has an effect, so the very scheme of implementing a project according to the EPC(M) model becomes contestable.

Under the insufficient legal framework, a shortage of integrator companies developed that use EPC/EPCM models for the implementation of high-tech projects. As of 2016, engineering services for project management amounted to 10%, which is including general contractors' and general designers' services associated with the integrated management of projects on construction of industrial or other facilities in the framework of EPC(M). Analysis of EPC-contracts revealed their main advantage for a purchaser that lies in reducing the risks of both budget overruns and increasing the execution period of a project [2]. In addition, signing an EPC-contract guaranties the best conditions for financing by banks or for development institutions in the framework of project financing [3]. The traditional Western structure of project implementation through the EPC-contract is shown in Fig. 2.

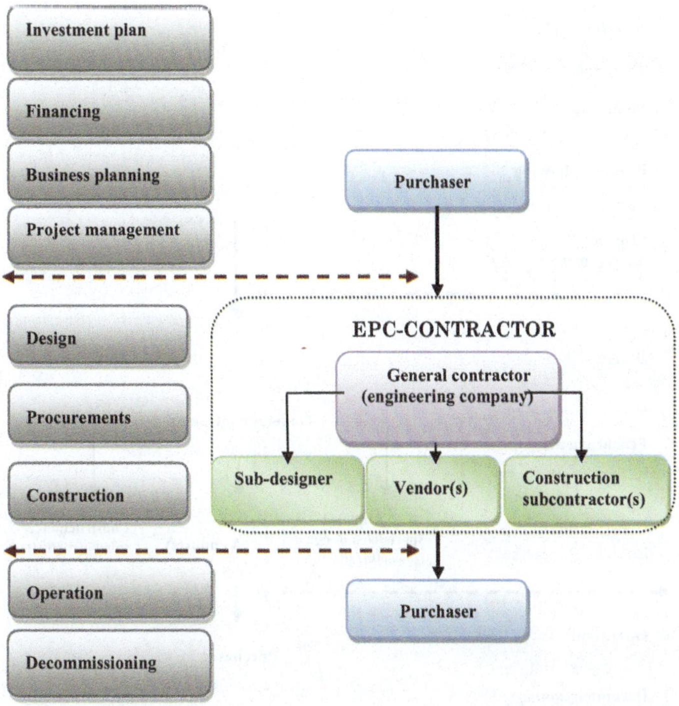

Fig. 2. The traditional Western model for the implementation of a project according to an EPC-contract.

When implementing large engineering projects in Russia, the functions of an EPC-contractor are usually performed by an affiliated, specialized design organization (Fig. 3). To date, most of specialized design institutes that were founded in the USSR have become the property of large companies implementing engineering projects. These design organizations either possess their own basic technology or take a third-party, mainly foreign technology and adapt it to Russian conditions.

This situation leads to the fact that an investment company has to accumulate project management competencies within itself instead of a project executor. Under this scheme, third-party contractors do not gain the experience that is so significant for engineering development, while the competencies of customers are constantly growing.

Technological factors that negatively affect the development of the Russian engineering services market are primarily determined by the strong dependence of projects and industries on foreign technologies. Despite an active policy of import substitution, there is a part of technologically complex equipment that Russia does not produce. Another interesting fact is that import substitution in Russia does not always ensure low prices for production, since domestic equipment often turns out to be more expensive than imported one.

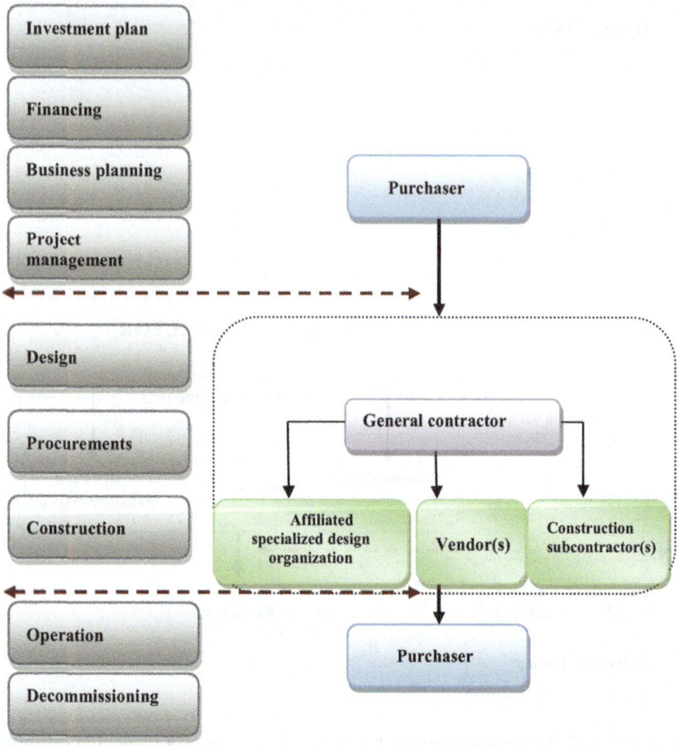

Fig. 3. The scheme for the implementation of engineering projects in Russia.

The problem of shortcomings in the domestic regulatory-and-technical base is also one of the factors that restrain the development of competitiveness of the engineering services market. The existing standards of industrial construction in Russia were primarily worked out in the 1960–1970s and now they significantly differ from those used in the rest of the world. Excessive requirements in the form of standards radically change the geometry of a project, which entails cost overruns at the construction stage, increased energy consumption at the operational stage, and higher expenses of the whole life cycle of a facility.

The final part of the research involved the study on new factors of the technological revolution implying the transition to a digital economy. In the context of the digital transformation of the economy, investment-and-construction engineering is becoming essential in terms of work organization, management and control during the life cycle of a capital facility that involves information modeling technologies [4]. Digital engineering may be considered as the combination of information modeling and making engineering and management decisions [5]. The research of digital engineering processes was performed using as the example some projects implemented by engineering organization “ViPS Group of companies” and their prime contractor STRABAG Russian branch office.

3 Results

The processes of digital engineering when bringing to life investment-and-construction projects implemented by ViPS Group of Companies were considered at the pre-investment stage of the life cycle of a permanent industrial facility. The technological audit of the project via building information modeling (BIM) of an industrial facility object showed that changing one of the components automatically changes all the regarding parameters: graphical images, specifications, work schedule, project budget, implementation deadlines. Three options for technological solutions were elaborated for the enterprise. Based on the BIM model and taking into account the existing infrastructure and the results of mathematical modeling performed with the use of Anylogic simulation software, a favorable option for a production line was chosen: option 2 that considers equipment loading and production output Fig. 4, Table 2.

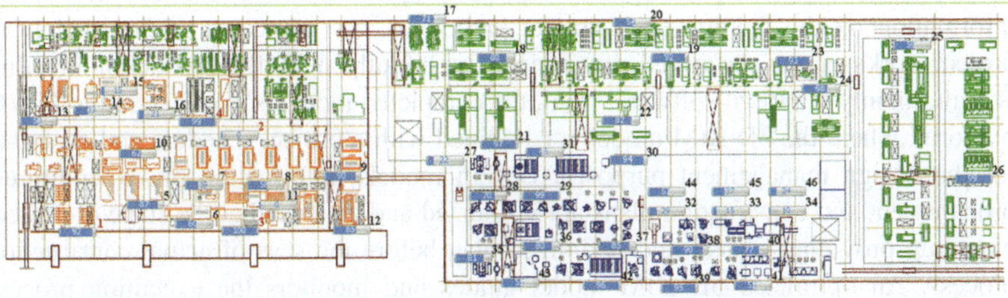

Fig. 4. Calculation of favorable option of production line via Anylogic simulation software.

Table 2. Output data when choosing the second option of technical solution for implementation of construction project of industrial facility designed to manufacture products of model A.

Product model	Totally manufactured, product A, items (%)	Average annual production
A-14	399 (88%)	133
A-15	162 (108%)	54
A-16	160 (106%)	53
A-17	782 (104%)	260
	1503 (100%)	

The use of digital engineering was also studied using the example of Russian branch office of STRABAG. ZAO Strabag, the Russian subsidiary company of the concern, is successfully implementing a number of projects on residential, industrial and civil construction in the central region of Russia. Since Russian market lacks large-scale infrastructure projects that are structured and financed according to the Western European model (EPC/EPCM), the company does not place a priority on such projects, yet it has certain competencies to implement them.

STRABAG started working on the idea of a "digital construction site" in the late nineties. Currently, a BIM.5D information modeling system has been created and implemented for working on a number of ISPs. 5D is considered by the companies as geometric model (3D) + time (4D) + process data (5D).

The use of information modeling in the company is carried out throughout the whole life cycle of a project. The maximum effect from using BIM can be obtained in projects, which were under BIM application from the earliest stages: design concept, feasibility study (working out options of project implementation). A model includes all information on the facility, each project participant has an access to the model, all project works can be performed simultaneously. Initially, a 3D BIM model of a facility is built, mainly via Revit software. Next, the completed model is exported to the ITWO platform, where the 3D BIM model is combined with a project schedule and cost data for the design management purposes. Design models, information on expenditures, procurement and construction plans are visualized and evaluated at the stage of project planning, providing a clear and optimized plan before the start of actual construction process. An optimized 5D BIM model guides and monitors the execution process during the stage of construction of a facility.

Work on a construction site is performed through remote access, which is obtained by the company with the use of remote servers, Revizto software and a tablet computer. According to the requirements of the concern, Strabag does not use open cloud servers; the work is managed from the local servers via telecommunication. Moreover, the company is currently working out its own cloud server. Revizto software serves as a tool for setting and tracking tasks online during simultaneous work on BIM projects. This software allows working with a most recent version of a project. Project participants monitor tasks in progress receiving instant notifications on any changes. Revizto software significantly accelerates and improves the process of field supervision: a task is created in Revizto just at a construction site, then it is instantly transferred via remote access to the office, where the changes are automatically displayed in detail design documents, particularly in drawings of a Revit model Fig. 5.

The use of Revizto software improves the quality of complaint management, especially in such a case when a contractor claims for payment for work beyond a contract, since the system is transparent so one can see the volume of work performed over time.

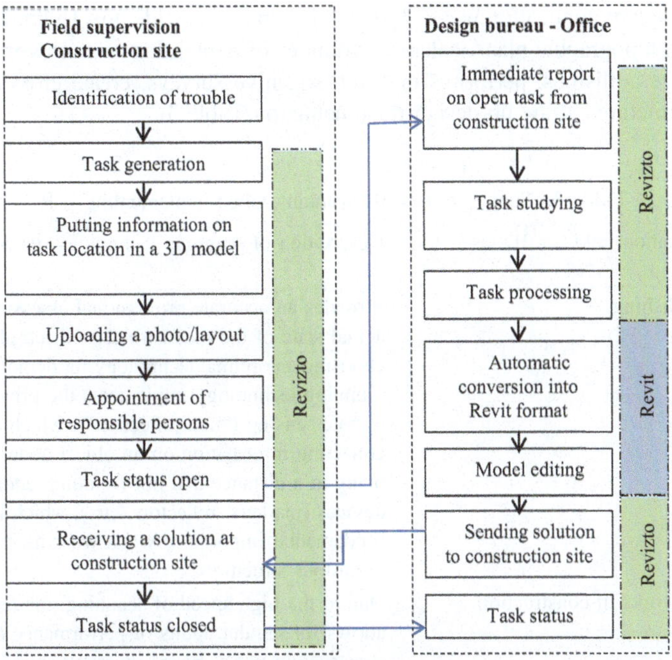

Fig. 5. Interaction of field supervision and design bureau in Revizto software.

BIM.5D system can also be used for the operation of facilities: visual navigation in the model allows accessing all types of saved and related data. To use BIM.5D comprehensively when managing an object during its operation, one should first compare a real object with a digital model. Further a facility can be reconstructed, demolished or disposed based on saved information models [6].

ZAO "Strabag" actively applies 3D laser scanning technology for information modeling of objects. Laser scanning is a technology that allows obtaining a maximum accurate three-dimensional image of a facility in a shortest time. Time spent for performing geodetic measurements via 3D laser scanner is several times lower than time spent for classical measurements via a total station theodolite or a level. When performing laser scanning, a laser 3D scanner Leica ScanStation P40 is used, which allows shooting at a distance of up to 270 m from an object with a measurement speed of up to 1 million points per second and measurement accuracy of up to 1 mm. 3D laser scanning technology consists in the following: a scanner measures all objects coming in its sight; scanning is performed from several sites; distance to an object, vertical and horizontal angles are measured, so finally spatial coordinates of points are found.

The result of a survey is a point cloud, which is an array of points, each of which has individual coordinates. These are usually millions or billions of points in one project, depending on its volume. Obtained scans are stitched together via Leica Cyclon software comprising a point cloud, which is further used as the basis for BIM.5D.

The range of application ways of 3D laser scanning is wide: this technology is used to perform topographic plans, calculate volumes of demolition or earthworks, monitor and analyze collisions, perform front and executive surveys, create cross-sections of existing structures, draw up design documentation Table 3.

Table 3. Scope of laser 3D scanning when implementing ICP.

№	Application field for 3D laser scanning	Description of scanning technology, its advantages
1.	Demolishing	Provides an accurate professional assessment of the actual state of the reconstructed object, allows choosing an optimal technology for demolishing. 3D scanning technology implements the principle of remote sensing ("non-reflection"), which allows collecting information on an object under survey, being at a distance and not installing additional devices (markers, reflectors, etc.), which is of fundamental importance when working on hazardous structures
2.	Civil works in construction	Due to the high speed of scanning, the technology allows for regular audits of performance of construction work. Based on results of scanning, measurements and as-built drawing are made without a need for a second site visiting
3.	Façade preparation	Based on the point cloud obtained when shooting an object under construction, when post-processing in the office, it is possible to plot 2D facade drawings without re-visiting an object. In addition, the density and completeness of shooting, as well as panoramic photography, give a detailed image of all facade elements without a need for an outline
4.	3D scanning of existing facilities and utilities	The use of a 3D scanner in the analysis of existing buildings and utilities significantly reduces the shooting time and labor costs. Using a resulting cloud, one can immediately build cross-sections; the completeness and density of points of these cross-sections allows seeing the entire object and its elements, which distinguishes 3D scanning from traditional shooting methods. Information models obtained on the basis of 3D scanning can be used to control a facility during the operation stage
5.	Local scanning of completed structures with maximum detailing	The technology has no alternatives when it is necessary to obtain an executive survey of a completed construction with a high density. For example, to create an executive geodetic diagram of a foundation slab for calculating the volume and making a polymer floor screed with an accuracy of 2 mm

4 Discussion

Goals and prospects of engineering in the context of digitalization should be clearly defined at the level of state policy [6]. In 2014, in order to develop engineering activities in Russia, the Ministry of Industry and Trade of the Russian Federation approved an Action Plan ("Road Map") in the field of engineering and industrial design. According to this Action Plan, the growth of the Russian engineering services market was supposed to be 8% per year excluding inflation, while the global engineering market annually grew averagely by 4%. The advance growth of engineering in Russia is reasoned by the low base effect: the current share of engineering in Russia's GDP is 2%, while in advanced countries this index can be up to 20%. However, according to expert estimates, the growth rate in developing countries should be at least 12%, otherwise the technological gap between advanced and developing countries will be unbridgeable.

One of the key aims of Road Map is to create an integrated monitoring system for the engineering services market in order to make reasoned management decisions, while the market is under state regulation and determination of support priorities. By the order of the Ministry of Industry and Trade of Russia NRU Higher School of Economics, a list definition of engineering services was elaborated for statistical monitoring of individual types such services. Distribution of engineering services by types as of 2016 is presented below, % (Table 4).

Table 4. Distribution of engineering services by types as of 2016, %.

№	Type of engineering service	Percentage of total provided services
1.	Engineering design	83.1
1.1	Engineering design of products	50.1
1.2.	Engineering design of technological (production) processes	5.5
1.3.	Engineering design of capital construction facilities	44.4
2.	Engineering consultations (not related to specific engineering design projects)	0.1
3.	Project management	10.2
4.	Other engineering services	6.8

It is worth noting that the engineering services market has not been monitored after 2016, which complicates the assessment of the effect of implementing measures of Road Map.

In order to increase the rate of investment in fixed assets, the Government of the Russian Federation is going to stimulate the country's economy growth through the implementation of National projects, particularly the Comprehensive Plan for the Modernization and Increasing of Main Infrastructure until 2024, which was developed in accordance with the Decree of the President of Russia of May 7, 2018 №204 "On national goals and strategic objectives of the development of the Russian Federation for the period until 2024". According to the Comprehensive Plan, the total amount of financing of transport infrastructure projects in the period from 2019 to 2024 should amount to 6.3 trillion rubles, 3 trillion of which are federal budget funds and the rest are from non-budgetary sources. The technological and organizational complexity of these projects implies the involvement of engineering companies, which in turn would contribute to the development of market of engineering services provided by domestic companies.

However, according to the results of a winter half of the year, expenses for national projects noticeably lag behind the planned ones. According to the report of Accounts Chamber of Russia, as of July 1, 2019, expenses for the implementation of national projects and Comprehensive Plan for the modernization and expansion of the main infrastructure amounted 558.8 billion rubles, which is only 32.4% of the sum recorded in the consolidated budget. It is 10.1% below the average level of expenditures of the federal budget, which for the recent years has been at its minimum of 42.5%. Thus, there is a lag in the financing of both national projects and a plan for modernization of transport infrastructure from other budget expenditures, as well as a significant failure to finance projects with the largest share of investments in the cost structure: safe and high-quality roads, digital economy, comprehensive plan for the development and modernization of main infrastructure. All this is caused by unavailability of investment project entailed among other reasons by the low development level of the engineering services market.

Modernization in most sectors of Russian economy is based on foreign equipment with the involvement of Western contractors, which also does not help domestic engineering companies get demanded competencies. In this regard, one of the main stages in the implementation of almost any modern industrial construction project in Russia is to find a foreign licensor, a technology donor. Moreover, a foreign licensor usually insists on working with already verified foreign suppliers of technological equipment, even if Russian counterparts are available. This choice is associated not only with a purpose to protect intellectual property, but primarily with an equipment quality issue. Licensors often control the production of critical equipment through certification and a variety of access systems for manufacturers, meanwhile creating additional barriers to the use of domestic solutions.

According to studies performed by RUPEC, Russia does not have any tools to attract funds for large-scale projects regarding industrial sectors that can compete with foreign lenders' project financing aimed at paying to foreign export credit agencies. Import of technological equipment is preferable even without taking into account the actual technological factors. If taking into account expenses for attracting credit resources to the Russian Federation, domestic equipment can be competitive in price if comparing to imported analogues only when its cost is 10–40% lower than a cost of an imported sample in rubles. As a result, import substitution is advisable only in those projects where it is economically justified.

The problem of dependence on foreign technologies can be partly solved through creating joint ventures with the involvement of domestic contractors. So, the preparation of construction project Arctic LNG-2 of NovaTEK company is carried out by a specially created engineering company LNG Novaengineering, a joint venture of Russian NIPIGaz, German Linde (this company has a license for the production of LNG through cascade process on the basis of mixed refrigerant, currently in use in Norway) - 15% and French Technip (international design institute) - 34.9%. NIPIGaz together with foreign engineering companies designs the technological part of the project, as well as the necessary infrastructure, adapts foreign design standards to Russian standards. Within the framework of Arctic LNG-2 project, a contract was signed by the All-Russian Vedeneev Hydraulic Engineering Research Institute, a subsidiary of RusHydro, and the Italian oilfield services company Saipem to carry out design works on gravity-type basement. Only Russian companies are planned to execute general contracting works on drilling fields for Arctic LNG-2, however, service will be provided by foreign organizations.

To date, companies solve the problem of shortcomings in the domestic regulatory framework, as one of the factors negatively affecting the development of engineering in Russia, by elaborating and obtaining special technical conditions for each project. However, to achieve high growth rates for industrial production in a long prospect, it is necessary to update the entire regulatory framework in the construction sector.

Currently, the demand for the integration of equipment, engineering infrastructure and digital systems in industrial production is coming to the fore in the context of the digital transformation of the economy [7]. In this regard, the industry is undergoing large-scale changes caused by the introduction of intelligent systems, characterized by the integration of production and communication networks. Technological changes are accompanied by the development of fundamentally new business processes at all levels. Experts believe that in order to increase the efficiency, productivity and competitiveness of the industrial sector in the world market, engineering companies should play the role of an integrator, as an integrated center of coordination in the currently forming digital industry environment [8]. In particular, it is necessary to include engineering companies in the marketing assessment of investment intentions and conduct thorough technological audits of projects at a pre-investment stage [10]. In addition, when bringing to life complex projects, it is necessary to implement world practices, including BIM, as a necessary element of digital engineering.

5 Conclusions

If comparing to foreign countries, the share of Russia's market of engineering services is small. Despite the positive dynamics, the pace of engineering development does not allow Russia to draw level with its main rivals: the USA and the EU countries. It would be advisable to work out a state policy in the field of systemic formation and development of engineering as a type of economic activity in digital environment.

All educed external, legal and technological factors that restrain the development of the engineering services market determine the priorities in elaboration and state regulation of engineering which is considered as a highly intellectual activity supported by implementation of digital technologies. These factors include updating the regulatory framework in accordance with international practices, legislative consolidation of EPC/EPCM-contracts, obligation of using these contracts in public procurement, activation of import substitution program with giving attention to both technical characteristics of domestic equipment and financial instruments, in order to reduce the cost of this equipment.

When bringing large-scale industrial or civil construction projects into life, engineering companies have shown high effectiveness in digital engineering, particularly in terms of product quality, reduction of execution terms and implementation of project budgets. The main users of digital technologies in the process of implementing ISPs (purchasers, designers, contractors) point out that BIM is not limited to visual representation of a facility under construction, it is a powerful analytical tool that allows managing a project and a life cycle of a facility online. A project begins with a concept, which already implies highly accurate calculation of its implementation options (feasibility study), then the design stage comes, the main task of which is to prevent conflicts, build an exact geometry of a facility and calculate a cost of a project, which will become a basis for tender documentation. Project management at the construction stage consists in tracking time of implementation and development of a project budget, making management decisions based on incoming data obtained from an information model. Stage of operation implies maintenance of a construction site on the basis of a created BIM model, subsequent overhauls, as well as decommissioning and disposal (demolishing) of a facility.

Therefore, introduction of digital engineering and particularly BIM-solutions to ISP implementation process should be accelerated with the support of the government, as long as it is the most effective information method for creating, managing and monitoring project implementation.

Acknowledgement. The proceedings were prepared with the supporting grant from Russian Foundation for Basic Research 18-010-01040 "Development of digital economy methods in the innovation system of managing investment-and-construction projects".

References

1. Kolbin, V.V.: Generalized mathematical programming as a decision model. Appl. Math. Sci. **8**(70), 3469–3476 (2014). https://doi.org/10.12988/ams.2014.44231
2. Konikov, A.I.: Perspective directions in the field of construction management information systems. Ind. Civil Eng. **6**, 64–69 (2019). https://doi.org/10.33622/0869-7019.2019.06.64-69. (in Russian)
3. Raychaudhuri, D.: Very small aperture terminal (VSAT) systems for digital satellite communication: an overview. IETE J. Res. **36**(1), 24–34 (1990). https://doi.org/10.1080/03772063.1990.11436
4. Wang, L., Liu, Y., Yin, Z.: A hybrid TDMA/CSMA-based wireless sensor and data transmission network for ORS intra-microsatellite applications. Sensors **18**(5), 1537 (2018). https://doi.org/10.3390/s18051537
5. Zhang, W., Jiang, L.: Algorithm analysis for big data in education based on depth learning. Wirel. Pers. Commun. **102**(4), 3111–3119 (2018). https://doi.org/10.1007/s11277-018-5331-3

6. Ivanov, N.A., Valpeters, M.L., Kireev, I.A.: Big data analytics and machine learning at risk management of contract obligations default in the construction industry. Ind. Civil Eng. **5**, 81–87 (2019). https://doi.org/10.33622/0869-7019.2019.05. (in Russian)
7. Stockinger, K., Bundi, N., Heitz, J., Breymann, W.: Scalable architecture for Big Data financial analytics: user-defined functions vs. SQL. J. Big Data **6**(1), 46 (2019). https://doi.org/10.1186/s40537-019-0209-0
8. Lyushinskiy, A.V., Fedorova, E.S., Roshan, N.R., Chistov, E.M., Golov, R.S.: Diffusion welding of 12Cr18Ni10Ti steel to palladium alloy foil. Weld. Int. **31**, 777–778 (2017). https://doi.org/10.1080/09507116.2017.1318505
9. Pinaev, S.A.: Application of polymer-cement corrosion protection for different strength concrete of reinforced concrete elements. In: IOP Conference Series, Materials Science and Engineering (2018). https://doi.org/10.1088/1757-899X/463/3/032012
10. Vladimirova, I., Kallaur, G., Bareshenkova, K.: Digital methods of real estate asset lifecycle management. Baltic J. Real Estate Econ. Constr. Manage. **6**, 165–174 (2018). https://doi.org/10.2478/bjreecm-2018-0013. (in Russian)

Economic Effect of the Renovation of Street Engineering Networks

Pavel Shatalov(✉), Anton Akopian, Vladimir Volokitin, and Andrey Eremin

Volonezh State Technical University, 20-letiya Oktyabrya Street, 84, Voronezh 394006, Russia
shatalovpavel@mail.ru

Abstract. The paper considers the problem of reconstruction of gas distribution networks using the example of the city of Voronezh. A developed innovative option is proposed to solve the growing problem of deterioration of gas distribution networks in the face of a shortage of funds for reconstruction. The gas supply schemes of various areas of the city and the problems occurred when connecting new consumers or during operation are presented. Modern methods of gas network reconstruction are considered. A cyclic algorithm for choosing a method of reconstruction of gas networks is proposed. A method is proposed for an integrated approach to the reconstruction process, in which all networks in a closed cluster change at once, combining work with the overhaul of buildings. The term "Renovation of street engineering networks" is proposed. The methodology for renovation (using the gas pipeline network as an example), the sequence of actions and the economic effect emerging in the process of construction and installation work and further operation are described. This method is applicable and recommended for use in all linear engineering networks of megalopolises. Its efficiency increases with an increase in the number of utilities and building density. The necessity of facilities for temporary accommodation during complex work is substantiated.

Keywords: Gas distribution network · Gas pipeline · Renovation · Algorithm · Renovation of street engineering networks · Voronezh · Economic effect

1 Introduction

Russian gas export in 2019 decreased relative to last year, which is caused by various factors, mainly non-economic ones. Domestic demand is stable and has a positive trend, despite the "European" winter of 2019–20, which was made possible thanks to the existing developed gas distribution system.

The state of the gas distribution system is significantly better than the condition of the other networks of the housing and communal services, which is caused by high safety requirements for gas transportation and increased danger of the physical properties of the transported substance. The danger is the loss of tightness with the release of gas into a confined space [1]. A serious problem of the gas industry in the Russian

V. Murgul and V. Pukhkal (Eds.): EMMFT 2019, AISC 1258, pp. 616–628, 2021.
https://doi.org/10.1007/978-3-030-57450-5_53

Federation is the significant deterioration of existing gas pipelines. From the report of the General Director of Gazprom Mezhregiongaz LLC, the total length of steel underground gas pipelines older than 40 years is 38.15 thousand km in the whole country. According to forecasts, by 2031, their length will grow by almost two, if the current trend will not be changed. Taking into account the service life of steel gas pipelines, it is now necessary to make decisions on their reconstruction and replacement [2]. Gas reduction points have a shorter period of safe operation (20 years). At the moment, more than 20 thousand units require reconstruction, forecast for 2031-50.74 units. To bring the entire designated gas distribution system in a safe condition by 2030, it is necessary to find 505.1 billion rubles. Moreover, the existing structure of sources of financing for reconstruction: depreciation, indexation of tariffs, special allowances, as well as the abolition of property tax, does not cover the entire necessary amount. It is necessary to find additional sources of almost 148 billion rubles (30%), or 15 billion rubles annually.

Despite this, the level of gasification in the Russian Federation is growing steadily, expenses for gasification of the regions are approaching 40 billion rubles, the total length of gas pipelines is increasing annually, which, in the absence of a fundamentally new approach to network reconstruction issues, will further aggravate the situation.

The order of reconstruction of networks is determined by the Russian State Standards GOST R 56290-2014 and SP-62.13330 and is carried out if:

- the service life indicated in the design documentation is over (for steel pipes, usually no more than 40 years);
- technical diagnosis gave a negative result;
- existing network capacity does not match demand.

In the Gazprom Innovation Development Program until 2025, the priority areas of R&D and innovative development are production and transportation along the main lines, as well as the development of LNG infrastructure. Blue fuel is the cheapest type of energy that can satisfy the needs of production and the population. Its availability for all consumers is the key to the investment attractiveness of the municipality (territory) in the eyes of investors, and also increases the standard of living of the population and their loyalty. Housing and communal services account for at least 25% of the country's energy balance.

In our study, we would like to propose an innovative version of the solution to the growing problem of deterioration of gas distribution networks in the face of a cash shortage. The purpose of the paper is to find an opportunity to reduce the cost of reconstruction of gas networks, to ensure the necessary level of safety and reliability of the system as a whole, with the maximum use of modern technologies and materials, to reduce operating costs, and as a result to free up additional funds for the reconstruction of networks in the long term. Our ideas correspond to one of the seven main areas of organizational innovation - the introduction of a life cycle management system for products (objects) based on modern digital technologies.

The development of efficient gas distribution systems has been reviewed by many authors. Their goal is to develop scientifically based concepts and design methods, efficient gas distribution systems in settlements, aimed at improving the reliability and cost-effectiveness of supplying consumers with natural gas [3]. We would like to consider the most acute problem of the reconstruction of gas distribution networks for large populated areas in crowded neighborhoods with a high building density.

2 Materials and Methods

In our work, the object of the study was the networks serviced by the gas distribution organization, the leader of the rating among the other GDOs for 2018 in the city of Voronezh. The city is powered by 5 gas distribution stations of the Voronezh Line Pipe Operation Center, located on both sides of the river, with a capacity of 900 thousand m^3/h, sufficient to satisfy all existing and prospective needs (the total load in the heating period does not exceed 60%). GDS are able to provide consumers with gas pressure up to 1.2 MPa, and the reliability of gas supply is provided by the loop of gas pipelines up to high pressure.

Voronezh has a diverse architecture of gas networks. Figure 1 shows various gas supply schemes for existing areas, depending on the type of their development: the private sector of low-rise buildings, multi-storey buildings and industrial areas.

a)

b)

c)

Fig. 1. Scheme of gas supply to various areas of the city of Voronezh: a) private sector b) multi-storey buildings c) industrial areas.

Figure 1b) shows the gas supply scheme of a new district with multi-storey buildings. In the central part, there are MSBs up to 9 floors, receiving gas in each apartment, and in the right part, MSBs are higher than 9 floors. In such buildings, gas is used in roof boiler rooms for heating. In new districts formed in undeveloped areas, designers can use all modern methods and approaches to ensure high reliability and security with minimal capital costs in accordance with the conditions of energy-saving policies in the gas industry. We need to achieve it when reconstructing old networks in the city center due to a successful layout of new areas, with minimal material characteristics of the network.

During the reconstruction of gas networks, we will encounter a large number of restrictions due to the increased danger of gas, the main ones:

Not allowed (with some exceptions): laying gas pipelines through and under the foundations of buildings, in tunnels, collectors and channels, violating the prescribed minimum horizontal distances to buildings, structures and utility networks (2–20 m in each direction), and also vertically to intersected communications.

The safety factor for all pipes is at least 2.0, and the thickness of steel pipes is at least 3 mm (for underground ones), with corrosion protection.

Ensuring uninterrupted gas supply ("uninterruptable clients").

The gas distribution system should be as reliable as possible, "looped".

In confined spaces of the city, with many underground linear communications, the cost of performing repair work on certain sections of the gas pipeline in an open way can exceed 35,000 rubles/m. The construction of a new underground gas pipeline through an uninhabited territory may cost less than 5,000 rubles/m. According to the enlarged standards of the construction prices NCC 81-02-15-2017, when laying polyethylene pipes in a trench from a stationary device and digging the soil into a dump, a Ø160 mm pipe laid to a depth of 1.5 m will cost 1630 rubles/linear meter. That is why in large cities it is recommended to carry out work in a trenchless manner. To the list of already traditional technologies for the reconstruction of individual sections of underground gas pipelines (Appendix A, GOST R 56290-2014), innovative ones (Primus Line, Polymer Armored Pipes) can be added. That is why the role of promising renovation technologies increases with the increasing population density of territories, the increasing number of intersected communications, roads, water barriers [4].

Modern methods for the reconstruction of gas networks in large cities include:

- The maximum use of polyethylene pipes instead of steel.
- The rational location of the gas pressure reduction stations, the abandonment of urban stand-alone gas distribution plants (GDP) and branched low-pressure networks, the installation of individual pressure regulating station for each consumer [5] in the standard version.
- The maximum use of high and medium pressure gas pipelines, such as a two-stage gas supply scheme with linear reduction points and house gas pressure stabilizers, reduces the annual discounted costs for the construction and operation of a village gas supply system by more than 23% [6].
- The use of trenchless technologies, both for reorganization and for laying new branches, is successfully developed and applied as part of an innovative strategy for the development of enterprises in the household and construction industries in dynamic conditions of organizational and economic changes [7], increasing its economic efficiency.
- Improving the reliability and stability of the system through the use of looping.

- Implementation of a "tie-in under pressure" in order to exclude blackouts of gas consumers and reduce greenhouse gas emissions for the period of the tie-in [8].

The main parts of the gas supply system have a long operational life. However, it is significantly less than the service life of residential buildings. That is why the methodology for renovating the gas pipeline system on an innovative basis [9] has a positive economic effect and, with due consideration, is able to ensure the achievement of the goals set in the work. It is proposed to reconstruct the gas supply system not in separate, the most worn-out parts, but in a complex, combining with the overhaul of buildings and the rest of the street infrastructure (roads and sidewalks, utility networks). The main goal of the methodology is to achieve compliance with the service life of the gas supply network and the rest of the infrastructure of the cluster and, at the expense of simultaneous repair of all street engineering parts of the selected area (cluster), save money and thereby fully implement the program for the reconstruction of gas networks. We propose not only to consider the life of the cluster, but, with a sufficient remaining life of the buildings, rely on the total wear of networks and roads. Thus, to reduce the total cost of reconstruction by dividing the additional costs (earthworks, landscaping, etc.) among all interested organizations.

When performing the replacement of all street engineering networks and highways in the cluster, serious difficulties will arise in the movement of residents living in the area, both by car and on foot, primarily due to the large amount of excavation, dismantling the roadway and sidewalks. It will be especially difficult for people with limited mobility. Access to emergency dispatch services is hindered. It is necessary to bear additional costs for the construction of temporary overpasses and side roads, the installation of fencing systems around all trenches, pits. Excavation on existing roads creates large traffic jams, especially in the first days of work, as most drivers are not aware of repair work and do not have time to find bypasses in advance. If one road from each communication line passes through the highway: gas supply, water supply, drainage, heating, power supply, telephone and Internet lines; when carrying out repair work in an open way, it is necessary to block traffic along it at least 7 times, open the asphalt and carry out expensive work to restore it.

When carrying out excavation work in the city, the problem is the storage of soil excavated from a trench, foundation pit. Usually it is not possible to dump the soil due to the high building density. When compiling estimates for the construction of gas pipelines, the aggregated prices took into account the work on the removal of soil to a distance of not more than 1 km. when carrying out work with loading into a motor vehicle. To find a place in the city center where it is possible to store several hundreds, and often thousands of cubic meters of soil, it is necessary to take it out of town to landfills, and upon completion of work, bring sand and gravel back. This increases the cost of reconstruction of gas pipelines in the central parts of the city.

Gas distribution networks were built gradually due to an increase in gas demand, growth and expansion of cities. Often, in cities you can meet such a town-planning phenomenon as "spot development", as well as a gradual increase in the city around the perimeter, without any prepared or planned infrastructure. Due to the lack of existing gas distribution system capacities, when connecting new large consumers, it is necessary to pull the gas pipeline over fairly long distances. Figure 2a shows that during the construction of a new residential area, it was necessary to lay a high-pressure gas pipeline with a length of more than 3.1 km. It was more economical than to route along existing gas pipelines, to which there is no more than 0.3 km (total length would be 2.4 km.). Figure 2b shows a similar situation, but with medium pressure pipelines, when connecting a boiler room, which is tied in a pipe located three times farther than the nearest gas pipeline (2 lines) of larger diameter. In the existing gas supply system, such routes are economically justified when connecting each consumer individually. Gas distribution organizations are ready to apply the necessary capacities to any territory at a much lower cost (laying gas pipelines along the shortest routes and a larger diameter), with a clear long-term urban development plan for the development of the territory of the settlement, and the absence of point unplanned development. There are negative examples when large consumers do not locate their enterprises in the region due to the high cost of connecting to gas distribution networks. This reduces the overall investment attractiveness of the region in the eyes of investors and does not contribute to improving the living standards of the population.

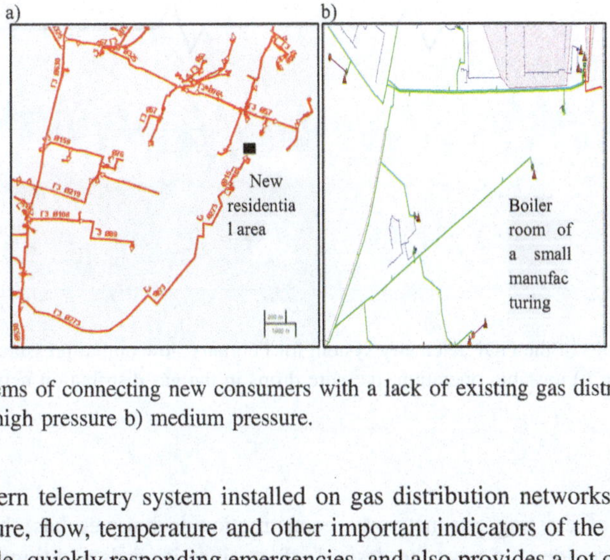

Fig. 2. Problems of connecting new consumers with a lack of existing gas distribution system capacities: a) high pressure b) medium pressure.

The modern telemetry system installed on gas distribution networks allows monitoring pressure, flow, temperature and other important indicators of the system in the "online" mode, quickly responding emergencies, and also provides a lot of information for analysis and decision making. Gas consumption during the year is not constant, there are both seasonal and daily fluctuations in transportation volumes. Consumption also changes on holidays and weekends. With the onset of winter, gas consumption is highly dependent on outdoor temperature. More than 10 years ago, Ivanov A.A. proved that with an unstable incoming pressure of gas to gas equipment, its efficiency decreases, and the greater the deviation from the optimal pressure, the lower the efficiency.

Figure 3 shows the data of the telemetry system (outlet pressure) of two gas control points supplying residential areas for February. Figure 3a shows stable operation throughout the entire period; the GDP power reserve and network capacity are sufficient to maintain the specified pressure among end users, which means that the efficiency of their gas equipment will be high. Figure 3b shows the operation of another GDP. Daily fluctuations in the low-pressure gas distribution network with an amplitude of more than 300 Pa are visible on it. At the GDP outlet, for the most distant consumers, the pressure drop can be even greater. Such areas must be monitored, pressure at the end users should be controlled, and, if possible, reconstruction should be done. Such a situation at low pressure is a fairly common situation. When the network was being built, there were much fewer consumers with less gas needs. As the city grew, many small houses turned into cottages, or even apartment buildings, the outskirts expanded, the streets lengthened, but the gas distribution system and the cross section of steel gas pipelines remained at the same level. The telemetry readings help identify areas that need reconstruction.

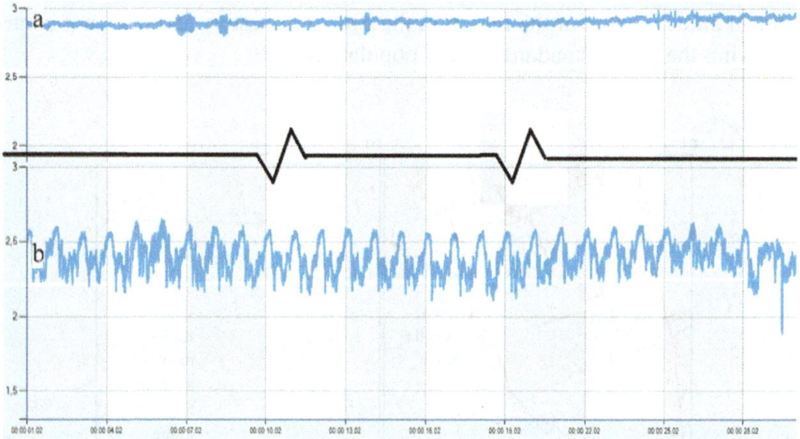

Fig. 3. Readings of the GDP telemetry system for February, low output pressure, kPa: a) stable daily operation; b) unstable operation, pressure drops in the gas distribution network.

3 Results

An analysis of existing gas network reconstruction techniques has shown that the current practice does not have the desired effect, since it needs additional financing in the amount of at least 30%, the sources of which are still unknown. To solve the problem of reconstructing the entire volume of worn out gas distribution networks without using additional financial resources, it is necessary to apply an innovative approach, according to the developed cyclic algorithm for choosing the method of reconstruction of gas networks shown in Fig. 4.

Renovation is interpreted by the Civil Code of the Russian Federation as the development of built-up territories. The problem of renovation of residential buildings and industrial territories over the past 5 years has been the subject of many works, the results of which have been successfully implemented in the Capital. The advantage of renovation is the formation of a housing fund that meets increased consumer qualities:

having a homogeneous social environment, high amenities, a new engineering infrastructure. Renovation resolves obsolete problems of deterioration of gas networks. In addition to the classical renovation of the entire district (cluster) with the demolition of buildings and structures, with a slight unrecoverable physical and moral deterioration of buildings, we propose the use of renovation of street engineering networks everywhere.

Renovation of street engineering networks is an intensive complex process of a synchronous complete replacement of all linear engineering networks of a separate cluster and structures on them, with the preservation (or partial preservation) of the existing buildings, aimed at creating a comfortable living environment that meets all modern requirements for reliability, economy, environmental friendliness, providing long trouble-free operation of engineering networks.

For the effective renovation of street engineering networks, it is necessary to:

- choose a cluster with the most deteriorated networks of all kinds;
- create facilities for temporary accommodation, where the residents of the cluster will live during the work implementation;
- develop a modern project that meets all modern requirements for reliability, cost-effectiveness, environmental friendliness and coordinate it with all services operating the underground and above-ground communications of the city, as well as those responsible for the maintenance and repair of buildings and structures, managing companies, public utilities organizations and services responsible for city landscaping and gardening, if necessary, plan a project for major repairs;
- relocate people to facilities for temporary accommodation;
- close free access to the area and prepare convenient bypass routes for the duration of the work;
- carry out overhaul and maintenance of buildings, carry out work on street engineering networks, starting with the most deep-seated communications and up to the surface, then above-ground ones and landscaping (including roads and sidewalks);
- return people to a new renovated house in a new modern area.

The economic effect of the renovation of street engineering networks will consist of many components (for example, gas networks) divided into two classes: savings in the construction and installation works "in the field", compared to work in separate areas using classical methods, and also savings during failure-free operation for the next 40 years.

When carrying out major repairs of buildings and structures, in addition to the use of modern safe pipe routing along the building façade and the use of modern materials, the coefficient of 1.5 will not be applied to the costs of work, which is used in works without resettlement.

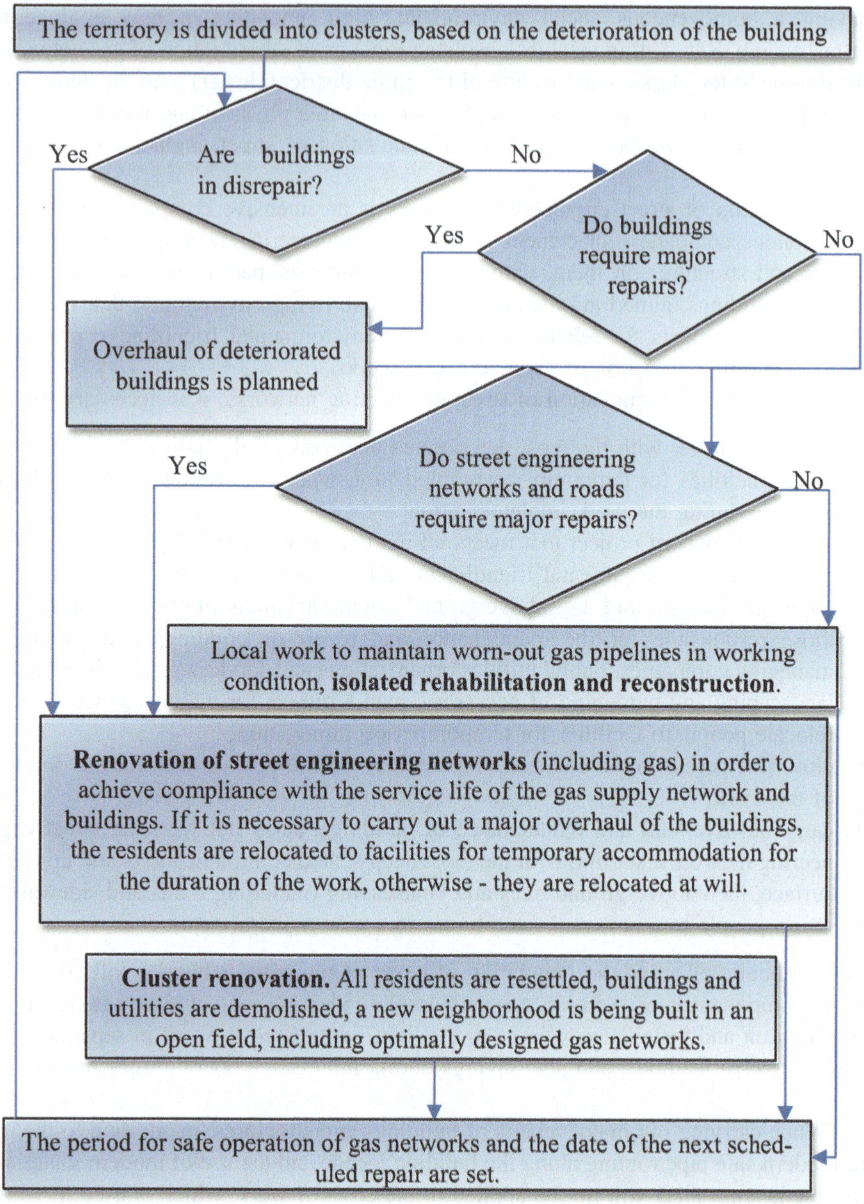

Fig. 4. The cyclic algorithm for choosing the method of reconstruction of gas networks.

There is an opportunity to plan a new route of gas distribution pipelines as short as possible, but able to satisfy all current and forecast gas demand. Make the most of high and medium pressure gas pipelines. It is rational to abandon the use of GDP in favor of individual pressure regulating station. All this will allow minimizing the material characteristics of the network, and, consequently, the cost of purchasing pipes.

The cost of work carried out "in the field" can do without additional costs for constraining the conditions, with the maximum use of polyethylene pipes, without the need to carry out subsequent work to improve the territory (primarily asphalt concrete pavement). All work will be carried out in an open way, without the need to use expensive trenchless technologies. All points of intersection with other communications will be known and easily identifiable, all communications with a depth deeper than the gas pipeline will be laid later and will not interfere with the work. Excavation work can be minimized by using a bar with a precisely controlled depth of excavation, which would be impossible in the city, conducting work in a classical way. Also, there is no need to carry out earthworks manually, which is usually done near other communications as prescribed by representatives of operating organizations.

To illustrate the effect of savings on construction and installation works, we consider an example of optimization of the gas network tracing for an individual cluster, with a given set of consumers. The most suitable optimization method is selected, which leads to the shortest distances when laying the pipeline route. Three methods can be applied: the least squares method, the Prim's algorithm, the Steiner method. It is proved that with the optimal choice of gas network tracing, a significant economic effect is achieved. Figure 5 shows the scheme obtained on the basis of the Prim's algorithm where: the vertices of the graph are consumers; graph edge - gas distribution pipe with a weight equal to the cost of construction.

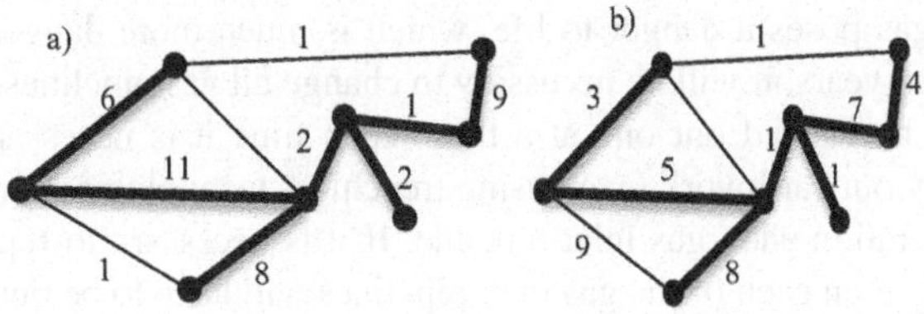

Fig. 5. A minimum spanning tree constructed using the Prim's algorithm a) a classical approach, b) an innovative approach.

The presented method and the result obtained (the minimum main tree) cannot be used as the final gas distribution network scheme, since there are no "loops", which reduces the overall reliability of the network, and long winding routes can form, which will lead to loss of pressure and the need to increase pipe diameters, as a result, investment is increased. Also, the possibility of adding intermediate vertices (constructing a Steiner tree) is not considered here, which often helps to significantly reduce costs. However, this will allow demonstrating the economic effect of the renovation of gas networks, and the effect increases with increasing building density, the area of asphalt concrete pavements and the number of lines of laid utility networks.

Additional income can be obtained after the sale of worn-out old gas pipelines, which will have a minimum cost of getting them out of the ground using this approach, and with the coincidence of old and new routes - without costs.

During the production process, it is possible to most reliably and accurately establish the place of pipe laying, which will update and clarify the data of geographic information systems (GIS). Recently, the accuracy of GIS has significantly improved, all inaccuracies, both on maps and on the profiles of laying old networks, will be eliminated. This will create reliable GIS (possibly even three-dimensional one) of all networks, establish owners and users of all networks (there will be no abandoned, no-man networks), which will give significant savings during operation.

During operation, it is possible to get even greater savings over the next 50 years. It is for such an operating period that modern underground gas pipelines are laid, specifically:

New networks have significantly less crashes! On a call “Smell of gas on the street”, according to the plan for localization and liquidation of accidents, a team consisting of a driver, two locksmiths under the guidance of a foreman in an equipped specialized car must leave for the place, carry out emergency recovery work (even minimal, if there is no real leakage), and return back. It lasts around 70–80 min and is quite expensive (5 person-hours, gasoline, and other overheads). If a leak is detected on the underground gas pipeline, then it will be necessary: to make a pit, repair the leak on the spot (if possible), if necessary, close the valves and shut off consumers, conduct welding work, restore the protective coating, backfill with soil and carry out landscaping work. If there is not enough money to change the trend in the number of gas pipelines older than 40 years by 2030, then there will be much more such on-site visits. The release of gas poses a danger to life, which is much more dangerous.

In the coming years, it will be necessary to change all gas pipelines in the cluster! If such work will be carried out one at a time, each time it is necessary to disconnect consumers, carry out earthwork (even using trenchless technologies), tie-in again, press and put into operation each gas inlet pipeline. If it is necessary to replace the old gas distribution pipe, then each of the gas inlet pipelines will have to be tied-in again, while producing at least 1 pit for each insert. If we take into account that the cost of replacing a single short section will cost about 4 times more expensive than similar ones, and all the pipes will have to be replaced in 40 years, then during this period, the costs of replacing the pipes alone will amount to 4 times more than what is needed now. If we take into account the time factor (taking into account the Central Bank’s inflation reduction target) of 8%, the savings will be 19% only for these works.

Replacing steel gas pipelines with polyethylene pipes will allow saving annually on electrochemical protection. The rate of overgrowth of the internal cavity of a PE pipe is much slower than steel one, which will also reduce energy costs due to lower pressure losses during operation. Improving the energy efficiency of the gas industry is a priority for gas distribution companies.

Since all networks in the cluster will be replaced, there will be almost no accidents on neighboring networks, earthworks will not be carried out, the number of accidents due to damage to gas pipelines and earthmoving equipment will decrease.

It is necessary to reduce losses during gas transportation, in case of accidents and scheduled works on gas pipelines.

It is worth noting the costs that have to be incurred in the implementation of the renovation program of street engineering networks.

It is necessary to carry out a large organizational work under the leadership of the city authorities in order to develop a unified strategy for the renovation of the entire city infrastructure in the future and each cluster individually; establish the order and terms of work, the area of responsibility of each service; prepare a plan for the organization of construction and work; provide a long-term forecast for the needs of the population and industry in household resources, as well as a plan for expanding the territory by at least 20 years.

It is necessary to create facilities for temporary accommodation of the required size for the temporary resettlement of residents.

It is necessary to ensure the readiness of each operating organization in terms of equipment, human resources, the availability of materials, blanks, products and consumables for such large-scale works.

4 Conclusions

Many scientists are on the path to finding effective ways to reconstruct gas pipeline networks; the same studies are being conducted on other engineering networks. We propose to combine efforts and, due to the synergistic effect, save money for intensifying the process of reconstruction of city networks. This problem is extremely urgent due to the strong deterioration and additional operating costs. Most of the developed technologies and approaches can be easily applied when working according to the proposed methodology, part of the work in the renovation of street engineering networks, part for local work to maintain worn-out gas pipelines in working condition. Our study shows that:

1) full use of cluster renovation and renovation of street engineering networks according to the developed cyclic algorithm will save the necessary missing funds for the full replacement of all worn-out gas pipelines, ensure timely updating of the entire gas network, annually reduce the number of emergencies, increase the reliability and stability of gas supply, reduce gas losses, reduce the cost of technical connection of new planned consumers, reduce the energy and material consumption of the gas network;
2) application of the proposed approach to updating engineering networks is fully suitable not only for gas supplying organizations, but also for all organizations operating underground linear facilities, the wear of which is more than 40%;
3) the economic synergy effect of the interconnected work of services of various departments will improve the living standards of the population, increase the profitability of utility organizations, reduce the growth of utility tariffs, increase the energy efficiency of the economy as a whole, reduce the cost of major repairs of buildings and street engineering networks, increase investment attractiveness of municipalities, and will not require large investment costs, since the work will be financed in the already planned volumes;

4) to implement the developed methodology, it is necessary to do a lot of legal and organizational work: it is necessary to develop mechanisms for temporary relocation of residents from clusters (someone will definitely be categorically against it), otherwise, the costs will increase significantly, and the safety of the remaining citizens will decrease; it is necessary to find ways to influence the owners of underground utilities that do not want to change old networks (no money, opportunity, time); indicate the dates and terms of the next major work in the cluster, taking into account the service life of buildings, structures and communications;
5) it is necessary to develop a methodology for calculating the life cycle of engineering communications (including gas), taking into account the cost of total expenses, and conduct calculations on each specific cluster separately for each of the participants in the process to demonstrate the resulting economic effect and to increase the interest and involvement of participants into the process.

References

1. Nezhnikova, E.: The use of underground city space for the construction of civil residential buildings. Proc. Eng. **165**, 1300–1304 (2016). https://doi.org/10.1016/j.proeng.2016.11.854
2. Romanova, T.N.: Current status of natural gasification. Bull. PNRPU. Constr. Architect. **10** (1), 80–90 (2019). https://doi.org/10.15593/2224-9826/2019.1.08
3. Potapov, Y., Polikutin, A., Panfilov, D., Okunev, M.: Comparative analysis of strength and crack resistance of normal sections of bent elements of T-sections, made of rubber concrete, cauton reinforcement and concrete. MATEC Web of Conf. (2016). https://doi.org/10.1051/matecconf/20167304018
4. Lyushinskiy, A.V., Fedorova, E.S., Roshan, N.R., Chistov, E.M., Golov, R.S.: Diffusion welding of 12Cr18Ni10Ti steel to palladium alloy foil. Weld. Int. (2017). https://doi.org/10.1080/09507116.2017.1318505
5. Safronova, N, Nezhnikova, E, Kolhidov, A.: Sustainable housing development in conditions of changing living environment. 2017 MATEC Web Conf. **106**, 08024 (2017). https://doi.org/10.1051/matecconf/201710608024
6. Pan, Z.H., Zhou, J.L., Jiang, X.: Investigating the effects of steel slag powder on the properties of self-compacting concrete with recycled aggregates. Constr. Build. Mater. **200**, 570–577 (2019). https://doi.org/10.1016/j.conbuildmat.2018.12.150
7. Nedeljkovic, M., Ghiassi, B., Laan, S., Li, Z.M., Ye, G.: Effect of curing conditions on the pore solution and carbonation resistance of alkali-activated fly ash and slag pastes. Cem. Concr. Res. **116**, 146–158 (2019). https://doi.org/10.1016/j.cemconres.2018.11.011
8. Yan, X.C., Jiang, L.H., Guo, M.Z., Chen, Y.J., Song, Z.J., Bian, R.: Evaluation of sulfate resistance of slag contained concrete under steam curing. Constr. Build. Mater. **195**, 231–237 (2019). https://doi.org/10.1016/j.conbuildmat.2018.11.073

Web-Based Power Management and Use Model

Vyacheslav Burlov[1], Oleg Uzun[1], Mikhail Grachev[2](✉), Sergey Faustov[1], and Dmitry Sipovich[3]

[1] Peter the Great St. Petersburg Polytechnic University, 29 Polytechnicheskaya str., 195251 St. Petersburg, Russia
[2] St. Petersburg University of the Ministry of Internal Affairs of Russia, 1 Pilot Pilyutov str., 198206 St. Petersburg, Russia
mig2500@mail.ru
[3] Russian State Hydrometeorological University, st. Voronezhskaya, 79, 192007 Saint-Petersburg, Russia

Abstract. The use and consumption of electricity occurs everywhere and constantly, including in social and economic systems. With the development of information technologies, including web technologies, they are being introduced into all spheres of human life, including those being introduced and used at electric power distribution facilities. Software and hardware systems are being introduced, allowing for control and operational monitoring of electricity distribution, as well as its consumption. Along with the development of software tools that help in everyday activities of a person to interact with power distribution plants, there is a destructive impact from outside that destroys the normal operation of the system. Power outages caused by the destructive effect on this hardware and software complex can cause great damage to the entire system of social and economic systems and only a correctly constructed model of an adequate response to emerging threats will allow the person exercising control to respond in a timely manner and eliminate the destructive impact in order to further normal distribution of electricity, including using web-based technologies online, wherever he is. This model will allow you to redistribute the time resource, which will increase the success of achieving the management goal.

Keywords: Web model · Power consumption · Model · Control · State of the system · Management decision

1 Introduction

Human development proceeds in parallel with the development of electrical networks. Together with them, the technical side and the software to it were developed. The advent of the Internet has given impetus to the development of innovative technologies such as web applications. Further development of the energy complex will undoubtedly increase, which will undoubtedly affect the business models that will introduce further development of technologies, including web-technologies of energy consumption management. Structural changes the enterprises (the plants, factories) happening in the

V. Murgul and V. Pukhkal (Eds.): EMMFT 2019, AISC 1258, pp. 629–641, 2021.
https://doi.org/10.1007/978-3-030-57450-5_54

power industry at the enterprises of the energy sector, including traditional are directed to automation as the enterprises in general, and to optimization of expenses in the power spent for work of the enterprise. The transition to autonomous networks with renewable energy resources, which implies a complex self-governing system, is made possible with the introduction of modern web technologies. As an example, we see projects such as controlled electrical networks or smart home.

Many business processes are based on energy and web technologies, and various threats of unauthorized access to personal data to users and databases of various web resources are common in this regard. Threats and hacking are common not because networks and web technologies are used, but because of imperfections in information technology, lack of software protection and other factors.

Web applications are objects that are constantly exposed to danger and vulnerability by intruders. Web applications are one of the most vulnerable elements in their operation, to which a hacker can access sensitive data placed on a computer and change its content, thus disrupting the integrity of the system, as well as disable the server with a distributed attack that will deny access to users.

This set of factors determines the relevance of the creation of a model for the management of energy networks in the global Internet on the basis of the development of an analytical and mathematical model for the management and use of energy consumption using web-technologies.

2 Threat Analysis for the Web Application

Like a conventional application or operating system, web pages may have security disadvantages. This is a serious problem because the web site uses sensitive data (passwords, credit card numbers, etc.). Even on a secure web server running a known secure operating system, security vulnerabilities can persist because they are mainly caused by programming errors of the application itself rather than the server.

The first step in eliminating existing hacker attacks and vulnerabilities is the identification of threats - the identification of all possible and current threats that occur to the Web site during its creation and operation [1, 2].

There are a wide range of vulnerabilities to web applications. However, some are better known and more dangerous than others. Let's highlight 6 vulnerability classes:

1. Authentication is a vulnerability associated with the functions of a web application to identify a user, service, or application.
2. Authorization - This class combines vulnerabilities associated with features to verify rights attached to a user, service, or application.
3. Client-side attacks are vulnerabilities that allow attackers to directly target web application users, such as providing them with illegal content, and pretending its information from the source site.
4. Executing Commands - This class contains vulnerabilities that allow remote execution of commands on web applications.
5. Confidential Disclosure - This class includes vulnerabilities that provide information about the system (operating system, version, and so on).

6. Logical and software errors: Vulnerabilities belonging to this class can lead to attacks that distract application implementation logic from illegal actions.

Web applications must manage user authentication and establish sessions to track requests because HTTP does not have this feature.

If all authentication information and session IDs are permanently unprotected by encryption and not protected from disclosure by other vulnerabilities such as XSS, the attacker can intercept the active session and impersonate another user.

In the event that a hacker detects a session from which the original user has not logged off, all account management functions and transactions must require new authentication, even if the user has a valid session ID. Two-factor authentication must also be considered for high-cost transactions [3].

Insufficient protection of the transport layer is due to the fact that web applications are not protected or poorly protected. If, for example, SSL/TLS is used only during the authentication phase, the data and session IDs can be represented in application network flows. In addition, this type of vulnerability can be caused by expiration or incorrectly configured certificates. One incorrect SSL configuration can contribute to phishing attacks * in particular. Such vulnerabilities expose user data and may cause it to spoof. *

- Phishing is a method used by scammers to obtain personal information to assign the identity of a customer, company, financial institution, or administration.
- Spoofing is creating Internet protocol packets containing a modified source address to hide the identity of the sender or impersonate another computer system.

Incorrect Security Configuration. An infrastructure that supports a web application includes many devices and software: servers, firewalls, databases, operating systems, and applications. All these elements must be configured and protected in a secure manner. One of the main reasons for poor system administration is the lack of adequate training for those responsible for managing web applications and basic infrastructure [4].

Some Approaches to Web Site Security

Developers and security administrators can use a variety of techniques to address web-based threats. The following are some of the main ways to ensure security, given the following objectives:

1. Vulnerability prevention - Prevent the introduction of vulnerabilities through rigorous design methods;
2. Fix vulnerabilities - Identify and resolve vulnerabilities using validation and testing methods.
3. Intrusion Prevention, Detection, and Resiliency: Protect your system from attacks and intrusions while you work by introducing security barriers (firewalls and detection systems) that enable your application to ensure the right service despite attacks.
4. Assessment: Assess the impact of vulnerabilities and attacks and the effectiveness of protection mechanisms.

Vulnerability prevention is typically based on rigorous processes, design, and operational rules to reduce the risk of vulnerabilities being deployed at different stages of the application lifecycle [5].

Whether the developer is new or already experienced in web application security, creating a new web site or securing an existing web site can be difficult, controlling all risks without applying appropriate processes and tools. For this reason, it is important to consider security aspects throughout the development phase of the application, especially in the early stages.

First, it is necessary to comply with the security requirements of applications, that is, to indicate what security means, because it is much more profitable to develop the application, protecting it from the beginning, rather than eliminating its shortcomings after its operation.

Second, standard security checks must be observed when writing application code. For example, a set of standard controls can greatly facilitate the development of Java, PHP, and Python applications [6].

The application of these rules can make a significant contribution to improving the security of web applications. However, given the complexity of applications and technologies used, these measures are insufficient to ensure that development is done without errors and should be complemented by other means [7].

Existing approaches allow web applications to be secured only in the case of a single request. Therefore, it is of scientific and practical interest to develop models and methods to ensure the security of web applications using web technologies in the context of complex requests.

For this purpose it is advisable to use the natural-scientific approach developed by the scientific school "Systematic integration of public administration processes." This scientific school is registered in the Register of Leading Scientific and Educational Schools of the Government of St. Petersburg. The security process will be based on a systematic integration of the following processes, namely, the targeted information process, the threat generation process, the threat detection process, the threat clearance process, and further action processes [8, 9].

As a result of system integration, the condition of existence of the process of ensuring system energy management is formed. This condition eventually allows one equation to be formed with two unknown:

- generalized characteristics of the threat identification process;
- is a generic characteristic of threat neutralization.

Formation of generalized characteristics of these two processes on the basis of the required level of security of web-applications allows to satisfy the condition of existence of uninterrupted operation process. In general, this result allowed to move to the development of an analytical mathematical model to ensure safe operation of the web-application of energy distribution under conditions of destructive effects of the environment.

Most Internet networks are based on a model such as a client-server. Especially recommended for networks requiring high levels of reliability. The main advantages of using the client-server architecture, since the server is located in the center of the network and accordingly can manage such resources as a centralized database and providc the best security, administer at the server level, make changes in the client network base without disrupting the network operation, and in itself is easy to configure. The negative fact is that the client-server architecture itself is very expensive in its cost.

Attackers can take advantage of different classes of web vulnerabilities. Today, however, no one can argue that there are robust computer security mechanisms, on the one hand, because not all vulnerabilities are known, on the other, because systems and technologies are rapidly evolving with many new vulnerabilities, so it is necessary to assess the effectiveness of new security tools or the improvement of existing technologies. Therefore, in order to ensure adequate security of the web application, it is necessary to have an adequate mathematical model of the human solution.

Therefore, it is proposed to solve the task on the basis of system integration.

The information security process is always based on human decision. The decision is made by the person on the basis of the model. An object model refers to the description or representation of the corresponding object, which gives you basic properties and characteristics about it. Therefore, the solution will be a model of the human process, and the process of the object is in action at a fixed purpose.

In order to ensure the achievement of the objective of the activity while ensuring safety, it is necessary to be able to form processes with well-defined properties. Without the conditions of these processes, we cannot shape them. This does not allow the formation of processes with well-defined properties. Without such an opportunity, we cannot be guaranteed to achieve the purpose of the activity. This is possible only on the basis of security methodology.

Without a methodological basis for solving the problems of the operation of a potentially dangerous object in the form of conditions for the existence of the process, we cannot guarantee the achievement of the objective of the activity. This situation gave rise to a fundamental problem: "The results of the security management system tasks do not meet the expectations of the decision-maker." The decision-maker acts on the basis of three categories. It is the System. Model. Purpose.

Two approaches are known for system development. System development based on analysis. In order to solve task 1, it is proposed to use for synthesis - the Law of Preservation of Integrity of the Object (RSPS), which ensures the achievement of the goal of operation of safety systems. (RSPA is a stable, objective, repetitive relationship between the properties of an object and the properties of its actions at a fixed purpose) [10].

The decision maker operates model-based security systems. To do this, you need to be able to synthesize adequate models. The goal of operating artificial intelligence systems is possible only on the basis of a properly constructed system (PPP) and an adequate model. There is no such approach to the operation of security systems in the prior art. Therefore, the entire PPP design is implemented on the basis of the RSPA, which confirms the feasibility of considering the RSPA as a condition for the existence of safety systems. The security model is based on the system integration of four processes. Target process. Formation of a problem. Process of recognition of a problem. Process of elimination of a problem [11].

To synthesize the management solution model, we will use a natural-scientific approach to form conditions that in turn will guarantee the achievement of the management goal. A natural-scientific approach is defined as the integration of the properties of human thinking, the surrounding world, and the connection of these two components through cognition, Fig. 1. The above components are reflected in the principles of the integrity of the world, the cognition of the world and the components of knowledge.

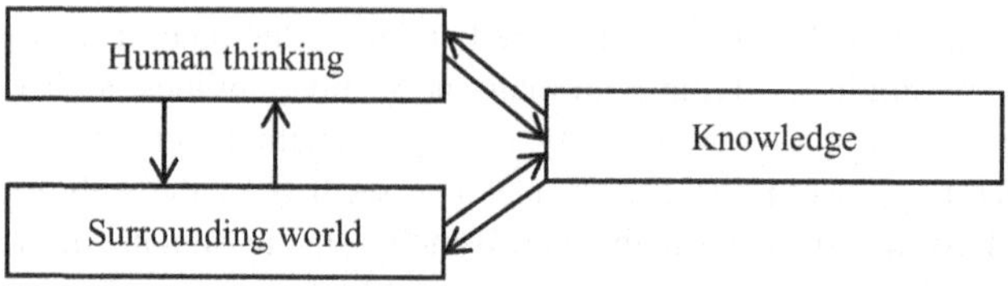

Fig. 1. Natural-science approach.

In turn, these principles are implemented by a stable, objective, repetitive relationship of object properties and actions at a fixed purpose or by another law of preservation of object integrity (RSPA) and the following methods, such as decomposition, aggregation and abstraction.

In his/her life activity, a person operates the following categories, such as "system," "model" and "purpose," in this regard, attention should be paid and these categories should be used. Two directions of system development (models) are known. Analysis-based design and synthesis-based solution. This approach is known from systems engineering [4]. Also academician of the Academy of Sciences of the USSR Anohin P. K. [5] pointed out and experimentally confirmed that for synthesis of the system it is necessary to identify the "main pattern" in the general theory of functional systems. It should be noted that on the basis of the above mentioned in the work for synthesis of the mathematical model of the management decision will be used, as well as one of the main points will be the conditions of its adequacy, for this we will apply the approach of completeness taking into account the main patterns of the subject area [12].

According to the management developed, each process will be represented by three components that correspond to the following properties "objectivity," "integrity" and "variability"(or the concepts of "object," "purpose," and "action," respectively). You will arrange these components horizontally. They can be interpreted in three various levels of knowledge of the world (abstract, abstract and concrete, concrete). Graphically, we will present this approach as three levels vertically, Fig. 2.

Let's enter a number of definitions:

Management decision is a condition for the realization of the purpose of the facility it manages in an appropriate environment in order to achieve the objective of management.

The environment is a set of factors and conditions in which activities are carried out.

Information and analytical work - continuous mining, collection, study, display and analysis of situation data.

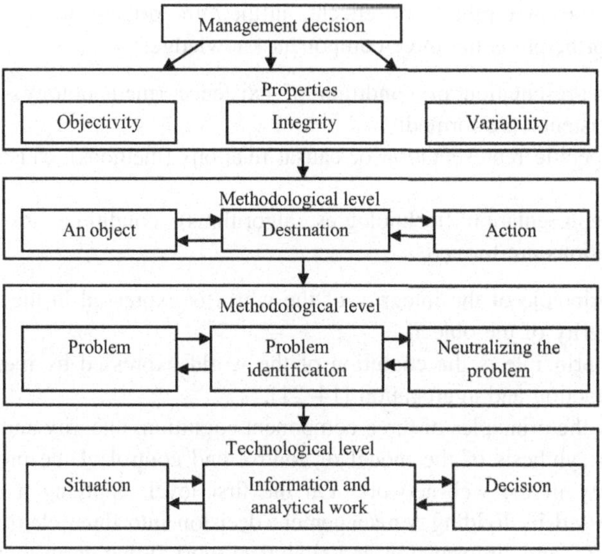

Fig. 2. Management decision as a process.

By dividing the concept of "management solution" into three basic elements - "environment," "information and analytical work" and "solution," it is necessary to move to synthesis of the solution model [13]. Figure 3 shows the design of the management solution model.

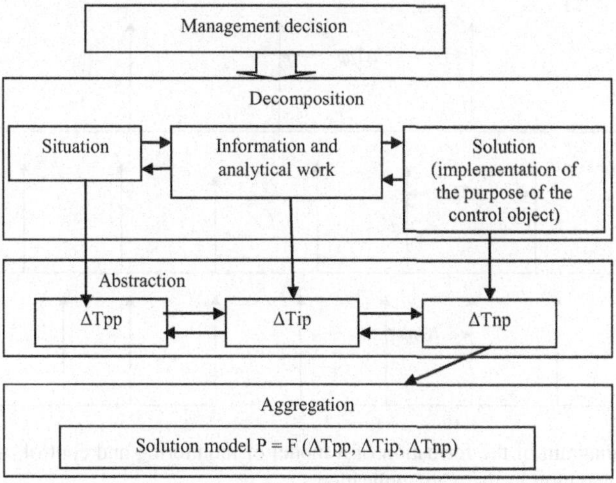

Fig. 3. Management solution model formation scheme.

As noted, the three pillars are reflected in the three principles.

The first principle is the three-component knowledge:

- Abstract representation or condition of existence (methodology), conditions of process existence are formed;
- Abstract-specific representation or causal relations (methods), causal relations are formed;
- Specific representation (technologies, algorithms), conditions for realization of causal relations are formed.

The second principle of the integrity of the world is expressed in the law of preservation of integrity of the object.

The third principle is the cognition of the world expressed by methods: decomposition, abstraction and aggregation [14–21].

Guided by the principles of three-component cognition, integrity and cognition, we will carry out synthesis of the model of control and control of the process of information security in the web-network. On the first level, applying a decomposition method expressed in dividing a management decision into three elements ("setting," "decision itself," and "information and analytical work") that correspond to "object", "purpose," and "action." Applying at the second level the method of abstraction, which is expressed in the separation of the "object" ("furnishings") with the frequency of manifestation of the problem in front of a person (ΔT_n). "Purpose" ("Solution") is identified with the frequency of neutralizing the problem (average time of adequate response to the problem) by a person (ΔT_p). "Action" ("informational and analytical work") is identified with the periodicity of problem identification (average time of recognizing a situation) (ΔT_{ip}) [13] (Fig. 4).

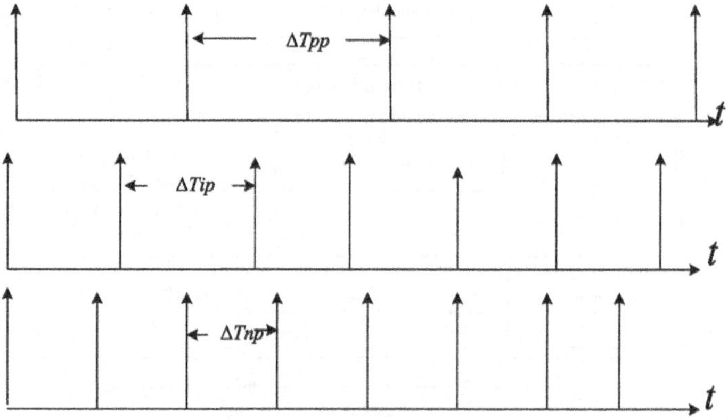

Fig. 4. Time diagram of the formation of a model of monitoring and control of the process of ensuring safe operation in the web-application.

As a result of the methods of decomposition, abstraction and aggregation considered, we have transformed the concept of "management solution" into a mathematical model of management solution of the following kind:

$$P = F(\Delta T_{PP}, \Delta T_{IP}, \Delta T_{NP}), \tag{1}$$

where ΔT_{pp} is the average time the problem occurred; ΔT_{ip} - average time of problem identification; ΔT_{np} is the average neutralization time of the problem;

P is the probability that an emerging problem will be recognized and solved before a decision is made. This is a prerequisite for the management process.

In order to implement this approach, we will need to draw up a system of differential equations of Kolmogorsk - Chapman. The characteristic of system transitions can be represented by Fig. 5.

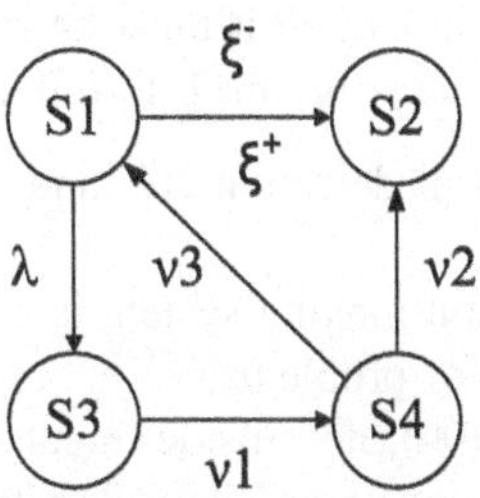

Fig. 5. Characteristic of transitions of a system.

Accordingly, in order to obtain this relationship, we will enter the following designations:

λ - size, the return to the average time of manifestation of a problem;
ν_1 - is the inverse of the average identification time of the problem;
ν_2 - is the inverse of the average neutralization time of the problem.

The person solution model characterizes four basic states:

A_{00} - person does not identify or neutralize;
A_{10} - the person identifies and does not neutralize;
A_{01} - the person does not identify and neutralize;
A_{11} - the person identifies and neutralizes.

According to the described feature of the management solution, it is necessary to introduce the probabilities of our management system in these four states. We, respectively, receive four probabilities of P_{00}, p_{10}, P_{01}, P_{11}, to finding of a system in conditions of A_{00}, A_{10}, A_{01}, A_{11}. The process of forming the solution can be considered as Markov chain, for example, in the work on safety research [11]. This approach does not sufficiently take into account the dynamics of the process, so it is advisable to use continuous Markov chains in the work [15].

The state of the A_{00} (00) system is in the initial state, and no effects occur. At emergence of a problem under the influence of intensity of λ the A_{00} (00) system passes into a condition of A_{10} (10), that is into a condition of recognition of a problem. From this state, the system, under the influence of the intensity v1, moves to the state A_{01} (01), in which the system starts the process of eliminating (neutralizing) the problem. In case of delay in resolution of the situation (insufficient qualification of SCL), under the effect of intensity v_3, the system changes from state A_{01} (01) to state A_{00} (00). This situation is possible if the problem is neutralized and another problem has not yet arisen. If the problem has not been resolved and the SCL needs more time to resolve the problem, the system changes to the A_{11} state (11) under v_2. Being in a condition of A_{11} (11) decision-makers involves resources and fixes a problem, under the influence of intensity ζ_2 the system passes into an initial condition of A_{00} (00). Return of a system from a condition of A_{00} (00) under the influence of intensity ζ_{1B} a condition of A_{11} (11) demonstrates failure of a task of elimination of a problem. Then the next problem comes to the input and it must be recognized [16].

The following assumptions must be made to reflect in mathematical form:

1. The time moments of problem detection are random and form a flow close to the Poisson flow;
2. The management and decision-making system is ready to solve problems aimed at recognition and elimination of problems;
3. When the system is at rest (00), no outside threats are received or processed;
4. For a system of differential equations, the following restriction is imposed:

$$P_{00}(t) + P_{10}(t) + P_{01}(t) + P_{11}(t) = 1. \tag{2}$$

On the basis of this we can apply the system of differential equations Kolmogorsk - Chapman and make the following equation [17–21]:

$$\begin{aligned}
\frac{d}{dt}P_{00}(t) &= -P_{00}(t)\lambda + P_{01}(t)v_2 \\
\frac{d}{dt}P_{01}(t) &= -P_{01}(t)(\lambda + v_2) + P_{11}(t)v_1 + P_{10}(t)v_1 \\
\frac{d}{dt}P_{10}(t) &= P_{00}(t)\lambda - P_{10}(t)v_1 + P_{11}(t)v_2 \\
\frac{d}{dt}P_{11}(t) &= P_{01}(t)\lambda - P_{11}(t)(v_1 + v_2)
\end{aligned} \tag{3}$$

3 Results

By solving the obtained system of linear algebraic equations we get final probabilities:

$$0 = (\zeta_2 + \lambda) \cdot P_{10} + \zeta_1 \cdot P_{11} + \nu_3 \cdot P_{01}; \quad (4)$$

$$0 = \zeta_2 \cdot P_{11} - \zeta_1 \cdot P_{11} + \nu_2 \cdot P_{01}; \quad (5)$$

$$0 = \lambda \cdot P_{00} - \nu_1 \cdot P_{10}; \quad (6)$$

$$1 = P_{00} + P_{01} + P_{10} + P_{11} \quad (7)$$

The solution of the system will be represented by the following expression:

$$P11 = \frac{(\lambda \cdot v1 \cdot v2 + v1 \cdot v2 \cdot \zeta2 + v1 \cdot \zeta2 \cdot v3)}{(\lambda \cdot v1 \cdot v2 + \lambda \cdot v1 \cdot \zeta1 + \lambda \cdot v3 \cdot v1 + v1 \cdot v2 \cdot \zeta2 + v1 \cdot v2 \cdot \zeta1 v1 \cdot v3\, \zeta2 + v1 \cdot v2 \cdot \zeta1)} \quad (8)$$

4 Conclusions

The resulting ratio shows us the effectiveness of the management process of ensuring safe operation in the web-application of energy distribution.

In this relationship, three parameters are related, which depend on the ability of the Web network. This example shows how we have established analytical dependence of the following characteristics: environment aimed at analysis and recognition of the threat and activity aimed at neutralization of the problem arising in the Web-application [18].

By considering the relationship (8) as a condition of the existence of the management process in the Web network and setting the necessary level of information security in the form of P_{11}, we can guarantee the achievement of the purpose of management, i.e. compliance with information security in the Web network. The model of information security management in the Web-network based on the system of differential equations Kolmogorsk - Chapman made it possible to implement a guaranteed approach to the management of the Web-application of distribution and control of energy [19–21].

References

1. Mesarovic, M.D., Takahara, Ya.: General Systems Theory: Mathematical Foundations. Academic Press, New York/San Francisco/London (1975)
2. Kaverzneva, T., Rumyantseva, N., Uljanov, A., Belina, N.: Use of the logical-statistical model as a procedure for assessing occupational risks in the OSH management. In: IOP Conference Series: Materials Science and Engineering, vol. 666, no. 1 (2019). https://doi.org/10.1088/1757-899x/666/1/012091

3. Burlov, V.G., Grobitski, A.M., Grobitskaya, A.M.: Construction management in terms of indicator of the successfully fulfilled production task. Mag. Civ. Eng. **3**, 77–91 (2016). https://doi.org/10.5862/MCE.63.5
4. Lu, M., Li, H.: Resource-activity critical-path method for construction planning. J. Constr. Eng. Manag. **129**(4), 412–420 (2003)
5. Antonenkov, D.V., Solovev, D.B.: Mathematic simulation of mining company's power demand forecast (by example of "Neryungri" coal strip mine). In: IOP Conference Series: Earth and Environmental Science, vol. 87, pp. 1–7 (2017). https://doi.org/10.1088/1755-1315/87/3/032003
6. Gud, G.H., Makol, R.E.: System Engineering. Sov. Radio, Moscow (1962)
7. Karlin, L., Gogoberidze, G., Abramov, V., Lednova, J.: Indicator method of estimation of human impact assessment for coastal local municipalities. In: 2014 IEEE/OES Baltic International Symposium, BALTIC 2014, Estonia (2014). https://doi.org/10.1109/baltic.2014.6887840
8. Gogoberidze, G., Karlin, L., Abramov, V., Lednova, J.: Indicator method of estimation of human impact assessment for coastal local municipalities, measuring and modeling of multi-scale interactions in the marine environment. In: IEEE/OES Baltic International Symposium. BALTIC 2014, Estonia (2014). https://doi.org/10.1109/baltic.2014.6887840
9. Andreev, A.V., Burlov, V.G., Grachev, M.I.: Information technologies and synthesis of the management process model in the enterprise. In: International Science and Technology Conference EastConf, pp. 371–376 (2019). https://doi.org/10.1109/eastconf.2019.8725428
10. Leonova, N., Avdeeva, M., Kaverzneva, T.: Developing individuals' professional qualities in the course of technosphere safety specialists training. E3S Web Conf. **140** (2019). https://doi.org/10.1051/e3sconf/201914008008
11. Byzov, A.P., Efremov, S.V., Gomazov, F.A.: Social and economic aspects of acceptable risk of ensuring ecological safety at management of municipal waste. In: Proceedings of the 2018 IEEE International Conference "Management of Municipal Waste as an Important Factor of Sustainable Urban Development". WASTE 2018, pp. 10–13 (2018). https://doi.org/10.1109/waste.2018.8554106
12. Avdeeva, M., Leonova, N., Gomazov, F., Strelcova, E.: The automated system of models of management of information resources of higher education institution. In: IOP Conference Series: Materials Science and Engineering, vol. 666, no. 1 (2019). https://doi.org/10.1088/1757-899x/666/1/012099
13. Borisova, M., Byzov, A., Efremov, S.: Assessment of the maximum possible number of victims of accidents at hazardous production facilities for insurance purposes. In: IOP Conference Series: Materials Science and Engineering, vol. 666, no. 1 (2019). https://doi.org/10.1088/1757-899x/666/1/012096
14. Salkutsan, V., Russkova, I., Faustov, S.: Methods for assessing safe seniority in high noise conditions. In: IOP Conference Series: Materials Science and Engineering, vol. 666, no. 1 (2019). https://doi.org/10.1088/1757-899x/666/1/012102
15. Burlov, V.G., Grachev, M.I.: Model of transport systems management, taking into account the possibilities of innovation. Tech. Technol. Prob. Serv. **42**(20), 34–38 (2017). https://doi.org/10.1016/j.trpro.2017.01.023, (in Russian)
16. Mankov, V., Efremov, S., Monashkov, V.: Grounding device for electrical networks and electrical installations in the arctic regions. In: IOP Conference Series: Earth and Environmental Science, vol. 302, no. 1 (2019). https://doi.org/10.1088/1755-1315/302/1/012066

17. Idrisova, I.I., Kaverzneva, T.T., Rumyantseva, N.V., Skripnik, I.L.: Neural network modeling of safety system for construction equipment operation in permafrost zone. In: IOP Conference Series: Earth and Environmental Science, vol. 302, no. 1 (2019). https://doi.org/10.1088/1755-1315/302/1/012128
18. Ferreira, E.F., Barros, J.D.: Faults monitoring system in the electric power grid of medium voltage. Proc. Comput. Sci. **130**, 696–703 (2018). https://doi.org/10.1016/j.procs.2018.04.123
19. Russkova, I., Dolgikh, N., Salkutsan, V., Logvinova, Y.: Russia's arctic is as an object of environmental monitoring. In: IOP Conference Series: Earth and Environmental Science, vol. 302, no. 1 (2019). https://doi.org/10.1088/1755-1315/302/1/012028
20. Burlow, V.G., Grachev, M.I., Shlygina, N.S.: Adoption of management decisions in the context of the uncertainty of the emergence of threats. In: Proceedings of 2017 XX IEEE International Conference on Soft Computing and Measurements (SCM), pp. 107–108 (2017). https://doi.org/10.1109/scm.2017.7970510
21. Tumanov, A., Venevsky, S.: Elaboration of theoretical methods for assessment of isolated and combined physical damage effects of technogenic accident while transporting radiological materials by multimodal transport. In: IOP Conference Series: Materials Science and Engineering, vol. 582, no. 1 (2019). https://doi.org/10.1088/1757-899x/582/1/012038

Analysis of Tools for Determining Professional Suitability to Perform Hazardous Construction Works

Liliia Kireeva(✉), Tatiana Kaverzneva, Regina Shaydullina, and Adel Farkhutdinova

Peter the Great St. Petersburg Polytechnic University, Polytechnicheskaya 29, Saint-Petersburg 195251, Russia
la.savchenkova@yandex.ru

Abstract. A person is an integral part of many processes, and his activities often directly affect the security of the entire system. Selection of workers whose professionally important qualities fully meet the profession requirements leads to a decrease in the risk of accidents and injuries. In this work, the recent investigations of issues on professional suitability and professional selection were studied for the purpose of analyzing them and possible adaptation for use in the current business of Russia when conducting professional selection for hazardous works in construction industry. It was found that the main direction of professional suitability study is focused on socially-oriented activities, and there are not enough works aimed at studying the links of professional selection to improve safety. It was established that personality tests are not applicable in determining the professional suitability of common working professions for hazardous work, in particular, in the construction industry. The need for development of new techniques that allow obtaining valid results for the formation of subsequent decisions on professional suitability with the possibility of their further integration into the existing enterprises has been identified.

Keywords: Suitability · Selection · Qualities · Hazard · Work · Safety · Construction

1 Introduction

Ensuring the required level of security is one of the most pressing issues that accompany human activities in developed and developing countries. Many areas of activity involve risks and dangers, either very significant in terms of damage, or having a significantly high frequency, which ultimately leads to damage. In many areas of activity, different countries develop and implement safety programs, tighten control and supervision, transform legislation, develop and implement new technical safety systems aimed at reducing injuries and accidents. According to the report of the Economic and Social Committee and the Committee of the regions on an EU Strategic Framework on Health and Safety at Work 2014–2020, a lot of countries are moving to a proactive risk-based approach, which is the most popular today [1]. This all applies to the construction industry, which remains one of the most traumatic areas of human activity.

V. Murgul and V. Pukhkal (Eds.): EMMFT 2019, AISC 1258, pp. 642–648, 2021.
https://doi.org/10.1007/978-3-030-57450-5_55

Ongoing research continues to develop new and improve existing tools, approaches and specific techniques aimed at increasing security [2–4]. The attention of researchers is very often directed to a person not only as an object of protection, but also as an integral part of any process, which mainly affects safety parameters. The so-called risk management based on the management of the “Human Factor” is becoming increasingly popular. In international literature, the methods of Human reliability analysis (HRA) are very common as a consequence of general recognition of the fact that the role of human operations in safety-critical systems is so important that they should be modeled as part of a risk assessment [5–8].

In order to reduce injuries in work associated with increased danger, provided that it is impossible to refuse the participation of a person in such work (ergatic systems), the concept is being developed. This concept implies that selection of workers whose professionally important qualities fully comply with the requirements of profession will result in reducing the risk of accidents and injuries, including that obtained at construction sites. By attracting the most relevant workers to hazardous work requiring increased responsibility, concentration of attention and the ability to quickly and correctly respond to the current situation (welders, special equipment operators, responsible executives of hazardous work), a significant part of hazardous work risks can be compensated. The main tool for implementing this approach is professional selection.

In most cases professional selection is a key element of modern systems and methods of personnel management, which allows one to increase the socio-economic efficiency of production and reduce occupational injuries [9]. Today, there are a large number of different methods for assessing professional suitability, with varying degrees of development or effectiveness. The methodology for personnel selection to perform certain tasks or work has a significant history, the first scientifically based methods for assessing professional suitability began to appear in the second half of the 19th century. The world practice of professional selection has been significantly enriched by American military developments and their orienting tests. At the end of the 20th century, the initiative was seized by the free labor market, which began to apply methods of professional selection for civilian positions, which are associated not only with danger, but with stress and responsibility. The business community actively embraced the practice of professional selection and continued to develop it outside of scientific institutions. Today, there are a sufficient number of detailed and approved test tasks and other activities that are used in the world with various intensities in order to identify the most suitable candidates for specific works.

This work is aimed at studying the recent publications devoted to the issues on professional suitability and professional selection with a purpose of analyzing them and possible adaptation for use in the current business of Russia when conducting professional selection for hazardous work in construction industry.

2 Materials and Methods

First of all, it is necessary to determine the general types and approaches to the professional selection procedure. Most often there are 4 main types of professional selection: Medical, Educational, Socio-psychological and Psycho-physiological. Since performing hazardous work with specific requirements is considered here, the authors did not consider Medical and Educational professional selections, as these directions are generally regulated at the state level and are obligatory in accordance with the requirements of the law with the aim of not allowing disabled or untrained personnel to do dangerous work. For the study and analysis, initially only the psycho-physiological and socio-psychological (or psychological) types of professional selection were left, which can be used to increase the general level of security and are of interest to the labor market.

To achieve this goal, the literature review was organized as follows:

1) the literature search method;
2) the selection of relevant publications;
3) retrieval and classification of data;
4) discussion of research.

Articles containing the following keywords were selected: professional suitability, professional selection, method, hazardous work, human factor, injury reduction, conformity, professional qualities, labor safety, safety measures, assessment, selection, analysis, quantitative, qualitative. The search was conducted in Russian and English by groups of keywords using "AND" or "OR" operators. The publications obtained in this way were used as the basis for studying the tools used for professional selection of personnel and the study of professional suitability. The analysis of names, keywords and theses was carried out [10]. The research included only scientific papers subjected to expert evaluation, research reports in the public domain. Thus, 36 peer-reviewed publications were analyzed. The main criteria for choosing professional selection methods are shown in Fig. 1.

Fig. 1. The relationship of the main criteria for the selection of tools for assessing professional suitability.

3 Results and Discussion

The works from 2007 have been analyzed. Most of the selected publications were devoted to professions of social orientation or that related to working with people (doctors, teachers, consultants, etc.). The main idea of these articles was the need to take into account the socio-psychological aspects of such activities, in particular the issue of conducting professional selection or predicting the success of candidates in such positions. Since these articles did not propose specific approaches or techniques in assessing significant qualities, but in fact, they simply stated the need for a psychological assessment of candidates, they were not further analyzed for the purposes of this article. Thus, 14 publications were left for analysis.

Conditionally, all processed data of the remaining publications can be divided into 3 blocks:

1) approaches to the definition of professionally important qualities;
2) approaches to conducting changes in indicators/values of professionally important qualities;
3) approaches to the method of decision making.

The papers of the first block devoted to the definition of professionally important qualities for specific professions are mainly dealing with the analysis of the activity itself, for which professional suitability is assessed with a parallel selection of the corresponding professionally important qualities. In fact, these are examples of partial formation of professiograms [9]. To conduct a suitability assessment procedure, one must first study the requirements for the job and identify the professionally important qualities that are necessary for more successful work. The relevance of this process is due to constantly appearing new types of work (in particular due to appearance of new types of equipment), changes in the nature of old professions, etc. Until now, the issues of choosing individual professionally important qualities have not been adequately studied, there is no single methodological approach, there is an incompleteness in solving the problem of professional selection of personnel for work related to danger [11]. It is worth noting that papers of this block were written mainly by authors from countries related to the developing type of economy (Russian Federation, Turkey, some countries of the Eastern Europe). Probably, this may be explained by, among other things, insufficient organizational component in the safety culture of these countries.

The papers of the second block are devoted to the collection of information, on the basis of which a conclusion is drawn on the degree of compliance with the requirements. The most popular way is testing. There are 4 categories of tests used in professional selection.

1. Intelligent and psychometric tests. This type of test was specially created to study the lower and middle personnel in industry. Such tests may have a relatively high validity for certain professions or may not have any validity at all. The success of their application largely depends on the correctness of identifying professionally important qualities. In recent years, no new techniques have been discovered. Nevertheless, it is a good, sufficiently valid and tested tool; further studies are required to increase its efficiency.

2. Psychomotor tests are aimed at assessing the speed and accuracy of motor coordination, the course of mental reactions. Most of these tests are passed on real equipment or special simulators. This is a fairly objective way of assessment, but it is greatly complicated by the need for special equipment, personnel with specific qualifications and serious time costs. For these reasons, the use of psychomotor tests in the current business environment is not widespread, but research continues to determine the relationship of indicators with safety [12, 13]. The development of methods is required to obtain data of sufficient accuracy subsequent with the possibility of their integration into existing enterprises.
3. Personality tests and interest tests. These tests are based on the assumption that the individual will be a more successful employee if the structure of his personality and interests is similar to that of a person who has reached heights in professional activity and is consistent with the culture and values of the company, etc. The main recognized drawback of this kind of tests is that many subjects can distort the answers when motivated to get a job. Despite this, personality tests are still very popular both in the market and among researchers. Nevertheless, based on works that analyze an impressive retrospective of research in this issue, the authors conclude that it is inappropriate to use personality tests as a tool for selecting personnel for hazardous work, due to their low validity, reliability and massive distortion of the results by subjects [14].
4. Projective tests. Test materials with a vague, undefined meaning are used, for example, black spots, unfinished sentence, plot pictures, etc. Initially, these tests were created by clinical psychologists to analyze human abnormalities. The validity of this kind of tests for the purpose of determining the ability to hazardous work was not evaluated due to inapplicability.

All works related to the 3rd block, namely 8 studies, propose the Multi-Criteria Decision-Making (MCDM) method or its variations for making a decision on staff selection, which indicates its great popularity among researchers and its high applicability in this area [12, 15–21].

4 Conclusions

Recently, the main attention in the field of study of professional suitability is directed to socially-oriented activities and professional selection is focused on the personality of the candidate. Today there are not many new studies in the field of professional suitability when performing hazardous work in order to reduce the risk of accidents and injuries [22, 23]. The basic principles of professional selection have already been formed, but the emergence of new technologies and a fundamental change in the filling of professions make us continue to study professional suitability with the aim of finding new approaches to studying professional suitability or increasing the effectiveness of the existing ones [23, 24]. The analysis showed that today there is still no ideal method for determining professional suitability, nevertheless, the developments available today can be used to improve and increase the reliability of professional suitability decisions when performing hazardous work, including that in construction. The results of this study demonstrate the need for further research in the field of reducing injuries during hazardous work on the basis of human factor management, in particular in the construction industry, as well as new methods to obtain valid results for the formation of subsequent decisions on professional suitability with the possibility of their further integration in the existing enterprises.

References

1. Guldenmund, F.: The Nature of safety culture: a review of theory and research. Saf. Sci. SAF SCI **34**, 215–257 (2000). https://doi.org/10.1016/s0925-7535(00)00014-x
2. Burlov, V., Andreev, A., Gomazov, F.: Mathematical model of human decision - a methodological basis for the realization of the human factor in safety management. Procedia Comput. Sci. **145**, 112–117 (2018). https://doi.org/10.1016/j.procs.2018.11.018
3. Glebova, E.V., Volokhina, A.T., Vikhrov, A.E.: Accidents and injury rates reduction in petroleum industry based on the development and implementation of the automated complex for the employees' professional competences assessment. In: IOP Conference Series: Earth and Environmental Science, vol. 272(3) (2019). https://doi.org/10.1088/1755-1315/272/3/032063
4. Onofrio, R., Trucco, P.: Human reliability analysis (HRA) in surgery: identification and assessment of influencing factors. Safety Sci. **110**, 110–123 (2018). https://doi.org/10.1016/j.ssci.2018.08.004
5. Boring, R.L., Hendrickson, S.M.L., Forester, J.A., Tran, T.Q., Lois, E.: Issues in benchmarking human reliability analysis methods: a literature review. Reliab. Eng. Syst. Saf. **95**(6), 591–605 (2010). https://doi.org/10.1016/j.ress.2010.02.002
6. Park, J., Arigi, A.M., Kim, J.: A comparison of the quantification aspects of human reliability analysis methods in nuclear power plants. Ann. Nucl. Energy **133**, 297–312 (2019). https://doi.org/10.1016/j.anucene.2019.05.031
7. Wang, Y., Ding, Y., Chen, G., Jin, S.: Human reliability analysis and optimization of manufacturing systems through bayesian networks and human factors experiments: a case study in a flexible intermediate bulk container manufacturing plant. Int. J. Ind. Ergon. **72**, 241–251 (2019). https://doi.org/10.1016/j.ergon.2019.05.001
8. French, S., Bedford, T., Pollard, S.J.T., Soane, E.: Human reliability analysis: a critique and review for managers. Saf. Sci. **49**(6), 753–763 (2011). https://doi.org/10.1016/j.ssci.2011.02.008
9. Svetlakova, A., Kaverzneva, T., Tarkhov, D., Belina, N.: Analysis of tools for assessing the terms of working environment of foreigners. In: MATEC Web of Conferences, vol. 245 (2018). https://doi.org/10.1051/matecconf/201824512004
10. Petukhov, I.V.: Methodological bases of assessment of profitability of the operator of ergatic systems modern problems of science and education. **2** (2013)
11. Chen, X., Wei, Z., Gao, L.: Professional driver suitability evaluation. Procedia Eng. **15**, 5222–5226 (2011). https://doi.org/10.1016/j.proeng.2011.08.968
12. Salkutsan, V., Russkova, I., Faustov, S.: Methods for assessing safe seniority in high noise conditions. In: IOP Conference Series: Materials Science and Engineering, vol. 666(1) (2019). https://doi.org/10.1088/1757-899x/666/1/012102
13. Morgeson, F.P., Campion, M.A., Dipboye, R.L., Hollenbeck, J.R., Murphy, K., Schmitt, N.: Reconsidering the use of personality tests in personnel selection contexts. Pers. Psychol. **60**(3), 683–729 (2007). https://doi.org/10.1111/j.1744-6570.2007.00089.x

14. Hashemkhani, Z., Banihashemi, S.: Personnel selection based on a novel model of game theory and MCDM approaches. In: 8th International Science Conference Business and Management (2014). https://doi.org/10.3846/bm.2014.024
15. Kabak, M., Burmaoğlu, S., Kazançoğlu, Y.: A fuzzy hybrid MCDM approach for professional selection. Expert Syst. Appl. **39**(3), 3516–3525 (2012). https://doi.org/10.1016/j.eswa.2011.09.042
16. Polyukhovich, M., Burlov, V., Mankov, V., Bekbayev, A.: Electric power supply management of the construction site in the interests of facilitating electrical safety. In: E3S Web of Conferences, vol. 140 (2019). https://doi.org/10.1051/e3sconf/201914008006
17. Kamisli Ozturk, Z., Toptancı, Ş.: An integrated MCDM model for occupational safety specialist selection. J. Bus. Res. Turk. **9**, 419–435 (2017). https://doi.org/10.20491/isarder.2017.339
18. Kilic, B., Ucler, C.: Stress among ab-initio pilots: a model of contributing factors by AHP. J. Air Transp. Manag. **80**(C) (2019). https://doi.org/10.1016/j.jairtraman.2019.101706
19. Afshari, A.R., Mojahed, M., Yusuff, R.M., Hong, T.S., Ismail, M.Y.: Personnel selection using ELECTRE. J. Appl. Sci. **10**(23), 3068–3075 (2010). https://doi.org/10.3923/jas.2010.3068.3075
20. Lin, H.T.: Personnel selection using analytic network process and fuzzy data envelopment analysis approaches. Comput. Ind. Eng. **59**(4), 937–944 (2010). https://doi.org/10.1016/j.cie.2010.09.004
21. Borisova, M., Byzov, A., Efremov, S.: Assessment of the maximum possible number of victims of accidents at hazardous production facilities for insurance purposes. In: IOP Conference Series: Materials Science and Engineering, vol. 666(1) (2019). https://doi.org/10.1088/1757-899x/666/1/012096
22. Leonova, N., Avdeeva, M., Kaverzneva, T.: Developing individuals' professional qualities in the course of technosphere safety specialists training. In: E3S Web of Conferences, vol. 140 (2019). https://doi.org/10.1051/e3sconf/201914008008
23. Avdeeva, M., Leonova, N., Gomazov, F., Strelcova, E.: The automated system of models of management of information resources of higher education institution. In: IOP Conference Series: Materials Science and Engineering, vol. 666(1) (2019). https://doi.org/10.1088/1757-899x/666/1/012099
24. Andreev, A.V., Burlov, V.G., Gomazov, F.A., Penner, Y.A.: Improving the system of higher education for enterprises of industrial and economic complex. In: Proceedings of 2018 17th Russian Scientific and Practical Conference on Planning and Teaching Engineering Staff for the Industrial and Economic Complex of the Region PTES 2018, pp. 86–88 (2019). https://doi.org/10.1109/ptes.2018.8604230

Offenses Prevention at Municipal Energy Facilities Under Geoinformation System Management

Vyacheslav Burlov[1], Aleksey Mironov[2](✉), Anna Mironova[3], Jamila Idrisova[1], and Irina Russkova[1]

[1] Peter the Great St. Petersburg Polytechnic University, St. Petersburg, Russia
[2] Russian State Hydrometeorological University, St. Petersburg, Russia
wakepolarbear@gmail.com
[3] ITMO University, Kronverksky Prospect, 49, 197101 Saint Petersburg, Russia

Abstract. In the conditions of narrow departmental management of the municipal infrastructure, energy security is being destroyed due to illegal connections to energy resources, thefts of equipment, and vandalism. The timely application of the geoinformation system in order to prompt prevention, reliable revealing and proper proving of offenses ensures a preventively guaranteed efficiency of energy supply. In this paper, proposed a synthesis the energy security management model, based on the natural-scientific approach to the adoption of managerial decisions. The condition for the existence of guaranteed management, which relates the characteristics of processes and performance indicators, is concretized by the Kolmogorov-Chapman equations system. Network models of interacting processes of the revealing and proving of offenses, the origin, and identification of violations, them neutralization through the managerial impact on the limited resources of ensuring energy security and of geoinformation system are built. Based on the simulation results, the requirements for the geoinformation system that ensure guaranteed management and maximum stability to the municipal energy sector within a reasonable time are determined.

Keywords: Energy security · Guaranteed management · Administrative practice · Geoinformation system · Natural-scientific approach

1 Setting Targets and Tasks

The huge latency, estimated up to 3/4 of the really committed misconducts punishable by law, indicates an unsatisfactory provision of the energy security in the municipal infrastructure [1]. The problem is related to opportunities for unauthorized connection to electric networks and pipelines of petrol or gas, for thefts of non-ferrous metals from energy equipment, for vandalism and hooliganism at municipal energy facilities. The destruction of the stability of energy security at municipal facilities is provoked by the random and highly departmental nature of the reasons used to initiate criminal and

V. Murgul and V. Pukhkal (Eds.): EMMFT 2019, AISC 1258, pp. 649–658, 2021.
https://doi.org/10.1007/978-3-030-57450-5_56

administrative affairs. The sporadic and post-factum response of law enforcement officials to the flow of offenses gives a vicious punitive and fiscal meaning to the production about them. The law enforcement agencies are focused on collecting fines for committed acts and are indifferent to the damage caused [2]. Overcoming the incompleteness, unreliability and untimely to the provision of energy security is especially relevant in relation to offenses, the traces of which are hidden behind the features of the terrain or covered by offenders. Prevention, revealing and proving within a reasonable time regularly requires the geographical coordinates with the signs of preparation or commission of offenses to law enforcement agencies in the territory under their jurisdiction, for carryout targeted inspections immediately and for prevent or suppress of harmful consequences.

To implement the geocoordinate support of preventive and operational-search activities, it is necessary to systematically analyse the cartographic information on tracking changes in the position, shape, and structure of anthropogenic or natural objects with an eye to the ascertainment of event and of elements on the offense [3]. Addresses for targeted inspections can be obtained using the geoinformation system during the monitoring of territories under the jurisdiction of the law enforcement agency [4]. Dozens of geoportal solutions at the federal, regional and local levels operate in government or municipal departments, commercial structures of Russia [5]. But because of the inability to timely prevent and suppress offenses, energy collapses and catastrophic consequences arise annually from the failure of point objects in the electric networks, pipelines of petrol or gas. Therefore, at the junction of energy security and geoinformatics, an adequate model for managing the prevention, detection, and proof about offenses using a geoinformation system is gaining scientific and practical interest [6].

According to the theory of functional systems, a decision-maker implements the management process based on the model, which is manifested in the chain of basic elements of its formation: “excitement” - “recognition” - “reaction to the situation” [7]. The relevance of this paper to the automation of management is determined by the lack of an adequate mathematical model for managerial decision making, based on ensuring a balanced unity of the functioning of social and geoinformation systems, the basic laws of the world order within the framework of a single approach to the universal formalized criterion.

In contrast to analysis, synthesis forms processes with predetermined properties [8]. The synthesis of energy security management will allow guaranteed to ensure the revealing and proving of offenses at municipal energy facilities within a reasonable time.

Thus, in this paper, the following tasks are solved:

- synthesis of a mathematical model for managerial decision making;
- substantiation of the criterion for guaranteed management of energy security using a geoinformation system;
- construction of structural-functional technology for automated management of prevention, detection, and proof about offenses at municipal energy facilities within a reasonable time.

2 Synthesis of a Model for Managerial Decision Making

In order to exclude arbitrariness in reasoning and conflicting conclusions, the axiomatic-deductive method is used. To ensure the adequacy of the model for managerial decision making, its synthesis is based on the law of object integrity conservation, as a stable repetitive relationship between the properties of the object and the properties of its actions for a fixed mission [9].

In accordance with the natural-scientific approach, integrating the properties of the world around us, consciousness and cognition, the process of managerial decision making is considered in the light of its three properties at each of the three levels of world cognition, as shown in Fig. 1 [10].

The formation of an adequate model for managerial decision making according to Fig. 2 is to establish formal analytical relations between the three technological components that are characterized by temporary resources, irreversible for the decision-maker:

- Situation (Object) is a set of current state characteristics of the energy security, factors, and conditions of decision-maker activity, which is identified with the period of the target process ΔT and the periodicity of offense occurrence (average time before the emergence of the problem) Δt;
- Procedural Decision (Mission) is the provision by the decision-maker of the condition for energy security implementation within a reasonable time in the current Situation to achieve the management goal, which is identified with an adequate periodicity of response to offenses (average time to neutralize problem) ΔN;
- Information-Analytical Work (Action) is the continuous extraction, accumulation, generalization, analysis of geodata about the current Situation, which is generalized into periodicity of offenses detection (average time to identify the problem) ΔI.

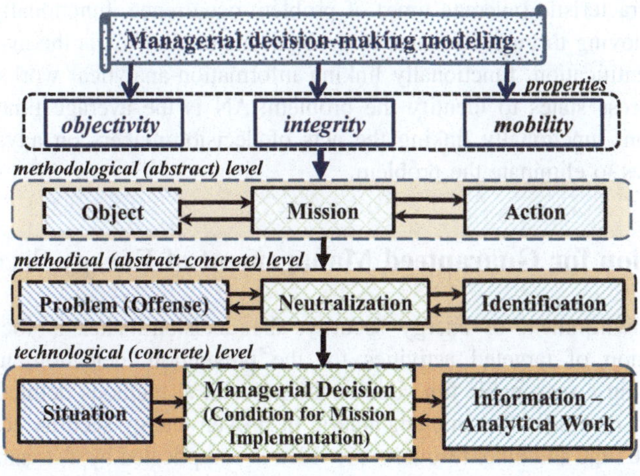

Fig. 1. Deployment of a natural-scientific approach to modeling managerial decision making.

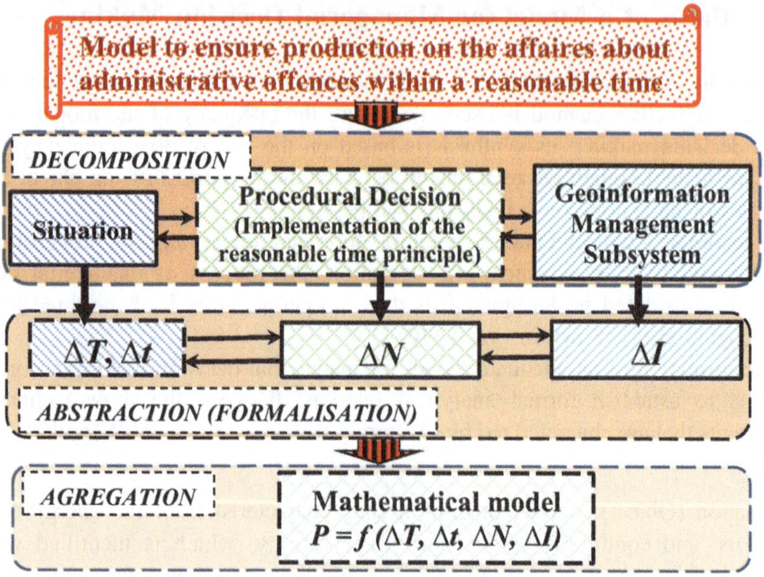

Fig. 2. Synthesis of a mathematical model for managerial decision making.

Using the methods of decomposition, abstraction, and aggregation, the process of managerial decisions making is formalized into the mathematical aggregate of the model, where:

$$P = f(\Delta T, \Delta t, \Delta N, \Delta I) \tag{1}$$

P is the probability of finding administrative proceedings during the management in each of its basic states: Initial, Target, Identification or Neutralization; Δt is the generalized characteristic (average time) of problem occurrence, functionally uniting the works for moving through its states to problem maturation; ΔI is the average time of problem identification, functionally linking information-analytical works on the passage of process states to identify the problem; ΔN is the average time of problem neutralization, functionally linking the acts of decision-makers on advance through process states to eliminate the problem.

3 Criterion for Guaranteed Management of Energy Security

In a real environment of energy security, the decision-maker is focused on the implementation of targeted activities for the prevention, detection, and proof of offenses within a reasonable time due to the procedural methods defined by law and departmental methods. The implementation of proven energy security schemes based on normative time and other resources provided is complicated by a stream of objective and subjective circumstances that make it difficult to quickly establish the geographical coordinates of the offense place where there are signs of the event and corpus delicti. Stochastically emerging problems can and should be proactively identified using a geoinformation system and guaranteed to be eliminated with the involvement of additional resources in the conditions of restrictions on their availability. If the goals of reliable fixing and proving the offense are not achieved within a reasonable time, the decision-maker is forced to delay, procedurally extend or terminate the affair investigation, which is tantamount to the target failure and to the breakdown of energy security management [11].

Considering the three-component nature of the basic model of managerial decision making to protect from the interference of problems (offenses) in the target activity, the energy security management includes four interacting processes, as shown in Fig. 3. The target process during the implementation of energy security in the normal mode is objectively accompanied by a Poisson stream of problems leading to failures of a reasonable time. When a reasonable period of implementation is delayed, the target process breaks back to the initial state by the extension or termination of the prevention, detection, and proof about offenses. In order to prevent the non-fulfillment of the target task and the failure of management, the identification and neutralization processes due to the works of the decision-maker in the automated management subsystem based on the geoinformation system for a comparable time diagnose problems occurrence and take remedial measures to eliminate them with a deficiency of resources. In cases of insoluble problems, part of the identification and neutralization processes can also be disrupted to the initial state.

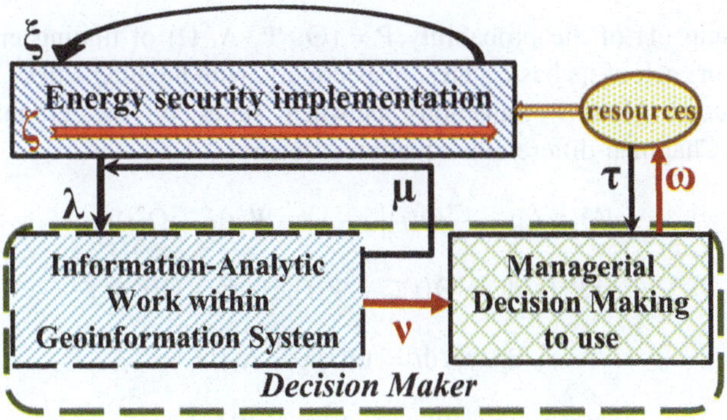

Fig. 3. Energy security management scheme.

In Fig. 4, the energy security management model in the form of a continuous Markov chain is characterized by the probabilities of being in one of four basic states associated with the intensities of target process $\zeta = 1/\Delta T$, of problem occurrence $\lambda = 1/\Delta t$, of its identification $\nu = 1/\Delta I$ and neutralization $\omega = 1/\Delta N$, as well as the failure rates of target process $\xi = f(\Delta T)$, of identification $\mu = f(\Delta I)$ and neutralization $\tau = f(\Delta N)$:

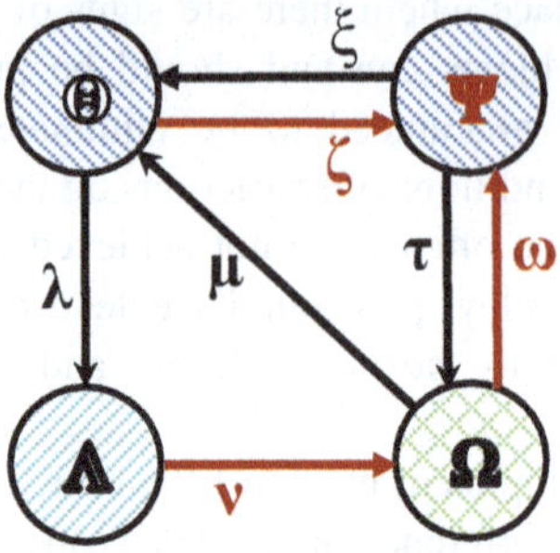

Fig. 4. Management states graph.

- Θ is the probability of initial state when the energy security management by the decision-maker is in the initial state of the prevention, detection, and proof of offenses, which does not require the identification or neutralization of problems;
- Ψ is the probability of target completing when the decision-maker fulfilled the target task of energy security management with the guarantee of neutralizing the associated problems (offenses) within a reasonable time;
- Λ is the probability of identification when the decision-maker identifies problems leading to a breakdown of a reasonable time for their subsequent neutralization;
- Ω is the probability of neutralization when the decision-maker, based on the results of identification, neutralizes problems that cause disruptions in ensuring a reasonable time.

The relationship (1) of the probability P = (Θ, Ψ, Λ, Ω) of finding energy security management in each of its basic states, respectively, with the intensities ζ, ξ, λ, ν, μ, ω, τ of continuous Markov transitions between states of the graph is specified Kolmogorov-Chapman differential equations system:

$$
\begin{aligned}
&d\Theta(t)/dt = -\Theta(t)[\zeta + \lambda] + \Psi(t)\xi + \Omega(t)\mu \\
&d\Psi(t)/dt = \Theta(t)\zeta - \Psi(t)[\xi + \tau] + \Omega(t)\omega \\
&d\Lambda(t)/dt = \Theta(t)\lambda - \Lambda(t)v \\
&d\Omega(t)/dt = \Psi(t)\tau + \Lambda(t)\nu - \Omega(t)[\mu + \omega] \\
&\Theta(t) + \Psi(t) + \Lambda(t) + \Omega(t) = 1
\end{aligned} \tag{2}
$$

When, over time, the Poisson streams of interacting processes tend to the limiting stationary mode, the Kolmogorov-Chapman differential Eqs. (2) are transformed into a system of linear homogeneous algebraic equations, the solution of which using the Cramer method is system-forming factors of management:

$$\Theta = \frac{\xi\nu\omega + \nu\mu(\xi + \tau)}{\xi\omega(\lambda+\nu) + \mu(\lambda+\nu)(\xi + \tau) + \nu(\zeta + \lambda)(\omega + \tau) + \nu(\zeta\mu + \xi\lambda)}$$

$$\Psi = \frac{\nu\omega(\zeta + \lambda) + \zeta\nu\mu}{\xi\omega(\lambda+\nu) + \mu(\lambda+\nu)(\xi + \tau) + \nu(\zeta + \lambda)(\omega + \tau) + \nu(\zeta\mu + \xi\lambda)} \quad (3)$$

$$\Lambda = \frac{\xi\lambda\omega + \lambda\mu(\xi + \tau)}{\xi\omega(\lambda+\nu) + \mu(\lambda+\nu)(\xi + \tau) + \nu(\zeta + \lambda)(\omega + \tau) + \nu(\zeta\mu + \xi\lambda)}$$

$$\Omega = \frac{\nu\tau(\zeta + \lambda) + \xi\lambda\nu}{\xi\omega(\lambda+\nu) + \mu(\lambda+\nu)(\xi + \tau) + \nu(\zeta + \lambda)(\omega + \tau) + \nu(\zeta\mu + \xi\lambda)}$$

Thus, in the current Situation, characterized by the intensities of target process ζ and of problem occurrence λ, at the normatively established levels of the maximum allowable disruption frequencies ξ, μ, τ and of the minimum sufficient efficiency Ψ, guaranteed criterion (3) of energy security management allows the decision-maker to control the sufficiency and to optimize the intensities of identification ν due to the Information-Analytical Work within geoinformation system and of neutralization ω due to the Managerial Decision by rationalizing the durations of transitions by their events.

4 Structural-Functional Technology for Automated Management

The intensity of the stream depends on the structure and duration of transitions by states (events) within the process. Therefore, it is advisable to obtain the intensity of target process $\zeta = 1/\Delta T$, the intensity of problem occurrence $\lambda = 1/\Delta t$, the intensities of its identification $\nu = 1/\Delta I$ and neutralization $\omega = 1/\Delta N$ by the structural-functional method. When using it, by clearly linking the works on transitions and the times spent on them in the network model of each process, network analysis makes it possible to evaluate periods ΔT, Δt, ΔI, ΔN through the critical path of each process [12]. According to the observations of administrative statistics regarding the timeliness of the issuance and execution of ordinances on the affairs about administrative offenses, $\xi \approx 0{,}25\zeta$ and $\tau \approx 0{,}05\omega$. The probability of technical errors in the functioning of the geographic information system does not exceed $\mu \approx 0{,}10\nu$.

Based on the average statistical durations of transition between process events, the following indicators are calculated in each network model:

- the earliest possible time of event j

$$T_e(j) = \max_{i \subset j}[T_e(i) + t_{ij}] \quad (4)$$

where $i \subset j$ occurs when event i precedes the next event j; t_{ij} is the duration of process transfer from event i to j;

- the latest allowed time of event i

$$T_l(i) = \min_{i \subset j}[T_l(j) - t_{ij}] \tag{5}$$

- the time reserve of event i

$$R_i = T_l(i) - T_e(i) \tag{6}$$

- the full reserve of time for transfer from event i to j

$$r(i,j) = T_l(j) - T_e(i) - t_{ij} \tag{7}$$

- the duration of the critical path of the process, considered as the maximum total duration to transfer the process from the initial to the final event of the network model or as the durations sum of work for which the full reserves of time are zero.

5 Findings

A numerical assessment of system-forming factors (3) on network models of the municipal energy infrastructures confirmed that, based on the average statistical duration of the process procedures, the proposed management model is capable of monitoring critical changes in the situation according to the conditions of energy security. It also helps to reconfigure the structure and functionality of prevention, detection, and proof of offenses within a reasonable time under the current situation.

In accordance with Fig. 5, the dependence of management efficiency Ψ on the failure rate of the target process ξ is close to a decreasing exponent. At the same time, a step-by-step twofold increase in the intensities of identification ν and neutralization ω causes an exponentially decaying increase of efficiency Ψ within identical slice ξ. Therefore, the failure rate of the target process ξ sets the status of compliance (suitability) of energy security by the signs of a reasonable time. It is within the framework of the suitability indicator ξ that the necessary management efficiency Ψ is adjusted by varying the intensities of identification ν and neutralization ω.

The leadership of the enforcement agencies was proposed a mechanism that, within the regulatory acceptable range of suitability for energy security, allows to maintain sufficient efficiency in the prevention, detection, and proof of offenses by attracting the necessary resources in a timely manner.

Figure 6 shows the significant decrease in management efficiency Ψ from the share of neutralization failure τ in comparison with the influence of a similar identification failure μ. Technical errors of the geographic information system can be partially

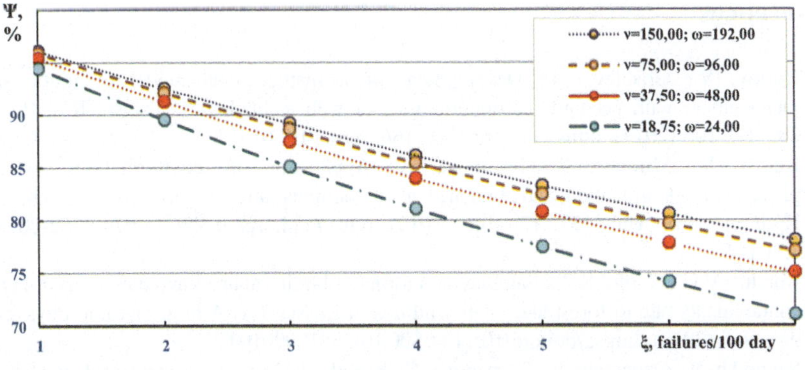

Fig. 5. Dependence of efficiency Ψ on the failure rate ξ during double extension ν, ω.

corrected by other means of information and analytical work. A limited set of procedural means of neutralizing offenses does not leave by the decision-maker the opportunity to compensate for irreparable losses of a reasonable time.

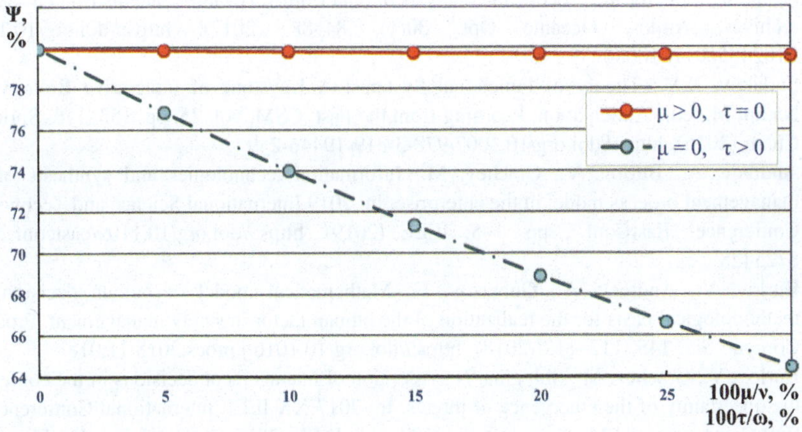

Fig. 6. Dependence of efficiency Ψ on the failure rates $\mu = f(\nu)$, $\tau = f(\omega)$.

When changes the situation, characterized by the intensity of the target process ζ and the problem flow λ, the intensities ν, ω, of activity of the geographic information system and of the decision-maker must be strengthened by the available reserves to identify and neutralize offenses within a reasonable time taking into account the failure frequencies μ, τ.

References

1. Burlov, V., Grachev, M.: Development of a mathematical model of traffic safety management with account for opportunities of web technologies. Transp. Res. Procedia **20**, 100–106 (2017). https://doi.org/10.1016/j.trpro.2017.01.023
2. Burlov, V., Mironov, A., Mironova, A.: Application of geoinformatiom system in prevention, identification and evidence about administrative offenses. Proc. Russ. State Hydrometeorol. Univ. **57**, 126–146 (2019). https://doi.org/10.33933/2074-2762-2019-57-126-146
3. Bondur, V.G., Gordo, K.A.: Satellite monitoring of burnt-out areas and emissions of harmful contaminants due to forest and other wildfires in Russia. Izv. Atmos. Oceanic Phys. **54**(9), 955–965 (2018). https://doi.org/10.1134/S0001433818090104
4. Savinykh, V., Oznamets, V., Selmanova, N., Tsvetkov, V.: System-categorical analysis used while monitoring of the land according to the remote sensing data. Geod. Aerophotosurveying **62**(1), 106–113 (2018). https://doi.org/10.30533/0536-101X-2018-62-1-106-113
5. Kuznetsov, A., Trusov, S., Baraboshkin, O., Bobrovskij, S., Romanov, A., Romanov, A.: Analysis of the results obtained over three years of operation of AIS vessel monitoring equipment based on the resurs-P No. 2 spacecraft. Rocket Device Eng. Inf. Syst. **5**(4), 80–87 (2018). https://doi.org/10.30894/issn2409-0239.2018.5.4.80.87
6. Kashnitskii, A.V., Lupyan, E.A., Balashov, I.V., Konstantinova, A.M.: Technology for designing tools for the process and analysis of data from very large scale distributed satellite archives. Atmos. Oceanic Opt. **30**(1), 84–88 (2017). https://doi.org/10.1134/S1024856017010080
7. Sudakov, K.V.: Theory of Functional Systems: A Keystone of Integrative Biology. In: Nadin, M. (ed.) Anticipation: Learning from the Past. CSM, vol. 25, pp. 153–173. Springer, Cham (2015). https://doi.org/10.1007/978-3-319-19446-2_9
8. Andreev, A., Burlov, V., Grachev, M.: Information technologies and synthesis of the management process model in the enterprise. In: 2019 International Science and Technology Conference "EastConf.", pp. 1–5. IEEE (2019). https://doi.org/10.1109/eastconf.2019.8725428
9. Burlov, V., Andreev, A., Gomazov, F.: Mathematical model of human decision - a methodological basis for the realization of the human factor in safety management. Procedia Comput. Sci. **145**, 112–117 (2018). https://doi.org/10.1016/j.procs.2018.11.018
10. Burlov, V., Grachev, M., Shlygina, N.: Adoption of management decisions in the context of the uncertainty of the emergence of threats. In: 2017 XX IEEE International Conference on Soft Computing and Measurements, pp 107–108. IEEE (2017). https://doi.org/10.1109/scm.2017.7970510
11. Byzov, A., Efremov, S., Gomazov, F.: Social and economic aspects of acceptable risk of ensuring ecological safety at management of municipal waste. In: 2018 IEEE International Conference "Management of Municipal Waste as an Important Factor of Sustainable Urban Development" (WASTE), pp 10–13. IEEE (2018). https://doi.org/10.1109/waste.2018.8554106
12. Burlov, V., Grobitski, A., Grobitskaya, A.: Construction management in terms of indicator of the successfully fulfilled production task. Mag. Civil Eng. **63**, 77–91 (2016). https://doi.org/10.5862/MCE.63.5

Mathematical Model for Managing Energy Sector in the Region

Vyacheslav Burlov, Oleg Lepeshkin, and Michael Lepeshkin(✉)

St. Petersburg Polytechnic University Peter the Great, Polytechnic Str., 29, St. Petersburg 195251, Russia
misha.lepeshckin@yandex.ru

Abstract. The paper discusses the methodological foundations and problems of modeling the energy sector management in the region. Approaches to the construction of a new dynamic mathematical model of the energy sector management in the region and the primary results of modeling are presented. Three main system-forming indicators of social, economic, energy systems, corresponding according to the law of preservation of integrity of object of V. G. Burlov. The model is based on an empirical approach based on the analysis of statistics of the main indicators number of population; number of jobs in the real economy and energy supply in the region. Calculations were performed using the obtained system of nonlinear dynamic equations, and the results are presented in the form of a phase portrait. In both direct and indirect assessment of these factors based on known statistics for the region and solutions of this system, we obtained the trajectories of the phase portrait, showing how it needs to change two measures that could be made to change the third one.

Keywords: Energy sector management · Balanced energy supply in the region · Produced fuel and energy resources · Consumed fuel and energy resources · Sustainable of socio-economic and energy development of the region

1 Introduction

The development of the energy industry is largely determined by such long-term processes such as industrialization, globalization, and urbanization. Increasing the amount of energy consumed and diversifying its sources, improving the efficiency of energy production and consumption is characterized by energy development. Globalization of energy and integration of energy markets increases the importance of research on the development of energy systems at the territorial level due to changes in production, distribution and consumption of energy in the context of integration, providing significant impact on the stability of local energy supply.

The use of energy resources is a prerequisite for the functioning of life-support and security systems, and has an impact on the quality and economic level of society. The modern energy complex includes the entire complex enterprises, installations and structures that link their economic relations that ensure the operation and development of production (production) of energy resources and all processes of their

V. Murgul and V. Pukhkal (Eds.): EMMFT 2019, AISC 1258, pp. 659–668, 2021.
https://doi.org/10.1007/978-3-030-57450-5_57

transformation, as well as end users. The fuel and energy complex represents it is a complex inter-industry system of production and production of fuel and energy, its transportation, distribution and use. For energy systems are largely characterized by the following specific features:

- close and growing interdependence of the development of the entire population energy systems, as a result of development in the direction of deepening the principle of consistency (along the way of forming a set of large systems based on a combination of production concentration, means transport of converted energy types and energy resources);
- materiality of connections of the main elements of systems;
- continuity, often enough, and continuity over time of processes production, distribution and consumption;
- complexity of purposeful equilibrium motion of systems, defined by the requirements of reliable operation, flexibility development and a number of other factors.

Method of modeling the energy sector management in the region based on system dynamics is especially important for Russia, which is living in conditions of external threats and is striving to achieve rapid growth in the development of regions. Balanced energy supply plays an important role in the socio- economic development of the region.

Therefore, the forecast of balanced energy supply and socio- economic development of the region within a particular region of Russia is very relevant. This would allow us to outline the limits of the possible and the impossible in order to build a strategy for our own development. From these considerations, it is necessary to create a new model that should reflect the trends of the main processes in the region and should be based on statistical data.

The model uses number of population, the number of jobs in the real economy, and energy supply as the main variables of the region. Various crises of recent times show the instability of the current model of development in the world. An important drawback of this model is absolutization economic growth and increased consumption of energy resources at the expense of solving social and environmental problems. The proposed model makes it possible to track through statistical indicators the negative impact of economic growth and increased consumption of energy resources on the social indicators of the region, in particular the birth rate, mortality, and migration balance. The negative impact of economic growth and increased consumption of energy resources leads to deterioration of environmental indicators and affects the health and well-being of people in the region. The model allows us to find the maximum of economic growth and increased consumption of energy resources that does not cause a significant deterioration of social indicators in the region.

2 Mathematical Model of the Energy Sector Management in the Region

The use of energy by mankind is growing. Given that energy is an essential element of sustainable development in the region, the goal of the office is to identify such methods of energy supply that would best ensure the development of the economy, improving the quality of life of people, while reducing to minimize the impact of human activities on human health and the environment.

A set of natural and industrial energy carriers, the stored energy of which at the current level of technology development and the technology available for use in economic activities is called fuel and energy resources. We proceed from the description of the movement of physical flows in general, i.e. production, production and consumption of fuel and energy resources in the region.

Energy is the fuel and energy complex of the country, covering receiving, transmitting, converting, and using different types of energy and energy resources. It is a point intersections of energy, economic and social components social development and regulatory factor in the ecological and economic space.

Since the second half of the XX century, in the conditions of scientific and technical revolution, the needs of human society in various types of energy, mainly electric, rapidly increasing. Power system it is a collection of energy resources of all types, methods for obtaining (extracting), converting, distributing, and use, as well as technical means and organizational complexes, ensuring the supply of consumers with all types of energy.

Dynamical systems theory comprises a broad range of analytical, geometrical, topological, and numerical methods for analyzing differential equations and iterated mappings. E. N. Lorenz's [1, 2] discovery in 1963 said that the solutions to his equations never settled down to equilibrium or to a periodic state instead they continued to oscillate in an irregular, aperiodic fashion. Moreover, if he started his simulations from two slightly different initial conditions, the resulting behaviors would soon become totally different. The implication was that the system was inherently unpredictable, tiny errors in measuring the current state of the atmosphere would be amplified rapidly, eventually leading to embarrassing forecasts. In 1971 Ruelle and Takens proposed a new theory for the onset of turbulence in fluids, based on abstract considerations about strange attractors. A few years later, R. M. May [3] found examples of chaos in iterated mappings arising in population biology, and stressed the pedagogical importance of studying simple nonlinear systems.

Modeling of management on the basis of synthesis and the law of preservation of integrity of object is presented in works of H. H. Goode, R. E. Machol, V. G. Burlov [4–14], O. M. Lepeshkin [13, 14] and other [4–14].

The dynamic mathematical model of the energy sector management in the region based on the synthesis is formalized as a system of nonlinear differential equations. Three main system-forming indicators of activity of the region to three basic interconnected properties:

- indicator of social system of the region "x" (number of population = birth rate-death rate + migration balance);

- indicator of economic system of the region “y” (number of jobs in the real economy = high-technology jobs + other jobs);
- indicator of technical and technological system of the region “z” (energy supply in the region = produced fuel and energy resources-consumed fuel and energy resources).

2.1 The Output of the First Differential Equation

We introduce the following notations.

x - demographic indicator, which is defined as

$$x = \frac{x^*(t_0) + x^*(t)}{x^*(t_0)} \tag{1}$$

$x^*(t)$, the current value of the demographic indicator at a time.

$x^*(t_0)$, the value of the demographic indicator of variability at a time.

The derivative of the indicator “x” $\frac{dx}{dt}$ is the rate of change in the human population. The rate of change “x” is proportional to the number of population in the region. That is, the larger the number of population, the greater its growth.

$$\frac{dx}{dt} = ax. \tag{2}$$

a, the coefficient of demographic activity.

Determine the impact on the rate of change of the demographic indicator of the other two system-forming indicators.

“y” is an indicator of economic development. Quantitatively estimated by the number of jobs in the real sector of the economy.

Determined based on the minimum number of people in the enterprise, necessary to produce a certain range of goods and services.

For a given value of the demographic indicator “x”, the number of “y” jobs, respectively, of working people, will reduce the rate of growth of the demographic indicator by an amount proportional to the value of “bxy”.

The differential equation describing the change in the demographic indicator “x” will take the following form:

$$\frac{dx}{dt} = ax - bxy \tag{3}$$

b, the coefficient of negative attitude to childbearing (1–2 children – acceptable; and 3–5 – already a lot).

For a given value of the demographic index “x”, the energy supply “z” index will increase the rate of growth of the human population, in proportion to the value of qxz. Therefore, the more energy enters the region, the faster the growth rate of the quantitative composition of the population of this region.

The differential equation is transformed to the following form:

The differential equation is converted to the following form:

$$\frac{dx}{d\,t} = ax - bxy + qxz \tag{4}$$

q, the coefficient of the level of energy supply of the region.

2.2 The Output of the Second Differential Equation

"y" is an indicator of economic development, which is defined as

$$y = \frac{y^*(t_0) + y^*(t)}{y^*(t_0)} \tag{5}$$

$y^*(t)$, the current value of the indicator of economic development at a time;
$y^*(t_0)$, the value of the economic development indicator at a time (number of jobs in the real sector of the economy).

The derivative of the indicator "y" $\frac{dy}{d\,t}$ is the rate of change in the indicator of economic development.

This indicator is associated with both the indicator of the development of the economy "y", and with the other two system-forming indicators. This "x" is a demographic indicator, and "z" is an indicator of the region's energy supply.

The rate of change in the indicator of economic development (the number of jobs in the real sector) is proportional to the number of jobs in the real sector of the economy with a minus sign. (The more jobs are created in the real sector of the economy, the more difficult it is to increase and increase the number of these places).

$$\frac{dy}{d\,t} = -py \tag{6}$$

p, the coefficient of development of the real sector of the economy.

Determine the impact on the rate of change in the economic development index of the other two system- forming indicators.

"y" is an indicator of the development of the economy. This is the number of jobs in the real sector of the economy, necessary.

Workplaces are determined based on the minimum number of people in the enterprise needed to produce a certain range of goods and services.

For a given indicator of the development of the economy "y", the demographic index "x" with the interest of people will increase the rate of change in the index of development of the real sector of the economy by an amount proportional to the value of "cxy". The manifestation of such a property is objective, since it is justified by the self-preservation of society. The reverse manifestation of this property will lead to the self-destruction of society. Where "c" is the coefficient of people's interest in the development of the economy.

The differential equation is converted to the following form:

$$\frac{dy}{dt} = -py + cxy \tag{7}$$

For a given indicator of the development of the economy "y", the number of "z" - energy units will contribute to an increase in the growth rate of the economic development indicator by an amount proportional to the value of "γyz". (The more energy is supplied for the development of the real sector of the economy, the higher the growth rate of the indicator of economic development).

The differential equation has the following form:

$$\frac{dy}{dt} = py - cxy + \gamma yz \tag{8}$$

γ, the coefficient of TS provision in the real sector of the economy.

2.3 The Output of the Third Differential Equation

"z" is the energy supply indicator. It is defined as

$$z = \frac{z^*(t_0) - z^*(t)}{z^*(t_0)} \tag{9}$$

$z^*(t_0)$, is the value of the energy supply at the initial time t_0;

$z^*(t)$, the value of the energy supply indicator at the current time (t).

The derivative of this indicator "z" $\frac{dz}{dt}$ is the rate of change in the energy supply. This speed is related to the energy consumption index "z". Also, two backbone indicators. This "x" is a demographic indicator and "y" is an indicator of the development of the economy. The rate of change in the energy consumption index "z" is proportional to the amount of energy consumed. That is, the more the energy is consumed in society, the higher the rate of its increase.

$$\frac{dz}{dt} = \mu z \tag{10}$$

μ, the coefficient of energy supply growth in the region.

Determine the impact on the rate of change in the energy consumption indicator of the region from the other two system-forming indicators.

"x" - the demographic indicator with increasing the quantitative composition of the region reduces the rate of increase in energy consumption.

"y" is an indicator of the development of the economy. Number of jobs in the real sector of the economy.

For a given energy consumption index "z" demographically, the "x" indicator will decrease the rate of change in the energy supply by an amount proportional to the value of τxz. (With the increase in the population, the rate of change in energy supply decreases). Where "τ" - is the ratio of the quantitative composition of the population to the level of energy supply in the region.

The differential equation takes the following form:

$$\frac{dz}{d\,t} = \mu z - \tau x z \tag{11}$$

For a fixed indicator of energy supply for the development of the real sector "z", an increase in the number of jobs in the real sector of the economy contributes to a decrease in the growth rate of the energy supply proportional to the value of δyz. That is, the more the real sector of the regional economy develops, the less energy "gets" to one workplace of the real sector of the economy.

A differential equation takes the following form:

$$\frac{dz}{d\,t} = \mu z - \tau x z - \delta y z \tag{12}$$

δ, the coefficient of correspondence between the level of development of the real sector of the economy and the level of energy supply.

2.4 System of Differential Equations

Formula (13) describes a system of differential equations of three systems of the region:

$$\begin{cases} \frac{dx}{d\,t} = ax - bxy + qxz; a, b, q > 0; \\ \frac{dy}{d\,t} = -py + cxy + \gamma yz; c, p, \gamma > 0; \\ \frac{dz}{d\,t} = \mu z - \tau xz - \delta yz; \delta, \mu, \tau > 0. \end{cases} \tag{13}$$

x, indicator of the number of population;
y, indicator of the number of jobs in the real economy;
z, indicator of the energy supply in the region;
a, the coefficient of demographic activity;
b, the coefficient of negative attitude of people to childbearing;
q, coefficient of provision of energy in the region;
c, coefficient of people's interest in economic development;
p, coefficient of development of the real sector of the economy;
γ, coefficient of energy supply of workplaces;
μ, coefficient of development of energy supply in the region;
τ, coefficient of compliance of the population with energy supply;
δ, coefficient of compliance of the economy's development with energy supply.

The backbone of the model is a system of differential equations and three dimensionless relative indicators: social, economic and energy supply. Nine coefficients of the system of differential equations implement mechanisms of energy sector management in the region.

Methods of nonlinear dynamics allow you to simulate fast, non-equilibrium processes (so-called phase transitions) in economic systems, related to the transition from one stable states in others. It is necessary to proceed to models aimed at describing non-equilibrium processes using the nonlinear dynamics apparatus.

This system of nonlinear differential equations does not have a purely analytical solution, it is possible only by methods of numerical integration, for example, such as Adams, Euler, Runge-Kutta in MathCAD, which allow you to build solutions in the form of smooth curves. The obvious drawback for the practical application of the method is the difficulty of perception for analysis. Since the values may be the same or in absolute value be unsuitable for viewing. The solution is to construction phase portraits, which sufficiently fully and succinctly reflect the properties of the function under consideration.

Considering that three systems are considered, the most complete is the consideration of phase portraits in three-dimensional space, where the solution is represented as a corresponding spiral.

A set of phase variables of a system is a minimal set of variables which fully describes the state of the system. Phase space is the space generated by the phase variables i.e. phase space is the space generated by the generalized moments of a system. A state of a system at any time is represented by a point in the system's phase space. Change of a system state over time is represented by a trajectory in the phase space. In other words, a trajectory is the path of an object in phase space as a function of time. A phase portrait is the collection of all possible trajectories of the system.

The phase portrait of a nonlinear system can contain isolated closed phase trajectories called limit cycles, which can be stable and unstable. They determine the stability of self- oscillations. If the limit cycle is stable, then the phase paths outside and inside are wound around it. The limit cycle is unstable if the phase paths outside and inside are reeled off from it (Fig. 1).

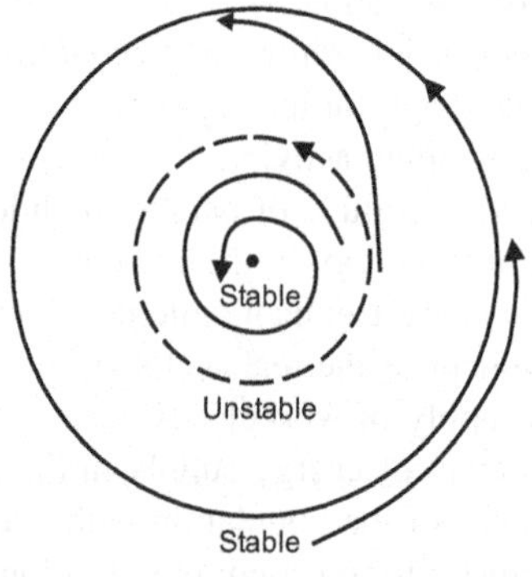

Fig. 1. Phase portrait a nonlinear system, stable and unstable cycle.

Phase portrait of the system in three-dimensional space ("x" -number of population; "y" - number of jobs in the real economy; "z" - energy supply in the Saint- Petersburg в 2010–2017 yeas) in MathCAD shown in Fig. 2.

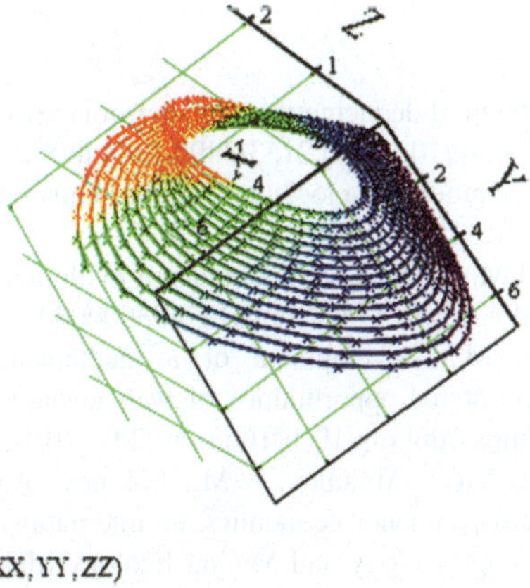

Fig. 2. Phase portrait of the system in three-dimensional space ("x" -number of population; "y" - number of jobs in the real economy; "z" - energy supply in the Saint- Petersburg в 2010–2017 yeas).

The above mathematical model of the energy sector management in the region, formalized as a system of three differential equations, the solution and the analysis of which is proposed through numerical integration, which allows to evaluate the behavior of the main characteristics of the long time interval and to generate proposals for the adjustment of certain parameters socio-economic development and energy supply of the region.

3 Conclusions

Thus, in the presence of huge reserves of fuel and energy resources in many regions of the Russian Federation, it is necessary to look for new scientific approaches to energy management in the region, taking into account the multidimensional nature of the problem of energy sector management in the region. Their goal is to solve the multi - criteria problem of energy supply, preservation of natural resources and the environment as a fundamental factor of public health in the region. We have developed a methodological approach that allows us to identify their positive or negative impact on the social and economic development of the region by modeling and setting the necessary values of the region's energy supply, and to determine the maximum energy supply at which the region's socio-economic development will be sustainable. It becomes possible to set values for the state of one of the three baselines at a certain point in time. In direct and indirect estimation of these factors, based on known statistical data for the region and solutions of this system, the phase portrait trajectories are obtained, showing how to change two measures that could be done to change the third.

References

1. Lorenz, E.N.: The problem of deducing the climate from the governing equations. Tellus **16**, 1–11 (2010). https://doi.org/10.1111/j.2153-3490.1964.tb00136.x
2. Lorenz, E.N.: Periodic solutions of a logistic difference equation. SIAM J. Appl. Math. **32**(1) (1977). https://doi.org/10.1137/0132005
3. Nowak, M.A., May, R.M.: The spatial dilemmas of evolution. Int. J. Bifurcat. Chaos **3**(1), 35–78 (1993). https://doi.org/10.1142/S0218127493000040
4. Burlov, V., Grachev, M.: Development of a mathematical model of traffic safety management with account for opportunities of web technologies. Transp. Res. Procedia **20**, 100–106 (2017). https://doi.org/10.1016/j.trpro.2017.01.023
5. Istomin, E.P., Burlov, V.G., Abramov, V.M., Sokolov, A.G., Bidenko, S.I.: Decision support model within environmental economics. In: International Multidisciplinary Scientific GeoConference Surveying Geology and Mining Ecology Management, SGEM, Sofia, vol. 19, no. 5.3, pp. 139–145 (2019). https://doi.org/10.5593/sgem2019/5.3/s21.018
6. Istomin, E., Abramov, V., Fokicheva, A., Sokolov, A., Burlov, V.: New approach to the assessment of geohazard in the management of the territories. In: International Multidisciplinary Scientific GeoConference Surveying Geology and Mining Ecology Management SGEM, Albena, Bulgaria (2017). https://doi.org/10.5593/sgem2017/21/s08.127
7. Istomin, E.P., Abramov, V.M., Lepeshkin, O.M., Baikov, E.A, Bidenko, S.I.: Web-based tools for natural risk management while large environmental projects. In: International Multidisciplinary Scientific GeoConference: SGEM, Sofia, vol. 19, no. 5.3, pp. 953–959 (2019). https://doi.org/10.5593/sgem2019/5.3/s21.120
8. Istomin, E.P., Abramov, V.M., Burlov, V.G., Sokolov, A.G., Popov, N.N.: Development of technology for environmental safety control based on geo-information systems. In: 17th International Multidisciplinary Scientific GeoConference SGEM, Sofia, vol. 17, pp. 859–866 (2017). https://doi.org/10.5593/sgem2017/21/s08.109
9. Burlov, V., Andreev, A., Gomazov, F., Somga-Bichoga, N.: System integration of security maintenance processes in knowledge management. In: Proceedings of the European Conference on Knowledge Management ECKM, vol. 1, pp. 112–122 (2018)
10. Burlov, V., Andreev, A., Gomazov, F.: Mathematical model of human decision - a methodological basis for the realization of the human factor in safety management. Procedia Comput. Sci. **145**, 112–117 (2018). https://doi.org/10.1016/j.procs.2018.11.018
11. Burlov, V., Andreev, A., Gomazov, F.: Development of a model for the management of environmental safety of the region taking into account of the GIS capacity. MATEC Web Conf. **193**, 1–8 (2018). https://doi.org/10.1051/matecconf/201819302038
12. Andreev, A.V., Burlov, V.G., Grachev, M.I.: Information technologies and synthesis of the management process model in the enterprise. In: International Science and Technology Conference "EastConf" (Vladivostok), Vladivostok, pp. 1–5 (2019). https://doi.org/10.1109/eastconf.2019.8725428
13. Burlov, V., Lepeshkin, O.: Modeling the process for controlling a road traffic safety system based on potentially active elements of space and time. Transp. Res. Procedia **20**, 94–96 (2017). https://doi.org/10.1016/j.trpro.2017.01.021
14. Burlov, V.G., Lepeshkin, O.M., Lepeshkin, M.O., Gomazov, F.A.: The control model of safety management systems. Conf. Ser.: Mater. Sci. Eng. **618**, 012088 (2019). https://doi.org/10.1088/1757-899X/618/1/012088

Improvement of the Tool of Strategic Management Accounting

Guzaliya Klychova[1], Alsou Zakirova[1(✉)], Shakhizin Alibekov[2], Aigul Klychova[1], Vitaly Morunov[3], and Ullah Raheem[1]

[1] Kazan State Agrarian University, Karl Marx, 65, 420015 Kazan, Russia
zakirovaar@mail.ru
[2] North Caucasus Institute (Branch) of the All-Russian State University of Justice (RPA of the Ministry of Justice of Russia), Agasieva, 87, 367008 Makhachkala, Russia
[3] Bugulma Branch of the Kazan National Research Technological University, Krasnoarmeyskaya 9, 423230 Bugulma, Russia

Abstract. Trends in increasing competition, the growing dynamics of business, manifested in the transience of strategic guidelines, lead to an objective need for a fundamental review of the main approaches to management in enterprises. In order to ensure the effective use of existing production resources and achieve sustainable economic development of enterprises, it is necessary to apply modern approaches and systems of management. One of such approaches is the concept of strategic management that allows connecting the corporate strategy of the enterprise with its organizational processes, considering problems of development in modern conditions of management. The purpose of the article is to justify the theoretical provisions and develop practical recommendations for the organization and practical application of tools of strategic management accounting. Objectives of the research: to study the content and role of strategic management accounting as an information base for making strategic management decisions; to substantiate the thesis about changing priorities of strategic management accounting at different stages of the life cycle of manufactured products; to develop a complex of unified forms of management reporting, necessary for the organization of strategic management accounting and segmental analysis. When writing this article, such methods as analysis of scientific and theoretical sources, system approach, method of comparative analysis, generalization were used. The results presented in the article allow estimating profitability of production and key clients of the enterprise, carrying out a general estimation, and raising efficiency of activity of economic entity within the limits of strategic management accounting.

Keywords: Strategic management accounting · Strategic management · Accounting and analytical information · Product life cycle · Strategic control · Management reporting · Customer profitability · Product profitability

V. Murgul and V. Pukhkal (Eds.): EMMFT 2019, AISC 1258, pp. 669–686, 2021.
https://doi.org/10.1007/978-3-030-57450-5_58

1 Introduction

In order to improve modern management concepts of enterprises based on the application of strategic approaches, there is a need in the system of accounting and analytical support of strategic management. Lack of systematic approaches to the implementation of management information systems and complex research in this area, taking into account the specifics of agricultural enterprises is one of the main problems.

The researches of modern aspects of agricultural enterprises functioning and construction of effective mechanisms of their strategic management require the formation of fundamentally new theoretical and methodological tools of their information support. In particular, there is a necessity of development of accounting system in the direction of substantiation of the essence and peculiarities of practical realization of strategic management accounting methodology on the basis of joint application of different models.

The modern stage of development of scientific researches in the sphere of strategic management is characterized by a considerable pluralism of approaches of scientists that can be explained by the following reasons:

- insufficient level of implementation of accounting and analytical support systems for strategic management at enterprises;
- sufficient novelty of the strategic management concept and existence of a great number of variants of its construction at the enterprises;
- lack of general rules for building a system of analytical support of management decisions;
- understanding of the system of analytical support of strategic management as a constituent subsystem of strategic management accounting;
- absence of the general concept of construction of subsystems of strategic financial and management accounting at the enterprise as one of the main information sources when making managerial decisions;
- absence of theoretical substantiation of system interconnections between accounting (financial and management) and analytical subsystems of strategic management system;
- theoretical and methodological tools of strategic analysis, economic analysis, investment analysis, planning and budgeting systems, which are used as tools providing analytical support of strategic management of an enterprise.

In this connection, there is a necessity to develop a general model of analytical support of the strategic management of agricultural enterprises with the application of the system approach, in particular, the construction of the strategic analytical system, which will promote the increase of efficiency of the strategic management system functioning by means of providing:

- establishment of interrelationships between strategic objectives and the system of analytical indicators;
- establishing relations between the motivation of employees and the chosen strategy
- quality assessment of the selected strategy [1–3].

To ensure the effectiveness of information support of strategic management of agricultural enterprises it is necessary to ensure a continuous monitoring of external parameters and internal performance of the enterprise with the observance of the main objectives of the chosen strategy of business development, which can be carried out only if a complex system of information support of strategic management of agricultural enterprises [4–6].

In order to develop an agricultural enterprise development strategy that ensures effective distribution and use of available resources (material, financial, labor, land and technologies) and promotes the establishment of a sustainable market position in a competitive environment, it is necessary to move to adaptive management, which includes structured analysis and calculation of forecast options for enterprise development [7–9]. This calls for the development and practical use of a system of analytical indicators on the internal and external environment of the enterprise, the complex construction of which is possible only if the strategic analytical system is implemented in the activities of agricultural enterprises [2, 10, 11].

Rational system of information support of strategic management of agricultural enterprises should take into account internal and external information spheres. Thus, the information support reflects complex consolidated information about the enterprise activity and its strategic prospects [12–14]. Rationalization of the model of information support of strategic management of agricultural enterprises should be based on a clear structuring of information flows to ensure prompt and reliable receipt of necessary data in case of such a necessity [15–17]. Strategic decision making should be based on current (reliable, accurate) and forecasting (calculation, orientation) information.

2 Materials and Methods

For maintenance of effective realization of process of strategic management at the enterprise the system of its accounting and analytical maintenance which would consider stages of realization of strategic management should be created. The following types of accounting and analytical information necessary to ensure the process of strategic management of the organization are distinguished (Fig. 1).

Specificity of the approach is integration of the accounting and analytical information (from the system of the financial analysis) for maintenance of realization of strategic management by the organization. At the same time the used analytical information is the result of processing of the accounting information, testifying about certain duplication of sources of information for maintenance of strategic management. Certain stages of strategic management provide continuity and cyclicity of this process.

An important role in the development of the system of accounting and analytical support of strategic management, except for procedural aspects of its implementation, also plays the classification of strategies, as the set of accounting and analytical tools, which should be created and adjusted at the enterprise for its formulation, introduction, evaluation and implementation, depends on the chosen type of strategy.

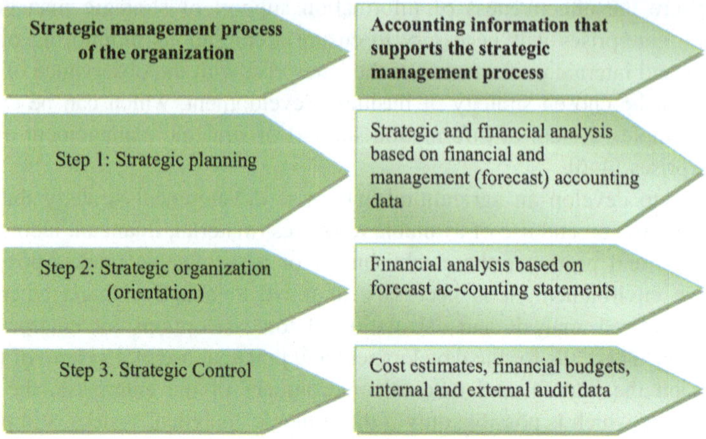

Fig. 1. Accounting and analytical information, providing the process of strategic management of the organization.

Comprehensive study of the place of elements of strategic management accounting in the system under consideration in the activities of the agricultural enterprise requires an explanation of the following condition: when implementing opportunities and strategic initiatives is extremely important and necessary:

- understanding of the nature of products manufactured and sold by an agricultural enterprise;
- the degree of market maturity of agricultural products;
- specifics of demand formation in the agricultural market.

The process of realization of agricultural production should be concluded in certain strategic framework. Evaluation of production in the considered branch can take place on the parameter of attractiveness for the consumer, with definition of competitive advantages for each segment of the market.

The complex analysis of questions of behavior of separate kinds of production in the market allows the concept of product life cycle (PLC). The life cycle of a product (PLC) involves a number of stages. For today various stages of life cycle of the goods are allocated, but as a whole, all of them are reduced to passing of following basic stages:

1) stage of conception of new products and their introduction into the market;
2) stage of demand growth;
3) stage of product maturity;
4) the stage of decline (fall) in demand for products.

The application of the concept of PLC allows structuring the organization of strategic management accounting at the enterprise of agrarian sphere of economy. In particular, the choice of strategic initiatives involves the optimization of various risk elements that inevitably accompany the activities of the enterprise in the market conditions. Let's consider the interrelation of the tasks of strategic management accounting with the stages of PLC [18–20].

At the first stage of PLC - a plan of new production (introduction of production on the market) - the strategy of development for a gain of a market share is defined, strategic reference points for reactions of competitors to change by the enterprise are developed, the information on expenses for development, the general information on the basic preferences of consumers is accumulated.

At the second stage of PLC - demand growth - relative levels of marketing expenses of all market participants are revealed, enterprise strategies are adapted to corresponding changes in the external environment, alternative variants of development strategy from positions of estimation of incomes on each strategy are developed.

At the third stage of PLC - maturity of production - relative expenses of competitors on the given production with comparison to own expenses of the enterprise, with corresponding working out of actions for decrease in expenses are estimated and analyzed, the information on volume of sales, calculation of an operational indicator of profitability of the invested capital (ROI) is formed.

At the fourth stage of PLC - decline (fall) of demand on production - perspective directions of development of the given kind of production are defined, the information on decisions on the further development of the enterprise is formed.

It should be noted that each stage of the PLC is based on the critical success factor and instruments of control by the strategic management accounting. Tools to control the implementation of the strategy at various stages of the PLC are shown in Fig. 2.

The main results obtained within the framework of the concept of PLC can serve as a basis for the implementation of strategic management accounting at an agricultural enterprise, and, accordingly, can be used as information support for the development of competitive strategies at the enterprise [21, 22].

Thus, the result of strategic management accounting functioning is the generation of information product in the form of strategic management reporting, which is a set of methods of obtaining the final information, provided to managers to ensure the functions of strategic management. Support of processes of environment scanning, formulation, introduction, evaluation and control of the strategy is a defining feature, which distinguishes strategic management accounting from all elements of the management information system.

Strategic management accounting serves as a tool of information support for strategic decision-making, assuming the organization of accounting procedures at various levels of the organizational structure of an agricultural enterprise. Strategic management distinguishes the level of a unified corporate strategy from the level of competitive strategies [23–25].

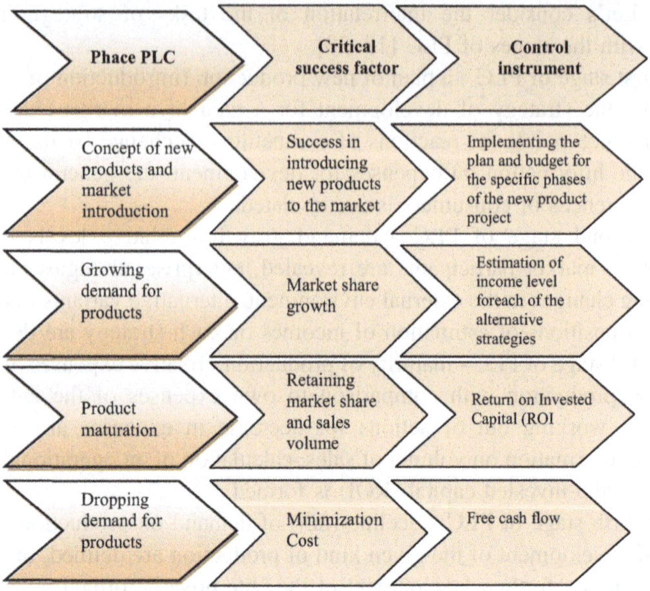

Fig. 2. Interrelation of strategic control tools at the stages of PLC as part of strategic management accounting.

When determining the competitive strategy and considering factors of external environment of agricultural enterprises, it is necessary to pay attention to competitive analysis.

Stage 1 - the level of competition at the branch market is analyzed: existing competitors are identified, potential competitors are identified, competitors are preliminary estimated using the method of expert estimations.

Stage 2 - competitors are evaluated: the expert opinion is analyzed, competitors are ranked, prospective analysis is performed (alternative strategies for leading competitors are analyzed).

Stage 3 - responses to the results of forward-looking competitive analysis are developed: strategic initiatives are developed, preventive measures are developed that provide a response to potential threats.

The result of the first stage is a preliminary assessment of the most important competitors (3 to 5 most important competitors are singled out) by their main characteristics. The assessment is done through expert analysis and scoring, and ranking of competitors by the following categories: "leaders" (from 21 to 25 points), "aggressive growth group" (from 16 to 20 points), "medium business" (from 10 to 15 points), "outsiders" (from 0 to 9 points).

The sources of information for experts can be data from mass media (for example, competitors' press releases), analytical reports on the industry, state and regional statistics.

Interrelation of the most important directions of strategic management accounting at competitive analysis with various stages of PLC is shown in Fig. 3.

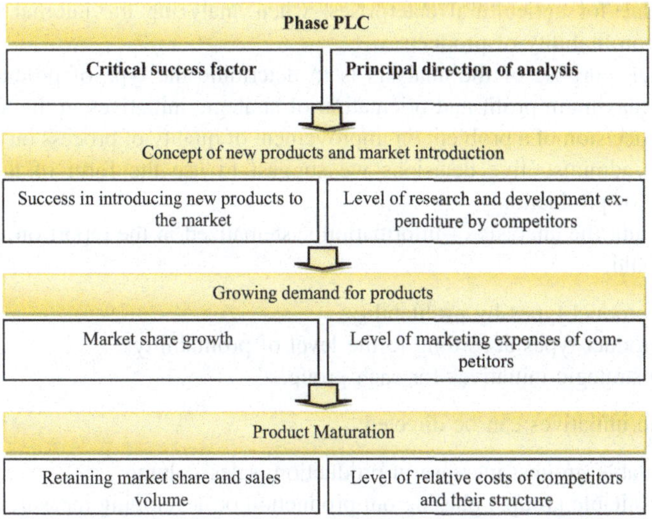

Fig. 3. Interrelation of the most important directions of strategic management accounting in competitive analysis with different stages of PLC.

We emphasize that the potential added value, according to the results of competitive analysis within the framework of strategic management accounting, is very significant. Questions of practical application of the basic toolkit are considered in modern practice of the account, when exactly rational selection and use of tools in practice of the agricultural enterprises gives the chance in realization of the complex qualitative analysis with granting of the relevant information taking into account not only internal resources of the enterprise, but also external environment [26–28]. A special role in this analysis acquire possible threats from competitors and the development of responses to them.

3 Results

The main catalyst for introducing strategic competitive analysis into the accounting practice is the introduction of information obtained from strategic management accounting into the system of regular management reporting. Consequently, in comparison with the traditional accounting there is a balance of the reporting information on the internal condition of the enterprise (financial indicators, the main business processes) with the information on the external environment of the enterprise (for example, comparison of financial indicators with the similar indicators on branch) [29, 30].

The internal condition of the enterprise is characterized by a set of indicators. One of the most significant is the relative indicator - profitability. Profitability can be analyzed by types of activity, by types of products, by groups of consumers. The greatest value for agricultural enterprises, when analyzing the internal state, is the analysis of profitability of products.

The main purpose of the analysis is to determine the type of products that can extract the maximum profit and orientation of strategic initiatives in this direction.

For the decision of a problem on improvement of quality of process on working out of strategic administrative decisions we suggest to use the form of the report on profitability of production.

As a result, the analysis of information systematized in the report on each type of product should:

- to rank product types by profitability;
- group product types according to the level of profitability;
- develop strategic initiatives for each group.

Strategic initiatives can be directed:

- for profitable groups: increase in production, sales volume;
- for unprofitable groups: phasing out production or developing measures to increase the level of profitability.

The main result of the analysis is qualitative information for two processes: planning and evaluation. However, besides the analysis of financial indicators (the main ones are costs, profit, profitability) it is necessary to analyze non-financial factors which can be investigated by means of drawing up a matrix of attractiveness of production.

Let's consider procedure of the analysis of profitability of production as one of directions of the strategic administrative account by an example of LLC "August-Muslum" of Muslumovsky area of Republic Tatarstan. From the point of view of identification of analytical segments of August-Muslum LLC of Muslumovsky District of the Republic of Tatarstan in relation to the products manufactured, it is reasonable to consider the following segment groups: winter grain, spring grain, sunflower, spring rape, perennial herbs (Table 1).

Table 1. Agricultural Product Profitability Report of August-Muslum LLC, Mu-Slum district, Republic of Tatarstan.

Product name	Volume of production, tons	Sale price, rub./tons	Direct costs per unit of production, rub./tons	Over-head production costs per unit of production, rub./tons	Selling expenses per unit of production, rub./tons	Total full cost per unit of production, thousand rubles	Profit per unit of production, thousand rubles	Product profitability, %
Winter cereals	15.72	11878	11247.66	181.32	4.90	11434.88	444.12	3.74
Spring cereals	48.40	3400	3132	134.35	2.38	3268.73	131.27	3.86
Sun-flower	30.47	19890	18927.86	276.45	5.37	19209.68	680.32	3.42
Spring rapeseed	41.80	23400	22455.73	159.15	159.15	22616.72	783.28	3.35
Perennial herbs	85.44	17504	16797.65	33.67	0.9	16832.23	671.77	3.84

As a result of calculations of profitability of production it is possible to draw a conclusion that the most profitable group is spring grain with the value of profitability 3.86%, further on a level of profitability we shall note perennial grasses 3,84%, winter grain with profitability 3.74%, sunflower – 3.42%, and the least profitable segment with the value of 3.35% is spring rape.

Based on the product profitability report of August Muslum LLC of the Muslumovsky District of the Republic of Tatarstan, a product attractiveness matrix has been compiled (Fig. 4).

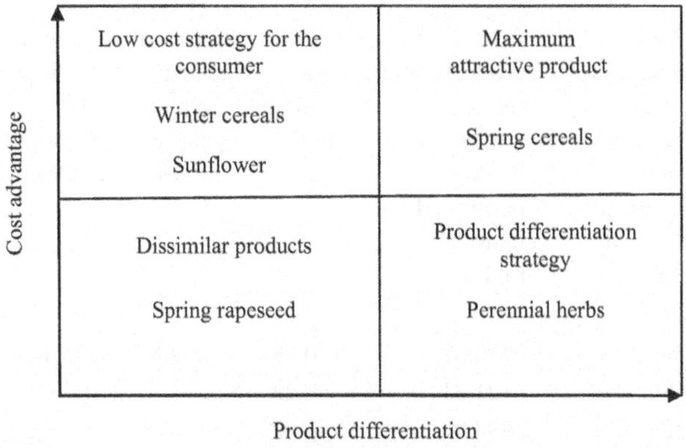

Fig. 4. Product attractiveness matrix of August-Muslum LLC of Muslumovsky District of the Republic of Tatarstan based on strategic management accounting data.

The management of August-Muslum LLC in the Muslumovsky District of the Republic of Tatarstan considers corn cultivation as a promising strategic direction. Last year, a good harvest was grown on cobs, but there was not enough capacity to dry grain. Corn is planned to be used as the main fodder crop in the future.

The data obtained from the product analysis are directly linked to the customer profitability analysis. Consumer Profitability Analysis or Strategic Client Analysis allows you to analyze the reasons for significant differences in profitability between different client groups.

For an objective assessment of the efficiency of management decisions it seems reasonable to carry out this analysis step by step.

On the basis of research and taking into account the specifics of agricultural enterprises, the following list of stages is proposed:

1. Definition of client groups.
2. Determination of the enterprise's costs for servicing certain (on the first stage) groups of clients.
3. Evaluation of intervals in the cost level.
4. Calculating the profitability of certain (first stage) customer groups based on actual data.
5. Ranking of client groups by cost and profitability.

6. Development of strategic decisions for each client group.
7. Analysis of the effectiveness of the proposed strategic alternatives.
8. Selecting the most attractive strategic initiative.

By analogy with the form of a report on the profitability of products, proposed a form of customer profitability report (Table 2). The structure of the form includes a summary of data on the client (client group):

Table 2. Profitability reports of customers of August Muslum LLC of Muslumovsky District of the Republic of Tatarstan. Reporting period January–August 2019.

Name of products sold, services rendered, works performed	Cost of products sold, services rendered, works performed, thousand rubles	Cost of products sold, services rendered and work performed, thousand rubles	Gross profit, thousand rubles	Costs of client activity, thousand rubles	Total profit per client, thousand rubles	Profitability client, %
LLC "New Style"						
Order 1	294.10	173.52	120.58	45.68	143.14	29.6
Order 2	189.54	121.31	68.23			
Total.	483.64	294.82	188.82			
LLC "KKV"						
Order 1	1105.00	729.30	375.70	111.72	552.12	28.98
Order 2	800.40	512.26	288.14			
Total.	1905.40	1241.56	663.84			
KES-Holding, LLC						
Order 1	980.40	696.08	284.32	192.44	718.24	34.24
Order 2	1117.00	826.58	290.42			
Order 3	600.45	468.35	132.10			
Order 4	784.00	580.16	203.84			
Total.	2097.40	2571.18	910.68			
LLC "Navruz"						
Order 1	408.10	342.80	65.30	87.22	284.38	26.38
Order 2	670.00	495.80	174.20			
Order 3	600.45	468.35	132.10			
Total.	1078.10	1306.96	371.60			
KFH "Vagapov"						
Order 1	298.00	202.64	95.36	25.68	129.97	26.72
Order 2	188.40	128.11	60.29			
Total	486.40	330.75	155.65			

- data for a certain period of time: order value, cost, gross profit;
- cost of client activity;
- profit of the client (client group);
- profitability of the client (client group).

As a result of the analysis the orders of the main clients (with the cost of orders for one month more than 200.00 thousand rubles) were analyzed. Thus, the most profitable client LLC "IES-Holding" with a value of profitability of 34.24%, LLC "Novy Style" with profitability of 29.60%, LLC "KKV" with profitability of 28.98%, KFH "Vagapov" with profitability of 26.72%, and the least profitable client LLC "Navruz" with profitability of 26.38%.

The information obtained as a result of filling out the form should be used in the practical activities of the agricultural enterprise and supplemented by a matrix of customer attractiveness.

Strategic initiatives in this case can be directed:

- for profitable groups: decisions to develop cooperation;
- for unprofitable groups: curtailing relationships or developing measures to revise these relationships.

Additional analysis of clients of August-Muslum LLC of the Muslumovsky District of the Republic of Tatarstan allowed to distinguish customer segments:

- segment A (focused on purchasing grain (winter and spring));
- segment B (focused on purchasing spring rapeseed and sunflower);
- segment C (focused on purchasing different types of products, including multiyear herbs).

For each of the segments of August-Muslum LLC in the Muslumovsky district of the Republic of Tatarstan, calculation of indicators: customer value (LTV) and customer profitability (LTP) (Table 3). This analysis is necessary for building a matrix of customer attractiveness.

Based on the results of value calculation, we can conclude that the most valuable segment is segment "B", the least valuable segment is segment "C". Analyzing customer profitability, we conclude that segment "B" is the most profitable segment, and the least profitable segment is segment "C". Therefore, on the basis of analytical data on the main customer segments of August-Muslum LLC of the Muslumovsky District of the Republic of Tatarstan it is possible to rank the main customer groups as follows: customer leaders (segment B), strong customers (segment C), problem customers (segment A).

In order to make rational decisions in "August-Muslum" LLC of the Muslumovsky district of the Republic of Tatarstan in terms of developing strategic initiatives in relation to key client groups, the "Profitability - Sales Volume" matrix was developed based on comparison of the analysis data (Fig. 5).

Based on the analysis of data presented in Fig. 5, the management and employees of August-Muslum LLC of the Muslumovsky District of the Republic of Tatarstan were able to rank the selected segments (segment "A", "B", "C").

Table 3. Calculation of indicators characterizing the value and profitability of the client. Reporting period: January–August 2019.

№	Indicator name	Segment "A"	Segment "B"	Segment "C"
Customer Value Statement (LTV)				
1	Total number of clients in the segment	6.00	18.00	12.00
2	Total number of orders in the segment	26.00	55.00	20.00
3	Total cost of orders, thousand rubles	896.40	2804.00	1108.00
4	The average cost of the order, thousand rubles. (page 3/page 2)	34.48	50.98	55.40
5	Average duration of client relationship, bottom. (n)	90.00	180.00	120.00
6	Number of clients attracted in the last n days	2.00	10.00	2.00
7	Total number of orders in the segment for the last n days	10.00	3.00	4.00
8	Estimated number of clients (page 1–page 6)	4.00	8.00	10.00
9	Calculated number of orders in a segment (page 2–page 7)	16.00	52.00	16.00
10	Average number of orders per customer (page 9/page 8)	4.00	6.50	1.60
11	Customer Value (LTV), thousand rubles. (page 4 × page 10)	137.91	331.38	88.64
Customer Profitability Report (LTP)				
1	Total number of clients in the segment	6.00	18.00	12.00
2	Total number of orders in the segment	26.00	55.00	20.00
3	Total profit from orders (total order value less total cost of goods and customer service expenses of the segment), thousand rubles	233.06	729.04	288.08
4	Average order profitability, thousand rubles. (page 3/page 2)	8.96	13.26	14.40
5	Average cost of attracting a segment customer (ratio of total marketing expenses for attracting segment customers to the total number of customers in a given segment), thousand rubles	0.21	0.19	0.16
6	Average duration of relations with the client, days (n)	90.00	180.00	120.00
7	Number of clients attracted in the last n days	2.00	10.00	2.00
8	Total number of orders in the segment for the last n days	10.00	3.00	4.00
9	Estimated number of clients (page 1–page 7)	4.00	8.00	10.00
10	Calculated number of orders in a segment (page 2–page 8)	16.00	52.00	16.00
11	Average number of orders per customer (page 10/page 9)	4.00	6.50	1.60
12	Client Profitability (LTP), thousand rubles. (page 4 × page 11)	35.86	86.16	23.05
13	Maximum allowable cost of attracting a new client in the segment, thousand rubles. (page 12 + page 5)	36.07	86.35	23.21

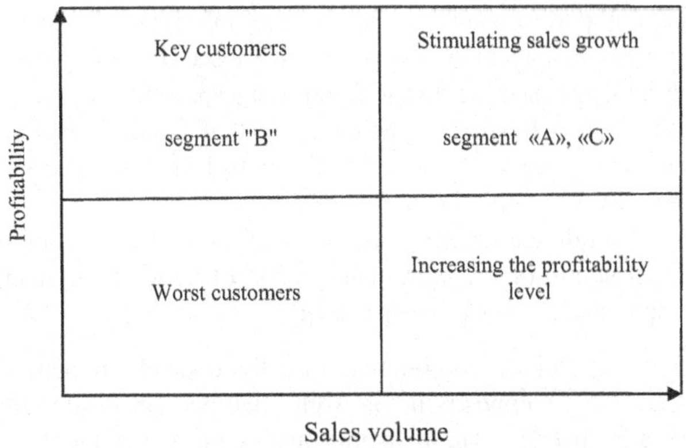

Fig. 5. Customer Attractiveness Matrix of August-Muslum LLC of Muslumovsky District of the Republic of Tatarstan based on strategic management accounting data.

The attractiveness matrix with corresponding cells is the basis for determining the strategic initiative: for segments "A" and "C" it is necessary to stimulate sales growth, therefore, the analyzed company should attract instruments of product promotion, and for segment "B" - to offer the best conditions of cooperation, as these customers are key.

Thus, there is a close relationship between the formation and content of analytical strategic reporting from the perspective of strategic management accounting and the overall development strategy of the enterprise "August-Muslum" LLC Muslumovsky district of the Republic of Tatarstan. In this context, analytical reporting is the basis for effective strategic management of the agrarian sector of the economy.

4 Discussion

Research of modern practice of activity of the agricultural enterprises testifies to presence of a complex of factors which reduce efficiency of acceptance of administrative decisions of the strategic character, directly concerning system of accounting and analytical maintenance of management system. Such factors can be referred to:

- imperfection of the current accounting and analytical model of forming information for making managerial decisions;
- unjustified dynamics of the correlation between the volumes of necessary and sufficient information, which leads to the phenomena of deficiency and oversaturation of the accounting and analytical system;
- lack of rational criteria for selecting necessary accounting and analytical information from the total volume of data;
- principal difference between classification features of accounting and analytical information required for strategic management needs;
- untimely preparation and submission of reports and formation of indicators necessary for making managerial decisions.

The considerable part of problems and lacks of information support of the process of making strategic management decisions is caused by the imperfection of the structure of the information base of agricultural enterprises.

To ensure the correct functioning of the system of accounting and analytical support of strategic management of agricultural enterprises it is necessary to take into account the specifics of specific business entities.

The process of building a strategic management accounting system as an element of information infrastructure of strategic management of agricultural enterprises is influenced by such specific features of their activity:

1. The results of agricultural production directly depend on natural and climatic conditions (quality composition of soils, temperature, humidity, etc.) And, accordingly, are unstable. These factors justify the expediency of forecasting in decision making and, accordingly, the need for rational accounting and information support of strategic management of agricultural enterprises.
2. Agricultural production is characterized by seasonality, which contributes to the uneven use of resources of the enterprise, the receipt of products only at certain intervals, the rhythmicity of sales and revenue from sales. This factor accents attention on expediency of strategic management of solvency of the agricultural enterprises.
3. Production processes in agriculture are characterized by a significant risk level of crop loss (heat, hail, frost, rain). The mentioned factors substantiate the expediency of strategic risk management application and the necessity to establish the mechanism of rational accounting and information support of strategic management of agricultural enterprises.
4. Agricultural business combines the influence of economic and biological laws of reproduction. The result of such synthesis is duration of production cycles (time limits of production of many products are longer than a calendar year and accordingly considerable specific weight of production costs of the reporting period is reflected as incomplete production).
5. Part of agricultural products produced by the enterprise in the next production cycle is consumed as means of production (feed, seeds). This feature indicates the specificity of the operating cycles of agricultural production and the need to take it into account in information support of strategic management of agricultural enterprises.
6. Agricultural production requires monitoring of external factors of influence: political, economic, social, legal. It is especially important to take into account national peculiarities, in particular concerning the range of products.
7. The main means of production in agriculture is land, on the quality and fertility of which the productivity of activities depends to a large extent. Characteristics inherent in land determine the expediency of specialization of agricultural enterprises and the structuring of strategic areas of activity.

Let's consider the influence of sectoral features on accounting and information support of management decisions of strategic nature in agricultural enterprises in Table 4.

Table 4. Specifics of agribusiness and its impact on accounting and information support for management decision-making of strategic nature.

Sectoral features of agriculture	Economic accounting option	Direction of accounting support of strategic management
Effect of natural and climatic conditions	Instability of production process	Forecasting strategy in decision making
The threat of natural disaster	High level of business risk	Strategic risk management
Mutual influence of economic and biological laws of reproduction	Production cycle time	Strategic monitoring of unfinished production dynamics
Seasonality of production	Uneven use of resources	Strategic solvency management
Use of finished products as a means of production	The finished product is consumed in the following cycle	Strategic forecasting of marketable product volume
Impact of national consumer preferences	Limited assortment	Strategic planning of the product range
Earth - main means of production	Specialization of agricultural enterprises	Strategic directions

Despite the existence of individual features of the strategic analysis, it is possible to formulate general organizational and methodical aspects of its implementation, in particular, regarding the calculations and interpretation of the results obtained, which significantly improve the process of implementation of this system in the practical activities of agricultural enterprises (Table 5).

In order to improve the analysis in the direction of ensuring its strategic orientation, the conceptual basis of its implementation has been developed, the practical use of which should contribute to the provision of such information: the reasons for the results obtained by the enterprise and their compliance with the selected strategy of the enterprise or strategic business unit, the impact of research on the possibility of further implementation of the selected strategy of the enterprise, the impact of the identified trends, industry and economy on the indicator of future cash flows, as well as the impact of research on the future of the enterprise.

Table 5. Conceptual framework for the analysis in the direction of its strategic orientation.

№ p/p	Implementation stage	Sources of information	Output results
1	Goal setting and forming the basis for strategic analysis	- information about the company's strategy - information on existing additional and specific requests of strategic management	- target map - the relationship between objectives and strategy - setting time and financial limits
2	Data collection	- information about the company's strategy - financial statements - accounting data - survey of managers, consumers, suppliers, competitors	- financial reporting and accounting data - processed survey results - current indicators - forecasting indicators
3	Analysis and interpretation	- financial reporting and accounting data - survey results - current figures - forecasting indicators	- analytical results
4	Conclusions and recommendations	- analytical results and reports - internal rules of analytical reporting	- strategic analytical reports - recommendations for selecting specific strategic alternatives - recommendations for adjustments
5	Monitoring	- information that affects the results, recommendations	- updated strategic analysis reports and recommendations

5 Conclusions

Thus, strategic accounting is one of the functional areas of accounting, which is aimed at informational support of making informed decisions. Strategic management accounting provides for gaining advantages when building a system of accounting and analytical support of strategic management.

In order to improve the order of strategic management accounting organization at agricultural enterprises, it is proposed to use the model of information links, which covers types of products, groups of clients, on the basis of which profitability reports can be formed. The information presented in the report is the basis for making managerial decisions concerning the improvement of the process of value formation at agricultural enterprises.

Analytical toolkit of strategic management accounting, used for research of products, clients (client groups), allows to obtain relevant, reliable, high-quality information, which contributes to the formation of a reliable basis for the development of rational, objective, and most important effective strategic management decisions for a long term perspective.

References

1. Rikhardsson, P., Yigitbasioglu, O.: Business intelligence & analytics in management accounting research: status and future focus. Int. J. Account. Inf. Syst. **29**, 37–58 (2018). https://doi.org/10.1016/j.accinf.2018.03.001
2. Fernandes, J.P.: Developing viable, adjustable strategies for planning and management – a methodological approach. Land Use Policy **82**, 563–572 (2019). https://doi.org/10.1016/j.landusepol.2018.12.044
3. Guesalaga, R., Gabrielsson, M., Rogers, B., Ryals, L., Cuevas, J.M.: Which resources and capabilities underpin strategic key account management? Ind. Mark. Manag. **75**, 160–172 (2018). https://doi.org/10.1016/j.indmarman.2018.05.006
4. Faizrakhmanov, Dz., Zakirova, A., Klychova, G., Yusupova, A., Klychova, A.: Formation and disclosure of information on social responsibility of agribusiness enterprises. In: E3S Web of Conferences, vol. 91, p. 06004 (2019). https://doi.org/10.1051/e3sconf/20199106004
5. Baumgartner, R.J., Rauter, R.: Strategic perspectives of corporate sustainability management to develop a sustainable organization. J. Clean. Prod. **140**(11), 81–92 (2017). https://doi.org/10.1016/j.jclepro.2016.04.146
6. Klychova, G., Zakirova, A., Sadrieva, E., Avkhadiev, F., Klychova, A.: Reporting in the area of sustainable development in agribusiness. In: E3S Web of Conferences, vol. 91, p. 06002 (2019). https://doi.org/10.1051/e3sconf/20199106002
7. Sultana, M., Ahmed, J.U., Shiratake, Y.: Sustainable conditions of agriculture cooperative with a case study of dairy cooperative of Sirajgonj District in Bangladesh. J. Co-oper. Organ. Manag. **8**, 100105 (2020). https://doi.org/10.1016/j.jcom.2019.100105
8. Yigezu, Y.A., Mugera, A., El-Shater, T., Aw-Hassan, A., Piggin, C., Haddad, A., Khalil, Y., Loss, S.: Enhancing adoption of agricultural technologies requiring high initial investment among smallholders. Technol. Forecast. Soc. Change **134**, 199–206 (2018). https://doi.org/10.1016/j.techfore.2018.06.006
9. Wei, M.M., Zhang, F.: Recent research developments of strategic consumer behavior in operations management. Comput. Oper. Res. **93**, 166–176 (2018). https://doi.org/10.1016/j.cor.2017.12.005
10. Klychova, A., Klychova, G., Zakirova, A., Sungatullina, R., Mukhamedzyanov, K., Philippova, E.: Social development mechanism of an agricultural enterprise formation. In: E3S Web of Conferences, vol. 110, p. 02072 (2019). https://doi.org/10.1051/e3sconf/201911002072
11. Klychova, G.S., Ziganshin, B.G., Zakirova, A.R., Valieva, G.R., Klychova, A.S.: Benchmarking as an efficient tool of social audit development. J. Eng. Appl. Sci. **12**, 4958–4965 (2017). https://doi.org/10.3923/jeasci.2017.4958.4965
12. Pavlatos, O., Kostakis, H.: Management accounting innovations in a time of economic crisis. J. Econ. Asymmetries **18**, e00106 (2018). https://doi.org/10.1016/j.jeca.2018.e00106
13. Klychova, G.S., Zakirova, A.R., Zakirov, Z.R., Valieva, G.R.: Management aspects of production cost accounting in horse breeding. Asian Soc. Sci. **11**(11), 308–312 (2015). https://doi.org/10.5539/ass.v11n11p308
14. Uyar, A., Kuzey, C.: Does management accounting mediate the relationship between cost system design and performance? Adv. Account. **35**, 170–176 (2016). https://doi.org/10.1016/j.adiac.2016.06.004
15. Klychova, G., Zakirova, A., Mannapova, R., Pinina, K., Ryazanova, Y.: Assessment of the efficiency of investing activities of organizations. In: E3S Web of Conferences, vol. 110, p. 02075 (2019). https://doi.org/10.1051/e3sconf/201911002075

16. Papulova, Z., Gazova, A.: Role of strategic analysis in strategic decision-making. Procedia Econ. Finan. **39**, 571–579 (2016). https://doi.org/10.1016/S2212-5671(16)30301-X
17. Bui, B., de Villiers, C.: Business strategies and management accounting in response to climate change risk exposure and regulatory uncertainty. Br. Account. Rev. **49**, 4–24 (2017). https://doi.org/10.1016/j.bar.2016.10.006
18. Liu, H., Chu, X., Xue, D.: An optimal concurrent product design and service planning approach through simulation-based evaluation considering the whole product life-cycle span. Comput. Ind. **111**, 187–197 (2019). https://doi.org/10.1016/j.compind.2019.07.008
19. Arnett, D.B., Sandvik, I.L., Sandvik, K.: Two paths to organizational effectiveness – product advantage and life-cycle flexibility. J. Bus. Res. **84**, 285–292 (2018). https://doi.org/10.1016/j.jbusres.2017.11.010
20. Mendes da Luz, L., Carlos de Francisco, A., Piekarski, C.M., Salvador, R.: Integrating life cycle assessment in the product development process: a methodological approach. J. Cleaner Prod. **193**, 28–42 (2018). https://doi.org/10.1016/j.jclepro.2018.05.022
21. Schwabe, O., Erkoyuncu, J.A., Shehab, E.: On the change of cost risk and uncertainty throughout the life cycle of manufacturing products. Procedia CIRP **86**, 239–244 (2019). https://doi.org/10.1016/j.procir.2020.01.004
22. Kassner, L., Gröger, Ch., Mitschang, B., Westkämper, E.: Product life cycle analytics – next generation data analytics on structured and unstructured data. Procedia CIRP **33**, 35–40 (2015). https://doi.org/10.1016/j.procir.2015.06.008
23. Klychova, G.S., Nizamutdinov, M.M., Safiullin, L.N., Mavlieva, L.M.: Priorities of agricultural credit cooperation development. Mediterr. J. Soc. Sci. **5**(18), 215–218 (2014). https://doi.org/10.5901/mjss.2014.v5n18p215
24. Oboh, C.S., Ajibolade, S.O.: Strategic management accounting and decision making: a survey of the Nigerian Banks. Future Bus. J. **3**(2), 119–137 (2017). https://doi.org/10.1016/j.fbj.2017.05.004
25. Klychova, G.S., Zakirova, A.R., Mukhamedzyanov, K.Z., Faskhutdinova, M.S.: Management reporting and its use for information ensuring of agriculture organization management. Mediterr. J. Soc. Sci. **5**(20), 220–224 (2014). https://doi.org/10.5901/mjss.2014.v5n24p104
26. Lebedev, P.: Getting insight into management accounting and control systems: a framework for survey-based research design for emerging markets context. Procedia – Soc. Behav. Sci. **213**, 293–298 (2015). https://doi.org/10.1016/j.sbspro.2015.11.540
27. Abdullah, N.H.N., Said, J.: Enhancing the governance of government linked companies via strategic management accounting practices and value creation. Procedia Econ. Finan. **28**, 222–229 (2015). https://doi.org/10.1016/S2212-5671(15)01103-X
28. Neumann, J., Robson, A., Sloan, D.: Monitoring and evaluation of strategic change programme implementation – lessons from a case analysis. Eval. Program Plan. **66**, 120–132 (2018). https://doi.org/10.1016/j.evalprogplan.2017.09.012
29. Bian, J., Li, K.W., Guo, X.: A strategic analysis of incorporating CSR into managerial incentive design. Transp. Res. Part E: Logist. Transp. Rev. **86**, 83–93 (2016). https://doi.org/10.1016/j.tre.2015.11.012
30. Creţu, R.C., Creţu, R.F., Muscănescu, A.: Comparative analysis of strategic and tactical decisions in agriculture under the IAS 41 Standard in the context of the emerging markets. Procedia Econ. Finan. **15**, 1641–1646 (2014). https://doi.org/10.1016/S2212-5671(14)00635-2

Information and Analytical System of Strategic Management of Activities of Enterprises

Alsou Zakirova[1(✉)], Guzaliya Klychova[1], Kamil Mukhamedzyanov[2], Zufar Zakirov[1], Almaz Nigmetzyanov[3], and Alfiya Yusupova[1]

[1] Kazan State Agrarian University, Karl Marx, 65, 420015 Kazan, Russia
zakirovaar@mail.ru
[2] University of Management "TISBI", Mushtari, 13, 420012 Kazan, Russia
[3] Kazan Branch of the Russian State University of Justice, 2nd Azinskaya Street, 7A, 420088 Kazan, Russia

Abstract. The main task of the enterprises is timely adaptation to dynamically changing conditions of the market in order to form an optimal development strategy. The given task is achievable only under condition of introduction of scientifically proved model of complex information-analytical maintenance of strategic management. The aim of the article is the development of organizational-methodical provisions and practical aspects of strategic management accounting at the enterprise as a system of information support of strategic management. Objectives of the research: to study modern concepts of strategic management accounting, in particular, the concept of alternative costs, a variety of cost drivers, construction of industry value chains; to identify areas of improvement of modern practice of organization and application of strategic management accounting at enterprises, to highlight the main principles of selection of key performance indicators of enterprises; to offer an algorithm of competitor rating analysis based on information support of strategic management. When writing this article, such methods as analysis of scientific and theoretical sources, system approach, method of comparative analysis, generalization were used. The research contains practical recommendations in the field of methodological support of strategic management accounting, aimed at improving the efficiency of processes of development and implementation of strategic management decisions.

Keywords: Strategic management · Information and analytical system · Management accounting · Value chain · Economic value added · Key performance indicators · Competitor analysis · Strategic business units

1 Introduction

In order to improve modern management concepts of enterprises based on the application of strategic approaches, there is a need in the system of information and analytical support of strategic management [1, 2]. Absence of systematic approaches to the introduction of management information systems and complex scientific research in

V. Murgul and V. Pukhkal (Eds.): EMMFT 2019, AISC 1258, pp. 687–707, 2021.
https://doi.org/10.1007/978-3-030-57450-5_59

this area, taking into account the specifics of agricultural enterprises is one of the main problems [3–5].

As a component of the accounting system, the strategic management account provides the performance of all its main functions and tasks, but as a specific instrument of information support, focused on monitoring of internal and external environment, formulation and implementation of the strategy of enterprise, it has its own special tasks, which concretize the accounting functions.

As the main function of accounting is the information one, then its realization strategic management accounting provides by granting to users retrospective and perspective information of strategic character about internal and external environment of enterprise or business unit for decision-making on introduction, realization or correction of strategies [6–8].

The financial result of the management decision is a comparison of income from the implementation of this decision and alternative costs. One of the actual concepts is the concept of cost factors, which solves the problem of in-depth cost analysis (in particular, the sources of occurrence).

Traditional management accounting considers mainly one expenditure-forming factor - the volume of production in combination with cost carriers within a certain production or commercial process. In CVP analysis ("cost-volume-benefit") costs are divided into variable and constant, average and marginal. However, strategic management accounting uses information from several cost drivers, based on the multilevel nature of their occurrence and the need to identify the underlying causes.

Cost drivers are classified into structural (scale, degree and range of vertical and horizontal integration, experience, technologies, assortment) and functional (labour force involvement, integrated quality management, use of production facilities, planning efficiency, configuration, use of relationships with suppliers and customers within the value chain) factors [9–12].

The next concept is the value-added concept. The main objective of an enterprise is to maximize value added [13, 14]. Many researchers point out the disadvantages inherent in this concept, which are essential for agricultural enterprises. It should be noted that for agricultural enterprises there is no general model for the construction of value chains, which is determined by the existence of significant differences in the functionality of agricultural enterprises [15, 16]. This applies both to the stages of the chain, providing direct agro-industrial production, and the stages of distribution and marketing of products. At the same time, researchers cite generalized or the most typical models of a value chain of agrarian enterprises to display the general interrelationships of the process of value creation (Fig. 1).

However, when developing strategic management accounting, not the above aggregated model should be applied, but the real and detailed value chain of the agricultural enterprise, which will allow to specify the subject of strategic management accounting based on value chains and develop a system of analytical indicators on the value of the enterprise necessary for the needs of strategic management.

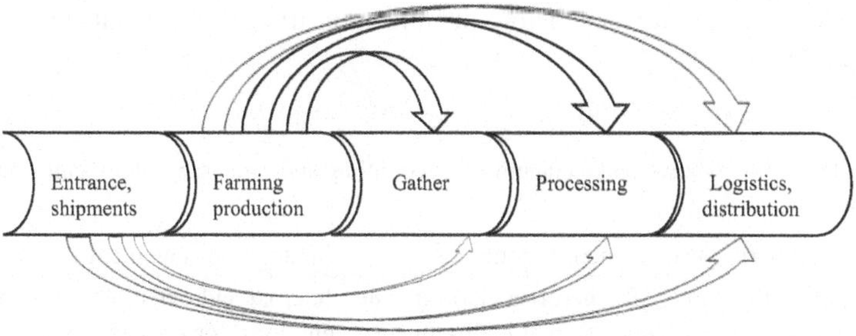

Fig. 1. Steps in the value chain.

Consequently, the improvement of the accounting system in the direction of ensuring compliance with the requirements of the strategic management system in the context of applying the concept of value chains acquires a significant relevance in the conditions of global value chains of agricultural enterprises.

2 Materials and Methods

The application of the value chain system concept differs from the traditional approach to cost analysis, which is used in management accounting and involves a focus on internal value generation factors. The differences between management accounting and strategic management accounting based on value chains are presented in Table 1.

The concept of value chain system as a mechanism of strategic cost management provides for the allocation of four spheres that allow to optimize the size of production costs: communications with suppliers (the first sphere), communications with consumers (the second sphere), technological communications within the value chain of one division (the third sphere), and communications between value chains within an enterprise (the fourth sphere).

The key role of the value chain idea from the point of view of strategic management accounting is orientation to the processes that take place, including those outside the enterprise. In this case, enterprises are considered separately in the context of a common chain of activities that create value.

In the context of information theory, the strategic management accounting subsystem on the basis of the value chain, on the one hand, uses input information about the supplier value chain, acting as an information recipient, and on the other hand, is an information donor, generating and providing information about its structural elements (activities). This approach not only increases the efficiency of enterprise value creation within the internal value chain, but also provides better interaction with external value chains, which is especially important in strategic decisions on vertical integration.

On the basis of the analysis of elements of system of value chains in a section of their cost-benefit or profitableness and on the basis of the analysis of possibilities of the enterprise the decision on realization of a strategic choice can be accepted, including on introduction of external concerning the enterprise of elements of a chain in structure of internal, for example, about replacement of market transactions by intrafirm deliveries, about transition from purchase before manufacture, from services outsourcing to their insourcing, etc.

Table 1. Differences between management accounting and strategic value-chain management accounting.

Comparison criteria	Management accounting	Strategic management accounting
1. Object of study (spatial orientation)	Internal enterprise environment, partially external environment	Internal and external environment of enterprise or strategic business unit, external environment of value chain system
2. Temporary orientation	Past, present	Past, present, future
3. Level of systemic coverage of the phenomena	System analysis of financial factors, partial accounting of non-financial factors	System analysis of financial and non-financial factors by type and type of activity
4. Basic approach	Added value	Value created in a chain built on related activities
5. Main role in the management system	Reducing costs at the enterprise level	Minimizing costs, product differentiation
6. Ways to reduce costs	Allocation of responsibility centers based on total cost determination	Allocation of responsibility centres based on types and activities that are elements of the value chain
7. Types of decisions that are focused on	Tactical	Strategic
8. Type of presentation of information	Internal management reporting	Internal (for value chains) and external (for the value chain system) strategic reporting
9. Users	Management staff	Analysts and consultants

In general, building a strategic management accounting subsystem on the basis of the value chain concept allows identifying factors that affect the long-term profitability of the enterprise and the process of strategy formation and implementation.

Each agricultural enterprise has its own value chain, but its components may vary significantly from enterprise to enterprise depending on the nature of its business activities, which are determined by the wide variety of business operations it undertakes and the strategic initiatives chosen by the enterprise. It was noted that the concept could be compared with the traditional method of accounting for product costs, with a significant increase in the range of information compared with financial reporting. Let us illustrate this conclusion with the help of Fig. 2.

The presented model is the basis not only for carrying out the analysis of expenses in a context of certain kinds of activity of the enterprise, but also for comparison of a position of the enterprise or its separate business unit in comparison with competitors from the point of view of expenses that also allows to reveal the basic factors of generation (or destruction) of competitive advantages and to provide differentiation of production and reception of a price premium.

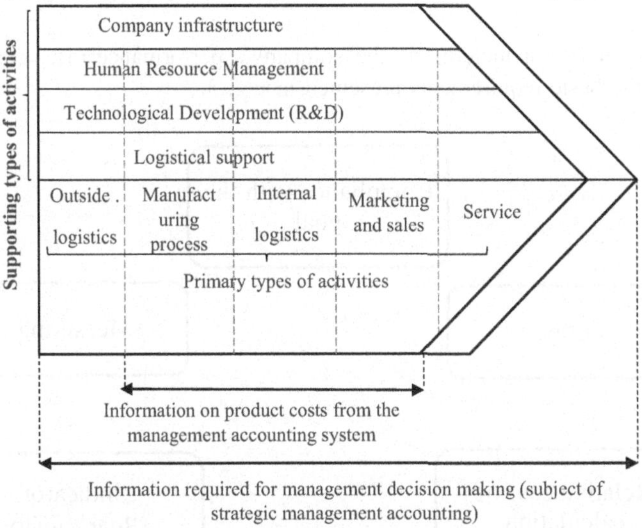

Fig. 2. Structure of the value chain by type of activity in the context of the subject of management and strategic management accounting.

The given Fig. 2 visually shows that for the purpose of acceptance of the correct strategic administrative decision monitoring and processing of the information on expenses of the enterprise at each stage of a chain is required. It can also be stated that traditionally reflected in the reporting costs (including production costs, raw materials stock assessment, EOR, etc.), are only a part of the amount of costs, which in reality is incurred by the enterprise in the agrarian sector.

At the current stage of competition and increasing integration of enterprises within the industry value chain, in addition to financial indicators, including profitability and profitability, the assessment of non-financial indicators, in particular, the performance of key business processes, also becomes essential for the development of the company's strategy.

Key performance indicators (in the practice of strategic management, enterprise management uses the abbreviated English name - KPI) to date is a relevant tool for assessing non-financial performance and intangible assets. In order to fully evaluate the effectiveness of the enterprise development strategy and its management, there is a need to use a wider analytical toolkit, namely, the concept of management by key performance indicators, and a balanced system of indicators based on it [17–19].

The purpose of the system of key performance indicators is to provide the management of the organization with comprehensive information on its status and possible problems to monitor the achievement of the set objectives.

The task of the KPI system and balanced indicators is to translate the organization's strategy into a comprehensive set of performance indicators that define the main parameters of the measurement and management system [20–22]. The set of indicators sets the basis for forming the company's strategy and includes quantitative characteristics to inform employees about the main factors of success in the present and future [23–25].

Each key indicator included in the company's performance measurement system must meet the basic requirements presented in Fig. 3.

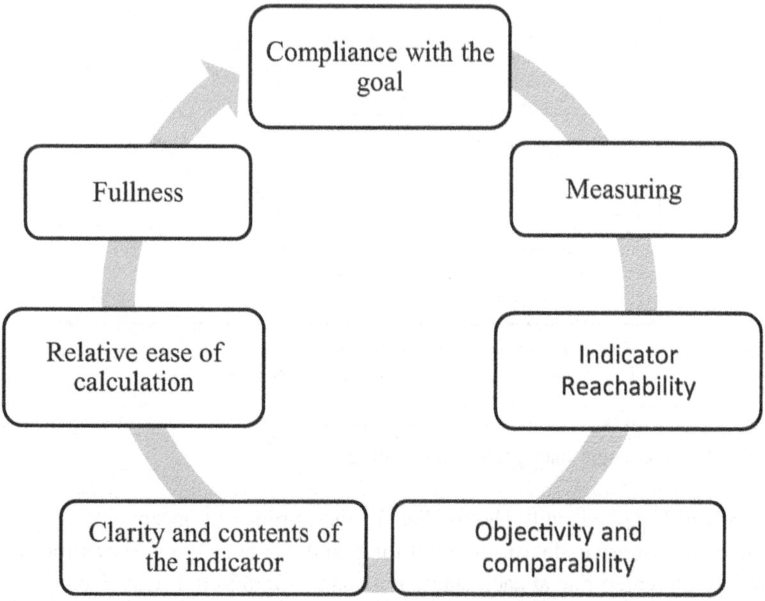

Fig. 3. Requirements for key performance indicators in strategic management accounting practice.

There are several classifications of key performance indicators. Researchers distinguish various classification features reflecting the specifics of the industry they are considering. The following ones have been singled out as classification signs:

- time factor;
- target designation;
- responsibility for achieving the target values;
- types of indicators;
- focus on internal and external environment of the enterprise.

Table 2 presents the types of key performance indicators depending on the above presented attributes.

The procedure for developing a system of key performance indicators is shown in Fig. 4.

Table 2. Classification of key performance indicators of the enterprise.

№	Character of classification	Type of key performance indicator	Feature
1	Time factor	Deferred key performance indicators ((lag) KPI)	Retrospectively, the main objective is to assess the results of actual events
		Key performance indicators ((lead) KPI)	Prognostic nature and act as factors in achieving the desired results
2	Targeted use	Strategic key performance indicators	Paired with the implementation of a set of strategic initiatives in critical functional areas of the enterprise
		Normative key performance indicators	Maintenance at a certain level within the selected time period
3	Responsibility for achieving target values	Indicator(s) key performance indicators	Are aligned with the purpose, objectives and processes of the respective management level, reflecting the degree of achievement of indicators and the effectiveness of the implementation of benchmarks
		Imperative (benchmark) key performance indicators	Delayed nature and are the main indicators by which the higher management level monitors the performance of the lower management level
4	By species	KPI Functionality	Relate to the performance indicators of business processes and allow you to assess how well the process meets the required algorithm of its execution
		KPI result	Show the quantity and quality of the result
		KPI costs	Shows the resources spent
		KPI performance	Derivative indicators characterizing the ratio of the result achieved to the time spent on it
		KPI efficiency	Derivative indicators that characterize the ratio of the result obtained to the cost of resources

(*continued*)

Table 2. (*continued*)

№	Character of classification	Type of key performance indicator	Feature
5	By focusing on the internal and external environment of the enterprise	Diagnostic (process-functional)	They reflect the state of the system "from within", show the degree of its orderliness, internal organizational efficiency
		Interactive (market)	Reflect the situation of contact with the external environment, the degree of impact on it, received changes in the system of superstructures

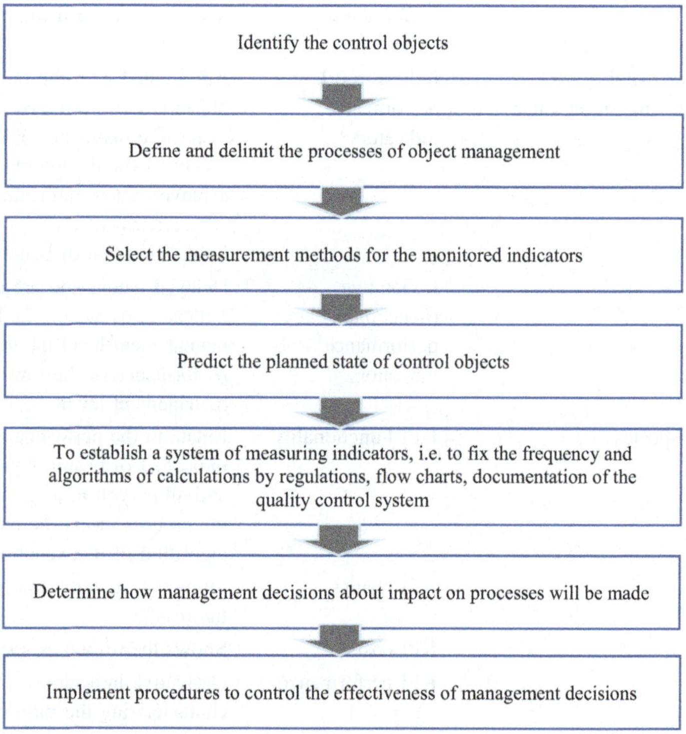

Fig. 4. Procedure for developing a system of key performance indicators.

Creation and introduction into strategic management accounting of the enterprise of the concept of management by key performance indicators, definition of KPI and allocation of responsibilities require compliance with the appropriate algorithm shown in Fig. 5.

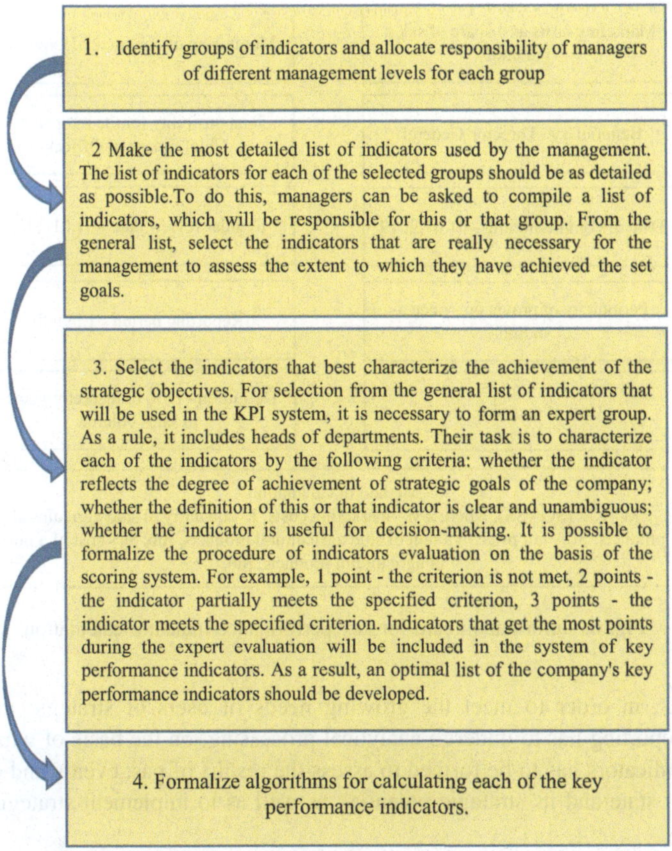

Fig. 5. Definition of KPI and allocation of responsibilities.

The algorithm for calculating each of the key performance indicators should be understandable to managers at different levels and unambiguously defined. In this regard, it is possible to develop an information card (Fig. 6) for each applicant, including the main parameters and description of methods of calculation of the indicator, the frequency of calculation and revision of the indicator, the responsible person, etc.

The main condition for implementation is the presence of clearly defined strategic goals and objectives. The essential value for effective introduction is the use of possibilities of information technologies which allow to simplify procedure on filling, the analysis of the information by means of automation of data input [26].

It is necessary to agree with opinion, that one of the basic lacks which the researchers-drivers who are engaged in questions of working out of an accounting and analytical maintenance of strategic management suppose, is granting of system of the strategic account by analytical functions and attributing to information product (the strategic reporting) received as a result of its functioning ability of full maintenance of all information requirements for acceptance of administrative decisions [27–29]. In

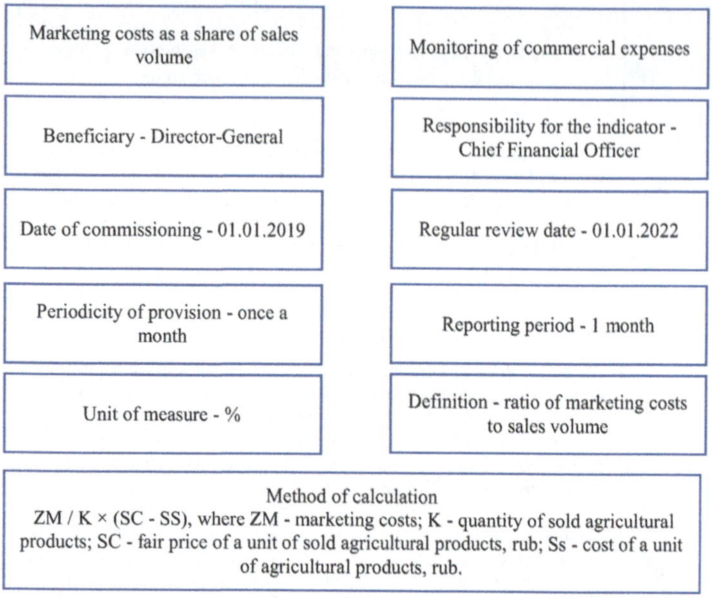

Fig. 6. Information card for key performance indicator calculation.

most cases, in order to meet the growing needs of users of strategic information, strategic reporting has to undergo analytical processing, on the basis of which a set of relevant indicators has to be formed to assess the results of past events and operations, the current state and its strategic potential, as well as to implement strategic elections [30].

The Readiness of the accounting strategic information for acceptance of administrative decisions is defined by its users who in the conditions of the economy globalization characterized by strengthening of competitive struggle, put forward new tasks to the system of accounting and analytical support of strategic management: expansion of an accounting subject; introduction of new objects of the account; specification of the basic principles of the account; increase of flexibility and relevance of accounting system.

3 Results

Let's consider practical aspects of the studied directions of application of tools of information-analytical tools of strategic management on an example of the agricultural enterprise "August-Muslum" LLC of Muslumovsky area of the Republic of Tatarstan.

"August-Muslum" agricultural enterprise of the Muslumovsky district of the Republic of Tatarstan was established within the framework of "August" agricultural project to create a commercially successful, profitable agricultural enterprise, on the basis of which it would be possible to practice and implement the most advanced agro-technologies, plant protection systems, to test the newest preparation mouths.

The main type of activity is mixed farming. "August-Muslum" LLC of the Muslumovsky district of the Republic of Tatarstan owns 28 thousand hectares of arable land. The structure of the arable land on which crops are cultivated is shown in Fig. 7.

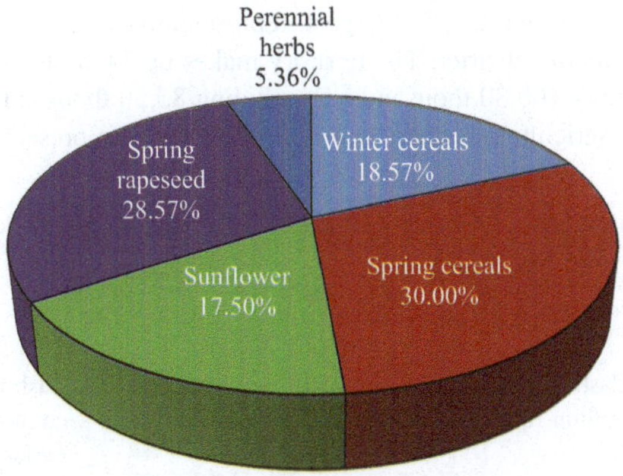

Fig. 7. Structure of the arable area of August-Muslum LLC, Muslumovsky District, Republic of Tatarstan, 2019.

The farm immediately made a bet on all the most modern and best - varieties, hybrids, se-men, machines, fertilizers and, of course, crop protection products. All crop varieties and hybrids introduced into production in August-Muslum LLC of the Muslumovsky District of the Republic of Tatarstan undergo mandatory tests, and numerous experiments on sowing norms, application of various norms and types of fertilizers, plant protection products, etc. are laid down here.

Analyzing the structure, we can conclude that winter cereals occupy 5.2 thousand ha (18.57%), spring cereals – 8.4 thousand ha (30.00%), including spring wheat – 3.4 thousand ha (12.14% of the total area), barley – 2.7 thousand ha (9.64% of the total area), peas – 2.3 thousand ha. (8.21% of the total area). Sunflower covered 4.9 thousand ha (17.50%), spring rape - 8.0 thousand ha (28.57%), perennial herbs - about 1.5 thousand ha (5.36%).

August-Muslum LLC, Muslyumovsky District of the Republic of Tatarstan, has been working with No-till technology since 2018.

The company implements the Cropio satellite monitoring and process control system (for example, sowing, chemical treatments), and drones are used to visually inspect fields. The farm has a modern fleet of equipment, including many agricultural machines that have not yet been used in the Republic of Tatarstan, for example, the Borgo sowing complexes with anchor or disc coulters, the John Deere tractors of the ninth series and various Klaas modifications. The enterprise cooperates in technology with the companies Kazan Agrohimservis, Agrimatko (Jacto sprayers), John Deere (they supply Maestro precision seeders for row crops), and the Pegasus Samara company (Tuman- self-propelled sprayers 2"), "Agromaster", "Amazone", representing a new modification of the DMC seeder, as well as the "Catros", "Liliani" aggregate, from which hopper reloaders are purchased, as well as equipment for laying and storing products in the sleeves.

The activities of the analyzed enterprise are carried out in a competitive market, where demand is formed by the largest representatives of agriculture, including Muslumovsky municipal district. The territory makes up 1464.30 sq. km, and the area of agricultural lands - 100.80 thousand ha, including 85.50 thousand ha of arable land. The indicators of agricultural enterprises for 2016–2019 are presented in Table 3.

Table 3. Performance indicators of agricultural enterprises of Muslumovsky municipal district of the Republic of Tatarstan for 2016–2019.

Reporting period		Indicators		
Years	Quarter	Cash receipts, million rubles	Cash proceeds per 1 hectare of arable land, thousand rubles	Cash receipts per 1 employee, thousand rubles
2016	I	178.8	2.1	136.0
	II	286.9	3.3	219.0
	III	234.6	2.8	179.0
	IV	555.5	6.5	583.0
	Total I–IV block	1255.8	14.7	1117.0
2017	I	257.2	3.0	251.0
	II	309.8	3.6	303.0
	III	236.8	2.8	231.0
	IV	796.2	9.3	777.0
	Total I–IV block	1600.0	18.7	1562.0
2018	I	210.8	2.5	138.0
	II	333.4	4.4	408.0
	III	266.7	2.6	246.0
	IV	1139.1	11.4	1199.0
	Total I–IV block	1950.0	20.9	1991.0
2019	I	247.5	2.9	300.0
	II	467.5	5.5	567.0
	III	399.0	4.6	483.0
	Total I–III block	1114.0	13.0	1350.0

According to the data presented in Table 3, we can conclude that the revenue growth for 2016 is stable, there is a slight decrease in the III quarter of 2016 compared to the II quarter of 2016 (−52.3 million rubles), which is associated with the features agricultural sector of the economy. Analyzing the growth rate of revenue in 2017 by 2016, we can conclude that the agricultural sector of the Muslyumovsky municipal district is developing, the annual growth rate was 127.4%.

In 2018, revenue increased (growth from the 1st to 4th quarter amounted to 928.3 million rubles, respectively, the growth rate was 540.4%). If we consider the growth rate in relation to 2017, we can say an increase of 121.9%, and in relation to 2016 - 155.28%.

Cash revenue during 2019, similarly to previous years, has a tendency to increase (an increase from I to III quarter amounted to 151.5 million rubles, respectively, the growth rate is 161.2%). If we consider the growth rate in relation to 2018, we can say an increase of 149.6%, in relation to 2017 - 168.5%, and by 2016 - 170.1%.

Based on the presented agricultural market development trends, it can be stated that in the Republic of Tatarstan the market is dynamic, has positive dynamics, which is associated, inter alia, with state initiatives (Table 4).

Table 4. Information on state support received by agricultural enterprises of the Muslumovsky District of the Republic of Tatarstan for 2016–2019.

Family farms		Beginning farmers		Cooperatives	
Units	Million rubbles	Units	Millions rubbles	Units	Millions rubbles
26	67.3	9	14.4	2	18

These factors predetermine the necessity to develop an effective strategy for the analyzed enterprise “August-Muslum” LLC of Muslumovsky region of the Republic of Tatarstan in order to achieve a sustainable competitive advantage in the agar market based on the analysis of the external environment. The main source of information is data obtained as a result of strategic management accounting.

Since agriculture is a complex, multi-stage process requiring a sufficiently large amount of investment, and the modern nomenclature is characterized by a high degree of diversity, it seems expedient to conduct a rating analysis of key competitors of August-Muslum LLC of the Muslumovsky District of the Republic of Tatarstan.

As existing key competitors of August-Muslum LLC of Muslumovsky District of the Republic of Tatarstan in the agricultural market, it seems necessary to distinguish the following ones: LLC Agrofirma Namus, LLC Agrofirma Rodnye Kray - Tugan Yak, KFH Islamov, KFH Avzalov, KFH Ayupov.

Within the framework of this research, an assessment of competitiveness has been made. The results of the comparative characteristics of the existing competitors of August-Muslum LLC in the Muslumovsky District of the Republic of Tatarstan based on expert assessments using key performance indicators are presented in Table 5.

Table 5. Results of comparative characteristics of existing competitors of August-Muslum LLC of Muslumovsky District of the Republic of Tatarstan based on expert assessments.

Competitors	Price	Product quality	Scope of activities	Production and technical capabilities	Sales activities	Ecology	Sum of points
LLC «A August-Muslum»	4	4	3	5	2	2	20
LLC «Agrofirma Namus»	4	4	5	5	4	3	25
LLC «Agrofirma Rodnye Krai - Tugan Yak»	5	4	5	5	5	4	28
KFH «Islamov»	3	3	3	3	3	3	18
KFH «Avzalo»	4	3	4	3	2	2	18
KFH «Ayupo»	3	3	3	3	2	2	16

Based on the competitors' rating analysis within the framework of strategic management accounting, based on expert assessments using key performance indicators, the following probable strategic initiatives of competitors were formulated in August-Muslum LLC, Muslumovsky District of the Republic of Tatarstan (Table 6).

Table 6. Results of comparative characteristics of existing competitors of August-Muslum LLC, Muslumovsky District of the Republic of Tatarstan.

Competitors	Probable strategic initiatives
LLC «Namus Agrofirm»	– expanding its presence on the market – sustainability of resources and energy – increasing its capacity and production volumes
LLC «Agrofirma Rodnye Krai - Tugan Yak»	– maintaining and strengthening market positions – reducing costs, introducing different varieties and hybrids of crops – reduction of negative impact of the enterprise's activity on the environment
KFH «Islamov»	– growing market share – cost leadership – high level of investment and modernization of production – implementation of internal environmental policies and resource conservation
KFH «Avzalov»	– growing market share – cost leadership – high levels of investment in equipment and personnel – increase in the share of resource-saving technologies
KFH «Ayupov»	– aggressive market share growth, cost leadership – the shortest terms of product development and production – high level of investment and modernization of production – optimization of energy and resource consumption

Research of modern practice of activity of the agricultural enterprises testifies to presence of a complex of factors which reduce efficiency of acceptance of administrative decisions of strategic character, directly concerning system of accounting and analytical maintenance of a control system:

- imperfection of the current accounting and analytical model of forming information for making managerial decisions;
- groundlessness of dynamics of correlation between the volumes of necessary and sufficient information, which leads to the phenomena of lack and saturation of accounting and analytical information;
- absence of rational criteria for selection of necessary accounting and analytical information from the total volume of data, which is connected with the absence of the criterion of orientation of construction of accounting system to ensure implementation of enterprise strategy;
- principal difference in classification features of accounting and analytical information required for strategic management;
- untimely preparation and submission of reports and formation of indicators necessary for making managerial decisions.

The considerable part of problems and lacks of information support of process of acceptance of administrative decisions of strategic character is caused by imperfection of structure of an information base of the agricultural enterprises.

4 Discussion

Another important issue, which concerns the organizational aspects of building a system of information and analytical support in agricultural enterprises, is the definition of subjects of strategic management accounting and implementation of economic analysis, focused on providing strategic management.

In our opinion, such variants of organization of strategic accounting at enterprises are possible.

- granting functions of strategic accounting to accountants of enterprises;
- creating a new accounting position for strategic accounting;
- attraction of external specialists for strategic accounting (auditors, representatives of consulting firms, financial analysts, etc.).

Each of the options has its advantages and disadvantages of practical implementation in the conditions of modern information-computer technologies and different degree of economic feasibility and information security.

Realization of advantages of strategic management accounting at the enterprise of agrarian sphere of economy is possible on condition of analysis of the enterprise through set of separate segments. Each segment, in this case, should be characterized by a certain type of product, which is sold to groups of clients in a competitive environment. For strategic segmentation, both the existing organizational structure and the structure of its responsibility centers can be used as basic elements. In particular, on the basis of structural units, strategic business units, whose activities cover the entire value chain, can be identified.

Strategic business units (SBU) are characterized by the availability of their own production and distribution channels.

In other words, SBU are characterized by presence of the certain nomenclature of production and unequivocally identified groups of clients. It is possible to allocate the following basic conditions which should satisfy SBU for the purposes of the strategic analysis (Fig. 8).

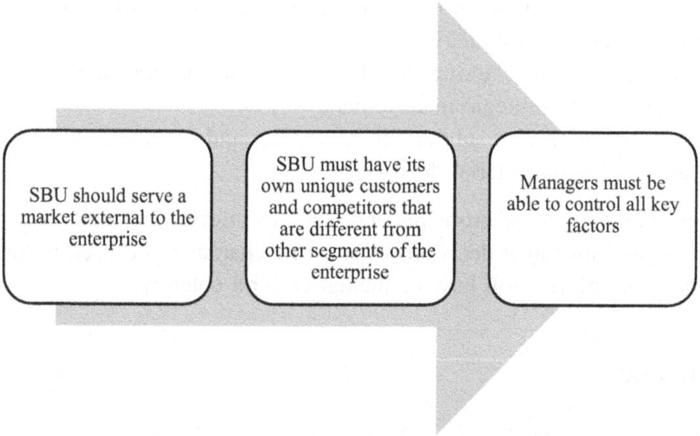

Fig. 8. Conditions for creating strategic business units.

Information obtained as a result of strategic management accounting should also be presented as part of segment reporting.

Taking into account the important role of strategic management for further development of agricultural enterprises, to provide its information support, we propose to apply a strategic analytical system using the strategic accounting system as an information basis. The use of the given system will allow to raise efficiency of development of strategies and carrying out of actions on their realization, creating adequate preconditions for reception of steady competitive advantages.

At many enterprises of agrarian sphere of economy the essential discrepancy between balance and market value of assets is observed. The main reason for the emergence of this situation is the increasing value in the activities of the enterprises considered intangible assets, as well as other factors that cannot be directly evaluated (for example, the assessment of reliability of relations with suppliers, with customers, brand).

For agricultural enterprises, the most rational use of the economic value-added model (abbreviated name EVA). On the basis of calculated cost indicators and value-added indicators in the context of strategically important types of activity as components of a value chain, derived analytical indicators can be calculated and used to make managerial decisions in this area. Therefore, the presence of the resulting EVA allows avoiding the difficulties associated with the calculation and analysis of many other unrelated indicators.

The main task of the management is to increase the value of the business, which corresponds to the main strategic task of the enterprise, which is expressed in ensuring the trend of positive value of EVA (Economic Value Added). Stable positive index assumes effective development of the enterprise, and, consequently, growth of its investment attractiveness.

Maximization of the EVA value can be achieved in the following ways (Fig. 9).

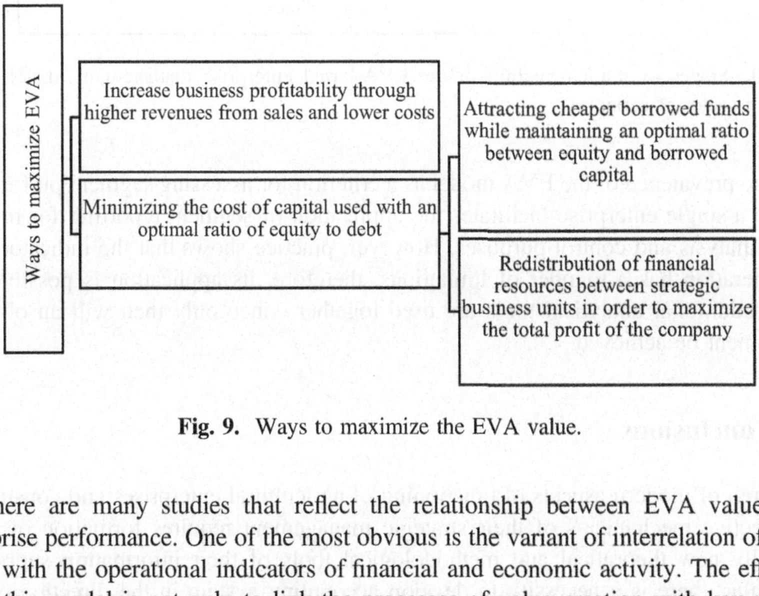

Fig. 9. Ways to maximize the EVA value.

There are many studies that reflect the relationship between EVA values and enterprise performance. One of the most obvious is the variant of interrelation of EVA value with the operational indicators of financial and economic activity. The effect of using this model is not only to link the processes of value creation with key performance indicators, but also the possibility of obtaining objective values of key performance indicators that contribute to the increase of economic value added by the enterprise.

On the basis of the above-mentioned facts, we will present the main stages of realization of the model under consideration (Fig. 10).

The object of the EVA calculation is:

- the enterprise as a whole;
- separate business lines;
- strategic business units.

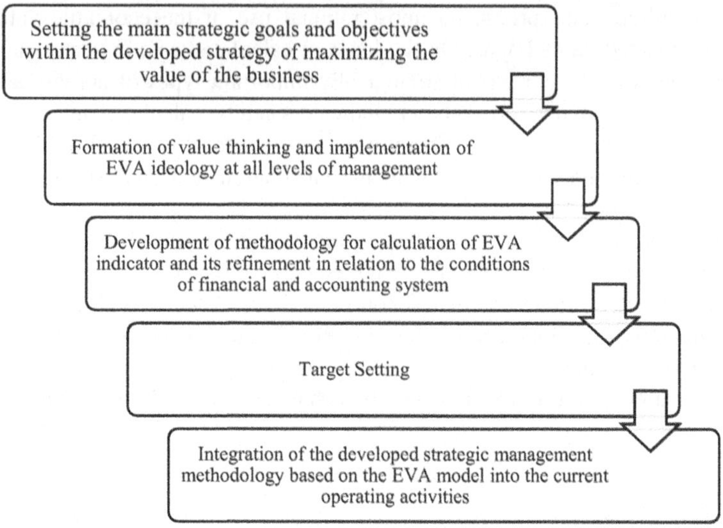

Fig. 10. Stages of implementation of the EVA-based enterprise management model for the agrarian sector of economy.

The prevalence of the EVA model as a criterion for assessing segment performance within a single enterprise facilitates the application of segment reporting for management analysis and control purposes. However, practice shows that the indicator under consideration has a number of limitations, therefore, its application is possible only when additional analytical tools are used together, since only then will an objective assessment be achieved.

5 Conclusions

Research of modern aspects of functioning of agricultural enterprises and construction of effective mechanisms of their strategic management requires formation of fundamentally new theoretical and methodological tools of their information support. In particular, there is a necessity to develop accounting system in the direction of substantiation of the essence and peculiarities of practical realization of strategic management accounting methodology on the basis of joint application of different models.

The rationalization of the model of information support of strategic management of agricultural enterprises should be based on the clear structuring of information flows to ensure prompt and reliable obtaining of necessary data in case of such necessity. Strategic decision making should be based on current (reliable, accurate) and forecasting (estimated, indicative) information.

The considered aspects of application of analytical tools within the framework of strategic management accounting at agricultural enterprises, in particular, strategic competitive analysis, are aimed at accounting and analysis of trends in the development of key value factors at two levels (the level of strategic segments of the enterprise and operational level) and should be regular.

To ensure correct functioning of the system of accounting and analytical support of strategic management of agricultural enterprises, it is necessary to take into account the specifics of specific business entities. Within the limits of the considered theme the questions on information support of strategic management accounting (use of a value chain), on the organization and practice of application of key efficiency indicators are analyzed; the rating analysis of competitors is carried out and the information base for acceptance of effective strategic management decisions for development of measures on strengthening of competitive advantages is formed. Implementation of strategic management aspects at agricultural enterprises allows creating prerequisites for ensuring survival in the competitive struggle in the long-term perspective.

References

1. Fernandes, J.P.: Developing viable, adjustable strategies for planning and management—a methodological approach. Land Use Policy **82**, 563–572 (2019). https://doi.org/10.1016/j.landusepol.2018.12.044
2. Lan, S., Yang, Ch., Tseng, M.-L.: Corporate sustainability on causal financial efficiency model in a hierarchical structure under uncertainties. J. Clean. Prod. **237**, 117769 (2019). https://doi.org/10.1016/j.jclepro.2019.117769
3. Klychova, G., Zakirova, A., Sadrieva, E., Avkhadiev, F., Klychova, A.: Reporting in the area of sustainable development in agribusiness. In: E3S Web of Conferences, vol. 91, p. 06002 (2019). https://doi.org/10.1051/e3sconf/20199106002
4. Su, Y., He, S., Wang, K., Shahtahmassebi, A.R., Zhang, L., Zhang, J., Zhang, M., Gan, M.: Quantifying the sustainability of three types of agricultural production in China: an emergy analysis with the integration of environmental pollution. J. Clean. Prod. **252**, 119650 (2020). https://doi.org/10.1016/j.jclepro.2019.119650
5. Klychova, A., Klychova, G., Zakirova, A., Sungatullina, R., Mukhamedzyanov, K., Philippova, E.: Social development mechanism of an agricultural enterprise formation. In: E3S Web of Conferences, vol. 110, p. 02072 (2019). https://doi.org/10.1051/e3sconf/201911002072
6. O'Sullivan, C.A., Bonnett, G.D., McIntyre, C.L., Hochman, Z., Wasson, A.P.: Strategies to improve the productivity, product diversity and profitability of urban agriculture. Agric. Syst. **174**, 133–144 (2019). https://doi.org/10.1016/j.agsy.2019.05.007
7. Hanrahan, L., McHugh, N., Hennessy, T., Moran, B., Shalloo, L.: Factors associated with profitability in pasture-based systems of milk production. J. Dairy Sci. **101**, 5474–5485 (2018). https://doi.org/10.3168/jds.2017-13223
8. Klychova, G.S., Ziganshin, B.G., Zakirova, A.R., Valieva, G.R., Klychova, A.S.: Benchmarking as an efficient tool of social audit development. J. Eng. Appl. Sci. **12**, 4958–4965 (2017). https://doi.org/10.3923/jeasci.2017.4958.4965
9. Yigezu, Y.A., Mugera, A., El-Shater, T., Aw-Hassan, A., Piggin, C., Haddad, A., Khalil, Y., Loss, S.: Enhancing adoption of agricultural technologies requiring high initial investment among smallholders. Technol. Forecast. Soc. Chang. **134**, 199–206 (2018). https://doi.org/10.1016/j.techfore.2018.06.006

10. Teirlinck, P.: Configurations of strategic R&D decisions and financial performance in small-sized and medium-sized firms. J. Bus. Res. **74**, 55–65 (2017). https://doi.org/10.1016/j.jbusres.2017.01.008
11. Faizrakhmanov, D., Zakirova, A., Klychova, G., Yusupova, A., Klychova, A.: Formation and disclosure of information on social responsibility of agribusiness enterprises. In: E3S Web of Conferences, vol. 91, p. 06004 (2019). https://doi.org/10.1051/e3sconf/20199106004
12. Rikhardsson, P., Yigitbasioglu, O.: Business intelligence & analytics in management accounting research: status and future focus. Int. J. Account. Inf. Syst. **29**, 37–58 (2018). https://doi.org/10.1016/j.accinf.2018.03.001
13. Wonglimpiyarat, J.: What is it about strategic implications of using financial models in the process of technology management? J. High Technol. Manage. Res. **30**, 82–90 (2019). https://doi.org/10.1016/j.hitech.2018.12.001
14. Sayari, N., Sari, R., Hammoudeh, S.: The impact of value added components of GDP and FDI on economic freedom in Europe. Econ. Syst. **42**, 282–294 (2018). https://doi.org/10.1016/j.ecosys.2017.03.003
15. Bui, B., de Villiers, C.: Business strategies and management accounting in response to climate change risk exposure and regulatory uncertainty. Br. Account. Rev. **49**, 4–24 (2017). https://doi.org/10.1016/j.bar.2016.10.006
16. Klychova, G.S., Safiullin, L.N., Zakirova, A.R.: Information-analitical support of cost management in horse breeding. Mediterr. J. Soc. Sci. **5**(18), 193–196 (2014). https://doi.org/10.5901/mjss.2014.v5n18p193
17. Elzahar, H., Hussainey, K., Mazzi, F., Tsalavoutas, I.: Economic consequences of key performance indicators' disclosure quality. Int. Rev. Financ. Anal. **39**, 96–112 (2015). https://doi.org/10.1016/j.irfa.2015.03.005
18. Li, H., Hong, T., Lee, S.H., Sofos, M.: System-level key performance indicators for building performance evaluation. Energy Build. **209**, 109703 (2020). https://doi.org/10.1016/j.enbuild.2019.109703
19. Banu, G.S.: Measuring innovation using key performance indicators. Procedia Manuf. **22**, 906–911 (2018). https://doi.org/10.1016/j.promfg.2018.03.128
20. Klychova, G.S., Zakirova, A.R., Kamilova, E.R.: The methodological instruments of social audit in the agricultural companies development. Int. Bus. Manage. **10**, 5254–5260 (2016). https://doi.org/10.3923/ibm.2016.5254.5260
21. Zhu, L., Johnsson, Ch., Varisco, M., Schiraldi, M.M.: Key performance indicators for manufacturing operations management – gap analysis between process industrial needs and ISO 22400 standard. Procedia Manuf. **25**, 82–88 (2018). https://doi.org/10.1016/j.promfg.2018.06.060
22. Badawy, M., Abd El-Aziz, A.A., Idress, A.M., Hefny, H., Hossam, S.: A survey on exploring key performance indicators. Future Comput. Inform. J. **1**, 47–52 (2016). https://doi.org/10.1016/j.fcij.2016.04.001
23. Bhadani, K., Asbjörnsson, G., Hulthén, E., Evertsson, M.: Development and implementation of key performance indicators for aggregate production using dynamic simulation. Miner. Eng. **145**, 106065 (2020). https://doi.org/10.1016/j.mineng.2019.106065
24. Joppen, R., von Enzberg, S., Gundlach, J., Kühn, A., Dumitrescu, R.: Key performance indicators in the production of the future. Procedia CIRP **81**, 759–764 (2019). https://doi.org/10.1016/j.procir.2019.03.190
25. Maté, A., Trujillo, J., Mylopoulos, J.: Specification and derivation of key performance indicators for business analytics: a semantic approach. Data Knowl. Eng. **108**, 30–49 (2017). https://doi.org/10.1016/j.datak.2016.12.004

26. Subaeva, A.K., Nizamutdinov, M.M., Mavlieva, L.M., Kalimullin, M.N.: Labor productivity in digital agriculture. In: BIO Web of Conferences, vol. 17, p. 00226 (2020). https://doi.org/10.1051/bioconf/20201700226
27. Pavlatos, O., Kostakis, H.: Management accounting innovations in a time of economic crisis. J. Econ. Asymmetries **18**, e00106 (2018). https://doi.org/10.1016/j.jeca.2018.e00106
28. Papulova, Z., Gazova, A.: Role of strategic analysis in strategic decision-making. Procedia Econ. Financ. **39**, 571–579 (2016). https://doi.org/10.1016/S2212-5671(16)30301-X
29. Klychova, G., Zakirova, A., Mannapova, R., Pinina, K., Ryazanova, Y.: Assessment of the efficiency of investing activities of organizations. In: E3S Web of Conferences, vol. 110, p. 02075 (2019). https://doi.org/10.1051/e3sconf/201911002075
30. Uyar, A., Kuzey, C.: Does management accounting mediate the relationship between cost system design and performance? Adv. Account. **35**, 170–176 (2016). https://doi.org/10.1016/j.adiac.2016.06.004

Technological Prospect of Innovative Development of the Processing Industry

Andrey Alekseev[1], Kirill Khlebnikov[2], Alexander Arkhipov[3](✉), and Alexander Schraer[1](✉)

[1] St. Petersburg State University of Economics, Sadovaya str., 21, St. Petersburg, Russia
sharier@rambler.ru
[2] Grozny State Oil Technical University named after Academician MD Millionshtchikov, pr. Isaeva 100, Grozny, Russia
[3] St. Petersburg State University of Industrial Technologies and Design, str. Bolshaya Morskaya, 18, St. Petersburg, Russia
aarhipov@list.ru

Abstract. The study is aimed at expanding the scientific discussion about the factors of investment efficiency in regional (Russia) and sectoral (processing industry) projections. The paper presents the results of the research on the process of investing into innovations of the Russian processing industry (sample of 2009–2017). The factors of economic efficiency have been identified for investment industrial projects. The theoretical statement was formulated that in today's economy, innovation investment efficiency is not determined that much by the internal potential of the enterprise and the infrastructure it is surrounded with (however, their significance in the balance of potentials is not denied), but rather by external, market factors - industry consolidation ratio on the platform of new technology innovation. The identified nature of innovation investment efficiency is considered as a new research statement, which has objective limitations: region- and industry-specific ones. The practical side of the obtained result can be formulated as a criterion for selecting industries (markets, lines of business) for investing into innovations by enterprises of the processing industry.

Keywords: Innovations · Investments · Entrepreneurship · Economic efficiency

1 Introduction

When an innovation is localized in a project, it is reasonable to draw upon the position of investment cycle with traditional indicators of (volume) effect and efficiency. Academically, the first one is expressed by net present value (NPV) and the second one by return on investment (ROI). An investment project is based on evaluating the economic formed by the innovation. The issues related to interpreting and assessing economic effects from innovations have been considered by Russian and foreign scientists: Puryaev A. [1]; Guan J., Chen K. [2]; Fagerberg J., Srholec M., Verspagen B. [3]; Carayannis E.G., Grigoroudis E., Goletsis Y. [4]; Dziallasa M., Blindab K. [5];

V. Murgul and V. Pukhkal (Eds.): EMMFT 2019, AISC 1258, pp. 708–717, 2021.
https://doi.org/10.1007/978-3-030-57450-5_60

Kharchenko E., Alpeeva E., Ovcharova O. [6]; Alekseev A., Khlebnikov K., Fomin E., Fomina N. [7]; Vasilyeva, E.U., Kudryavtseva T.Y., Ovsyanko D.V. [8]; Polder M., Leeuwen G. [9]; Kuczmarski T. D. [10]; Drake M.P., Sakkab N., Jonash R. [11]; Koskinen Y., Maeland J. [12] et al. Consolidating the results of the scientific discussion, the unanimously accepted statements can be formulated:

1. The result of an innovation project is expressed with local effects from business activity of the enterprise: growth of price, cost-cutting, growth of productivity, etc.
2. Local effects are determined and economically interpreted by innovation areas or classes: product, process, marketing, and organizational classes;
3. Economic models are described for interpreting local effects of innovations through profit and profitability indicators of projects.

Thus, in scientific literature methods have been formed for planning economic effects and efficiency of investing capital into innovation projects.

1.1 Discussion on the Nature of Innovation Effectiveness

On the other hand, the authors point out that the scientific discussion about the nature and the factors of innovation efficiency is incomplete. Most scientific works investigate investment in R&D. Searching for the interrelations between this factor and the effects is very popular in economic studies because the information on R&D costs is available as well as the data on economic activity at the micro, meso and macro levels. This information is contained in the public reports of enterprises and state statistics. The most commonly recognized effect of investing in R&D is defined as "productivity" (the performance indicator): [13], Lim E.S. et al. [14], Drake M.P. et al. [11] et al. But it is important to understand the limitations of the results obtained in the research of this interrelation, objectively formulated by the authors of the publications. For instance (the first limitation), R&D costs amount from 0.3 to 23% (OECD, 2015) from the total investment budget of an innovation project, which also includes capital investments of the production, operation and disposal cycle. Formally, the investment cycle of an innovation project [15] is determined by the total amount of costs from the "investment memorandum" to "disposal". The effects and efficiency of innovation can be objectively measured only in application to the implemented investment cycle and the total amount of costs of an innovation project. The second limitation: innovation projects do not always affect the main line of business of an enterprise. They can be local in nature. In particular, it is true if the scale of a project is insignificant in the financial turnover of the enterprise. This statement is fully fair for medium- and low-tech industry, whose level of research intensity is below 7%. The third limitation concerns the popular paradigm of "open innovation" and the network model used for organizing innovation process. Innovation projects can be organized as a consortium (network, distributed form of an innovation project), set up by several specialized enterprises. Consequently, including a single enterprise in managements accounts will be incorrect in terms of assessing the growth of productivity. So, when productivity of an enterprise is correlated within the main line of business and investment into R&D, there are many limitations and assumptions in terms of economic modeling of efficiency achieved through investing into innovations. Correspondingly, in the authors' opinion, objective assessment of innovation efficiency must rely on investigating individual projects and investment cycles. If the problem is stated like this, the statistics is hard to form. There is need for management accounting of investment projects, aimed at the implementation of an innovative solution. Such information has to be collected "manually", through interpreting management accounts and accounting statements of individual innovation projects.

1.2 The Nature of the Effectiveness of Innovation Projects

Nonetheless, we would like to focus on individual scientific research studies, which made an effort to define the nature and factors of economic efficiency of innovation projects within an entire investment cycle. It is unanimously understood that the economic effect and efficiency of an innovation project is a balance of three potentials (Rothwell R., [16]): the innovation (research and technology) potential of the enterprise; the innovation (meso- and macro-) potential of the infrastructure; and the innovation potential of the market. As of today only empirical models of the component interrelation have been proposed. These are not justified by econometric validation and multi-factor regression models. But it is possible to point out some investigations into individual factors of efficiency achieved from investing into innovation projects. The analyzed publications make it possible to conclude that the identified factors concentrate in the internal (potential, infrastructure) and external (market) environments. The internalities were researched by: Morris L. [28], Tsai C.-L., Chang H.-C. [18] - innovation infrastructure; Bueche A.M. [19] - the size of intellectual capital; Vega-Jurado, J. et al. [20] presence of their own research departments in enterprises; Quintella, V. et al. [21] level of financial risks; and others. The externalities. Tajeddini K. et al. [22] point out at the dependence on the external marketing factors, and subsidiarity of the impact created by potential and infrastructure. A Caggese A. [23] suggested a model for evaluating through integral level of risk. Zahra S., Das S. [24] formulate the dependence on the "level of leadership", interpreted through market share. Nanda R., Rhodes-Kropf M. [17] determine efficiency through market growth rates. The above interrelations are not universal. The researchers localize the result by individual industries, lines of business, and types of economies.

In the context of the mentioned analysis it seems reasonable to expand the scientific discussion about the factors of investment efficiency in the regional (Russia) and industry-specific (processing industry) projections.

2 Methods

The target of the research is the nature and factors of investment efficiency in the national processing industry.

The method of the research is correlation analysis. In terms of this analysis, an objective is set to identify the factors which affect the performance characteristic - the level of return on investment (ROI) into innovation projects. The authors point out 8 internalities and 4 externalities (Fig. 1), which hypothetically determine the profitability of investing into innovation pojects. Hypothetical factors are selected based on the following criteria: frequency of accentuation in scientific literature; measurability; isomorphism, relative independence on the scale of projects; possibility of comparison in projects; availability from management accounts documents (business plans, investment projects, forms 1–4 of national accounts, marketing research reports).

The sample of the research is based on the investigation of the retrospective of 92 innovation projects of processing industry enterprises, whose investment cycle was implemented in the period of 2009–2017. The statistics on the internal investment indicators of the projects are supplemented by the data on the external indexes of the industry (market), obtained from the business plans of the innovation projects, marketing research studies and official statistics (in particular, FAS of Russia). In order to refine some quantitative assessments and justify the nature of economic efficiency, in-depth interviews were conducted with the managers of 17 innovation projects.

3 Results

The results of investigating into the sample of innovation projects are presented through correlation coefficients (r2, Fig. 1) of hypothetical factors with the return on investment (ROI) indicator – performance characteristic. The analysis allows us to formulate the conclusions:

1. Externalities have a comparatively greater influence on the economic efficiency of investing into innovations;
2. The "highest" level of correlative interrelation is determined with the factor of economic concentration of an industry, market, innovation project (coefficient by the sum of three, CR3).

The results of the research are consistent with the scientific conclusions about the dominance of external factors in the formation of economic efficiency of investments into innovative activities: Tajeddini K. et al. [22], Caggese A. [23], Zahra S., Das S. [24], Okrepilov V. [25], Nanda R., Rhodes-Kropf M. [17]. Attention should be paid to the "average" level of correlation with the performance characteristic and other externalities. The market share of the industrial enterprise for the main line of business (r 0.447) characterizes a considerable marketing potential for promoting and selling innovative products. I.e. the innovative product can be distributed via the sales channels of the enterprise that have been established within the main line of business. At the same time, the average market growth rate (r 0.571), characterizes the demand dynamics and is explicitly related to the economic concentration ratio of the industry, which is objectively shown in the model by A.T. Kearney (Schuh C. et al., [26]). The presented conclusions are also congruent with modern methodological statements of the theory of industrial organization (Patchell J. [27]).

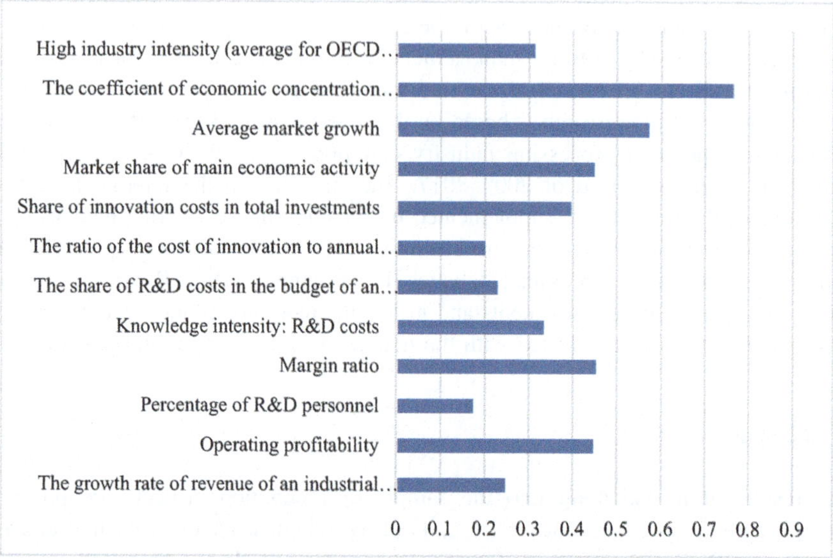

Fig. 1. Correlation coefficients (r2) of hypothetical factors with the indicator of return on investment (ROI) for an innovation project (performance indicator).

In order to develop the formal evaluation of the results of correlation analysis, the authors conducted in-depth interviews with the managers of some innovation projects. One of the issues was about obtaining expert evaluation of the balance of influence of internal and external factors on economic results of the investment project. The obtained results (Fig. 2) showed that external, market conditions dominate in forming the efficiency of the project with respect to the internal potential. The only exception was the biotechnology sector. But in this case the exception proved the rule. The economic concentration ratio of this sector is very low (<20%). There is virtually no competition and the success of the innovation project depends entirely on the scientific potential of the enterprise, which is pointed out by the experts in the survey.

The level of correlation of externalities, the results of the in-depth interview and consistency of the methodology justify the first theoretical conclusion about the dominance of externalities in forming economic efficiency of innovative entrepreneurship.

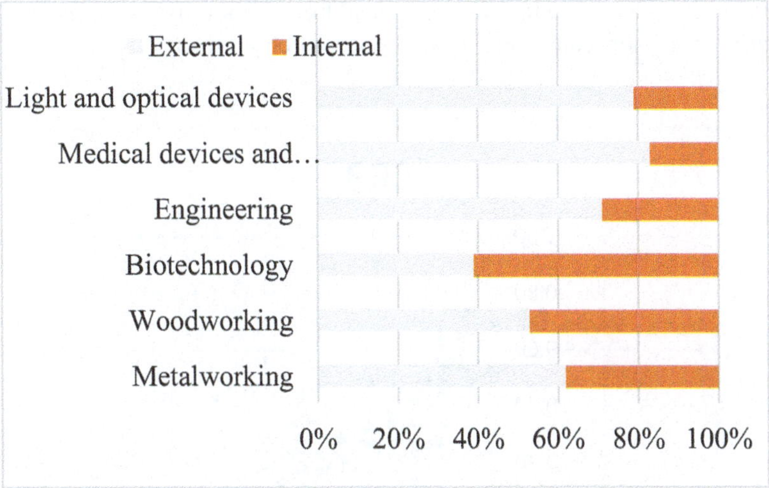

Fig. 2. Expert evaluations of the influence balance (survey scale) of externalities and internalities on the economic result of the innovation project. Interpreted according to the data of in-depth interviews with the managers of 17 innovation projects.

The identified formal interrelation between the economic concentration factor and innovation efficiency (conclusion 2) needs to be explained from the perspective of its nature. It is namely what will help justify the objectivity of the phenomenon. The trend line can be found out in application to the correlation field (Fig. 3)[1]. It separates 2 zones (zone A and B), which have the following economic logic. Markets with a low (<0.45) ratio of economic concentration (zone A) have a low level of competition intensity. There are no big players ("leaders" identified in the study by Zahra S., Das S., [24]) on the market in terms of market share and competition is close to being "perfect". The market of innovative product is in deficit, the demand is not satisfied and there are no formed standards of consumption. Entry barriers are virtually absent.

The situation is reverse in the zone of high concentration (zone B). The entry barriers are formidable and formed by 3–4 leaders. The agents, large by their market share, (belonging to CR3) implement strategies of "cost leadership" thanks to the "scale effect", which prevent new products from entering the market. It is especially clear in the zone of projects (Fig. 3, zone B), launching innovative product on highly consolidated markets (CR3 > 0.5). More than half innovation projects in this zone of consolidation (B) have negative ROI indicators. In the in-depth interviews (3 projects with negative ROI), the project managers synonymously pointed out at high market barriers as a cause of failure. Barriers, built primarily due to ".... a critically low price

[1] Regretfully, the size of the sample does not give a confident result in terms of regression analysis for approximating the interrelation function. The authors put forward a hypothesis that the sample is highly inhomogeneous because it contains innovation projects with a various level of technological advancement and from different industries. The size of the sample is not sufficient for differentiated investigation of industries. This area is defined by the authors as an area for further research.

of the market leaders…., which prevents small lots of innovative products from being created in an economically efficient way" (citation of an expert).

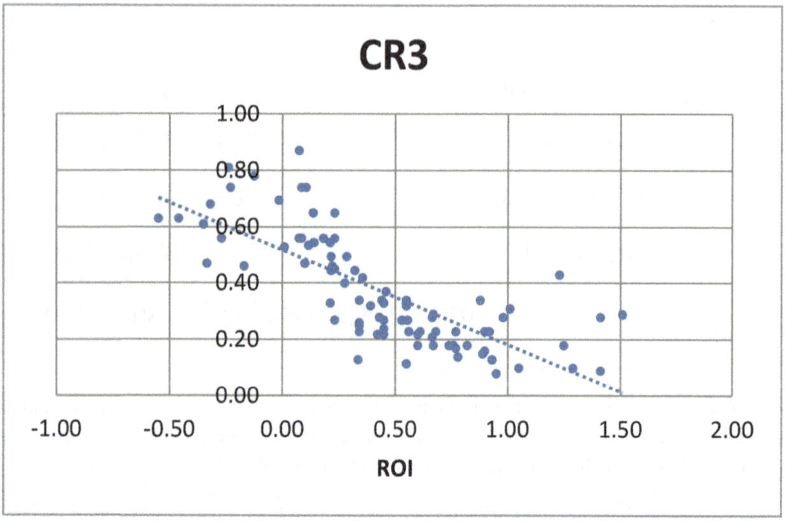

Fig. 3. The trend of the economic concentration ratio affecting innovation efficiency.

The proposed view of the nature of connection between the factors is revealed in consistency with the model of industry consolidation by A.T. Kearney (Schuh C. [26]). The well-known diagram of the industry consolidation dynamics (Fig. 3) shows 4 stages of industry development from the perspective of change of the economic concentration coefficient. The "opening" stage characterizes the "birth" of the industry, initiated by a new technology platform - radical innovation. The "scale" stage is a process of evolutionary development of the market with low intensity of competition and deficit of a morally new product. At the "focus" stage the market becomes highly consolidated. It develops through mergers and acquisitions (M&A) and high barriers are created to entering the market. Innovation projects of small innovation enterprises (SIE) virtually do not have any potential for investment scaling. They are either blocked or acquired by the leaders.

The authors superposed[2] the trend of economic concentration ratio affecting the innovation investment efficiency (Fig. 3) on the industry consolidation diagram by A.T. Kearney [26]. The superimposition gives understanding of 2 analogous zones (Fig. 3, A, B) for projecting the innovation investment efficiency. The zone of the "opening-scale" stages (CR3 15–45%) is potentially favorable for investing into innovative projects due to the expected low entry barriers. The zone of the "focus" stage (CR3 > 45%) is determined

[2] This superimposition is empirical in terms of references to the stages. For strict quantitative referencing of trends, it is necessary to build a superimposed dynamics of investment projects and economic concentration. This task is seen as an area of further research on the part of the authors.

by a high-risk zone of investing into innovations. The project can have a high research and development potential, be supported by infrastructure, but high economic concentration blocks marketing opportunities of the industrial enterprise. Of cause, at the "focus" stage there is an exception for the leaders of the market included in the evaluation of sum CR3-4. Their innovation projects have a marketing advantage, which provides them with a prospect of investment efficiency. It should be said that at the focus stage the market leaders use only incremental innovation projects, which have only insignificant moral and technology-operation novelty. Incremental innovations of the leaders strengthen the market barriers, which cannot be overcome by "ordinary" industrial enterprises, even if they invest in radical novelties. So the focus stage is seen as stagnation, where innovative development of industry and investing even in incremental innovations is economically inefficient. However, by the end of the focus stage and by the start of the balance stage, a new technology platform (radical innovation) appears, with the potential of investment efficiency being characterized by zone A - opening stage. So, the two revealed zones of investment efficiency (Fig. 3) are consistent with the relevant strategy methodology and concepts of the genesis of development of manufacturing industries.

The accounted genesis of innovation investment efficiency of the national processing industry is the research result and contribution of the authors to the theory of innovations, industrial economy and organization. The practical side of the presented research study can be formulated as a basis for deciding upon investing into innovation projects, criterion for the formation of the technology innovation portfolio by enterprises of processing industry.

4 Conclusion

The authors formulate the theoretical statement that in today's economy, innovation investment efficiency is not determined that much by the internal potential of the enterprise and the infrastructure it is surrounded with (however, their significance in the balance of potentials is not denied), but rather by external, market factors - industry consolidation ratio on the platform of new technology innovation.

The practical side of the obtained result can be formulated as a criterion for selecting industries (markets, lines of business) for investing into innovations by enterprises of the processing industry. Formulating the criterion: those sectors are determined as best, where the economic concentration ratio amounts to 15–45% at the moment the innovative product starts selling, which characterize the scale stage on the diagram by A.T. Kearney.

Further research. The identified nature of innovation investment efficiency is considered as a new research statement, which has objective limitations: region- and industry-specific ones. Expanding statistics and sampling will make it possible to elaborate on the following positions within further research studies.

1. Regression analysis will allow us to select a descriptive model, a trend of interrelation between return on investment and economic concentration ratio;
2. Multi-factor regression analysis (Table 1) will allow us to consider a number of internalites and externalities with a high and average level of correlation;

3. To consider the applicability and specifics of the identified interrelation within individual sectors and levels of technological advancement (knowledge content);
4. To expand the research study in terms of referencing the dynamics of projects to consolidation of the industry, in particular, to find out the best way to assess concentration (HHI, CR3, CR4 and other coefficients).

References

1. Puryaev, A.: Evaluating of innovative projects' effectiveness at industrial enterprises. SHS Web Conf. **35**, 01102 (2017). https://doi.org/10.1051/shsconf/20173501102
2. Guan, J., Chen, K.: Modeling the relative efficiency of national innovation systems. Res. Policy **41**(1), 102–115 (2012). https://doi.org/10.1016/j.respol.2011.07.001
3. Fagerberg, J., Srholec, M., Verspagen, B.: Innovation and economic development. In: Handbook of the Economics of Innovation, vol. 2, pp. 833–872 (2010). https://doi.org/10.1016/s0169-7218(10)02004-6
4. Carayannis, E.G., Grigoroudis, E., Goletsis, Y.: A multilevel and multistage efficiency evaluation of innovation systems: a multiobjective DEA approach. Expert Syst. Appl. **62**, 63–80 (2016). https://doi.org/10.1016/j.eswa.2016.06.017
5. Dziallasa, M., Blindab, K.: Innovation indicators throughout the innovation process: an extensive literature analysis. Technovation **80–81**, 3–29 (2019). https://doi.org/10.1016/j.technovation.2018.05.005
6. Kharchenko, E., Alpeeva, E., Ovcharova, O.: Innovative potential of Russian regions: methodological aspects of analysis and development trends. Procedia Econ. Finance **14**, 313–319 (2014). https://doi.org/10.1016/S2212-5671(14)00718-7
7. Alekseev, A., Khlebnikov, K., Fomin, E., Fomina, N.: Factors of economic efficiency of innovative entrepreneurship in the manufacturing industry. IOP Conf. Ser. Mater. Sci. Eng. **618**, 012089 (2019). https://doi.org/10.1088/1757-899X/618/1/012089
8. Vasilyeva, E.U., Kudryavtseva, T.Y., Ovsyanko, D.V.: Evaluation of the effectiveness of investment in innovation in industry. Bul. Altai Acad. Econ. Law **9**(1), 13–18 (2019). https://doi.org/10.17513/vaael.693
9. Polder, M., Leeuwen, G.: Product, process and organizational innovation: drivers, complementarity and productivity effects. Publications 2010s-28, 23719, pp. 1–46 (2010). https://doi.org/10.2139/ssrn.1626805
10. Kuczmarski, T.D.: Measuring your return on innovation. Mark. Manag. **9**(1), 24–32 (2000). https://doi.org/10.1111/1540-5885.1960455-i24
11. Drake, M.P., Sakkab, N., Jonash, R.: Maximizing return: on innovation investment. Res. Technol. Manag. **49**(6), 32–41 (2006). https://doi.org/10.1080/08956308.2006.11657406
12. Koskinen, Y., Maeland, J.: Innovation, competition, and investment timing. Rev. Corp. Finance Stud. **5**(2), 166–199 (2016). https://doi.org/10.1093/rcfs/cfw002
13. Bobillo, A.M., Sanz, J.A.R., Gaite, F.T.: Innovation investment, competitiveness and performance in industrial firms. Thunderbird Int. Bus. Rev. **48**(6), 867–890 (2006). https://doi.org/10.1002/tie.20126
14. Lim, E.S., Lee, C., Nagaraj, S.: The relationship between innovation and firm performance: evidence from Malaysian manufacturing firms. In: Conference: The Asian Business and Management Conference 2012At: Osaka, JapanVolume: The Asian Business and Management Conference Official Conference Proceedings (2012). https://doi.org/10.1016/j.sbspro.2013.04.026. ISSN 2186-5914

15. Domnina, S.V., Savoskina, E.V., Shekhova, N.V.: On innovative decisions in the investment-construction cycle. Procedia Eng. **153**, 741–746 (2016). https://doi.org/10.1016/j.proeng.2016.08.236
16. Rothwell, R.: Successful industrial innovation: critical factors for the 1990s. R&D Manag. **22**(3), 221–240 (1992). https://doi.org/10.1111/j.1467-9310.1992.tb00812.x
17. Nanda, R., Rhodes-Kropf, M.: Investment cycles and startup innovation. J. Financ. Econ. **110**(2), 403–418 (2013). https://doi.org/10.1016/j.jfineco.2013.07.001
18. Tsai, C.-L., Chang, H.-C.: Evaluation of critical factors for the regional innovation system within the HsinChu science-based park. Kybernetes **45**(4), 699–716 (2016). https://doi.org/10.1108/K-02-2015-0059
19. Bueche, A.M.: Investment in innovation. J. Technol. Transf. **3**(2), 1–13 (1979). https://doi.org/10.1007/BF02174845
20. Vega-Jurado, J., Gutiérrez-Gracia, A., Fernández-de-Lucio, I., Manjarrés-Henríquez, L.: The effect of external and internal factors on firms' product innovation. Res. Policy **37**(4), 616–632 (2008). https://doi.org/10.1016/j.respol.2008.01.001
21. Quintella, V., Silva Jr., A.F.A., Almeida, J.R.U.C., Embiruçu, M.: Financial exposure and technology innovation investment: measuring project results in Brazilian commodity industries. Academia Revista Latinoamerica de Administracion **30**(4), 547–564 (2017). https://doi.org/10.1108/arla-06-2016-0165
22. Tajeddini, K., Trueman, M., Larsen, G.: Examining the effect of market orientation on innovativeness. J. Mark. Manag. **22**(5–6), 529–551 (2006). https://doi.org/10.1362/026725706777978640
23. Caggese, A.: Entrepreneurial risk, investment, and innovation. J. Financ. Econ. **106**(2), 287–307 (2012). https://doi.org/10.1016/j.jfineco.2012.05.009
24. Zahra, S., Das, S.: Innovation strategy and financial performance in manufacturing companies: an empirical study. Prod. Oper. Manag. **2**(I), 15–37 (1993). https://doi.org/10.1111/j.1937-5956.1993.tb00036.x
25. Okrepilov, V.V., Kuzmina, S.N., Makarov, V.L., Bakhtizin, A.R.: Application of supercomputer technologies for simulation of socio-economic systems. Econ. Reg. **2**, 340–350 (2015). https://doi.org/10.15826/recon.2015.2.016
26. Schuh, C., Raudabaugh, J., Kromoser, R., Strohmer, M.F., Triplat, A., Pearce, J.: The Purchasing Chessboard: 64 Methods to Reduce Costs and Increase Value with Suppliers, pp. 49–206 (2017). https://doi.org/10.1007/978-1-4939-6764-3
27. Patchell, J.: Industrial organization. In: International Encyclopedia of Human Geography, pp. 237–242 (2020). https://doi.org/10.1016/b978-0-08-102295-5.10083-6
28. Morris, L.: The innovation infrastructure. Int. J. Innov. Sci. **1**(1), 41–49 (2009). https://doi.org/10.1260/175722209787951215

Pandeconomic Crisis and Its Impact on Small Open Economies: A Case Study of COVID-19

George Abuselidze[1(✉)] and Anna Slobodianyk[2]

[1] Batumi Shota Rustaveli State University, Ninoshvili, 35, 6010 Batumi, Georgia
george.abuselidze@bsu.edu.ge

[2] National University of Life and Environmental Science of Ukraine, Heroiv Oborony, 11, Kiev 03041, Ukraine

Abstract. In the beginning of 2020, prevalence of coronavirus pandemic in the global world has raised an issue of replacing the existing economy with an alternative economy – pandeconomic (pandemic + economy), which unlike traditional economies will reduce environmental risks and losses and will focus on raising public welfare and social equality. Issues such as recession phase in which the economy will find itself, the increase in GIG economy and entire societies are discussed in the article. The article explores the problems associated with the onset of the economic crisis and the outbreak of COVID-19 pandemic in countries with small open economies (SOE). It was proved that the key challenges the small open economies countries will face are primarily the devaluation of the national currency, the massive closure of small and micro businesses, the growth of social problems, etc. The review of the literature, as well as reports from financial institutions, show that the economies of individual countries are entering a phase of stagnation and recession. In the coming years, the economy will be affected by the coronavirus pandemic. Freelancing will change nature of employment and working places will shift to GIG-economy. Exactly these factors will expectedly provoke the growth of socio-economic crises in the country.

Keywords: Business fluctuations · Forecasting · Economic growth of open economies · Open economy macroeconomics · SOE

1 Introduction

Historically, the world economy goes through a certain process of renewal every 10 years. In this regard, it is interesting to remember other historic stock market/economic crises and to compare it with current one. It is meant here the Great Depression of 1930–1933 (−19% decline), Black Monday of 1987 (−12%), the 2000s Recession (−11%), and the Great Recession of 2007–08 (−38.5%). Trends contributed to the intensive development of certain processes, the market is being adjusted, the economy and markets are returning to a more balanced state. Black Monday (March 9, 2020 The

V. Murgul and V. Pukhkal (Eds.): EMMFT 2019, AISC 1258, pp. 718–728, 2021.
https://doi.org/10.1007/978-3-030-57450-5_61

Dow Jones has declined by about 28%) in the stock markets pushed the world into a new crisis – a sharp drop of oil prices, within last 30 years, has become a shock that coronavirus weakened markets will not manage to tolerate (see Fig. 1).

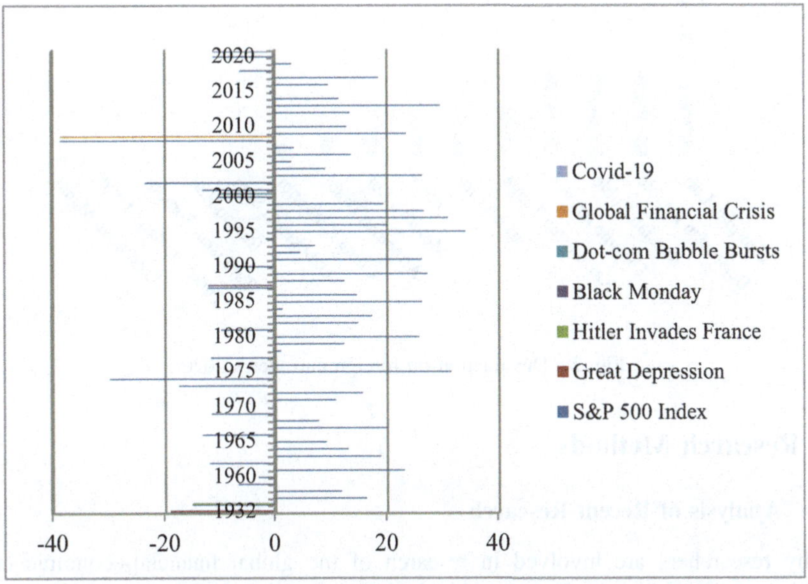

Fig. 1. How the covid-19 crash compares to historic market falls.

If the oil prices had fallen without coronavirus case, then this could have been considered a pleasant surprise. Now probability that markets will survive coronavirus without financial crisis and recession is minimal. The situation is aggravated by the fact that everyone has long been waiting for the crisis. And now the effect of the forecast, which is coming true, may work.

Georgia and Ukraine are always suffering from turbulences in the global economy. The Georgian and Ukrainian small open economy are focused on the sale of agricultural products. Any financial crisis in the world means a crisis in the country. On the one hand export revenues are falling, on the other hand - the outflow of investment capital from the country begins. Eventually, the country will receive an exchange rate strike on the currency (see Fig. 2), as a result, devaluation will possibly other problems.

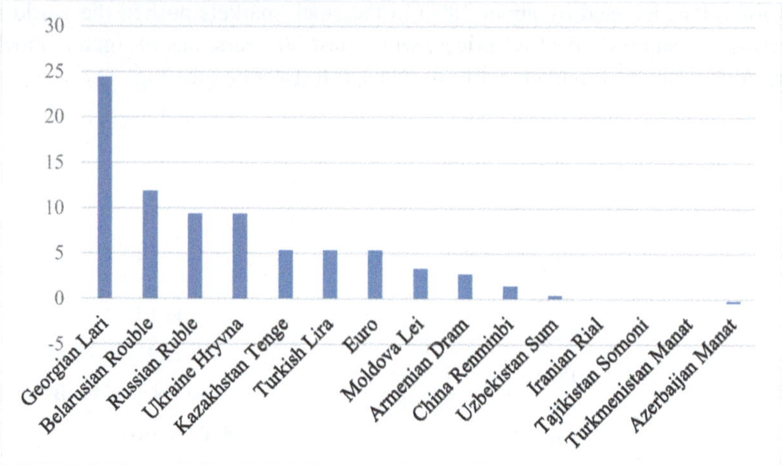

Fig. 2. Devaluation on foreign currencies 2020.

2 Research Methods

2.1 Analysis of Recent Research

Many researchers are involved in research of the global financial-economic crisis problem. Among them are P. J. Arnold, N. V. Beketov, S.R. Benatar, S. Gill, I. Bakker, T. Ciro, K. Hawtrey, R. Johnson. The research highlights issues such as the nature of world economic crisis in the context of civilizational development of world [1–6], the consequences of world crises, for individual economic systems of the world in general and particularly for SOE [7–13], the development of anti-crisis measures, although fragmented [14–18]. As for pandemic crisis researchers, the greatest attention to these issues was paid by L. Alfaro, A. Chari, B. K. Baker, C. W. Calomiris, S. Claessens, K. Forbes, N. J. Gormsen, R. S. Koijen, P. O. Gourinchas, H. Hafiz, S. Y. Oei, D. M. Ring, N. Shnitser, R. Peckham, S. Ramelli, A. F. Wagner, A. Renda, R. J. Castro, F. Stephany, N. Stoehr, P. Darius, L. Neuhauser, O. Teutloff, F. Braesemann, Ł. Sułkowski and others [19–31].

However, the problems caused by coronavirus pandemic have become not only a medical problem, but also an economic one and have become global. Closed borders and quarantine measures will bring enormous losses to the global economy, while governments, IMF and international financial institutions are forced to allocate billions of dollars to fight the epidemic and support their economies. According to Bloomberg experts, as a result of the COVID-19 pandemic, the global economy could lose about 27 trillion dollars. However, it is important to outline the issues of adaption and support to the economies of Ukraine and Georgia during the global crisis.

2.2 Research Methodology

For solving outlined tasks, the following research methods were used: the grouping and graphic methods, the abstract logical method, the monographic and dialectical method – in the process of collecting, systematizing and processing information for research, as well as the economic and statistical method, synthesis and comparison method – for processing and analysis of mass statistic data, that were necessary for assessing the state, studying variations, dynamics and making comparisons of the indicators of the problem, graphic method – for visual interpretation study's results.

The task of this study is to analyze the indicators of economic and social activity by small open economies in the context of the rapid spread of coronavirus, the fall of stock indexes, the outflow of investors and a decrease in demand in world markets for raw materials, the oil war and other crisis phenomena in the world threaten the Ukrainian and Georgian economies and their financial systems.

3 Results and Discussion

As a result of significant reforms in 2012–2019, Georgia has made significant progress in creating a favorable business environment and an efficient market, resulting in the following form of economic growth decomposition (see Fig. 3).

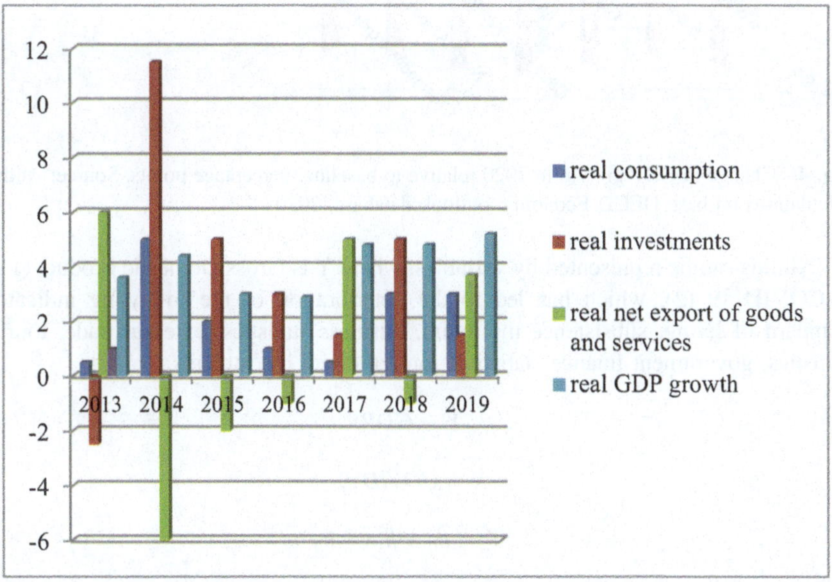

Fig. 3. Decomposition of GDP Growth (2013–2018). Source: Compiled by the authors based on data from the National statistics office of Georgia [32].

According to the International Monetary Fund, Georgia should have had the highest economic growth (5.0%) in the region in 2019–2024. However, the spread of coronavirus at the end of 2019 has radically changed expectations. The whole world, including Georgia and Ukraine has being faced severe challenges, in reality, economic relations are in the process of disruption that might result in economic collapse. According to the World Bank forecast, global economic growth is about to decline and in the first half of this year economic growth will shrink to 2.5%, while economic growth in developing countries in East Asia and the Pacific will drop to 2.1% this year instead of 5.8% growth, which was observed in 2019 (see Fig. 4).

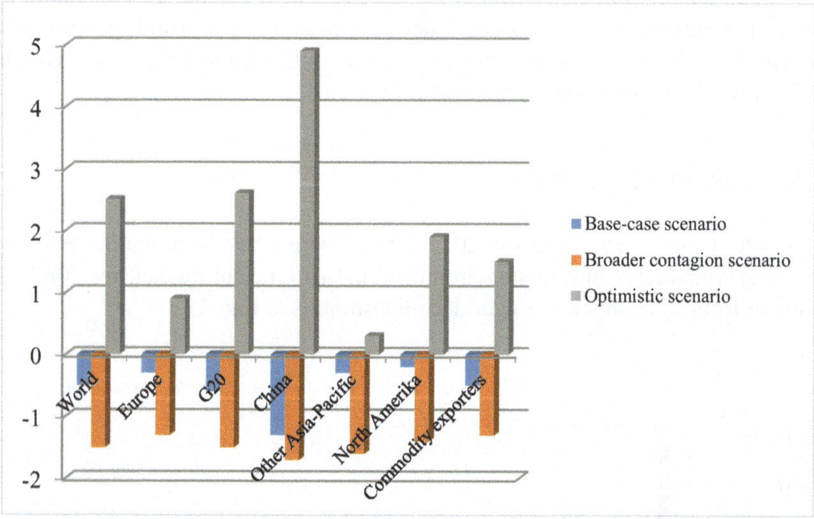

Fig. 4. Change in GDP growth in 2020 relative to baseline, percentage points. Source: Authors calculations on base OECD Economic outlook database, 2020.

All this can be represented by variables, where Y is Gross Domestic Product (1), X is COVID 19 (2), which has led to the deterioration of the following indicators: Standard of living, subsistence minimum, business statistics, external trade, tourism statistics, government finance statistics, foreign direct investments.

$$Y - GDP \tag{1}$$

$$X - COVID19 \tag{2}$$

$$Y = \beta_0 + \beta_1 x + u \tag{3}$$

$$Y = \beta_0 + \beta_a Covid19 + u \tag{4}$$

$$\Delta Y = \beta_1 \Delta covid19 \tag{5}$$

For different countries

$$Y_i = \beta + \beta_1 covid19_i + U_i \tag{6}$$

$$Y_i = \beta + \beta_1 covid19_1 + \beta_2 covid19_2 + \ldots \beta_k covid19_k + \widehat{u_i} \tag{7}$$

As can be seen from the formula (7), current world events have impact on Georgia as a country with small open economy. Accordingly, several economic scenarios are expected in 2020–2021: Optimistic scenario (2020 – (2.4%), 2021 – (5.0%)), Mild scenario (2020 – (−2.3%), 2021– (2.4%)) and Pessimistic scenario (2020 – (−5.0%), 2021 – (−0.5%)) (see Fig. 5).

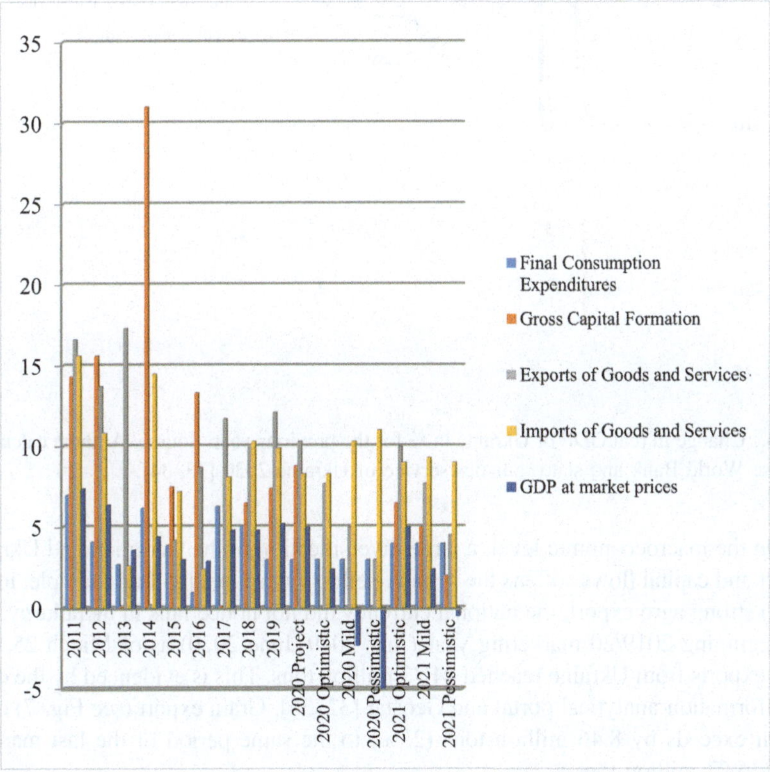

Fig. 5. Change in GDP growth in 2020–2021 relative to baseline, percentage points. Source: Compiled by the authors based on data from the National statistics office of Georgia [32].

And the Ukrainian economy may shrink by 4–9%, the national currency will devalue, because the COVID-19 pandemic will slow down the growth of Ukraine's GDP by half (see Fig. 6).

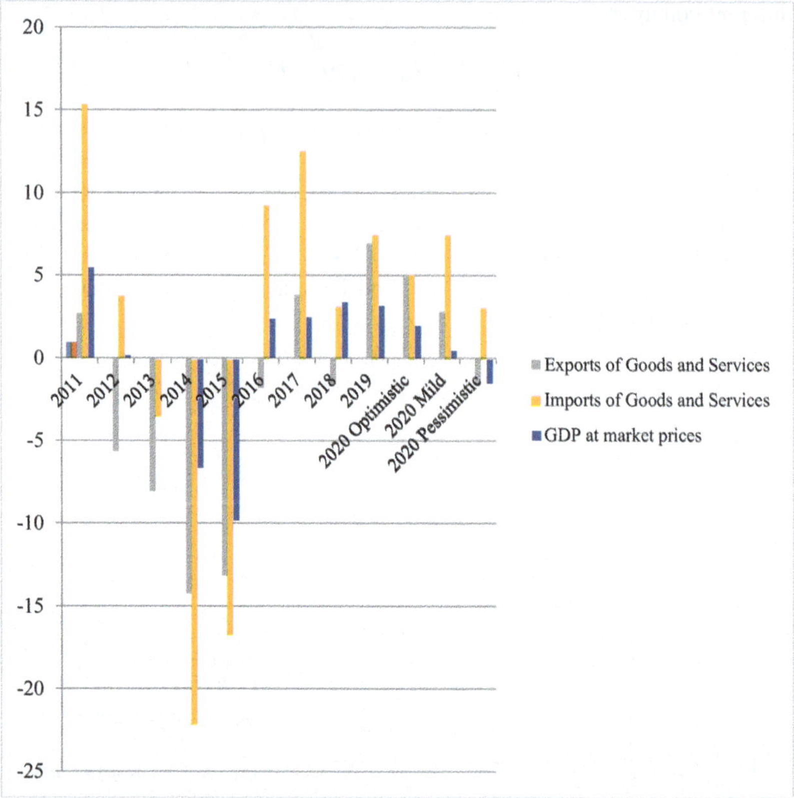

Fig. 6. Change in real GDP of Ukraine in % for the previous year. Source: Authors calculations on base World Bank and state statistics service of Ukraine 2020 [33, 34].

On the macroeconomic level, a large diversification of the Georgian and Ukrainian export and capital flows softens the negative effects substantially. For example, in 2019 due to strong agro export, the national currency did not notice falls in metallurgy. From the beginning 2019/20 marketing year (July 2019–June 2020) as of March 25, 2020, grain exports from Ukraine reached 44.73 million tons. This is evidenced by the data of the information-analytical portal and Geostat [32, 35]. Grain export (see Fig. 7) on this season exceeds by 8.46 million tons (23%) to the same period of the last marketing year (36.27 million tons).

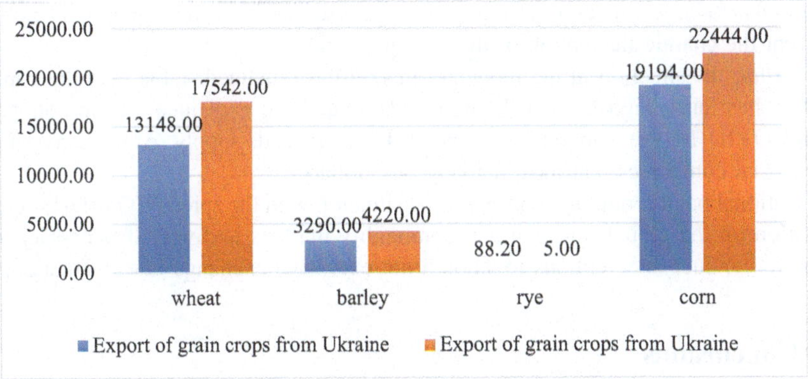

Fig. 7. Export of grain crops from Ukraine [35].

Price reduction of ore and metals will have a negative impact on the Ukrainian metallurgy and related industries. The beginning of 2020 for steelmakers is still successful: they significantly increased their production in January after a significant decline in the 4th quarter of 2019.

Consequently, the impact on the budget will be slightly different. Decreased prices for metals and ore will potentially reduce the taxable profits of exporters and their rental payments. Reduced fuel prices, on the on hand, will reduce the import VAT revenues to the budget, and on the other hand, will increase the taxable profits of farmers and other energy-intensive producers. In general, the impact of commodity market will be multi-directional, but negative will most likely weigh out.

The impact on Georgian and Ukrainian exports in the first quarter and throughout the year will be tangible, but moderate. China is not Ukraine's largest trading partner, it is only one of the five largest countries of destination of our exports after Turkey, Italy and Poland and occupies about 5% of Ukrainian exports, while for Georgia it is the largest trading partner and occupies about 12% of Georgian exports [32, 36]. In general, Georgia and Ukraine naturally trades more with European countries than Asian. The direct impact may be noticeable in the market of iron ore, corn, sunflower and colza oil, turbines and grist – exactly these are the goods producers export mainly to China.

As for the prospects, most companies admitted that they cannot evaluate themselves, as the situation is unpredictable for many factors. Among the greatest risks is the unstable exchange rate (Lari, Hryvnia, Lira, ruble, etc.) and the inability to predict its prospects. The global economic crisis and the further spread of coronavirus in the EU and in the associated states (Georgia and Ukraine) can drastically reduce the demand for products. Considering the uncertainty, entrepreneurs simply act on the circumstances and have not yet developed any anti-crisis plans.

It is a well-known that Georgia and Ukraine will receive a significant financial assistance from international financial institutions in 2020. Ukraine has in principle managed to repay its credit amnesty. And the National Banks has been able to accumulate a large amount of foreign exchange reserves, by using which will be able to prevent the chaotic fluctuation of the exchange rate.

So far, the situation in the monetary market has remained stable, and the banking system has fairly high level of liquidity. National bank explained that in connection with COVID-19, the demand for currency (USD) exceeds supply. At the same time, it is external companies that most actively buy dollars.

National banks continue to monitor the situation with the spread COVID-19 and the situation on the global financial and commodity markets and will, if necessary, continue to conduct interventions to smooth out excessive exchange rate fluctuations.

4 Conclusions

The key challenges that the population will face first of all will be an increase in the exchange rate, which will significantly affect inflation, because prices that are of foreign origin will rise. Respectively there will be a massive closure of small and micro businesses, wage growth will slow down or even stop; increase of fiscal pressure on medium and large businesses, we are talking about tax reform, which will make it impossible to optimize taxes for the abolition of a single social contribution for small and micro businesses at the time of quarantine and employers are required to pay all taxes. The National Bank's lending rate can increase, thereby increasing the cost of money for business and as a result, worsening its condition, the budget will be financed from National bank (it is possible to issue bonds and print money to buy them back), which is exactly what will provoke inflation growth. In our opinion, in such situation, it is especially necessary for states to move to MP plus (MP+) policy [37, 38], i.e. it is necessary to cohabit and coordinate orthodox and heterodox monetary methods in financial practice.

References

1. Arnold, P.J.: Global financial crisis: the challenge to accounting research. Acc. Organ. Soc. **34**(6–7), 803–809 (2009). https://doi.org/10.1016/j.aos.2009.04.004
2. Avčin, B.A., Kučina, A.U., Sarotar, B.N., Radovanović, M., Plesničar, B.K.: The present global financial and economic crisis poses an additional risk factor for mental health problems on the employees. Psychiatr Danub **23**(1), 142–148 (2011)
3. Beketov, N.V.: Globalization of the world economy and the global financial crisis. Finan. Credit **25**(361), 2–13 (2009)
4. Benatar, S.R., Gill, S., Bakker, I.: Global health and the global economic crisis. Am. J. Public Health **101**(4), 646–653 (2011). https://doi.org/10.2105/AJPH.2009.188458
5. Ciro, T.: The Global Financial Crisis: Triggers, Responses and Aftermath. Routledge, Abingdon (2016)
6. Hawtrey, K., Johnson, R.: On the atrophy of moral reasoning in the global financial crisis. J. Relig. Bus. Ethics **1**(2), 4 (2010)
7. Bilorus, O.: World structural crisis and transformations of the global financial system. Econ. Ann. XXI **07–08**(1), 4–7 (2014)

8. Cheung, L., Tam, C.S., Szeto, J.: Contagion of financial crises: a literature review of theoretical and empirical frameworks. Hong Kong Monet. Auth. Res. Note **2**, 1–18 (2009). https://www.hkma.gov.hk/media/eng/publication-and-research/research/research-notes/RN-02-2009.pdf. Accessed 28 Mar 2020
9. Jordà, Ò., Singh, S.J., Taylor, A.M.: Longer-run economic consequences of pandemics. Federal Reserve Bank of San Francisco Working Paper 2020-09, pp. 1–17 (2020). https://doi.org/10.24148/wp2020-09
10. Kattel, R.: Financial and economic crisis in Eastern Europe. J. Post Keynes. Econ. **33**(1), 41–60 (2010). https://doi.org/10.2753/PKE0160-3477330103
11. Korowicz, D.: Catastrophic shocks through complex socio-economic systems: a pandemic perspective, pp. 1–10 (2012)
12. Kotz, D.M.: The financial and economic crisis of 2008: a systemic crisis of neoliberal capitalism. Rev. Radic. Polit. Econ. **41**(3), 305–317 (2009). https://doi.org/10.1177/0486613409335093
13. Rechel, B., Suhrcke, M., Tsolova, S., Suk, J.E., Desai, M., McKee, M., Semenza, J.C.: Economic crisis and communicable disease control in Europe: a scoping study among national experts. Health Policy **103**(2–3), 168–175 (2011). https://doi.org/10.1016/j.healthpol.2011.06.013
14. Abuselidze, G.: Optimal tax policy-financial crisis overcoming factor. Asian Econ. Financ. Rev. **3**(11), 1451–1459 (2013)
15. Abuselidze, G.: Fiscal policy directions of small enterprises and anti-crisis measures on modern stage: during the transformation of Georgia to the EU. Sci. Stud. Account. Financ. Probl. Perspect. **12**(1), 1–11 (2018). https://doi.org/10.15544/ssaf.2018.01
16. Favero, C.A., Giavazzi, F.: Is the international propagation of financial shocks non-linear?: Evidence from the ERM. J. Int. Econ. **57**(1), 231–246 (2002). https://doi.org/10.1016/S0022-1996(01)00139-8
17. Fetisov, G.: On measures for overcoming the global crisis and forming sustainable financial and economic system (proposals for G20 on financial markets and world economy). Voprosy Ekonomiki **4**, 31–41 (2009). https://doi.org/10.32609/0042-8736-2009-4-31-41
18. Goldin, I., Vogel, T.: Global governance and systemic risk in the 21st century: lessons from the financial crisis. Glob. Policy **1**(1), 4–15 (2010). https://doi.org/10.1111/j.1758-5899.2009.00011.x
19. Alfaro, L., Chari, A., Greenland, A., Schott, P.: Aggregate and firm-level stock returns during pandemics, in real time. NBER Working Paper 26950, pp. 1–23 (2020). https://doi.org/10.3386/w26950
20. Baker, B.K.: The impact of the international monetary fund's macroeconomic policies on the AIDS pandemic. Int. J. Health Serv. **40**(2), 347–363 (2010). https://doi.org/10.2190/HS.40.2.p
21. Calomiris, C.W.: The great depression and other "contagious" events. In: The Oxford Handbook of Banking, 1st edn. (2012). https://doi.org/10.1093/oxfordhb/9780199640935.013.0027
22. Claessens, S., Forbes, K.: International financial contagion: an overview of the issues and the book. In: International Financial Contagion, pp. 3–17. Springer, Boston (2001). https://doi.org/10.1007/978-1-4757-3314-3_1
23. Gormsen, N.J., Koijen, R.S.: Coronavirus: impact on stock prices and growth expectations. University of Chicago, Becker Friedman Institute for Economics Working Paper 22 (2020)
24. Gormsen, N.J., Koijen, R.S.: The corona virus, the stock market's response, and growth expectations. Becker Friedman Institute, Working Paper 22 (2020)
25. Gourinchas, P.O.: Flattening the pandemic and recession curves. mitigating the COVID economic crisis: act fast and do whatever 31 (2020)

26. Hafiz, H., Oei, S.Y., Ring, D.M., Shnitser, N.: Regulating in pandemic: evaluating economic and financial policy responses to the coronavirus crisis. Boston College Law School Legal Studies Research Paper 527 (2020). https://doi.org/10.2139/ssrn.3555980
27. Peckham, R.: Economies of contagion: financial crisis and pandemic. Econ. Soc. **42**(2), 226–248 (2013). https://doi.org/10.1080/03085147.2012.718626
28. Ramelli, S., Wagner, A.F.: Feverish stock price reactions to Covid-19 (March 2020). CEPR Discussion Paper DP14511 (2020)
29. Renda, A., Castro, R.J.: Chronicle of a pandemic foretold. CEPS Policy Insights No 2020-05/March 2020 (2020)
30. Stephany, F., Stoehr, N., Darius, P., Neuhauser, L., Teutloff, O., Braesemann, F.: Which industries are most severely affected by the COVID-19 pandemic? A data-mining approach to identify industry-specific risks in real-time 2003, 12432 (2020)
31. Sułkowski, Ł.: Covid-19 pandemic; recession, virtual revolution leading to de-globalization? J. Intercult. Manag. **12**(1) (2020). https://doi.org/10.2478/joim-2020-0029
32. National Statistics office of Georgia. Gross Domestic Product, Standard of Living, Subsistence Minimum, Business Statistics, External Trade, Tourism Statistics, Government Finance Statistics, Foreign Direct Investments (2020). https://www.geostat.ge/en/modules/categories/93/regional-statistics. Accessed 28 Mar 2020
33. State Statistics Service of Ukraine. Gross Domestic Product, Standard of Living, Subsistence Minimum, Business Statistics, External Trade, Tourism Statistics, Government Finance Statistics, Foreign Direct Investments (2020). http://www.ukrstat.gov.ua/. Accessed 28 Mar 2020
34. World Bank Group. Exports-Imports of goods and services (annual % growth) - Ukraine. https://data.worldbank.org/indicator/NE.EXP.GNFS.KD.ZG?locations=UA&name_desc=false. Accessed 25 Mar 2020
35. Latifundist. Grain export 2019/20: corn shipment went up by 17% (2020). https://latifundist.com/en/novosti/49287-eksport-ukrainskoj-kukuruzy-vyros-na-17. Accessed 28 Mar 2020
36. Better Regulation Delivery Office. BRDO's analysis: the impact of the economic crisis on key areas and markets (2020). https://en.brdo.com.ua/analytics/brdo-s-analysis-the-impact-of-the-economic-crisis-on-key-areas-and-markets/. Accessed 23 Mar 2020
37. Abuselidze, G.: Modern challenges of monetary policy strategies: inflation and devaluation influence on economic development of the country. Acad. Strateg. Manag. J. **18**(4), 1–10 (2019)
38. Abuselidze, G.: Optimality of tax policy on the basis of comparative analysis of income taxation. Eur. J. Sustain. Dev. **9**(1), 272–293 (2020). https://doi.org/10.14207/ejsd.2020.v9n1p272

Functional and Spatial Development of Agricultural Subregional Localities

Oksana Kolomyts[1], Inna Ivanova[2], and Emil Velinov[3](✉)

[1] Kuban State Technological University, Moskovskaya Str., 2, Krasnodar 350042, Russia
[2] Kuban State Agrarian University, Kalinina Str., 13, Krasnodar 350004, Russia
[3] Skoda Auto University, Na Karmeli 1457, 293 01 Mladá Boleslav, Czech Republic
emil.velinov@savs.cz

Abstract. The paper sheds a light on spatial development of agricultural subregional localities in Krasnodar Region in the Russian Federation. Nowadays, there is a high need of creating so called "agro-cities" in the regions across Russia in order to sustain the regional economy and to open more possibilities for the local entrepreneurs. Furthermore, the paper suggests combination of approaches in order to identify dynamic changes in the spatial organization of rural areas, considering their spatial and sectoral potential. The study explains how the agricultural business in Krasnodar region has been developing and what factors have influenced it, which triggers high need of transformation from economic, environmental, and social perspectives. Paper results show that agro-cities contribute to the prosperous development of urban and rural areas of national economy through higher cooperation among cities and rural small business ecosystems. The study concludes that this synergy would decrease the socio-economic difference between the urban and rural areas in Russia.

Keywords: Spatial development · Agricultural · Agro-cities · Small business

1 Introduction

The transformation of the economy and the development prospects of rural municipalities have recently been the subject of debate and discussion not only in the scientific community, but also at various levels of government bodies and local self-government of the Russian Federation.

To achieve the goals of sustainable economic growth of the whole country, it is necessary to ensure conditions for the development of each territory, included in its composition, including rural municipalities, the structure of which is currently quite heterogeneous. This is partly due to the fact, that the reforms gave the "push" for the economic development of some municipalities, while others on the contrary have become depressed and operate at the expense of higher-level budgets. Against this

V. Murgul and V. Pukhkal (Eds.): EMMFT 2019, AISC 1258, pp. 729–737, 2021.
https://doi.org/10.1007/978-3-030-57450-5_62

backdrop, disparities in the socio-economic status of the intraregional territories in terms of the standard of living and the quality of life of the population are increasing.

The rising unemployment rate, mass poverty, the moral and cultural degradation of rural labor potential and the outflow of the working population are proof of the crisis in rural areas and are leading to their dehumanization.

The consequence of these is dew both the decline in the number of rural settlements and the decrease in the number of people living in rural areas. According to the monitoring of the development of the system of local self-government in the Russian Federation only in 2016 and the first half of 2017 the number of rural settlements decreased by 251 settlements in Russia. Even more indicative is the change in the rural population, which has declined by 1.4 million over the past 17 years. Decline in the size of the population was on average 87.5 thousand per year (see Fig. 1).

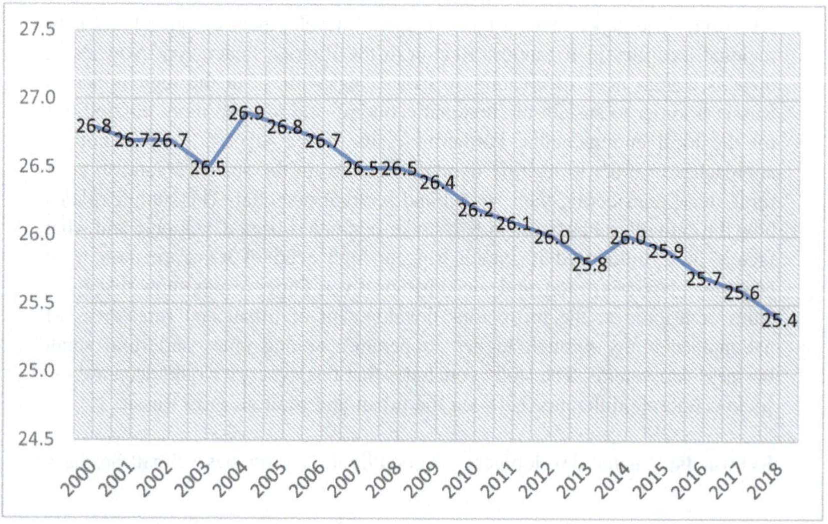

Fig. 1. Rural population as a percentage of the total population of the Russian Federation, %. Source: own elaboration.

A defining attribute for the half of Russian regions is signs of depression, which exacerbates the already unfavorable rural development conditions.

In comparison with other countries, Russia can now be easily attributed to a country with the prerequisites of agricultural orientation, because ¾ of its spaces are rural areas, which account for about 90% of the total area developed by the farm [1].

The social and economic progress of the rural areas urgently requires the development of new effective forms of spatial organization based on rural territories of agricultural regions - agricultural towns [2], which can solve a whole range of problems: social, economic, environmental, cultural, ethnic, etc.

The vector of the functional-spatial development of agro-cities should coincide with the leading positive trends in modern society [3]: Its mobility, efficiency, technological rearmament, a desire for innovative innovations, a synthesis of scientific and practical human activity, the humanistic direction of the ecumenical development and the development of the human being.

In recent years, the attention of domestic and foreign scientists to the problems of rural development has increased. In the previous literature review on regional agricultural business have been explored modern economic, social and demographic features of rural development, and problems related to the systems of livelihood of rural areas [4].

The economic aspects of the sustainable development of the agro-industrial complex of modern Russia were studied before [5].

Current scientific research on agrarian business aims at determining the sectoral and market characteristics of the development of agriculture in the forecast period [6].

Previous work of numerous researchers from Central and Eastern Europe covered the issues and developed strategic tools for the sustainable development of the agro-industrial complex [7].

In the field of the establishment and development of the cooperative movement of the Russian agro-industrial complex should be included the works of numerous foreign and Russian researchers and practitioners of the concept of entrepreneurial ecosystems [8].

The idea of an agricultural town arose in modern Russia, now it is known abroad. Until recently, in the English language there was no analogue of the word "agricultural city" and there wasn't the definition of this concept. Now we can find the word «agrotown» or «agro-town». Its insufficient scientific development and the existence of its various interpretations, readings and understanding of the use of an integrated approach to solving the problems of the Russian agro-industrial complex and the participation of a wide range of specialists from various branches of science in the design of a modern agricultural city that can satisfy the needs of modern society [9].

In the Russian Federation, however, the phenomenon under consideration is more at the stage of scientific research and justification than of practical use.

Despite a lot of research on various aspects of this issue, they don't give complete a picture of the entrepreneurial ecosystem - the foundation for the creation of agro-cities in territorial entities; in economic science, hasn't been developed a mechanism for the interaction of entities of entrepreneurial ecosystems with respect to rural territories [10].

2 Methodology

The complexity and multi-dimensionality of the above-mentioned problems necessitate the use of a combination of systems, monographic, structural, economic, statistical and cluster analysis methods, as well as the need to apply an interdisciplinary approach. As a methodological basis for the study is the theory of sustainable development, in which territorial formation is a complex, multi-component and multifunctional system.

Empirical and factual data for the study are taken from official data of the Federal State Statistics Service (Rosstat) and its regional bodies, specifically the Krasnodar Territory.

Grouping of municipalities according to concentration of agricultural production by multi-dimensional statistical procedure - cluster analysis, which involves determining the distances between the compared objects, combining them into relatively homogeneous groups according to a set of diverse indicators.

Euclidean distance is used as a measure of proximity:

$$(x, y) = \|x - y\| = \sqrt{\sum_{p=1}^{n} \left(x_p - y_p\right)^2} \quad (1)$$

where x, y € R^n.

The rule of association (communication) - full communication (method of the most distant neighbors) [11].

The combination of approaches is aimed at identifying dynamic changes in the spatial organization of rural areas, taking into account their spatial and sectoral potential, which will allow to identify prospects for the formation and use of the concept of agricultural cities and entrepreneurial ecosystems, their characteristics and strategic results for the conservation and sustainable development of rural territories.

3 Findings and Discussion

Analysis of the current state of Russian agriculture shows that, new ways are needed to ensure intensive and sustainable development in order to successfully realize the objectives of the chosen import substitution policy [12]. In the agricultural sector, the economies of the country must be properly elected and placed right priorities, which, first of all, would contribute to the technological modernization of the industry, its restructuring, the optimal distribution of productive forces and the expansion of domestic demand for agricultural products.

At the same time, it will be necessary to find optimal ways of social development of rural territories, which requires considering a large complex of factors to maintain their potential, which is strategically important for the stability of the state.

The current priority in Russia for the creation of large agrarian associations and firms with capital contributes to the development of agricultural products, but for agricultural societies it has mixed consequences. Domestic practice shows that, often, possible monopolization in the countryside is a direct route to the degradation of rural settlements [13].

The regions of the Russian Federation are, for the most part, large populated centers with peripheries that are differentiated to varying degrees, which causing the process of urbanization and leading to the decay of large rural areas, and consequently, the regional center is overcrowded and unable to function effectively in the face of many negative factors (high unemployment, rising crime, etc.).

The analytical phase of the study involves the analysis of the concentration of agricultural production and related branches of the agro-industrial complex at the level of rural territories of a specific subject of the Russian Federation. Based on the available statistical information in Rosstat, the key indicators for benchmarking on agricultural business are agricultural products in all categories of farms, the average number of employees in agricultural organizations and the value of the fixed assets of agricultural organizations [14].

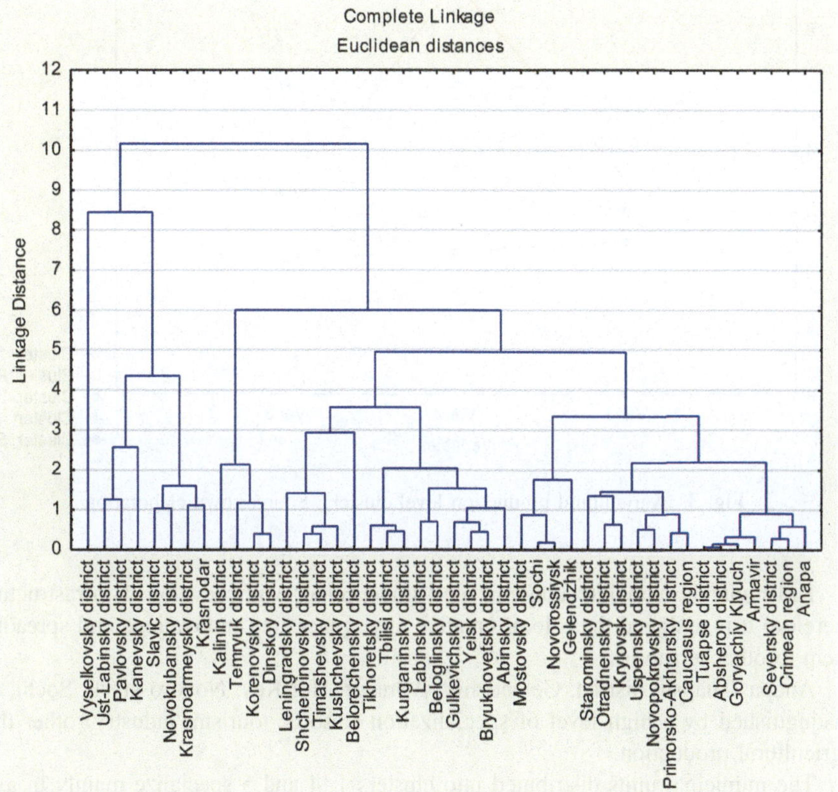

Fig. 2. Dendrogram of classification of cities and districts of Krasnodar Region according to concentration of agricultural production. Source: own elaboration.

The results obtained by clustering are illustrated in Fig. 2, which shows the objects of research (cities and districts of Krasnodar Region) on axis the X, and the distance between districts or groups of districts in nominal units (k) on axis the Y.

The analysis makes it possible to highlight the established center of agricultural production in the Vyselkovsky district (cluster 2).

Kanevsky, Slavic, Novokubansky, Ust-Labinsky, Pavlovsky districts (cluster 3) can be considered as emerging points of growth in agricultural production (Fig. 3).

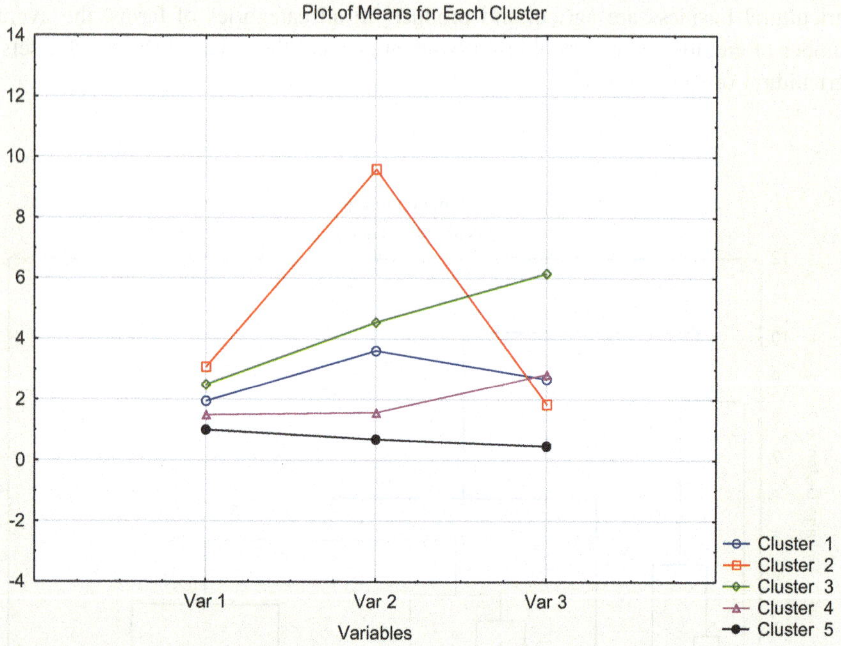

Fig. 3. Agricultural production level clusters. Source: own elaboration.

Krasnodar is a major economic and trade center with developed infrastructure, therefore it is assigned the role of creative city, generating innovations and spreading them to other territories.

Anapa, Tuapsin district, Gelendzhik, Armavir, Hot Key, Novorossiysk, Sochi, are distinguished by a high level of specialization (resorts, tourism, industry) other than agricultural production.

The municipal units distributed into clusters 1, 4 and 5 specialize mainly in agriculture, processing of agricultural raw materials and partly in industrial production [15].

The analysis suggests that a promising vector for the transformation of the economic space for the effective life of the considered rural territories of the Krasnodar Territory, their reformatting into modern, world-class spatial objects, can be an idea the creation of «agro-cities» - well-established settlements with urban infrastructure and small population, whose main occupation will be agriculture, processing of products from local raw materials in small enterprises, handicrafts. The core of the agro-city model will be entrepreneurial ecosystems - alternative forms of business organization based on the principles of cooperation and considering the uniqueness of local specifics.

4 Conclusion

The main goal of entrepreneurial ecosystems is to increase the efficiency of the functioning of small businesses, based on new forms of organizing production, advanced modern technologies, solutions, and creating comfortable conditions for the development of small businesses in rural areas [12].

This is possible with such an organization of entrepreneurial activity, which will be aimed at creating new forms of labor organization and the formation of an innovative development system for the region. The business ecosystem will be the link between sector-wide and regional development, considering the environmental dimension [7]. Of great importance is the idea of an environmental focus on the development of entrepreneurial ecosystems, which implies a significant reduction in the level of man-made environmental stress in agro-cities compared to that of modern industrial cities.

At the same time, the presence of an industrial component in the city of this type is not only not being replaced but is generally based on the agricultural component and contributes to its growth (for example, Vyselkovsky district).

The concept of agro-cities is not a direct extrapolation of the idea of urban development with regard to the organization of new institutional forms of agricultural production, but rather a set of specific characteristics, the spatial organization of the territory, the specificity of infrastructure and communication development, etc. The idea of establishing agro-cities with the status of more advanced territories is a promising one.

Agro-cities have the advantage of having large economic resources, of being able to use them efficiently, of attracting investment, and of being attractive for life and work. Agro-cities are also needed in modern Russia in order to solve problems of primary processing, storage and marketing of agricultural products, support and development of entrepreneurship. In the current conditions, agriculture is a high-tech industry that employs not only physical labor and endurance is required, but also requires specific knowledge and skills to strategically think and effectively use the achievements of scientific and technological progress [2] which is possible only in large-scale production.

The transformation of agrarian regions into a mode of proactive development is possible by implementing an agrarian policy focused on the creation of agro-cities, the effective functioning of which is possible based on entrepreneurial ecosystems as mechanisms for generating and diffusing innovations [5].

The State agricultural policy of the Russian Federation is currently aimed at ensuring the strategic development of rural areas, increasing the competitiveness of products produced in its territories and the reproduction and conservation of natural resources used in agriculture.

Natural growth and the migration factor hinder a significant increase in the number and level of development of rural areas. Rural population tends to decline. Therefore, the emergence of agro-cities can lead to fundamental changes in the demography of rural populations.

However, the demographic situation in rural areas is very difficult. This is due to the massive exodus of young people from rural areas and the pattern of migration. The outflow of young people increases the demographic burden on the working population and reduces the Territory's labor potential. Moreover, the new technologies needed by rural enterprises could not be introduced without the young labor force.

Accordingly, the orientation of rural agro-urban centers will make it possible to solve the problems of the social and economic development of the territories in a comprehensive manner, and will help to increase their sustainability, reduce social tension and ensure the growth of employment, improvement of the demographic situation, etc.

The proposed innovation will contribute to increase the efficiency of the agricultural sector, levelling the socio-economic, organizational, and technological relations between urban and rural areas. Moreover, it will contribute to the development of innovative forms of agricultural production aimed at the introduction of the latest scientific and technological developments.

The proposed agribusiness project is possible in the absence of strategic alternatives to agricultural development. All previously proposed rural development concepts have proven to be inadequate and irrelevant. Agro-cities should become a national priority and have an advantage over the next 5 to 10 years not only in lending and taxation but also in the acquisition of land ownership.

Acknowledgement. The research was carried out with the financial support of RFFI and Krasnodar Region within the framework of the scientific project 19-410-230041.

References

1. Mason, C., Brown, R.: Entrepreneurial Ecosystems and Growth Oriented Entrepreneurship. The Hague, Netherlands (2014)
2. Kolomyts, O.N., Ivanova, I.G., Popov, M.N.: Regional economic development through the effective functioning of the business sphere. Azimuth Sci. Res. Econ. Adm. **8**(1(26)), 176–179 (2019). https://doi.org/10.26140/anie-2019-0801-0037
3. Feld, B.: Startup Communities: Building an Entrepreneurial Ecosystem in Your City. Wiley, Hoboken (2012)
4. Apishev, A.A., Khakhuk, B.A.: Socio-economic assessment of natural (land) resources as the basis of modeling of the mechanism of paid land tenure. Bull. Adyghe State Univ. Ser.: Econ. **20**, 96–203 (2011)
5. Trukhachev, V.I., Mazloev, V.Z., Sklyarov, I.Y., Sklyarova, Y.M., Kalugina, E.N., Volkogonova, A.V.: The strategic directions of innovative economy development in Russian agribusiness. Montenegrin J. Econ. **12**(4), 97 (2016)
6. Prokhorova, V.V., Klochko, E.N., Kolomyts, O.N., Gladilin, A.V.: Prospects of the agro-industrial complex development: economic diversification, business development, mono-industry town strengthening and expansion. Int. Rev. Manag. Mark. **6**(6), 159–164 (2016)
7. Stam, E.: Entrepreneurial ecosystems and regional policy: a sympathetic critique. Eur. Plan. Stud. **23**(9), 1759–1769 (2015). https://doi.org/10.1080/09654313.2015.1061484
8. Prokhorova, V.V., Kolomyts, O.N., Zakharova, E.N., Bailagasov, L.V.: Development of strategic benchmarks for the creation and operation of digital regional ecosystem industry profile. Indo Am. J. Pharm. Sci. **6**(3), 6877–6880 (2019)
9. Isenberg, D.: How to start an entrepreneurial revolution. Harvard Bus. Rev. **88**, 41–50 (2010)

10. Kolomyts, O.N., Prokhorova, V.V., Ivanova, I.G.: Advantages and prospects of creating regional high-tech agrarian clusters. Indo Am. J. Pharm. Sci. **6**(3), 6881–6885 (2019)
11. Velinov, E., Ponomarev, M.: Organizational development of Russian SMEs: current trends. J. Eastern Eur. Central Asian Res. (JEECAR) **3**(2), 10 (2016). https://doi.org/10.15549/jeecar.v3i2.142
12. Kolomyts, O.N., Stepanets, L.Yu.: Ecosystem approach to the development of regional agribusiness entrepreneurship as a factor of sustainable rural development. Azimuth Sci. Res. Econ. Adm. **8**(4(29)), 347–350 (2019). https://doi.org/10.26140/anie-2019-0804-0079
13. Armington, C., Odle, M.: Small business: how many jobs? Brookings Rev. **1**(2), 14–17 (1982)
14. Jackson, D.J.: What Is an Innovation Ecosystem? National Science Foundation, Arlington (2011). http://urenio.org/wp-content/uploads/2011/05/What-is-an-Innovation-Ecosystem.pdf. Accessed 11 Mar 2020
15. Ivanova, I.G., Makhnyreva, O.A.: The role of small businesses in the current environment. Small business prospects in Russia. Mod. Eur. Res. **1**, 56–63 (2017)

Internal Management Reporting on Efficiency of Budget Funds Use

Guzaliya Klychova[1], Alsou Zakirova[1](✉), Regina Nurieva[1], Rashida Sungatullina[2], Elena Klinova[2], and Evgenia Petrova[3]

[1] Kazan State Agrarian University, Karl Marx, 65, 420015, Kazan, Russia
zakirovaar@mail.ru
[2] Vyatka State Agricultural Academy, October Avenue, 133, 610017, Kirov, Russia
[3] Kazan Innovative University named after V.G. Timiryasova, Moskovskaya, 42, 420111, Kazan, Russia

Abstract. Adoption of effective and justified management decisions depends on the quality of preparation of detailed and reliable information that most fully reflects the effectiveness and targeted use of state aid funds received by organizations during the reporting period. Reasonable and detailed disclosure of information on the state aid in reporting is important, because these funds are an additional source of financing for the organization and an investment by the state in the development of this organization. The purpose of the article is development of organizational and methodical provisions and practical aspects of formation of management reporting on target use of budgetary funds. Objectives of the study: to determine the principles and criteria for evaluation of subsidies to agricultural organizations based on the study of sectoral peculiarities; to develop a form of internal management reporting on the state of budgetary funds as an information base and the basis for monitoring the target use and evaluation of efficiency of spending of state aid in organizations. In this article, such methods as analysis of scientific and theoretical sources, system approach, method of comparative analysis, generalization were used. In the process of research on the basis of generalization of peculiarities of formation and use in management of information on budgetary funds, the form of internal management report on the state of budgetary funds, allowing estimation of the system of economic, financial and operational indicators, characterizing the state of subsidizing and efficiency of use of budgetary funds, was developed.

Keywords: Management reporting · Government aid · Subsidies · Budget funds · Government support · Efficiency · Subsidies · Control

1 Introduction

In modern conditions, the main purpose of agrarian policy in Russia is continuous targeted support of enterprises, their price support by using intervention and collateral prices and creating price monitoring; protection of domestic agricultural producers in the international market. Besides, the models of effective management, product

V. Murgul and V. Pukhkal (Eds.): EMMFT 2019, AISC 1258, pp. 738–758, 2021.
https://doi.org/10.1007/978-3-030-57450-5_63

associations, introduction of high technologies; development of personnel specialization; provision of conditions for attraction of additional investments into agrarian sector of Russia are being worked out [1–4].

Provision of state aid is an important element of agrarian policy of the Russian Federation, which aims to significantly mitigate the consequences of the non-equivalence in the commodity exchange of agriculture with other sectors of the economy and ensure the effective functioning of the industry as a whole [5–7].

Over the past decade, events have been organized in the Russian Federation aimed at forming a multi-channel system of state support for agriculture, and the volume of budget financing has increased significantly. As a result of these measures, agricultural production has been stabilized and increased, but this has not made it possible to bring the industry and social sphere of rural areas to a high level of development [8–10]. To solve this problem, a new approach is needed in the organization of agrarian policy of the state, focused on improving the efficiency of agricultural production, competitiveness of products, significant growth of incomes of rural population and development of rural areas, which should be carried out simultaneously with the development of accounting and control over the targeted use of budgetary funds [11–13].

In our opinion, the development of the assessment of the effectiveness of the state aid should be carried out by addressing the following issues:

1) determination of the role and place of state support for agriculture and processing of agricultural products, taking into account the development of the industry in order to justify the importance and priority of recommendations to improve accounting and control [14, 15];
2) evaluation of the priority of production of certain types of agricultural products in order to substantiate the need to finance these sectors. Justification of the need for public financing of certain sectors should be defined as one of the criteria for subsidizing. In the future, when budget funds are accepted for accounting, this fact can be considered as one of the conditions for subsidizing, which will increase the informativeness of the recognition of funds, as well as will serve as an information base for determining the effectiveness of subsidizing and internal management reporting [16, 17];
3) ensuring receipt of budget funds sufficient for financial rehabilitation of agrarian enterprises. In most cases, the amounts allocated under financing are insignificant and cannot even cover part of the costs incurred by the organization. At the same time, additional costs often arise in the course of preparatory activities and collection of documents necessary to obtain subsidies. Therefore, it is often not profitable to justify the measures aimed at developing the accounting system, and their application by agricultural organizations may lead to higher costs in relation to the income received from the subsidy. Therefore, it is necessary to ensure that organizations have sufficient budget funds to cover both the costs of targeted activities and the operations related to their receipt, recognition, accounting and reporting.

In addition, it is necessary to carefully control the receipt and spending by the organization of the full amount of state subsidies, as these amounts may affect the change in financial results, increase in the value of the company, which, accordingly, will affect the data of internal management reporting on the state of budget funds.

2 Materials and Methods

The system of state regulation is a general concept that includes various ways of interaction for the development of agro-industrial complex. If we consider the problem in the theoretical sense, in the most general sense state regulation, apart from state support, includes administrative impact, legal impact, restrictive measures, etc.

The main directions of financing agricultural organizations of the Republic of Tatarstan from the federal and regional budgets will be discussed in Table 1.

Table 1. Main directions of financing agriculture in the Republic of Tatarstan from the federal and regional budget for 2019.

№ p/p	Main directions of financing agriculture in the Republic of Tatarstan	For 2019		
		From the federal budget, thousand rubles	From the regional budget, thousand rubles	Total, thousand rubles
1	Grants to support elite seed production	88219.0	110382.80	198601.80
2	Grants to cover part of the cost of laying and maintaining perennial fruit and berry plantations	42093.70	30481.60	72575.30
3	Grants for reimbursement of part of expenses of agricultural producers for payment of insurance premium accrued under the agricultural insurance contract in the field of crop production	63149.70	45729.10	108878.80
4	Compensation to agricultural producers who in accordance with the established procedure provided insurance protection of their property interests related to the production of agricultural products, damage caused as a result of natural emergency situations in 2019 in the territories of the constituent entities of the Russian Federation, from the federal budget funds	120154.90	–	120154.90
5	Grants to help achieve the development targets of crop and livestock sectors	617142.70	446896.40	1064039.10
6	Subsidies to finance a portion of costs associated with the acquisition of mineral fertilizers	–	1209268.60	1209268.60

(*continued*)

Table 1. (*continued*)

№ p/p	Main directions of financing agriculture in the Republic of Tatarstan	For 2019		
		From the federal budget, thousand rubles	From the regional budget, thousand rubles	Total, thousand rubles
7	Subsidies to compensate for part of the expenses of legal entities for the purchase of barley for processing for the production of food products	–	1500000.00	1500000.00
8	Grants for untied support to agricultural producers in developing seed potato and open field vegetable production	453915.80	328697.60	782613.40
9	Subsidies for calcification of acid soils	–	250000.00	250000.00
10	Grants to support crop production	–	2237.70	2237.70
11	Grants to agricultural producers for activities aimed at plant protection	–	30000.00	30000.00
12	Subsidies to increase the productivity of dairy cattle	408084.80	826999.20	1235084.00
13	Grants to support livestock breeding	194880.00	417781.80	612661, 80
14	Subsidies for reimbursement of part of agricultural producers' expenses for payment of insurance premiums under agricultural insurance contracts in the field of animal husbandry	8560.30	6198.90	14759.20
15.	Grants for veterinary and sanitary improvement activities	–	158700.00	158700.00
16	Subsidies to compensate for part of the cost of raising the brood stock of sheep and goats	1740.00	1260.00	3000.00
17	Livestock support subsidy	–	228601.60	228601.60
18	Grants for prevention of animal diseases and protection of the population from diseases common to humans and animals	–	72867	72867
19	Grants for agricultural machinery fleet renewal	–	1880056.2	1880056, 2
20	Grants for agricultural land reclamation development	75895.00	462946.30	538841.3
21	Total government support	2073835.9	8009104.8	10082940.7

* The table is based on the data of consolidated annual reports of agricultural organizations of the Ministry of Agriculture and Food of the Republic of Tatarstan.

As can be seen from Table 1, the activities of agricultural organizations of the Republic of Tatarstan were financed from the regional budget funds 8009104.9 thousand rubles, which is 79.4% of the total amount of state support.

In 2019 in the Republic of Tatarstan special attention was paid to the development of farming, personal subsidiary farms and consumer cooperatives (Tables 2, 3, Figs. 1, 2 and 3).

Table 2. Grants to cooperatives in 2019.

Directions	Number of applicants, e.g.	Amount of subsidy, million rubles
Buying products from members of cooperatives	27	37.8
Acquisition of property for transfer to members	6	5.4
Acquisition of machinery and equipment	28	55.6

Table 3. Stimulation of private subsidiary farming activities in 2019.

Areas of state support	Amount, million rubles
Reimbursement of mini-farm construction costs	89.9
Reimbursement for the purchase of young birds (broiler chickens, ducks, geese, turkeys)	26.5
Reimbursement of dairy cows, goats, veterinary service of cows	365.4
Reimbursement of expenses for purchase of commodity and pedigree heifers and first heifers	12.9
Credit subsidies	33.3
Total	528.0

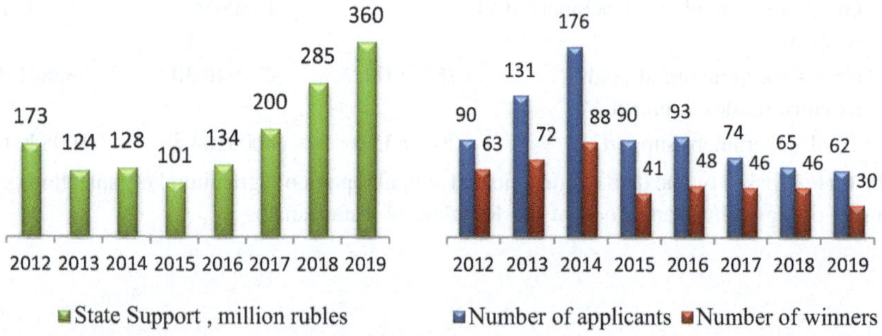

Fig. 1. Implementation of grant support programs for beginner farmers for 2012–2019.

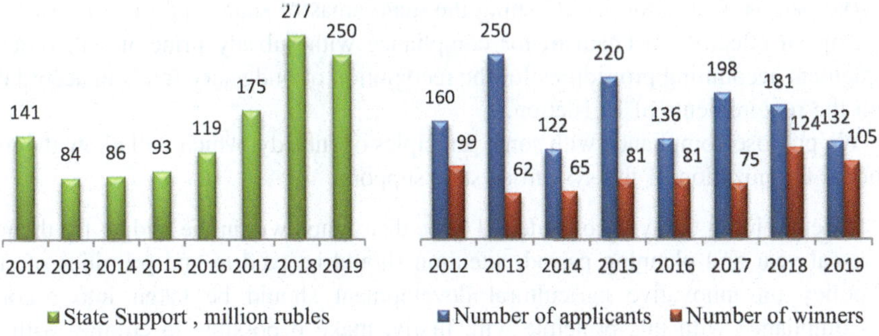

Fig. 2. Implementation of grant support programs for family livestock farms for 2012–2019.

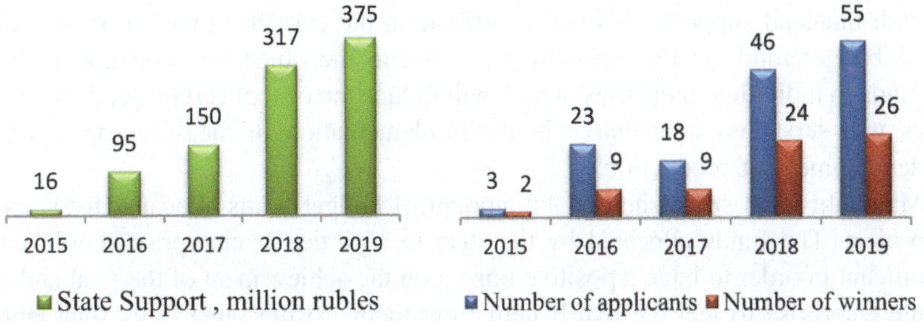

Fig. 3. On the development of the material and technical base of agricultural consumer cooperatives, 2015–2019.

Analysis of Tables 2, 3 and Figs. 1, 2 and 3 leads to the conclusion that the Republic of Tatarstan pays much attention to the development of farms, agricultural consumer cooperatives and personal subsidiary farms. Thus, the amount of grant support provided to novice farmers in the period from 2012 to 2019 increased 1.8 times, family livestock farms- 2.1 times, and the amount of state support aimed at developing the material and technical base of agricultural consumer cooperatives from 2015 to 2019 increased 24 times.

Under the federal project "Creation of a system of support of farmers and development of cooperation" 418.8 million rubles were spent. Under the "Agrostartup" project, 103 KFHs won grants worth 312.7 million rubles. For the current year the grant amount per farmer increased to 5 million rubles. The terms and conditions of the Beginning Farm and Family Livestock Farm grant programs allow farmers to increase their stock and production many times over.

As can be seen, the government has taken serious measures to stimulate small and medium businesses in the agricultural sector of the economy to sustainable development, increase the number of small businesses, modernize the material and technical base of agricultural consumer cooperatives and improve the competitiveness and quality of their products, and improve the efficiency of land use from agricultural land.

We believe that before establishing the main areas of state support, it is crucial to develop an effective mechanism for compliance with subsidy principles in order to implement accounting procedures for the recognition of budgetary funds in accordance with the requirements of legislation.

We propose compliance with some principles of subsidy, which will allow the most rational organization of the system of state support.

1. In determining the types of state aid to be distributed within the budget for the next fiscal year and planning period, attention should be paid to and priorities of state policy on innovative agricultural development should be taken into account. Compliance with this principle will, firstly, make it possible to comply with the national policy on innovative development of the economy. Secondly, it will make it possible to form federal budget obligations on priority directions of development based on clear objectives set by the RF Government. Thirdly, it will allow agrarian organizations to understand which projects and to what extent they will be provided with financial support. This will contribute to the correct formation of information on budget funds in the accounting system, to the objective allocation of budget funds to individual industries, which will reduce the cost of certain products, as well as will serve as an incentive in the implementation of measures on innovative development of production.
2. Materiality and sufficiency of the amount of budget funds allocated for a specific project. The funds allocated by the state to a particular enterprise should be significant in order to have a positive impact on the achievement of the goal and allow the enterprise to feel the real benefit from them. At the same time, such amounts should be economical from the point of view of the state budget, especially in the current crisis conditions.
3. Strengthening of control measures over the rational use of allocated funds in the distribution by the RF subjects and local authorities. This will make it possible to control the movement of state aid, assess the rationality of its distribution among specific organizations, and facilitate control procedures when planning external state control.

3 Results

Modern conditions are characterized by the presence of a high concentration of information flows. The management structures of agricultural organizations need more and more skills to make quick but effective decisions, as the economy of resources used, the quantity and quality of products produced, and the growth of competitiveness depend on these decisions. Equally important is the management of information flows related to the receipt and use of budgetary funds, as such funds are an additional source of financing costs on the one hand, the state investment in the development of a company on the other, and the obligation of the organization to the state on the third [18].

It is known that the management accounting system is designed to provide operational information to the organization, managers and managers of the company to manage financial flows [19, 20]. On the basis of the data formed in the management accounting system, an internal management reporting should be prepared, which combines information on all structural subdivisions, responsibility centers and even detailed information on certain types of products [21, 22].

The format and structure of internal management reporting depend on the industry affiliation of the organization; the adopted technological processes; the scale of activities and product range; organizational structure; various external and internal factors affecting development, and, of course, the personal and professional skills and abilities of management personnel [23, 24].

The information formed in the management accounting system and presented in the management reporting is the basis for the functioning of the management system in the organization.

In the domestic practice of agricultural organizations the management accounting system is formed for quite a long time, and many farms do not consider it at all, based on the fact that it is optional. In addition, the organization of management accounting and presentation of internal financial statements requires the investment of a certain level of material, financial and labor costs, the payback of which may come after a long enough time [25–27].

In our opinion, in agricultural organizations the management accounting system should be organized as a subsystem of accounting and internal control, because it can ensure the effective functioning of these two systems [28]. We believe that the results obtained in the management accounting system and the information disclosed in the internal management reporting of the organization can serve as a basis for exercising control functions in order to improve the efficiency of the company [29, 30].

Competently drawn up internal management reporting in an agricultural organization is the main fundamental element of the entire management structure and the most important tool of management and control function, the structure of which consists of internal management documentation containing a system of economic, financial and operational indicators necessary for making informed and effective management decisions.

Three types of management reports are generated in Russian companies:

1) management reporting, which characterizes the financial situation of the organization, financial results of activity, factors and dynamics of change of its financial situation. Estimates of income and expenses, production plan, business plan, investment plan, etc. serve as a basis for building reports of this nature.
2) management reports that characterize key performance indicators, reflect operational performance indicators and contain information on types of products manufactured in the organization;
3) management reports on budget execution of the organization, which contain information on cash flow, information on income and expenses of the organization, etc.

Research in the field of management reporting among the works of various modern scientists allowed us to identify some features of internal management reporting (Fig. 4).

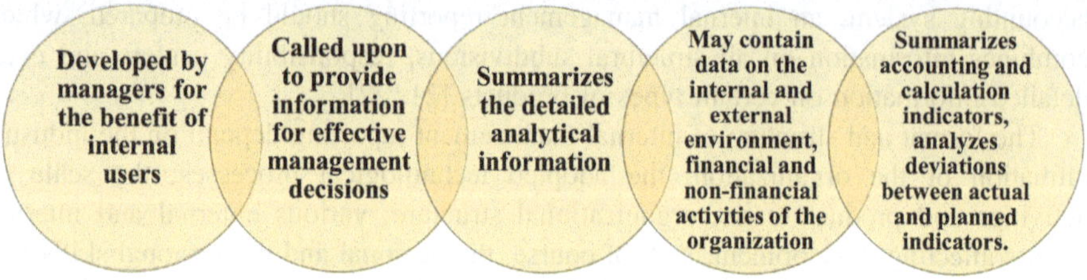

Fig. 4. Features of internal management reporting.

Thus, the internal administrative reporting of the agricultural organization is a system of the interconnected accounting and settlement indicators which are a result of gathering, processing and systematization of the information of financial and nonfinancial character about internal and external factors of activity of the organization, called to detail the information on all components of activity of the company (kinds of production, the centers of responsibility, divisions and branches and other segments interesting internal users), with a view of acceptance of the manual.

We believe that in order to achieve positive and effective results of management reporting, it is necessary to systematize the stages of its construction depending on the procedures. With regard to the formation of an internal management report on the movement and use of budget funds, we suggest systematizing procedures for the following stages (Fig. 5).

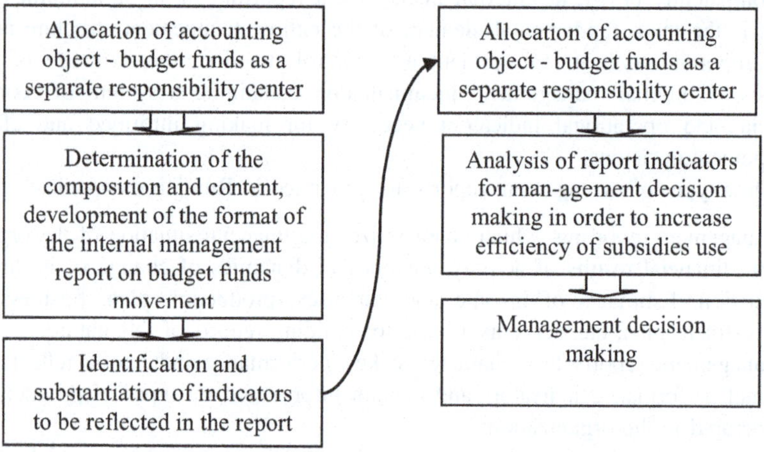

Fig. 5. Stages of forming internal management report on budget funds movement.

When developing the management report on the state of budget funds, we recommend to follow the following principles:

1. objectivity, which means that the prepared report should be useful and objective, contributing to making effective managerial decisions;

2. addressability, defining specific addressees-managers for whom the report is intended and by whom it is prepared;
3. completeness, due to which the information and indicators disclosed in the report are fully sufficient for making managerial decisions;
4. compactness, which means that the report should contain only those indicators, which may be useful for making decisions and do not distract attention of managers and managers;
5. analyticity, meaning that the data presented in the report are the main basis for further analysis of the efficiency of budget funds use;
6. reliability, i.e. the data contained in the reports should be true and fair.

The main purpose of developing an internal management report on the state of budgetary funds is to provide management with the fullest possible information on the movement of budgetary funds from the moment of their receipt in the accounts of the organization, which is necessary to assess their efficiency and target use and to make effective management decisions aimed at improving such efficiency, at attracting additional sources and preventing possible risks of their improper use.

We nominate as a center of financial responsibility a subsidy center, i.e. a responsibility center, the head of which is responsible for maximizing revenues from the use of budget funds and increasing the efficiency of subsidy. As can be seen, in order to build a management report on the state of budget funds it is necessary not only to introduce a management accounting system in the company, but also to nominate a separate center of responsibility, which will require a lot of time, cost and effort and may not immediately produce positive results.

The process of management reporting system formation should be considered from two points of view:

- systematization and justification of the format of internal management reporting;
- definition and formation of indicators to be reflected in the reporting.

If we consider the system of indicators of internal management reporting as a basis for information disclosure, the following stages should be adhered to when defining and justifying them:

1. Identifying the subjects involved in internal reporting and the subjects for whom the information disclosed in it is intended. The nature, completeness and objectivity of information on the selected indicators depend on the skills, abilities and professional abilities of the first group of subjects. The needs and interests of the second group of subjects depend on the goals and objectives of the first group, in particular, the definition of the group of indicators disclosed in the reports. If we consider the principles of management report formation, which have already been listed, at this stage the fundamental principles are objectivity, targeting and completeness.
2. Determination of accounting objects, information about which will be disclosed in the management report. In our case it is information about the movement of budget funds from the moment of their receipt to the accounts of the organization and till their full use. Therefore, we set a goal to develop and substantiate the indicators disclosing the operations on budget funds movement, which in future may serve as

an informative basis for assessment of subsidizing efficiency. At this stage it is important to follow the principles of objectivity, similarity and reliability.
3. Identification of sources of information for the establishment and evaluation of indicators, in order to create a database of accounting and analytical information for the development of an internal management report on the movement of budget funds. We believe that the main sources of information for the definition of indicators and further preparation of this report are the Explanatory Notes to the Balance Sheet and Statement of Financial Performance and Statement of Targeted Funding of the Organization. Besides, if necessary, primary documents and registers of analytical and synthetic accounting on accounting of target-financing funds and working documentation prepared by the internal control service can be used. The main principle to be followed at this stage is the sufficiency principle, which means that the information considered in the sources is sufficient to fully achieve the objective.
4. Identification of consolidated indicators that are necessary to substantiate the main indicators and calculate the indicators characterizing the efficiency of budget funds application. The main summary indicators are: the sum of profit (uncovered loss), the sum of proceeds, the sum of total production costs, the cost of fixed assets, the sum of the organization's equity capital.
5. Evaluation of efficiency of budget funds use and decision-making on measures of its increase, as well as decisions on minimization or prevention of risks of inefficient and improper use of funds.

Based on this algorithm we have defined the main indicators of the internal management report on the state of budget funds, which are presented in Table 4.

When forming an internal management report on the state of budget funds, it is necessary to simultaneously analyze operational data, which will allow to calculate their effectiveness in a less laborious way. However, it is necessary to take into account the sectoral specifics of agricultural organizations, which may significantly affect the level of efficiency and profitability of subsidies.

Tables 5 and 6 propose form of internal management report on the state of budgetary funds (for example, LLC "Tashkichu" of the Arsky district of the Republic of Tatarstan), formed taking into account the above principles, the stages of constructing management reporting and presented to managers or the leadership of an agricultural organization for further relevant decisions.

Table 4. Main indicators of the internal management report on the state of budget funds.

No p/p	Indicator name	Comments
1	RGS - state subsidy amount	Main quantitative indicator, it is recommended to reflect the amount for 2 reporting periods in the report to analyze the growth rate and growth rate
2	Ratio of the amount of the state subsidy to the financial performance of the organization	Qualitative indicator that characterizes the share of budget funds in sales revenue, net profit (uncovered) loss
3	Ratio of total expenses incurred by the organization for regular activities to expenses compensated from budget funds	Qualitative indicator that characterizes the level of compensability of production costs at the expense of the state budget
4	EIS - efficiency of investment subsidies	Qualitative indicator that characterizes the ratio between the level of renewability of fixed assets and the amount of subsidies that compensate for such expenses
	EOS - efficiency of operating subsidies	Qualitative index characterizing the change in the cost of material and production expenses in the reporting period to the amount of subsidies compensating such expenses
5.	ESFIA - effectiveness of subsidies aimed at financing intellectual development.	Qualitative indicator characterizing the ratio between the level of renewal of intangible assets and the amount of budget funds financing expenditures for their acquisition
6.	Share of non-monetary budget funds in total cost of fixed assets	Qualitative indicator that characterizes the share of fixed assets received as a government subsidy in the total value of fixed assets

Based on this report, as well as on the statements of financial results and cash flows, it is possible to monitor and evaluate the effectiveness of the use of budgetary funds in agricultural organizations, using the income approach. In this regard, the following methods of control and evaluation of budgetary funds have been developed and proposed as part of the study:

$$KD = \frac{\mathrm{DO}_1}{\mathrm{DO}_0} \times \frac{\mathrm{DP}_1}{\mathrm{DP}_0} \times \frac{\mathrm{DB}_1}{\mathrm{DB}_0}, \quad \text{or}, or \tag{1}$$

$$KD = \frac{DT_1}{DT_0} \times \frac{DI_1}{DI_0} \times \frac{DF_1}{DF_0}, \tag{2}$$

where KD is the growth rate of return from the use of budget funds;

Table 5. Form of internal management report on state of budget funds (proposed form). Operational Data

Indicators	At the end of the previous period, thousand rubles	At the end of the reporting period, thousand rubles	Absolute deviation, thousand rubles	Growth rate, %
Overall figures:				
Amount of net profit (uncovered loss)	2682	56736	54054	2115.4
Amount of revenue	268966	152469	−116497	56.7
Cost of fixed assets	621559	625812	4253	100.7
Value of intangible assets	–	–	–	–
Amount of main production costs	543414	232516	−310898	42.8
Budget funds	65600	57922	−7678	88.3
For the development of the "Plant Industry", among other things: (detailing depending on the subsidies received in a particular organization)	51365	57168	5803	111.3
For the development of the livestock industry, among other things: (detailing depending on the subsidies received in a particular organization)	14235	754	−13481	5.3
Subsidies to finance capital expenditures (detailed description by groups of fixed assets is possible)	–	140	140	–
Grants for financing current expenses (detailed description by type of expenses is possible)	65600	57781	− 5819	88.1
Grants for financing intellectual activity (detailed elaboration by groups of intangible assets is possible)	–	–	–	–
Non-monetary government subsidies	–	–	–	–

Table 6. Form of internal management report on state of budget funds (proposed form). Calculation part

Indicators	Method of calculation	At the end of the previous period, thousand rubles	At the end of the reporting period, thousand rubles	Growth rate, %
The ratio of the size of the budget funds to the financial results of the organization:	X	X	X	X
Amount of revenues per 1 ruble of budget funds	Amount of revenue/Budgetary funds	4.10	2.63	64.1
Amount of profit per 1 ruble of budget funds	Amount of profit/Budgetary funds	0.04	0.98	2450
The ratio of the total amount of expenses incurred by the organization in ordinary activities to expenses compensated by state subsidies	X	X	X	X
Share of budget funds provided for cost recovery in total cost of expenditures	Amount of budgetary funds/Amount of expenses by usual types of activity	0.12	0.25	208.3
EIS - Efficiency of investment subsidies	Amount of capital expenditure subsidies/Change in fixed assets value for the reporting year * 100%	–	3, 3	–
EOS - Efficiency of operating subsidies	Amount of subsidies for financing current expenses/Amount of expenses by ordinary activities * 100%	12.1	24.9	205.8
ESfIA - Efficiency of subsidies aimed at financing intellectual activity	Amount of grants for intellectual activity financing/Change in the value of intangible assets for the reporting year * 100%	–	–	–
Share of non-monetary government subsidy in total fixed asset value	Amount of non-monetary state subsidy/Total cost of fixed assets	–	–	–

DO_1, DO_0 - respectively, income from ordinary activities in the reporting and base years, thousand rubles;

DP_1, DP_0 - respectively, income from other operations in the reporting and base years, thousand rubles;

DB_1, DB_0 - respectively, deferred income in the reporting and base years, thousand rubles;

DT_1, DT_0 - respectively, income from current activities in the reporting and base years, thousand rubles;

DI_1, DI_0 - respectively, income from investment activities in the reporting and base years, thousand rubles;

DF_1, DF_0 - respectively, income from financial activities in the reporting and base years, thousand rubles.

Thus, the coefficient of growth of profitability from the use of budget funds is equal to:

$$KD = \frac{\mathrm{DO}_1}{\mathrm{DO}_0} \times \frac{\mathrm{DP}_1}{\mathrm{DP}_0} \times \frac{\mathrm{DB}_1}{\mathrm{DB}_0} = \frac{152469}{268966} \times \frac{58601}{68194} = 0.476 \quad (3)$$

$$KD = \frac{DT_1}{DT_0} \times \frac{DI}{DI_0} \times \frac{DF_1}{DF_0} = \frac{109769}{326183} = 0.336 \quad (4)$$

This form of the report is standard, the report includes the main indicators characterizing the efficiency of budgetary funds in general. For more detailed analysis of efficiency of subsidies by types of costs, fixed assets, branches of subsidizing, etc., organizations can independently disclose additional indicators. For example, the efficiency of budget funds financing the purchase of mineral fertilizers can be calculated using the same methodology. For this purpose, it is necessary to determine the share of subsidies, financing purchases of mineral fertilizers, in the total cost of mineral fertilizers used in the reporting period, etc. Detailed indicators depend on requests from managers and types of government subsidies received by the organization in the reporting year.

After calculating all presented indicators, managers should determine the growth rate for each indicator, which will visually represent the growth or decline in efficiency for the periods under review.

4 Discussion

The mechanism of distribution, procedure and conditions of budget funds provision to agrarian organizations are determined by regional executive and legislative bodies and are regulated by relevant regulatory and legislative acts. Some criteria for optimization of subsidy directions for efficient and targeted use of budget funds are identified:

1. priority of the industry (products) for the region;
2. maximum attraction of funds from the federal budget for 1 rub. of the regional budget funds;
3. efficiency of use of state support funds.

In our opinion, the efficiency of subsidies is most significantly affected by the sectoral features of agriculture: the climatic conditions and geographic location of this or that farm. The fact is that in regions with difficult and unfavorable conditions for agricultural activities, the level of material and labor costs is much higher than in other regions with favorable climatic conditions.

Efficiency of subsidizing agricultural production should be assessed through internal control of targeted use or performance audit of such funds.

In our opinion, the whole system of substantiation of the subsidizing assessment criterion in agricultural organizations should be based on the principle of efficiency, which provides for reasonableness, rationality, practical and theoretical substantiation of the necessity of subsidizing agricultural organizations. Efficiency can be manifested from different aspects - in economic, technological, social, ecological and even political understanding. From an economic point of view, the very concept of "economic efficiency" provides for achieving an optimal balance between the costs of production of an organization and the positive effect that can be obtained from such production. Minimizing the risks of unforeseen costs, achieving the goal with minimum cost and maximum return are also indicators that characterize the cost-effectiveness of something.

Based on the analysis of literature sources, systematization of known principles and pre-filling with new ones, we propose the following criteria for assessing subsidies in agricultural organizations for accounting and reporting purposes.

1. Evaluation of the manner and conditions of budget funds provision. Grants are provided to legal entities on a non-reimbursable basis as budgetary allocations. It is known that budget appropriations are allocated by the budget of this or that level in the form of a ceiling on the amount of cash resources provided in the respective financial year for the performance of certain budget obligations. Consequently, it is true that the receipt of subsidies entails certain obligations, which organizations usually fulfil by fulfilling the conditions of the State subsidy.
2. Impact of the amount of budget funds on the revenue and expenditure part of the budget. Providing agrarian organization with state subsidy, the government carries out direct transfer of funds from the budget of the country and has a financial impact on the business entity in order to obtain positive results for itself. The state subsidy is a gratuitous and non-refundable form of state aid. Any amount provided to agricultural organizations for the budget is a negative financial flow, so it is fair to say that subsidies only increase budget expenditures and have no positive impact on it. However, we believe that government agencies, by providing subsidies, not only provide financial assistance to economic entities, but also hope to implement certain goals in their own interests. We believe that government agencies, by providing state subsidies to agricultural organizations, hope that these funds will have a positive impact on the activities of the entity. Here we can consider two positive aspects for the state: first, the amounts of budget funds can have a positive impact on the increase in financial results, which means an increase in the taxable base; second, state subsidies can have a positive impact on the increase in gross output, which means an increase in gross output produced within the state.

3. The sufficiency of the amount of budget funds to offset the costs of the subsidized project. Depending on the direction of development of industries, the goals and objectives of the subsidized project, the scale and volume of financing will vary. For example, activities related to improving the stimulation of growth in livestock production in a particular economic entity may require less financial and labor costs than activities aimed at introducing new breeds of animals and caring for them. For these purposes, a rational definition of the priorities of measures, their directions and the ultimate goal is important. This allows you to more accurately determine the amount of budget funds allocated to each business entity so that they do not have a shortage or funds are not used to finance less priority projects.
4. The degree of confidence that after the implementation of financing the expected benefit will be significantly higher than the amount of budget funds. In our opinion, this is the most important criterion by which it is necessary to evaluate budget funds. An assessment by this criterion makes it possible to determine the profitability of state investments in agriculture. Initially, it is not always possible to assess the profitability and payback of a subsidized project, since we are talking about organizations operating in a specific industry related to climatic conditions and living organisms. For example, the full introduction and adaptation of a new crop or the creation of one hundred heads of pedigree cattle will require several years. Such conditions require a forward-looking analysis, and if in the current year the company receives positive results from a particular project, this does not mean that the same results can be achieved next year as, we note again, the activities of agricultural organizations are closely related to nature and living organisms.
5. The absolute advantage of subsidies, i.e. whether this type of budget funds is absolutely indispensable for a given project or whether there is an alternative to their replacement. It should be noted that along with government subsidies, subsidies, subventions and budget loans can also be allocated as budget funds. Receiving subsidies, from the economic point of view, is a more complicated process, as they entail obligatory fulfillment of certain conditions by the agricultural organization for accounting purposes. The legislation does not regulate the allocation of subsidies or subventions with or without certain conditions. Subsidies, for example, can be allocated without specifying the directions and conditions for their use. Subventions usually focus only on the intended purpose without the existence of additional accompanying conditions. The advantage of grants and subventions is that they are less subject to control by both the State and the business entity, as they do not have any conditions for granting. However, in our opinion, this very advantage is the main disadvantage, as uncontrolled funds may be accompanied by the least economic efficiency or lost in financial flows.

We believe that it is expedient to allocate budgetary credits for the most large-scale projects, the payback period of which can be shown in years. We explain this by the fact that from an accounting point of view, government subsidies can be accounted for as either capital or profit. However, if the amount of subsidies is large enough, it can create unreasonable sources of funds in the balance sheet of the organization. In cases when an organization has been granted a budget loan, they can be accounted for as long-term or short-term accounts payable.

6. Assessment of the applicability of this form of resource support in the implementation of similar projects. An important point at this stage is the evaluation and analysis of the results of already completed projects. We believe that this is the most important criterion, as it allows us to assess the effectiveness of the use of subsidies received by the organization in previous reporting periods.

We believe that this procedure should be implemented in several directions:

- an analysis of the commercial efficiency of subsidies, which makes it possible to assess the level of positive effect from the use of government subsidies in a given project, i.e., to assess the economic profitability of subsidies;
- budget efficiency assessment, which makes it possible to analyze the impact of government subsidy on the revenue and expenditure parts of the budget;
- assessment of the national economic efficiency, which makes it possible to assess the level of impact of the subsidy amount on the increase in gross output and increase in the productive potential of the country, as well as on the decrease in the consumer price of agricultural products.

The assessment of the economic efficiency of subsidizing agricultural organizations should be based on a comparative analysis of the means used and the result achieved under this project. At such analysis it is expedient to compare the amounts of used budgetary funds and the result achieved after the use of these funds. Here one can consider such categories as increase in the volume of agricultural production, reduction of material and monetary expenses for production, etc. Such analysis makes it possible to assess the effectiveness of direct subsidy from the budgets of various levels based on the study and calculation of the ratio between the volume of aggregate support and the final results.

5 Conclusions

Evaluation of the efficiency and targeted use of budget funds and further effective management decisions aimed at improving such efficiency, attracting additional sources and preventing possible risks of their improper use are possible only by providing management with the fullest possible information on the movement of these funds from the moment they reach the accounts of the organization. The main element of the information base at such an assessment can be the internal management reporting of the agricultural organization, which is the result of collection, processing and systematization of information of financial and non-financial nature on internal and external factors of the organization, designed to detail information on all components of the company. At the current stage of development, most agricultural organizations do not form such reporting, so it is not possible to assess the system of economic, financial and operational indicators, including indicators that characterize the effectiveness of subsidies in the organization. In this connection, the procedures and stages of formation of internal management reporting on the state of budgetary funds have been systematized; the indicators disclosed in it have been justified, the methodology of their calculation has been presented; the format of such reporting reflecting the efficiency of subsidizing in agricultural organizations has been proposed.

We believe that the formation of an internal report on the movement of budgetary funds illustrates the effectiveness of subsidy and will allow managers to identify the main effective and inefficient areas of subsidy, to identify the causes of inefficiency or low efficiency and the areas to which the greatest attention should be paid in order to prevent the risk of total inefficient use of budgetary funds. The analysis of this nature will make it possible to determine the tactics and strategy of further measures aimed at attracting additional amounts of budget funds and achieving maximum efficiency in their use.

References

1. Subaeva, A.K., Nizamutdinov, M.M., Mavlieva, L.M., Kalimullin, M.N.: Labor productivity in digital agriculture. BIO Web Conf. **17**, 00226 (2020). https://doi.org/10.1051/bioconf/20201700226
2. Klychova, G., Zakirova, A., Dyatlova, A., Klychova, A., Zaugarova, E., Zalyalova, N.: Methodological tools to ensure economic security in the personnel management system of enterprises. In: E3S Web of Conferences, vol. 135, p. 04008 (2019). https://doi.org/10.1051/e3sconf/201913504008
3. Stupak, N.: Adaptation of Russian agriculture to climatic variability: the role of federal and provincial policies. Environ. Sci. Policy **68**, 10–19 (2017). https://doi.org/10.1016/j.envsci.2016.10.003
4. Uzun, V., Shagaida, N., Lerman, Z.: Russian agriculture: growth and institutional challenges. Land Use Policy **83**, 475–487 (2019). https://doi.org/10.1016/j.landusepol.2019.02.018
5. Chen, Y.-H., Chen, M.-X., Mishra, A.K.: Subsidies under uncertainty: modeling of input- and output-oriented policies. Econ. Model. **85**, 39–56 (2020). https://doi.org/10.1016/j.econmod.2019.05.005
6. Bayramoglu, B., Copeland, B.R., Jacques, J.-F.: Trade and fisheries subsidies. J. Int. Econ. **112**, 13–32 (2018). https://doi.org/10.1016/j.jinteco.2018.01.009
7. Sibande, L., Bailey, A., Davidova, S.: The impact of farm input subsidies on maize marketing in Malawi. Food Policy **69**, 190–206 (2017). https://doi.org/10.1016/j.foodpol.2017.04.001
8. Klychova, A., Klychova, G., Zakirova, A., Sungatullina, R., Mukhamedzyanov, K., Philippova, E.: Social development mechanism of an agricultural enterprise formation. E3S Web Conf. **110**, 02072 (2019). https://doi.org/10.1051/e3sconf/201911002072
9. Klychova, G.S., Nizamutdinov, M.M., Safiullin, L.N., Mavlieva, L.M.: Priorities of agricultural credit cooperation development. Mediterr. J. Soc. Sci. **5**(18), 215 (2014). https://doi.org/10.5901/mjss.2014.v5n18p215
10. Tian, Q., Holland, J.H., Brown, D.G.: Social and economic impacts of subsidy policies on rural development in the Poyang Lake region, China: insights from an agent-based model. Agric. Syst. **148**, 12–27 (2016). https://doi.org/10.1016/j.agsy.2016.06.005

11. Varela-Candamio, L., Calvo, N., Novo-Corti, I.: The role of public subsidies for efficiency and environmental adaptation of farming: a multi-layered business model based on functional foods and rural women. J. Clean. Prod. **183**, 555–565 (2018). https://doi.org/10.1016/j.jclepro.2018.02.109
12. Chen, J.-Y., Dimitrov, S., Pun, H.: The impact of government subsidy on supply chains' sustainability innovation. Omega **86**, 42–58 (2019). https://doi.org/10.1016/j.omega.2018.06.012
13. Liang, L., Wang, Y., Ridoutt, B.G., Lal, R., Zhao, G.: Agricultural subsidies assessment of cropping system from environmental and economic perspectives in North China based on LCA. Ecol. Ind. **96**, 351–360 (2019). https://doi.org/10.1016/j.ecolind.2018.09.017
14. Klychova, G.S., Ziganshin, B.G., Zakirova, A.R., Valieva, G.R., Klychova, A.S.: Benchmarking as an efficient tool of social audit development. J. Eng. Appl. Sci. **12**, 4958–4965 (2017). https://doi.org/10.3923/jeasci.2017.4958.4965
15. Lopez, R.A., He, X., De Falcis, F.: What drives China's new agricultural subsidies? World Dev. **93**, 279–292 (2017). https://doi.org/10.1016/j.worlddev.2016.12.015
16. Huang, J., Wang, X., Rozelle, S.: The subsidization of farming households in China's agriculture. Food Policy **41**, 124–132 (2013). https://doi.org/10.1016/j.foodpol.2013.04.011
17. Ginn, W., Pourroy, M.: Optimal monetary policy in the presence of food price subsidies. Econ. Model. **81**, 551–575 (2019). https://doi.org/10.1016/j.econmod.2018.06.012
18. Huang, W., Zhou, W., Chen, J., Chen, X.: The government's optimal subsidy scheme under manufacturers' competition of price and product energy efficiency. Omega **84**, 70–101 (2019). https://doi.org/10.1016/j.omega.2018.04.006
19. Combes, J.-L., Kinda, T., Ouedraogo, R., Plane, P.: Financial flows and economic growth in developing countries. Econ. Model. **83**, 195–209 (2019). https://doi.org/10.1016/j.econmod.2019.02.010
20. Sasaki, T.: Financial cash flows and research and development investment. Pac.-Basin Financ. J. **39**, 1–15 (2016). https://doi.org/10.1016/j.pacfin.2016.05.002
21. Klychova, G., Zakirova, A., Sadrieva, E., Avkhadiev, F., Klychova, A.: Reporting in the area of sustainable development in agribusiness. E3S Web Conf. **91**, 06002 (2019). https://doi.org/10.1051/e3sconf/20199106002
22. Van Erp, W., Roozen, F., Vosselman, E.: The performativity of a management accounting and control system: exploring the dynamic relational consequences of a design. Scand. J. Manag. **35**, 101077 (2019). https://doi.org/10.1016/j.scaman.2019.101077
23. Faizrakhmanov, Dz., Zakirova, A., Klychova, G., Yusupova, A., Klychova, A.: Formation and disclosure of information on social responsibility of agribusiness enterprises. E3S Web Conf. **91**, 06004 (2019). https://doi.org/10.1051/e3sconf/20199106004
24. Prajogo, D., Toy, J., Bhattacharya, A., Oke, A., Cheng, T.C.E.: The relationships between information management, process management and operational performance: internal and external contexts. Int. J. Prod. Econ. **199**, 95–103 (2018). https://doi.org/10.1016/j.ijpe.2018.02.019
25. Shad, M.K., Lai, F.-W., Fatt, C.L., Klemeš, J.J., Bokhari, A.: Integrating sustainability reporting into enterprise risk management and its relationship with business performance: a conceptual framework. J. Clean. Prod. **208**, 415–425 (2019). https://doi.org/10.1016/j.jclepro.2018.10.120
26. Dowdell, T.D., Herda, D.N., Notbohm, M.A.: Do management reports on internal control over financial reporting improve financial reporting? Res. Acc. Regul. **26**, 104–109 (2014). https://doi.org/10.1016/j.racreg.2014.02.011
27. Klychova, G., Zakirova, A., Khusainov, Sh., Dyatlova, A., Raheem, U., Sagadeeva, E.: Accounting and analytical support of internal management reporting on reclamation works. E3S Web Conf. **157**, 04022 (2020). https://doi.org/10.1051/e3sconf/202015704022

28. Chalmers, K., Hay, D., Khlif, H.: Internal control in accounting research: a review. J. Acc. Lit. **42**, 80–103 (2019). https://doi.org/10.1016/j.acclit.2018.03.002
29. Liew, A.: Enhancing and enabling management control systems through information technology: the essential roles of internal transparency and global transparency. Int. J. Acc. Inf. Syst. **33**, 16–31 (2019). https://doi.org/10.1016/j.accinf.2019.03.001
30. Lo, K., Ramos, F., Rogo, R.: Earnings management and annual report readability. J. Account. Econ. **63**, 1–25 (2017). https://doi.org/10.1016/j.jacceco.2016.09.002

The Concept of Anthropotechnical Safety of Functioning and Quality of Life

Ruben Kazaryan(✉)

Moscow State University of Civil Engineering, 26 Yaroslavskoe Shosse, Moscow 129337, Russia
r.kazarian@mail.ru

Abstract. Purpose: There are many methods, techniques and technologies for the formation of directive (prescribed for use) standards of integrated security: • fundamentally new (original); • developed "by analogy with world experience" (adapted); • borrowed. Methods: The elemental base of the methodology of these norms extends to all known spheres of life: • "environmental" or the sphere of human existence in nature; • "social" or the sphere of collective relationships and interactions in the community; • "technogenic", recultivated, or the sphere of all material created in processes of thought and subsequent productive activities of a person; • "informational" or the sphere of origin, perception, processing, accumulation, transmission and impact of information, as well as reactions to it; • "energy" or the sphere of detection (generation) of energy. Results: Modern theoretical concepts (not related to business and employment problems) combine the last two areas into one "informational and energy" sphere based on the well-known functional dependence of the "bit" and "joule" units (according to Felker). Conclusion: For each of these five spheres of life, as appropriate, legitimate (authorized for use by the official authorities) averaged standards of integrated security that have an officially regulated period of their functioning are created.

Keywords: AFAS concept · LCL · "Person-home-activity" · "Anthropotechnics of integrated life safety" · OTRO · Qualimetric instrumental engineering assessment · COP · WEP · WOP · SNiP · PPC

1 Introduction

Preconditions for the development and main provisions of the concept of the advanced formation of anthropotechnical safety of functioning and quality of life (AFAS concept)

ANTHROPOS (Greek: άνθρωπος - "man") is a part of nature, the most complex and incompletely known living organism. This is a personal (individual) unique object of study in many natural sciences (biology, physiology, psychology, pedagogy, medicine, etc.).

As a result of studying the properties and manifestations of man, it was revealed that in almost all of his diagnosed parameters and characteristics, each person displays his individual adaptive norm of state (physiological and psychological), behavior

V. Murgul and V. Pukhkal (Eds.): EMMFT 2019, AISC 1258, pp. 759–767, 2021.
https://doi.org/10.1007/978-3-030-57450-5_64

(individual and social) and activity (mental activity, creative activity, productive activity, vital activity, etc.).

The mental activity (thinking) of an individual is the most important type of human activity for ensuring the safety of functioning and quality of life. It must certainly be the initial stage of any other type of human activity. Here, a person creates a project of any subsequent applied (and then materially realized) activity, which he needs, and then develops the technology for its implementation, the required resource support, and outlines the desired results. In this project, one of the most important (if not the most important!) aspect is the "human factor", the technique of working with a person (anthropos), i.e. anthropotechnics.

The aspect of anthropotechnical safety of functioning and quality of life occurs.

2 Methods

In this aspect, two fundamentally different approaches to solving problems can be distinguished.

The first approach implies unconditional compliance with the current directive (prescribed for use) integrated security standards. A law-abiding performer of these standards (for example, a builder who erects or reorganizes a building or structure) can be sure that there will be no complaints about this part, and his products will be legitimate on the market. Everything is good in the system "construction business - integrated security policy guidelines".

It is believed that the future consumer-operator of construction products (buildings or structures) is formally and anonymously taken into account in this system when forming a comprehensive safety policy based on averaged statistical data. No one has ever erected buildings and structures in national mass housing construction of economy class objects, oriented to adaptive norms of the state, behavior and activity of a specific person.

The first approach is averaged (like the development of a construction organization project (COP) based on SNiP data for entering a tender). Then, the general contractor identified at the tender develops a work execution plan (WEP) (and, if necessary, a work organization plan (WOP), which considers the picture of the construction of a particular object by a specific construction organization in specific realities. That is, even with the first approach, when the future specific consumer-operator of construction products is not known, construction goes through two phases: generalized ideas about the processes and technologies for their implementation (COP), and the most specific data representations (WEP and WOP).

The second approach immediately implies the presence of two stages:

- the first stage requires compliance with current directive (prescribed for use) integrated safety standards at the stage of creating primary construction products with a view to their entry into the market; the stage corresponds to the current idea of the customer who formed ("built") the building or structure and sold it to the owner or operator on the construction market; the latter, in turn, may repeatedly resell or lease its property in whole or in part to other owners or operators. This stage of ownership change can last forever;

- the second stage involves taking into account the individual adaptive norms of the state, behavior and activities of a particular person or group of persons (team) in a building or structure pre-erected or rebuilt at the first stage in compliance with the current comprehensive security guidelines. Without denying the need and the prevailing understanding of the feasibility of the first stage, it becomes possible to take into account and anthropotechnically implement the individual adaptive norm of the state, behavior and activities of a particular person-operator of the building.

The usual way of reacting of modern Russian society to incidents when implementing the first approach to ensuring the anthropotechnical safety of functioning and quality of life is the so-called "post factum": something happens and only after that the reaction to such an emergency event begins.

3 Results

In connection with the foregoing, the concept of the advanced formation of anthropotechnical safety of functioning and quality of life, the AFAS concept, seems more promising:

- individual adaptive norms of the state, behavior and activities of a particular person or group of persons (team) in a particular building or structure can be identified in advance;
- pathological interactions of the components of the "person-home-activity" system and building processes and technologies that compensate them, anthropotechnical techniques, and medications can be unambiguously determined and implemented in advance;
- diagnostics and monitoring of safety parameters for functioning and quality of life in the "person-home-activity" system can be performed in the necessary cycles in order to identify and timely compensate for violations of the volume and quality of human health.

Two main approaches to assessing situations are known:

- the most ancient one, taking into account the generalized results of private opinions (expert approach), which is very impressive to the average person - you don't need to study anything, just express your opinion – that's all!
- the one based on quantitative instrumental estimates of the current values of the normalized parameters of the object of study according to the approved scientifically sound methodology in the framework of computer information technology. This approach requires certain efforts, expenditure of resources (time, finances, etc.), discipline to comply with the standards of the information technology used in instrumentation quality assessment; and most importantly, it ensures the objectivity of the assessment (its independence from the properties and personal manifestations of a series of changing operators, and its compatibility and comparability with the assessments performed for other objects of research using the same express technology).

Therefore, the original method and its implementing engineering instrument anthropotechnical express technology for quantifying the level of comfortable living (LCL) of a particular person in a particular environment are selected as the technique for diagnosis and monitoring the safety parameters of functioning and quality of life in the "person-home-activity" system in AFAS concept. The basic characteristics of this position are:

- an infographic approach as an anthropotechnical method;
- a "human-technology-environment" system as an object of research and design in anthropotechnics;
- methods, techniques and means of express diagnostics in anthropotechnics;
- a person as an object of research in diagnostic and monitoring processes, as well as the only objective criterion for comfortable living;
- individual and subjective nature of LCL;
- technology and organization of engineering monitoring of LCL [1–4].

4 Discussion

The concept of the advanced formation of the anthropotechnical safety of functioning and quality of life, the AFAS concept, in its essence, can be taken as the basis for functioning:

- housing and communal complex of market services, since it relates to the sphere of consumer services, and not to the construction industry, as it is now;
- a new separate area of the construction business, implemented at the second stage of the second approach to solving the problems of anthropotechnical safety of functioning and quality of life;
- a new separate area of market business services for the diagnosis and monitoring of individual adaptive norms of the state, behavior and activities of a particular person or group of people (team) in a particular building or structure.

The combination of the approaches listed in this section form the basis of a new scientific direction "Anthropotechnics of integrated life safety", which was put forward by a prominent modern scientist, professor, doctor of technical sciences, Chulkov V.O.

5 Conclusions

Implementation of the basic principles of the AFAS concept in the residential complex "Iris" (Moscow region, Odintsovo district, the village of Lesnoy Gorodok).

The residential complex "Iris-2" (Fig. 2), with all the wealth of choice of new buildings in the Odintsovo district of the Moscow region, advantageously outstands among them, first of all, due to its extraordinary appearance, an informal creative approach to solving architectural tasks. The complex is small, it consists of two buildings. The first is an eight-sectional house of varying number of storeys. It looks like a closed triangle from a bird's eye view, and when viewed from a height of human, it becomes clear that this small "fortress" has an easy entrance and passage into the interior of a cozy courtyard. At little way away is a five-story building with a slightly different facade design. Roof architecture compositionally combines the Iris complex of buildings into one whole. It is reminiscent of unknown birds with a rather large wingspan, which is extremely unusual. Exterior finish: facing brick, two colors: "Milk Chocolate" RAL 8025 (dark), "Ivory" RAL1015 (light). Painting of reinforced concrete surfaces of two colors for the selected brick is approved by the chief architect of the project together with the painter at the construction site [5–11].

Base-lining with concrete slabs "Eurokam" for artificial stone in the color "Chocolate" - Brega product. Stained glass filling in two colors RAL8019 (belt along reinforced concrete slabs, ceilings and insertion on a stained glass column) and RAL1015 (lower filling of stained glass).

The design of the monolithic-brick structure is extraordinary and is clearly distinguished by grace against the background of long-bored rectangular high-rises. The sections of the buildings are located around the perimeter of the triangle, forming a complex line of the facade with smooth curves, alternating with the angled ledges of the loggias. The footage of apartments is from 40.4 to 96.5 m^2. Note that domestic developers have recently been increasingly using the term "economy comfort", putting in the head of the homeowner the idea that half of this term enclosed in the word "economy" refers to value, and the other half, "comfort", is all the rest that the buyer of the home decides to evaluate for himself. A person, buying a home for himself, is able to dream up anything within the framework of "comfort". But, as a rule, he cannot specifically and unequivocally determine whether the housing (as the environment for its future habitat) will be pathogenic for him (causing damage to him in terms of quality and volume of health), neutral or recreational (providing him with restoration and improvement of quality and volume of health). He does not even think about the possibility of such a prospect. Usually a person tries to find a compromise between economy and comfort. There are not that much three-bedroom apartments in the housing complex "Iris" based on the potential audience. However, the area of one-room (40–46 sq.m) and two-room (62–75 sq.m) apartments can be considered quite decent. The unusual sensations are caused by the lack of simple lines in the layouts that are unusual for the average tenant (a consequence of the imagination of architects), which leads to the appearance of apartments with a studio layout and a small kitchen. These are apartments on the tops of triangularly planned buildings, and here the residents are likely to show extraordinary thinking. In order to be able to compare the parameters for assessing functional zones, it is necessary to switch from their absolute values to relative values. To this end, in 2010, Gazaryan R.K. and Chulkov V.O. substantiated the construction of an infographic model of the actual level of organizational and technological reliability of operation (OTRO) "… for each indicator, the minimum and maximum numerical value of the indicator is revealed. The value of the interval from the minimum (taken as zero) to the maximum value (taken as unit) of the indicator is an estimated interval for this indicator. Each current value of the indicator in this interval is related to the value of this interval, getting the relative value of this current indicator in fractions or in percent. The maximum value of the indicator value taken per unit is considered the reference (maximum possible) value of this indicator. Next, a proportion is built, where they take the reference value of the parameter as 100%, and the real level for X and, in accordance with the accepted indicators, reveal the actual level of organizational and technological reliability of operation of the studied indicator". Using the above-described proportions, for each of the functional zone assessment indicators, the level of OTRO of the assessment indicators is determined, and a star-shaped infographic model is constructed, in which the number of axes originating from the "zero" point corresponds to the number of parameters characterizing a specific functional zone. Evaluation parameters are ranked in accordance with their tendency as centripetal and centrifugal ones. To determine the level of OTRO of one functional zone, it is necessary to build two infographic star-shaped models, one of which is constructed by centripetal indicators, and the other - by centrifugal ones. Therefore, the assessment of design decisions requires:

- introducing a step-by-step modular review (discretization) of organizational and technological processes (works);
- taking into account the specific features of the implementation of organizational and technological work of the preparatory phase of construction (PPC);
- ranking of technological processes as their degree of automation increases.

As an example, Gazaryan R.K. and Chulkov V.O. examined and built models of the production zone of the Krasnaya Roza weaving factory (Fig. 1) with the corresponding evaluation parameters (see table). On each of the axes of the model, the reference values of the indicator of the evaluation criterion (unit) and the actual values of the evaluation criterion in relative units are plotted. As a result, we obtain a polygonal star-shaped infographic model (Fig. 1), where the hatched polygon is a broken line connecting the actual values of the level of the OTRO zone for specific parameters, and the outer contour is the maximum possible value of the indicators for evaluating the functional zone [1–4] (Table 1).

Table 1. Some indicators for assessing the level of OTRO.

No.	Functional zones of an industrial enterprise	Indicators for assessing the level of OTRO of the study zone
2	Production zone	Centrifugal trend of indicators Nomenclature of buildings and structures included in the composition of the study area; Total area of the zone, m^2; Total area of buildings and structures, m^2; Building density, %; Construction volume, m^3; Production capacity (in kind); Annual output of products (in kind); Utilization rate of production capacity; Utilization rate of the area of buildings and structures; Utilization ratio of the area of the zone; Profitability of products, %; Profitability of production, %; Labor productivity per one worker (in kind and in value terms); Centripetal trend of indicators The total number of rejects in manufactured products (actual size); Depreciation of equipment, %; The loss ratio of the functional zone, %; Number of equipment malfunctions, pcs

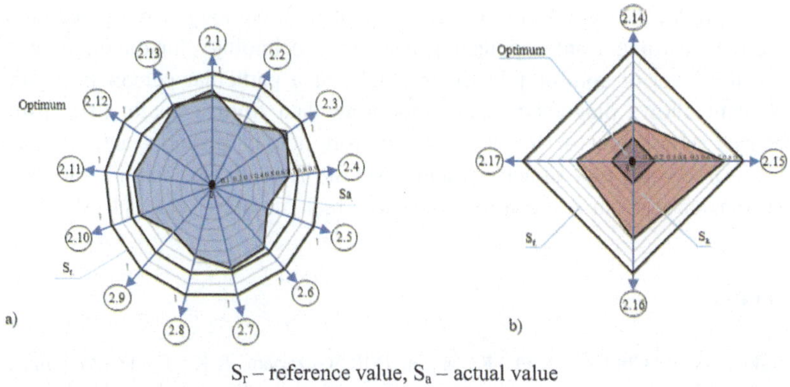

Fig. 1. The star-shaped model of the level of OTRO of the production zone: a) centrifugal trend of indicators, b) centripetal trend of indicators.

The number of axes in the star-shaped infographic model corresponds to the number of evaluation parameters.

It is in such rooms of the residential complex "Iris" that the above AFAS concept has been successfully applied, using the enormous potential of a star-shaped infographic model to compare parameters for assessing optimal functional zones of a person's living comfort, which ultimately will provide an opportunity to study actual indicators of the level of organizational and technological reliability of operation (OTRO), and, apparently, will ensure the profitability of home sales and contribute to improving the quality of life of its purchasers [1–4].

Fig. 2. General view of the territory of the residential complex "Iris-2".

Using AFAS concept based on an original methodology and implementing its engineering instrument anthropotechnical express technology for quantitative assessment of the level of comfortable living (LCL) of a particular person in a particular environment allows diagnosing and monitoring data from a qualimetric instrument engineering assessment of the level of comfort of any person's activity upon his request, as a separate type of activity and business. Thus, the residential complex "Iris" can be transferred from an economy class to a real "comfort" class [1–4].

References

1. Volkov, A.A., Chulkov, V.O., Kazaryan, R.R., Gazaryan, R.K.: Cycle reorganization as model of dynamics change and development norm in every living and artificial beings. Appl. Mech. Mater. **584–586**, 2685–2688 (2014). https://doi.org/10.4028/www.scientific.net/AMM.584-586.2685
2. Volkov, A.A., Chulkov, V.O., Chulkov, G.O., Kazaryan, R.R., Kyzina, O.N.: Qualities of documentation management chain (part 1). Adv. Mater. Res. **1065–1069**, 2401–2404 (2015). https://doi.org/10.4028/www.scientific.net/AMR.1065-1069.2401
3. Volkov, A.A., Chulkov, V.O., Chulkov, G.O., Kazaryan, R.R., Kyzina, O.N.: Qualities of documentation management chain (part 2). Adv. Mater. Res. **1065–1069**, 2405–2408 (2015). https://doi.org/10.4028/www.scientific.net/AMR.1065-1069.2405
4. Volkov, A.A., Chulkov, V.O., Chulkov, G.O., Kazaryan, R.R., Kyzina, O.N.: Qualities of documentation management chain (part 3). Adv. Mater. Res. **1065–1069**, 2409–2412 (2015). https://doi.org/10.4028/www.scientific.net/AMR.1065-1069.2409
5. Almazova, N., Rubtsova, A., Krylova, E., Barinova, D., Eremin, Y., Smolskaia, N.: Blended learning model in the innovative electronic basis of technical engineers training. In: Katalinic, B. (ed.) Proceedings of the 30th DAAAM International Symposium, pp. 0814–0825. DAAAM International, Zadar, Croatia (2019). https://doi.org/10.2507/30th.daaam.proceedings.113
6. Almazova, N., Barinova, D., Ipatov, O.: Forming of information culture with tools of electronic didactic materials. In: Katalinic, B. (ed.) Annals of DAAAM and Proceedings of the International DAAAM Symposium, vol. 29(1), pp. 0587–0593. Danube Adria Association for Automation and Manufacturing, DAAAM, Zadar; Croatia (2018). https://doi.org/10.2507/29th.daaam.proceedings.085
7. Almazova, N., Bylieva, D., Lobatyuk, V., Rubtsova, A.: Human behavior as a source of data in the context of education system. In: SPBPU IDE 2019: Proceedings of Peter the Great St. Petersburg Polytechnic University International Scientific Conference on innovations in digital economy. ACM, Saint – Petersburg (2019). https://doi.org/10.1145/1234567890
8. Bylieva, D., Lobatyuk, V., Safonova, A., Rubtsova, A.: Correlation between the practical aspect of the course and the e-learning progress. Educ. Sci. **9**, 167 (2019). https://doi.org/10.3390/educsci9030167
9. Bylieva, D.S., Lobatyuk, V.V, Nam, T.A.: Academic dishonesty in e-learning system. In: Soliman, K.S. (ed.) Proceedings of the 33rd International Business Information Management Association Conference IBIMA 2019: Education Excellence and Innovation Management through Vision 2020, pp. 7469–7481. International Business Information Management Association, IBIMA (2019)

10. Pozdeeva, E.G., Shipunova, O.D., Evseeva, L.I.: Social assessment of innovations and professional responsibility of future engineers. In: IOP Conference Series: Earth and Environmental Science, p. 012049 (2019). https://doi.org/10.1088/1755-1315/337/1/012049
11. Pokrovskaia, N.N., Ababkova, M.Y., Fedorov, D.A.: Educational services for intellectual capital growth or transmission of culture for transfer of knowledge—consumer satisfaction at St. Petersburg universities. Educ. Sci. **9**, 183 (2019). https://doi.org/10.3390/educsci9030183

Aspects in Managing the Life Cycle of Construction Projects

Ruben Kazaryan(✉)

Moscow State University of Civil Engineering, 26 Yaroslavskoe shosse, Moscow 129337, Russia
r.kazarian@mail.ru

Abstract. Goal: The problem of accounting and reporting net assets and the procedure for their formation, taking into account the specifics of the economic and legal status of property, is one of the cardinal aspects of managing the life cycle of construction projects. Methods: The purpose of the study is to substantiate the rules for reflecting net assets in accounting using method statements for the construction and installation work in the form of 3D images. Results: The principles, methods and means of visualization of organizational and technological solutions developed as part of method statements for the construction, installation and special works are proposed. An alternative valuation of assets to be returned to the owner upon the liquidation of the facility is proposed. Conclusion: The most important factor in modeling the key performance indicators of a system target approach to assessing the sustainability level of net assets based on IPSAS is the multi-criterion of assessing the fundamentals of the management strategy of the elements of quality systems for the operational and strategic planning of the life cycle of construction projects, which will apparently provide their multimedia visualization.

Keywords: IPSAS (International Public Sector Accounting Standards) · KPMG · NanoCAD · Organizational and Technological Documentation (OTD)

1 Introduction

In conditions of risk and uncertainty, there is a need for continuous management of economic processes of operational and strategic planning of the life cycle of construction projects. Changes in recent decades have led to a shift in the financial and managerial accounting of accents from expense management and financial flows to the management of economic processes in the life cycle of construction projects (financial situation, risks, reserve system of the enterprise, reorganization processes, control of added value) based on the use of financial engineering tools (monitoring, financial, hedging or other reports). The net assets ratio in conjunction with net liabilities is one of the most important indicators for assessing economic processes, efficiency and sustainable development of the life cycle of construction projects. However, for entities that are not focused on generating income or on satisfying the needs of the population in services as the results of investment development, this indicator is difficult to determine. Therefore, many proven models and methods of financial monitoring of net

V. Murgul and V. Pukhkal (Eds.): EMMFT 2019, AISC 1258, pp. 768–776, 2021.
https://doi.org/10.1007/978-3-030-57450-5_65

assets cannot be directly accepted for use in this area. This situation determines the discrepancy between the urgent need for the application of the scientific and methodological apparatus in the field of information and analytical support for management and performance assessment using the net asset indicator in the life cycle management of construction projects.

IPSAS (International Public Sector Accounting Standards) - a series of financial monitoring standards issued by IASB (International Accounting Standards Board) on August 31, 2015 are used by public sector enterprises around the world in the preparation of financial statements. The application of these standards allows improving budget control and supervision, as well as creating communication tools for dialogue and synchronizing the work of state institutions in different countries (European Commission [2]; European Commission [1]). IPSAS have actually become international benchmarks for assessing public sector accounting practices around the world. For these reasons, IPSAS deserves the attention of both policy makers in financial monitoring and practitioners and academics (KPMG [3]). Today, the development of national accounting and reporting standards is based precisely on this group of international standards, which are actively developed and are used worldwide [1–3].

This poses a problem associated with the need to develop new and adapt existing national and international models and instruments for accounting for net assets in the field of construction production and build effective methods based on them with the aim of forming elements of quality systems in managing the life cycle of construction projects.

2 Methods

Modern software systems “Hector-builder”, NanoCAD, “Builder” and others allow improving the design and development of method statements for the construction and installation work [4–7].

The basic principle for streamlining the development of organizational and technological documentation (OTD) is the maximum reduction in the manual labor of designers, increasing labor productivity in design and construction organizations, reducing the duration of design and improving the quality of OTD. The improvement tool should be a set of methods and technical means based on the use of computers, peripheral devices, reprography and organizational equipment, allowing to combine the technological process and visualization of documents into a single system [8].

The main conditions for rationalization are: the use of planar prototyping and documents - blanks, electrographic apparatus for producing originals, microfilms of enlarged copies of documents, as well as aperture cards as a means of automating copying and storage of documents [5–7].

One of the modern areas in the development of computer graphics is the construction of photorealistic images of objects modeled on a computer. This is true for computer aided design, film industry, advertising, design, computer games, etc. The increase in the quality of computer graphics products is rapidly spreading in the field of practical application, and the requirements of users for realistic 3-D images are constantly growing [9].

3 Results

The growth of computing power allows solving increasingly complex tasks of visualization of three-dimensional scenes (the number of objects, light sources, image resolution). The increase in complexity of the tasks is ahead of it. There are two main approaches to the visualization of 3-D models:

- creation of pseudo-three-dimensional fragments with their subsequent processing according to the principles of two-dimensional graphics, with little use of the simplest laws of the third dimension;
- creation of a model of a three-dimensional world with its further projection on the plane of the screen. Thus, all modern three-dimensional games are performed.

Creating an animation is a complex, requiring attention to detail, frame-by-frame arrangement of objects of the building master plan, setting the trajectory of their movement and interaction with the rest of the 3D model. Depending on the power of the computer, the quality of animation and visualization can vary from photorealistic to a schematic demonstration of the capabilities of the software package.

From the foregoing, we can conclude that the creation of any animation begins with the calculation of time and number of frames. So, in order to create a five-minute animation, it should be taken into account that the optimal number of frames per second is 30. Using simple arithmetic calculations, we get a finite number of frames equal to 9000. It takes about 30-33 s to calculate each frame on a modern computer, which implies that video with 9000 frames is created in about 75 h.

It is recommended that the creation of a visualization of the process of building a hydroelectric power station be carried out in the reverse order. That is, the building on the animation should not be built, but disassembled. This is the right approach to visualizing the construction process. By using this method, it is possible to avoid a heap of objects and, most importantly, correctly display the construction process.

The construction process chart is not compiled without the location and calculations of the used construction equipment. In animation, its presence is also necessary, since without technology it is impossible to realistically display the process of constructing a building.

An important stage in creating animation is a frame-by-frame arrangement of objects interacting with a building model. It is necessary to set the interval of occurrence of each object and each part of the model at the exactly specified time. For example, the first object is a crane, which begins the installation of a crane beam. It is given a trajectory of movement, and the disappearance of the elements and building structures, which interact with it, is set on the key frame. Thus, after processing and inverting the visual range of the video file, the effect of editing an industrial building is achieved. This technology is one of the first steps to improve the design of method statements for the construction and installation works [8–11].

4 Discussion

3D presentation of information is limited to simple and unambiguously perceived colors (red, green, yellow, white, black, gray). Passion for dynamic pictures (for example, animation) will distract and tire the designer of the concept design. The perception and use of photorealistic images of objects will worsen. It is proposed to carry out work on the creation of a method statement for the installation of beams under a bridge crane in the machine room of the HPP building in the bundle of Archicad 19 and Artlantis 6 programs. In Archicad 19, a model is created and subsequent layout is done, and visualization of 3d models using standard Archicad tools is done in Artlantis 6:

- overlapping with cutouts for the turbines were performed using the "overlap" tool;
- columns were made using a combination of tools "column" and "complex profile";
- floor beams were made with the tool "beams";
- the crane was exported as "pdf" - element from the library;
- the beams under the overhead crane were created by the combination of tools "beam of complex profile" (for the main body of the beam), "solid modeling operations" to create cutouts in the upper part of the beams; mounting outlets were created from "cylindrical primitives" from the standard library, rails were created from beams of a special profile "corner";
- scaffolding was created from a combination of elements "overlap", "columns of cylindrical section", "cylindrical beams";
- the mounting bolt is created by combining a "cylindrical column" and "overlap" with a hexagonal profile;
- gasket is created by "overlapping" with a cut;
- mounting holes for bolts are made by combining a "cylindrical primitive" from the standard library of elements and "solid modeling operations";
- the terrain to indicate the location of the crane was created using the 3D mesh tool. During the practical test, it was found that the proposed method is workable and suitable for practical use by quality services in the field of information and analytical support for management and performance evaluation using the net assets indicator in the life cycle management of construction projects both in peacetime and during restoration of facilities in natural and man-made emergency situations. This technique allows you to objectively evaluate the effectiveness of the QMS (quality management systems) as a whole and its individual elements.

5 Conclusions

The most urgent is the complex of problems that arise in the formation of financial monitoring and assessing the value of net assets in managing the life cycle of construction projects. The most significant relationships of system elements, factors determining the effectiveness of the element base are presented in Fig. 1.

The figure shows that the effectiveness of the quality management system as a whole is determined through the effectiveness of its individual components

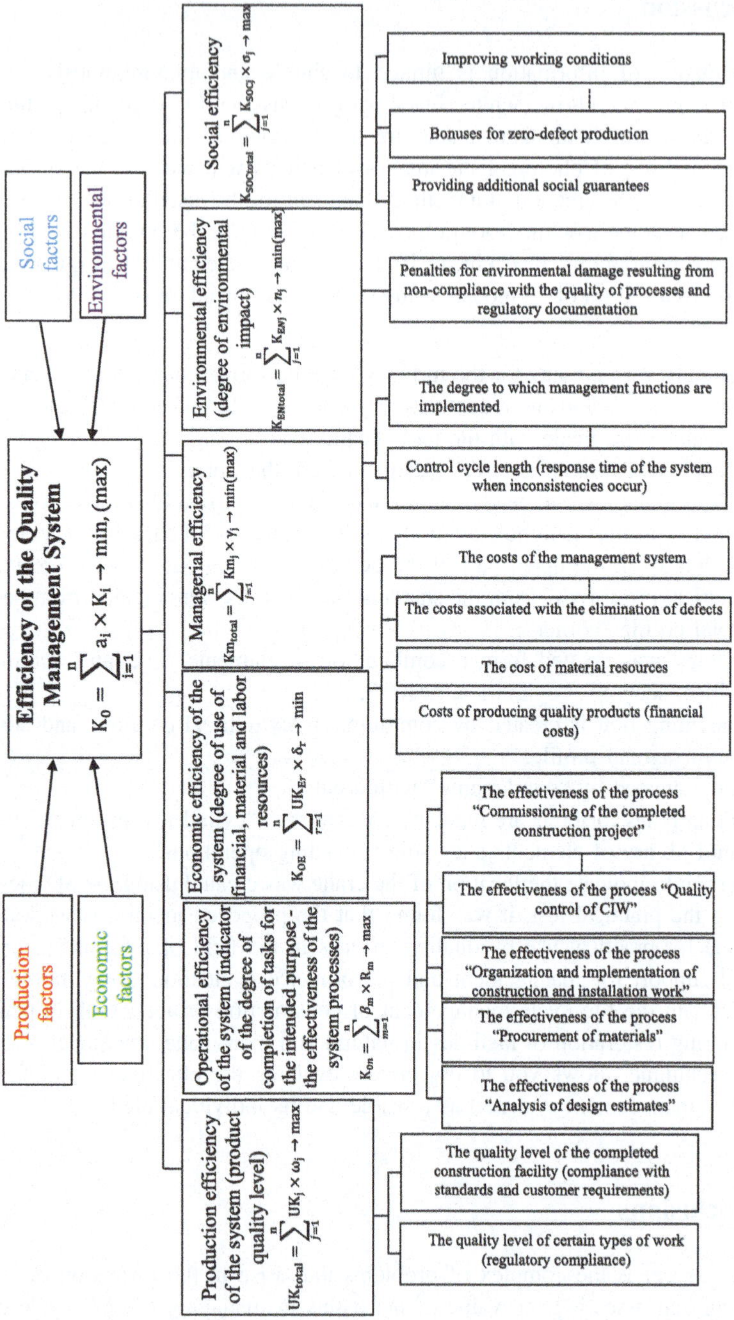

Fig. 1. System target approach to assessing the efficiency of the quality management system and the level of sustainability of net assets.

(production, economic, operational, social efficiency and others). These types of efficiency, in turn, will be characterized by the efficiency of solving lower goals (for example, economic efficiency is determined by the efficiency of using financial, material and labor resources, etc.) [7, 11–14].

At this stage, it is possible to identify the performance indicators of the quality management system and its individual elements. Indicators may be quantitative. In this case, it is necessary to distinguish between indicators of effect, which are measured in physical quantities. For example, the use of financial resources is estimated in real costs of financial resources or their savings (economic effect), and the ratio of these costs to allocated financial resources, measured in fractions of a unit, is an indicator of economic efficiency. A number of performance indicators relate to qualitative (or ranking) indicators that do not have a quantitative measure. These include indicators of operational, managerial effectiveness. Evaluation of the effectiveness of the QMS (quality management system) of construction products was carried out according to the scheme shown in Fig. 2.

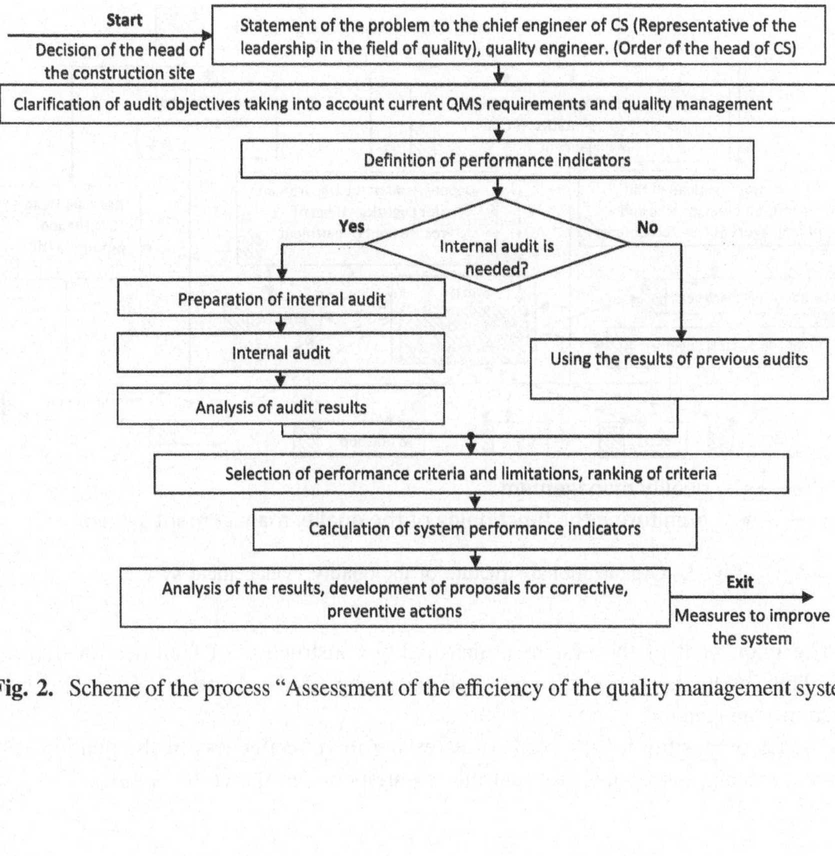

Fig. 2. Scheme of the process “Assessment of the efficiency of the quality management system”.

Within the framework of the research, the proposed methodology was tested and its implementation was carried out in the ICC "Zapad" and IC "Garant". The results are shown in the Table 1.

Table 1. Efficiency calculation results.

No.	Company	R_{Syst}	K_{OE}	U_e	K_{soc}	K_o
	Significance of the criteria	0.29	0.29	0.29	0.13	
1	ICC "Zapad"	0.61758	1.28227	0.76	0.62	0.85196
2	IC "Garant"	0.62959	0.59129	0.72	0.7	0.65385

The organizational structure of the quality management system ICC "Zapad" and IC "Garant" is presented in Fig. 3.

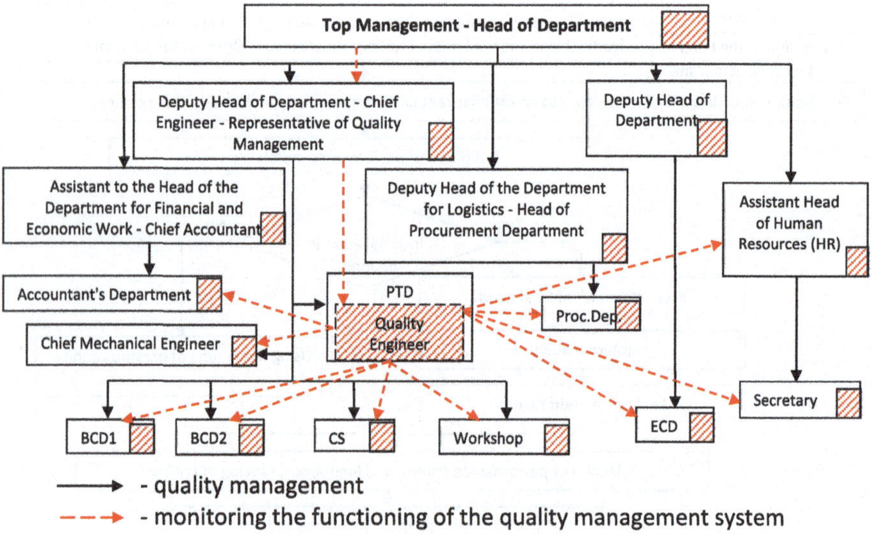

Fig. 3. Organizational structure of the quality management system.

The main goal of the system is the quality construction of facilities on time, in accordance with the estimated cost that meets the needs of the customer and construction management.

The decomposition of the goals of assessing the effectiveness of the quality management system, taking into account the requirements, is shown in Fig. 4.

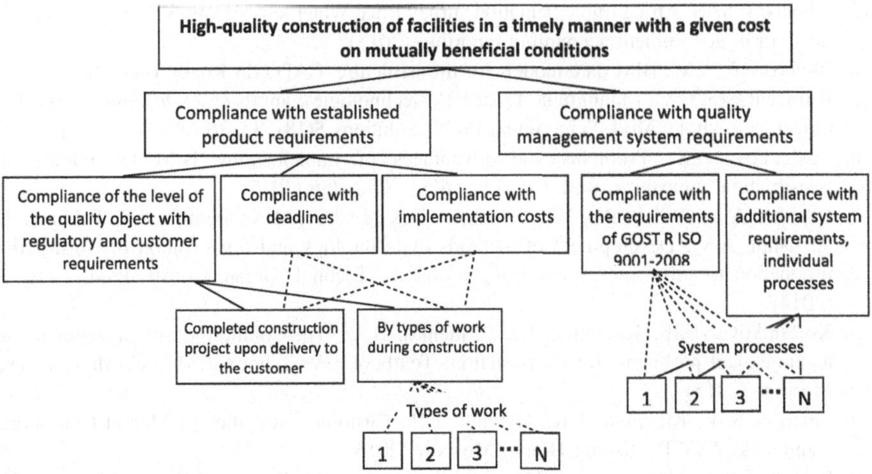

Fig. 4. Decomposition of goals in assessing the efficiency of a quality management system.

Given the volume of requirements for products and processes, the figure shows the goals of the upper levels. The objectives of the processes are indicated in the relevant process documentation. Product requirements are indicated in the design estimates and regulatory documentation. Analysis of existing factors of management and performance assessment using the net asset indicator in the life cycle management of construction projects shows the following [7, 11–14]:

1. the lack of a clear separation of the concepts of "quality management efficiency" and "quality management system";
2. efficiency is reduced, as a rule, to economic efficiency and overall effectiveness, while the concept of efficiency is broader than economic efficiency and effectiveness;
3. there are no unified approaches to solving the issue of assessing the effectiveness of quality management systems for construction organizations;
4. there is no experience in assessing the effectiveness of quality management systems in the organizational structures of construction enterprises.

References

1. European Commission: Assessment of the suitability of the International Public Sector Accounting Standards for the member states—Public consultation (2012a). http://www.epp.eurostat.ec.europa.eu/portal/page/portal/public_consultations/documents/Privacy_statement_feb_2012_EN.pdf. Accessed 02 Apr 2020
2. European Commission: Public consultation paper. Document accompanying the public consultation on the suitability of the International Public Sector Accounting Standards for EU member states (2012b). http://www.epp.eurostat.ec.europa.eu/portal/page/portal/public_consultations/documents/IPSAS_stakeholders_consultation_paper.pdf. Accessed 02 Apr 2020

3. KPMG: IPSAS, Africa Implementation survey Price Waterhouse coopers, (2013); Towards a new era in government accounting reporting (2013)
4. Skvortsovб, A.V.: BIM data models for infrastructure. CAD GIS Roads **1**(4), 16–23 (2015)
5. Rumyantseva, E.V., Manukhina, L.A.: BIM-technologies: approach to designing a building object as a whole. Mod. Sci.: Actual Probl. Solutions **5**(18), 33–36 (2015)
6. Lushnikov, A.S.: Problems and advantages of implementing BIM technologies in construction companies. Bull. Civ. Eng. **6**(53), 252–256 (2015)
7. Lapidus, A.A., Telichenko, V.I., Tumanov, D.K., Ershov, M.N., Oleinik, P.P., Feldman, O. A., Ishin, A.V.: Development of methods of technology and organization of construction production to solve energy efficiency problems. Technol. Organ. Constr. Prod. **2**, 10–16 (2014)
8. Kolesnikova, E.B., Kuzmina, T.K., Sinenko, S.A.: The solution of organizational and technological problems. In: Construction: Textbook (Workshop). ASV Publishing House, Moscow (2015)
9. Sinenko, S.A., Kuzmina, T.K., Oleynik, P.P.: Customer Activities in Market Conditions: Handbook. ASV Publishing House, Moscow (2015)
10. Sinenko, S.A., Ginzburg, V.M., Sapozhnikov, V.N., Kagan, P.B., Ginzburg, A.V.: Automation of Organizational and Technological Design in Construction. Higher Education, Saratov (2019)
11. Sinenko, S.A., Danilova, E.D.: To the question of improving the forms of displaying norms and standards for the organization and technology of construction. BST: Bull. Constr. Equip. **5**(1005), 54–55 (2018)
12. Oleynik, P.P.: Organization of Planning and Management in Construction. Scientific Publication. ASV Publishing House, Moscow (2017)
13. Oleinik, P.P., Brodsky, V.I.: Organizational Forms of Construction. Textbook. Scientific Publication (2015)
14. Oleynik, P.P., Shirshikov, B.F.: The Composition of the Sections of Organizational and Technological Documentation and the Requirements for Their Content. Scientific Publication. Publishing House of Moscow State University of Civil Engineering, Moscow (2013)

Method for Determining the Reliability Indicators of Elements in the Distribution Power System

Madina Plieva[1(✉)], Maret Madaeva[2], Aslanbek Khadzhiev[2], Soslan Marzoev[1], and Oleg Kadzhaev[1]

[1] North-Caucasian Institute of Mining and Metallurgy (State Technological University), 44, Nikolaeva Street, Vladikavkaz 362021, Russia
madosya80@mail.ru

[2] Grozny State Oil Technical University named after Academician MD Millionshchikov, 100, Isaeva Avenue, Grozny 364051, Russia

Abstract. The article presents the results of developing a program that allows determining the reliability indicators of individual elements of the electric power system. Distribution electric power systems perform an important function in the overall energy structure. That is, the reliable operation of individual elements of the power system depends on the uninterrupted transmission of electricity over air lines to end users. An example of calculation is given for Sevkavkazenergo IDGC of the North Caucasus. The summary table shows the values of reliability indicators of the elements of the energy system under study, distributed over the years and used for reliability analysis and forecasting. Using the data from this table, regression models can be built, and the adequacy and reliability of the forecast are estimated based on known methods of factor-regression analysis. To do this, it is necessary to develop appropriate programs that meet the tasks set. It was found that the distribution of the failure rate among them is not the same, so about 20% of the lines on average were disconnected 2–3 times a year for different reasons and for different durations, including the time when the automation was switched off. More than 30% of air lines were disconnected no more than 2–3 times during the entire study period of 2011–2017.

Keywords: Distribution · Power · System

1 Introduction

The problem of reliability in the modern world is a key one that determines the safe conditions for the existence of mankind. The problem of reliability of electric power systems should be highlighted as one of the main aspects of ensuring energy security in the world.

One of the main consumers of electricity is industrial enterprises that contain a large amount of electric load. Reliability of power supply of industrial enterprises is determined by the probability and frequency of occurrence of events in the power supply system, in which the functioning of electric motors that drive technological installations is disrupted. At the same time, violation of technological processes is possible not only

V. Murgul and V. Pukhkal (Eds.): EMMFT 2019, AISC 1258, pp. 777–790, 2021.
https://doi.org/10.1007/978-3-030-57450-5_66

with long power interruptions, but also with voltage dips, which are usually not taken into account when analyzing reliability.

Ensuring high reliability of electric power facilities and systems is the main requirement for them. Reliability of power supply systems of enterprises is one of the main tasks, both in their design and operation. Violation of the power supply system of many technological processes can lead to a complete shutdown of the enterprise, as well as loss of life [1–5].

The accumulation of failure statistics for electric power system (ES) elements and the calculation of reliability indicators is performed on the basis of a database. A large number of relational database management systems (DBMS) have been developed for personal computers, but most of them are designed to perform specific tasks. In practice, unified DBMS such as FOXPRO, CLIPPER, PAPADOX, WORKS, etc. are often used. However, the use of universal DBMS has not only well-known advantages, but also disadvantages, such as reducing the performance and increasing the required amount of RAM when using DBMS and application programs for calculating power supply reliability indicators.

In the energy sector, element failures are rare events and average values for a given period of operation are calculated to quantify the time to failure. This requires the accumulation of failure statistics for similar elements (power lines, transformers, high-voltage switches, bus sections, power supplies) over the study period. The longer this time period, the greater the volume of statistical data and the higher the compliance of the calculated data with the reliability indicators of the EPS elements with the actual conditions of their operation. However, when constructing regression models of reliability indicators for EES elements, it is necessary to take into account the aging of information and distortion of trends in their predicted values. Calculations carried out by many researchers show that the optimal period for studying the reliability of elements can be considered a period of 6–8 years.

2 Methods and Approaches for Developing a Program for Registering Failures of Power System Elements

The Department of "Power supply of industrial enterprises" of SKSMI (GTU) has collected a large number of statistics on the reliability of individual elements of electric power systems [6–9].

With the development of computer technology, it became possible to register every failure of a multi-element system, store and accumulate a database for calculating reliability indicators. The Department "Power supply of industrial enterprises" has developed a program ("Hermes") for registering failures of elements of electric power systems up to 1 kV and higher. For the convenience of users who have little experience working on a computer, working with the program is built in the form of a dialog. When starting the program, the service line offers the main menu for selecting the operating mode (Fig. 1): working with the database or determining reliability indicators for any period of time defined by the user.

1. Data base 2. Reliability index 3. Exit

1. Registering a new device
2. Add data
3. To edit the data
4. General viewing

Fig. 1. Operating mode selection Menu.

1. Registration of a new device is necessary to generate the code for each element, so that the program can find similar elements in the future (for example, VL-110 kV, VL-35 kV, TR-110 kV, etc.). When you register an item indicates the date of commissioning, in any climatic area of operation, which sub-station is connected, the voltage level, item type, availability of devices and seed number among elements of the same type (Fig. 2).

The parameters of the element

Area code: ____________

Substation number: ____________

Element type: ____________

Availability of automation: ____________

Sequential number of the element: ____________

< 1. Yes > < 2. No >

Fig. 2. Registration of an element of the studied electric power system.

2. The "Add data" Command is executed when you enter the date of the current failure and recovery, or the date when the element was intentionally disabled, for example, for conducting preventive tests, and put it into operation. In this case, a menu similar to Fig. 2. When you call the element appears in the database for this item.
3. The "Edit data" Command may be necessary for erroneous entries of element parameters and for viewing the failure time and recovery time of this element.
4. The "General view" Command is necessary to visually view the number of registered items and their codes, the date of commissioning, and the reasons for failures.

3 Method of Distribution of Power System Elements by Climate Regions

When registering an element, the program offers to fix the selected climate area code. This area code can be represented as any single digit that the user wants, so that when analyzing the reasons for failures, the climatic conditions of geographical areas can be distinguished. Similarly, enter the number of the substation that this element is connected to. In particular, for RSO-Alania and PJSC IDGC of the North Caucasus «Sevkavkazenergo» it is proposed to use the following codes for the regions of the Republic (Table 1).

Table 1. The codes used by the administrative districts of North Ossetia-Alania.

№	Administrative region of RSO-Alania	Area code
1	Vladikavkaz	1
2	Alagirsky district	2
3	Iraf district	3
4	Digor district	4
5	Ardon district	5
6	Kirovsky district	6
7	Mozdok district	7
8	Prigorodny district	8
9	Pravoberezhny district	9

The area code is defined by a single digit (set at the user's request for climate or administrative reasons). To identify climate impacts on the reliability of an element, zones with similar climatic conditions are identified on the map and each of them is assigned a number (code) to distinguish them. The substation numbers correspond to the serial number that is being maintained in the power system. The voltage level is selected from the menu (Fig. 3). The line Number corresponds to the voltage code.

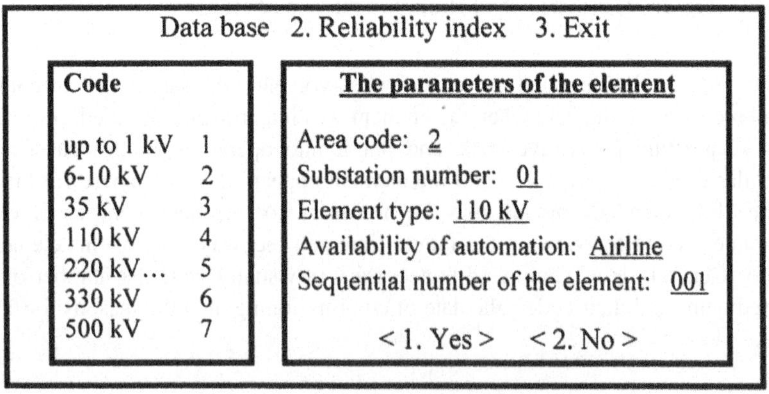

Fig. 3. Generating the voltage level code.

The item type is also selected from the menu (Fig. 4). The line Number corresponds to the item type code.

	Code
Cable line	1
Airline	2
Consumer	3
Transformer	4
Switch	5
Disconnector	6
The section of the busbar	7
The automation device	8
Power plant generator	9

Fig. 4. Typical elements of an electric power system.

The presence of automation is registered as an answer to the question: Yes - 1 or no - 0. The sequence number of the element is entered by the user in accordance with the accepted number on the electrical diagram of the studied EES. If you do not enter the district and substation number, the element is registered in the zero climate region, which is attached to the zero substation. In the future, you can edit the specified numbers by entering the menu (Fig. 1). There are two ways to represent the area code.

In the first case, each administrative region is assigned an ordinal number (Table 1). In this case, the selected code will be convenient due to the strictly administrative division by districts of the RSO-Alania. Another option is to code areas by climate factors (wind or ice load). This code is more consistent with the cause-and-effect relationships of assessing the reliability of the EPS elements (Table 2).

Table 2. Proposed codes for the RSO-Alania climate regions.

№	Climate the district of North Ossetia-Alania	Area code
1		1
2		2
3		3
4		4
5		5
6		6
7		7

The boundaries of districts by climate factors are indicated on the physical map of RSO-Alania. For example, let's register the failure of the VL-2 connecting the PS «V-1» and «S-Z», voltage 110 kV. Date of commissioning 2005, the last failure was 26.06.2013 12:01 (included in 12:49).

Initially, when registering VL-2, the 2nd climatic region was selected (Table 2) and the substation of connection B-1, code 01 (you could specify the PS «S-Z», code 07.

Sets the level of the rated voltage of the element to be entered into the data Bank (Fig. 2). The code corresponds to the menu bar, for example, for 110 kV 4th line, that is, the RA-ven 4 code. When you press «Enter» (Enter), and any attempt to record or to move to the next item, the program offers an extra menu (Fig. 3).

The type of item is offered from the menu (Fig. 4). Selecting the appropriate item type, e.g. «Air line» press on PC keyboard "Enter" and automatically-fixed item type and code (for VL code = 2, Fig. 4).

To increase the reliability of electric power system elements, automatic devices (APV, AVR, ARCH) are usually installed, so you can determine the average reliability indicators separately for elements with automatic devices and without them. This allows us to quantify the effectiveness of measures to improve the reliability of elements. For VL-2, the presence of an automation device characterizes code 1.

For elements of the Sevkavkazenergo power plant, three-digit numbers are assigned, analogous to the numbers on the electrical diagram of the studied power system, for example, for VL-2, the code of the serial number 002 (Table 3).

Table 3. An example of the numbering sequence of the elements of the EPS.

№	Element type	Item number
1	VL-I	001
2	VL-2	002
3	VL-3	003
4	VL-4	004
5	Transformer T-1 PS «B-1»	001

The greatest efficiency of ASDCPC is achieved when it is operated jointly with ASCME - an automated system for commercial metering of electricity (Fig. 2).

4 Method for Calculating the Time to Failure of Power System Elements

For each element, according to the program «Hermes», the result of calculating the time to failure is offered. To record the date of the current emergency failure or intentional shutdown, for example, for preventive tests, you need to enter the «Add data» menu (Fig. 1). To find any program item in the main menu, you can also select the «Add data» line and the program offers a menu (Figs. 1, 2, 3 and 4). Just like when registering an item, the program offers a form for recording the date of failure (if there is one) or deliberate disconnection.

If there were no previous failures before the current failure, the time to failure is determined from the installation date. The element repair time for the current emergency failure is calculated as the difference between the date of the current activation and the emergency shutdown [10–12]. Similarly, intentional shutdowns are recorded for preventive testing and scheduled maintenance (s_m) (t_{smi}). As indicators of element reliability, the time between two intentional shutdowns is calculated, as is the time between failures:

$$t_{smi} = t_{sm1} - t_{sm2}, \quad (1)$$

where: t_{sm1}, t_{sm2} – date of the current and previous scheduled repair or test.

These calculated values can be viewed by going to the «General view» menu, and the entire database is displayed, where each element is represented by a 9 – digit numeric code:

digit 1 – represents the conditional area code (ice, wind, flood, etc.).

digits 2, 3 – specifies the number of the substation.;

figure 4 – determines the voltage level (Fig. 3), for example, the number «4» corresponds to a voltage of 110 kV;

figure 5 identifies the type of the element (Fig. 4);

digit 6 – can only be 1 or 0, i.e. there is or is not an automatic device;

the numbers 7, 8, and 9 correspond to the item's ordinal number.

For example (Fig. 4), Table 4 shows the output information during the «General view» of the main menu (Fig. 1).

Table 4. Example of coding of elements of the EPS.

Element code	Installation date	Comments
101431120	21.01.77	Line break
201421001	01.01.51	MTZ
503431102	01.01.78	Is unknown

When viewing such a table, you need some experience, so, for example, in the second line, the code «201421001» is not very familiar to a person, but the program allows you to group elements by region, by connection, by element type, and taking into account the presence of automation devices. For the specified code, climate zone 1, substation 01 (V-1), 110 kV (4), overhead line (2), without automation devices (0), serial number 001 (VL-1).

Table 5. Example of entering statistical data.

№	A	B	C	D	E	F	I	J	K		
1	Number	Year of entry	Length, km	Number of supports	Mark the wires	Support material	Cable	Insulator	Is the wire	Located m/d phases	h dewlap's
2	VL - 1	1934 1971	35.84	206	AS-120 AS-185	Metal 83	c-50	PF-65 5298	Δ	3.0	12.5
3	VL - 2										
4	VL - 3										
5	VL - 4										
6	VL - 7										
7	VL - 8										
8	VL - 9										
9	VL - 10										
10	VL - 11										
11	VL - 12										

The proposed program for creating a database on the elements of the EES allows you to store and use passport data and statistics on both reliability and operating modes.

In accordance with the research goals, the developed database management systems (DBMS) are constantly being developed, that is, supplemented and debugged. Improving the program used leads to excessive complexity of the DBMS being developed and reduces their main advantage in performing specific tasks, so now they are trying to use universal programs. The main advantage of the developed program «Hermes» is the simplicity of the dialogue with the PC when entering statistical data by a specialist (Table 5), who has little experience working with a computer. The calculation results presented in tables are used for further analysis and development of measures to improve the reliability of power supply to consumers [13–15].

The disadvantage of this program, like any other non-universal program, is the complexity of the program and, consequently, leads to a lot of work. In addition, DBMS of this type are designed to perform only certain functions, in particular, the program «Hermes» is designed to store the passport data of elements and register emergency failures and intentional shutdowns for the entire period of operation. On the basis of these retrospective data on the reliability of EES elements, complex indicators of the reliability of power supply to its consumers are determined. Work is currently underway to further develop this program to analyze reliability in various climatic regions and the impact of operating modes on the reliability of power supply.

The level of computer technology development is currently quite high and the proposed programs allow performing mathematical calculations and various studies of any complexity. In particular, Excel software allows you to immediately enter both the passport data of elements and statistics on their failures in ready-made tables and each cell is used for further calculations and research. The cell format can be any with text and numeric and in the form of a date.

Passport data of 110 kV overhead lines and 110 kV transformers are entered in the cells of the spreadsheet. For example, Tables 6 and 7 show the distribution of cells for storing information and using it in further calculations. Only the initial recording of the elements ' passport data is performed manually, cells are Called, reliability indicators are calculated, predicted, and graphs are generated automatically.

Table 6. Creating a database on failures of EES elements in EXCEL.

№	*A*	*B*	*C*	*D*	*E*	*F*	*I*	*J*	*K*
1	Type								
2									
3									
4									
5									

Table 7. Data on the mean time to failure.

Years							
2011	2012	2013	2014	2015	2016	2017	2018
17520	3838	12622	25618	4523	13283	22068	38246
	0.006252	0.001901	0.000937	0.005305	0.001807	0.001088	
	0.0182577	0.438128					

For example, the database table (Table 6) in EXCEL. According to the researchers, in this table, the first column A is assigned for writing textual information by element types.in the future, all rows with such letters can be found by characteristic features (for example, VL-1). In columns B-F, enter dates. The remaining columns are reserved for getting calculated values.

The average time to failure for the entire period of operation in 2011–2017 is set by the formula for calculating the average value in cells (Table 8).

Table 8. Calculation of time to failure in EXCEL.

Year	Calculation formula	Computing
2012	«=C5-B2»	(2012-25.07.2011)·8760 = 3838 (hour)
2013	«=D5-B2»	(2013-25.07.2011)·8760 = 12598 (hour)
2014	«=E5-B2»	(2014-25.07.2011)·8760 = 21358 (hour)

5 Calculation and Forecasting of Reliability Indicators of the Elements of the ES (for Example, the 110 KV Electric Network) in the Program «Hermes»

The main indicators of the structural reliability of the electric network are the reliability indicators of the elements of the studied EES. At the Department of «Power supply of industrial enterprises» of SKSMI (GTU), statistics of failures of elements of 110 kV EES from 2011 to 2017 have been accumulated. Currently, «Sevkavkazenergo» operates 81 110 kV overhead lines and 33 PS with 61 transformers installed.

To quantify the reliability of elements of the 110 kV «Sevkavkazenergo» power plant, the failure flow parameter and recovery time were used. In the original database table, each line is searched for the number of failures of each line for each year and entered in a new table. Part of it is shown as an illustration (Table 9).

Using this table, you can calculate the relative failure rate of each line for the entire period of operation and the average value of the failure rate parameter for years for the 110 kV overhead line (number of lines N = 81).

Table 9. Failure rate of overhead line-110 kV of «Sevkavkazenergo».

№	*A*	*B*	*C*	*8*	*D*	*E*	*F*	*I*	*J*	K	L
1	Failure rate of overhead lines-110 kV by year									Average	Probability of failure
2	LEP	2011	2012	2013	2014	2015	2016	2017	General		
3	VL - 1	5	6	8	7	6	6	2	40	5.7143	0.6596
4	VL - 2	1	0	0	2	0	0	0	3	0.4286	0.0526
5	VL - 3	1	1	1	0	3	0	1	7	1	0.122
6	VL - 4	0	4	8	2	1	3	1	19	2.7143	0.3246
7	VL - 5	2	0	15	1	1	1	0	20	2.8571	0.3411
8	VL - 6	0	1	1	0	0	0	0	2	0.2857	0.0352
9	ω, 1/г										

For each line, the probability of failures is determined, assuming the simplest flow of failures:

$$Q = \frac{(\omega \cdot t)^k}{k!} \cdot exp(-\omega \cdot t) \tag{2}$$

where: t – is the study period, t = 1 year;

k – is the number of intervals, k = 7.

The disadvantage of calculating average reliability indicators is that the individual characteristics of each element and the forecast will not be objective enough. If there is a database for each element, a more objective assessment of reliability is the time to failure for each element.

To calculate the time between failures for each line, you must determine the time between two failures for each year. If there were no refusals in the current year, the difference between the date of the current year and the previous refusal is determined.

For an example of calculating the time to failure of VL-1, you must specify where to enter the calculation result (in this case, in cell F3). The failure time for this element will be equal to the date of the last failure (cell B3) minus the date of the penultimate failure (cell B2). Then it is enough to specify the calculation formula «=B3-B2» and the result will be entered automatically in the same date format or numeric format, for example:

1. pitch ("=B3-B2») = 26.06.2014 12:01-25.07.2011 01:30 = 01.11.2012 10:30
2. pitch ("=B3-B2") = 26.06.2014 12:01-25.07.2011 01:30 = 2,92 year (25550 h).

When registering all elements of the electrical network, you can calculate the current operating time for failure of each of them and the recovery time.

The time to failure of each element in EXCEL spreadsheets is determined by the specified formula in the cell where the reliability indicator is entered (Table 9). Line 5 is reserved for registering the current year of operation. In line 6 for the corresponding year, the calculation formula is entered, so, for example, for VL-2, the operating time for failure in 2011 is determined by the previous date of failure in 2009. Then enter the formula «=B5-2009» in cell B6, so that the time to failure is calculated in hours. Can change this formula by multiplying the number of calculated years by 8760 h. Then calculations are made automatically in this cell: (2011–2009).8760 = 17520 (hour).

The time spent on failure in other years is determined by the date of the previous failure, entered in cell B2 (25.07.2011 1:30: 00) and in 3 (26.06.2014 12:01:00). The recovery dates are entered in cells C2 and C3 to calculate the duration of the element's failure. The time to failure in L2 in the current years 2012, 2013, 2014 is determined by the specified formulas in cells C6, D6, E6.

When analyzing circuit reliability, the average values of elements are usually used regardless of their installation location, so for the same type of elements, the average annual reliability indicators are determined [16–18]:

- average time to failure:

$$\bar{t} = \frac{1}{N} \cdot \sum_{i=1}^{N} t_i; \tag{3}$$

- average recovery time:

$$\bar{t}_в = \frac{1}{N} \cdot \sum_{i=1}^{N} t_{вi} \tag{4}$$

The calculations typically use the inverse of the:

- average failure rate:

$$\lambda = \frac{1}{\bar{t}} \tag{5}$$

- the average intensity of the recovery:

$$\mu = \frac{1}{\bar{t}_{в}} \tag{6}$$

The calculated reliability indicators for each year are summarized in a General table, which can be viewed in the main menu "reliability Indicators".

In the summary table (Table 10), the reliability indicators are distributed by year, for the analysis and forecasting of the reliability of the elements of the studied EES. Using the data from this table, you can build regression models and assess the adequacy and reliability of forecasting based on known methods of factor-regression analysis. To do this, it is necessary to develop appropriate programs that meet the tasks set.

Table 10. Summary table of reliability indicators for 110 kV elements.

Years	Element type	Average reliability indicators			
		crash		deliberate disconnections	
		λ, 1/year	μ, 1/h	$\lambda_{в}$, 1/year	$\mu_{в}$, 1/h
2011–2017	Airline	0.23649	0.05579	0.863	0.1022
	Transformer	0.03996	0.01187	0.08635	0.0325
	Switch	0.01843	0.08865	2.092	0.2266
	The automation device	0.00188	0.223	–	–
	Power plant generator	0.05833	0.0312	0.442	0.0634

6 Conclusion

The distribution of failure rates among them is not the same, so about 20% of lines on average were disconnected 2–3 times a year for different reasons and for different durations, including the time of automatic operation. More than 30% of overhead lines for the entire study period 2011–2017 were switched off no more than 2–3 times.

References

1. Heylen, E., Deconinck, G., Hertem, D.: Review and classification of reliability indicators for power systems with a high share of renewable energy sources. Renew. Sustain. Energy Rev. **97**, 554–568 (2018). https://doi.org/10.1016/j.rser.2018.08.032
2. Rexhepi, V.: An analysis of power transformer outages and reliability monitoring. Energy Proc. **141**, 418–422 (2017). https://doi.org/10.1016/j.egypro.2017.11.053
3. Davidov, S., Pantoš, M.: Optimization model for charging infrastructure planning with electric power system reliability check. Energy **166**, 886–894 (2019). https://doi.org/10.1016/j.energy.2018.10.150
4. Afzali, P., Keynia, F., Rashidinejad, M.: A new model for reliability-centered maintenance prioritisation of distribution feeders. Energy **171**, 701–709 (2019). https://doi.org/10.1016/j.energy.2019.01.040
5. Rebello, S., Yu, H., Ma, L.: An integrated approach for system functional reliability assessment using dynamic bayesian network and hidden markov model. Reliability Engineering & System Safety **180**, 124–135 (2018). https://doi.org/10.1016/j.ress.2018.07.002
6. Klyuev, R.V., Fomenko, O.A., Gavrina, O.A., Sokolov, A.A., Sokolova, O.A., Plieva, M.T., Kabisov, A.A., Ikoeva, E.Yu.: Ensuring the consumer reliability based on retrospective analysis. IOP Conf. Series: Materials Science and Engineering 663, (2019). https://doi.org/10.1088/1757-899x/663/1/012033
7. Kortiev, L.I., Klyuev, R.V., Kulumbegov, R.P., Kortiev, A.L., Bosikov, I.I., Gavrina, O.A., Madaeva, M.Z.: Technical support of the power lines design - as a linear structure in difficult mountain conditions. IOP Conf. Ser.: Mater. Sci. Eng. **663** (2019). https://doi.org/10.1088/1757-899x/663/1/012034
8. Plieva, M.T., Gavrina, O.A., Kabisov, A.A.: Analysis of technological damage at 110 kV substations in JSC IDGC of the North Caucasus-«Sevkavkazenergo». In: International Multi-Conference on Industrial Engineering and Modern Technologies (FarEastCon), Vladivostok, p. 19229305 (2019). https://doi.org/10.1109/fareastcon.2019.8934076
9. Buryanina, N.S., Korolyuk, Y.F., Maleeva, E.I., Lesnykh, E.V.: Power transmission lines with a reduced number of wires in mountain territories. Sustain. Dev. Mt. Territ. **10**(3), 404–410 (2018). https://doi.org/10.21177/1998-4502-2018-10-3-404-410
10. Yssaad, B., Abene, A.: Rational Reliability Centered Maintenance Optimization for power distribution systems. Int. J. Electr. Power Energy Syst. **73**, 350–360 (2015). https://doi.org/10.1016/j.ijepes.2015.05.015
11. Heylen, E., Ovaere, M., Proost, S., Deconinck, G., Hertem, D.: A multi-dimensional analysis of reliability criteria: From deterministic N − 1 to a probabilistic approach. Electr. Power Syst. Res. **167**, 290–300 (2019). https://doi.org/10.1016/j.epsr.2018.11.001
12. Shukla, D., Arul, A.: A smart component methodology for reliability analysis of dynamic systems. Ann. Nuclear Ener. **133**, 863–880 (2019). https://doi.org/10.1016/j.anucene.2019.07.027
13. Klyuev, R.V., Bosikov, I.I., Mayer, A.V.: Complex analysis of genetic features of mineral substance and technological properties of useful components of Dzhezkazgan deposit. Sustain. Dev. Mt. Territ. **11**(3), 321–330 (2019). https://doi.org/10.21177/1998-4502-2019-11-3-321-330
14. Bosikov, I.I., Klyuev, R.V., Egorova, E.V.: Assessment of oil and gas potential prospects of the north eastern unit of the South Khulym deposit. Sustain. Dev. Mt. Territ. **11**(1), 7–14 (2019). https://doi.org/10.21177/1998-4502-2019-11-1-7-14

15. Zhukovskiy, Y., Batueva, D., Buldysko, A., Shabalov, M.: Motivation towards energy saving by means of IoT personal energy manager platform. J. Phys.: Conf. Ser. **1333**(6) (2019). https://doi.org/10.1088/1742-6596/1333/6/062033
16. Girón, C., Rodríguez, F., Urtasum, L., Borroy, S.: Assessing the contribution of automation to the electric distribution network reliability. Int. J. Electr. Power Energy Syst. **97**, 120–126 (2018). https://doi.org/10.1016/j.ijepes.2017.10.027
17. Tong, Y., Tien, I.: Analytical probability propagation method for reliability analysis of general complex networks. Reliab. Eng. Syst. Saf. **189**, 21–30 (2019). https://doi.org/10.1016/j.ress.2019.04.013
18. Cadini, F., Agliardi, G., Zio, E.: A modeling and simulation framework for the reliability/availability assessment of a power transmission grid subject to cascading failures under extreme weather conditions. Appl. Energy **185**, 267–279 (2017). https://doi.org/10.1016/j.apenergy.2016.10.086

E-trading: Current Status and Development Prospects

Olga Sushko[1(✉)] and Alexander Plastinin[2]

[1] Bauman Moscow State Technical University, 2nd Baumanskaya St., 5, bld. 1, Moscow 105005, Russia
osushko@mail.ru
[2] Northern (Arctic), Federal University named after M.V. Lomonosov, embankment of the Northern Dvina, Arkhangelsk 17163002, Russia

Abstract. The relevance of the research topic is associated with the active influence of the Internet on all areas of economy, politics and social life, emergence and development of a new electronic economy. Along with fundamental factors, the development of e-commerce is accelerated by the current negative trends in Russia and in all countries of the world connected with the pandemic of the coronavirus infection, which negatively influences many industries and entire sectors of the economy. The object of this study is e-trading in Russia and the key aspects affecting its functioning. The main goal of the first stage of the comprehensive study is to analyze the current state of the e-trading market, assess risks and potential development opportunities of models and the market as a whole. The article presents the results of the statistical analysis of the time series of e-trading, as well as some data of the econometric analysis. The results of the correlation and regression analysis help us determine the relationship between the GDP, retail and e-trading; correlation, elasticity and β-coefficients were calculated. The article shows preliminary calculations of the forecast dynamics of e-trading. The concept and content analyses were carried out to systematize international best practices and modern domestic practices of developing information technologies in the field of e-commerce. Due to the factor analysis the article discusses factors affecting e-trading. The author conducted the primary factor selection in relation to the development conditions of the e-commerce market and its segments.

Keywords: E-commerce · E-trading · Content analysis · Statistical analysis · Factor selection · Expert analysis of factor matrices · Forecasted dynamics of e-trading

1 Relevance, Subject and Object of the Research

The relevance of the research topic is significantly increasing in connection with the growing Internet needs of subjects in all areas of the economic and social life. The role of the Internet in all areas of the economy, politics and social life is increasing, and is the source of the emergence of a new electronic economy, which is determined by the creation of innovative opportunities for all entities.

V. Murgul and V. Pukhkal (Eds.): EMMFT 2019, AISC 1258, pp. 791–805, 2021.
https://doi.org/10.1007/978-3-030-57450-5_67

According to analysts from Statistica and the Eurasian Economic Commission [1], e-trading in the global economy accounts for 15% of traditional retail with annual growth rates of more than 20% per year. Short-term forecasts of Statistica analysts are positive: in 2021, e-trading will reach the level of 17.5%, and of 22% in 2023 [1, 2]. These analytical forecasts, in our opinion and in the opinion of many analysts, can be considered rather conservative due to the emergence of the new drivers. The catalyst was the current negative events associated with the coronavirus pandemic, adversely affecting many industries and entire sectors of the economy in Russia and in all countries of the world. The object of this study is e-trading in Russia and the key aspects affecting its functioning and development. The main goal of the first stage of the comprehensive study was to assess the factors and potential development opportunities of e-trading based on the retrospective analysis and modeling of the development directions.

The recent negative events associated with the global coronavirus pandemic affecting all countries have negatively influenced many industries and entire sectors of the economy. There is already a large-scale decline in revenues in the sectors most affected by the pandemic and quarantine measures such as passenger transportation, consumer services (food, tourism, trade, entertainment services) and manufacturing. The negative situation revealed promising directions for the development of government-to-consumer, business-to-consumer, consumer-to-consumer, peer-to-peer, consumer-to-business-to-consumer segments of e-business, and entrepreneurs received a development driver in the form of efficient application of IT. If earlier IT played only a supporting role for many industries and the state and was used optionally, now it is becoming a vital necessity. Internet technologies make it possible to continue working, doing sport, providing oneself with food and other essential goods, communicating and relaxing. With its help you can both continue working even during quarantine and restrain the development of the epidemic. IT innovations can help withstand the economic downturn and even fight the pandemic. Especially significant growth rates are observed in the e-trading segment. Thus, according to the National Association for Distance Selling of the Russian Federation, in the first half of 2020, the Russian e-trading market, especially the food segment, was growing at an accelerated pace due to the coronavirus pandemic.

According to the experts of Enterprise Technology Research [3], a change in the structure of spending on information technology is expected in the near future. We agree that expenses for capital expenditures on IT equipment, pilot projects, and the services of service providers can be reduced. On the other hand, those industries that are now demonstrating the ability to function even in the conditions of remoteness of employees and remote work of counterparties will increase the expenditures on Virtual Private Network, which provides a logical network, other IT infrastructure and information security tools necessary to ensure the protection of communications for remote workers, remote users.

Government agencies and enterprises have stepped up in the development of Internet services. So, the Ministry for Digital Development, Communications and Mass Media of the Russian Federation launched the All. Online portal in March this year with Internet services for work, study, entertainment and leisure. Currently, a state project Access to All is being implemented in conjunction with the commercial organizations, which also offers free or cheap entertainment services, food delivery. These are just a few projects that have great development prospects.

Thus, e-commerce is not a new subject of research, although, unfortunately, it has not been studied enough. The relevance of this topic is currently growing. The rapid growth of e-commerce is predicted, which can be explained by the fact that it is considered to be one of the ways to increase the efficiency of small and medium-sized businesses, large corporations and government agencies, and it acquires strategic importance in many aspects under conditions of the pandemic.

2 Tasks and Research Methods

The topic under consideration is relatively new from the point of view of coverage in the scientific literature. Thus, analytical reviews and sources of the leading analytical Russian and foreign companies are mainly taken for the analysis. With regard to statistics, the work provides estimates of the international organizations (UNCTAD, IMF) and statistical agencies (eMarketer, AliResearch, Ecommerce Foundation, Euromonitoring) [3–5]. Objective data on the status, trends and development prospects of the Internet in Russia are presented in the annual industry reports of the Federal Press and Mass Communications Agency [6] and the Russian Association for Electronic Communications (RAEC).

This study is at its initial stage. At the current stage, the main theoretical objectives of this study are as follows:

- to clarify the contours of the field of research, conduct a discursive analysis of the concepts on the research topic,
- to conduct a concept analysis and content analysis of the regulatory framework for the provision of e-commerce and e-trading,
- to determine the current state of e-trading in Russia, show the difference from global trends,
- to conduct an initial factor selection in relation to the development conditions of the e-commerce market and its segments, develop matrices for selecting priority factors, conduct a quantitative assessment of factors,
- to highlight the preliminary prospects for the development of e-trading in Russia.

The study is relevant and useful to people involved in tasks related to the development of digital products and those who provide their services in the B2B, B2C and C2C market segments, as well as to entrepreneurs and owners of small and medium enterprises. Since it is these e-commerce models that are rapidly changing today, they make up most of the transactions.

In the process of carrying out this study, the methods of empirical and theoretical research were used: a discursive analysis of concepts on the research topic, a content analysis of normative legal acts and standards that regulate the scope of activity; a monographic survey and content analysis with systematization of publications of best foreign practices and modern domestic practices in the field of IT development; an organizational and economic analysis and modeling of the competitiveness of Internet services. To solve the tasks of the scientific analysis, special methods were used:

registration of facts and events in the process of observation, expert interviewing and questioning of users; data collection based on the results of continuous and selective observation; processing and analyzing e-database data with its description and classification; system analysis.

A comprehensive statistical analysis was carried out, from calculating absolute and relative indicators to correlation and regression analysis, as well as preliminary forecasting using the analysis of Excel tabular data and the "PREDICTION" and "TREND" functions in Excel based on existing numerical values, etc. Determining the relationship between the GDP, retail and e-trading was carried out on the basis of correlation and regression analyses; the correlation coefficient was calculated. Elasticity and β-coefficients were determined to assess the bond strength of the factor with the result. Mandatory procedures were performed to check the quality of the regression equation; t-statistic and Student's t-test were applied. Darbin-Watson and Spearman tests were used to verify autocorrelation of excess and heteroskedasticity. In the study, the SPSS and Statistica software packages were applied to process data arrays and conduct the statistical analysis.

3 Preliminary Results of the Concept and Content Analyses

Active economic activity on the Internet or using its capabilities has led to the emergence of a new concept of e-commerce. Currently, e-commerce is a multifaceted phenomenon that requires close attention from researchers. E-commerce appeared in the 1960s, when the e-trading with airline tickets began. In the first publication called E-commerce [7], Doctor of Science D. Cozier from the USA mentioned that "e-commerce began with the sale, purchase and transfer of funds through computer networks". D. Cozier's definition is supported in a manual edited by the Minister of Communications and Information Technology L. Reiman stating that it is "the technology that provides a complete closed loop of operations… with the help of electronic means and information technology". However, today this concept has expanded significantly and includes trade with fundamentally new types of goods, for example, information in the electronic form. In Russia, the emergence of e-commerce is linked with two dates. The first Internet banking system Internet Service Bank was launched in 1998. A little earlier, in 1994 the first domain was registered, which is considered to be the date of the Internet creation in Russia. The following year, an automatic Internet terminal was created at the Moscow Interbank Currency Exchange to process applications for the purchase and sale of securities. Since the beginning of the 2000s, high growth rates of online stores and the development of the e-trading market was observed.

E-commerce and e-trading as new concepts have many controversial copyright and several official definitions. The official definition took shape in 1988 in the work of the World Trade Organization [8]: it is "… production, distribution, marketing, sale or delivery of goods and services using electronic means of communication" [8]. Later, many organizations submitted their definitions. Thus, the Commission of Trade and Industry Councils of the United States and Japan presented the following definition: "Entrepreneurship through electronic means of communication" [9]. The Ministry of

Economy, Finance and Industry of France gave its own interpretation: "The totality of operations for the electronic exchange of data related to the implementation of entrepreneurial activity: transfer of information and transactions with goods or services" [10]. The government of Canada believes e-commerce to be "buying, selling, and other transactions through telecommunications and computer technology. This includes transactions concluded using telephone, fax, ACM, credit cards, debit cards, EDI and the Internet" [10]. In the definition of the European Commission [11], e-commerce is presented "as remote services usually provided for remuneration by electronic means and at the individual request of the recipient of services" [11].

So, today there is no single standardized definition of this concept, but a generalization of numerous copyright and official definitions and approaches to identifying e-commerce shows that it is one of the types of commerce and, in a broad aspect, is an economic activity using information technology [10].

Another problematic issue is the synonymous use of the concepts of "e-commerce", "e-business" and "e-trading", which, according to IT experts, denote various phenomena. The broadest of the presented is the concept of "e-business", which includes "e-commerce", and, in turn, "e-trading". The last concept is the narrowest of the given ones.

When summarizing the official definitions of e-trading (the European Commission, the World Trade Organization, the Organization for Economic Cooperation and Development), the essence of this trade consists in organizing the process of commodity-money exchange in the form of sale and purchase based on electronic technologies. This concept is more widely interpreted by the UN Commission on International Trade Law (UNCITRAL) [10]: "e-trading is the organization and technology of selling goods and services electronically using telecommunication networks and electronic financial and economic instruments". The common similarity of multiple definitions is that they contain the main essence: purchase and sale, which is carried out electronically in a virtual mode, in contrast to e-commerce, i.e. e-trading does not cover the entire process of commodity-money exchange, but only the part related to purchase and sale.

In the legislation of the Russian Federation, the concepts of "e-commerce" and "e-trading" are distinguished. At the same time, the draft Federal Law of the Russian Federation "On Electronic Trading" provides another definition, which, in our opinion, is broader and is suitable as a definition of the term "electronic commerce". With regard to the regulatory framework of e-commerce and e-trading, the Russian legislation has not been fully developed. The regulation of the activities of entities in the field of e-commerce is carried out through the norms and rules of the Constitution of the Russian Federation, the Civil Code and some federal laws designed to regulate other areas of activity unrelated to electronic activities. So, e-trading is regulated by the main federal laws and orders:

- Federal Law No. 44 "On the Contract System in the Field of Procurement";
- Federal Law No. 94 "On the Placement of Orders for the Supply of Goods";
- The Order of the Ministry of Economic Development of the Russian Federation No. 54 "On Approval of the Procedure for Conducting Open Tenders".

To create a common base for conducting e-trading, a separate draft law of 2000, which has not been adopted yet, is being developed and the initiatives and discussions on the development of new laws are being actively undertaken. The main areas of normative regulation in recent years have been in the field of regulation of data turnover, social networks and content, anti-piracy and requirements for telecom operators. In recent years, many legislative initiatives have been proposed. Thus, the Association of E-trading Companies developed the "E-Trading Development Strategy for 2017–2018 and until 2025" and proposed to consider the main provisions concerning the activities of online trading platforms and aggregators, adopt a law on online retail and create a non-bank payment system for B2B.

Such type of e-commerce as social commerce is developing quite confidently due to the use of social networks, social and online media. This concept was introduced in 2005 by an American company that owns Yahoo Internet portal [9] to describe online means of user work on commercial websites. This term has been studied to a lesser extent, but generally it is the use of social networks for e-trading transactions. D. Biesel developed the concept of social commerce to designate users of advertising content on sites. Today, social commerce includes social applications and social marketing, reviews and ratings, recommendations and tools for social purchases, communities and forums, etc.

Thus, the question of the need to unify the conceptual apparatus in the field of e-commerce continues to be relevant and debatable, which is associated with national and international rule-making processes and effective legal regulation of e-commerce, requiring a common understanding of this phenomenon. Therefore, the legislators in many countries have codified only the most necessary terms that do not cause strong disagreements, such as an "electronic document", "electronic signature" and "electronic transaction", until a common understanding of the term "electronic commerce" is developed [9, 11, 12].

4 The Results of the Analysis of the E-Trading State in Russia

Despite certain barriers to the development of e-commerce and all its segments, Russian Internet industry occupies the leading positions in Europe in many respects. So, the experts of the analytical company GfK [13, 14] published that the number of Internet users in Russia in 2018 amounted to 90 million people and in 2019 to 93 million people" [1, 6] (Fig. 1), which significantly exceeds the number of users and the growth rate in Europe and some countries of the world. This positive indicator against the background of the negative demographic situation in Russia and the decline in the population over the past two years encourages some optimism. With a share of the Russian population of 1.9% of the total world population, the proportion of Internet users is higher and makes up 2.3% of the global Internet community.

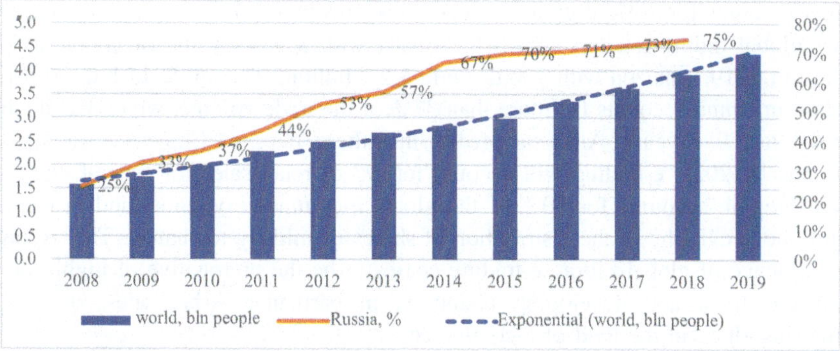

Fig. 1. Dynamics of the number of internet users in Russia and the world, source: euromonitoring, 2020.

Positive indicators of the growth of the Internet audience are confirmed by the annual industry report of the Federal Press and Mass Communications Agency of the Russian Federation [1, 6]: the number of Internet users in 2018 amounted to 93 million people (76% of the total population) with an increase of 6% to the level of 2017. A similar 6% increase in Internet users in 2018 was recorded in the world with a total of 4.38 billion people (67% of the world's population). The same data on the Internet audience is provided in the report on the global state of digital technologies We Are Social and Hootsuite for 2019 (4.39 billion people), but user growth was recorded at 9% compared to last year. Positive growth indicators (9%–10%) are observed in social networks and mobile Internet. The reports of various organizations indicate the number of mobile Internet users of 3.26 billion people in 2018 and 3.986 billion people in 2019. According to the Association of E-Trading Companies [6], the volume of e-trading in Russia in 2018 amounted to 1.66 trillion rubles, which is 59% more compared to 2017 [1, 6]. In previous years, the market showed moderate growth rates – not more than 20% per year. Such positive dynamics is explained by the contribution of small and medium-sized online sellers. The structure of Russian Internet users is dominated by women, whose share in the total number of Internet users is 52.5%. The age structure has a high level of Internet penetration in the group of 16–29 years old (97%) and 30–54 years old (82%). The smallest group consists of people over 55 years old (28%) [9]. The largest regions in terms of the share of e-trading are Moscow, Moscow Region and St. Petersburg.

The pace of the development of world and Russian e-trading shows growth rates that are faster than traditional trade, as recorded by the World Trade Organization. The benefits of the Internet as a sales channel are obvious to entrepreneurs around the world. According to the German analytical company Statistica [14], the global volume of e-trading (B2C) in 2019 amounted to more than 14% of sales, and according to forecasts it will reach 17.5% in 2021 and 22% in 2023 respectively. Given that a quarter of the world's population is currently shopping online, these forecasts seem reasonable, and even a bit conservative. The global volume of e-trading (B2C) according to the last year data is estimated at 1,850 billion US dollars [14, 15]. Today,

China accounts for 40% of global online trading revenue ($740 billion), and it has the main global online retailers [15–18]: Alibaba with a gross value of goods at $768 billion in 2019, Amazon with goods worth $239 billion, ID.com ($215 billion) [6]. E-trading in China accounts for more than 33% of the trade volume, which has doubled for the last five years. Analysts predict that these rates of Chinese e-trading will continue by 2023, e-trading will account for 2/3 of retail sales. In general, the Asia-Pacific region accounts for 80% of the global e-commerce volume, and in the near future the geography of the distribution of shares is unlikely to change. That is, under the influence of globalization, e-trading ceases to be the prerogative of highly developed countries, and developing countries, in particular Asian ones, occupy an increasing share in the field of e-commerce.

The United States ($561 billion) and England ($93 billion) are also among the three leaders [6, 14]. Significant online turnovers take place in other European countries: Germany ($77 billion) and France ($55 billion). Japan ($87 billion) and South Korea ($69 billion) also have large market shares in global e-trading [6, 18, 19].

According to the Federal State Statistics Service of the Russian Federation [6], online sales have grown significantly in recent years. The volume of e-trading in 2019 amounted to 19 billion US dollars. This indicator, of course, is not comparable with the Chinese or European volume of e-trading. So, in Germany e-trading makes up 9.5%, and excluding food – 15.3% [6]. Nevertheless, the share of e-trading in Russia is currently approaching 5% of the total retail turnover, although for a long time, given the population, it has been less than 2%. Thus, the current trends in the development of e-trading and their extrapolation will help determine positive results in the near future and predetermine the "digital potential" in the projected future of Russia. In addition, in Russia, the e-trading market is at the stage of formation and at the present stage of development it is possible to identify the Russian leading Internet retailers.

There are many conservative and optimistic short-term forecasts of various statistical organizations and analytical agencies, which were mentioned above, but unexpected long-term forecasts of e-commerce have appeared as well. So, Nasdaq has predicted a complete transition to e-trading by 2040 [1, 6].

5 Market Development Dynamics

For analysis, a database of retail and e-trading indicators in Russia has been compiled from 2006 to the present. To determine the general trends, a comparison with GDP dynamics over the same period was carried out (Fig. 2). In order to process and analyze the data, the frequency analysis and statistical calculation of the studied indicators (arithmetic mean, variance, mean linear deviation, quadratic deviation, coefficient of variation, etc.) were performed. At the stage of quantitative assessment of the dynamics of retail and e-trading, different types of statistical analysis (correlation, variance, regression), econometric analysis and modeling were used. In the study, the SPSS and Statistica software packages were applied to process data arrays and conduct the statistical analysis.

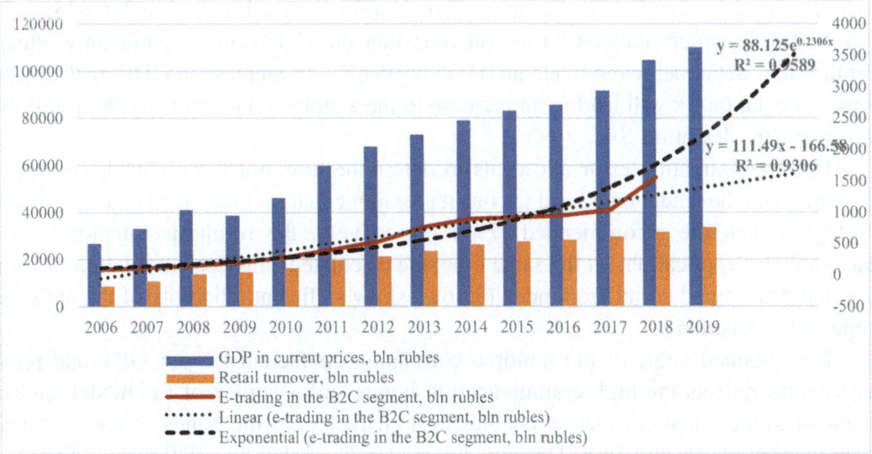

Fig. 2. Dynamics of the GDP, retail and e-trading, source: Rosstat, 2020.

The graphical analysis of the dynamics of the GDP, retail and e-trading shows unidirectional trends, which led to the formation of a hypothesis about the close relationship of the studied trends, as well as the impact of GDP on the development of retail and e-trading.

The author studied the dependence of the development of retail and e-trading on the GDP dynamics. At the specification stage, paired exponential regression was chosen. Its parameters were estimated by the least squares method. The statistical significance of the equation was verified using the determination coefficient and Fisher's ratio test. Elasticity and β-coefficients were determined to assess the bond strength of the factor with the result. Mandatory procedures were performed to check the quality of the regression equation; t-statistic and Student's criterion were calculated. Darbin-Watson and Spearman's tests were used to verify autocorrelation of excess and heteroskedasticity.

The analysis of the field of correlation of the values of the general populations in GDP and retail trade in Russia from 2006 to 2019 allowed us to put forward a hypothesis about the exponential nature of the relationship between all possible values of the population. For our GDP and retail values, the system of empirical regression coefficients b = 1.5E−5, a = 8.9032 of the equations has the form of:

$$14a + 942013 \cdot b = 138 \cdot 887 \qquad (1)$$

$$942013 \cdot a + 72734905243 \cdot b = 9486568.209 \qquad (2)$$

The regression equation (empirical regression equation) is as follows:

$$y = e^{8.9032247111139} e^{1.5E-5x} = 7355.6552 e^{1.5E-5x} \qquad (3)$$

where x is the GDP value,

y is the retail value.

The elasticity coefficient was more than 1 and amounted to E = 67286.643 (1.5E −5) = 1.017, which allowed us to conclude that the GDP value significantly affects retail trade. Beta-coefficient made up 0.966 or 96.6%. An increase in GDP by the mean square deviation Sx will lead to an increase in the average retail value by 96.6% of the mean square deviation Sy.

The approximation error allows us to determine how much the calculated values deviate from the actual ones, and for our regression equation this error is 9.07%, which is higher than the recommended (7%), respectively, the resulting equation is not advisable for application. At the same time, the calculated empirical correlation relation and the determined correlation index (95.6%) show sufficient reliability of the obtained regression equation.

The obtained value of the multiple correlation coefficient for our GDP and retail aggregates reflects the high communication bonds and accuracy of the model, and in contrast to the linear correlation coefficient it characterizes the bonds of the nonlinear coupling but not its direction. The data obtained indicate that the GDP factor affects the size of retail trade significantly.

Evaluation of the parameters of the regression equation, the calculation of the regression error variance (S^2 = 5957157.276) and the standard estimation error as a measure of the dispersion of the observation data from the simulated values (S = 2440.73) allow us to better determine the quality of the obtained model.

Autocorrelation and testing for heteroskedasticity ($H_0 = 2.179 > 0.6$) showed their absence, which is one of the conditions for the effectiveness of the regression model, and that it cannot complicate the interpretation of the regression results, since the value of the output variable depends not only on the magnitude of the input change, but also on of relative to what magnitude this change occurs.

Next, the correlation analysis of the values of the GDP aggregates and e-trading was carried out. Also, based on the correlation field for the general set of two time series of values determined at the specification stage, a paired exponential regression was selected, and the empirical regression equation was made:

$$y = e^{3.6811970975268} e^{3.8E-5x} = 39.69388 e^{3.8E-5x} \tag{4}$$

where x is the GDP value,

x is the importance of e-trading.

The calculated elasticity coefficient E = 60611.417 (3.8E−5) = 2.307 is greater than 1, which shows that the GDP value significantly affects the volume of e-trading. Beta-coefficient made up 98.4%, showing that an increase in GDP by the mean square deviation will predetermine an increase in the average value of e-trading of the same period by 98.4% of the mean square deviation.

The approximation error showing the deviation of the calculated values from the actual ones was 13.08%, which also indicates that it is undesirable to apply the obtained equation and other types of regression equations should be developed. At the same time, the correlation index amounted to 96.1%, which indicates a close relationship of the considered dynamic series and high reliability of the regression equation. The multiple correlation coefficient we obtained when constructing a one-factor correlation model, reflecting the bond of the nonlinear coupling and the accuracy of the model, is close to the pair correlation coefficient and varies within [1].

The determination index, which shows the accuracy of the selection of the regression equation, is slightly lower than in previous calculations and is equal to 92.28%. The remaining 7.72% of the change in the value of e-trading is explained by factors not taken into account in the model (as well as specification errors).

The calculated confidence intervals of the regression coefficients with a reliability of 95% will be as follows:

$$\begin{gathered}(b - t_{crit}\,S_b;\; b + t_{crit}\,S_b)\\(3.8\,E-5-2.228*0.00142;\; 3.8\,E-5+2.228*0.00142)\\(-0.00313;\; 0.0032)\end{gathered} \tag{5}$$

That is, it can be expected that with a probability of 95%, the e-trading value will be within the found interval.

Based on the determination of the value of F statistics (F = 119.609) and Fisher's criterion and their comparison with the table value (F_{table} = 4.96) with degrees of freedom $k_1 = 1$ and $k_2 = 10$, the actual value $F > F_{table}$. Thus, the coefficient of determination is statistically significant, and the estimation of the regression equation is statistically reliable.

A similar analysis was carried out for the general populations of retail and e-trading. The specification and calculation of the regression equation was carried out. Its parameters are estimated by the least squares method. The statistical significance of the equation is determined using the determination coefficient and Fisher's criterion. It was found that 92.3% of the total variability of the dynamics of e-trading from 2006 to 2019 is associated with changes in retail trade. It was also established that the model parameters are not statistically significant, and therefore it is necessary to further determine the statistically significant model.

Thus, the results of regression and correlation analyses allow us to judge the unidirectional nature and stability of changes in the initial time series. Even the exponential regression equations of the studied dynamics have close coefficients. However, it was revealed that there are differences between the growth rates of the GDP, retail and e-trading. So, with moderate growth or a slowdown in the GDP in certain periods from 2006 to 2019, the growth rate of e-trading remained at a high level, which demonstrates the influence of other factors and the need for a thorough factor analysis, the first stage of which included the determination of the main factors.

6 Factor Analysis Results

The development of the Internet and e-commerce leads to significant changes in the economy and the traditional tenets of economic theory and practice. The development of e-commerce and all its segments is associated with fundamental macroeconomic indicators. At the same time, an analysis of the dynamics of retail and e-trading shows that along with unidirectional movement there are significant differences in growth rates and cyclicality, which is determined not only by macroeconomic factors, but also by new determinants. Information and communication technology is the main driver of globalization. Technologies and standards for data transfer via the Internet have become a universal means of exchanging commercial information and have largely determined the principles of doing business in the field of e-commerce. Using the Internet transforms the processes of economic interaction between companies and their customers, partners and suppliers. Methods and means of conducting commercial transactions in the field of e-commerce are somewhat different from those in the real economy. Significant factors in the development of e-commerce and the dynamics of e-trading are similar in all countries: technical innovations, growth of professionalism, computerization and mobilization of the population with the growth of technical opportunities for customers to access web shops, scaling of Internet networks and market penetration. According to GfK, more than 75% of Russians have access to the Internet, and about 60% of users go online using smartphones. Thus, the spread of smartphones has led to the fact that approximately 60% of purchases are made by Russians using mobile applications.

We can distinguish the main groups of factors for functioning and prerequisites for the favorable development of e-commerce and its segments: socio-economic, technical and technological, information and communication, demographic ones. All factors are interrelated, which leads to a synergistic effect. Such results as the expansion of Internet networks, computerization and "mobilization" of the population are due to the general interaction of all these groups of factors.

The market environment for e-commerce and all its segments is changing under the influence of many factors, the number of which can be significantly limited by means of logical analysis. Five main groups of factors were selected (Table 1), for each of

Table 1. Fundamental factors.

Group of factors	Fundamental factors
Technical and technological	Implementation of new equipment and technologies Equipping organizations with specialized digital equipment Internet infrastructure development Lower infrastructure access costs
Socio-economic	GDP growth Real income growth Formation and structuring of the e-commerce market Increased competition
Demographic	Population growth Life extension Urbanization increase Better computer literacy
Information and communication	Increase in activity and mobility of the population in social networks Greater activity of organizations in online space
Uncertain	Pandemics Environmental and climate issues

them matrices were made to determine external priority factors. At this stage, the intercorrelation coefficients were determined, which made it possible to exclude dependent factors from the model, since the independence of factors is an important condition for the development of the model. The expert analysis of the matrices shows the planned probability of the appearance of external strategic factors and the degree of their potential impact on the development dynamics.

To quantify the influence of the factors, the hierarchy analysis method was used. As a result, weight coefficients were obtained, the sum of which is equal to one (or 100%) [11]. Due to the different size of the factor groups, the weight coefficients were reduced to a single scale and were multiplied by the intergroup weight coefficients [11]. As a result of the analysis, the parity of one group of factors was not revealed, which is explained by their high significance and a direct impact on the effectiveness of Internet retailers (Fig. 3) [11].

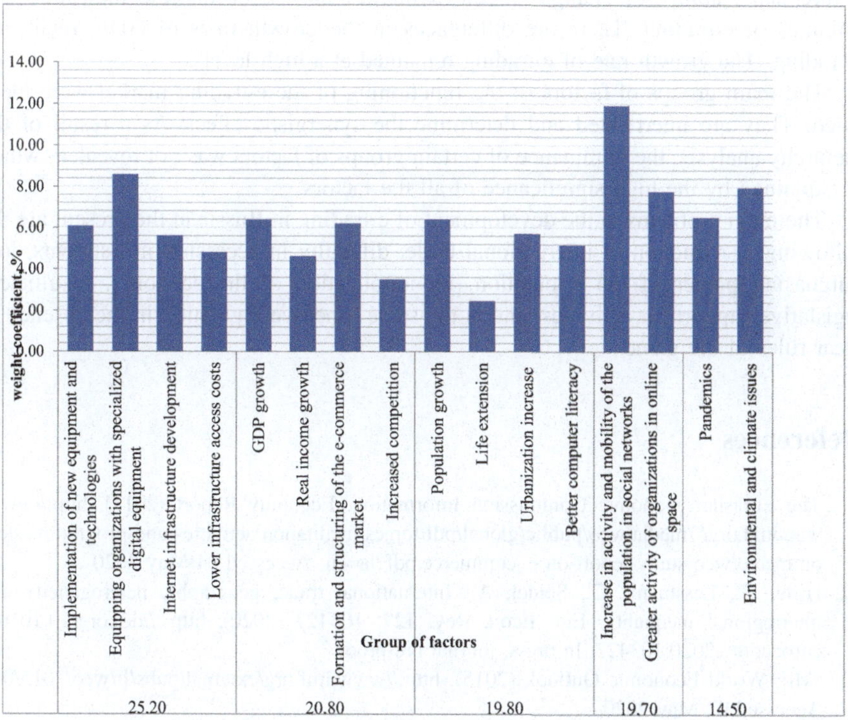

Fig. 3. E-trading development rankings.

The main tasks of the next stage of work is the formation of an integrated socio-economic model, which will allow for a systematic analysis to assess and predict the state of the main factors in the development of B2B, B2C and C2C models of e-trading.

7 Conclusion

Digital support of any activity today shows exceptional importance for business, for the social life of families and society as a whole. Organizations and the population operate in the context of globalization and integration of industries and countries, development of digital technologies, transformation of social networks and increase of population mobility. All this creates favorable conditions for the development of e-trading, and at the same time transforms the economy and society.

As a result of fulfilling the tasks of the first stage of the comprehensive study, the following main results were obtained, necessary for the next stage of the study.

Russia is the largest country in Europe by the number of Internet users, and the number of users is growing steadily due to areas where network coverage is still relatively small, including rural population [5].

Regression and correlation analyses made it possible to judge the unidirectional nature and stability of changes in the dynamics of macroeconomic indicators and volumes of e-trading. There are differences in the growth rates of GDP, retail and e-trading. The growth rate of e-trading remained at a high level.

The main groups of factors of the functioning of the e-trading market were identified. They are interrelated and determine the synergistic effect. As a result of the hierarchy analysis, the dominance of certain groups of factors was not revealed, which is explained by the high significance of all the factors.

The main problems in the development of e-trading in Russia at the present are the following: a reduction in international trade, difficulty in accessing investments, low purchasing power of the population, monopolization of the economy, insufficient legislative support for activities, and at the same time overregulation in the absence of clear rules of the game.

References

1. The Eurasian economic Commission Information. Economy Report (2017). http://www.wcoomd.org/-/media/wco/public/global/pdf/topics/facilitation/activities-and-programmes/ecommerce/wco-study-report-on-e_commerce.pdf?la=en. Accessed 04 May 2020
2. Hirte, G., Lessmann, C., Seidel, A.: International trade, geographic heterogeneity and interregional inequality. Eur. Econ. Rev. **127**, 103427 (2020). https://doi.org/10.1016/j.euroecorev.2020.103427, In press, journal pre-proof
3. IMF: World Economic Outlook (2015). http://www.imf.org/external/pubs/ft/weo/2015/01/. Accessed 02 May 2020
4. UNCTAD Information Economy Report (2015). http://unctad.org/en/PublicationsLibrary/ier2015_en.pdf. Accessed 29 Apr 2020
5. Emarketer Asia-Pacific Is Home to Majority of World Retail Ecommerce Market (2015). http://www.emarketer.com/Article/Asia-Pacific-Home-Majority-of-World-Retail-EcommerceMarket/1013352. Accessed 05 May 2020
6. Federal Agency for press and mass communications. Internet in Russia. Forward Print, Moscow (2019)
7. Kozye, D.E.: Commerce: TRANS. from English-Moscow. Russian Edition (1999)

8. Measuring electronic business definitions, underlying concepts, and measurements plans. Government of the United States of America (1999)
9. Directive 2000/31/EC of the European Parliament and of the Council of 8 June 2000 on certain legal aspects of information society services, in particular electronic commerce, in the Internal Market (Directive on electronic commerce). Official J. L **178**, 0001–0016 (2000)
10. Official site Canadian e-Policy Resource Centre (CePRC) (1998). http://www.ic.gc.ca/eic/site/ceprc-ccrcp.nsf/eng/00025.html. Accessed 27 Apr 2020
11. Joint Statement on Electronic Commerce, Committees of U.S. – Japan Business Council, Inc. and Japan. Business Council, U.S (1999)
12. Cybersecurity Law Cyber Security Law of the People's Republic of China (Draft) (2015) . http://www.npc.gov.cn/npc/xinwen/lfgz/flca/2015–07/06/content_1940614.htm. Accessed 18 Mar 2020
13. Frost, S.: The Global B2B E-commerce Market Will Reach 6.7 Trillion USD by 2020 (2015). http://ww2.frost.com/index.php?cID=10746. Accessed 18 May 2020
14. Une nouvelle donne pour les consommateurs, les entreprises, les citoyens et les pouvoirs publics, Rapport Du Groupe De Travail Preside Par M. Francis Lorentz, Ministere de l'Economie, des Finances et de l'Industrie (1998)
15. China Daily China Completes Drafting E-commerce Law, 3 October 2016. http://europechinadaily.com.cn/business/2016-03/10/content_23812316.htm. Accessed 18 Feb 2020
16. Lichtenstein, F.: New framework for online sales in China – two sets of regulations take effect. E-Commerce Law and Policy (2014)
17. Schaub, M., Bing, C., Huckerby, M.: New challenges in China cross-border e-commerce. Lexology (2016)
18. Ali Research Global Cross Border B2C E-commerce Market 2020: Report Highlights & Methodology Sharing (2016). http://unctad.org/meetings/en/Presentation/dtl_eweek2016_AlibabaResearch_en.pdf. Accessed 15 Apr 2020
19. GSPANN Europe and Asia Continue to Fuel E-Commerce Growth in 2015, 1 June 2015. http://blog.gspann.com/europe-and-asia-continue-to-fuel-e-commerce-growth-in-2015/. Accessed 03 May 2020

Model of Sustainable Economic Development in the Context of Inland Water Transport Management

Svetlana Borodulina(✉) and Tatjana Pantina

Admiral Makarov State University of Maritime and Inland Shipping,
5/7, Dvinskaya Str., Saint-Petersburg 198035, Russia
piter00000@mail.ru

Abstract. The present paper has worked out logical structures about ways to achieve goals of sustainable development of the Russian Federation in terms of branch administration. The study attempted to complete the analysis and integration of goals of sustainable development in accordance with the UN's methodology, and development priorities elaborated by the Analytical Center for the Government of the Russian Federation for the purposes of Russia's economy. Analysis of tasks of national economy development was performed in terms of the role and value of inland water transport development on the basis of current branch-wise and socioeconomic trends of 2018–2019 on their way to 2024 and then to 2030. The present study is aimed to investigate several urgent issues of advancement of inland water transport sub-branch in conjunction with solving problems of Russian economy development. Another aim is to give a broad outline of a model of Russia's sustainable development as a unity of social, economic and ecological realms on the basis of priority of achieving strategic aims of national economy development. Strategic advantages of inland water transport in terms of socioeconomic and environmental parameters were described. Nowadays, transport is becoming both a tool for implementing state interests and an object of national development projects. The study defined factors of formation and components of the effect of measures implemented by the government to support projects of inland waterways development. The paper gives an image of the outline of Russia's sustainable development in terms of management of inland water transport facilities.

Keywords: Model · Economy · Inland water transport · Industry management

1 Introduction

In the modern world, raising population's welfare, employment rate and quality of social sectors of the economy is one of the most urgent tasks for all countries, the fulfillment of which is associated with facing a number of such problems as poverty, technical and technological backwardness, low level of education and consequently low qualification of personnel at enterprises. The key goal for governments of many countries is to effectively implement the prescribed direction to the development of economy and social sphere, which is clearly demonstrated by the course for stabilization and sustainable development of economic systems taken by the UN.

V. Murgul and V. Pukhkal (Eds.): EMMFT 2019, AISC 1258, pp. 806–819, 2021.
https://doi.org/10.1007/978-3-030-57450-5_68

Recent changes in the composition and structure of the Government of Russia are reasoned by the need for accelerated development, modernization and digitalization of the country's economy and for the increase in responsibility for implementing the May Decrees of the President of Russia. These changes are aimed at ensuring the sustainability of development of Russia in the context of urgent economic tasks arising when implementing national projects under conditions of national legislation priority. In this regard, within the next 2–3 years, changes in vital socio-economic areas are expected which will involve significant structural changes in the country's economy.

Sergey Glazyev believes that the key criterion for choosing the priorities of domestic economy restructuring consists in determining the areas of relative advantages of the national economy, which can become points of economic growth [1, 2]. Taking into account the geographic location of Russia, the transport sector can become such a point of economic growth, partly due to the use of capabilities of the national transport system and its expanded reproduction through the development of transportation and related branches including the following: industrial engineering, construction of roads, ports, berths, terminals of cargo transshipment and all kinds of related facilities, informational support of the industry and its digitalization. Such development will help create new workplaces and solve a number of problems concerning sustainable development [1, 3].

The authors attempted to complete the analysis and integration of goals of sustainable development in accordance with the UN methodology [4], of development priorities elaborated for Russian realities by the Analytical Center for the Government of the Russian Federation [5], of tasks of Russia's national economy development. The analysis was performed in terms of the increasing value of water transport development on the basis of current branch-wise and socioeconomic trends of 2018–2019 on their way to 2024 and then to 2030. The paper is aimed at studying a number of relevant aspects that allow outlining a model of sustainable development of Russia as a unity of social, economic and ecological realms on the basis of priority of achieving strategic aims of national economy development in the context of the growth of the inland water transport sub-branch.

The hypothesis of the present study consists in the idea that within the framework of transport component (particularly, inland water transport) a number of problems of sustainable development of the Russian economy can be solved, which can be described via a certain model. The present paper presents the outline of this model as a system of interrelated tasks and vectors of industry development in terms of its direct and indirect impact on the fulfillment of goals set for the country's economy.

2 Materials and Methods

In 2015, the UN set up the 2030 Agenda for Sustainable Development [4]. The program presented ($\sum$Pi, i = 1… 17) includes 17 goals [of the UN]:

- Eliminate Poverty (P1)
- Erase Hunger (P2)
- Establish Good Health and Well-Being (P3)

- Provide Quality Education (P4)
- Enforce Gender Equality (P5)
- Improve Clean Water and Sanitation (P6)
- Grow Affordable and Clean Energy (P7)
- Create Decent Work and Economic Growth (P8)
- Increase Industry, Innovation, and Infrastructure (P9)
- Reduce Inequality (P10)
- Mobilize Sustainable Cities and Communities (P11)
- Influence Responsible Consumption and Production (P12)
- Organize Climate Action (P13)
- Develop Life Below Water (P14)
- Advance Life On Land (P15)
- Guarantee Peace, Justice, and Strong Institutions (P16)
- Build Partnerships for the Goals (P17)

Each of the goals presented contains a number of parameters that should be achieved by the global economy in 15 years by means of governments, industry companies and the society. These goals are of comprehensive nature which is peculiar to global recommendations, so they can be customized in accordance with any national realities and specifics. According to the UN methodology [4] and the interpretation [5], sustainable development consists in maintaining the results achieved in terms of the aspects listed above, so that performance of the modern economy would not worsen them.

The research conducted by the Analytical Center [5] distinguished the priority goals for sustainable development (PPi) taking into account socioeconomic and ecological aspects. For instance, social development sphere has the following priority goals: P1–P5, P10–P12, P16, P17, P6 and P8. Sustainable economic development is to be ensured through the implementation of P7, P9, P12, as well as P1, P11, P13, P14 and P17. Stability of ecological parameters should be ensured through implementing P6, P8, P13–P15, as well as P2, P7, P12.

Features of the national economy and the development of Russia described in work [6] primarily include the parameters of the social sphere and are focused mainly on the content and adjustment of the wording of goals above, such as P1, 2, 5, 3, 7, 14, 17. For instance, poverty reduction and elimination of hunger (D1), ensuring affordability of education (D2), ensuring gender equality (D3), reducing maternal mortality and mortality of children under the age of 5 (D4), combating diseases (D5), ensuring environmental sustainability (D6), engaging in global cooperation which would meet Russian national interests (D7).

The authors of works [5, 7] denote that in Russia, the successful implementation of the main goals of sustainable development was reasoned by significant growth of the economy and welfare of the population in the early 2000s. Particularly, cash incomes increased 13.3 times per capita during 2000–2015 [8], average life expectancy became longer by 6 years. By 2020, the same indicator increased by an average of 8.4 years [9]. However, the geopolitical situation of 2014–2016 revealed the instability of the Russian development model, exacerbated several problems of the economy associated with number of economic goals and objectives unfulfilled. This fact causes the need to find new ways of development, taking into account the previous stage destabilization, which entailed the latest changes, first of all, in the economic frame of the Government of the Russian Federation.

L. V. Kharlamova [3] states that today Russia's technological lag behind economically developed countries is especially noticeable and leads to weakening of Russia in the field of foreign policy and trade. "In implementation of large geostrategic plans of Russia, the transport industry has an advantage peculiar to powerful corporate structures with great opportunities in elaboration and implementation of a number of geo-economic projects, which directly influence the establishment of stable trade and economic relations. While transiting to a production-and-investment model of international economic relations, which will cause the extraordinary growth of foreign trade, Russia will need a qualitative review of the role and value of the entire country's transportation component, which would ensure the implementation of promising large geo-economic transport projects regulated by the society".

Under current conditions, transport is becoming a tool for implementing national interests in many life sectors of Russia. A reasonable strategy for the development of the transport complex is able to inspire the development of international economic relations, transport mobility of the population, which determines social mobility, as well as contribute to solving urgent socio-economic problems of the country [10]. Industry experts point out that the strategic advantages of inland waterway transport (IWT) are transparent. They include the low cost of bulk cargo transportation, the possibility of carrying bulky and heavy cargo, low energy intensity (Fig. 1), relatively low costs for the development and maintenance of inland waterway infrastructure (Fig. 2), the possibility of saving costs for warehousing and storing goods as part of logistics costs of Russian and foreign business, the possibility of delivering goods to areas that do not have any other options of transport accessibility [11].

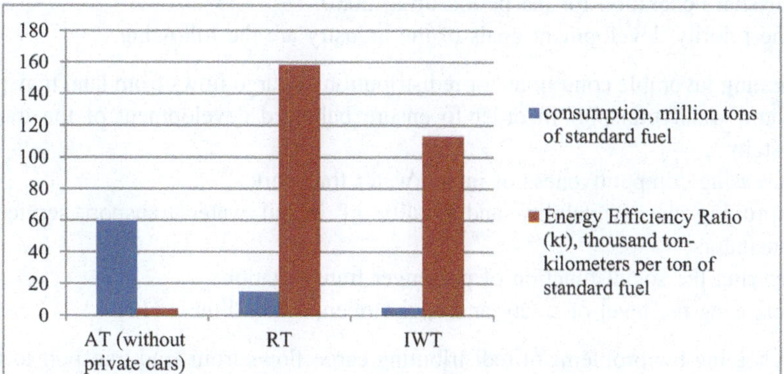

Fig. 1. Comparison of energy efficiency rates by modes of transport (AT – road transport (automobiles), RT – railroad transport).

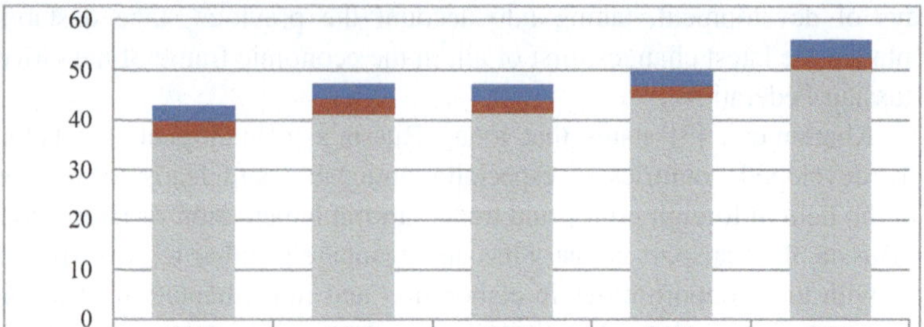

	2012	2013	2014	2015	2016
IWT	3.1	3	3.3	3.2	3.5
RT	3.2	3.1	2.5	2.3	2.4
AT	36.7	41.3	41.5	44.6	50.2

Fig. 2. Amount of financing the infrastructure of inland water and rail transport, federal and regional public highways from both budgetary and extra-budgetary sources, billion rubles per 10 ton-kilometers of cargo.

Today, the capacity of inland waterways is not fully utilized. However, the growing quality of inland water transport infrastructure allows for faster cargo delivery and ensures the transportation of increasing volumes while compressing and complicating the movement in supply chains.

The experts of The Federal Agency for Maritime and River Transport [12] consider that ensuring the sustainable development of inland water transport is associated with overcoming a set of restrictions that include a number of infrastructural, organizational, legislative, technical and technological ones. Sustainable development also demands creating certain conditions and developing a model of stable, balanced and efficient development of water transport as a part of the whole transport system of Russia, which is partially reflected in the Strategy for the Development of Inland Water Transport of the Russian Federation for the period up to 2030.

The priority development goals of the industry are the following:

- creating favorable conditions for redistribution of cargo flows from land transport to inland water transport in order to ensure balanced development of the transport system;
- increasing competitiveness of inland water transport;
- improving the availability and quality of inland water transport services for consignors;
- ensuring the social function of passenger transportation;
- increasing the level of safety and environmental friendliness [12].

When solving the problems of redistributing cargo flows from land transport to inland water transport, the geographical and climatic features of Russia should be taken into account, as they determine the peak values of bulk cargo transportation by road (AT) and rail transport (RT), which appear in a summer season and coincide with peak values of passenger traffic. Moreover, in many regions of Siberia and the Far East, inland waterways are the only routes of communication. Thus, in summer, the obvious advantages of inland water transport can be used for shipping bulk cargos (non-metallic building materials, coal, chemical goods and fertilizers, grain, oil products), increasing the pass-through and carrying capacity of other modes of transport and reducing the budgetary expenses for repairs and reconstruction of land transport communications. However, in order to implement such a scenario, the infrastructure of inland water transport should be adequate in terms of ensuring the regular cargo delivery and the reliability in new supply chains that are being formed or adjusted. Timely financing of works on adaptation of the infrastructure of inland water transport to the increased volumes of cargo will enable the implementation of measures aimed to recover the lost dimensions of the shipways through executing the necessary track works and, first of all, dredging operations in order to increase the carrying capacity of the fleet by means of full load and reduction in unproductive costs. Measures for ensuring twenty-four-hour ship traffic will also be enabled.

The implementation of measures of state support for the redistribution of relevant cargo flows from land roads to inland waterways should involve legal-and-financial regulation aimed at restricting the transportation of non-metallic building goods by roads, restricting the use of heavy vehicles in large cities having internal water transport. These restrictions should be implemented at the legislative level of constituent entities of the Russian Federation. Some measures on development of cargo flows, implicitly tending towards inland waterways should be worked out with the involvement of the key enterprises. Industrial clusters oriented to the transportation of products and raw materials by river transport also should be developed. In addition, governmental support for fleet renewal will revitalize the business climate in shipbuilding and related industries, and increase employment rates.

The Russian transport industry, which involves large-scale projects, is a driver for both the growth of people's well-being and reduction of unemployment. Transport creates conditions for the permanent functioning and development of all other sectors of the economy, and helps increase sectoral and social efficiency. The studies that were carried out earlier [13] proved that large investment projects applied in the field of transport attract fund inflows to the country's budget, have a high socio-economic effect, which causes the generation of an industry-specific (direct and indirect) effect and non-transport effect (effect of related industries).

The results of the published literature sources that were examined by the authors indicate a significant interest in the methodological and practical aspects of determining such effects, as multiplier effects in various economy sectors, including the transport industry and some separate transport types, effects on growth of country's gross domestic product (GDP) and other macroeconomic indicators. The works of Richard Kahn (1931) and John Maynard Keynes (1936; 1973) should be noted as the most significant ones, as well as the papers belonging to the following contemporary authors working the field of assessing the impact of industry development on the economy of states: A.A. Shirova [14], L.Y. Titova [15].

In general, when analyzing modern publications, the authors found neither estimates of the effects caused by projects of managing the development of inland water transport facilities nor beneficiaries of those effects. However, this is a rather important element of the system of managing the country's economy in the context of achieving goals of its sustainable development through the management of transport industry facilities.

3 Results and Discussion

When assessing the prospects of the transport complex of the Russian Federation in terms of national interests, the guidelines of the state should not be aimed at short-term profit, but at the overall strategic socio-economic effect, complemented by the advantages of a geopolitical nature. At the same time, the effects occurring during the implementation of transport projects arise at different times, which means that they differ by horizons of obtaining diverse benefits, they can be time-dispersed (for example, the effect of investing in infrastructure), and contribute to solving a number of national problems and ultimately to achieving sustainable development goals.

The extension of transport infrastructure directly and indirectly affects the economic development of the country and the growth of gross domestic product, therefore it is advisable to consider the created transport infrastructure as a form of assets that can be economically and socially profitable. Erection of transport infrastructure facilities is an important factor contributing to strengthening markets through improving access to financial, land and labor resources. It leads to increased labor productivity, generation of scale effects, directly affecting the rate of economic growth.

The development of inland waterways enables increasing the volumetric indicators of cargo transporting by inland water transport. In the case of redirecting some parts of cargo traffic from roads to inland waterways during the navigation season, socio-economic effects will manifest due to:

- reduction of the cost of finished products in macro-logistic systems due to the decrease in a transportation component of the final price of goods, works or services, which will increase the population's purchasing power in the economy of Russia and its individual regions, as well as the level of wages;
- increase in energy efficiency (reduction of energy intensity) of the transport industry in the country's economy;
- relative reduction (saving) of the expenditures for road repair works, including roads of regional and local importance, and consequently, increase in the efficiency of using financial resources coming from budgetary funds;
- increase in tax payments due to changes in their configuration because of increase in taxes collected from constituent entities according to the general taxation system (shipping companies), compared to the collection of taxes according to the simplified system or applying UTII (motor transport enterprises, self-employed entrepreneurs);
- reduction of road traffic, and consequentially of traffic jams (decrease in time spent on the road), reduction of accidents and deaths on roads, which is a component of the social effect;
- improvement of the environmental situation (in terms of pollution of air, soil and water, climate impact, noise level, etc.).

The decrease in damage from atmospheric pollution is manifested in the reduction of the morbidity of the population and of negative consequences of polluting natural resources, while the bio-productivity of natural-economic complexes and the recreational potential of territories are increasing. An assessment of the effect of a decrease in the level of environmental pollution can be performed based on the calculation of the prevented damage that occurs within the contaminated area. A recipient has to pay for eliminating pollution, reclamation of nature, compensation for risking people's health, as well as to attract additional natural resources to neutralize the flow of pollutants, etc.

In case of redirecting a part of cargo traffic from railway to water transport, the effects formation is due to:

- reduction of the cost of finished products in macro-logistic systems due to the reduction of the transport component in the final price of goods, works and services, which will increase the population's purchasing power in the economy, as well as the level of real wages of the population;
- reduction of railway traffic during peak periods (summer season, vacations), which will create favorable conditions for increasing the speed of distribution of goods and for organizing a larger number of passenger train runs and coordination of schedules;
- decrease in wear of the railroad track during transportation of heavy and bulky cargos and reduction of expences for repair and maintenance of railway infrastructure;
- increase in the export potential of the country, including agricultural producers, due to an increase in the volume of cargo for export transported by inland waterways under the existing pass-through capacity limits of roads accessing seaports.

Creation of competitive transport corridors based on technically and technologically integrated transport and logistics infrastructure in the Russian Federation will allow generating effects by taking advantage of each type of transport and increasing the capacity of the transport system as a whole.

The composition of beneficiaries in relation to the type of effect they receive when implementing measures of state support (MSS) is presented at Fig. 3 and Table 1.

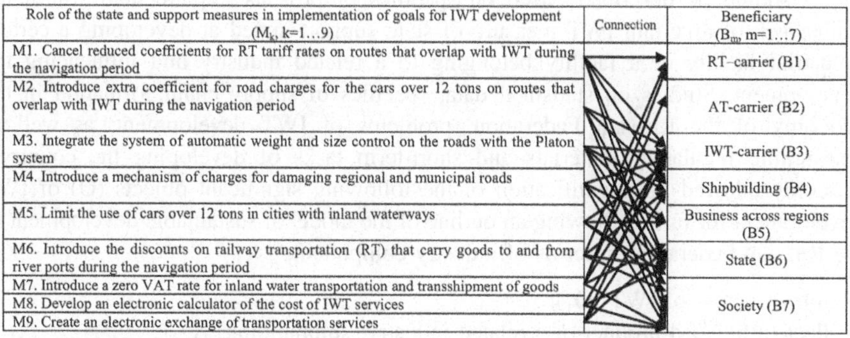

Fig. 3. Beneficiaries from the implementation of measures of state support for the development of inland water transport (IWT)

Table 1. Beneficiaries in regard with types of the effect that has received when implementing measure of state support for the development of inland water transport (IWT).

MSS	Beneficiary/effect							
	Economic effect (EcE)		Social effect (SE)		Environmental effect (EE)		Budgetary effect (BE)	
	Direct	Indirect	Direct	Indirect	Direct	Indirect	Direct	Indirect
M_1	B1, B3, B5	B4, B5, B7		B6, B7	B6, B7			
M_2	B3, B4, B5	B4		B6, B7	B6, B7		B6	
M_3		B4, B6		B6, B7		B7		B6
M_4		B3, B4	B7	B6	B6, B7		B6	
M_5	B3	B1, B4, B5, B6	B7		B6, B7			
M_6	B1,	B3, B4, B5		B6, B7	B6, B7			
M_7	B1, B2, B3	B4		B7	B6, B7		B6	
M_8	B3	B4, B7		B7		B6, B7	B6	
M_9	B3	B4, B7		B7		B6, B7		

Table 2 shows factors and components of the socio-economic effect of implementing the proposed measures of state support (MSS) aimed at creating conditions for the reorientation of cargo flows to inland water transport.

Therefore, on the basis of the materials studied, the outline of the model of sustainable development of Russia in the field of studying the strategic advantages and potential of the inland water transport sub-branch can be represented as follows (Fig. 4).

Today, transport development is regulated with the use of project management methods based on the Comprehensive Plan for the modernization and expansion of the main infrastructure for the period until 2024. The effects' indicators which are shown in Fig. 3 should be described based on specifics of projects implemented within the framework of a certain IWT measure of state support aimed at developing a certain industrial facility or a facility belonging to a related industry and influencing the development. Studies on statistical data, specifics of supply chain formation in the economy of the Russian Federation, problems of IWT development, as well as researching regulatory materials and short-term tasks of developing the country's economy allowed the identification of the following significant objects (O) of IWT development for further drawing an outline of the model of sustainable development of the Russian Federation in terms of industry component:

- infrastructure of IWT (O_1);
- fleet (a finished product of a related industry: shipbuilding) (O_2);
- redistribution of cargo flows in the Russian economy (governmental regulation of transportation in terms of redirection of certain cargo types to energy-efficient and environmentally friendly modes of transport (O_3).

Table 2. Forming factors and components of social-and-economic effect of implementation of measures of state support for shifting the cargo traffic to inland water transport.

MSS	Effect forming factors	Effect components
1	2	3
M_1	Increasing the country's export potential due to the growth of the cargo traffic on inland waterways with a limited capacity of feeder road to sea ports; Growing of investment opportunities for shipping companies to upgrade the fleet; Reducing deterioration of the railroad track when carrying bulky and heavy goods; Reducing the load of railways in the summer months, creating conditions for increasing the speed of goods distribution and optimizing the schedule of passenger trains; Decreasing pollution.	*Economic and budgetary effects*: - GDP growth because of investments in construction of the fleet; - reduction of expenses for repair and maintenance of railway infrastructure; *Social and environmental effects:* - increase in transport accessibility for people during the holidays; - lower accident rate; - diminishing the transport influence on the environment.
M_2	Reducing the cost of the finished product in macro-logistics networks due to a decrease in the transport component in the final price of goods; Growing investment opportunities for shipping companies to upgrade the fleet; Improving the quality of roads and providing new ways to compensate the repair works due to the damage caused by heavy goods vehicles; Improving the environmental parameters of the transport network when redistributing part of the cargo flows to inland water transport; Diminishing the transport influence on the environment; Improving the energy efficiency of cargo transportation; Lower accident rate.	*Economic and budgetary effects*: - GDP growth due to higher effective demand of the population; - GDP growth because of investments in construction of the fleet; - increase in "Platon" payments; - reduction of energy intensity for cargo transportation; - increase in taxes when applying the general taxation system by shipping companies in comparison with the application of the simplified tax system (STS), UTII by a road carrier; *Social and environmental effects*: - increase of actual wages; - increased transport accessibility; reduced downtime due to traffic jams; - Lower accident rate; - reduction of harmful emissions into the atmosphere; - increase in the level of environmental safety; - increase in energy efficiency of transport;
M_3	Elimination of violations of weight parameters in terms of gross weight and (or) axle load and improving road safety; Improving the environmental parameters of the transport network when redistributing part of the cargo flows to inland water transport; Diminishing the transport influence on the environment; Improving the energy efficiency of cargo; Lower accident rate.	*Economic and budgetary effects*: - increase in "Platon" payments; - reduction of energy intensity for cargo transportation; - increase in taxes when applying the general taxation system by shipping companies in comparison with the application of the simplified tax system (STS), UTII by a road carrier; *Social and environmental effects*: - increased transport accessibility; reduced downtime due to traffic jams; - Lower accident rate; - reduction of harmful emissions into the atmosphere; - increase in the level of environmental safety; - increase in energy efficiency of transport;
M_4	Reducing the cost of the finished product in macro-logistics networks due to a decrease in the transport component in the final price of goods; Improving the quality of the regional and municipal roads and providing new ways to compensate the repair works due to the damage caused by heavy goods vehicles; Improving the environmental parameters of the transport network when redistributing part of the cargo flows to inland water transport; Diminishing the transport influence on the environment; Improving the energy efficiency of cargo transportation; Lower accident rate.	*Economic effect (regional level)*: - increase in payments to compensation the damage to public roads of regional and municipal level caused by vehicles with a permitted maximum weight of over 12 tons; - reduction of energy intensity for cargo transportation; - increase in taxes when applying the general taxation system by shipping companies in comparison with the application of the simplified tax system (STS), UTII by a road carrier; *Social and environmental effects:* - increased transport accessibility; reduced downtime due to traffic jams; - Lower accident rate; - reduction of harmful emissions into the atmosphere; - increase in the level of environmental safety; - increase in energy efficiency of transport;
M_5	Restrict to move freight vehicles in cities; Automation of the process of monitoring and detecting freight vehicles, that drive in cities; Improving the environmental parameters of the transport network when redistributing part of the cargo flows to inland water transport; Diminishing the transport influence on the environment; Improving the energy efficiency of cargo transportation; Lower accident rate.	*Economic effect (for metropolises and cities)*: - GDP growth due to investments in the infrastructure of inland water transport, in the construction of the fleet, in the system of access, monitoring and control of motor vehicles in cities; - reduction of energy intensity for cargo transportation; *Social and environmental effects*: - increased transport accessibility; reduced downtime due to traffic jams; - Lower accident rate; - reduction of harmful emissions into the atmosphere; - increase in the level of environmental safety; - increase in energy efficiency of transport;
M_6	Reducing the cost of the finished product in macro-logistics networks due to a decrease in the transport component in the final price of goods; Government subsidies to railway operators in order to reimburse lost income as a result of discounts on rail transportation.	*Economic effect*: - GDP growth due to investments in the construction of the fleet; *Social and environmental effects:* - increased transport accessibility; reduced downtime due to traffic jams; - Lower accident rate; - increase in the level of environmental safety;
M_7	Reducing the cost of the finished product in macro-logistics networks due to the development of direct mixed cargo transportation capitalizing on the advantages of each type of transport; reduction of the transport component in the final price of goods, works and services; Improving the environmental parameters of the transport network when redistributing part of the cargo flows to inland water transport; Diminishing the transport influence on the environment; Improving the energy efficiency of cargo transportation; Lower accident rate.	*Economic and budgetary effects*: - GDP growth due to investments in the construction of the fleet, in the infrastructure of inland water transport; - GDP growth due to higher effective demand of the population; - reduction of energy intensity for cargo transportation; - increase in taxes when applying the general taxation system by shipping companies in comparison with the application of the simplified tax system (STS), UTII by a road carrier; *Social and environmental effects*: - increase of actual wages; - increased transport accessibility; reduced downtime due to traffic jams; - Lower accident rate; - reduction of harmful emissions into the atmosphere; - increase in the level of environmental safety; - increase in energy efficiency of transport;
$M_{8,9}$	Sale promotion of inland water transport services, broadening the opportunities to be included in cargo transportation network; Improving the environmental parameters of the transport network when redistributing part of the cargo flows to inland water transport; Diminishing the transport influence on the environment; Improving the energy efficiency of cargo transportation; Lower accident rate.	*Economic and budgetary effects*: - GDP growth due to investments in the construction of the fleet, in the infrastructure of inland water transport; - reduction of energy intensity for cargo transportation; - increase in taxes when applying the general taxation system by shipping companies in comparison with the application of the simplified tax system (STS), UTII by a road carrier; *Social and environmental effects:* - increased transport accessibility; reduced downtime due to traffic jams; - Lower accident rate; - reduction of harmful emissions into the atmosphere; - increase in the level of environmental safety; - increase in energy efficiency of transport;

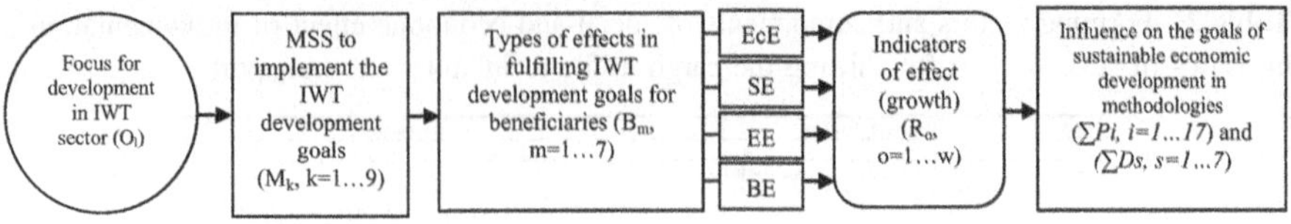

Fig. 4. Outline of sustainable development of the Russian Federation in the context of development of IWT.

Further, a more detailed elaboration of the object of study and its decomposition, as well as distinguishing other relevant objects for controlling this system, is possible for creating a model of the outline of development.

A general view of the model for O_3 object—the redirection of cargo flows to IWT in the Russian economy—which generates direct effects, is shown in Fig. 5.

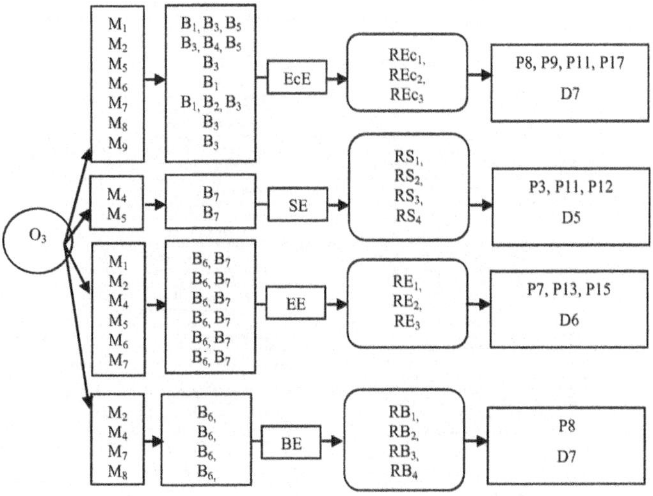

Fig. 5. Outline of the model of sustainable development of the Russian Federation in the context of IWT development (for object O3, direct effects).

Indicators Ro are partially shown in Table 3 without specifying the control object.

For this outline, the following Ro indicators can be distinguished that are characteristic for conditions given (the object is the redistribution of cargo flows in the Russian economy (O_3), direct effects), which describe the occurrence of possible effects from the implementation of Mk (Table 3).

Table 3. Indicators of effects for the management of the focus for development - inland water transport infrastructure.

Type of effect	MSS	Indicator R_o	R_o change characteristic
EcE	M_1, M_2, M_5, M_6, M_7, M_8, M_9	REc_1	GDP growth (in terms of value added of IWT) due to improved use of transport infrastructure
		REc_2	GDP growth (in terms of added value of IWT) due to increased load on transport infrastructure
		REc_3	GDP growth (in terms of added value of IWT) due to an increase in volume indicators of IWT
EE	M_1, M_2, M_4, M_5, M_6, M_7	RE_1	Reduction in energy intensity
		RE_2	Reduction of environmental emissions
		RE_3	Increased environmental safety
BE	M_2, M_4, M_7, M_8	RB_1	Increase in paid taxes by shipping companies in comparison with motor vehicles
SE	M_4, M_5	RS_1	Increased transport accessibility
		RS_2	Reduced downtime in traffic jams
		RS_3	Lower accident rate
		RS_4	Employment growth

The given outline of the model of sustainable development of Russia within the segment of the transport component of the country's economy will allow seeing the strategic potential of taking advantage of inland water transport. For further research, it is advisable to study the foreign experience of regulating cargo flows and redirecting them to inland water transport. At the same time, further elaboration is required for issues related to the description of management objects in the transport industry. It is also important to take into account the strategic position of each type of transport (inland water, road, rail), the current state and expected growth points of the economy and transit capacity of Russia, which will allow drawing up clear outlines of industry management based on the principles of transport balance.

4 Conclusions

Therefore, the paper states that a number of problems of sustainable development of the Russian economy can be solved through the process of inland water transport advancement. This process can be described via the relevant model, the outline of which is presented in the paper in shape of a system of interrelated tasks and directions for industry development. The model is based on taking into account national projects and projects for the development of inland water transport that are directly or indirectly influencing the achievement of goals set for the country's economy by certain effects occurring. In terms of sub-branch advancing, the given outline of the model of sustainable development of Russia will allow perceiving the strategic potential of using inland water transport.

The authors have substantiated the idea, that today transport is becoming a tool for implementing national interests in many life aspects. It contributes to solving urgent socio-economic problems of the country, and also is an incentive for the development of international economic relations. The authors note the following strategic advantages of inland water transport: low cost of bulk cargo transportation, high energy efficiency, ability to transport bulky and heavy cargo, low energy intensity, relatively low costs for the development and maintenance of inland waterway infrastructure, possibility of reducing logistics costs of Russian and foreign business due to saving of expenditures for warehousing and storing goods, ability to deliver goods to areas that do not have alternative transport accessibility.

The paper provides the results of the study on directions of development of inland waterways, which will allow increasing the volume of cargo transportation by inland water transport. In case of redirecting part of cargo traffic from roads to inland waterways within the navigation season, the socio-economic and environmental effects will occur. The article gives a list of beneficiaries depending on types of effects that occur when implementing measures of state support for the development of inland water transport in the Russian Federation.

For the further research, the authors have set a task to study foreign experience of regulating cargo flows and redirecting them to inland waterways. In addition, there has arisen a question of further elaboration of controlling objects in the transport industry on the basis of strategic positions of different modes of transport and expected growth points of the economy and predicted transit capability of Russia. This will allow drawing up a clear outline of sub-branch management with taking into account of goals of sustainable development.

References

1. Glazyev, S.Y.: Progressive forms of international economic integration. Finan.: Theory Pract. **1**, 64–68 (2015). https://doi.org/10.26794/2587-5671-2015-0-1-64-68
2. Ubushiev, E.V.: Economic security in various technological systems. Theor. Appl. Econ. **3**, 1–21 (2018). https://doi.org/10.25136/2409-8647.2018.3.27119
3. Kharlamova, Y.A.: The struggle for Eurasia in the focus of transport geostrategies: a monograph. INFRA-M, Moscow (2020). https://doi.org/10.12737/1033110
4. Valencia, S.C., Simon, D., Croese, S., Nordqvist, J., Oloko, M., Sharma, T., Buck. N.T., Versace, I.: Adapting the sustainable development goals and the new urban agenda to the city level: initial reflections from a comparative research project. Int. J. Urban Sustain. Dev. **11**(1), 4–23 (2019). https://doi.org/10.1080/19463138.2019.1573172
5. Bobylev, S.N., Chereshnya, O.Y., Kulmala, M., Lappalainen, H.K., Petäjä, T., Solov'eva, S. V., Tikunov, V.S., Tynkkynen, V.: Indicators for digitalization of sustainable development goals in peex program. Geogr. Environ. Sustain. **11**(1), 145–156 (2018). https://doi.org/10.24057/2071-9388-2018-11-1-145-156
6. Bobylev, S., Goryacheva, A.: Identification and assessment of ecosystem services: the international context. Int. Organ. Res. J. **14**(1), 225–236 (2019). https://doi.org/10.17323/1996-7845-2019-01-13

7. Bobylev, S.N., Solovyeva, S.V.: Sustainable development goals for the future of Russia. Stud. Russ. Econ. Dev. **28**(3), 259–265 (2017). https://doi.org/10.1134/S1075700717030054
8. Shulgin, S.G., Zinkina, Y.V., Scherbov, S.Y.: Life expectancy of the elderly in Russia, depending on educational status. Demograph. Rev. **5**(1), 25–38 (2018). https://doi.org/10.17323/demreview.v5i1.7708
9. Vishnevsky, A.G., Andreev, E.M., Timonin, S.A.: Mortality from cardiovascular diseases and life expectancy in Russia. Demograph. Rev. Engl. Sel. (2019). https://doi.org/10.24411/2411-8621-2019-12003
10. Pantina, T.A., Saveliev, M.N.: Human resources as a factor in the strategic development of inland water transport. J. Vestnik State Univ. Sea River Fleet Named After Admiral S. O. Makarov **1**(20), 194–199 (2013). https://doi.org/10.21821/2309-5180-2013-5-1-194-199
11. Pantina, T.A., Borodulina, S.A.: Criteria and growth factors for the competitiveness of inland water transport. Bull. Astrakhan State Tech. Univ. Ser.: Econ. **3**, 68–77 (2018). https://doi.org/10.24143/2073-5537-2018-3-68-77
12. Zaitsev, A.A., Sokolova, Y.V., Pantina, T.A.: Innovative development of transport system using magnetic levitation technology. World Transp. Transp. **17**(4), 36–45 (2019). (in Russian) https://doi.org/10.30932/1992-3252-2019-17-4-36-45
13. Pantina, T.A., Borodulina S.A.: Formation of monitoring system for federal program «Development of transport system in Russia (2010–2020)» (Direction «Inland Water Transport»). Vestnik Gosudarstvennogo universiteta morskogo i rechnogo flota imeni admirala S.O. Makarova **3**(31), 124–132 (2015). https://doi.org/10.21821/2309-5180-2015-7-3-124-1329
14. Ivanter, V.V., Shirov, A.A., Belkina, T.D., Belousov, D.R.: Recovery of economic growth in Russia. Stud. Russ. Econ. Dev. **27**, 485–494 (2016). https://doi.org/10.1134/S1075700716050105
15. Titov, B., Shirov, A.: Strategy of growth for Russia. Voprosy Ekonomiki **12**, 24–39 (2017). (in Russian) https://doi.org/10.32609/0042-8736-2017-12-24-39

The Impact of Transport Costs on Sales in Supply Chains

Valery Mamonov(✉) and Vladimir Poluektov

Novosibirsk State Technical University, Karl Marx Avenue, 20,
630073 Novosibirsk, Russia
v.mamonov@corp.nstu.ru

Abstract. Supply chains uniting related market participants are characterized by a number of parameters. One of the keys is the value of transport costs, which, in turn, has an impact on such indicators as: price, sales volumes, profits of participants in the supply chain. For this reason, the study of the impact of transport costs and their structure on the economic performance of participants in the supply chain is an urgent task, both in scientific and in practical aspects. The presented paper contains a consideration of the parametric dependence of the value of transport costs on a number of variables, using the method of fragmentation of the supply chain at the level of the link "supplier (focal company) - first-level consumer", and its relationship with the final profit of the participants. According to the degree of accounting and assignment of transport costs, the following options are being studied: inclusion of transport costs in the costs of the supplier (focal company); the inclusion of transport costs in the costs of a first-level consumer in the case of both non-integrated and integrated supply chains. The results obtained made it possible to formulate the problem of maximizing the profit of each of the participants in the logistics chain in the link "supplier (focal company) - first-level consumer" and get its solution. Estimates of the impact of transport costs on supply volume and final price are also given.

Keywords: Logistics · Supply chains · Transportation costs

1 Introduction

The development of sustainable long-term relationships between companies in the supply chain, which are based on economically sound mechanisms for managing relationships between partners, is an important emphasis and a modern trend in the development of research in the theory and practice of supply chain management (SCM) [1–4]. Managing the relationship of companies in the supply chain provides the opportunity to create significant competitive advantages and strengthen their market positions, which is especially true in conditions when consumers are more value-oriented and less loyal to a particular brand or supplier [5].

The SCM concept [6–9] is based on the principles of interaction between all participants in the supply chain in order to improve the performance of each of them and the entire chain as a whole, to ensure the delivery of products on time and with minimal total logistics costs throughout the chain. In this regard, it should be noted that

V. Murgul and V. Pukhkal (Eds.): EMMFT 2019, AISC 1258, pp. 820–829, 2021.
https://doi.org/10.1007/978-3-030-57450-5_69

logistics costs constitute one of the significant cost items associated with doing business in modern conditions [10], second only to raw materials and materials in production or cost of goods sold in wholesale or retail trade. Therefore, reducing logistics costs to the optimal level and its further maintenance is one of the priority tasks of the participants in the supply chain [11]. Obviously, reducing logistics costs is impossible without developing a scientifically based methodology for their identification, accounting and analysis. The analysis of scientific publications on this issue [10–13] showed that in the economic literature, the total costs of logistics are considered to consist of the costs of transportation of products, transport packaging, warehousing, stockpiling, logistics management, as well as indirect costs of logistics. At the same time, the costs associated with the transportation of products from producer to consumer comprise the bulk of the logistics costs [14]. Based on this, the study of the impact of transport costs and their structure on the economic performance of participants in the supply chain is an urgent task both in scientific and in practical aspects, which determines the feasibility of this study. The practical significance of this study lies in the development of recommendations for modeling supply chain management processes, the use of which will create conditions for increasing the coordination of processes for planning sales volumes and delivery time.

2 Materials and Methods

Let's consider a fragment of the network structure of supply chains, including a supplier of finished products (usually a focal company) and first-level wholesale consumers who bring finished products to the final consumer. Finished products are delivered through the transport network. In the network structure of the supply chain, product transportation can be organized by both participants in the supply chain and a specialized transport company. If the participants in the supply chain do not use outsourcing, the transportation of products is carried out by the participants themselves and at the same time, transportation costs can be included in their own costs by both the focal company and wholesale consumers.

In fact, the organization of the supply chain can be implemented in one of many possible options. And first of all, it is necessary to take into account an important property when modeling the supply chain: is it an integrated chain or not. It is proposed to first consider a non-integrated supply chain with two options for accounting for transportation costs by participants in the chain:

- the case when the wholesale consumer includes transport costs in their own costs;
- when transport costs are included in the cost of manufacturing products by the focal company.

Before moving on to the general case, we consider the interaction of a focal company with one wholesale consumer and describe the main business processes. Let the wholesale consumer include transportation costs in his own expenses and sell the products on the consumer market, taking into account the statistically observed regularity - the dependence of the value of the goods on the value of the offer. Without loss of generality, we assume that the inverse demand function is described by a linear function:

$$P(Q) = a - b \cdot Q, \tag{1}$$

where P – the price of the goods with a supply volume equal to Q; a and b – constant parameters.

We denote by s the transportation costs for the delivery of a conventional unit of finished products from the focal company to the wholesale customer. In a non-integrated supply chain structure, the wholesale customer will solve the problem of maximizing profits relative to such a sales volume of the product that will provide the maximum amount of profit depending on the price of the product at which the focal company will sell him this product.

Thus, the wholesale consumer determines the sales volume of the product, taking into account the demand function determined by the consumer market, and depending on the price P_1 set by the focal company. Then the amount of profit of the wholesale consumer is determined by the expression:

$$\pi(q) = (a - bq - P_1 - s) \cdot q \to \max_q . \tag{2}$$

The solution to the problem of unconditional extremum (2) determines the expression for the volume of sales of products on the consumer market, which depends on the price of goods of the focal company, the parameters of the inverse demand function and transport costs:

$$q = \frac{a - s - P_1}{2b}. \tag{3}$$

With this sales volume, the wholesale customer's profit will be maximum when the focal company sets the price at P_1.

Let's consider the behavior of a focal company. Suppose that in the manufacture of a conventional unit of finished goods, there are costs equal to c, which are the price of the entire set of components necessary for the production of finished products, at which first-level suppliers supply them to the focal company. The focal company will maximize its own profit by determining the price so that its profit would be maximum for any volume of product delivery to the wholesale customer. Given that transportation costs are included in the final price of the consumer, the task of maximizing profits by the focal company will be written in the form:

$$\pi(P_1) = (P_1 - c) \cdot q = (P_1 - c) \cdot \frac{(a - s - P_1)}{2b} \to \max_{P_1} . \tag{4}$$

Solving the problem, we find the expression for the unit price of the finished product, which maximizes the profit of the focal company with the volume of product delivery to the wholesale consumer in the amount determined by expression (3):

$$P_1 = \frac{a + c - s}{2}. \tag{5}$$

Note that the price in expression (5) contains the transport component with a minus sign. Thus, despite the fact that transportation costs are included by the wholesale customer in the final price of the consumer, finished products are sold to the wholesale customer at a lower price due to the transport component. At the same time, participants' profits are equal to:

$$\begin{aligned} \pi(P_1) &= \frac{2(a - c - s)^2}{16b}, \\ \pi(q) &= \frac{(a - c - s)^2}{16b}. \end{aligned} \tag{6}$$

Total profit is determined by the expression:

$$\pi = \frac{3(a - c - s)^2}{16b}. \tag{7}$$

We determine the volume of supply and the final price for the consumer:

$$q = \frac{a - c - s}{4b}, \tag{8}$$

$$P = \frac{3a + c + s}{4}. \tag{9}$$

Note that transportation costs reduce the supply volume and, consequently, according to the law of demand, increase the final price of products.

Let's consider the situation when transport costs are included in the price of finished products of the focal company. In this case, the wholesale customer solves the problem of maximizing profit in the form of:

$$\pi(q) = (a - bq - P_1)q \rightarrow \max_q \tag{10}$$

and obtains:

$$q = \frac{a - P_1}{2b}. \tag{11}$$

The focal company, including transport costs in the price of products, solves the problem of maximizing profits in the form of:

$$\pi(P_1) = (P_1 - c - s)q \rightarrow \max_{P_1} \tag{12}$$

and relative to the price, it gets the following expression:

$$P_1 = \frac{a + c + s}{2}. \tag{13}$$

Comparing the expression obtained with the value (5), we note that the selling price of products to the wholesale consumer increased by the value of specific transport costs. All other characteristics of the supply chain (participants' profits, total profits, the number of products on the consumer market and the price of products) remain as in the previous case and are quantified in accordance with expressions (6), (7), (8), (9). Thus, the identity of the results is obvious, which allows us to draw the following conclusion: from the point of view of the characteristics of the consumer market (price, quantity of offered goods), it does not matter which of the participants includes transport costs in its own costs. The difference is observed in the interaction of the focal company with the wholesale customer: in the first case, the price of the products of the focal company is lower by the amount of transport costs, and in the second, respectively, higher by the same amount of transport costs. Therefore, without affecting the final results, the inclusion of transport costs by one of the participants in the supply chain in their own costs is only an agreement in the process of interaction between the participants.

Let's consider the case of an integrated supply chain, when the economic interaction of the participants is expressed in the creation of a single profit pool, which is then distributed among the participants in accordance with the established regulations (quotas). The task of maximizing profit is formulated as:

$$\pi(q) = (a - bq - c - s) \cdot q \to \max_q \tag{14}$$

and its solution is the sales volume of products in the consumer market:

$$q = \frac{a - c - s}{2b} \tag{15}$$

and corresponding price:

$$P = \frac{a + c + s}{2}. \tag{16}$$

In this case, the profit in the supply chain will be equal to:

$$\pi = \frac{4(a - c - s)^2}{16b}, \tag{17}$$

which is more than the total profit in the supply chain in the absence of integration defined by expression (7). At the same time, the volume of supply of goods in the consumer market (8) is greater than in a non-integrated system, and the price of the final consumer is lower (9).

After considering the interaction of two participants in the supply chain, it is logical to consider the general case when the number of wholesale consumers is m. The different locations of wholesale consumers are expressed in different transport costs and in different parameters of the demand curves for the same product. In fact, this means that the focal company, producing the same or homogeneous products, is dealing with spatially differentiated goods.

We assume that the focal company does not implement a discriminatory policy regarding the price set by an individual wholesale customer. The mechanism of interaction between the focal company and wholesale customers involves a single price for manufactured products and the inclusion of transportation costs by wholesale customers in their own costs. Based on previous results, each wholesale consumer determines such a volume of delivered products from a focal company that maximizes his profit, taking into account transport costs and parameters of the demand curve for products in his consumer market:

$$q_i = \frac{a_i - s_i - P_1}{2b_i}, \quad i = 1, \ldots, m. \tag{18}$$

Two situations are possible here:

- focal company has no restrictions on the volume of production, and the volume of sales of products in the consumer market is regulated by demand;
- the volume of production of products of the focal company is limited by the conditions of production, and the volume of sales of products in the consumer market is regulated by demand.

In the first situation, the focal company solves the problem:

$$\pi(P_1) = (P_1 - c) \cdot \sum_{i=1}^{m} q_i \to \max_{P_1} \tag{19}$$

and as a result, determines the price:

$$P_1 = \frac{\sum_{i=1}^{m} \frac{1}{b_i}(a_i - s_i + c)}{2\sum_{i=1}^{m} \frac{1}{b_i}}. \tag{20}$$

It follows from expression (20) that the price at which the focal company sells products to wholesale suppliers depends not only on the parameters of the inverse demand functions in consumer markets, but also on the values of all transport costs and their structure.

Finally, we consider a situation where, according to the production conditions, the focal company can produce the quantity of finished products no more than Q. In this case, the focal company finds the optimal selling price of finished products to wholesale consumers, solving the problem:

$$\begin{aligned} &\pi_n = (P_1 - c)\sum_{i=1}^{m} Q_i \rightarrow \max \\ &\sum_{i=1}^{m} Q_i = Q \end{aligned} \tag{21}$$

For a given value of the volume of manufactured products Q by the focal company, the price of the product is the solution to problem (21) and is determined by the expression:

$$P_1 = \frac{\sum_{i=1}^{m} \frac{a_i - s_i}{2b_i} - Q}{\sum_{i=1}^{m} \frac{1}{2b_i}}. \tag{22}$$

Analyzing the obtained dependences, we establish that if there is a restriction on the volume of manufactured products according to the conditions of production and their reduction, the sale price of finished products to wholesale customers increases.

3 Results and Discussion

Let's consider the impact of changes in transport costs on the volume of supply of products and sales using the obtained formulas. Since the nature of the change in the transport component in the logistic transport network may be different, any assumption about an increase or decrease in transport costs for an individual wholesale consumer will be private, i.e. match one of the many possible implementations. With this approach, it is not possible to identify patterns or dependencies common to the supply chain. We introduce the coefficient $k > 0$, which indicates the change in all values of transport costs in the system and the cases corresponding to this change:

- if there is an increase in transport costs, then they increase $(1 + k)$ times at all wholesale customers;
- if there is a decrease in transportation costs, then they decrease $1/(1 - k)$ times at all wholesale customers.

We obtain the dependencies of the selling price of the goods to wholesale customers and the volume of goods for the first case, when all values of the transport cost increased $(1 + k)$ times. With such a change in transportation costs in the system, the selling price of the product by the focal company to the wholesale consumer decreases and will be equal to:

$$P_1(k) = P_1 - k \cdot \left(\sum_{i=1}^{m} \frac{s_i}{b_i}\right) \cdot \left(2\sum_{i=1}^{m} \frac{1}{b_i}\right)^{-1}. \tag{23}$$

Let us turn to the expression that determines the profit of the wholesale customer with an increase in transport costs:

$$\pi_i(k) = \frac{(a_i - (1+k)s_i - P_1(k))^2}{4b_i} = \frac{(a_i - s_i - P_1 - kd_i)^2}{4b_i}. \quad (24)$$

In this case, the volume of sales of goods to wholesale consumers will be:

$$q_i(k) = \frac{(a_i - s_i - P_1 - kd_i)}{2b_i}, \quad (25)$$

where constant d_i is:

$$d_i = \left(\sum_{j=1}^{m} \frac{2s_i - s_j}{b_j}\right) \cdot \left(2\sum_{i=1}^{m} \frac{1}{b_i}\right)^{-1}. \quad (26)$$

The obtained dependences make it possible to draw the following conclusions for this case about the relationship between changes in transport costs in the system and the economic performance of supply chain participants. Moreover: if $d_i > 0$, then the profit and sales volumes of products from the i-th wholesale customer are reduced; if $d_i < 0$, then both profit and sales volumes increase.

Let us illustrate the dependence of relative values on the parameter k (Fig. 1, 2):

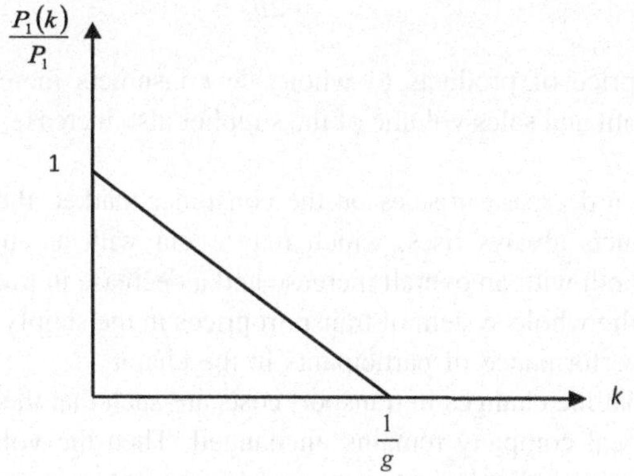

Fig. 1. Dependence of the relative value $\frac{P_1(k)}{P_1}$ on the parameter k, where $g = \left(\sum_{i=1}^{m} \frac{s_i}{b_i}\right) \cdot \left(2P_1 \sum_{i=1}^{m} \frac{1}{b_i}\right)^{-1}$.

For the case when all transport costs are reduced $1/(1-k)$ times, these dependencies have the following form:

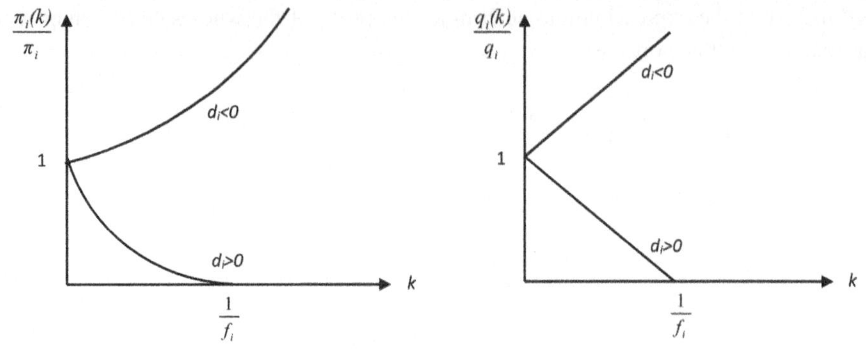

Fig. 2. Dependence of relative values $\frac{\pi_i(k)}{\pi_i}$, $\frac{q_i(k)}{q_i}$ on the parameter k, where $f_i = \frac{d_i}{a_i - s_i - P_1}$.

$$P_1(k) = P_1 + k \cdot \left(\sum_{i=1}^{m} \frac{s_i}{b_i}\right) \cdot \left(2\sum_{i=1}^{m} \frac{1}{b_i}\right)^{-1}, \tag{27}$$

$$\pi_i(k) = \frac{(a_i - (1-k)s_i - P_1(k))^2}{4b_i} = \frac{(a_i - s_i - P_1 + kd_i)^2}{4b_i}, \tag{28}$$

$$q_i(k) = \frac{(a_i - s - P_1 + kd_i)}{2b_i}. \tag{29}$$

Then the selling price of products to wholesale consumers increases. Moreover, if $d_i > 0$, then the profit and sales volume of the supplier also increase; when $d_i < 0$, profit and sales decrease.

Note that with a decrease in sales on the consumer market, the price for the final consumer of products always rises, which may occur with an individual wholesale seller of products both with an overall increase and a decrease in transportation costs in the system. Thus, the whole system of transport prices in the supply chain is influenced by the economic performance of participants in the chain.

It is possible that the changes in transport costs are such that the selling price of the products by the focal company remains unchanged. Then the volume of profits and sales increases for the wholesale customer who reduces transport costs included in their own costs.

4 Conclusion

Depending on the objectives of the study, the obtained analytical dependencies allow for a multilateral analysis of the results of economic activity of participants in the supply chain. The presence of transport costs in the formulas explicitly provides useful information for improving the system of organization of transport logistics.

References

1. Gualandris, J., Golini, R., Kalchschmidt, M.: Do supply management and global sourcing matter for firm sustainability performance? An international study. Supply Chain Manag. **19** (3), 258–274 (2014). https://doi.org/10.1108/SCM-11-2013-0430
2. Kamalahmadi, M., Parast, M.M.: A review of the literature on the principles of enterprise and supply chain resilience: major findings and directions for future research. Int. J. Prod. Econ. **171**(1), 116–133 (2016). https://doi.org/10.1016/j.ijpe.2015.10.023
3. Ralston, P.M., Richey, R.G., Grawe, S.J.: The past and future of supply chain collaboration: a literature synthesis and call for research. Int. J. Logist. Manag. **28**(2), 508–530 (2017). https://doi.org/10.1108/IJLM-09-2015-0175
4. Um, K.-H., Kim, S.-M.: The effects of supply chain collaboration on performance and transaction cost advantage: the moderation and nonlinear effects of governance mechanisms. Int. J. Prod. Econ. **217**, 97–111 (2019). https://doi.org/10.1016/j.ijpe.2018.03.025
5. Christopher, M., Gattorna, J.: Supply chain cost management and value-based pricing. Ind. Mark. Manag. **34**(2), 115–121 (2005). https://doi.org/10.1016/j.indmarman.2004.07.016
6. Lambert, D.M., Cooper, M.C.: Issues in supply chain management. Ind. Mark. Manag. **29** (1), 65–83 (2000). https://doi.org/10.1016/S0019-8501(99)00113-3
7. Bowersox, D., Closs, D., Cooper, M.B.: Supply Chain Logistics Management, 4th edn. McGraw-Hill Education, New York (2012)
8. Christopher, M.: Logistics & Supply Chain Management, 5th edn. Pearson Education, New York (2016)
9. Islam, D.M.Z., Fabian Meier, J., Aditjandra, P.T., Zunder, T.H., Pace, G.: Logistics and supply chain management. Res. Transp. Econ. **41**(1), 3–16 (2013). https://doi.org/10.1016/j.retrec.2012.10.006
10. Engblom, J., Solakivi, T., Töyli, J., Ojala, L.: Multiple-method analysis of logistics costs. Int. J. Prod. Econ. **137**(1), 29–35 (2012). https://doi.org/10.1016/j.ijpe.2012.01.007
11. Iida, T.: Coordination of cooperative cost-reduction efforts in a supply chain partnership. Eur. J. Oper. Res. **222**(2), 180–190 (2012). https://doi.org/10.1016/j.ejor.2012.03.029
12. Pettersson, A.I., Segerstedt, A.: Measuring supply chain cost. Int. J. Prod. Econ. **143**(2), 357–363 (2013). https://doi.org/10.1016/j.ijpe.2012.03.012
13. Hu, J., Hu, Q., Xia, Y.: Who should invest in cost reduction in supply chains? Int. J. Prod. Econ. **207**, 1–18 (2019). https://doi.org/10.1016/j.ijpe.2018.10.002
14. Minken, H., Johansen, B.G.: A logistics cost function with explicit transport costs. Econ. Transp. **19**, 100116 (2019). https://doi.org/10.1016/j.ecotra.2019.04.001

Author Index

V. Murgul and V. Pukhkal (Eds.): EMMFT 2019, AISC 1258, pp. 831–833, 2021.
https://doi.org/10.1007/978-3-030-57450-5

The manufacturer's authorised representative in the EU is Springer Nature Customer Service Centre GmbH, Europaplatz 3, 69115 Heidelberg, rmany. If you have any concerns regarding our products, please ct ProductSafety@springernature.com

d bound by CPI Group (UK) Ltd, Croydon, CR0 4YY

30/09/2026

02225495-0020